PERIODIC TABLE OF THE ELEMENTS*

Group	1 I IA	2 II IIA		3 IIIB	4 IVB	5 VB	6 VIB	7 VIIB	8	9 VIIIB
Period										
1										
2	3 **Li** 6.941	4 **Be** 9.0122								
3	11 **Na** 22.990	12 **Mg** 24.305								
4	19 **K** 39.098	20 **Ca** 40.078		21 **Sc** 44.956	22 **Ti** 47.867	23 **V** 50.942	24 **Cr** 51.996	25 **Mn** 54.938	26 **Fe** 55.845	27 **Co** 58.933
5	37 **Rb** 85.468	38 **Sr** 87.62		39 **Y** 88.906	40 **Zr** 91.224	41 **Nb** 92.906	42 **Mo** 95.94	43 **Tc** 98.91	44 **Ru** 101.07	45 **Rh** 102.91
6	55 **Cs** 132.91	56 **Ba** 137.33		71 **Lu** 174.97	72 **Hf** 178.49	73 **Ta** 180.95	74 **W** 183.84	75 **Re** 186.21	76 **Os** 190.23	77 **Ir** 192.22
7	87 **Fr** (223)	88 **Ra** (226)		103 **Lr** (262)	104 **Rf** (261)	105 **Db** (262)	106 **Sg** (263)	107 **Bh** (262)	108 **Hs** (265)	109 **Mt** (266)

1
H
1.0079

Lanthanides 6

57 **La** 138.9	58 **Ce** 140.1	59 **Pr** 140.9	60 **Nd** 144.2	61 **Pm** (145)

Actinides 7

89 **Ac** (227)	90 **Th** 232.04	91 **Pa** 231.04	92 **U** 238.03	93 **Np** (237)

*Atomic masses quoted to the number of significant figures given here can be regarded as typical of most naturally occurring samples. Parentheses around an atomic mass indicate the most stable isotope of a radioactive element.

			13 III IIIA	14 IV IVA	15 V VA	16 VI VIA	17 VII VIIA	18 VIII VIIIA
								2 **He** 4.0026
			5 **B** 10.811	6 **C** 12.011	7 **N** 14.007	8 **O** 15.999	9 **F** 18.998	10 **Ne** 20.180
10	11 IB	12 IIB	13 **Al** 26.982	14 **Si** 28.086	15 **P** 30.974	16 **S** 32.066	17 **Cl** 35.453	18 **Ar** 39.948
28 **Ni** 58.693	29 **Cu** 63.546	30 **Zn** 65.39	31 **Ga** 69.723	32 **Ge** 72.61	33 **As** 74.922	34 **Se** 78.96	35 **Br** 79.904	36 **Kr** 83.80
46 **Pd** 106.42	47 **Ag** 107.87	48 **Cd** 112.41	49 **In** 114.82	50 **Sn** 118.71	51 **Sb** 121.76	52 **Te** 127.60	53 **I** 126.90	54 **Xe** 131.29
78 **Pt** 195.08	79 **Au** 196.97	80 **Hg** 200.59	81 **Tl** 204.38	82 **Pb** 207.2	83 **Bi** 208.98	84 **Po** (209)	85 **At** (210)	86 **Rn** (222)
110 **Uun**	111 **Uuu**	112 **Uub**		114 **Uuq**				

Metals ←   → Nometals

Semimetals

62 **Sm** 150.36	63 **Eu** 151.96	64 **Gd** 157.25	65 **Tb** 158.93	66 **Dy** 162.50	67 **Ho** 164.93	68 **Er** 167.26	69 **Tm** 168.93	70 **Yb** 173.04
94 **Pu** (244)	95 **Am** (243)	96 **Cm** (247)	97 **Bk** (247)	98 **Cf** (251)	99 **Es** (252)	100 **Fm** (257)	101 **Md** (258)	102 **No** (259)

The PRACTICE
of CHEMISTRY

Donald J. Wink
University of Illinois at Chicago

Sharon Fetzer Gislason
University of Illinois at Chicago

Sheila McNicholas
Truman College

W. H. Freeman and Company • New York

Publisher: Susan Brennan
Editors: Jessica Fiorillo, Clancy Marshall
Development Editors: Kathleen Civetta, Guy Copes
Marketing Manager: Mark Santee
Project Editor: Vivien Weiss
Text Designer: Diana Blume
Cover Designer: Victoria Tomaselli
Illustration Coordinator: Bill Page
Illustrations: Fine Line Illustrations
Photo Editor: Vikii Wong
Production Coordinator: Susan Wein
Supplements and Multimedia Editors: Charles Van Wagner, Rebecca Pearce, Amanda
    McCorquodale
Composition: TechBooks
Layout: Jerry Wilke Design, TechBooks
Manufacturing: RR Donnelley & Sons Company

Library of Congress Control Number: 2002117801

ISBN-13: 978-0-7167-4871-7
ISBN-10: 0-7167-4871-1

Printed in the United States of America

Third printing

W. H. Freeman and Company
41 Madison Avenue
New York, NY 10010
www.whfreeman.com

Brief Contents

Contents

PREFACE

The Practice of Chemistry is written for a one-semester introductory chemistry course for students with little or no background in science and math and who plan to continue with the regular course in general chemistry. This is a text that prepares students to succeed in chemistry by building skills in problem solving, mathematics, and scientific reasoning, and then applying these skills to real-world situations. It is the result of seven years of curriculum development, initiated by a National Science Foundation grant to the authors.

This book is designed to help students build a strong foundation of basic chemical principles and skills to carry with them as they continue in general chemistry and beyond. As such, we do not feel it necessary to cover every topic that might be seen in general chemistry. Instead, we emphasize those skills that students need to master in order to succeed in a general chemistry course.

Organization

The organization of the book stems from our understanding of students who have limited experience in understanding chemistry, chemical substances, and the calculation of chemical quantities. It is important for students to see that they are learning a set of connected skills and concepts, and that each new chapter is related to those around it. We hope that we have created a topical sequence that allows students to understand important concepts and techniques fully in the context of the complete preparatory course. That is why we have generated a sequence that reflects the three key components of most preparatory chemistry courses. Each part has a unifying theme:

- **Part 1: Characterizing Chemical Substances and Chemical Reactions** builds a student's ability to look at a chemical substance or a chemical reaction and work with the information contained in the formula or equation. Chapters 1–5, "Elements and Compounds," "Molecular Substances and Lewis Structures," "Ionic Compounds," "The Mole and Chemical Equations," and "Chemical Reactions," comprise Part 1. Students are shown how to derive important information from three key chemical "statements"—chemical formulas, chemical structures, and balanced chemical reactions. This helps them develop the conceptual knowledge they must have before carrying out meaningful quantitative problem solving in the remainder of the course and in subsequent general chemistry courses.

- **Part 2: Chemical Quantities** concentrates on measurement and proportional reasoning, and how these apply to chemical stoichiometry and the gas laws. Students revisit many of the concepts developed in Part 1 but now put these concepts to use in calculations. We find that students with a good conceptual understanding of chemical statements from Part 1 are able to move easily to numerical calculations at this point. Here, students gain a complete understanding of how chemical calculations are used. While we expect that their further general chemistry studies will delve deeper into these topics, we find that students using our material have a complete set of tools for most later chemical calculations. In Part 2, a key aspect of our approach is also apparent: the presentation and the use of a single, unified problem-solving strategy for chemical calculations. This builds upon the algebraic principles behind proportional reasoning, conversion factor methods, and dimensional analysis. Included are Chapters 6–10, "Quantitative Properties of Matter," "Counting and Measurement in Chemical Experiments," "Measurement of Chemical Substances," "Chemical Stoichiometry," and "Discovering the Gas Laws."

- **Part 3: Chemical Systems** delves deeper into explanations for chemical behavior, including energy and heat, solutions, reactions of acids and bases, equilibrium, organic chemistry, and nuclear chemistry. Again, the foundations of conceptual and quantitative ideas from the first two parts are the basis for further explorations. According to our understanding of preparatory chemistry classes, not all of these chapters will always be covered, but those that are discussed will be the basis for richer and more complex elaboration in a subsequent general chemistry class. This part contains Chapters 11–17, "Chemical Systems and Heat," "The Atomic Nucleus," "Electrons and Chemical Bonding," "Solutions, Molarity, and Stoichiometry," "Acids and Bases," "Equilibrium Systems," and "Organic Chemistry and Biochemistry."

Meaningful Chemistry

This organization allows students to experience some success with analyzing chemical formulas, reading chemical equations, and predicting products of reactions—in essence, "doing" some real chemistry up front, before concentrating on more quantitative topics. The advantage of this approach is that it gives students who have not taken a chemistry class an extended experience with thinking and talking about chemical systems before working on calculations. This approach matches educational research, which emphasizes that deep learning is best done in a context where students carry out *meaningful* operations. If we push quantitative aspects of chemistry before students have a meaningful understanding of the chemical substances they are working with, then weak algorithmic strategies—not thoughtful ones—are developed. Instead, our text allows students to work with chemical formulas, structures, and reactions long before they must look at these for the information needed for calculations.

Integration of Biochemistry and Organic Chemistry

Of course, the meaning that students attach to chemistry will also reflect what they want to learn from the course. For the vast majority of these students, this will involve either further study of organic chemistry, biochemistry, or both. Therefore, examples from both areas are used throughout the text. For instructors who wish to add more coverage of organic chemistry and biochemistry, Chapter 17 introduces major concepts about functional groups and the building blocks of life—amino acids, nucleic acids, and carbohydrates.

Flexibility and Incorporating the Laboratory

There are cases where the course is best taught with more quantitative material earlier in the semester. Although we have often used this text in the order presented for a "stand-alone" preparatory course, we also have taught this material in courses that have a laboratory and that are linked to an algebra course. In those cases, we rearrange the material in a more "conventional" manner. The book is flexible enough to allow for moving quickly into quantitative operations on chemical substances. Here, briefly, is how this is done: After Chapter 1, we go to the nomenclature sections of Chapters 2 and 3, and then introduce the mole concept in Chapter 4. This is sufficient for work in Chapters 6, 7, and 8. We then go back to Chapter 5 and to Chapter 9 for more concentrated discussions of chemical reactions and chemical stoichiometry. After this, the remaining material in Chapters 2 and 3 can be presented. Chapter 10 and (depending on the needs of the course) Chapters 14 and 15 complete the preparation for students to perform most further chemical calculations. Should the instructor expect students to master some concepts and skills relevant to thermochemistry and atomic structure, then Chapters 12 and 13 can be presented.

Practical Chemistry

Work presented by cognitive scientists at Vanderbilt University underlies our approach to the inclusion of "Chapter Practicals" in every chapter of the book. This research uncovers how a "semantically rich macro-context" is helpful in learning. In the case of preparatory chemistry, we have interpreted this finding as an indication that students are able to learn better when they can see and discuss a practical use for the skills and concepts being presented. Chapter Practicals provide real-world context and applications for the material in the chapter and ask students to put their new skills to use. The Practicals provide a unifying theme for each chapter and connect chemical concepts to a problem or situation outside of the preparatory chemistry classroom. Students are able to answer the questions posed in the Practicals by using the chemical concepts and problem-solving methods presented in the chapter.

PRACTICAL **D**   What kinds of chemical reactions are used to test for metal ions?

In Practical A, we discussed using color changes to identify metal ions during chemical analysis. If a color change occurs as the result of a chemical reaction, the reaction is often an oxidation-reduction reaction. Color depends on electronic energy. The change in oxidation number that occurs in an oxidation-reduction reaction changes the number of valence electrons and often the color of the substance. The most highly colored substances contain transition metal ions. These are easy to recognize when a colored solution results after mixing two colorless solutions or when one of the reacting solutions is colorless and the color of the other changes after the mixing process. Show that the following reactions are oxidation-reduction reactions by finding the elements that undergo a change in oxidation number.

Figure 5-8 ▲ This photo shows the precipitation of either PbS, CdS, or Sb₂S₃. Which of these compounds has been formed here? How do you know? (Richard Megna/Fundamental Photographs.)

$$18 \ H^+ \ (aq) + Bi_2O_5 \ (s) + 4 \ Mn^{2+} \ (aq) \longrightarrow 4 \ MnO_4^- \ (aq) + 10 \ Bi^{3+} \ (aq) + 9 \ H_2O \ (l)$$
pale pink   purple

(a) What color change indicates a chemical reaction has occurred?
(b) Which element is oxidized in this reaction?
(c) Which element is reduced in this reaction?

A test used to determine alcohol in breath uses an oxidation-reduction reaction involving chromium.

$$2 \ Cr_2O_7^{2-} \ (aq) + C_2H_5OH \ (l) + 16 \ H^+ \ (aq) \longrightarrow 4 \ Cr^{3+} \ (aq) + 11 \ H_2O \ (l) + 2 \ CO_2 \ (g)$$
orange   green

(a) What color change indicates a chemical reaction has occurred?
(b) Is Cr being oxidized or reduced in this reaction?
(c) What gas is evolved in this reaction?

It is easier to be sure that a chemical reaction has occurred when an insoluble solid is formed from the mixing of two clear solutions. The chemical properties of the reacting substances determine whether or not a reaction occurs and a new substance is formed. But a *physical* property of the product determines whether or not the new substance is an insoluble solid. This physical property is the solubility. Figure 5-8 shows the precipitation of an insoluble solid from a solution. Predict possible products for the following reactions. Then, using the solubility rules, determine whether or not an insoluble solid is formed.

1. H₂S (aq) + Pb(NO₃)₂ (aq) →
   (a) Write the formula for any solid formed.
   (b) What is the name of this compound?
   (c) What is the color of this compound? (See Practical A.)
2. H₂S (aq) + Cd(NO₃)₂ (aq) →
   (a) Write the formula for any solid formed.
   (b) What is the name of this compound?
   (c) What is the color of this compound?
3. H₂S (aq) + Sb(NO₃)₃ (aq) →
   (a) Write the formula for any solid formed.
   (b) What is the name of this compound?
   (c) What is the color of this compound?
4. Use solubility rules to predict a compound containing sulfide ion that is soluble in water solution.

PRACTICAL B   How much sand do you need to build a patio?

If you visit a construction site, such as the one shown in Figure 7-7, you will probably see huge amounts of sand in use. In the United States, sand is usually sold by the cubic yard. Of course, in the metric system we use the cubic meter instead. This is a *lot* of sand, but it takes a lot to build even simple things, as we will discover here. We can use what we've learned about the metric system and significant figures to calculate the amount of sand needed to build a patio.

According to the plans for a townhouse community, each of 180 townhouses is to have a patio, made either of concrete or paving bricks. Here are the specifications given to the building contractor:

Plan A: A concrete patio

The depth of the concrete patio is to be 0.015 m, the length is to be 4.5 m, and the width is to be 3.5 m. Calculate the volume of concrete needed to fill this space. How many significant figures did you include in your answer?

Concrete is a mixture of cement, water, sand, and crushed stone. If the concrete mixture is 22% sand by volume, calculate the volume of sand needed to give 90 townhouses a concrete patio. How many significant figures are in your answer? Why?

Plan B: A brick patio

A second option for the prospective home-owners is a patio made with paving bricks. Sand is still needed as it provi...

Figure 7-7 ▲ Measurement is important at a construction site. (Geoff Tompkinson/Science Photo Library/Photo Researchers.)

patio of the same length and width as the concrete patio is to be constructed. But instead of concrete, it needs a bed of sand that is 5.0 cm deep.

Convert 5.0 cm to m. Next, determine the volume of sand needed for one brick patio. How many significant figures are in your answer?

How much sand should be ordered to provide 90 townhouses with brick patios? Give your answer in cubic meters and in kilograms, assuming that the ...

PRACTICAL B   How much carbon is "fixed" in a living plant?

We learned earlier that phototrophs (green plants) use simple molecules like CO_2 and H_2O as the building blocks of their cells. But where do these molecules come from? Plants absorb carbon dioxide from the atmosphere. The carbon is converted into plant material, and oxygen is released. Some plants absorb carbon from the soil as well. Converting carbon into plant material is known as "fixing" carbon.

Plants also need water. Most plants obtain water from the soil, but certain plants absorb most of the water they need directly from the air around them. An excellent example of such a plant is the orchid, which grows among rocks in the wild, and can be grown in a pot with little moisture or soil. An orchid, shown in Figure 9-6, grows by fixing carbon and absorbing water from the air. Let's assume that an orchid has a "dry weight" (the weight of just the solid, carbon-containing material after all free water is removed) of 10.0 grams. The majority of this weight is the carbohydrate known as starch, which has an empirical formula of $C_6H_{10}O_5$. We can calculate that 10.0 grams of starch is equal to 0.0618 mole of starch, and contains 0.370 mole of carbon. Mole/mass calculations allow us to determine how many moles of carbon dioxide and water are used to make a particular amount of plant material.

We start by writing a balanced chemical equation for a simplified chemical reaction for the formation of starch:

Figure 9-6 ▲ Orchids are plants that obtain all their components, including water, from the air alone. (David Stone/Rainbow.)

$$6\,CO_2 + 5\,H_2O \longrightarrow C_6H_{10}O_5 + 6\,O_2 \qquad (1)$$

We can use this equation to calculate the number of grams of carbon dioxide and water needed to make the 10.0 grams of carbohydrate we assume to be in an orchid. If we find that the air in the room where the orchid grows contains 0.025 gram of carbon dioxide per liter of air, then how many liters of this air are needed to make the orchid?

We vary the focus of the Practicals considerably so that students with different goals are engaged. For example, in Chapter 3, students examine labels from vitamin supplements and food packages, and identify the ionic compounds in these products. In Chapter 4, they discover the answer to the question "How is a safety match put together?" and in Chapter 7, they calculate "How many moles of sand are on your favorite beach?" The Chapter Practicals are not simpleminded boxes that students can read passively. Rather, they ask students to apply their new-found knowledge and skills in different ways. Students are motivated to master the chapter material in order to develop answers to the Practical questions.

The Chapter Practicals are organized around the following topics:

- Chapter 1: Chemistry in Our Lives
- Chapter 2: Power Molecules
- Chapter 3: Ionic Compounds—Our Source of Dietary Minerals
- Chapter 4: The Match
- Chapter 5: What Is in that Metal and How Do We Prove It?
- Chapter 6: Interpreting Chemical Experiments

Chapter 7: The Measure and Pleasure of Sand

Chapter 8: Medicinal Drugs—Molecules with Special Uses

Chapter 9: Metabolism and the Chemical Budget of Life

Chapter 10: Mathematical Models in Chemistry

Chapter 11: Portable Fuel

Chapter 12: What Happens When the Identities of Atoms Change?

Chapter 13: Electrons and the Properties of Molecules and Atoms

Chapter 14: Solution Concentrations in Medicine

Chapter 15: Titration

Chapter 16: What's in Our Water Supply?

Chapter 17: How Do Molecules Work?

Developing Problem-Solving Skills

The key to student success in chemistry is the ability to solve problems. A wealth of worked-out example problems appears in every chapter. We take great care in these examples to provide students with a consistent method of first *thinking through the problem*, then setting up and *working through the problem*, and finally stating the *answer* with the correct number of significant figures. When key ideas

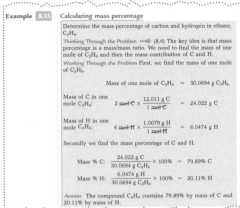

Example 8.15   Calculating mass percentage

Determine the mass percentage of carbon and hydrogen in ethane, C_2H_6.

Thinking Through the Problem (8.6) The key idea is that mass percentage is a mass/mass ratio. We need to find the mass of one mole of C_2H_6 and then the mass contribution of C and H.

Working Through the Problem First, we find the mass of one mole of C_2H_6.

$$\text{Mass of one mole of } C_2H_6 = 30.0694 \text{ g } C_2H_6$$

$$\text{Mass of C in one mole } C_2H_6: \quad 2 \text{ mol C} \times \frac{12.011 \text{ g C}}{1 \text{ mol C}} = 24.022 \text{ g C}$$

$$\text{Mass of H in one mole } C_2H_6: \quad 6 \text{ mol H} \times \frac{1.0079 \text{ g H}}{1 \text{ mol H}} = 6.0474 \text{ g H}$$

Secondly we find the mass percentage of C and H.

$$\text{Mass \% C: } \frac{24.022 \text{ g C}}{30.0694 \text{ g } C_2H_6} \times 100\% = 79.89\% \text{ C}$$

$$\text{Mass \% H: } \frac{6.0474 \text{ g H}}{30.0694 \text{ g } C_2H_6} \times 100\% = 20.11\% \text{ H}$$

Answer The compound C_2H_6 contains 79.89% by mass of C and 20.11% by mass of H.

PROBLEMS

26. List some physical properties of metals.

27. List some physical properties of nonmetals.

28. Concept question: Review the description of the periodic law. Would it have been possible to formulate this law if there were not a clear difference between an element and a compound?

29. Group II metals (M) will react with group VII nonmetals (X) to form compounds with the general formula MX_2. Predict formulas for compounds containing (a) Ca and Cl (b) Ba and I (c) Mg and F.

30. Concept question: The periodic table consists of groups of elements with similar—but not identical—properties. This means we expect the same behavior down a group, but that we aren't surprised when differences appear among elements in the same group. Show that this is true for the metal/nonmetal/semimetal classification of elements in groups II, IV, and VII of the periodic table.

31. Indicate whether the following statements are true or false. Write corrected statements for those you marked false.

2.1   Molecules in Chemistry

SECTION GOALS

✓ Predict whether or not a compound is likely to be a molecular substance.
✓ Know what information is given by the chemical formula of a substance.
✓ Understand and use numbering schemes for naming molecular substances.
✓ Apply the rules for naming molecular substances.

Molecular Substances

In Chapter 1 we discussed how pure chemical substances can be composed of a single element or a compound of two or more elements. In this chapter, we discuss an important class of pure substances, where the atoms are held together in certain fixed arrangements

from the text are needed to solve the problem, a small key icon ( ) appears within the example.

The following features work together to build student confidence in their problem-solving abilities:

◄ **Examples** An average of twenty examples per chapter, with titles for easy reference, take students step by step as they think through the problem, work through the problem, and arrive at the answer. Key icons link problem solving to the key ideas in the chapter.

► **How Are You Doing?** After each set of related problems, these practice problems ask students to solve similar problems without help. These problems give immediate feedback as to whether or not the student has understood the preceding examples. Answers are provided in the back of the book.

Na: 32.37 g Na $\times \dfrac{1 \text{ mol}}{22.990 \text{ g Na}} = 1.41 \text{ mol Na}$ $\dfrac{1.41 \text{ mol}}{0.70 \text{ mol}} = 2$

S: 22.53 g S $\times \dfrac{1 \text{ mol}}{32.066 \text{ g S}} = 0.70 \text{ mol S}$ $\dfrac{0.70 \text{ mol}}{0.70 \text{ mol}} = 1$

O: 45.05 g O $\times \dfrac{1 \text{ mol}}{15.9994 \text{ g O}} = 2.82 \text{ mol O}$ $\dfrac{2.82 \text{ mol}}{0.70 \text{ mol}} = 4$

Answer There are 2 moles Na, 1 mole S, and 4 moles O in the formula. The empirical formula of the compound is Na_2SO_4, sodium sulfate.

How are you doing? 8.10

A compound is analyzed and found to contain 62.0% C, 10.4% H, and 27.5% O. Determine the empirical formula of this compound.

▼ **Section Problems** Problem sets appear at the end of each chapter *section*, again giving immediate problem-solving reinforcement. This is an unusual placement for chemistry text problems, but it is in fact the norm in mathematics texts. It allows students and instructors to determine reliably "what can be done right now," often after a lecture that focuses on that section.

● **Chapter Problems** The Section Problems build pieces of student learning, but there is still the need for additional problems that summarize understanding at the end of each chapter, drawing from the chapter as a whole.

◄ **Concept Questions** These appear in many problem sets, and ask the student to think more deeply about the concepts presented in the chapter. These can be used as collaborative learning and as essay exercises.

Pedagogical Features

We have thought long and hard about what sorts of learning aids best serve the beginning chemistry student. The following features are carefully designed to build student confidence and to motivate students to work through each chapter:

◄ **Section Goals** A list of goals at the start of each chapter section alerts students to the skills and concepts that should be mastered. Margin notes, called *Meeting the Goals,* appear as each of the goals is discussed, recapping the important points. All section goals from the chapter are restated in the chapter Summary as questions and answers.

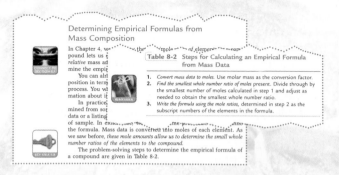

Determining Empirical Formulas from
Mass Composition

In Chapter 4, we ⋯⋯⋯ the ⋯⋯ mole ⋯ of elemen⋯ ⋯ com⋯
pound lets us ⋯⋯⋯⋯⋯⋯⋯⋯⋯⋯⋯⋯⋯⋯⋯⋯⋯⋯⋯⋯
relative mass an
mine the empi⋯

You can al⋯
position in term
process. You w
mation about it

In practice
mined from so⋯
data or a listing
of sample. In eith⋯⋯ ⋯⋯ ⋯⋯ ⋯ ⋯ ⋯ ⋯⋯ ⋯⋯⋯
the formula. Mass data is converted into moles of each element. As
we saw before, *these mole amounts allow us to determine the small whole
number ratios of the elements to the compound.*

The problem-solving steps to determine the empirical formula of
a compound are given in Table 8-2.

**Table 8-2 Steps for Calculating an Empirical Formula
from Mass Data**

1. *Convert mass data to moles.* Use molar mass as the conversion factor.
2. *Find the smallest whole number ratio of moles present.* Divide through by
   the smallest number of moles calculated in step 1 and adjust as
   needed to obtain the smallest whole number ratio.
3. *Write the formula using the mole ratios,* determined in step 2 as the
   subscript numbers of the elements in the formula.

◀ **Study Icons** Special icons flag three types of
content crucial to problem solving in chemistry.
Remember icons mark tables, equations, and
facts that should be memorized. *Connect to*
icons appear where material from other chapters is referenced and used. *Key Idea* icons
appear next to concepts that are vital to understanding the chapter and to solving its problems.
These icons allow students to easily review each
chapter.

▶ **Study Tips** Boxed
study hints appear in
some chapters. These
provide students
with ideas for good
study habits and help
to develop problem-
solving skills.

STUDY TIPS Thinking About Chemical Substances

We encourage you to study with another, more comprehensive,
goal in mind: to gain the ability to look at a name, a formula, or
a structure and immediately start thinking about what kind of substance it represents. This means that you examine the substance
to see what kinds of atoms are in it, and how the atoms are bonded
together.

This is a goal that will take many more courses in chemistry
before it is perfected. But even the beginning chemist can start to
develop a habit of studying for *familiarity*. What kinds of things
can you look for? You should already be able to read the words
"carbon tetrachloride" and think that this is an OK name for a compound. What if you heard "carbon quadchloride"? That should
sound odd to you, as it would to a more highly trained chemist,
because the prefix "quad" is not used in chemistry, even though it
means the same thing as "tetra."

◀ **Chapter Summaries** Each chapter summary consists of a glossary of the key terms that are highlighted
in the chapter, as well as a summary of section goals
presented in a question-and-answer format.

Chapter 6 Summary and Problems   ▣ Review with Web Practice

VOCABULARY

quantitative property	A property that has a numerical measurement.
extensive property	A property that depends on the size of the sample; the measured value of an extensive property will change as the size of the sample changes.
intensive property	A property that is intrinsic to a sample of material; the measured value of an intensive property does *not* depend on the size of the sample.
metric system	The numerical system of measurement used by scientists; it is based on powers of ten.
mass	The measure of the amount of matter present.
weight	A measure of the amount of matter present in a sample as it is acted upon by gravity.
temperature	A measure of the hotness or coldness of a substance compared to some temperature scale.
melting/freezing temperature	The temperature at which a substance melts/freezes, i.e., converts between the solid and liquid phases.
boiling temperature	The temperature at which a substance boils/condenses, i.e., converts between the liquid and gaseous phases.
vapor	The gas formed when a liquid evaporates.
normal boiling point	The boiling point temperature measured at sea level.
physical constant	A measurement of a physical property of a substance that does not change when measured at the same conditions of temperature and pressure.
density	A physical property determined by dividing the mass of a substance by its volume; density is an intensive property that can be used (with other criteria) to identify a substance.
proportionality	A proportionality occurs when we form the ratio of two extensive variables.
constant of proportionality	The k in the equation $y = kx$, where the variable y is directly proportional to the variable x.
proportional reasoning	A problem-solving technique based on the use of proportionality. One ratio in the proportion gives some known relationship; the second ratio contains the same quantities in the numerator and denominator as does the known ratio but in this case, one of the values is not known.

SECTION GOALS

What is the difference between an intensive and an extensive property?	Extensive properties depend on the size of the sample; intensive properties do not. These properties are used in the laboratory to characterize how substances exist or act; they are also used to make predictions.
What common metric units are used to measure length, mass, and volume?	Common scientific units that measure length, mass, and volume are centimeter, gram, and cubic centimeter, respectively.
How does temperature influence the physical state of a pure substance?	The general trend is for substances to be solids at lower temperature, liquids at intermediate temperatures, and gases at higher temperature. However, not all substances have a liquid phase under normal conditions.

Each of these features is essential to using this text
well in the learning of chemistry. But a second component of these features is how they support students in
the essential review of material. We have placed the
icons in positions that allow students to go back
through a chapter to quickly and easily find information that should be memorized ("Remember" icon) and
understood ("Key Idea" icon) before a test. Finally, the
chapter summaries can be used as a quick review of
key ideas and vocabulary. We find that placing these at
the end of the chapter is very useful to students in their
reviews, far superior to an end-of-book glossary. Chapter summaries are also linked to online Web Practice
assignments discussed below.

Building Math Skills: The Basis of Chemical Calculations

This textbook grew out of a National Science Foundation MATCH
program grant to integrate mathematics and chemistry education
(coauthor Sheila McNicholas is a mathematician). The MATCH program, which continues to be taught, was the basis of our engagement

with one of the major challenges in preparatory and general chemistry: getting students to successfully use math, especially algebra, in problem solving.

Special *Making it Work with Math* sections reinforce basic math concepts as they are needed. These focus on the math alone, without the chemistry application, and take the place of a separate math appendix. By appearing conveniently within the chapter where the math concepts will be applied, the material is more likely to be noticed and used by the students. For example, the sections "Graphing in the Cartesian Coordinate System" and "Linear Relationships" appear in Chapter 10, "Discovering the Gas Laws." We also have a section on the algebra of base-10 logarithms that, we find, demystifies much of what we teach about pH.

One of our key insights is that students rarely see that a very large fraction of chemical calculations are linked to a simple mathematical principle: the proportional relationship of two extensive quantities. In most texts the use of such proportional reasoning occurs under the guise of "conversion factors," "dimensional analysis," or "unit factors." All are valid representations in one sense or another. However, we find that a clear presentation of the proportional mathematics and repeated reference to it allows a host of unit conversion problems (metric conversions, molar mass, molarity, and mole ratios, to name a few) to be linked to a student's single, unified understanding of algebra.

We have studied this approach extensively, and it has been presented in the chemical education literature ("The MATCH Program: A Preparatory Chemistry and Intermediate Algebra Curriculum," *Journal of Chemical Education*, August 2000). An important finding of this paper is that, when students are taught the mathematics of chemistry in a mathematically rigorous way, they find it much easier to use math in chemistry. (It also helps their understanding of math!) Students from the MATCH program performed better in their follow-up general chemistry classes, at a statistically significant level, when compared with those from disconnected courses using traditional preparatory chemistry and algebra materials. Aside from the full research study we employed in the MATCH paper, we believe we have seen the same transfer occur when students come from a stand-alone preparatory chemistry class using the *Practice of Chemistry* text. In the final analysis, we are proud and

---

MAKING IT WORK WITH MATH

Working with Exponents

Exponents are used as a type of mathematical shorthand for multiplication. If you have the factored form of a number such as $81 = 3 \cdot 3 \cdot 3 \cdot 3$ it is easier to write the expression in exponential form, $81 = 3^4$. In this case 3 is the base and 4 is the exponent. *The exponent indicates the number of times the base appears as a factor.*

Remember that an exponent operates only on the factor *immediately below and to the left of the exponent. If the exponent appears beside a parentheses, it operates on everything inside the parentheses.* Thus $(-2)^3 = -2 \times -2 \times -2 = -8$.

Example 7.7 Simplifying numbers with exponents

Simplify the following numbers.
(a) 2^3 (b) $(-5)^2$ (c) -4^2

Thinking Through the Problem (7.6) We know that the exponent "works" only on the factor immediately preceding it.

Working Through the Problem Part (a) is straightforward as it has only one digit with the exponent: $2^3 = 2 \cdot 2 \cdot 2$. Part (b) shows the

---

6.3 Proportions in Chemical Measurement

SECTION GOALS

✓ Understand that extensive measurements can be used to determine an intensive measurement, using a ratio called a proportionality.

✓ Recognize a constant of proportionality.

✓ Find and write a proportionality using a relationship among extensive units given in a problem.

✓ Write and use ratios that involve percentage, parts per million, and parts per billion.

MAKING IT WORK WITH MATH

Constants of Proportionality

✔ Meeting the Goals

Whenever two variables are directly proportional, their ratio gives a constant value called a constant of proportionality. For example, density is the constant of proportionality between mass and volume.

The ratio of two extensive variables is called a **proportionality.** If this ratio is the same whenever we compare these variables, the constant is called the **constant of proportionality.** When we use proportionalities to help us solve problems, the process is called **proportional reasoning,** which is a very important skill in chemistry and other sciences. For example, you can measure the distance from your house to a bus stop and the amount of time it takes to jog this distance. These are both simple, extensive measurements. The ratio of distance to time gives an intensive measurement—the rate—which tells us how quickly or slowly we cover a particular distance. See Figure 6-11.

Mass and volume are both extensive variables. When we relate these two extensive variables, we get a proportionality called the density. Proportionalities like density free us from measuring every extensive property in every sample: if we know the proportionality, we need to measure only one extensive variable. The other variable can be calculated.

Constants of proportionality are found whenever one measurement changes with respect to another by direct variation. In mathe-

Figure 6-11 ▲ Although both distance and time are extensive properties, their ratio (the rate) is an intensive property. (Larry Brownstein/Rainbow.)

confident in pointing out that there is a direct link between our text and a rigorous research study, something we think is unique in the market today.

Integrated Web-Based Media

A Web site has been specially designed to accompany this textbook, consisting of three components that are tied directly to the text. It can be found at www.whfreeman.com/practiceofchemistry.

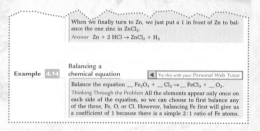

When we finally turn to Zn, we just put a 1 in front of Zn to balance the one zinc in $ZnCl_2$.
Answer $Zn + 2 HCl \rightarrow ZnCl_2 + H_2$

Example **4.14**   Balancing a chemical equation   ◄ Try this with your Personal Web Tutor

Balance the equation __ Fe_2O_3 + __ $Cl_2 \rightarrow$ __ $FeCl_3$ + __ O_2.
Thinking Through the Problem All the elements appear only once on each side of the equation, so we can choose to first balance any of the three, Fe, O, or Cl. However, balancing Fe first will give us a coefficient of 1 because there is a simple 2 : 1 ratio of Fe atoms.

◄ **Personal Web Tutor** Every chemistry student must tackle certain key types of problems, such as Lewis structures and multi-step stoichiometry. An interactive example of each of these types of problems appears on the Web, providing a step-by-step walk-through of the problem, with feedback along the way. Worked-out examples in the text that are included in the *Personal Web Tutor* are identified with a special media icon.

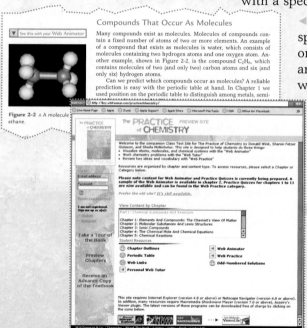

Compounds That Occur As Molecules

▼ See this with your Web Animator

Many compounds exist as molecules. Molecules of compounds contain a fixed number of atoms of two or more elements. An example of a compound that exists as molecules is water, which consists of molecules containing two hydrogen atoms and one oxygen atom. Another example, shown in Figure 2-2, is the compound C_2H_6, which contains molecules of two (and only two) carbon atoms and six (and only six) hydrogen atoms.
Can we predict which compounds occur as molecules? A reliable prediction is easy with the periodic table at hand. In Chapter 1 we used position on the periodic table to distinguish among metals, semi-

Figure 2-2 A molecule of ethane.

◄ **Web Animator** Molecular level animations of specific molecules and their reactions are included on the Web site. Where possible, stills from media animations are included in the text, and marked with a *Web Animator* media icon.

◄ **Web Practice** Students can use the Web to review material in each chapter. Included are flashcards of key terms, practice tests, and hyperlinks to text material. A *Web Practice* icon in each chapter reminds students to take full advantage of these study aids.

A Complete Supplements Package

The following supplements are provided to enhance the textbook and to help students and instructors alike. Please contact us at chemistry@whfreeman.com about availability of supplements for class testing.

For students

The Practice of Chemistry Web Site is where students can review material from the text, including all photos and art, take practice quizzes, use flashcards, and explore Web links to interesting chemistry sites. Included is the unique *Personal Web Tutor*, which presents the step-by-step solution of examples from the text and then guides the student through the solution of similar problems, providing feedback at each step. Just go to www.whfreeman.com/practiceofchemistry.

Study Guide/Solutions Manual, by Pam Mills of Hunter College and Amina El-Ashmawy of Collin County Community College, provides a chapter-by-chapter guide to the text, which helps students to contemplate and practice the important skills and concepts. Worked-out solutions to all odd-numbered problems from the text are also provided in the *Study Guide*. ISBN: 0-7167-4875-4

Laboratory Experiments in Introductory Chemistry, by Philip Reedy of San Joaquin Delta College, provides a selection of 19 Preparatory Chemistry laboratory experiments, available as one complete manual or through our custom publishing services. Labs include post-lab exercises that reinforce the skills and chemical concepts practiced in the laboratory experiments. Available now. ISBN: 0-7167-4975-0

For instructors

Instructor's Manual, by Donald Wink of the University of Illinois at Chicago and Amina El-Ashmawy of Collin County Community College, provides chapter-by-chapter insight into teaching the material and working with the text, as well as discussions of dynamic teaching situations, including collaborative learning. Worked-out solutions to the even-numbered problems are also available here for instructors.

Test Bank (print and electronic), by Juliette Lecomte of Pennsylvania State University, contains over a thousand multiple-choice questions, divided by chapter. The *Test Bank* is also available as a dual-platform CD-ROM, allowing professors to edit questions and create exams. Available now.

Overhead Transparencies represent approximately one hundred illustrations from the text, enlarged for use in the classroom and lecture halls.

Acknowledgements

There are only three authors listed on the cover of this book, but in the seven years that we have been working on this project and its predecessor, we have been aided, influenced, and directed by dozens of vital friends and colleagues. The MATCH program was aided by several colleagues at the University of Illinois at Chicago (UIC) and at the nearby community colleges and universities that form part of our network of collaborators. We recall fondly the way the late Herb Alexander of the UIC math department guided our understanding of how students really experienced learning the key concepts in algebra. John Baldwin of UIC's math department and Julie Ellefson Kuehn and Mercedes McGowan of William Rainey Harper College were also important in helping us think about how these materials might translate to other campuses, and Joseph Young of Chicago State University helped to frame our thinking about fitting our material into a full chemistry curriculum.

As a research-based curriculum development program, MATCH could not have proceeded without significant input from the findings of others. Our quantitative and qualitative research program was led by Barbara J. Zusman of the UIC Office for Data Resources and

Institutional Analysis and by Robert C. Mebane of the Department of Chemistry of the University of Tennessee at Chattanooga. In developing the actual curriculum, we are very thankful to Mary O'Brien and David Chalif of Edmonds Community College in Lynwood, Washington, who have shared their well-established CHEMATH program with us from the beginning of the MATCH program. William Sweeney, Pamela Mills, and their colleagues at Hunter College also shared seminal results of their two-year chemistry-physics-math program. Similarly, the more advanced math and chemistry integration efforts based at Rose Hulman Institute (Edward Mottell) and at the University of Alabama (David Nickles) also kept us abreast of other developments. As suggested, this group helped us bring out a custom-published text for the MATCH program. The General Chemistry Secretary at the University of Illinois at Chicago, Mrs. Victoria Scates, hand-produced this through her duplicating service, often on the morning of the lecture. Later a more formal production was done, and we thank the workers at Hayden-McNeil publishing in Michigan for providing this service.

As the text moved into development by W. H. Freeman and Company, another equally important set of colleagues lent their expertise in different ways. We begin with Brian Coppola, who first introduced our materials to Freeman at his "Day 2-to-40" conference at the University of Michigan in May 1997. That conversation essentially initiated the process of moving the results of the MATCH program into the present text. Another essential group gave us their frank and very helpful reviews of manuscript in one or more drafts, or participated in developmental focus groups.

Henry Abrash	California State University, Northridge	John Cullen	Ricks College
Gul Afshan	MSOE	Son Do	University of Southern Louisiana, Lafayette
Edward Alexander	San Diego Community College District	David Dollimore (deceased)	University of Toledo
Ramesh Arasasingham	University of California, Irvine	Liz Dorland	Mesa Community College
		Jerry Driscoll	University of Utah
Danny R. Bedgood, Jr.	Arizona State University	Doris Eckey	University of Iowa
William F. Berkowitz	Queens College	Amina El-Ashmawy	Collin County Community College
Steve Borick	Scottsdale Community College	Lisa Fridman	Glendale Community College
Joe Brundage	Cuesta College		
Maureen Burkhart	Georgia Perimeter	Dennis Fujita	Santa Rosa Junior College
Joe Burnett	Iowa State University	Stanley Grenda	University of Nevada, Las Vegas
C. J. Carrano	Southwest Texas State University	Carol Handy	West Virginia University
Karen Creager	Clemson University	James Hardcastle	Texas Women's University
James Cress	California State University, Sacramento	C. Alton Hassell	Baylor University
		Susan Holladay	Purdue University

Jeffrey Hurlbut	*Metropolitan State College*	Diann Schneider	*University of Arkansas, Fayetteville*
Marina Ionina-Prasov	*University of Michigan, Flint*	Jeffrey Seyler	*University of Southern Indiana*
Charles Jaffe	*West Virginia University*	Carl Shepherd	*Coconino Community College*
Richard Jones	*Sinclair Community College*		
Helen Kemp	*San Diego State University*	Phil Silberman	*Scottsdale Community College*
Ann Kamppainen	*University of Toronto, Scarborough*	Thomas E. Sorensen	*University of Wisconsin, Milwaukee*
Kevin Klausmeyer	*Baylor University*	Spencer Steinberg	*University of Nevada, Las Vegas*
Juliette Lecomte	*Pennsylvania State University*	Uni Susskind	*Oakland Community College*
John Long	*Henderson State University*		
Peggy McClure	*Midlands Technical College*	Paris Svoronos	*Queens Community College*
Claude Mertzenich	*Luther College*		
Deborah Miller	*Albuquerque Technical Vocational Institute*	William Sweeney	*Hunter College*
Barbara Mowery	*Thomas Nelson Community College*	John M. Toedt	*Eastern Connecticut State University*
Karl Muller	*Pennsylvania State University*	Marcy Towns	*Ball State University*
Glenn Nomura	*Georgia Perimeter*	Tod Treat	*Parkland College*
Raymond O'Donnell	*State University of New York, Oswego*	Petra Van Koppen	*University of California, Santa Barbara*
John Paparelli	*San Antonio College*	Ann Verner	*University of Toronto, Scarborough*
James Petrich	*San Antonio College*		
Phil Reedy	*San Joaquin Delta College*	Tracey Whitehead	*Henderson State University*
Erwin Richter	*University of Northern Iowa*	Linda Wilson	*Middle Tennessee State University*
Sue Roper	*Sacramento Community College*	James Wright	*Saddleback College*
Douglas Sawyer	*Scottsdale Community College*	David Zellner	*California State University, Fresno*

The task of pulling a project of this type together is only partly the authors' responsibility. We owe an uncounted debt to the people at W. H. Freeman and Company who saw, in our self-published MATCH text, the beginnings of a full preparatory chemistry effort. Michelle Julet was the key initial contact, with Jessica Fiorillo later joining as Chemistry Editor, and Kathleen Civetta working many hours as Developmental Editor. Charlie Van Wagner and Rebecca Pearce coordinated the media and supplements, and Roy Tasker of CADRE Design provided more than technological support. The give-and-take that occurred as he started designing Web Tutors was invaluable in our thinking about problem-solving strategies, both for those problems and in general. Diana Blume created the intricate book design, and Vivien Weiss and Susan Wein managed the production process. As the project moved into its final stages, UIC graduate students Kathleen Mandell and Anna Sromek were able to endure countless proofreading hours at our side.

    We also acknowledge the National Science Foundation Division of Undergraduate Education, which allowed us to start this project. We proudly include their logo as an indication of our gratitude to them. This material is based on work supported by the National Science Foundation under Grant No. DUE-9354526. Any opinions, findings, and conclusions or recommendations expressed in this material are those of the authors and do not necessarily reflect the views of the National Science Foundation.

    Finally, we owe everything to our families who have put up with many late-night and early-morning sessions on our computers as this textbook was put together.

Elements and Compounds: The Chemist's View of Matter

An ice cube is an example of a pure substance, the chemical compound water. (J & M Studios/Liason.)

PRACTICAL CHEMISTRY   Chemistry in Our Lives

Chemistry is the study of matter and the changes that matter undergoes. Our bodies change, as do many of the things around us. Ice melts, iron rusts, flowers grow, food spoils, and paper burns. Chemistry, then, must be a big part of our lives! When we see a candle burn, we are witnessing chemistry in action. When we cook a hamburger, we are "doing" chemistry.

It may be easy to see that chemistry is important, but being able to master its concepts requires discipline and hard work. The definition of chemistry as the study of matter and the changes that matter undergoes includes everything you will learn in this course. As you read this book, you will learn to discuss, describe, and think about matter and its changes. You will also need to learn how chemists classify matter and its changes in specific ways. For example, matter is classified as anything that occupies space and has mass. Our bodies are matter, and everything around us is made of matter.

This chapter introduces some very fundamental concepts of chemistry. Here, and in every chapter, we will organize our thoughts around questions that relate to how you use and encounter chemistry in your day-to-day life. These questions are more deeply explored in *Chapter Practical* boxes that go beyond the basics. They will help you to develop a feel for how chemistry matters in your everyday life. Each practical will present certain questions that you will be able to answer as you work through the chapter.

This chapter will focus on the following questions:

PRACTICAL A How does chemistry let us look at the world in an organized way?

PRACTICAL B What are the basic building blocks of matter?

PRACTICAL C What are chemical elements?

PRACTICAL D What are the important differences among chemical substances?

PRACTICAL E What systematic trends does the chemical viewpoint let us see?

1.1 The Vocabulary and Organization of Chemistry

SECTION GOALS

✔ Understand the difference between an element and a compound.

✔ Understand the difference between a pure substance and a mixture.

Elements and Compounds

Let's look again at our definition of chemistry. If we are to understand chemistry, we need to understand two words that have special meaning to chemists: **matter** and **change.** We will look first at matter. Matter is anything that occupies space and has mass. This general definition of matter includes almost everything we call "our world"; this book, your desk, and your chair are all examples of matter. You probably recognize that there must be more than one kind of matter. Your textbook is very different from, say, an ice cube. To try to make sense of all the matter around us, scientists classify it, or sort it out, in various ways.

Chemists generally consider the simplest kind of matter **elements.** There are a little over one hundred different elements now known to exist. Gold is an example of an element. So is aluminum. Oxygen in the air is an element.

Chemistry would be a very simple science if the elements were the only kind of matter. But the chemical elements combine with other elements in specific and constant patterns to form **compounds.** Most of the matter around us consists of compounds. Water, for example, is a compound made of the elements hydrogen and oxygen. A compound is a substance that contains two or more different elements that are chemically combined.

Both elements and compounds are pure substances. A **pure substance** contains only one kind of matter, either one element (such as gold) or one compound (such as water), and cannot easily be broken down into other things.

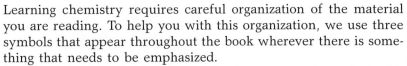

STUDY TIPS   ## What to Think about in Reading This Material

Learning chemistry requires careful organization of the material you are reading. To help you with this organization, we use three symbols that appear throughout the book wherever there is something that needs to be emphasized.

Anything that should be committed to memory is marked with a "memorization icon" when it first appears in this book and should be thoroughly learned. This icon reminds us that we should put this information "in storage" for later use. Some of these items are important to remember because they are part of the basic vocabulary of chemistry. When vocabulary is first introduced, the word will appear in **bold** type.

Sometimes we introduce a key concept, or a way of "thinking like a chemist." Key concepts and thought processes are marked with a "key idea" icon. You will see that some of these keys are used in the worked examples.

Finally, a third icon, the "linking icon," is present in places where we are linking ideas from one point in the book to another. This icon will assist you in connecting the material throughout the book.

These three "tools"—stored information, key concepts and techniques, and links—are essential to solving different chemistry problems. When you encounter a problem and you think, "I need to remember something in order to do this problem," you may want to flip back for memorization icons. On the other hand, when you are in need of a concept or technique, the key concept icon may be the place to look. Finally, if a problem in a chapter requires knowledge from earlier chapters, consider looking for the "link" icon.

Most samples of matter are combinations of pure substances. Combinations of pure substances are called **mixtures.** Examples of mixtures are sand (a mixture of solid materials), air (a mixture of gases), and salt water (a mixture of salt and water). Each pure substance in a mixture retains its individual identity and can be physically separated from the rest of the mixture. You are probably familiar with many mixtures already. Beach sand is a mixture. When you look closely at a handful of sand, you see that the grains vary in color and texture. Sand is a mixture of several different kinds of minerals. You have probably created some mixtures yourself. Brewed coffee, cream, and sugar form a familiar mixture in your coffee mug. The classification of matter into pure substances and mixtures is summarized in Figure 1-1.

Figure 1-1 ▶ Chemists classify matter as composed of pure chemical substances and the mixtures of these substances. Elements and compounds are the two kinds of pure chemical substances.

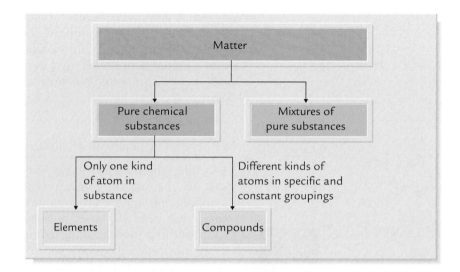

Figure 1-2 ▲ Granite (a) and beach sand (b) are good examples of heterogeneous mixtures. Lemonade (c) is a heterogeneous mixture that is a liquid. Notice the pieces of lemon pulp in some parts of the mixture. (a: John D. Cunningham/Visuals Unlimited; b: L. West/Photo Researchers; c: Erv Schowengerdt.)

Figure 1-3 ▼ Clear apple juice is an example of a homogeneous mixture. (Erv Schowengerdt.)

Many mixtures have obvious differences in their appearance. Mixtures that differ in properties from point to point are called **heterogeneous** mixtures. (The word *heterogeneous* means "having different kinds.") Beach sand is a good example of a heterogeneous mixture. Grains of sand in the same handful have different colors. Examples of heterogeneous mixtures are shown in Figure 1-2.

Sometimes the components of a mixture are well blended and the mixture appears the same throughout. For example, tea with sugar appears the same and tastes the same throughout. Such mixtures are said to be **homogeneous.** Homogeneous mixtures are called **solutions.** Another example of a homogeneous mixture is saline solution (a homogeneous mixture of salt and water). The air we breathe is a solution because it is a homogeneous mixture of several different kinds of gases. An example of a homogeneous mixture is shown in Figure 1-3.

Figure 1-4 expands our classification of matter to include the very important characteristics of homogeneous and heterogeneous appearance.

Figure 1-4 ▶ The homogeneous and heterogeneous appearance of matter is an important clue to its chemical composition.

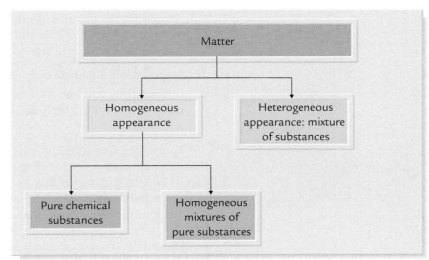

▼ See this with your **Web Animator**

Figure 1-5 ▲ A representation of one molecule of water, containing two hydrogen atoms (light gray) and one oxygen atom (red).

KEY IDEA 1.1

✔ **Meeting the Goals**

Both elements and compounds are pure substances. An element contains only one kind of atom. A compound contains two or more elements (kinds of atoms) in a fixed arrangement.

Atoms and Molecules

Now that we have classified matter as consisting of pure substances and mixtures of pure substances, let's look again at our definition of a pure substance. We said that a pure substance consists of only one kind of matter—either one element or one compound. If we were to break down a sample of an element into smaller and smaller parts, eventually we would be left with the smallest example of that element—the **atom.** An atom cannot be broken down into simpler components by chemical means.

There are many different kinds of atoms. *Atoms of different elements are different. Atoms of the same element are the same in their chemical identities.*

If we were to break down a compound into smaller and smaller parts, eventually we would be left with the smallest part of that compound—either a molecule or a formula unit. A molecule consists of two or more atoms that are chemically joined. For example, the pure substance water is a compound consisting of molecules that are formed by the combination of hydrogen atoms and oxygen atoms in a two-to-one ratio. There are two atoms of hydrogen (shown in Figure 1-5 as light gray) and one atom of oxygen (shown in Figure 1-5 as red) in every molecule of the compound water. Compounds are very different from a simple mixture of the elements. In a mixture each element retains its own identity, whereas a compound has an identity that is usually quite different from that of the elements it contains.

STUDY TIPS

How To Solve Problems in This Book

Throughout this book, you will see examples of the kinds of questions you can answer using what you have learned. You should always begin by **Thinking Through the Problem.** Sometimes the question can be answered very simply. At other times, the answer will require more detailed thinking. It is important to know which information and which skills will enable you to answer the question.

The tools you should use to solve problems are:

1. Information given directly in the problem.

2. Information you can look up or derive from the wording of the problem. This includes definitions and mathematical equations.

Your collection of tools will constantly increase as we progress through chemistry. Learning chemistry is very much like learning a new language. New words and phrases must be clearly defined in your mind so that you can use them properly. The same can be said for mathematical equations. You will only need a limited number of mathematical equations in this course.

How then do you proceed toward the answer? First you have to know what you are looking for. If you know what the answer must look like, then you are more confident that you can **work through the problem** to get the answer.

Often after working through a set of examples, you will want to know how you are doing. A similar problem that you should work out for yourself appears in a **How are you doing?** box. These problems are a good check of whether or not you are ready to proceed. If you cannot solve this kind of problem, study the preceding example until you understand the thought process involved. Then try again.

Example **1.1** **Determining that a sample is heterogeneous**

Indicate how it is apparent that the following samples are heterogeneous.

(a) a piece of rock
(b) ice cubes floating in a glass of iced tea
(c) fog

Thinking Through the Problem "Heterogeneous" means that different parts of a sample have different properties. So in each case, we need to think of a way to tell that some parts of the sample are different from others.

Answer

(a) A rock contains very fine variations from one point to the other; these may be differences in color or the presence of air pockets in the rock. Therefore, the rock is heterogeneous.

(b) Ice is composed of a single substance (water in the solid state), but the liquid is a solution (water and sugar plus material from the tea itself). Therefore, the sample is heterogeneous.

(c) In a fog there appears to be one continuous cloud. But close examination indicates that there are very fine droplets of water in the air. So this sample is heterogeneous.

Indicate whether the following samples are homogeneous or heterogeneous.

(a) soil   (b) filtered water   (c) automobile exhaust

PROBLEMS

1. Define the terms "homogeneous" and "heterogeneous."

2. What does a chemist mean by the word "mixture"? Give several examples of mixtures.

3. In this section we classified air as a mixture. List some substances in a sample of air whose presence would result in the samples being homogeneous.

4. What is the chemical meaning of "pure substance"? Give several examples of pure substances.

5. Give one way in which you can tell the difference between a mixture and a pure substance.

6. We can use our everyday knowledge to make some reasonably accurate decisions about whether a particular substance is homogeneous or heterogeneous. Do this for each of the following substances that can be found in most kitchens. Then decide whether the substance is a mixture or a pure substance.

(a) white cake mix
(b) table salt
(c) a gelatin dessert
(d) an apple
(e) mayonnaise

7. Define: (a) element (b) compound. Give an example of a substance that fits each definition.

1.2 Chemical Elements and the Periodic Table

SECTION GOALS

✔ Identify and name the first twenty elements on the periodic table.

✔ Identify the different elements in a chemical formula.

✔ Find an element in the periodic table when given its symbol or name.

Symbols of the Elements

All of the known elements can be organized into a table called the **periodic table.** Each element is represented in the table by a symbol, which is an abbreviation of its name. Figure 1-6 shows the periodic table.

Note that the periodic table is the ultimate "key idea" for chemists.

Many symbols of the elements begin with either the first letter (H = hydrogen), the first two letters (He = helium), or the first and third letter (Mg = magnesium) of the element's name in English. A few of the elements have symbols based on other languages (for example, Na = natrium, from Latin). Four elements have three

H																	He
Li	Be											B	C	N	O	F	Ne
Na	Mg											Al	Si	P	S	Cl	Ar
K	Ca	Sc	Ti	V	Cr	Mn	Fe	Co	Ni	Cu	Zn	Ga	Ge	As	Se	Br	Kr
Rb	Sr	Y	Zr	Nb	Mo	Tc	Ru	Rh	Pd	Ag	Cd	In	Sn	Sb	Te	I	Xe
Cs	Ba	Lu	Hf	Ta	W	Re	Os	Ir	Pt	Au	Hg	Tl	Pb	Bi	Po	At	Rn
Fr	Ra	Lr	Rf	Db	Sg	Bh	Hs	Mt	Uun	Uuu	Uub		Uuq				

	La	Ce	Pr	Nd	Pm	Sm	Eu	Gd	Tb	Dy	Ho	Er	Tm	Yb
	Ac	Th	Pa	U	Np	Pu	Am	Cm	Bk	Cf	Es	Fm	Md	No

Figure 1-6 ▲ Periodic table of the elements.

letters, because their names are actually Latin words for the atomic number. Figure 1-7 shows some common elements.

There are rules for writing the symbols of the elements. Only the first letter in an element's symbol is capitalized. For example, the symbol for helium is He and the symbol for magnesium is Mg. When a symbol contains a second or third letter, those letters are written in lowercase. A symbol for an element always contains a *single* uppercase letter.

Figure 1-7 ▶ Some common chemical elements: liquid bromine, liquid mercury, solid iodine, cadmium, red phosphorous, and copper. (Ken Karp.)

REMEMBER

See this with your **Web Animator** ▶

A compound, on the other hand, consists of more than one element, and therefore is represented by combining more than one element's symbol into what is called a **chemical formula.** The chemical formula of a compound includes the symbols for each of the elements in the compound and shows the number of atoms of each element that occurs in the basic arrangement of that compound. For example, H_2O is the chemical formula of the compound called water, which contains two elements, hydrogen and oxygen. The number 2 indicates that there are two hydrogen atoms present for every one atom of oxygen, as we saw in Figure 1-5.

The names and symbols of the first twenty elements, along with bromine (Br) and iodine (I), will be used extensively in this text and should be memorized. These are contained in Table 1-1. This chapter also contains practice problems with some other elements so that you gain familiarity with them, too. A complete alphabetical listing of all the known elements and their symbols is found inside the cover of this book.

Table 1-1 Symbols and Names of the First Twenty Elements

Atomic Number	Atomic Symbol	Name	Atomic Number	Atomic Symbol	Name
1	H	Hydrogen	11	Na	Sodium
2	He	Helium	12	Mg	Magnesium
3	Li	Lithium	13	Al	Aluminum
4	Be	Beryllium	14	Si	Silicon
5	B	Boron	15	P	Phosphorus
6	C	Carbon	16	S	Sulfur
7	N	Nitrogen	17	Cl	Chlorine
8	O	Oxygen	18	Ar	Argon
9	F	Fluorine	19	K	Potassium
10	Ne	Neon	20	Ca	Calcium

Example 1.2 **Identifying symbols of elements**

✔ **Meeting the Goals**

The first twenty elements and their symbols are contained in Table 1-1. When symbols contain two letters, only the first letter is capitalized. Because every element begins with an uppercase letter, it is a simple task to find all the different elements contained in a chemical formula.

How many different symbols are there in each of the following formulas?

(a) HBr (d) Es_2O_3
(b) $(NH_4)_2Cr_2O_7$ (e) SI_4
(c) OsO_4 (f) $Ir(NO_3)_3$

Thinking Through the Problem We know that each symbol begins with an uppercase letter, so we will scan each formula for uppercase letters to find the different symbols for the elements.

Answer

(a) HBr: H and Br = 2 symbols = two different elements
(b) $(NH_4)_2Cr_2O_7$: N, H, Cr, and O = 4 symbols = four different elements

PRACTICAL A — How does chemistry let us look at the world in an organized way?

Concrete is used in large and small objects, such as a concrete block, a high-rise building, a dam, or a statue. Regardless of how it is used, it is always formed by the careful mixture of materials.

The interesting examples of art, architecture, and industry in Figure 1-8 are examples of chemistry. Part of the enticement of learning a new subject is that you begin to look at the world through different eyes.

Even a mundane material like concrete becomes interesting as it is examined more closely through the eyes of a practicing chemist. Right now, the primary difference between a practicing chemist and you is how you each see the world. Try using what you've just learned to see concrete the way a chemist would.

Concrete is made by mixing cement, water, air, and both coarse and fine filler materials such as

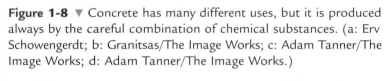

Figure 1-8 ▼ Concrete has many different uses, but it is produced always by the careful combination of chemical substances. (a: Erv Schowengerdt; b: Granitsas/The Image Works; c: Adam Tanner/The Image Works; d: Adam Tanner/The Image Works.)

(c) OsO_4: Os and O = 2 symbols = two different elements
(d) Es_2O_3: Es and O = 2 symbols = two different elements
(e) SI_4: S and I = 2 symbols = two different elements
(f) $Ir(NO_3)_3$ = Ir, N, and O = 3 symbols = three different elements

How are you doing? 1.2

How many symbols are there in each of the following formulas?
(a) HCN   (b) K_2SO_4 (c) SO_3 (d) $Mg(ClO_4)_2$

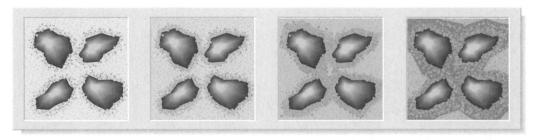

Figure 1-9 ▲ The setting of concrete involves the growth of cemented regions around filler particles. It takes the right materials, and it takes time. Gaps may remain, even in a strong concrete.

pebbles and sand. But simply mixing these ingredients together is not enough to make a good concrete. What binds the mixture together? What makes the concrete block so hard? The answer to both of these questions involves chemistry. The dry cement and the water undergo a chemical reaction and form strong connections between the pebbles and sand.

When dry cement mix is first added to water, the powdered mix spreads through the water, forming a solution. As more mix is added, the solution becomes paste-like because not all of the cement will dissolve. Gradually, the water reacts with the cement to form hydrated (that is, water-containing) solids that build up on the surface of the filler particles. At this point we notice that the concrete is "drying," or solidifying. With the passage of more time, the hydrated solids eventually form bridges between the original grains of mix and the concrete is "set." As shown in Figure 1-9, the cement hardens as a network of hydrated solids builds.

A chemist realizes that concrete is the result of chemical reactions taking place within a mixture. These reactions occur in the mixture of cement and water, and result in the formation of solid compounds that hold together the other components of the mixture. Chemistry gives us an organized way of looking at concrete, and at the rest of the world as well.

Think About It . . .

There is a striking parallel between this process and the process of acquiring new knowledge. At first we have small, isolated facts—definitions, pictures, formulas, and rules. As we add more facts to our "mix," we begin to see relationships between some of the pieces until, gradually, bridges are formed between chemical concepts. It is the relationships in chemistry that are important. Look for them. Use them.

Finding Elements in the Periodic Table

The first thing we notice about the periodic table, shown again in Figure 1-10, is its shape. *There are seven horizontal rows, called* **periods,** *and eighteen vertical columns, called* **groups.** Two columns on the left (marked I and II) and six on the right (marked III through VIII) bracket a longer section in the center. Part of the center is also taken up with two sets of fourteen elements, shown below the table. These elements actually lie between elements in the main body of the table—think of them as being located in the narrow areas indicated by the double lines. We can find any element if we know both its group and its period number. For example, the element with group number I and period number 1 is H, hydrogen.

I												III	IV	V	VI	VII	VIII
1																	18
H 1	**II** 2											13	14	15	16	17	**He** 2
Li 3	**Be** 4											**B** 5	**C** 6	**N** 7	**O** 8	**F** 9	**Ne** 10
Na 11	**Mg** 12	3	4	5	6	7	8	9	10	11	12	**Al** 13	**Si** 14	**P** 15	**S** 16	**Cl** 17	**Ar** 18
K 19	**Ca** 20	**Sc** 21	**Ti** 22	**V** 23	**Cr** 24	**Mn** 25	**Fe** 26	**Co** 27	**Ni** 28	**Cu** 29	**Zn** 30	**Ga** 31	**Ge** 32	**As** 33	**Se** 34	**Br** 35	**Kr** 36
Rb 37	**Sr** 38	**Y** 39	**Zr** 40	**Nb** 41	**Mo** 42	**Tc** 43	**Ru** 44	**Rh** 45	**Pd** 46	**Ag** 47	**Cd** 48	**In** 49	**Sn** 50	**Sb** 51	**Te** 52	**I** 53	**Xe** 54
Cs 55	**Ba** 56	**Lu** 71	**Hf** 72	**Ta** 73	**W** 74	**Re** 75	**Os** 76	**Ir** 77	**Pt** 78	**Au** 79	**Hg** 80	**Tl** 81	**Pb** 82	**Bi** 83	**Po** 84	**At** 85	**Rn** 86
Fr 87	**Ra** 88	**Lr** 103	**Rf** 104	**Db** 105	**Sg** 106	**Bh** 107	**Hs** 108	**Mt** 109	**Uun** 110	**Uuu** 111	**Uub** 112		**Uuq** 114				

La 57	**Ce** 58	**Pr** 59	**Nd** 60	**Pm** 61	**Sm** 62	**Eu** 63	**Gd** 64	**Tb** 65	**Dy** 66	**Ho** 67	**Er** 68	**Tm** 69	**Yb** 70
Ac 89	**Th** 90	**Pa** 91	**U** 92	**Np** 93	**Pu** 94	**Am** 95	**Cm** 96	**Bk** 97	**Cf** 98	**Es** 99	**Fm** 100	**Md** 101	**No** 102

See this with your Web Animator ▲

Figure 1-10 ▲ Periodic table of the elements with group numbers, symbols, and atomic numbers. The elements in the shaded squares are among the most important elements. The textbook website has a periodic table feature that lets you investigate the properties of the elements.

Notice in Figure 1-10 that there is a number under each symbol. The elements are numbered in order, beginning with hydrogen and pro-

Example 1.3 **Distinguishing between periods and groups**

The symbol C stands for the element carbon. Find carbon in the periodic table. What period contains carbon? Which group? ◄► (1.3)

Thinking Through the Problem This exercise helps you to distinguish between period number and group number. Periods are horizontal rows; groups are vertical columns.

Answer Carbon is in the second period and the fourth group. Another way to phrase this is that carbon is a second period element in Group IV.

Example 1.4 **Determining the period and group of an element**

What is the symbol and name of the element found in the first group of period 5 in the periodic table?

Thinking Through the Problem The first group is labeled I. What is the element in the first column that begins the fifth row?

Answer The symbol of this element is Rb. Its name is rubidium.

PRACTICAL B   What are the basic building blocks of matter?

Did you know that concrete becomes stronger with age? For example, five-year-old concrete may be as much as nine times stronger than it was on the day it was poured. What appears to be an unchanging material is actually a beehive of activity, as many of the substances present in the concrete undergo changes to more stable forms. As the concrete ages, chemical reactions continue to occur within it, producing new compounds that build on and around each other. The concrete becomes stronger as this network of interlacing crystals continues to grow and develop.

We can begin to think about the compounds in concrete by first examining their chemical formulas. The original mix begins with three compounds: $CaCO_3$, SiO_2, and $Ca_3Al_2O_6$. How many different elements are represented in the formulas of these compounds? What part of the formula do we look at to determine this answer?

One of the compounds that forms has a formula written as $(CaO)_3 \cdot Al_2O_3$. The raised dot in this formula signifies a weak connection or attraction between the substance appearing before the dot and the substance appearing after the dot. Compare this formula, $(CaO)_3 \cdot Al_2O_3$, with the formula, $Ca_3Al_2O_6$. What is the same about these two formulas? What is different? Identify and count the elements present in each formula.

A compound that forms within the concrete during the first month of hardening is $(CaO)_3 \cdot SiO_2$. Over longer periods of time, the structure of this compound changes to $CaSiO_3 \cdot H_2O$. Identify and count the elements present in each of these formulas.

✔ Meeting the Goals

A row in the periodic table is called a period. There are seven periods. The major *groups* refer to the first two and the last six columns on the periodic table. Every element in these eight columns appears at the intersection of a period and a group. Thus, you can locate an element if you know both its period and its group number.

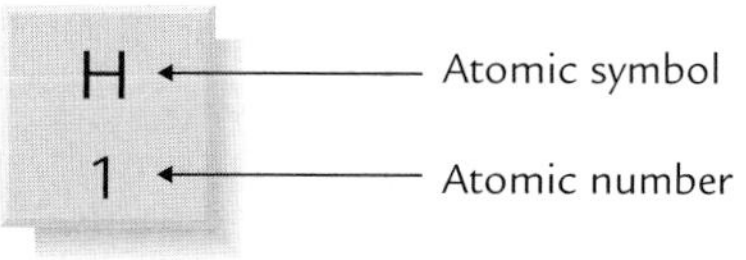

Figure 1-11 ▶ The symbol of the element and its number are included in the periodic table.

How are you doing?   1.3

Find the group number for: At, As, and Ar. What is the name of each of these elements? In which period will each of these be found?

ceeding across each row in turn. Figure 1-11 shows the entry for hydrogen on the periodic table.

The number of the element is called its **atomic number.** We can identify any element given either its symbol or its atomic number. You will learn more about the chemical meaning of an atomic number in the following sections.

H ◀——— Atomic symbol

1 ◀——— Atomic number

PROBLEMS

8. How many symbols are there in each of the following formulas?

(a) HAt   (b) H_2 (c) HI

9. Determine whether the following represent compounds or elements.

(a) P_4 (b) P_4O_{10} (c) O_3 (d) H_2O_2

10. Give the name of each of the following elements.

(a) Mg   (b) Na   (c) Si   (d) B   (e) K

11. Give the symbol for:

(a) beryllium   (b) oxygen   (c) helium

12. Determine the number of symbols present in the following compounds.

(a) $(NH_4)_2HPO_4$
(b) $Ca(C_2H_3O_2)_2$
(c) $Al(ClO_4)_3$

13. Give the name of the elements you identified in the previous problem.

1.3 Elements, Atoms, and Atomic Numbers

SECTION GOALS

✔ Know the relative mass and relative charge for each of the three subatomic particles.

✔ Know where protons, neutrons, and electrons are located in the atom.

✔ Define the terms "atomic number" and "mass number."

✔ Know how to write and interpret an atomic symbol.

✔ Determine the number of protons, neutrons, and electrons in a neutral atom when given its atomic symbol.

Inside the Atom

You have probably seen drawings of atoms and have heard of atomic bombs, atomic energy, and nuclear waste. But what do you really know about atoms? What do you think atoms are like? Could you draw a picture of an atom? Some ancient Greek and Roman philosophers believed that atoms were the smallest bits of matter that could exist. In fact, the name *atom* comes from the Greek word *atomos* which means "not divisible." To the Greeks, there was nothing smaller than an atom.

Our understanding of the atom has changed over the years into what we call "modern" atomic theory. The originator of modern atomic theory is generally considered to be the nineteenth-century English scientist, John Dalton. Dalton thought that atoms were small, hard spheres that could combine with one another to somehow form compounds. Dalton's theory of the atom evolved slowly as new experiments produced data that his theory of the atom could not explain.

The heart of modern atomic theory lies in the idea that there are certain building blocks of all matter, called atoms; that atoms of the same element are chemically the same; and that atoms of different elements are different from one another. Only a limited number of elements are known.

Although scientists believed for many years that atoms were truly the smallest particles that existed, today we know that the atom has "insides"! We call the particles contained inside the atom *subatomic particles.* How many subatomic particles have you heard about? Scientists know of many subatomic particles with names such as quarks, muons, mesons, etc. Most of these particles exist free for a very short time, usually less than one second. Although these particles are interesting, we will not discuss them in this course. Chemists typically need to study just three subatomic particles: the **proton,** the **neutron,** and the **electron.** The masses and charges of these three subatomic particles account for almost all the properties we will discuss in a general chemistry course.

The discovery of the existence of subatomic particles was extremely exciting and challenging for the scientific world. Ironically, the first subatomic particle to be discovered was the smallest—the electron. The electron was found to have a negative charge, which posed a problem for the scientific community because they knew that the atom was electrically neutral. How could a neutral atom contain negatively charged particles? The most plausible explanation was that the atom must also contain positively charged particles. Scientists proposed the existence of a particle that possessed a positive charge of equal magnitude to the charge of the electron. Years later, this particle, called the proton, was proven to exist. But there was yet another problem to solve. Experiments showed that the mass of an atom was larger than the sum of the masses of that atom's electrons and protons. To account for that "extra" mass, there had to be yet *another* undiscovered particle in the atom! Scientists postulated the existence of a third subatomic particle that had to be neutral and possess a mass similar to the mass of the proton. They named this particle the neutron. When the neutron was discovered in 1932, it did, in fact, possess properties very similar to those proposed by scientists.

The masses of these three subatomic particles have been experimentally determined and are listed in the second column of Table 1-2. Protons and neutrons have almost identical masses but the mass of an electron is approximately 2,000 times smaller. According to the table, the mass of the proton is 0.0000000000000000000000001672623 grams. This is a very small mass indeed. Such small numbers are usually written in what is called scientific notation, where the number is written as a larger number multiplied by a negative power of ten (here, 1.672623×10^{-24}). Scientific notation is discussed in more detail in Chapter 7.

Column three of the table lists the *relative* mass of each subatomic particle. The word *relative* means *compared to* something else. In this case, the mass of each subatomic particle is compared to the mass of a proton. The proton is a very important particle because the number of protons is linked to the identity of the element. The relative masses listed in the last column are obtained by dividing each of the experimental masses in the middle column by the mass of the proton. Thus, we say these masses are "relative" to the mass of the proton.

Table 1-3 summarizes our discussion of the major subatomic particles. In this table, protons are symbolized by p^+, and electrons by e^-, where the superscripted positive or negative signs remind us of the particle's electrical charge. The charges of the electron and the proton are experimentally measured quantities, but these charges are also very small numbers. It is conventional to use *relative* charge values of one positive unit for a proton and one negative unit for an

✔ **Meeting the Goals**

The relative masses for the proton, neutron, and electron are 1, 1, and 0, respectively. The relative charges for the proton, neutron, and electron are +1, 0, and −1, respectively.

Table 1-2 Masses of Subatomic Particles

Subatomic Particle	Mass (g)	Relative Mass
Proton	1.672623×10^{-24}	$1.672623 \times 10^{-24} \div 1.672623 \times 10^{-24} = 1$
Neutron	1.674929×10^{-24}	$1.674929 \times 10^{-24} \div 1.672623 \times 10^{-24} = 1.001379$
Electron	9.109390×10^{-28}	$9.109390 \times 10^{-28} \div 1.672623 \times 10^{-24} = 0.0005446170$

Table 1-3 Relative Masses and Charges
of Subatomic Particles

Particle	Relative Mass	Relative Charge
Proton, p^+	1	$+1$
Neutron, n	1	0
Electron, e^-	0	-1

Figure 1-12 ▲ Size comparison of the nucleus and the atom. (Walter Schmid/Tony Stone Worldwide.)

✔ **Meeting the Goals**

The protons and neutrons are located in the nucleus of the atom. All the electrons are located outside the nucleus in a large, diffuse cloud. The outermost electrons are called *valence electrons*.

electron. The relative mass and charge of these three subatomic particles are used frequently in general chemistry discussions and should be memorized.

We know from experiment that all the neutrons and protons are packed together in a small, dense **nucleus** at the center of the atom. If an atom were the size of the large baseball stadium shown in Figure 1-12, then the nucleus would be the size of a golf ball located in the center of the stadium. The distribution and location of subatomic particles within the atom are also very important. This is shown in Figure 1-13, which shows the nucleus greatly magnified relative to the electrons. Neutrons and protons never leave the nucleus during an ordinary chemical reaction. Thus, they can be considered a fixed part of the atom. Because protons are positively charged and neutrons are neutral, the nucleus itself is positively charged. The total charge on the nucleus is the sum of the charges of all the protons contained within that atom. Also, because electrons have a relative mass of zero, the mass of the atom is concentrated primarily in its nucleus. Thus, the mass of the nucleus is a good estimate for the mass of the whole atom.

In contrast to the small, compact nucleus, the electrons are present in a large, diffuse "cloud" outside the nucleus. Some may be considered very mobile and on the outer edge of the atom. These are called **valence electrons.** They easily move away from the atom to form chemical bonds and ions when atoms interact in chemical systems. (Valence electrons are discussed further in Chapter 2.) The other

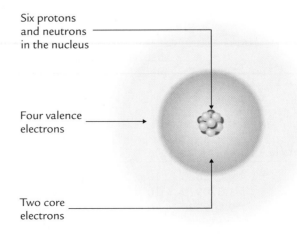

Six protons
and neutrons
in the nucleus

Four valence
electrons

Two core
electrons

Figure 1-13 ▶ Simple diagram of the nuclear and electronic structure of an atom of carbon. Electronic structure is discussed in more detail in Chapter 2.

electrons are considered part of the atom's **core electrons.** The size of an atom is determined by the extent of its electron cloud.

Atomic Numbers and Mass Numbers for Elements

KEY IDEA 1.5

The chemical elements differ from each other by the number of protons contained in their atoms. *The* **atomic number** *of an element is defined as the number of protons contained in one atom of that element.* In a very real sense, the atomic number of an element serves as an identity number.

Let's consider how the atomic number of an element tells us something unique about that element. Hydrogen has one proton; its atomic number is one. We locate H on the periodic table by finding element 1. Because the nucleus of an atom does not change during an ordinary chemical reaction, the elemental identity of an atom remains the same before, during, and after an ordinary chemical reaction.

Example 1.5 **Determining an element from its atomic number**

What is the name of the element that has atomic number 52?

Thinking Through the Problem We know the atomic number corresponds to the numbering on the periodic table. Looking at the periodic table, we see that the element with an atomic number of 52 has a symbol Te. This is the symbol for tellurium.

Answer Tellurium is the element with the atomic number 52.

Example 1.6 **Determining an element from the number of protons**

What is the name of the element that contains 27 protons in its nucleus?

Thinking Through the Problem We know that the atomic number equals the number of protons; number of protons = atomic number = 27. The element with an atomic number of 27 on the periodic table has the symbol Co.

Answer Cobalt is the element with 27 protons.

How are you doing? 1.4

Fun with atomic numbers! Match the descriptions in column A to the word formed from the symbols of the elements whose atomic numbers are given in column B.

Column A	Column B
(1) What not to do before exams	(a) #46
(2) A famous Donny	(b) #20 + #10
(3) Nice to have printed on a bill	(c) #24 + #95
(4) A useful aid in walking	(d) #76 + #42 + #60

✔ Meeting the Goals

The atomic number of an element is the number of protons contained in the nucleus of an atom of that element. The mass number of an element is the number of protons and neutrons contained in the nucleus of an atom of that element.

The mass of an atom is the sum of the masses of its protons, neutrons, and electrons. Since the relative mass of the electron is zero, and the relative mass of both the proton and the neutron is one (see Table 1-3), we can get a close approximation of the mass of an atom simply by counting the number of neutrons and protons it contains. In doing so, we ignore the very real but relatively small mass contributed by the electrons present in the atom. (See Table 1-2 for the actual mass of an electron.) The sum of the number of protons and neutrons contained in the nucleus of an atom is a very useful number called the **mass number.** For example, the nucleus of the atom shown in Figure 1-13 contains six protons and six neutrons. This carbon atom has an atomic number of 6 and a mass number of 12.

Consider the information contained in Table 1-4. Helium, He, has an atomic number two. This means that a helium nucleus contains two protons. The most common form of helium also contains two neutrons. Therefore, helium has a mass number of four. (The determination of the mass number is shown in the shaded area of the table.) Lithium (Li) has an atomic number three. A typical lithium nucleus contains three protons and four neutrons, giving it a mass number of seven. Beryllium (Be), atomic number four, has four protons and most commonly five neutrons and a mass number of nine.

Table 1-4 **Atomic Numbers and Mass Numbers of Common Types of the First Four Elements**

Element	H	He	Li	Be
Atomic number	1	2	3	4
# Protons, p^+	$1\,p^+$	$2\,p^+$	$3\,p^+$	$4\,p^+$
# Neutrons, n	$0\,n$	$2\,n$	$4\,n$	$5\,n$
Mass number ($p^+ + n$)	1	4	7	9

Atomic Symbols of Elements

✔ Meeting the Goals

An atomic symbol gives the symbol of the element preceded by its mass number (superscripted) and its atomic number (subscripted). An atomic symbol has the form

$$\begin{matrix} \text{mass number} \\ \text{atomic number} \end{matrix} \text{X,}$$

where X is the symbol of the element.

Atomic symbols are a shorthand summary of the element's symbol, atomic number, and mass number. The atomic number is written as a subscript number and the mass number is written as a superscript number, both preceding the symbol of the element. An atomic symbol has the form shown below:

$$\begin{matrix} \text{mass number} \\ \text{atomic number} \end{matrix} \text{X}$$

where X is the symbol of the element.

For the elements in Table 1-4, the atomic symbols are ^1_1H, ^4_2He, ^7_3Li, and ^9_4Be. The format of an atomic symbol is very useful. It shows, at a glance, both the number of protons *and* the number of neutrons contained in the nucleus of an atom. The number of neutrons = mass number − atomic number.

$$
\begin{array}{rl}
\text{mass number} & \#p^+ + \#n \\
- \quad \text{atomic number} & \#p^+ \\
\hline
\text{number of neutrons} & \#n
\end{array}
$$

The following definitions show the relationships that exist among the atomic number, the mass number, the number of protons, the number of neutrons, and the number of electrons in an atom.

atomic number = number of protons

mass number = number of protons + number of neutrons

number of protons = number of electrons (*for a neutral atom*)

These relationships among the components of an atom allow us to determine a lot about an atom from very little information. Taking a small amount of information and expanding upon it using logical rules is an important skill in science. It often seems quite difficult, but it is actually something you may do every day. For example, say you are told that behind a door is a large mammal with stripes. What can you say about the animal? In this case, you would have to *speculate* about its identity. But then if you were told "it is in the cat family," you would be certain that it was a tiger. And then you could say a lot of other things, such as the color of its stripes and even that it was probably not a good idea to open the door!

We can use the same process in determining the composition of an atom. Let's say we know that a particular neutral atom has a mass number of 24. This, as with the stripes, tells us something, but not enough to know what the atom is. But if we are also told it has twelve electrons, then we can start to reason systematically about the identity of the atom.

1. Twelve electrons in a neutral atom → twelve protons in the nucleus

2. Twelve protons in the nucleus → atomic number 12

3. Atomic number 12 → element is Mg

4. Mass number − atomic number = number of neutrons → 24 − 12 = 12 neutrons

In this way, the two statements, "12 electrons and a mass number 24," are linked to four other statements.

The following example shows that it is possible to do the same thing with different starting information.

Example **1.7**   **Determining an element from the number of electrons**

What is the name of the element whose atoms contain 17 electrons?

Thinking Through the Problem An atom is neutral, and therefore must have an equal number of protons and electrons. In a neutral atom, the number of electrons = the number of protons = atomic number.

Answer The element with atomic number 17 is Cl, chlorine.

Example `1.8`   Determining the number of electrons in an atom

How many electrons are there in an atom with the atomic symbol, $^{27}_{13}Al$?

Thinking Through the Problem Vocabulary is very important in solving these questions. For this question, we need the above relationships: the subscript number 13 represents the atomic number of aluminum. The atomic number is the number of protons. In a neutral atom, the number of protons equals the number of electrons.

Answer This atom has 13 electrons.

Example `1.9`   Determining the composition of atoms

Complete the following table by filling in the missing numbers.

Element	Number of Protons	Number of Neutrons	Number of Electrons	Atomic Number	Mass Number
Potassium			19		39
	17	19			
Sulfur		17			
	13	14			

Thinking Through the Problem We must find the missing number of electrons, protons, and neutrons in this table. We must also find the atomic number and mass number when missing.

The necessary equations are given above example 1.8. It is not necessary—in fact, it is usually impossible—to fill in the boxes beginning on the left and proceeding straight across each row. The order in which the blanks are filled in depends on the information given.

For each row in the table, use the given information to determine the missing values. For example, in the first line, you are given a neutral atom with 19 electrons. The easiest way to proceed is to remember that the number of electrons must equal the number of protons in a neutral atom (fill in the number of protons) and then that the number of protons is the atomic number (fill in the atomic number). For a neutral atom, these three numbers will be the same. The remaining box makes use of the fact that the mass number (given) is the sum of the number of protons plus neutrons.

Answer

Element	Number of Protons	Number of Neutrons	Number of Electrons	Atomic Number	Mass Number
Potassium	**1.** Neutral atom, $\#e^- = \#p^+ = 19$	**3.** Mass # − atomic # = 39 − − 19 = 20	19	**2.** # Protons = atomic # = 19	39

PRACTICAL C How do we recognize chemical elements?

We owe much of our understanding of a special group of elements, called radioactive elements, to the work of Marie Curie, her daughter Irene, and their husbands. At a time in history when women had few rights, both Marie and Irene Curie contributed greatly to the revolutionary changes in scientific thought. Marie and Pierre Curie performed pioneering experiments that resulted in the discovery of radioactive elements. Their dramatic experiments have had a tremendous impact that reaches beyond the realm of scientific thinking. A photograph of the Curie family is shown in Figure 1-14.

The discovery and isolation of radioactive elements and the development of nuclear energy have a lot to do with an understanding of elements. Radioactivity is associated with atoms that undergo non-chemical changes and become atoms of a different element.

They studied naturally-occuring radioactivity in uranium salts. What is the atomic number of uranium? How many electrons are there in a neutral atom of uranium?

Two other naturally occurring radioactive elements studied by the Curies are polonium and radium. What is the atomic number of polonium? What is the atomic number of radium?

Irene Curie and her husband Frederic Joliet later pioneered the study of artificial elements. Artificial radioactive elements can be made by bombarding atoms with high energy particles. For example, $^{10}_5B$ can be bombarded to form an atom of $^{14}_7N$ that is radioactive. We can discover the identity of the high-energy particle used in the bombardment of boron if we compare the atomic number of boron with that of the resulting atom. What is the difference in the

Figure 1-14 ▲ Marie and Pierre Curie with daughter Irene. (Mary Evans Picture Library/Photo Researchers.)

two atomic numbers? What element has this atomic number? This element is the particle used in the bombardment process.

Suppose we use the same high energy particle to bombard $^{27}_{13}Al$ atoms. What element might result from the bombardment if the same type of process occurs as it did with the boron?

At this point you cannot make predictions about radioactivity. That will be possible when we revisit these elements in Chapter 12.

2. Element from periodic table = Cl	17	19	**1.** Neutral atom, $\#p^+ =$ $\#e^- = 17$	**3.** Atomic number = $\#p^+ = 17$	**4.** Mass # = #p + #n = 17 + 19 = 36
Sulfur	**2.** Protons = atomic # = 16	17	**3.** Neutral atom, $\#p^+ =$ $\#e^- = 16$	**1.** Atomic # from periodic table = 16	**4.** Mass # = #p + #n =16 + 17 = 33
2. Element from periodic table = Al	13	14	**3.** Neutral atom, $\#p^+ =$ $\#e^- = 13$	**1.** Atomic number = $\#p^+ =$ 13	**4.** Mass # = #p + #n = 13 + 14 = 27

How are you doing?   **1.5**

Supply the missing information for the following neutral atoms.

Element	Number of Protons	Number of Neutrons	Number of Electrons	Atomic Number	Mass Number	Atomic Symbol
Oxygen		8				$^{16}_{8}O$
Copper					65	Cu
	11	12				
			13		27	

PROBLEMS

14. Concept question: Why is the number of protons equal to the atomic number for an element? Why not use the number of electrons, or the number of neutrons, or the number of neutrons + protons? Express your answer by thinking through the effect of the charge in an atom's nucleus on the charge of an atom.

15. Write the number of protons, the symbol, and the name of the elements with the following atomic numbers:

Atomic Number	Number of Protons	Symbol	Name of Element
14			
94			

16. Match the descriptions in column A to the word formed by combining the symbols of the elements whose atomic numbers are given in column B.

Column A	Column B
(1) A parakeet's home	(a) 71 + 52
(2) Spoils a picnic	(b) 20 + 32
(3) A misbehaving child	(c) 16 + 8 + 39
(4) A musical instrument	(d) 88 + 49
(5) A type of bean	(e) 35 + 85

17. Complete the table below by filling in the missing information for these neutral atoms.

Element	Symbol	Number of Protons	Number of Neutrons	Number of Electrons	Atomic Number	Mass Number
Lithium				3		7
		48	65	48		
				53	53	127
	W		111	74		

18. Determine the atomic symbol for a neutral atom from the information given in this table.

Element	Number of Protons	Number of Neutrons	Number of Electrons	Atomic Number	Mass Number	Atomic Symbol
Potassium		21				
Oxygen					17	
	78	117				
			50		119	

19. Web Animator: Use the periodic table feature on the website to answer this and the following question. Find the symbols for the stable atoms found for the following elements. Write the full symbol of each atom, and indicate the number of protons, and electrons for that element.

(a) B   (c) F   (e) Sb

(b) O   (d) Mg   (f) Hg

20. Web Animator: It is possible for two different elements to have atoms with the same mass number. For the following atoms, find a stable atom that has the same mass number. Comment on whether we can tell the identity of an atom from *just* the mass number.

(a) $^{3}_{2}\text{He}$ (c) $^{40}_{20}\text{Ca}$

(b) $^{40}_{18}\text{Ar}$ (d) $^{70}_{30}\text{Zn}$

1.4 Physical Properties of Elements and Compounds

SECTION GOALS

✔ Understand the word *property*. Distinguish between a physical property and a chemical property.

✔ Identify several physical properties.

✔ Know what is meant by the physical states of matter.

✔ Know the names of the processes that convert matter from one physical state to another.

What Are Physical Properties?

KEY IDEA 1.6

The term **property** is very important in chemistry and in all the sciences. *A property is a fixed characteristic of a substance that can be used to distinguish that substance from another substance.* In chemistry, we use the term *property* to refer to the characteristics of a sample of the substance large enough to be handled in the laboratory. Chemists generally distinguish between two kinds of property, physical and chemical.

A **physical property** tells us about the physical characteristics and appearance of a substance. Examples of physical properties are color, odor, hardness, and whether we can easily change the shape of the substance by hammering it into sheets (malleability) or drawing it out into thin wire (ductility). The physical state of the substance—whether it is a solid, liquid, or gas under ordinary conditions—as well as the temperatures at which the substance will **boil** (normal boiling point) or **freeze** (normal freezing point) are also physical properties. Physical properties can vary greatly from substance to substance and can be determined without destroying or changing the identity of the substance. For example, the physical properties of water include the fact that it is a colorless, odorless liquid at room temperature, it boils when heated at 100 °C, and freezes when cooled at 0 °C.

A **chemical property** tells us whether a substance is transformed to a new substance when it is combined with other chemical substances or exposed to heat or light. We will talk more about chemical properties when we study chemical reactions in Chapter 4. For now, we will concentrate on the physical properties of elements and compounds.

✔**Meeting the Goals**

A property is a fixed characteristic of a substance that can be used to distinguish one substance from another. A physical property describes the physical characteristics of the substance. Some physical properties are physical state of matter, color, odor, and normal boiling and freezing temperatures. A chemical property tells us how the substance acts when it is combined with other substances.

A substance is said to undergo a **physical change** when it changes without losing its original identity. For example, when a block of gold is hammered into a flat sheet, the substance remains gold. A **chemical change,** however, occurs when a substance combines with another substance to form an entirely new substance with different properties.

Example **1.10** Determining if a property is physical or chemical

A substance floats on water. Is this a physical or a chemical property?

Thinking Through the Problem First we think about what "floating" means. When something floats, does it change into a new substance?

Answer Floating is a physical property. (I am not changed into a different substance when I float in the pool!)

Example **1.11** Determining if a property is physical or chemical

As shown in Figure 1-15, a strip of magnesium metal burns with a white flame. Is "burning" a physical or a chemical property?

Thinking Through the Problem You may not be familiar with what happens when magnesium burns. This may make you feel very uneasy about this question. However, you have had a lot of experience with other things burning and the process is the same. Think about a familiar example (burning paper or trash). Then think about what you have at the end of the burning process. Is it the same substance that you had before the burning started?

Answer Burning is a chemical property. The substances you have at the end of the burning process (usually contained in ashes and gases) are not the same as what you had in the beginning.

Figure 1-15 ▲ The burning of magnesium metal can be seen here. (Richard Megna/ Fundamental Photographs.)

How are you doing? 1.6

Classify the following properties as either chemical or physical properties.

	Chemical property	Physical property
A metal is silvery white in color.		
Titanium metal will not react with water.		

Physical Properties and Physical States

Table 1-5 lists some of the physical properties of the elements carbon, hydrogen, and oxygen, and three compounds that contain these elements. Compare the properties of the elements with the proper-

Table 1-5 Physical Properties of Some Related Substances

Substance	Formula	Appearance	Density	Melting Point	Boiling Point
Carbon (graphite)	C	Black powder	2150 g L^{-1}	Sublimes at 3652 °C	4827 °C
Oxygen	O$_2$	Colorless gas	1.43 g L^{-1}	−218.4 °C	−182.9 °C
Hydrogen	H$_2$	Colorless gas	0.090 g L^{-1}	−259.2 °C	−252.8 °C
Methane	CH$_4$	Colorless gas	0.717 g L^{-1}	−182.6 °C	−161.4 °C
Water	H$_2$O	Colorless liquid	998 g L^{-1}	0 °C	100 °C
Carbon monoxide	CO	Colorless gas	1.25 g L^{-1}	−205 °C	−191.5 °C

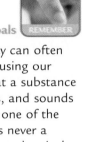

✔ Meeting the Goals

A physical property can often be determined by using our senses. It tells what a substance looks, feels, smells, and sounds like. (Taste is also one of the five senses but it is never a good idea to taste a chemical substance. Many chemical substances are poisonous.) We can determine physical properties without changing the identity of the substance.

ties of the compounds. Do the compounds have the same physical properties as the elements they contain? You will see that they do not.

The most common physical property of a pure substance is its **physical state,** whether it is a solid, a liquid, or a gas. These physical states of matter differ from one another in several basic ways. For now, let us only consider the requirements imposed by volume and by shape. Materials that are solid have both a definite shape and a definite volume. Solid samples do not require a container. A liquid sample, however, must be contained. Liquids have a definite volume, and they take on the shape of whatever container they occupy. Gases have neither definite shape nor definite volume. A sample of gas must be contained on all sides and therefore takes the shape and the volume of its container. Table 1-6 summarizes the shape and volume requirements for the three states of matter.

Often, the temperature determines whether a substance is a solid, a liquid, or a gas. Consider ordinary water, for example. Most of us think of water in its liquid phase, but we are equally familiar with ice, which is the solid phase of water, and steam or water vapor, which is the gaseous phase of water. Liquid water can be **evaporated** to form a gas or frozen to form a solid. There are some solids that can

Table 1-6 Shape and Volume Requirements for the Three Physical States of Matter

Physical State	Shape	Volume	Example
Solid	Definite	Definite	Ice cube
Liquid	*Indefinite*, takes the shape of the container	Definite	Pool of water
Gas	*Indefinite*, takes the shape of the container	*Indefinite*, expands to fill all available space	Water vapor

See this with your **Web Animator** ▼

Figure 1-16 ▶ Liquid water can be evaporated to form a gas, or frozen to form a solid. Solid water, liquid water, and gaseous water are drawn here at the molecular level.

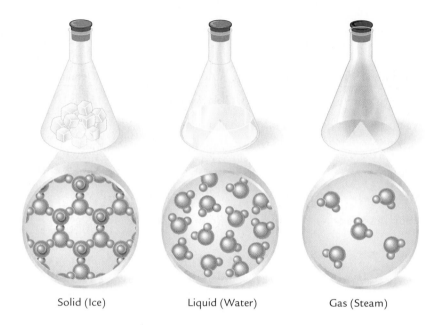

Solid (Ice)   Liquid (Water)   Gas (Steam)

change directly from the solid phase to the gaseous phase in a process called **sublimation.** Changes of state are physical changes because the identity of the substance does not change. Figure 1-16 shows solid water, liquid water, and gaseous water at the molecular level. Only the physical state of the water is changed.

A physical change is reversible. Solids melt to form liquids and liquids freeze to form solids. Other processes responsible for physical phase changes are given in Figure 1-17.

You will notice from Figure 1-17 that we use two words to describe the transitions between two phases. The choice of words is dependent on the direction of the change. We say that solids **melt** to form liquids but that liquids **freeze** to form solids. *Melt* and *freeze* are words that give more information about what is happening. For a

See this with your **Web Animator** ▼

Figure 1-17 ▶ Phase changes for chemical substances.

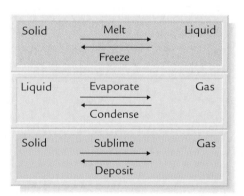

✔ **Meeting the Goals**

The words *melt* and *freeze* indicate a physical change of state between solid and liquid. It is important to realize that the freezing point temperature equals the melting point temperature. The different words are used to indicate the direction of the change. The words *evaporate* and *condense* indicate a physical change of state between liquids and gases. The words *sublime* and *deposit* indicate a physical change of state between a solid and a gas.

pure substance, melting and freezing occur at a single temperature—that is, the temperature at which a substance melts is the same as its freezing temperature. But the temperature is called either the melting point or the freezing point, depending on which transition is occurring. The melting (or freezing) temperature is a physical property for a pure substance. The substance will be a solid for temperatures lower than the melting point temperature and a liquid for all the temperatures between the melting point and the boiling point. Above the boiling point, the substance will be a gas.

Each one of the physical changes shown in Figure 1-17 occurs at a specific temperature. The temperatures at which these changes occur vary from substance to substance. For example, under ordinary conditions water freezes at 0 °C; ice melts at 0 °C. Water boils at 100 °C; steam **condenses** at 100 °C. Bromine has a melting point temperature of −7.2 °C and a boiling point temperature of 58.8 °C. Different substances generally have different boiling and melting points. We can predict the physical state of a substance at a specified temperature if we know its normal boiling point and melting point.

Example 1.12 **Determining if a property is physical or chemical**

Underline all words that indicate a physical property. Circle the words that indicate a physical state. Bracket the words that indicate a chemical property or process.

(a) Nitrogen is a colorless, odorless gas at room temperature.
(b) Nitrogen gas can be converted to liquid nitrogen by lowering its temperature and increasing its external pressure.
(c) Nitrogen gas does not burn in air.
(d) In the Haber process, nitrogen gas reacts with hydrogen gas to form ammonia gas.

Thinking Through the Problem This is mainly a vocabulary question. Each of the processes must be shown to be either a chemical or physical process. We look through the descriptions of the properties and try to tell if the nitrogen is converted into a different chemical substance in each case. As soon as we detect a change in the identity of the chemical substances, then we know we have a chemical property.

Answer (a) This is a *description* of a physical state. *Colorless* and *odorless* denote *physical properties* of nitrogen. (b) This is a *physical change* of state. Nitrogen gas is changed to nitrogen liquid, but it is still nitrogen, the *same* chemical substance before and after the change. (c) This is a *description* of how nitrogen gas *acts;* it is a *chemical property.* (d) Nitrogen reacts to form a new substance, ammonia. This is a *chemical property.*

PRACTICAL D   What are the important differences between different chemical substances?

Chemists must carefully distinguish between the properties of an element, the properties of a compound, and the properties of a compound containing that element. Classify the following statements according to how they apply to one of these three categories. Note that some statements fit more than one category.

Statement	Describes Property of a Compound	Describes Property of an Element	Describes Property of an Element in a Compound
Water, **H₂O**, will extinguish a fire.			
The **aluminum** coating of a pot will chip if tomato sauce is added.			
Bleaching wood pulp with **chlorine dioxide** works as well as elemental chlorine.			
Iron is needed as a component of the molecules that transport oxygen in the blood.			
The **iron** in the ship's hull is starting to rust.			
Since **gold** is used in jewelry, and gold is used in arthritis drugs, we wonder if we can prevent arthritis by wearing lots of jewelry.			

Example 1.13   Determining the state of a substance

Bromine has a melting point temperature of −7.2 °C and a boiling point temperature of 58.8 °C. Will bromine be a solid, liquid, or gas at −0.12 °C?

Thinking Through the Problem The relationship between melting points, boiling points, and physical states of matter can be summarized as:

$$\text{SOLID} \quad \underset{\text{freezing point temp.}}{\overset{\text{melting point temp.}}{\longleftrightarrow}} \quad \text{LIQUID} \quad \underset{\text{condensing point temp.}}{\overset{\text{boiling point temp.}}{\longleftrightarrow}} \quad \text{GAS}$$

Answer The temperature, −0.12 °C, falls between the melting point and the boiling point of bromine. It will be a liquid at this temperature.

How are you doing?   1.7

Copper has a melting point of 1083 °C and a boiling point of 2567 °C. What is the physical state of copper at 750 °C?

PROBLEMS

21. Concept question: What information can be directly read from a chemical formula of a compound? What information can be inferred?

22. Classify the following as a chemical change, a physical change, or neither of these.

(a) Fuel in a camp lantern is ignited with a spark.
(b) A liquid is red.
(c) A piece of iron is attracted by a magnet.
(d) Grapes ferment to wine.
(e) Gold does not burn.
(f) A Ping-Pong ball bounces 5 times.

23. Discussion question: Indicate whether or not the evidence in the following sentences allows us to tell if the substance is an element or a compound.

(a) Heating a substance causes it to melt into a thick liquid. It returns to the solid state when cooled.
(b) A gaseous substance is compressed into a smaller volume and then explodes, liberating oxygen gas.
(c) A metallic-looking solid cannot be separated into simpler substances, but it can be made to burn in the air.

Which sentences describe chemical or physical properties? Does that affect whether we can distinguish a compound from an element?

24. The melting point of sodium is 97 °C and its boiling point is 883 °C. Predict a temperature at which sodium will be a solid. Predict a temperature at which sodium will be a liquid and another temperature at which sodium will be a gas.

25. Ozone is a form of oxygen that has the formula O_3. Ozone is a light blue gas at room temperature and pressure. It has a boiling point of -111.9 °C and a melting point of -193 °C. What is its physical state of matter (solid, liquid, or gas) at the following temperatures and under the same conditions of pressure? (Hint: What happens at the boiling point? What is the physical state of matter at temperatures below the boiling point? Above the boiling point?)

(a) 0 °C
(b) 100 °C
(c) 200 °C
(d) -100 °C
(e) -150 °C

1.5 The Periodic Table: How Chemists Discuss Matter

SECTION GOALS

✔ Identify an element as a representative, transition, or inner transition element.

✔ Classify an element as a metal, nonmetal, or semimetal.

✔ Identify an element as an alkali metal, alkaline earth metal, a chalcogen, a halogen, or a noble gas.

The Periodic Table—An Example of Scientific Thinking

The development of the periodic table of elements is an interesting example of the way in which scientific thought develops. Scientists learn about the world by observing natural phenomena and by doing experiments. Results of experiments are recorded and shared with other scientists who may seek to repeat and verify the experimental results. Similarities and trends are looked for in the results. The accumulation of chemical knowledge is a long process involving many observations and experiments.

The development of the periodic table is a good example of the way in which scientists think. First they performed experiments, carefully recording their observations. All their observations were examined in hopes of discovering similarities or trends in properties. When trends were found, a general statement was formulated and predictions about other behavior were made on the basis of these general statements. Then the predictions were tested with a new series of experiments. This process continued, including more and more systems of matter, and has created the large body of chemical knowledge we have today.

The development of chemistry in the nineteenth century gave rise to important observations about trends among the elements. Scientists used trends to try to put some order into what was known about the elements. For instance, they knew that different elements had different relative masses, so they used the atomic mass as a guide to arrange the elements in a systematic way.

A more convincing and effective organization came about later in the nineteenth century, when a new chemical law emerged. This law, called the **periodic law,** is still used by virtually all chemists today. It states that all of the known elements, though unique, have properties that repeat *periodically,* allowing us to construct the periodic table as shown in Figure 1-10. The periodic table, with its wealth of information about the elements, was formulated because of knowledge gained through chemical experiments. Chemists at that time did not understand the behavior of elements. That occurred later, after physicists made discoveries about the nature of the atom.

Regions in the Periodic Table

Chemists identify regions of the periodic table that resulted when the elements were grouped according to similar properties. For example, elements contained in the first two and in the last six columns of the periodic table are called the **representative elements** because they are the most important elements represented in living things. Elements in the center of the periodic table are called the **transition elements.** Transition elements contained within the body of the table are sometimes called outer transition elements, whereas those listed under the table are called **inner transition elements.** Often the inner transition elements are referred to as the lanthanide or actinide elements in connection with the names of the first elements in each set, lanthanum (La) and actinium (Ac). The lanthanide elements are also called rare earth elements because they are found in geological deposits called rarefied earth. Positions of representative and transition elements are shown in Figure 1-18.

The numbering across the top of the periodic table is important. Roman numerals are used to label the columns of the representative elements, as shown in Figure 1-19. Thus, we see that group I contains H, Li, Na, K, Rb, Cs, and Fr. Group VIII contains He, Ne, Ar, Kr, Xe, and Rn. The transition elements (in this scheme) are then treated as an unnumbered group. A newer scheme numbers the groups with Arabic numerals from 1 to 18 (and here the twenty-eight inner transition elements at the bottom are treated as an unnumbered group).

✔ **Meeting the Goals**

Elements in the periodic table are called representative elements when they appear in the first two or last six columns of the table. Transition elements are those elements in the center of the periodic table that lie between the representative elements. The inner transition elements appear in the two rows beneath the periodic table.

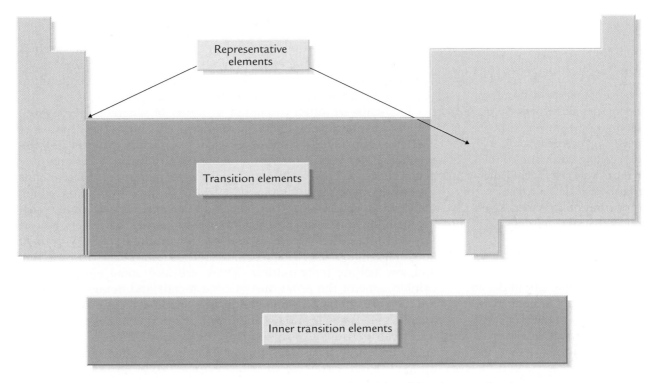

Figure 1-18 ▲ Periodic table of the elements by element type.

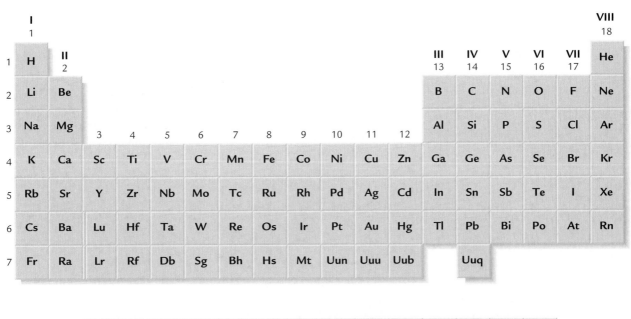

Figure 1-19 ▲ Periodic table of the elements with group numbers (top) and period numbers (side).

Properties and the Periodic Table

The periodic table groups many different elements together according to their properties—hence the basis for the name "periodic" table. In other words, two elements in the same group (vertical column) of the periodic table are expected to have similar properties.

✔ Meeting the Goals

We can separate metals from nonmetals in the periodic table by drawing a staircase-like line beginning at boron and descending down and to the right toward astatine. Elements that border this line have properties of both metals and nonmetals and are called semimetals or metalloids. Elements to the left of the line have properties of metals; elements to the right of the line have properties of nonmetals. The majority of the elements have properties that classify them as metals.

Elements are often discussed in terms of whether they exhibit metallic or nonmetallic behavior. Figure 1-20 shows the division of elements into metals, nonmetals, and semimetals. **Metals** are elements that are very good at conducting electricity and heat and they generally have high melting and boiling temperatures. Most metals are easy to bend (as shown in Figure 1-21) or draw into wires, which makes them good at absorbing shock in many cases but poor at resisting bending under great stress. These properties of metals are all physical properties. Elements that are **nonmetals** are very poor at conducting either electricity or heat and generally have low melting and low boiling temperatures. There are also some elements that exhibit some of the properties of both metals and nonmetals. These are the **semimetals,** or metalloids. The semimetals are found in the area between metals and nonmetals on the periodic table.

According to the periodic law, the properties of elements in a group have significant similarities. This is why some of the groups

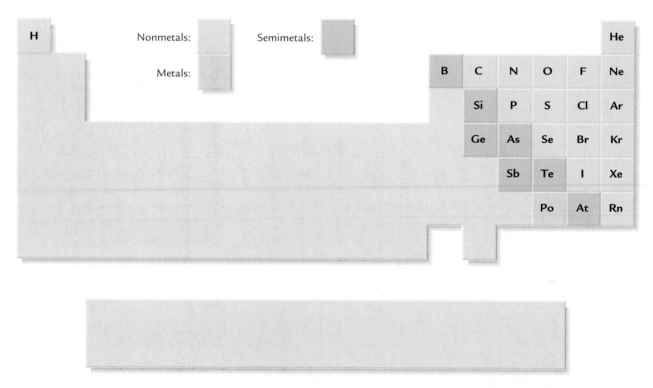

Figure 1-20 ▲ Periodic table of the elements showing classifications as nonmetals, metals, and semimetals.

Figure 1-21 ▲ Malleability is an important physical property of metals that contributes to its usefulness. (a) Aluminum foil. (b) Copper covering the surface of the Statue of Liberty. (a: IFA/eStock Photography/Picture Quest; b: Jeff Greenbergh/Photo Edit.)

have their own names. The most commonly used group names are shown in Figure 1-22; some of the characteristics of these groups are summarized in Table 1-7.

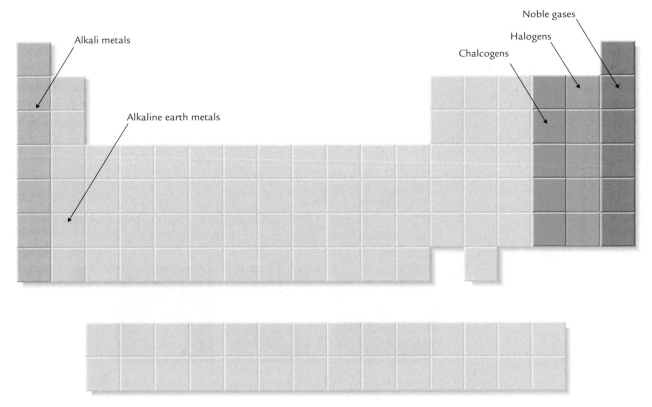

Figure 1-22 ▲ Periodic table of the elements by group name.

Table 1-7 Characteristic Properties of Groups I, II, VI, and VII

Group Number	Group Name	Group Characteristics
I	Alkali metals	Very reactive with water; very soft; combine rapidly with oxygen, chlorine, and hydrogen to form a salt structure; most salts very soluble in water
II	Alkaline earth metals	Reactive with water, though some elements require heating or some other initiation; combine easily with oxygen and halogens
VI	Chalcogens	Form compounds easily; S and O very reactive with many metals to form salts
VII	Halogens	"Salt-forming" elements that are gases at room temperature or that form gases easily; often found in nature in saltlike structures with metals; very reactive with most other substances

Example 1.14   **Properties of the groups**

What properties do the chalcogens and the halogens have in common?

Thinking Through the Problem We will need to list common properties described in Table 1-7.

Answer All the halogens and chalcogens form salts with metals. The halogens, and the chalcogens S and O, are very reactive. These are both chemical properties.

✔ **Meeting the Goals**

Elements in group I are called alkali metals and are shown in Figure 1-23. These are very reactive metals, as can be seen in Figure 1-24. Group II elements are called alkaline and are almost as reactive as the alkali metals. Chalcogens are represented in group VI. Many elements in this group form soft, chalky compounds (nonmetallic properties). Group VII elements are called halogens (salt-forming) and group VIII elements are the noble gases.

How are you doing? **1.8**

What properties do the alkali and the alkaline earth metals have in common? Are these chemical or physical properties?

Figure 1-23 ▲ The alkali metals of Group 1: (a) lithium; (b) sodium; (c) potassium; (d) rubidium and cesium. (W. H. Freeman photos by Ken Karp.)

Figure 1-24 ▲ Reactions of alkali metals with water. (a) Lithium reacts quietly. (b) Sodium reacts vigorously, releasing enough heat to melt the unreacted metal. (c) Potassium reacts so vigorously that the hydrogen produced by the reaction is ignited by the heat of the reaction. (W. H. Freeman photos by Ken Karp.)

STUDY TIPS   How Can You Study Well for Chemistry?

We close this part of the chapter with a section that is not about chemistry itself, but about learning chemistry. Suppose you were going to learn a new language before going on a trip—a long trip that will impact your whole life. You would need the same depth of studying that you will need to succeed in this course, for here you will learn to use chemistry at a level that prepares you for a course in general chemistry, and later, perhaps, for other courses, such as organic chemistry or biochemistry. Just as with a language course, you will have to do some memorization, some practice in reading, and, of course, a lot of talking to others in the new tongue.

But you may ask, "Studying is one thing, but why should I talk to my classmates about this?" Learning to communicate about your learning is, for most people, what makes success possible. In addition, should you one day take a job that requires chemistry, you will be expected to work with others and to communicate using chemistry. So, while taking this course, you might find it helpful to form a study group.

What study groups do:

1. Reinforce, clarify, and deepen learning by providing the opportunity to teach.
2. Provide feedback.
3. Provide insights into other ways to solve problems and understand material.
4. Help you become articulate in talking about a subject.

How to form one:

1. Get to know your classmates. Determine who is serious about the course and who is interested in forming a study

group. Don't worry if they have different abilities than you: strong and weak students together in the same group can give the maximum benefit to all.

2. Start by comparing notes or problem solutions. This will help you get a feel for styles of learning that should match.

3. Find classmates with a compatible schedule. If you never come to campus before 9 A.M., don't expect to participate easily in a 7:30 A.M. group.

4. If you like to communicate by email, then make sure your group members do, too! It can be very handy to post a message to one of them at 1 A.M. and have an answer when you wake in the morning.

How to manage one:

1. Each person should study individually before the group meets.

2. Begin with a few minutes to discuss what is going on in the course; make sure you all have a similar view of the "big picture" for the course at that point in time.

3. If you have many problems to work out, consider giving each person responsibility for leading a discussion of one problem.

4. Find an appropriate area to study. Don't try to make informal meeting places serve a formal study purpose. But do make sure the space you use is comfortable and has tables for you to write on, if that will be important.

Pitfalls and problems:

1. Keep in mind that everyone is there to learn, but some may have to learn more than others. Refrain from critical comments, so you are all able to show your weaknesses and strengths without worry.

2. Some socializing is OK, even normal, but if your group deteriorates into merely a social group, then begin to use a strict agenda. Also, know when it is time to call it quits for the evening.

3. Study groups are a great place to talk about larger college issues, but consider moving such topics to the end of the session.

We would like to thank the University of Illinois at Chicago Academic Center for Excellence for permission to adopt their study group materials for this Study Tip.

PRACTICAL E What systematic trends do chemical substances follow?

The following table describes some observations of six different sodium compounds. Each compound has been reacted in four different ways, and observations have been recorded. Look down each vertical column to see how the different sodium compounds behave. Note when the observations are similar and when they are different.

Compound	Heat in Air	Dissolve in Water	React with Vinegar	React with Bleach
Na_2S	A sour smell results, and a white solid remains	A slightly alkaline solution forms	A "rotten egg" odor is emitted from the solution; the vinegar is neutralized	The solid dissolves
NaBr	No reaction	A neutral solution results	The solid dissolves; the vinegar is not neutralized	The solid dissolves to give a brown solution with a pungent odor
NaF	No reaction	A slightly alkaline solution forms	The solid dissolves; the vinegar is neutralized	The solid dissolves
Na_2O	No reaction	A strongly alkaline solution forms	The solid dissolves; the vinegar is neutralized	The solid dissolves
NaI	The solid discolors to a purple-brown color	A neutral solution results	The solid dissolves; the vinegar is not neutralized	The solid dissolves; a pale yellow solution forms
Na_2Se	A pungent sour smell results	A slightly alkaline solution forms	The solid dissolves and the vinegar is neutralized; a *really* nasty smell results	The solid dissolves

List the compounds that are most similar to NaBr. Now list those that are most similar to Na_2S.

For each of the similar compounds, mark with a circle the properties that they will have in common, and mark with an "X" the observations where they differ.

PROBLEMS

26. List some physical properties of metals.

27. List some physical properties of nonmetals.

28. Concept question: Review the description of the periodic law. Would it have been possible to formulate this law if there were not a clear difference between an element and a compound?

29. Group II metals (M) will react with group VII nonmetals (X) to form compounds with the general formula MX_2. Predict formulas for compounds containing (a) Ca and Cl (b) Ba and I (c) Mg and F.

30. Concept question: The periodic table consists of groups of elements with similar—but not identical—properties. This means we expect the same behavior down a group, but that we aren't surprised when differences appear among elements in the same group. Show that this is true for the metal/nonmetal/semimetal classification of elements in groups II, IV, and VII of the periodic table.

31. Indicate whether the following statements are true or false. Write corrected statements for those you marked false.

(a) All of the group I elements are metallic.

(b) Some group VI elements are metallic.

(c) No group VII elements are nonmetals.

32. Discussion question: Chemical periodicity means that the elements contained in the same group on the periodic table have similar properties and form similar compounds. We know that the following compounds of fluorine (called fluorides) are stable under normal conditions:

AlF_3 SiF_4 PF_5 SF_6

Other elements in the same groups as these should form similar fluoride-containing compounds, i.e., they should have the same ratio with fluorine atoms. From their positions on the periodic table, predict the likely formulas for stable fluoride-compounds that are formed by the following elements:

In   Sn   As   Te

You have just predicted some new chemical formulas. Does this mean these are the only formulas these elements can form?

33. Concept question: The properties of an element in a compound may or may not be as important as the properties of the compound itself. Answer the following questions using your understanding of the difference between an element and a compound.

(a) When sugar is treated with concentrated sulfuric acid, steam is given off and a mass of pure carbon is left behind. This suggests that

sugar contains carbon. Why doesn't carbon taste sweet?

(b) Since water is made of hydrogen and oxygen, and hydrogen burns, why can't water be used as a fuel?

(c) There is oxygen in water, so why isn't it surprising that putting water on a fire makes the fire go out, because the oxygen should help the fire burn?

34. Each element on the periodic table lies at the cross section of a period and a group. (This is similar to finding an ordered pair of numbers on a coordinate graph system.) Give the period and group number that determine each of the following. The first one has been done for you.

Symbol	Name	Period	Group
Mg	Magnesium	3	II
P			
Ar			
Li			
H			
He			

35. Determine whether the following formulas represent the oxide of a metal or a nonmetal.

(a) CO   (b) P_4O_{10} (c) N_2O (d) SO

36. In the table below, check off all of the categories that can be used to describe the elements listed in the first column.

Element	Metal	Nonmetal	Semimetal	Chalcogen	Halogen	Transition Metal	Representative Group Element	Alkali Metal
Ge								
Y								
Sr								
Sn								
Fe								

37. A closed, flexible vessel is filled with a pure gas. When the top of the vessel is opened, the gas immediately bursts into flame and a new gas is formed. What can you say about the physical and chemical properties of the *product* substance?

Chapter 1 Summary and Problems

 Review with **Web Practice**

CHAPTER SUMMARY

There are two parts to the chapter summary: a list of key vocabulary introduced in the chapter (where the terms are shown in bold in the chapter) and the chapter goals. We present each goal again, as a question and answer.

Each part of the summary is presented as a list where you can move down the list with a piece of paper covering the "answers." You can cover the definition of the key word or the answers to the goal question and test yourself, or read through the list to review.

VOCABULARY

matter	Anything that occupies space and has mass.
change	A new arrangement of matter.
element	The simplest kind of chemical substance, which cannot be broken down into any parts; atoms of an element are the same chemically.
compound	A chemical substance that is formed from more than one element.
pure substance	A pure substance contains only one kind of matter; it is the simplest component of a mixture.
mixture	A combination of pure substances.
heterogeneous	Heterogeneous samples have different regions apparent to observers.
homogeneous	Homogeneous samples have a uniform appearance throughout.
solution	Homogeneous mixtures are called solutions.
atom	The smallest part of a substance; atoms cannot be broken down into smaller components by ordinary chemical means.
periodic table	An organized table of all the known elements.
chemical formula	The chemical formula indicates the elements that compose a chemical substance.
period	A period of the periodic table corresponds to a single row; periods are referred to as first, second, third, etc.
group	A group of the periodic table corresponds to a single column; groups are numbered from left to right.
atomic number	An integer that uniquely identifies an element; it is equal to the number of protons in the nucleus of all atoms of that element.
proton	The positively charged particle in the nucleus of the atom.
neutron	A neutral particle in the nucleus of the atom that has a mass very close to that of the proton.
electron	Light, negatively charged particles that surround an atom's nucleus.
nucleus	The innermost part of the atom that contains most of the mass of the atom in its protons and neutrons.
valence electrons	Among the outermost and most accessible electrons in the atom, valence electrons are involved in the atom-atom interactions most important in chemistry.
core electrons	Any electrons in an atom that are not valence electrons.
mass number	The sum of the number of neutrons and protons in an atom; it can be different for different atoms of the same element because the number of neutrons in the atom can vary.
atomic symbol	A shorthand summary of an element's symbol, atomic number, and mass number.
property	A fixed characteristic of a substance that can be used to distinguish that substance from another.

physical property	Physical properties tell us about the appearance of a substance or mixture; each chemical substance has a unique set of physical properties.
chemical property	A characteristic about how a chemical substance engages in chemical change, where it is transformed into a new substance.
physical change	Physical change occurs without altering the chemical substances present.
chemical change	Chemical change causes changes in the identity of the chemical substances present, to form an entirely new substance with different properties.
physical state	Refers to whether a certain sample is a liquid, solid, or gas.
boil	A change in phase wherein a liquid turns into a vapor, accompanied by the formation of bubbles within the liquid.
freeze	A change in phase wherein a liquid becomes a solid.
evaporate	A change in phase wherein either a liquid or a solid forms a vapor.
sublimation	A change in phase wherein a solid turns directly into a vapor.
melt	A change in phase wherein a solid becomes a liquid.
condense	A change in phase wherein a vapor turns into either a liquid or a solid.
periodic law	Describes how chemical elements have properties that are similar in a repeating, or periodic, fashion.
representative elements	The elements in the two "tall" sections of the periodic table; these comprise groups 1, 2, and 13–18 in the Arabic numbering of the groups, and I to VIII in the Roman numbering.
transition elements	The elements in the central portion of the periodic table, comprising groups 3–12 in the Arabic numbering of the groups.
inner transition elements	The elements in the sections of the periodic table usually separated from and positioned below the table; these include twenty-eight elements.
metals	Elements that are very good at conducting electricity and heat; they generally have high melting and boiling temperatures.
nonmetals	The nonmetals are hydrogen and the elements in the top right area of the periodic table; with rare exceptions, the nonmetals never exhibit metallic properties.
semimetals	The semimetals are found at the border between the metals and nonmetals; they often exhibit properties, especially when pure, that are like the metals.

SECTION GOALS

Do you know the difference between a pure substance and a mixture?

A pure substance contains only one kind of matter. A mixture is a physical combination of two or more pure substances.

Do you know the difference between an element and a compound?

Both elements and compounds are pure substances. An element contains only one kind of atom. A compound contains two or more elements in a fixed arrangement.

Can you identify and name the first twenty elements on the periodic table?

The first twenty elements and their symbols are contained in Table 1-1. Symbols contain a maximum of three letters. Only the first letter is capitalized.

Can you identify the different elements in a chemical formula?

Because every element begins with an uppercase letter, it is a simple task to find all the different elements contained in a chemical formula.

Can you find an element in the periodic table when given its symbol or name?

Many symbols of the elements begin with either the first letter, the first two letters, or the first and third letter. A few elements have symbols based on other languages.

What are the relative masses and relative charges for each of the three subatomic particles?	The relative masses for the proton, neutron, and electron are 1, 1, and 0, respectively. The relative charges for the proton, neutron, and electron are $+1$, 0, and -1, respectively.
Do you know where protons, neutrons, and electrons are located in the atom?	The protons and the neutrons are located in the nucleus of the atom. All of the electrons are located outside of the nucleus in a large, diffuse cloud.
Can you define the atomic number and the mass number of an element?	The atomic number of an element is the number of protons contained in the nucleus of that element. The mass number of an element is the sum of the number of protons and the number of neutrons contained in its nucleus.
Can you write and interpret an atomic symbol?	An atomic symbol gives the symbol of the element preceded by its mass number (superscripted) and its atomic number (subscripted). An atomic symbol has the form, $$\begin{smallmatrix}\text{mass number}\\\text{atomic number}\end{smallmatrix}X,$$ where X is the symbol of the element.
Can you determine the number of protons, neutrons, and electrons in a neutral atom when given its atomic symbol?	The number of protons in an atom is given by the atomic number. For a neutral atom, the number of protons equals the number of electrons. The number of neutrons in the atom can be found by subtracting the atomic number from its mass number.
Can you define property? Can you distinguish between a physical and a chemical property?	A property is a fixed characteristic of something that can be used to distinguish one substance from another. A physical property describes the physical characteristics of the substance. Some physical properties are physical state of matter, color, odor, and normal boiling and freezing temperatures. A chemical property tells us how the substance reacts.
How can you know if a property is a physical or a chemical property?	A physical property usually can be determined by using our senses. It tells what a substance looks, feels, smells, sounds like. We can determine physical properties without changing the identity of the substance. Chemical properties tell us how a substance reacts with other substances.
Do you know what is meant by the physical states of matter?	The physical states of matter refer to whether the substance is a solid, a liquid, or a gas.
Do you know the names of the processes that convert from one physical state and another?	The words melt/freeze indicate a physical change of state between solid and liquid. It is important to realize that the freezing point temperature is the same as the melting point temperature. The different words are used to indicate the direction of the change. The words evaporate/condense indicate a physical change of state between liquids and gases. The words sublime/deposit indicate a physical change of state between a solid and a gas.
Can you identify an element as a representative, transition, or inner transition element?	Elements in the periodic table are called representative elements when they appear in the first two or last six columns of the table. Transition elements are those elements in the center of the periodic table that lie between the representative elements. The rare earth elements (sometimes called inner transition elements) are those elements that appear in the two rows beneath the periodic table.
Can you classify an element as a metal, nonmetal, or semimetal?	We can divide metals from nonmetals in the periodic table by drawing a staircase-like line beginning at boron and descending down and to the right toward astatine. Elements that border this line have properties of both metals and nonmetals and are called semimetals or metalloids. Elements to the left of the line have properties of metals; elements to the right of the line have properties of nonmetals. The majority of the elements have properties that classify them as metals.
Can you identify an element as an alkali metal, alkaline earth metal, a chalcogen, a halogen, or a noble gas?	Elements in group I are called alkali metals. These are very reactive metals. Group II elements are called alkaline earth metals and are almost as reactive as the alkali metals. Chalcogens, or "chalk-forming" elements, are the elements in group VI. Group VII elements are called halogens (salt-formers) and group VIII elements are the noble gases.

PROBLEMS

38. Determine whether the following represent metals, nonmetals, or semimetals.

(a) Ca   (b) Ge   (c) F   (d) Cs

39. Decide, from the formulas alone, whether the substance is an element (E) or a compound (C).

(a) C   (c) CO   (e) Na   (g) NH_3 (i) No

(b) Co   (d) CF_4 (f) F   (h) N   (j) NO

40. Name each of the elements present in:

(a) H_2SO_4 (c) NH_4Cl
(b) $NaC_2H_3O_2$ (d) $CaCO_3$

41. Decide which of the following are physical changes and, if there is a physical change in state, name the process involved.

	Physical Change?	Physical Change of State?
A board is sawed in two		
Water is decomposed into hydrogen and oxygen gases		
Mothballs "disappear" with time		
A tennis ball is bounced		
Wax melts		

42. **Discussion question:** From the following descriptions, decide whether the original sample was a mixture or a pure substance.

(a) A white solid is added to a beaker of water. A white solid remains undissolved on the bottom of the beaker. The solution is filtered and the clear filtrate (the water solution) is collected in a second, clean beaker and is evaporated. When all of the filtrate has evaporated, a white powdery substance remains in the second beaker. Was the original substance probably a mixture or a pure substance?

(b) A clear, yellow liquid is put into a flask and slowly heated while recording the temperature readings every minute. The temperature of the liquid continues to rise until the boiling point is reached. The temperature remains constant until all of the liquid has boiled away. Was the original substance a mixture or a pure substance?

(c) A clear, brown liquid is put into a flask and slowly heated while recording the temperature readings every minute. The temperature of the liquid continues to rise until the liquid begins to boil. Then the temperature remains constant for a while as the brown liquid boils away. After a time the flask is found to contain a light yellow liquid and the temperature readings begin to rise again, becoming constant for a second time at a higher temperature. Was the original substance a mixture or a pure substance?

(d) A magnet is held over a small pile of black powdery particles. Some of the black particles cling to the magnet while others remain in the pile. Was the original substance a mixture or a pure substance?

In cases where the original sample was a mixture, can we determine if the components of the mixture were elements or compounds?

43. Indicate if the evidence in the following sentences indicates if a substance is an element, a compound, or a mixture.

(a) A clear liquid is boiled away, leaving a white powder behind.
(b) When a solid is heated, it forms a gas and a different solid.
(c) Cooling a liquid causes it to solidify. It returns to a liquid when it is gently warmed.

44. From the descriptions in the table below, decide whether the experiment describes a physical or chemical change.

Description	Physical Change?	Chemical Change?
Solid silver chloride is formed when two clear liquids are mixed.		
Solid iodine changes to gaseous iodine.		
Candle wax burns.		
Liquid ether evaporates at room temperature.		
Soft metals can be hammered into sheets.		

45. How many neutrons are contained in a neutral atom of $^{60}_{27}Co$?

46. How many protons are contained in a neutral atom of $^{197}_{78}Pt$?

47. Write the atomic symbol for a neutral atom of cadmium (Cd) that contains 67 neutrons.

48. Write the atomic symbol for a neutral atom of tungsten (W) that contains 106 neutrons.

49. Classify the actions in the table below as physical or chemical changes by putting an X in the correct column.

Action	Physical Change?	Chemical Change?
A solid dissolves in a liquid and produces heat.		
Bubbles of gas are formed when metal is added to a solution of acid.		
Copper metal can be drawn into copper wires.		
Phosphorus burns when put into water.		

50. How many symbols are there in each of the following formulas?

(a) $TiCl_3$ (b) XeO_4 (c) O_3

51. Determine whether the following represent compounds or elements.

(a) S_8 (b) Co (c) CO (d) HCN

52. Give the name of each of the following elements.

(a) F (b) Be (c) P (d) C (e) Cl

53. Give the symbol for:

(a) sodium (b) phosphorus (c) fluorine

54. Complete the table below by filling in the missing information for these neutral atoms.

Elements	Symbol	Number of Protons	Number of Neutrons	Number of Electrons	Atomic Number	Mass Number
		28	31			
	Sn		70			
Arsenic						75

55. Determine the atomic symbol for a neutral atom from the information given in this table.

Name	Number of Protons	Number of Neutrons	Number of Electrons	Atomic Number	Mass Number	Atomic Symbol
				17	35	
Barium		81				
	88	138				

56. Classify the following as a chemical change, a physical change, or neither of these.

(a) Baking soda reacts with acid to liberate a gas.
(b) An iron nail rusts.
(c) An opening is cut in a brick wall.
(d) The pine freshener hanging in the closet "disappears" over time.

57. Each element on the periodic table lies at the cross section of a period and a group. (This is similar to finding an ordered pair of numbers on a coordinate graph system.) Give the period and group number that determine each of the following.

Symbol	Name	Period	Group
K			
N			
O			
Cl			
Ne			
S			

58. Check off all of the categories that can be used to describe the elements listed in the first column below.

Element	Metal	Nonmetal	Semimetal	Chalcogen	Halogen	Transition Metal	Representative Group Element	Alkali Metal
Pb								
Te								
Si								
Au								
Br								

59. From the following descriptions, decide whether the experiment describes a physical or chemical change.

Description	Physical Change?	Chemical Change?
Liquid ether can be frozen.		
Candle wax melts.		
A solid is formed when two clear liquids are poured together.		
Sodium carbonate dissolves in acid, producing heat.		
Hydrogen gas is formed when metal is added to a sodium of acid.		
Methane gas burns to form CO_2 and water.		

60. Classify the following as chemical or physical changes by putting an X in the correct column.

Action	Physical Change?	Chemical Change?
Water vapor condenses on cold surfaces.		
Aluminum melts at high temperatures.		
Solid iodine sublimes.		
Butane gas burns.		

61. (a) Distinguish between a molecule and a formula unit. Give an example of each.
   (b) Distinguish between a compound and a mixture. Give an example of each.

62. (a) Write the symbol and name of the element in group II and period 3.
   (b) Write the symbol and name of the 3rd period element in group VI.
   (c) Write the symbol and name of the group V element in period 3.
   (d) Name all of the elements in the second period.

63. (a) Identify and count the elements present in Na_3AsO_4.

(b) Identify and count the elements present in $Na_2B_4O_7 \cdot 10\ H_2O$.

64. (a) Which of the subatomic particles—protons, neutrons, or electrons—have appreciable mass?
   (b) Which of the subatomic particles—protons, neutrons, or electrons—carry a charge?

65. (a) A neutral atom contains 15 protons and 16 neutrons. How many electrons does it have?
   (b) Write the atomic symbol for a neutral atom that contains 20 electrons and 21 neutrons.
   (c) Write the atomic symbol for a neutral atom that contains 35 electrons and 45 neutrons.

66. (a) Write the atomic symbol for a neutral atom of tin that contains 70 neutrons.

(b) Write the atomic symbol for a neutral atom of iodine that contains 74 neutrons.

67. Name two physical properties of gold that make this metal easy for jewelers to work with.

68. Steel is a homogeneous mixture of metals. Explain how this can be possible when the metals combined in the steel are solids at room temperature.

69. Of the following properties, the one that is not a physical property is (a) freezing, (b) burning in air, (c) malleability.

70. (a) Of the following properties of H_2O_2, the one that is not a physical property is (i) decomposition into O_2 and H_2O, (ii) ductility, (iii) evaporation.
   (b) Of the following properties, the one that is not a physical change of state is (i) condensation (ii) sublimation (iii) decomposition

71. (a) What physical property of aluminum makes the production of aluminum foil possible?
   (b) What physical property of copper makes it a good metal to use in the production of wires?

72. Liquid A has a melting point of -12 °C and a boiling point of 48 °C. Liquid B has a melting point of 22 °C and a boiling point of 89 °C. Solid C has a melting point of -2 °C and a boiling point of 122 °C. Solid D has a melting point of -12 °C and a boiling point of 48 °C. Which two of these substances might be the same substance?

73. Using the information in Table 1.5, give the physical state of:
   (a) carbon at 3675 °C.
   (b) oxygen at 100 °C.
   (c) carbon monoxide at -220 °C.
   (d) methane at -175 °C.

74. Concept question: What does the periodic law have to do with the periodic table?

75. A substance that has a high boiling point but is a poor conductor of heat and electricity is expected to be a

   (a) metal.   (b) nonmetal.   (c) semimetal.

76. The element most likely to have a low melting point is

   (a) strontium.   (b) antimony.   (c) sulfur.

77. The element most likely to be a poor conductor of electricity and to have a low boiling point is

   (a) scandium.   (b) silicon.   (c) selenium.

78. A chemical property of sulfur is that it

   (a) reacts with metals to form salts.
   (b) is very reactive with water.
   (c) is a gas at room temperature.

79. (a) Name the alkali metal in period 6.
   (b) Name the chalcogen in period 4.
   (c) Name the halogen in period 5.

CHAPTER 2

Molecular Substances and Lewis Structures

The controlled demolition of this hotel in Hartford, Connecticut, illustrates the power of certain molecular substances.

(Jim Zipp/Photo Researchers.)

PRACTICAL CHEMISTRY Power Molecules

In this chapter we will see how chemists, and we hope you, too, can recognize certain chemical substances as being safe to handle while others have clear warning signs associated with their arrangement of atoms. We call the molecules of these more dangerous substances "powerful" because, in one way or another, they can have a rapid and profound effect on their surroundings—including us!

Reading this chapter will not qualify you to be a chemical safety officer, but you will be able to make some reliable statements about some safety-related properties of chemical substances based on what you've learned about molecular structures.

This chapter will focus on the following questions:

PRACTICAL A How do we recognize small molecules that are likely to be gases and need plenty of room for their storage?

PRACTICAL B What important air pollutants have structures that suggest that they may be very reactive?

PRACTICAL C What are the components of the nitrates, and why can they be so dangerous to handle?

2.1 Molecules in Chemistry

SECTION GOALS

✔ Predict whether or not a compound is likely to be a molecular substance.

✔ Know what information is given by the chemical formula of a substance.

✔ Understand and use numbering schemes for naming molecular substances.

✔ Apply the rules for naming molecular substances.

Molecular Substances

In Chapter 1 we discussed how pure chemical substances can be composed of a single element or a compound of two or more elements. In this chapter, we discuss an important class of pure substances, where the atoms are held together in certain fixed arrangements called molecules. A **molecule** is a combination of a specific number and kind of atoms held in a specific arrangement that cannot be broken down without creating a different substance. This means that if we split a molecule we will no longer have a molecule of the same substance. Instead we will have either different molecules, charged fragments, or individual atoms.

A substance that is normally found with its atoms arranged as molecules is called a **molecular substance.** The chemical formula of a molecular substance is the same as the formula of its individual molecules. Most molecular substances retain their molecular character in the gas, liquid, and solid phases. In the gas phase the molecules are far apart, while in liquids and solids they are very close together. A common and well-known molecular substance is water. In Chapter 1, Figure 1-16 shows a drawing of water molecules in water vapor, in liquid water, and in solid water (ice). Note that the water molecules are rigidly packed against one another (solid), are randomly arranged but still touching (liquid), or are well separated (gas). But in all cases the molecule of water, H_2O, is the basic building block of the sample.

Figure 2-1 ▲ From left to right, chlorine is a gas, bromine is a liquid, and iodine is a solid at room temperature. (Chip Clark.)

Elements That Occur As Molecules

There are certain elements that usually exist as molecules. The most important examples of these elements are the **diatomic** elements, with two atoms of the same element held together in a molecule. These are listed in Table 2-1.

Because these seven elements are found as diatomic molecules under standard conditions, when we say the name of the pure element we mean the diatomic molecule. For example, when we speak of the element hydrogen, we mean H_2.

The last four diatomic elements in Table 2-1 are halogens. Under normal conditions, two of the halogens are gases (F_2 and Cl_2), one is a liquid (Br_2), and one is a solid (I_2). See Figure 2-1. But in all cases the

Table 2-1 Diatomic Elements

Hydrogen	H_2
Nitrogen	N_2
Oxygen	O_2
Fluorine	F_2
Chlorine	Cl_2
Bromine	Br_2
Iodine	I_2

basic building block is a molecule of two atoms bonded to one another.

Compounds That Occur As Molecules

Many compounds exist as molecules. Molecules of compounds contain a fixed number of atoms of two or more elements. An example of a compound that exists as molecules is water, which consists of molecules containing two hydrogen atoms and one oxygen atom. Another example, shown in Figure 2-2, is the compound C_2H_6, which contains molecules of two (and only two) carbon atoms and six (and only six) hydrogen atoms.

Can we predict which compounds occur as molecules? A reliable prediction is easy with the periodic table at hand. In Chapter 1 we used position on the periodic table to distinguish among metals, semimetals, and nonmetals. Usually, when a compound contains only nonmetals, or nonmetals and semimetals, it will exist as molecules. Don't forget that hydrogen is a nonmetal. For example, because only nonmetals appear in the compounds CH_4, SO_3, and IF_7, we expect that these compounds are molecular. If nonmetals and semimetals combine with one another, then the result is usually, but not always, a molecular compound. Examples of molecular compounds that contain a nonmetal and a semimetal are BF_3, SiH_4, and As_2O_5.

If a metal is present in a compound, then the compound typically does *not* have a molecular structure. The compounds ZnO, Na_3P, and $CrCl_3$, for example, each contain a metal and therefore do not have a molecular structure. We will discuss these types of compounds in Chapter 3.

▼ See this with your **Web Animator**

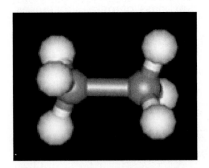

Figure 2-2 ▲ A molecule of ethane.

✔ Meeting the Goals

If we recognize that a compound contains only nonmetals or nonmetals and semimetals, then we can assume that it is a molecular substance.

Example 2.1   **Determining whether or not compounds are molecular**

Indicate whether or not the following compounds are likely to be molecular.

(a) HF   (b) NaH   (c) $SiCl_4$ (d) SCl_4

Thinking Through the Problem If we do not find a metal, then we expect a molecular compound.

Answer (a) H and F are nonmetals. We expect HF to be molecular. (b) Na is a metal. We expect NaH not to be molecular. (c) Cl is a nonmetal, but Si is a semimetal. We expect this compound to be molecular. (d) S and Cl are nonmetals. We expect this compound to be molecular.

How are you doing? 2.1

Determine whether or not we expect the compounds KBr and PBr_3 to be molecular.

Naming Molecular Compounds

In order to communicate clearly, it is important that scientists agree on the names of the compounds that they are studying. Indeed, many breakthroughs in chemistry and biology have been linked to new forms of classifications and to new ways of naming things. **Nomenclature** refers to any systematic way of naming things. In this textbook, we will present systematic rules for naming compounds. These rules give the **systematic names** of compounds. Knowing how to write the systematic name of a compound when given its formula is an important technique in this course. Systematic naming of molecular compounds is based on a simple idea: we need to specify the type and number of the elements in the compound.

We will generally avoid the use of **trivial names,** which are short names applied to specific compounds to make it easier to refer to them. We will use only two—*water* for H_2O and *ammonia* for NH_3—but you should be aware that other trivial names do exist. Also, many carbon-containing compounds are named using the systematic nomenclature of organic chemistry.

In this chapter, we will only work on naming molecular compounds that have two elements in them. Such compounds are called **binary** compounds, to indicate that there are two—and only two—elements in the compound. To name a binary molecular compound:

- The element to the left in the periodic table (in an earlier group) is written and named first.
- If two elements are from the same group, the element in the lower period is written and named first.
- Name the first element using its elemental name. Use a prefix from Table 2-2 to indicate the number of atoms of that element in the molecule if it is two or more ("mono" is never used on the first element in the name).
- Alter the name of the second element by applying the suffix *-ide* to the root of the element name. Thus, oxygen becomes oxide, chlorine becomes chloride, and sulfur becomes sulfide. To this altered name, we add the appropriate counting prefix.

Thus, SF_4 is named sulfur tetrafluoride while CS_2 is carbon disulfide and SO_3 is sulfur trioxide.

REMEMBER

Table 2-2 Prefixes for Use in Counting Atoms
in Molecule Names

Count	Prefix
1	mono
2	di
3	tri
4	tetra
5	penta
6	hexa
7	hepta
8	octa
9	nona
10	deca

Example 2.2   **Naming molecular compounds**

Name the molecular compounds CBr_4, SO_2, N_2O, and ClF_5.

Thinking Through the Problem The formula is the starting point. With our knowledge of the periodic table, we can name each element in turn. We use the prefixes in Table 2-2 to indicate the count of the elements.

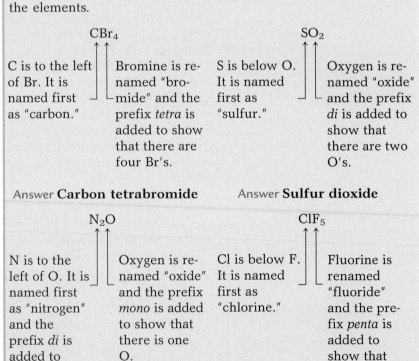

CBr_4

C is to the left of Br. It is named first as "carbon." | Bromine is re-named "bro-mide" and the prefix *tetra* is added to show that there are four Br's.

Answer **Carbon tetrabromide**

SO_2

S is below O. It is named first as "sulfur." | Oxygen is re-named "oxide" and the prefix *di* is added to show that there are two O's.

Answer **Sulfur dioxide**

N_2O

N is to the left of O. It is named first as "nitrogen" and the prefix *di* is added to show there are two N's. | Oxygen is re-named "oxide" and the prefix *mono* is added to show that there is one O.

Answer **Dinitrogen monoxide**

ClF_5

Cl is below F. It is named first as "chlorine." | Fluorine is renamed "fluoride" and the pre-fix *penta* is added to show that there are five F's.

Answer **Chlorine pentafluoride**

✔ Meeting the Goals

To name a molecular compound, name each element giving the second element an -*ide* ending. Use prefixes to "count" the number of times each element appears in the formula. (An exception to this is that the prefix *mono*- is not used if the element appears first in the formula. Thus, CO is simply carbon monoxide, not monocarbon monoxide.)

Example 2.3   Naming molecular compounds

Name the molecular compounds P_4O_{10}, S_2O, SF_2, and CCl_4.

Thinking Through the Problem This is done in the same way as the previous problem. In two cases we have more than one of the first element. The first element is still named first, but a prefix is needed to give the count.

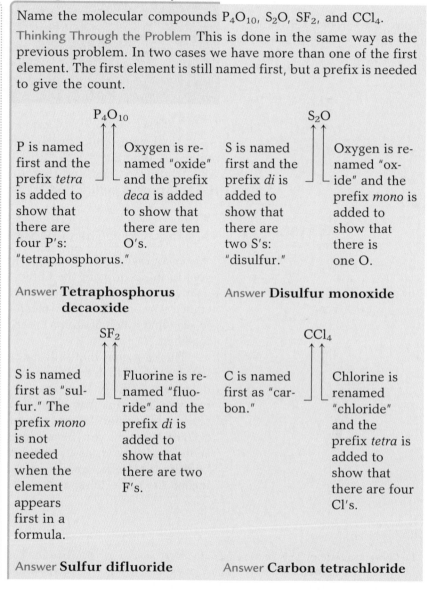

P_4O_{10}

P is named first and the prefix *tetra* is added to show that there are four P's: "tetraphosphorus."

Oxygen is renamed "oxide" and the prefix *deca* is added to show that there are ten O's.

Answer **Tetraphosphorus decaoxide**

S_2O

S is named first and the prefix *di* is added to show that there are two S's: "disulfur."

Oxygen is renamed "oxide" and the prefix *mono* is added to show that there is one O.

Answer **Disulfur monoxide**

SF_2

S is named first as "sulfur." The prefix *mono* is not needed when the element appears first in a formula.

Fluorine is renamed "fluoride" and the prefix *di* is added to show that there are two F's.

Answer **Sulfur difluoride**

CCl_4

C is named first as "carbon."

Chlorine is renamed "chloride" and the prefix *tetra* is added to show that there are four Cl's.

Answer **Carbon tetrachloride**

How are you doing? 2.2

Name the compounds SO_3, PBr_3, and NCl_3.

It is also important that we be able to look at a representation of certain molecules and simply count the atoms present. This gives us another way to determine the formula. For example, in Figure 2-2 we see a molecule containing C and H. With dark gray for carbon and light gray for hydrogen, we can see two carbons and six hydrogens. The formula of this molecular substance is C_2H_6.

Figure 2-3 shows three molecules made of phosphorus and chlorine. Their formulas are, from left to right, PCl_3, P_2Cl_4, and PCl_5.

Figure 2-3 ▶ Molecules formed by phosphorus (orange), and chlorine (green).

A final point concerns the naming of compounds with the formula H_xA, where $x = 1, 2 \ldots$ and A is a nonmetal. These are named with "hydrogen" first and then the unnumbered name of "A." Thus, H_2S is hydrogen sulfide and HF is hydrogen fluoride.

STUDY TIPS   **Thinking About Chemical Substances**

We encourage you to study with another, more comprehensive, goal in mind: to gain the ability to look at a name, a formula, or a structure and immediately start thinking about what kind of substance it represents. This means that you examine the substance to see what kinds of atoms are in it, and how the atoms are bonded together.

This is a goal that will take many more courses in chemistry before it is perfected. But even the beginning chemist can start to develop a habit of studying for *familiarity*. What kinds of things can you look for? You should already be able to read the words "carbon tetrachloride" and think that this is an OK name for a compound. What if you heard "carbon quadchloride"? That should sound odd to you, as it would to a more highly trained chemist, because the prefix "quad" is not used in chemistry, even though it means the same thing as "tetra."

PROBLEMS

1. Concept question: This section introduces systematic nomenclature. We follow rules for naming, so others may know exactly what we mean. Sometimes the systematic name might sound strange to you. Can you think of three "official" names from outside of chemistry that are systematic, but that also seem strange?

2. Indicate whether or not the following compounds are molecular.

(a) CaH_2 (c) PCl_3 (e) XeO_4
(b) H_2S (d) $TiCl_3$ (f) U_3O_8

3. Underline the compounds in the following list that are expected to have a molecular structure.

SiH$_4$ PCl$_3$ AlN
NaI   HI   SF$_6$
ZrI$_4$ UF$_4$

4. Give the systematic names for the following compounds. You may have to look up the names of elements below the third row of the periodic table.

(a) ClO_2 (b) SF_6 (c) I_2O_7

5. Determine the formulas for the following compounds.

(a)   Nitrogen triiodide
(b)   Diphosphorus pentaoxide
(c)   Phosphorus tribromide
(d)   Dichlorine monoxide
(e)   Sulfur tetrafluoride

6. Give the systematic names for the following compounds. You may have to look up the names of elements below the third row of the periodic table. (a) IF$_7$ (b) ICl (c) NCl$_3$

PRACTICAL A How do we recognize small molecules that are likely to be gases, and that need plenty of room for their storage?

We have looked at molecules and considered how their properties are, in part, derived from their specific arrangement of atoms. Molecular substances have weak connections between molecules (intermolecular connections), compared to the bonds within the molecule (intramolecular connections) which are stronger. You may think that it is necessary to know the exact properties of different substances before you can tell if they will form a gas. But you have already learned the most essential point: you can recognize whether a substance contains a metal, nonmetal, or semimetal from its position in the periodic table. In many cases, then, you know if a substance is likely to be molecular and that it might be a gas or vapor. Some molecular substances are solids, such as waxes or ice.

Others are liquids, like oils and liquid water. Almost all small molecular substances (with usually fewer than 30 atoms) exist as gases at or near room temperature. We only have to heat water to its boiling point to see the formation of huge amounts of water vapor. Substances like ammonia (NH_3) and oxygen (O_2) are familiar and important gases.

The formation of a gas from a liquid can have disastrous consequences on the surroundings. Consider, for example, the prohibition against heating aerosol cans and butane lighters. These devices contain a liquid that easily turns into a vapor. When the vapor forms at room temperature, the walls of the container are strong enough to hold the vapor. But when the liquid is heated, much more of it is converted into a gas exerting greater pressure on the can. Explosions occur when the metal container can no longer withstand the pressure of the confined gas. Figure 2-4 shows gases under pressure in an aerosol can, a large tank, and on a tanker truck. If the escaping gas burns easily, then it can cause a fire. If it is a poison, especially if it is denser than air, the gas can cause great harm.

Other substances are gases under ordinary conditions of pressure and temperature. To keep such gases contained requires very strong containers. To efficiently transport these substances often requires cooling to very low temperature or the use of high pressure equipment.

As these examples show, it is important to know when a substance will form a gas under different conditions. You can then take precautions to stop the substance from forming a high-pressure gas, or you can make sure that it is held in a very strong container.

Even in cases where a substance does not form a gas capable of bursting its container, some molecular substances form a vapor that is toxic. Therefore, an open container of a gas-forming substance can be dangerous.

Look at the following list of compounds. Which are ones that you would expect to be gases or vapor-forming substances? Which are not likely to form a gas under normal conditions? For those that are likely to be gases, name them based on the principles developed in this section.

$$H_2S \quad CaS \quad KF \quad CF_4 \quad Br_2$$
$$CO \quad LiCl \quad ICl \quad NCl_3 \quad SO_3$$

Your choices probably show that you already know some important things about molecules. Now let's add another "wrinkle." It is generally known that molecular substances containing a single atom bonded to oxygen, a halogen, or hydrogen (for example, SO_2 or SiH_4) form gases readily, and often completely, at room temperature. Molecular substances containing just hydrogen and another element are gases, too, if the number of atoms is below fifteen.

Which of the following compounds are likely to be gases at room temperature? Except for the first substance, name all of the molecular substances.

$$C_2H_6 \quad HCl \quad P_2O_5 \quad P_2H_4$$
$$LiH \quad BrF_3 \quad NO_2 \quad CO_2$$

Figure 2-4 ▲ Some substances are almost always gases at room temperature. To keep them contained requires very strong containers. (Left: Antman/The Image Works; center: David Wells/The Image Works; right: Joseph Nettis/Photo Researchers.)

7. Discussion question: Nitrogen forms five compounds with oxygen. Name each of these compounds.

(a) N_2O (c) N_2O_3 (e) N_2O_5
(b) NO (d) N_2O_4

If someone said, "All nitrogen and oxygen compounds are the same," how would you use these to show their error?

8. Complete each of the following phrases with the number that corresponds with each prefix.

tetra- means _____
deca- means _____
hepta- means _____

tri- means _____
hexa- means _____

9. Name each of the compounds below.

(a) NH_3 (b) NF_3

10. Web question: For the illustrations in Web Animator Problem 2-10, determine the formula of the molecules shown. The dark gray indicates carbon and the light gray hydrogen.

11. Web question: For the illustrations in Web Animator Problem 2-11, determine the formula of the molecules shown. Besides C and H the red atoms are oxygen and the blue atoms are nitrogen.

2.2   Lewis Structures

SECTION GOALS

✔ Draw the Lewis structure of a representative element or its ion.

✔ Draw the Lewis structure for binary compounds.

✔ Draw the Lewis structure for polyatomic ions.

Bonding in Molecules: Valence Electrons and Chemistry

In the preceding section we learned to identify molecular substances. A very important point was that the forces between atoms in a molecule are much stronger than the forces between the molecules themselves. What accounts for these strong forces between atoms in a molecule? The answer lies in the nature of the **chemical bonding** in the substance. Chemical bonding is the way that atoms are connected to one another.

A molecule is formed when atoms build chemical bonds with other atoms by sharing electrons. The electrons are not transferred from one atom to another but instead are shared. The electrons that are shared are called valence electrons. The valence electrons are the outermost electrons of the atom, as illustrated in Figure 1-13. They are available for atom-to-atom interactions. Only a certain number of the electrons in an atom are valence electrons. The quickest way to determine the number of valence electrons in an atom of a particular element is to note the column of the periodic table (labeled with a Roman numeral) in which the element appears. *This number, the group number, is the same as the number of valence electrons for that atom.* For example, an atom of carbon, which is in Group IV of the periodic table, contains four valence electrons. Other electrons, if present, make up the atom's less available core electrons.

Most of our study in this book will concern the representative group elements in Groups I through VIII of the periodic table. The

Figure 2-5 ▶ Valence electrons in Period 2 elements. Note that the number of valence electrons increases as you move across the period.

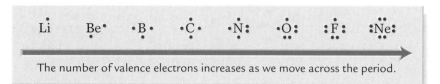

The number of valence electrons increases as we move across the period.

✔ **Meeting the Goals**

We begin by looking at the group number of the element. The Roman numeral we use to indicate the group number of a representative element is the same as the number of valence electrons in a neutral atom of the element. If we have an ion with a positive charge, we know that valence electrons have been removed. If we have an ion with a negative charge, we know that valence electrons have been added.

neutral atoms of these elements have from one to eight valence electrons. The group number tells us the number of valence electrons in a neutral atom. (The only exception to this rule is He, which has two valence electrons despite its presence in Group VIII.)

For example, calcium (Ca) and magnesium (Mg) are both in Group II. As neutral atoms, they each have two valence electrons. On the other side of the periodic table, chlorine and fluorine are both Group VII elements. When neutral, they each have seven valence electrons.

An important part of working with valence electrons is our ability to draw them in ways that help us analyze the structure of atoms and molecules. When working with valence electrons, we commonly use a scheme first developed by the chemist G. N. Lewis over eighty years ago. We represent each valence electron with a dot, known as a **Lewis dot** or an electron dot. When we draw the **Lewis structures** of the second period elements, shown in Figure 2-5, we use one dot for lithium's valence electron and eight dots for neon's eight valence electrons.

When there are unequal numbers of electrons and protons in an atom, we no longer have a neutral atom, but instead have an **ion**. In Chapter 1 we learned that electrons have a negative charge, so an ion that has more electrons than protons has a negative charge. We call such an atom an **anion**. An ion that has more protons than electrons has a positive charge and is called a **cation**. Many compounds are composed of anions and cations. The overall compound is neutral because the charges on the cations and the charges on the anions add up to zero. We will discuss such compounds, called ionic compounds, in more detail in Chapter 3.

Example 2.4   **Determining the number of valence electrons**

How many valence electrons are on a chlorine atom? How many valence electrons are on a chloride ion, Cl^-?

Thinking Through the Problem We must determine the number of valence electrons on both an atom and an ion of chlorine. ➡️ (2.1) The key idea is that chlorine is in Group VII of the periodic table. Therefore, we know that a chlorine atom has 7 valence electrons. The chloride ion has one negative charge. Therefore, Cl^- must have one more electron than Cl, or 8 valence electrons.

Working Through the Problem $:\dot{Cl}: + 1\ e^- \longrightarrow :\ddot{Cl}:^-$

Answer Cl has 7 valence electrons. Cl^- has 8 valence electrons.

Example 2.5   **Determining the number of valence electrons**

> How many valence electrons are on sulfide ion S^{2-}?
>
> Thinking Through the Problem We must find the number of valence electrons on the sulfide ion. Sulfur is in Group VI of the periodic table so the neutral atom has 6 valence electrons. The ion, S^{2-}, has a double negative charge. Therefore, S^{2-} has $6 + 2 = 8$ valence electrons.
>
> Working Through the Problem $\cdot\ddot{S}: +\ 2\ e^- \longrightarrow :\ddot{S}:^{2-}$
>
> Answer   The S^{2-} ion has 8 valence electrons.

Counting the number of valence electrons present on a metal ion follows a similar procedure except the valence electrons of metals are lost during ion formation, leaving a positive charge.

Example 2.6   **Determining the number of valence electrons**

> How many valence electrons are on an ion of Mg^{2+}?
>
> Thinking Through the Problem Since magnesium is in Group II of the periodic table, the neutral Mg atom has two valence electrons. The cation, Mg^{2+}, has a double positive charge. This means it has lost two electrons. Mg^{2+} has $2 - 2 = 0$ valence electrons.
>
> Working Through the Problem $\cdot Mg\cdot -\ 2\ e^- \longrightarrow Mg^{2+}$
>
> Answer Mg^{2+} has zero valence electrons.

How are you doing?   **2.3**

How many valence electrons are present in an atom of nitrogen? How many valence electrons are on a nitride ion, N^{3-}?

Covalent Bonding in Diatomic Molecular Substances

The sharing of valence electrons in molecules is known as **covalent bonding** (literally "shared valence bonding"). If we think for a minute about what happens in a covalent bond, we get an insight into why covalent bonds are so strong and stable. To break a covalent bond means that we must "unshare" the electrons in it. This means that the stability gained by sharing is lost, and (if we do manage to break the bond) the resulting fragments are often very unstable.

When electrons are shared, chemical bonds form and a molecule is created. We can describe the bonding in a molecule with Lewis structures. Lewis structures are drawn using the valence electrons in an atom. In examining the structures of atoms that share electrons, we can see that the covalent bonding is arranged to give eight electrons around each atom. This is sometimes known as the **octet rule:** that the main group elements form bonds so that each atom has eight

KEY IDEA 2.2

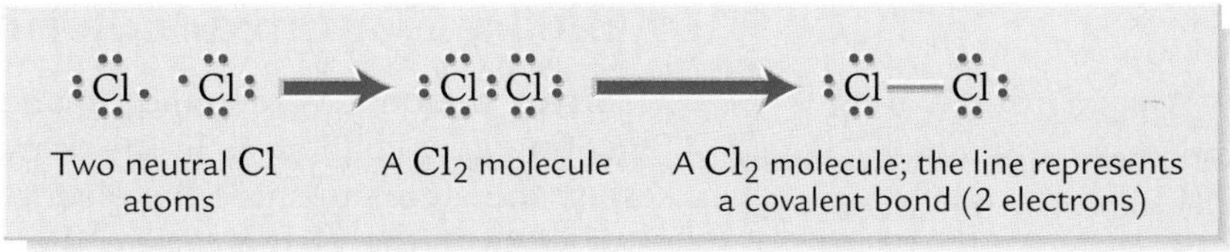

Figure 2-6 ▶ The formation of a Cl—Cl bond by octet formation.

$$\ddot{\text{:Cl}}\cdot \quad \cdot \ddot{\text{Cl}}\text{:} \quad \longrightarrow \quad \ddot{\text{:Cl}}\text{:}\ddot{\text{Cl}}\text{:} \quad \longrightarrow \quad \ddot{\text{:Cl}}\text{—}\ddot{\text{Cl}}\text{:}$$

Two neutral Cl atoms   A Cl$_2$ molecule   A Cl$_2$ molecule; the line represents a covalent bond (2 electrons)

electrons, or an octet, around it. This is the same number of valence electrons found in the noble gases, which are very stable (unreactive) elements. One important exception is hydrogen, which is stable in molecules where it has two electrons.

We can draw Lewis structures to show the bonding in almost any molecule. We draw the valence electrons as dots, and then arrange the atoms and the dots so that each atom has around it a stable octet of eight electrons.

Let's begin by drawing the Lewis structure for a chlorine molecule. A neutral chlorine atom (Group VII) has seven valence electrons, and so we draw the atom with seven dots. The chlorine atom needs one more electron to achieve a stable octet (eight valence electrons). We know from Table 2-1 that a chlorine molecule contains two atoms of chlorine. Because we have two chlorine atoms, we draw the picture shown on the left in Figure 2-6.

How do we analyze the bond formed in the Cl$_2$ molecule? It won't help to transfer one electron from one of the Cl atoms to the other, because that would leave one Cl atom with only six electrons. Instead, the two chlorine atoms can each share one electron. This sharing is shown in the center of Figure 2-6, where each Cl atom has eight electrons around it. This electron sharing explains why Cl$_2$ is a stable molecule.

The two shared electrons in the chlorine molecule, which are located between the chlorine atoms in the Lewis structure, are called the **bonding pair.** The remaining 6 pairs of electrons are called **lone pair electrons.** Each pair of bonding electrons is said to make up one **covalent chemical bond.** We indicate a covalent bond by drawing a line between the atoms as shown on the right in Figure 2-6. The line represents a pair of electrons in a chemical bond.

It is possible to have more than two shared electrons (a single covalent bond) between two atoms in a molecule. An excellent example of a molecule that has four shared electrons (a double bond) between the same atom is oxygen. Neutral oxygen (Group VI) has six valence electrons, and therefore needs two electrons to achieve an octet. Each oxygen atom must share two electrons with the other oxygen atom. This places four electrons in the space between these two atoms. An oxygen molecule thus has a double bond between the O atoms, as you can see in Figure 2-7. Notice how the Lewis dots of the single atoms of oxygen are redrawn to make the arrangement of the bond electrons clearer.

REMEMBER

See this with your **Web Animator** ▼

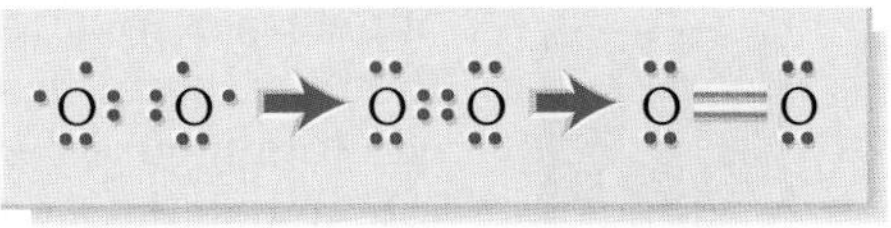

Figure 2-7 ▶ The formation of an O═O bond.

$$\cdot\ddot{\text{O}}\text{:} \quad \text{:}\ddot{\text{O}}\cdot \quad \rightarrow \quad \ddot{\text{O}}\text{::}\ddot{\text{O}} \quad \rightarrow \quad \ddot{\text{O}}\text{═}\ddot{\text{O}}$$

Binary Diatomic Molecular Compounds

✔ **Meeting the Goals**

We form a proper skeleton for the molecule and account for the valence electrons around each atom. Then we form bonds with adjacent atoms to create a structure with eight electrons (two for H) around each atom.

Binary diatomic compounds, made from two different elements, present no new challenges in the drawing of Lewis dot structures. We allow the atoms to share the electrons needed to bring each to a count of eight if the element is in the second period or below. A special case is the element hydrogen. It requires only two electrons (this gives the H the same number of valence electrons as the nearest noble gas, helium). Consider HF, made from hydrogen (Group I, one valence electron, needs one more to reach its required pair) and fluorine (Group VII, seven valence electrons, needs one more to reach its required octet). Both atoms achieve their goals by sharing one electron from each. A single bond results, as shown in Figure 2-8.

Figure 2-8 ▶ The formation of an H—F bond.

$$\text{H} \cdot \ddot{\underset{\cdot\cdot}{\text{F}}} : \longrightarrow \text{H} : \ddot{\underset{\cdot\cdot}{\text{F}}} : \longrightarrow \text{H} - \ddot{\underset{\cdot\cdot}{\text{F}}} :$$

Example 2.7   **Drawing Lewis structures**

Draw the Lewis dot structure of carbon monoxide, CO.

Thinking Through the Problem We can find the valence electrons for both C and O by locating them on the periodic table. We arrange the valence electrons on each atom to give the other atom access to eight. Note that the Lewis dots of the neutral carbon atom have been regrouped in order to clarify the way the bond arises from the three pairs of electrons.

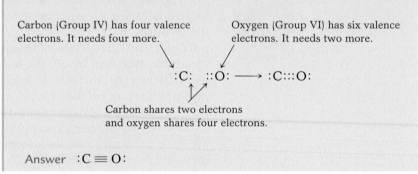

Carbon (Group IV) has four valence electrons. It needs four more.

Oxygen (Group VI) has six valence electrons. It needs two more.

$$:\text{C}: \quad ::\text{O}: \longrightarrow :\text{C}:::\text{O}:$$

Carbon shares two electrons and oxygen shares four electrons.

Answer $:\text{C} \equiv \text{O}:$

How are you doing?   **2.4**

Draw the Lewis dot structure of chlorine monofluoride, ClF.

Polyatomic Ions

REMEMBER

Sometimes a cation or anion in an ionic substance consists of more than one atom. For example, the substance NaClO, sodium hypochlorite (see Figure 2-9), is composed of a sodium cation (Na$^+$) and an anion, hypochlorite, ClO$^-$, which consists of two different atoms, Cl and O. We call an ion that contains different atoms a **polyatomic ion.**

Figure 2-9 ▶ Sodium hypochlorite is the active ingredient in chlorine bleach, which is commonly used as a disinfectant. (David Young–Wolff/Photo Edit.)

KEY IDEA 2.3

When we draw a Lewis structure for a polyatomic ion we must account for the charge by changing the number of valence electrons. *A positive charge on the ion means that there are more protons than electrons in the ion.* This means that one or more of the atoms in the ion has lost some of its original electrons. *A negative charge on the ion means that there are more electrons present than in the neutral molecule.* We account for this by adding valence electrons to the Lewis structure of the ion.

Example 2.8   **Drawing Lewis structures**

Draw the Lewis structure for NO^+.

Thinking Through the Problem The neutral nitrogen atom has five valence electrons. The neutral oxygen atom has six. We know that one or the other atom lost an electron in order to get the positive charge on the polyatomic ion. It does not matter which atom we take the electron from; let's say it is nitrogen. Then we construct our molecule from a nitrogen atom with four electrons and an oxygen atom with six electrons.

(a) Nitrogen atom has four electrons ⇒ Needs four more ⇒ Oxygen must share four.

(b) Oxygen atom has six electrons ⇒ Needs two more ⇒ Nitrogen must share two. These two statements tell us how to lay out the molecule: the nitrogen shares two electrons with oxygen and the oxygen atom shares four electrons with nitrogen, as shown below. The result is six shared electrons, a triple bond, between N and O.

Answer $\ddot{N}:\ \ :\!:\!\ddot{O} \longrightarrow \dot{N}\!:\!:\!:\!\ddot{O} \longrightarrow [\ddot{N}\!\equiv\!\ddot{O}]^+$

Again, the Lewis dot arrangements of electrons in the neutral atoms have been redrawn to show more clearly how the bond can be pictured.

Note that we check our result by counting to be sure there are eight electrons around both the nitrogen and the oxygen. We place brackets around the polyatomic ion. Because the charge is really spread over the whole molecule, we indicate its presence by a positive sign above and to the right of the molecule.

Example 2.9   Drawing Lewis structures

✔Meeting the Goals

To draw Lewis structures for polyatomic ions, follow the same procedure given for neutral molecules, adjusting the electron count to accommodate the charge on the ion. The charges are not truly assigned to any one atom, but we need to add (for a negative ion) or subtract (for a positive ion) the electrons from an atom.

Draw the Lewis structure for ClO^-.

Thinking Through the Problem For an anion with a -1 charge we add an extra electron. Here we can add it to the oxygen. The oxygen atom (group VI = 6 valence electrons) now has seven electrons and the chlorine seven, also. Both atoms need one more electron to reach eight, so both share one electron with the other atom. Because the charge is really spread over the whole polyatomic ion, its presence is indicated by a negative sign above and to the right of the ion.

Answer $:\ddot{\underset{..}{Cl}}\cdot \quad \cdot\ddot{\underset{..}{O}}: \quad \longrightarrow \quad :\ddot{\underset{..}{Cl}}:\ddot{\underset{..}{O}}: \quad \longrightarrow \quad \left[:\ddot{\underset{..}{Cl}}-\ddot{\underset{..}{O}}:\right]^{-}$

How are you doing?   2.5

Because it does not matter which element gives up the electron, we ought to get the same final structure for the NO^+ molecule if we remove the electron from oxygen. Determine the electron dot structure for NO^+ starting from a nitrogen atom with five electrons and an oxygen atom with five electrons.

Triatomic and Higher Polyatomic Molecules

We have presented a method for drawing Lewis structures that is quite simple: have each atom share enough electrons to allow each atom to have a stable octet of electrons (or in the case of H, 2 electrons). The same idea works for molecules and polyatomic ions that have three or more atoms. However, it is helpful if we follow a series of steps to do this, so we can make sure that all atoms have the number of valence electrons they need. These steps are summarized in Table 2-3.

Table 2-3   How to Draw Lewis Structures of Binary Molecules

1. Lay out the atoms in the proper skeleton.
2. Draw the proper number of valence electrons on each atom.
3. Connect electrons so that each of the *outer* atoms has the proper number of electrons. This is eight electrons (second period or below) or two electrons (H).
4. Evaluate the central atom to be certain that there are at least eight electrons around it. If there are not, then move one or more lone pairs from the outer atoms to a bonding position.
5. Convert each bonding pair into a neat line between the atoms.

Here is how this method works for a typical molecular substance, SiH_4:

H
H Si H
H

1. **Lay out the atoms in the proper skeleton.** This can be a complicated procedure when there are different ways to put the atoms together. But here we will deal with systems that have a single atom in the center and multiple atoms on the outside. Generally, **the first atom in the formula is in the center** (an exception is H_2O). In addition, **the other atoms are attached only to the central atom, not to each other.** This is summarized for the molecule silane, SiH_4, in the picture at left.

2. **Draw the proper number of valence electrons on each atom.** For silane this means the Si gets four electrons and each H gets one. Note that we are thinking ahead: we know that each H will need one more electron, so we arrange the four Si electrons to "point at" the H atoms.

3. **Connect electrons so that each of the *outer* atoms has the proper number of electrons.** In this case, we share one electron from Si with each of the H atoms. We can draw this by circling the electrons to make pairs for the H atoms.

4. **Evaluate the central atom to be certain that there are at least eight electrons around it.** This is already true in this case.

5. **Convert each bonding pair into a neat line between the atoms.** The bonding pairs are converted into lines. Our drawing is now complete.

Example 2.10   Drawing Lewis structures

Draw the Lewis structure for ammonia, NH_3.

Thinking Through the Problem According to the plan we have:

1. The skeleton consists of a central nitrogen and three hydrogens around the outside.

2. The nitrogen, in Group V, has five electrons and the hydrogens each have one electron.

3. The hydrogen atoms *each* need one more electron to have a stable pair. They will accept one apiece from the central nitrogen to form a bond.

4. The structure we have after satisfying the H's is also satisfactory for the N.

5. The final drawing has one lone pair on N and three bonding pairs, one between each H and the N.

Working Through the Problem   Answer

H N H	H··N̈··H	H⊙N̈⊙H	H—N̈—H
H	H	H	H
Step 1: Skeleton	Step 2: Valence electrons	Steps 3/4: Satisfy outer and inner atoms	Step 5: Convert bonding pairs

Example 2.11   **Drawing Lewis structures**

Draw the Lewis structure for carbon dioxide, CO_2.

Thinking Through the Problem According to the plan we have:

1. CO_2 consists of a central carbon and two outer oxygen atoms.
2. Each oxygen has six valence electrons (and needs two), and the C has four valence electrons (and needs four more).
3. We can obtain eight electrons around each O by sharing two electrons from the central C with each O.
4. In this case, step 3 leaves a structure with just four electrons around C. We give the C four more electrons by "moving" a lone pair on each O into a bonding pair.
5. The final structure has two C=O double bonds and two lone pairs on each O.

Working Through the Problem

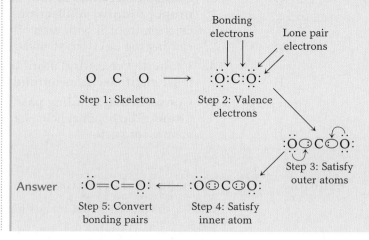

Example 2.12   **Drawing Lewis structures**

Draw the Lewis structure for NCl_3.

Thinking Through the Problem According to the plan we have:

1. As with ammonia, the skeleton consists of a central nitrogen atom and three chlorine atoms around the outside.
2. The nitrogen has five electrons, and it will share one with each chlorine atom. The chlorine atoms each have seven electrons.
3. Each Cl will need one electron. They will share one apiece with the central nitrogen. The drawing then has a satisfactory number of electrons around N.
4. There is a single bond between nitrogen and each chlorine. There are also three lone pairs on each chlorine and a single lone pair on nitrogen.

Answer $:\overset{..}{\underset{..}{Cl}}-\overset{..}{N}-\overset{..}{\underset{..}{Cl}}:$

$\overset{|}{\underset{..}{\underset{:Cl:}{\vphantom{x}}}}$

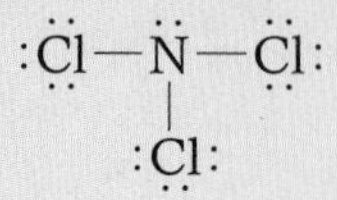

2.6

Draw the Lewis dot structure for water, H_2O, and carbon tetrachloride, CCl_4.

The Lewis structure is built from the number of valence electrons that we expect for the atoms in the molecule. That means that if we *count* the number of electrons in the final structure we can check to see if the structure is correct. For example, in the structure of NCl_3 we have ten lone pairs and three bonding pairs. That means we have drawn a structure with $2 \times (10 + 3) = 26$ electrons. Is this correct? The structure started with one N (5 electrons) and 3 Cl's ($3 \times 7 = 21$ electrons), for a total of 26 electrons. A structure that, as we see here, has the correct number of valence electrons is very, very likely to be correct.

Example 2.13   **Checking Lewis structures**

In the three structures below, count the valence electrons and compare them to the expected number.

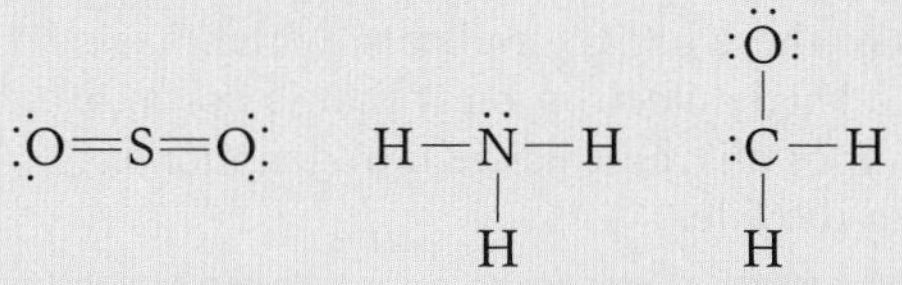

$$:\overset{..}{O}=S=\overset{..}{O}: \qquad H-\overset{..}{\underset{|}{N}}-H \qquad :\overset{..}{\underset{|}{C}}-H$$
$$\qquad\qquad\qquad\qquad\quad H \qquad\qquad H$$

Which structures are incorrect?

Thinking Through the Problem We count the electrons in each, then count the electrons that *should* be in the structure.

Working Through the Problem First structure: 4 lone pairs + 4 bonding pairs = 8 pairs. There are 16 electrons present. One S (6 electrons) and 2 O's (12 electrons) should have 18 electrons. *Too few electrons in this structure.*

Second structure: 1 lone pair + 3 bonding pairs = 4 pairs. There are 8 electrons present. One N (5 electrons) and 3 H's (3 electrons) should have 8 electrons. *Structure is correct.*

Third structure: 4 lone pairs + 3 bonding pairs = 7 pairs. There are 14 electrons present. One C (4 electrons), 2 H's (2 electrons) and 1 O (6 electrons) should have 12 electrons. *Too many electrons in this structure.*

Answer   Only the second structure is correct.

PRACTICAL B — What important air pollutants and related ions have structures that suggest that they may be very reactive?

The existence of lone pairs of electrons on Lewis structures can be very, very important in describing the stability and the hazards associated with certain molecules. When atoms do not have a stable arrangement of electrons, they tend to react rapidly (and sometimes violently) with other atoms in order to attain the stability they lack. This means that if the Lewis structure of a molecule does not show a stable arrangement of electrons around each of its atoms, then we can (correctly) guess that the molecule may be dangerous.

Radicals

The Lewis structures that you have learned about all have the same principle "aim": the formation of a group of eight electrons around a representative element, or two electrons around a hydrogen. But structures that don't meet this criterion can be found. One important class of these are the radicals, which have one atom with a single electron on it that is *not* in a lone pair. Two important radicals are nitrogen monoxide and nitrogen dioxide.

$$:N\!\!=\!\!\ddot{O}:   \qquad :\ddot{O}\!\!-\!\!N\!\!=\!\!\ddot{O}:$$

These are both major air pollutants, in part because they need to react with other molecules or atoms to rectify their unpaired electron problem.

Verify that both of these structures have the correct number of valence electrons.

Another important radical in biological systems is the molecular ion "superoxide," which has the formula O_2^-. Determine the Lewis structure of superoxide (but beware that one atom will not have an octet of electrons!).

Radicals do not just occur in the gas phase. They also occur in condensed phases. One very important radical is OH. Another is CH_3. Draw their Lewis structures. Do you notice the unpaired electron in each case?

An interesting thing occurs when radicals react with one another. Consider how two OH molecules can combine to make hydrogen peroxide, H_2O_2. The NO_2 radical itself forms a compound, N_2O_4, that contains a N—N single bond. How does this bond "take care" of the unpaired electron?

PROBLEMS

12. Give a Lewis dot structure for the following atoms and ions. Indicate how many electrons are needed to make an octet of electrons around the atom.

  (a) Ge
  (b) I and I^{3+}
  (c) P and P^{3-}

13. Draw the Lewis dot structures of Br_2, HBr, and BrF.

14. Draw the Lewis dot structures of SF^-, O_2^{2-}, and NO^-. Note the number of bonds between the atoms.

15. Draw the Lewis dot structures of PH_3, NH_2^-, BH_4^-, and CH_3^-.

16. Discussion question: Draw the Lewis dot structures of $COCl_2$, which has carbon at the center, and $SOCl_2$, which has S at the center.

How do these molecules differ? What effect does moving from C to S have?

17. Give the Lewis dot structure of the following molecules. Include all lone pairs.

  (a) CO
  (b) H_2S
  (c) HCN (C in center of H and N)

Verify that each of your answers has the correct number of valence electrons present.

18. Draw the structures you expect for the following molecules. A single atom (the first) is in the center.

  (a) NOCl
  (b) CH_2Br_2

2.3 Additional Features of Lewis Structures

SECTION GOALS

✔ Draw structures of larger charged polyatomic ions in a systematic way.

✔ Draw resonance Lewis structures in appropriate cases.

✔ Draw Lewis structures of elements with more than eight electrons around certain atoms.

Lewis Structures of Polyatomic Ions

Now we are ready to deal with larger polyatomic ions that contain *more than two* atoms. An excellent example is the chlorate ion, ClO_3^-, a negatively charged ion formed by one atom of chlorine and three atoms of oxygen. The central atom in ClO_3^- is the chlorine and the three oxygen atoms are on the outside, bonded *only* to the chlorine as shown in Figure 2-10. The single negative charge on this ion means there is one more electron than we would have if all the atoms were neutral.

The only question for us to worry about is where to put the extra electron. *For convenience, we add it to the electron count of the central atom.* (The charge is considered to be spread out over the entire molecular ion but, to draw a Lewis structure, we must place the electron dot somewhere in the drawing. It is easiest to put it on the central atom.) Then, we note that each oxygen requires one pair of additional electrons from the chlorine. These electrons form the Cl—O bonds. The chlorine, with eight electrons in the drawing, can share. In addition, the eight electrons on chlorine means that it already satisfies the octet rule. So our picture is complete with one bond between the chlorine and each of the oxygen atoms and one lone pair on chlorine and three lone pairs on each of the oxygens. Since the charge is really spread over the whole molecule, we indicate its presence by a negative sign above and to the right of the molecule.

It is very important that we remember that there is a real charge on the whole molecule. Therefore, if we look at the total number of valence electrons in a Lewis structure, it should match the number of

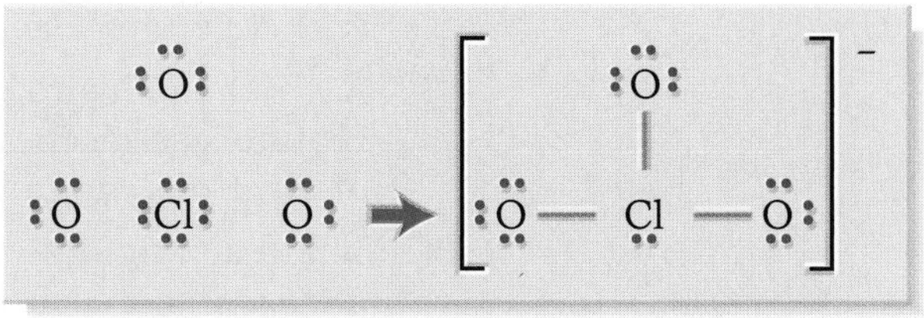

Figure 2-10 ▲ The Lewis structure of chlorate.

electrons we expect for the atoms in the structure, with additional electrons in the case of negative ions and fewer electrons in the case of positive ions.

For example, in the structure of chlorate above, the final structure has 10 lone pairs and 3 bonding pairs, for a total of 26 electrons. This is what we expect for one chlorine (7 valence electrons), three oxygens ($3 \times 6 = 18$ electrons) and one negative charge (1 valence electron), so that $7 + 18 + 1 = 26$. As we saw in the last section, counting electrons is a very effective way to check the final result in a Lewis structure problem.

The step-by-step procedure, shown in Table 2-4, for the Lewis structure of polyatomic ions is a simple modification of the one we used for binary molecules themselves. The modification occurs in step 2.

Table 2-4   A Scheme for Lewis Structures of Polyatomic Ions

REMEMBER

1. Lay out the atoms in the proper skeleton.
2. Draw the proper number of valence electrons on each atom. *If we have a polyatomic ion, adjust the number of valence electrons on the central atom to reflect the charge: Remove electrons if the ion has a positive charge and add electrons if there is a negative charge.*
3. Connect electrons so that each of the *outer* atoms has the proper number of electrons.
4. Evaluate the central atom to be certain that there are at least eight electrons around it. If there are not, then move one or more lone pairs from the outer atoms to a bonding position.
5. Convert each bonding pair into a neat line between the atoms.

Example 2.14   **Drawing Lewis structures for polyatomic ions**

Draw the Lewis structure for the ammonium ion, NH_4^+.

Thinking Through the Problem We lay out this molecule with one electron missing from the nitrogen. Removing electrons from the central element in a positively charged ion (a cation) is the best approach to handling positive charge in larger molecules.

1. The central atom will be N in this case.
2. When we remove the valence electron, the central nitrogen will have four electrons.
3. Each of the hydrogen atoms will have one electron and will require one electron. Doing this leaves a satisfactory number of electrons around N.
4. The resulting drawing has one single bond between the nitrogen and each hydrogen and no lone pairs. Since the charge is really spread over the whole molecule, we indicate

See this with your **Web Animator** ▶

its presence by a positive sign above and to the right of the molecule.

Working Through the Problem			Answer

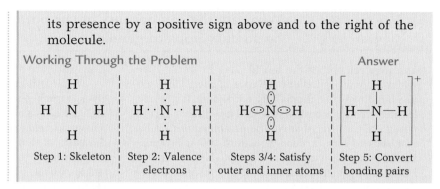

| Step 1: Skeleton | Step 2: Valence electrons | Steps 3/4: Satisfy outer and inner atoms | Step 5: Convert bonding pairs |

How are you doing? 2.7

Draw the Lewis dot structures for hydronium ion, H_3O^+, and PCl_4^+.

Molecules with More than One Acceptable Structure: Resonance

Our survey of Lewis dot structures has focused on molecules where one—and only one—structure is acceptable, but there are some molecular compounds and polyatomic ions where two or more equivalent structures can be drawn. In such a case, **resonance structures** can be drawn. *Resonance structures are equivalent Lewis structures that differ from one another in the placement of single, double, or triple bonds.* Resonance structures occur when the electrons are more delocalized (not confined to one place or locality) than our Lewis drawings can show.

It is not necessary that you learn to predict when resonance will be found. When present, resonance structures will result as a natural outcome of drawing Lewis dot structures. The pivotal step in the drawing process occurs when we seek to ensure that all atoms have satisfactory numbers of electrons. Whenever we have a choice about *which* lone pairs to convert into a bond, we know that we will have two or more resonance structures!

As an example, let's draw the Lewis structure for the nitrate anion, NO_3^-. The structure has an N in the center and three O's on the outside (step 1). Next we assign the negative charge to the N, so it has six electrons. Each of the three oxygen atoms around the outside of the ion has six electrons (step 2). For step 3 we make sure that each O gets two electrons from the nitrogen. This gives us the following structure:

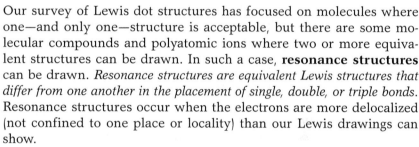

Structure after step 3

For step 4, we need to get two more electrons to the N atom. We could have the nitrogen atom take one electron from two different O

PRACTICAL C — What are the components of the nitrates, and why can they be so dangerous to handle?

You may remember the terrible day in April, 1996, when a bomb ripped out the core of the Federal Office Building in Oklahoma City, Oklahoma. It was later learned that the bomb was fueled by nitrates, a component of fertilizer that can be combined with other materials to form a deadly mixture. What few may know, however, is that this was not the first time in history that Americans experienced a disaster caused by a nitrate-fueled explosion. Shortly after the end of the Second World War, in April 1947, a container ship in Texas City, Texas, caught fire. Soon after, as firefighters rushed to the scene, the ship exploded as its load of ammonium nitrate, NH_4NO_3, detonated. Photographs of the destruction at Oklahoma City and Texas City are shown in Figures 2-11(a) and 2-11(b).

What is it about the polyatomic ion nitrate that gives it such potential? The answer lies in examining its Lewis structure, and that of related powerful nitrogen based explosives like TNT and nitroglycerin, the basis of dynamite.

Compare the structures below to the relatively stable radicals NO and NO_2 discussed in Practical B. Do they share features with those molecules that suggest they also form radicals?

For these structures only one resonance structure is provided. Which of the atoms are likely to be sites of resonance? Draw at least one resonance structure where appropriate.

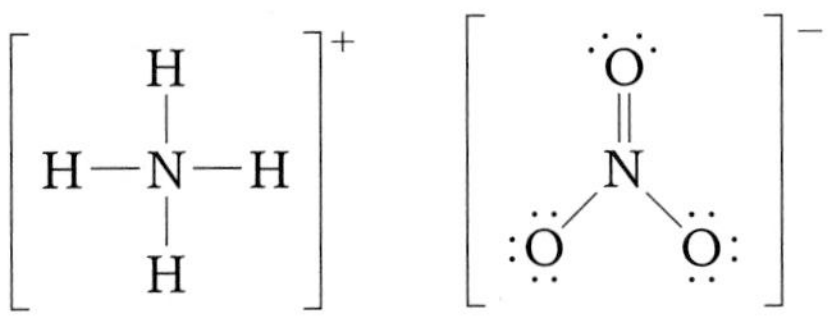

Ammonium nitrate

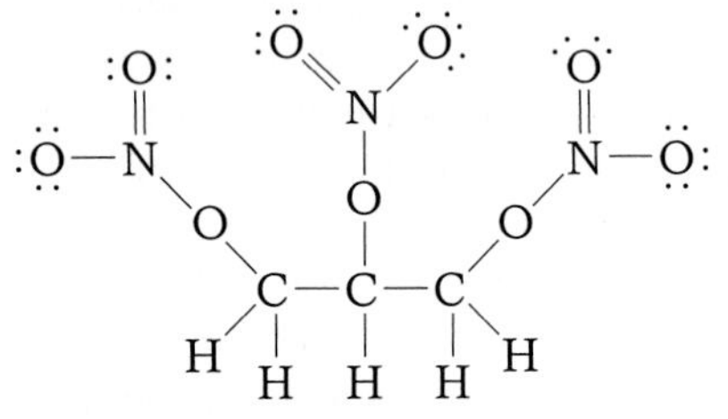

Nitroglycerine

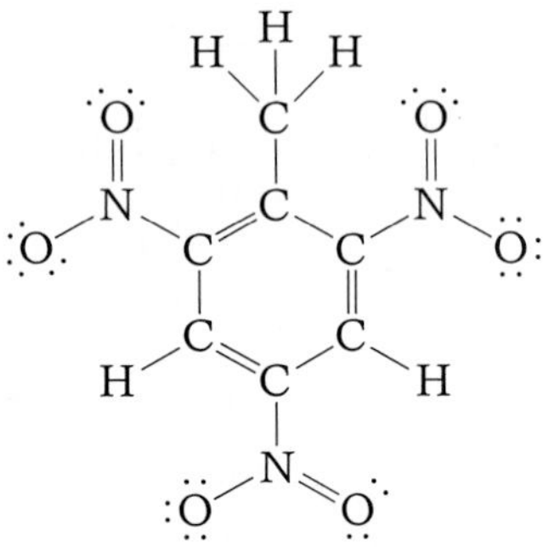

TNT

atoms, but this would leave electrons *unpaired*, and this should be avoided whenever possible. Thus, it is better if the nitrogen atom gets two electrons from any one of the oxygen atoms. Which one? It doesn't matter, so let's pick the bottom oxygen atom:

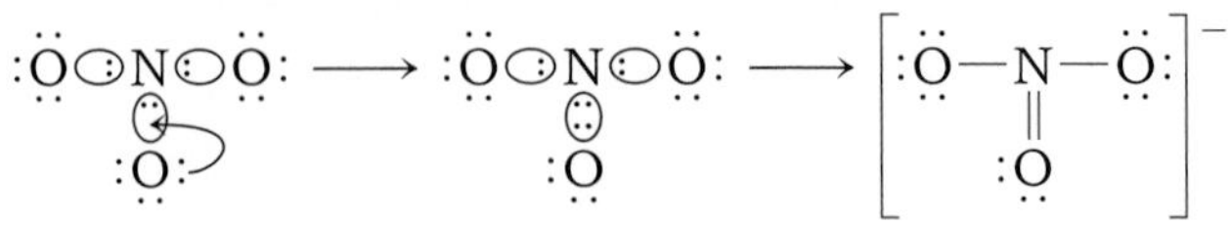

We made a choice that the bottom oxygen atom would share two electrons, creating a double bond between it and the nitrogen atom. The resulting Lewis structure satisfies the octet rule. We might have chosen either the oxygen atom on the left or the oxygen atom on the

Figure 2-11 ▲ (a) The Federal Building was devastated by a nitrogen-based explosion in the criminal bombing at Oklahoma City, Oklahoma, in 1996. (b) A container ship exploded in Texas City, Texas, in 1947, after an accidental fire detonated the ship's cargo of ammonium nitrate. (a: B. Daemmrich/The Image Works; b: UPI/Corbis-Bettmann.)

Assuming that the atoms in these molecules are connected properly, how could you efficiently determine that these structures have been drawn correctly?

Why do these molecules cause such dangers? The answer lies in the potential for all of them to decompose with the release of energy and the formation of N_2 gas. In the process, oxygen atoms and radicals are created.

There is, however, a molecule that is an interesting exception to the rule that compounds containing N bonded to O pose great hazards. That molecule is dinitrogen monoxide, N_2O:

$$2 \;:\!N\!\equiv\!N\!-\!\ddot{O}: \longrightarrow 2 \;:\!N\!\equiv\!N\!: + \;:\!\ddot{O}\!=\!\ddot{O}:$$

This can decompose under *mild* conditions to give N_2 and O_2 gas.

right. This leads to *three equivalent choices* of which oxygen atom will share two electrons with nitrogen. *We must draw all three structures:*

$$\begin{bmatrix} :\ddot{O}-N-\ddot{O}: \\ \| \\ :O: \end{bmatrix}^{-} \longleftrightarrow \begin{bmatrix} :O=N-\ddot{O}: \\ | \\ :\ddot{O}: \end{bmatrix}^{-} \longleftrightarrow \begin{bmatrix} :\ddot{O}-N=O: \\ | \\ :\ddot{O}: \end{bmatrix}^{-}$$

Taken together, we have three resonance structures for nitrate. All three are equally valid, and we symbolize this with the double-headed arrows among the drawings.

Resonance structures can be very worrisome to the beginner. When can you expect them? How can you be sure that you have them all? One thing to recognize is that most resonance structures differ in the positions of single and double bonds. Resonance structures are present when we have a choice of where to put a double bond.

How are you doing?   **2.8**

Draw the Lewis structure for the nitrite anion, NO_2^-. Show both of the possible structures.

Example 2.15   **Drawing resonance structures**

◀ Try this with your **Personal Web Tutor**

Draw the Lewis structure of ozone, O_3.

Thinking Through the Problem We know that oxygen is in group VI, so each oxygen has six valence electrons. We draw the skeleton with the oxygen atoms connected in a line.

Working Through the Problem Place the oxygen atoms next to one another with one oxygen atom in the center of the other two. The outer oxygen atoms take two electrons each from the central oxygen. The center oxygen also needs two electrons, and we make a *choice* of taking them from the oxygen on the right, first. This gives a single bond to the oxygen on the left, and a double bond to the oxygen on the right. Since the structure we developed for ozone was made with an arbitrary choice, we expect resonance. The second structure has the double bond on the left. This is the other resonance structure for ozone.

Answer

$$:\ddot{O}-\ddot{O}=\ddot{O} \longleftrightarrow \ddot{O}=\ddot{O}-\ddot{O}:$$

✔ Meeting the Goals

A resonance structure exists when you can draw two or more equivalent Lewis dot structures. Resonance structures differ from one another in the location of double and single bonds.

How are you doing?   **2.9**   ◀ Try this with your **Personal Web Tutor**

Draw the Lewis resonance structures for carbonate ion, CO_3^{2-}.

Octet Expansion

We have developed rules based on the octet rule, where we expect eight electrons to have a stable structure for the atom's valence electrons. Do these rules apply to all compounds? For example, how can we explain such odd molecules as SF_6, a very stable compound that must have six S—F bonds, requiring twelve electrons on the S? Or we may be confronted by the fact that oxygen and fluorine both react under the correct conditions with the noble gas elements Kr and Xe to make compounds and ions such as KrF_2, H_2XeO_4, and XeF_4O.

The answer to such puzzling molecules is to accept their existence as clear evidence that the *third and higher period elements can expand to accommodate more than eight electrons.* This is sometimes called **octet expansion.** With this in mind, we can now explore a host of interesting compounds, based on the principles we have already developed.

Octet expansion does not occur in the second period elements (B, C, N, O, F) and H is always stable with two electrons (the helium configuration). For other elements, we expand the octet only when needed: if we can draw a structure that has just eight electrons on an element in the third or higher periods, we accept it. Octet expansion, like resonance, is something we find is the inevitable consequence of drawing Lewis structures of certain molecules.

Let's see how it works with the example of sulfur tetrafluoride, SF_4. We carry out the same steps as we did earlier, but after step 3 we find that there are ten electrons on S:

F F S F F	:F̈ :F̈: S :F̈ :F̈:	:F̈ S F̈: :F̈ F̈:
Step 1	Step 2: Orient 4 dots on S at the F's	Step 3

The S at this point (step 4) will have to have ten electrons around it: eight in bonding pairs, two in a lone pair. While we would be concerned if S had *fewer* than eight electrons, more than eight is fine. The structure is finished with four S—F single bonds:

$$:\ddot{F}\,S\,\ddot{F}: \quad \longrightarrow \quad :\ddot{F}\!-\!S\!-\!\ddot{F}:$$

Example 2.16   **Drawing Lewis structures with octet expansion**

Sketch the Lewis structure of the $XeOF_4$ molecule.

Thinking Through the Problem This molecule has eight valence electrons from Xe, six from O, and $4 \times 7 = 28$ from F, for a total of 42 valence electrons. We expect the Xe at the center. Xe will have one single bond to oxygen (and the oxygen will have three lone pairs). We expect one single bond between Xe and each of the F's (with three lone pairs on each F). At this point, we find the structure contains forty valence electrons, so the extra two electrons must be on the Xe in a lone pair.

Answer

$$\begin{array}{c} :\ddot{O}: \\ | \\ :\ddot{F} \diagdown \diagup \ddot{F}: \\ Xe \\ :\ddot{F} \diagup \diagdown \ddot{F}: \end{array}$$

How are you doing? **2.10**

Sketch the Lewis structures for KrF_2 and for SF_6.

PROBLEMS

19. Draw all possible resonance structures of SO_3.

20. Determine the Lewis structure of NF_4^+.

21. Draw the Lewis dot structures of SiO_4^{4-}, PO_4^{3-}, SO_4^{2-}, and ClO_4^-. Note any trends that you observe in the structures and in the number of electrons on each structure.

22. Give the Lewis dot structure of the polyatomic ion BrO_4^-.

23. Give the Lewis dot structure of the following compounds or ions.
(a) H_2Se (b) SO_2

24. Draw the Lewis structure for each of the following molecules.
(a) Cl_2O (O in center)
(b) PF_3
(c) $NOCl$ (N in center)

25. Draw the Lewis structure for each of the following ions.
(a) ClO_3^- (b) PF_4^+ (c) NO_2^-

26. Draw the Lewis structure for the following molecules. The carbons are in the center.
(a) C_2H_6 (b) C_2H_4 (c) C_2H_2

27. Draw a scheme for the assembly of BrO^-, BrO_2^-, BrO_3^-, and BrO_4^-. Because Br is a fourth period element, it can have more than an octet of electrons. Yet there is no known BrO_5^- ion. Examine your work and suggest why we do *not* expect BrO_5^- ion to exist.

28. (a) Draw the Lewis structure for the molecules IF_3 and IF_5.
(b) The molecule IF_7 exists. Can you predict its Lewis structure?

29. Web question: For the illustrations in Web Animator Problem 2-35, determine the formula of the compounds shown. Redraw the molecules as Lewis structures. C, H, and O are present. Use the rules for common bonding patterns to determine the location of any lone pairs.

30. Web question: For the illustrations in Web Animator Problem 2-36, determine the formula of the molecules shown. C, H, and N are present. Redraw the molecules as Lewis structures. Use the rules for common bonding patterns to determine the location of any lone pairs.

Chapter 2 Summary and Problems

◀ Review with **Web Practice**

VOCABULARY

molecule	An arrangement of atoms with a fixed formula and set of connections.
molecular substance	A substance made up of molecules as its basic building blocks.
diatomic	A diatomic group of atoms contains only two atoms; usually refers to a molecule with only two atoms.
nomenclature	A system of naming things by definite rules; in chemistry this includes recognizing if a substance is likely to be molecular.
systematic name	A name given to a substance according to a general naming system.
trivial name	A nonsystematic name given to a substance; a trivial name is often the most common way to refer to certain substances, like water.
binary compound	A compound that contains two and only two elements.
chemical bonding	The way atoms are connected to one another in a chemical substance.
Lewis dot	A representation of an electron in an atom or molecule by a dot that represents one electron.

Lewis structure	A representation of the electrons and the bonding in molecules and ions where the valence electrons are drawn as dots or (in bonds) as lines.
ion	An atom or group of atoms that have an electrical charge.
anion	An ion that carries a negative charge.
cation	An ion that carries a positive charge.
ionic	Pertaining to interactions between charged particles, especially in substances formed through ionic bonding.
covalent	Pertaining to interactions involving sharing of valence electrons, particularly for molecular substances whose intramolecular interactions are covalent in character.
octet rule	A rule whereby molecules and polyatomic ions are represented so that each non-hydrogen atom has eight valence electrons.
bonding pair	Electrons, in pairs, associated with bonding two atoms.
lone pair electrons	Electrons, in pairs, associated with just one atom.
covalent chemical bond	A connection between two atoms involving the sharing of valence electrons.
polyatomic ion	An ion that contains two or more atoms.
resonance structures	Two or more acceptable Lewis structures for a molecule where the two structures have the same number of bonds and lone pairs.
octet expansion	Formation of valid Lewis structures where an element in the third period or below has more than eight valence electrons.

SECTION GOALS

What information is given by a chemical formula of a substance?

A chemical formula includes the symbols of the elements and subscript numbers that indicate the number of times the element appears in the formula.

What are the numbering schemes that we use in naming molecular substances?

The elements in a molecular compound are named using prefixes that correspond to the number of times that element appears in the formula.

What are the rules for naming molecular substances?

To name a molecular compound, name each element giving the second element an *-ide* ending. Use prefixes to count the number of times each element appears in the formula. (An exception to this is that the prefix *mono-* is not used if the element appears first in the formula. Thus, CO is simply carbon monoxide, not monocarbon monoxide.)

How do we draw the Lewis structure of an atom of a representative element or ion?

We begin by looking at the periodic table to find the group number of the element. The Roman numeral we use to indicate the group number of a representative element is the number of valence electrons in a neutral atom of the element. If there is a positive charge, we remove valence electrons and if there is a negative charge we add valence electrons.

What steps do we use to draw the Lewis structure for binary compounds?

We form a proper skeleton for the molecule and account for the valence electrons around each atom. Then, starting from the outermost atoms, we form bonds with adjacent atoms to create a structure with eight electrons (two for H) around each atom.

What is needed to draw the Lewis structure for polyatomic ions?

We need to add (for a negatively charged ion, or anion) or subtract (for a positive ion, or cation) the electrons from an atom to assemble an ion with the right charge.

What changes do we make to draw structures of larger polyatomic ions?

We can work the same way that we do with diatomic ions. But for the sake of consistency, we make adjustments to the central atom when we start to construct the Lewis structure of a large polyatomic ion.

How do we know when to draw resonance Lewis structures?

Resonance structures arise when we have two or more possible equivalent Lewis structures, usually involving more than one O atom. We know we need to use these if we have to make an arbitrary choice about drawing a single or double bond to two atoms on a central atom.

What do we do when we find more than eight electrons on an atom?

Atoms of elements in the third and higher periods of the periodic table can have more than eight valence electrons. We do not set out to give such an atom more than eight, but in certain cases it will be inevitable. There is no problem with such a structure.

PROBLEMS

31. Determine the number of valence electrons in the following atoms and ions.

(a) N^+ (c) N^- (e) H^- (g) B^{3-}
(b) C^{2-} (d) Si^{4-} (f) Br^+

32. In the first chapter, you were asked to remember which elements are found in normal conditions as diatomic elements. Draw the Lewis structure of each of these.

33. A few years ago chemists determined that the anion AlF_4^- had a notable structure, something *you* can predict. What do you expect is the structure of this anion?

34. Draw the Lewis structures of the following molecules. (a) AsH_3 (b) ClF_5

35. Evaluate the structures in the table below to see if they have a reasonable number of bonds and electrons.

Structure	Skeleton OK?	Electrons in Structure	Electrons from the Atoms	Electrons on Outer Atoms	Electrons on Central Atom

36. For those Lewis structures from Problem 35 that are "wrong," redo them properly here.

37. One of these three structures is not a valid Lewis structure. Which one is it, and why is it invalid? Redraw the structure correctly.

SF₄   SO₂   OF₂

38. Underline the compounds in the following list that are expected to have a molecular structure.

$TiCl_4$ SeH_2 $NiCl_2$ BF_3
BaH_2 Cs_2O_2 SO_3

39. Give the systematic names for the following compounds. You may have to look up the names of elements below the third row of the periodic table.

(a) OF_2 (b) N_2O_4 (c) CS_2

40. Give a Lewis dot structure for the following atoms and ions. Indicate how many electrons are needed to make an octet of electrons around the atom.

(a) O and O^+
(b) C and C^{2-}
(c) F and F^-

41. Give the Lewis dot structure of the following molecules. Include all lone pairs.

(a) CBr_4 (b) $SiCl_4$ (c) SO_2

Verify that each of your answers has the correct number of valence electrons present.

42. Give systematic names for (a) XeO_3 and (b) SeN_2.

43. Write correct formulas for (a) antimony pentafluoride and (b) diantimony trioxide.

44. (a) What is the total number of atoms in one molecule of iodine heptafluoride?

(b) What is the total number of atoms in one molecule of dichlorine heptoxide?

45. Name Cl_2O_7, ClO_2, and Cl_2O.

46. What is the difference between valence electrons and core electrons?

47. How many valence and core electrons are there in a neutral atom of selenium? How many are there in a neutral atom of iodine?

48. Why it is dangerous to heat aerosol cans?

49. The substance not expected to form a gas under normal conditions is

(a) NCl_3.   (b) KF.   (c) SO_3.

50. Molecular substances containing a single atom bonded to oxygen, a halogen, or hydrogen are expected to

(a) explode at room temperature.
(b) form gases at room temperature.
(c) conduct heat readily.

51. What are the components of a Lewis structure?

52. What is a bonding pair of electrons? What is a lone pair of electrons?

53. Carbon monoxide has three bonding pairs and two lone pairs of electrons. How many electrons is this?

54. Nitrogen monoxide has three bonding pairs and two lone pairs of electrons. Is this number of electrons consistent with the formula, NO?

55. How many valence electrons are there in ClO^- and in ClO_4^-?

56. How many valence electrons are there in NO?

57. Draw the Lewis structure for superoxide, O_2^-. What is different about this structure?

58. What is a resonance structure?

59. Draw the Lewis structure for N_2O in which N is in the center of the drawing.

CHAPTER 3

Ionic Compounds

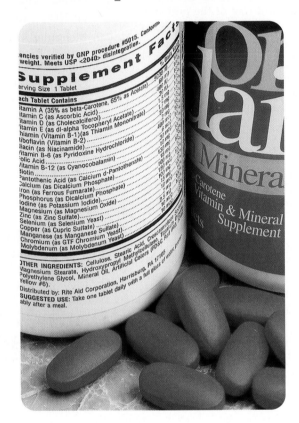

The nutrition labels from these containers of mineral supplements show the names of some ionic compounds.

(Richard Megna/Fundamental Photographs.)

PRACTICAL CHEMISTRY Ionic Compounds— Our Source of Dietary Minerals

In this chapter, we will concentrate on recognizing ionic compounds both from their names and from their formulas. You may notice that some of these same names and formulas appear on the labels of vitamin and mineral supplements and other food nutritional labels. After reading this chapter, you will be able to recognize many of these names and formulas as representing ionic compounds.

This chapter will focus on the following questions:

PRACTICAL A The table salt controversy—which alkali metal halides are beneficial to health?

PRACTICAL B Is calcium important for bones only?

PRACTICAL C Is the oxidation number of an element related to its nutritional benefit?

3.1 Ionic Compounds with Two Elements

SECTION GOALS

✔ Determine the charge and the number of electrons on each element in the formula of an ionic compound.

✔ Write the formula for binary ionic compounds by using the preferred charge of their elements.

Ionic Compounds

In Chapter 2 we discussed compounds where the bonding between atoms involves the sharing of electrons. Such sharing, which gives rise to covalent bonds, is the key to the stability of molecular compounds. To break a covalent bond requires that we break a bonding pair, leaving atoms with fewer electrons than they need for a stable configuration.

Another kind of bonding involves the very strong forces associated with ions (charged atoms or groups of atoms). Positively charged ions and negatively charged ions attract one another to form an **ionic bond.** Compounds that are formed as a result of ionic bonding are called **ionic compounds.** Ionic compounds contain a cation with a positive charge and an anion with a negative charge. Together, these charges add up to zero. A general formula for an ionic compound is A_aX_x. Convention indicates that "A" is a cation and "X" is an anion.

The presence of charges on the cations and anions of ionic compounds means that each cation or anion is a stable structure in itself. *Unlike the atoms in a covalent bond, ions do not share electrons to obtain a stable octet, and they do not form bonds in this way. Instead, ionic compounds are held together by the electrical attraction between their positive and negative charges.*

Ionic compounds are generally found in solid form, where the structure is held together by the attractions between many cations and anions. The attractions between cations and anions allows ionic compounds to form very large structures called **extended structures.** For example, sodium chloride is a solid that contains huge numbers of Na^+ and Cl^- ions (Figure 3-1). We do not need to count how many cations and anions are present in the extended structure of sodium chloride. We simply write the formula for the simplest whole number ratio of anions and cations. Because there is one Na^+ ion for every one Cl^- ion, sodium chloride has the formula NaCl. This formula exhibits the *simplest whole number ratio* of sodium to chloride ions. An important requirement of any compound is that it be electrically neutral. That is, all the charges must add up to zero. This means that in an ionic compound the sum of the charges on the anions must be equal and opposite to the sum of the charges on the cations:

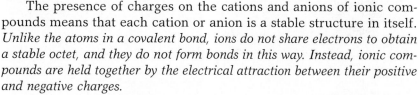

charges on cations + charges on anions = 0

charges on anions = −charges on cations

This simple rule is called the **principle of charge balance,** which states that *the charges of the ions in an ionic compound must add up to zero.*

Example **3.1**

Applying the principle of charge balance

Determine which of the following formulas satisfy the requirements of charge balance.

(a) Mg^{2+} S^{2-} (b) Na^+ S^{2-} (c) Mg^{2+} OH^- (d) Na^+ OH^-

Thinking Through the Problem ◀▷(3.1) We add the charges of the ions in each case, and see what the total charge is. If it is not zero, then the compound does not exhibit charge balance and the formula is, therefore, incorrect.

Working Through the Problem

(a) $1 \times (+2) + 1\,(-2) = 2 - 2 = 0$: charge balance OK
(b) $1 \times (+1) + 1\,(-2) = 1 - 2 = -1$: charge balance *not* OK
(c) $1 \times (+2) + 1\,(-1) = 2 - 1 = +1$: charge balance *not* OK
(d) $1 \times (+1) + 1\,(-1) = 1 - 1 = 0$: charge balance OK

Answer   Charge balance is preserved in (a) and (d). These formulas are correct.

▼ See this with your **Web Animator**

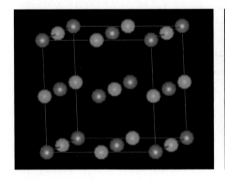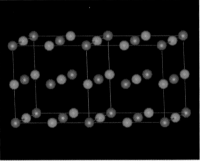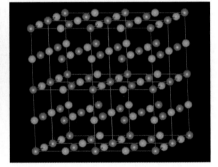

Figure 3-1 Structure of Sodium Chloride ▲

Whether we have (a) one, (b) two, or (c) more units of this compound, the structure is always the same. We write the formula for sodium chloride in terms of the simplest ratio of Na and Cl atoms, 1 : 1. The formula, then, is NaCl.

We do not need to count how many cations and anions are present in the extended structure of an ionic compound. We simply write the formula for the *simplest whole number ratio* of anions and cations.

Example **3.2**

Writing formulas for ionic compounds

For the following ionic compounds, rewrite the formula in the simplest whole number ratio

(a) Na_2Cl_2 (c) Pb_2Cl_4 (e) Ga_3Cl_9
(b) K_4O_2 (d) Al_4S_6

Thinking Through the Problem The subscript numbers are the essential item for this exercise. We look to see if the two numbers share a common factor. If so, we divide both numbers by that factor.

Answer

(a) 2 and 2 have a common factor, 2. The correct formula is NaCl.
(b) 4 and 2 have a common factor, 2. The correct formula is K_2O.
(c) 2 and 4 have a common factor, 2. The correct formula is $PbCl_2$.
(d) 4 and 6 have a common factor, 2. The correct formula is Al_2S_3.
(e) 3 and 9 have a common factor, 3. The correct formula is $GaCl_3$.

Valence Electrons and Charge

Later in this chapter we will discuss ionic compounds where A and/or X is a polyatomic ion. But first let's look more closely at the simplest ionic compounds, where both A and X are ions of a single element. Such compounds are an important example of binary compounds, which are compounds containing two elements only. NaCl, MgF_2, Al_2O_3, and SiO_2 are all examples of binary ionic compounds. Figure 3-2 shows the structure for SiO_2.

Let us examine how the atoms of elements usually change their number of valence electrons when they combine to form a binary ionic compound. Locate each of the following elements on the periodic table and determine whether it is a metal or a nonmetal. Put a check in the last column if the element is a metal.

See this with your **Web Animator**

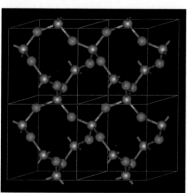

Figure 3-2 Structure of Quartz, SiO_2 ▲

The Si atoms are shown in gray and the O atoms are shown in red.

	Neutral atom	Valence electrons		Preferred ion	Valence electrons	Metal?
Magnesium	Mg	2	⇒	Mg^{2+}	0	——
Sulfur	S	6	⇒	S^{2-}	8	——
Fluorine	F	7	⇒	F^-	8	——
Rubidium	Rb	1	⇒	Rb^+	0	——

We see that the number of valence electrons is different for each of the neutral atoms of these four elements. However, when we look at the number of valence electrons present in the ion with preferred charge of each of the above elements, there is a *trend* in the numbers as each element changes from a neutral atom to an ion. Two of the ions have 0 valence electrons, and two have 8 valence electrons. Can you see a relationship between these numbers and the checks you made in the last column? Notice that the nonmetals have 8 valence electrons in their ions, while the metals have 0 valence electrons in their ions. Semimetals typically act as metals in ionic compounds, having 0 valence electrons.

This trend also applies to atoms when they are in ionic compounds. Both silicon (Si) and sulfur (S) appear in the formula SiS_2.

Silicon is in group IV; its neutral atom has 4 valence electrons. As silicon is a semimetal, it tends to lose electrons to preferably form the ion Si^{4+}, which has 0 valence electrons. Sulfur is in group VI, so it has 6 valence electrons in its neutral atom. Sulfur is a nonmetal; it gains electrons to form the ion S^{2-}, which has 8 valence electrons.

	Neutral atom	Valence electrons		Ion	Valence electrons
S in SiS_2	S	6	$\Rightarrow$	S^{2-}	8
Si in SiS_2	Si	4	$\Rightarrow$	Si^{4+}	0
Al in Al_2O_3	Al	3	$\Rightarrow$	Al^{3+}	0
O in Al_2O_3	O	6	$\Rightarrow$	O^{2-}	8

In Al_2O_3, Al is a metal and O is a nonmetal. When we examine the number of valence electrons in the neutral atoms and in the preferred ions of Al and O, we see that the trend continues. The values we observe as the "preferred" number of valence electrons for the ions in binary ionic compounds are the values chemists associate with a **closed shell** of valence electrons. In other words, if an ion has either 0 valence electrons or 8 valence electrons it will no longer tend to either gain or lose electrons. In effect, its valence electron shell is "closed."

KEY IDEA 3.3

We can now answer questions about the formulas of binary ionic compounds based on the identity of the elements. *The formulas for binary ionic compounds of metals and nonmetals usually contain the corresponding anions and cations that have achieved a closed shell of electrons.* These observations can be summed up in two rules, shown in Table 3-1, that you can use with almost all binary ionic compounds.

✔ **Meeting the Goals**

Metals lose all of their valence electrons to make a positive ion that has a closed shell of 0 valence electrons, as shown in Figure 3-3(a). The resulting charge on the metal ion is positive and is determined by the number of electrons lost. Figure 3-3(b) shows how nonmetals gain valence electrons to achieve a closed shell of 8 valence electrons. The resulting charge on the nonmetal ion is negative and is determined by the number of electrons gained.

Table 3-1 Rules for Charges on Ions of Metals and Nonmetals

While there are exceptions, the closed shell of valence electrons generally causes two things to happen in a binary ionic compound:

Metals lose all of their valence electrons to make a positive ion (cation) that has a closed shell of 0 valence electrons. Since the number of valence electrons on a metal atom is equal to the group number (in the Roman numeral scheme), the preferred charge on a metal in an ionic compound is $+m$, where m is the group number. Thus, the metals in Group I are found as $+1$ ions, those in Group II as $+2$ ions, and those in Group III as $+3$ ions.

Nonmetals gain valence electrons to make a negative ion (anion) that has a closed shell of 8 valence electrons. Since the number of valence electrons on the neutral atom is equal to the group number (in the Roman numeral scheme), the preferred charge on a nonmetal in a binary ionic compound is $-n$, where $n = 8 - m$. Thus, the nonmetals in Group VII are found as -1 ions, those in Group VI as -2 ions, and those in Group V as -3 ions.

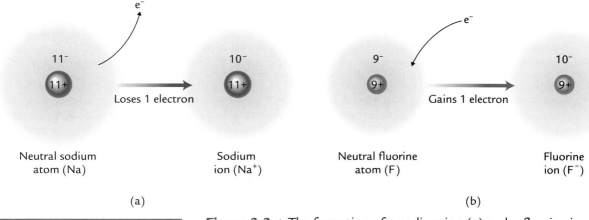

Neutral sodium atom (Na) — Loses 1 electron — Sodium ion (Na^+)

Neutral fluorine atom (F) — Gains 1 electron — Fluorine ion (F^-)

(a)   (b)

See this with your **Web Animator** ▶

Figure 3-3 ▲ The formation of a sodium ion (a) and a fluorine ion (b). The neutral sodium atom loses one electron to become a positively charged sodium ion. The neutral fluorine atom gains one electron to become a negatively charged fluorine ion.

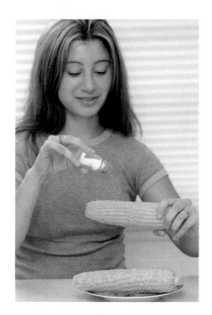

Table salt (NaCl) is an ionic compound that adds flavor to many of our foods. (David Young-Wolff/PhotoEdit.)

How are you doing? 3.1

Determine the charge and the number of valence electrons on each element in the following list of ionic compounds.

(a) Al_2O_3 (c) $MgCl_2$ (e) PbO_2 (g) BaO
(b) Li_3P (d) NaF (f) K_2Se

We can now begin to write correct formulas for any binary ionic compound that contains elements in groups I to VII. A summary of preferred ionic charges is given in Table 3-2. Rules for predicting the formula of common binary ionic compounds are given in Table 3-3.

Table 3-2 Preferred Charges on Ions of Representative Elements

Group Number	I	II	III	IV	V	VI	VII	VIII
Valence electrons in neutral atom	1	2	3	4	5	6	7	8
Valence electrons in ion	0	0	0	0	8	8	8	8
Preferred charge on ion	+1	+2	+3	+4	−3	−2	−1	0

Table 3-3 Predicting the Formula of Common Binary Ionic Compounds

REMEMBER

1. Determine the group number of the elements from the periodic table.
2. Determine the preferred charge on each of the ions. (See Table 3-2.)
3. Apply the principle of charge balance to determine how many of each ion are present to give an overall zero charge.

Example 3.3   Determining formulas for ionic compounds

Write a correctly balanced formula for the compound we expect to be formed by sodium and chlorine.

Thinking Through the Problem Sodium (Na) is in Group I; chlorine (Cl) is in Group VII. A sodium ion has a charge of $+1$ and is written Na^+. Chlorine is most commonly found as a -1 ion, Cl^-. Because the charges are equal but opposite in sign, both subscript numbers are the same, resulting in a 1:1 ratio of Na to Cl and an understood subscript number 1.

Answer   NaCl

Example 3.4   Determining formulas for ionic compounds

Write a correctly balanced formula for the compound made of boron and oxygen.

Thinking Through the Problem The preferred charge on boron is $+3$; on oxygen it is -2. The formula then starts as:

$$(B^{3+})_a(O^{2-})_x$$

Working Through the Problem ◀ (3.2) The principle of charge balance reminds us that the values of a and x must make an overall neutral compound. Therefore:

$$(\text{Charges on cations}) + (\text{Charges on anions}) = 0$$
$$(3 \times a) + (-2 \times x) = 0$$
$$(3a) + (-2x) = 0$$
$$3a = 2x$$

Solving for the ratio of a to x, $\dfrac{a}{x} = \dfrac{2}{3}$

This tells us that the value for a is 2 and the value for x is 3; B_aO_x becomes B_2O_3.

Answer B_2O_3

✔ **Meeting the Goals**

To predict the formula of a common binary ionic compound, determine the group number of the elements from the periodic table, determine the preferred charge on each of the elements, and write the formula in a manner that balances the electrical charge and leaves the compound electrically neutral.

How are you doing?   3.2

Write a balanced formula for the compound of potassium and phosphorus and the compound of beryllium and fluorine.

KEY IDEA 3.4

There is an obvious relationship between the charges of the cation (A^{m+}) *and the anion* (X^{n-}) *and their subscript numbers in a correctly written formula.* Binary compounds of the form $A_a^{m+}X_x^{n-}$ have a formula A_aX_x. For ionic compounds, the subscript numbers a and x must be reduced to lowest terms. Thus, when $m = n$, the formula is always AX, as in MgO. When $m = 2n$, the formula is AX_2, as when Sn_2O_4 is reduced to SnO_2. Recall that we do not reduce the formula for molecular compounds.

Table 3-4 Common Formulas for Ionic Compounds

Charges	Initial Formula	Correct Formula	Charges	Initial Formula	Correct Formula
A^+X^-	A_1X_1	AX	A^+X^{2-}	A_2X	A_2X
$A^{2+}X^{2-}$	A_2X_2	AX	$A^{2+}X^-$	AX_2	AX_2
$A^{3+}X^{3-}$	A_3X_3	AX	$A^{3+}X^{2-}$	A_2X_3	A_2X_3
			$A^{4+}X^{3-}$	A_3X_4	A_3X_4
			$A^{4+}X^{2-}$	A_2X_4	AX_2

Some other examples of this kind of reasoning are shown in Table 3-4.

Example 3.5

Determining formulas for ionic compounds

Determine the formula of the compound that will form for the following cation/anion pairs.

(a) Pb^{2+} and O^{2-}
(b) Fe^{3+} and S^{2-}

Thinking Through the Problem We can write an initial formula by interchanging the ionic charges, removing the positive and negative signs, and using these numbers as subscripts. Then, if necessary, we simplify to the lowest whole number ratio.

Answer

(a) Pb^{2+} and $O^{2-} \rightarrow Pb_2O_2 \rightarrow PbO$
(b) Fe^{3+} and $S^{2-} \rightarrow Fe_2S_3$

Example 3.6

Determining formulas for ionic compounds

Determine the formula of the compound you expect for the following pairs of elements.

(a) barium (Ba) and iodine (I)
(b) magnesium (Mg) and sulfur (S)
(c) aluminum (Al) and phosphorus (P)
(d) cesium (Cs) and nitrogen (N)

Thinking Through the Problem We begin by noting the expected charges on the ions. We then use these charges to get a formula, which is simplified to the lowest whole number ratio if needed.

Answer

(a) Ba and I; Ba^{2+} and I^-; BaI_2
(b) Mg and S; Mg^{2+} and S^{2-}; Mg_2S_2; MgS
(c) Al and P; Al^{3-} and P^{3-}; Al_3P_3; AlP
(d) Cs and N; Cs^+ and N^{3-}; Cs_3N

Until now we have seen problems where we write the formula of a compound from known elements. We can use this same process to determine whether or not a formula that we are given is correct. To do this, we determine whether the ions in the formula occur in the correct ratio. For example, if someone told us that the formula of an ionic compound was KCl_3, we would pause to note that if K is K^+ and Cl is Cl^- then the formula should be KCl, *not* KCl_3. We could also reason that one K^+ and three Cl^- ions would have a total charge of -2, not the neutral charge given in the incorrect formula.

Example 3.7

Determining formulas for ionic compounds

Determine if each of the following compounds has a reasonable formula. If the formula is wrong, correct it.

(a) K_3N (b) Al_3S_2 (c) BaO

Thinking Through the Problem   We can do this kind of problem efficiently by breaking the formula into the expected ions, then seeing if the result gives a neutral charge.

Working Through the Problem   (a) Here K_3N, with three K^+ and one N^{3-}, has a neutral charge. The formula K_3N is reasonable. (b) Al_3S_2 would have three Al^{3+} and two S^{2-}, for a charge of $+9 - 4 = +5$. Al_3S_2 is *unreasonable*. Instead, we expect the formula Al_2S_3. (c) BaO has Ba^{2+} and O^{2-}, giving a neutral charge. The formula BaO is reasonable.

Answer   The formulas in parts (a) and (c) are reasonable. The formula in (b) is not, and should instead be written as Al_2S_3.

Naming Binary Ionic Compounds

In order to name a binary ionic compound we must name both the cation and the anion in the compound. It is fairly simple to name ions that have a fixed expected charge. First, when naming the cation of a metal, we use the simple metal name alone. For example, sodium is a metal and thus sodium ion, Na^+, is simply called "sodium." Similarly, aluminum ion, Al^{3+}, is called "aluminum." In cases where we want to talk about the pure elemental substance, we can also use the atomic name alone, but for clarity we often explicitly call it a metal. Therefore, the element Ca is called "calcium metal" to distinguish it from Ca^{2+}, which, as the cation in an ionic compound, is called simply "calcium," or, even more unmistakably, "calcium ion."

CONNECT TO
SECTION 2.1

Next, we name the anion in the binary ionic compound. We then take the root of the element's name and add *-ide*. This is the same suffix that we use to name molecular substances in Chapter 2. So, the anion of sulfur, S^{2-}, is named "sulfide" and the anion of nitrogen, N^{3-}, is "nitride."

Each of the anion names refers to a single ion. Therefore, we also can omit any information about the number or the charge of the ion. $BaCl_2$, for example, is composed of Ba^{2+} and Cl^- ions and is named barium chloride, not barium dichloride.

| PRACTICAL | A | The table salt controversy—which alkali halides are beneficial to health? |

Ionic compounds readily release their ions into solution when dissolved in water. The ions that result from this process are available for immediate use. It is not surprising, then, that ionic compounds supply us with many of the minerals needed for a healthy and productive life. Although both sodium and chloride are cited as essential dietary elements, NaCl (table salt) has been the focus of close medical attention because of its possible connection to hypertension (high blood pressure). It is estimated that 36 million people in the United States suffer from hypertension, a condition that may worsen when there is too much sodium in the diet. A certain amount of sodium is essential for life, but too much sodium may lead to fluid retention and high blood pressure. Salt substitutes often contain potassium rather than sodium compounds.

Write balanced formulas for the ionic compounds in the following table.

Name of Compound	Ions Formed	Formula of Compound	Nutritional Value of Compound
Sodium chloride	Na^+ and Cl^-		Sodium compounds are important to the exchange of fluids between cells and plasma.
Potassium chloride	K^+ and Cl^-		Potassium compounds ensure proper muscle and nerve function and can affect the blood pressure.
Potassium iodide	K^+ and I^-		Iodide is needed for proper thyroid function. Salt containing supplementary iodine in the form of KI is called iodized salt.

REMEMBER

These rules for naming a binary ionic compound apply only if there is one and only one cation formed by the metal (that is, the ion of the cationic element can possess only one charge). For our purposes, all of the metals belonging to groups I and II, Al (in group III), zinc (Zn), cadmium (Cd), and silver (Ag) are characterized by having only one possible charge in their compounds. Group I elements and Ag *always* have a charge of +1. Group II elements, Zn, and Cd *always* have a charge of +2. Al (group III) *always* has a charge of +3. Compounds containing these metals are the easiest to name.

Example 3.8 **Naming binary ionic compounds**

Name AgF.

Thinking Through the Problem Ag is a transition metal, but we know that it only forms important compounds with a +1 charge. The name of the metal, Ag, is silver. We name the nonmetal using the root of the name and adding an *-ide* ending (F = fluorine; with the binary ending, it is fluor*ide*).

Answer AgF is silver fluoride.

How are you doing? **3.3**

Name the following compounds.

(a) CaO   (f) SrI_2
(b) MgF_2 (g) CsBr
(c) Ba_3N_2 (h) BeSe
(d) Al_2O_3 (i) NaF
(e) LiCl   (j) RbCl

Example 3.9   **Writing binary ionic formulas**

Write the formula for zinc sulfide.

Thinking Through the Problem Zinc, Zn, is a transition metal that forms compounds only with the Zn^{2+} ion. Sulfur, S, is an element in group VI that forms the preferred ion, S^{2-}. As the two ions have equal but opposite charges, they are combined in a one-to-one ratio into the correct formula, using the principle of charge balance.

Answer The formula of zinc sulfide is ZnS.

How are you doing? **3.4**

What are the formulas for the following compounds?

(a) magnesium nitride
(b) potassium phosphide
(c) cesium sulfide

PROBLEMS

1. Concept question: Many ionic compounds are important in electronics. These include GaAs, InP, and CdS. All have structures similar to pure silicon, perhaps the best known semiconductor. Look at the positions of Ga and As, In and P, and Ca and S in the periodic table. What pattern explains why the ionic compounds they form are similar to Si?

2. Complete the following statement: The ions of metals tend to have_____valence electrons.

3. Complete the following statement: The ions of nonmetals tend to have_____valence electrons.

4. Determine the preferred charge on the atoms in the following table when they form ionic compounds. Each ion should have either eight or zero valence electrons. The first one is done for you.

Atom	Preferred Charge on Ion	Valence Electrons in the Ion
Al	+3	0
Mg		
C		

Atom	Preferred Charge on Ion	Valence Electrons in the Ion
S		
P		
N		
I		
O		
Br		

5. Indicate the number of valence electrons on the ions in the following table. The first one has been done for you.

Ion	Valence Electrons in the Ion
Sn^{2+}	$4 - 2 = 2$
Mg^{2+}	
Ca^{2+}	
C^{3-}	
S^{2-}	
P^{3+}	

6. Discussion question: Write correct formulas for the compounds that result by combining each metal (first column) with each nonmetal listed across the top of the table. For example, combine Na with Se, Na with Br, etc. (Hint: determine the charges of the ions first.)

	Se^{2-}	Br	O	P	I
Na^+	Na_2Se				
Ge^{4+}					
Al^{3+}					
Li^+					
Mg^{2+}					

A useful but sometimes misapplied rule of thumb in writing ionic formulas suggests that the number of cations, a, is equal to the charge on the anion, n, while the number of anions, x, is equal to the charge on the cation m. Use the results of this table to show and discuss when this rule is valid, and when it does not directly apply. Restate the rule to take care of the cases where it needs to be modified.

7. Determine if each of the following formulas is reasonable. If not, correct the formula.

(a) Li_2Se (d) Al_3O_3
(b) $SrCl$ (e) AgI_2
(c) KBr_2

8. Name the following compounds.

(a) K_2S (c) $BeCl_2$
(b) CaO (d) ZnF_2

9. Name the following compounds.

(a) K_3N (c) CdO
(b) BaS

10. For the following compounds, (1) determine if the formula is reasonable, (2) write the new formula consistent with these two elements, and (3) name the compound.

(a) KBr_2 (c) Li_2S_2
(b) $CaCl_3$

3.2 Polyatomic Ions in Ionic Compounds

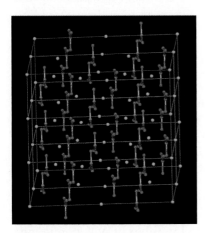

▼ See this with your **Web Animator**

Figure 3-4 Structure of $MgCO_3$ ▲

Magnesium atoms are shown in green, carbon atoms are shown in gray, and oxygen atoms are shown in red.

SECTION GOALS

✔ Recognize and name the common polyatomic ions.
✔ Write a correctly balanced formula containing a polyatomic ion and a single element with only one possible charge.
✔ Name ionic compounds containing polyatomic ions.

Ionic Compounds with Polyatomic Ions

In Chapter 2 we saw that polyatomic ions are important examples of a group of atoms bonded in a fixed ratio and arrangement. Because polyatomic ions like NO_3^- and NH_4^+ have a charge, they are never found alone in a substance, something we know by the principle of charge balance, as explained in the last section. But these ions, which contain more than one kind of atom, can be the building blocks of ionic compounds. In an ionic compound the cation, the anion, or both can be polyatomic ions. When ionic compounds containing polyatomic ions form, they follow the same rules we discussed earlier.

Consider the compound $MgCO_3$. This is composed of the cation Mg^{2+} and the anion CO_3^{2-}. As with the compound NaCl, the Mg^{2+} and CO_3^{2-} combine in a 1:1 ratio. An extended structure, shown in Figure 3-4, results.

In the last section, we learned how to write the formula of a binary ionic compound using an expected number of valence elec-

Table 3-5 Names, Formulas, and Charges of Some Common Polyatomic Ions

REMEMBER

Key Element Present	Formula	Name of Ion
Nitrogen	NO_3^-	Nitrate ion
	NO_2^-	Nitrite ion
	NH_4^+	Ammonium ion
Sulfur	SO_4^{2-}	Sulfate ion
	SO_3^{2-}	Sulfite ion
	HSO_4^-	Hydrogen sulfate ion
	HSO_3^-	Hydrogen sulfite ion
Phosphorus	PO_4^{3-}	Phosphate ion
	HPO_4^{2-}	Hydrogen phosphate ion
	$H_2PO_4^-$	Dihydrogen phosphate ion
	PO_3^{3-}	Phosphite ion
Carbon	CO_3^{2-}	Carbonate ion
	HCO_3^-	Hydrogen carbonate ion
	$C_2H_3O_2^-$	Acetate ion
	CN^-	Cyanide ion
	$C_2O_4^{2-}$	Oxalate ion
Chlorine	ClO_4^-	Perchlorate ion
	ClO_3^-	Chlorate ion
	ClO_2^-	Chlorite ion
	ClO^-	Hypochlorite ion
Boron	BO_3^{3-}	Borate ion
Hydrogen	H_3O^+	Hydronium ion
	OH^-	Hydroxide ion
Metals	MnO_4^-	Permanganate ion
	CrO_4^{2-}	Chromate ion
	$Cr_2O_7^{2-}$	Dichromate ion

trons on an ion of a metal or a nonmetal. That procedure does not apply to ionic compounds containing polyatomic ions. The composition of a particular polyatomic ion is fixed. So is the charge on the entire polyatomic ion. *This* charge—not the individual preferred charge of the atoms in the polyatomic ion—determines how the ion combines in ionic compounds.

KEY IDEA 3.5

In practice, the most effective way to work with the formulas of ionic compounds containing polyatomic ions is to know the formula of common polyatomic ions beforehand. Only a limited number of polyatomic ions are important. The formulas, charges, and names of some of these are given in Table 3-5 and should be memorized. Your instructor may have others as well.

A word on parentheses is in order. We use parentheses to indicate if there is *more than one* occurrence of a *group* of atoms in a compound. Parentheses are *not* used if there is only one of the group of atoms, or if the anion or cation in question is not polyatomic but contains a single atom, like Ca^{2+} or P^{3-}. Thus, the formula for the compound formed by calcium, Ca^{2+}, and perchlorate, ClO_4^-, is written as $Ca(ClO_4)_2$. On the other hand, the compound formed by ammonium ion, NH_4^+, and chlorate, ClO_3^-, is written as NH_4ClO_3, not $(NH_4)(ClO_3)$. Likewise, the formula for calcium chloride is written as $CaCl_2$, not $Ca(Cl)_2$.

Example 3.10 **Determining formulas for ionic compounds with polyatomic ions**

Determine the formula of the compounds formed from the following ions.

(a) calcium and hydroxide ion
(b) potassium and phosphate ion
(c) magnesium and sulfate

Thinking Through the Problem As with the binary ionic compounds, we write an initial formula to balance charge, then rewrite to make the formula have the simplest whole number ratio of the ions.

Answer

(a) Ca^{2+} and OH^-: $Ca(OH)_2$
(b) K^+ and PO_4^{3-}: K_3PO_4
(c) Mg^{2+} and SO_4^{2-}: $Mg_2(SO_4)_2$, rewritten correctly as $MgSO_4$

Example 3.11 **Checking formulas for ionic compounds with polyatomic ions**

Note where these formulas have an error in the simplest whole number ratio, the use of parentheses, or both.

(a) $K(ClO_2)$
(b) $(Ca)_2(ClO_4)_4$
(c) $(NH_4)_4S_2$

Thinking Through the Problem We can break each formula into the expected anion and cation, and then reassemble them.

Answer

(a) $K(ClO_2)$ should have K^+ and ClO_2^- ions, with a 1 :1 ratio and no parentheses. The formula should be $KClO_2$.
(b) $(Ca)_2(ClO_4)_4$ has Ca^{2+} and ClO_4^- ions. The formula should be $Ca(ClO_4)_2$. Note that there is *never* a case where a metal ion should be written in a formula surrounded on its own by parentheses.
(c) This formula should be reduced to a simple 2 :1 ratio; the formula should be $(NH_4)_2S$.

Our discussion so far has focused on compounds where we know the formula of both the anion and the cation. Suppose we have an ionic compound of Na, S, and O. Can we predict its formula? The answer is no, for the following reasons. We can say for certain that the Na will be present in the form of Na^+ because that ion has sodium with zero valence electrons. But the S and O are *not* present independently as S^{2-} and O^{2-}. Instead, they may be present as either SO_4^{2-} or SO_3^{2-}. Both ions have a -2 charge, so both form 2:1 salts with Na^+. The compound will be either Na_2SO_4 or Na_2SO_3, depending on which polyatomic ion of S and O we have.

You may find that some formulas, like $KClO_2$, seem confusing. Is this a salt of chloride and oxide with K^+? Absolutely not! Table 3-5 guides us to recognize ClO_2^- as the chlorite ion with a negative charge. A Cl atom and two O atoms are combined in a single polyatomic ion, and we do not separate them.

Writing Formulas and Names of Compounds Containing Polyatomic Ions

Polyatomic ions may require some time and effort to learn but, once mastered, you will find it easy to name and to write formulas for compounds containing them. *To name ionic compounds with polyatomic ions, we name the cation followed by the name of the anion.* The polyatomic ion is named just as it appears in Table 3-5. Thus, NH_4Cl is ammonium chloride, $NaNO_3$ is sodium nitrate, and NH_4NO_3 is ammonium nitrate. To write a formula for a compound that contains a polyatomic ion, we follow the same rules that we developed for binary ionic compounds, using the principle of charge balance.

Example 3.12   Naming compounds containing polyatomic ions

Name $MgCO_3$.

Thinking Through the Problem We recognize Mg as the magnesium ion, Mg^{2+}. The carbon-oxygen grouping is carbonate ion, CO_3^{2-}. As Mg^{2+} has a fixed oxidation number, we simply need the name "magnesium."

Answer   Magnesium carbonate

Example 3.13   Naming compounds containing polyatomic ions

Name the following compounds.
(a) NH_4Cl (b) $KClO$ (c) $Mg_3(PO_4)_2$

Thinking Through the Problem We identify the names of the cation and the anion, then write their names in sequence.

Answer

(a) The cation is NH_4^+, ammonium, and the anion is Cl^-, chloride. The compound is ammonium chloride.
(b) The cation is K^+, potassium, and the anion is ClO^-, hypochlorite. The compound is potassium hypochlorite.
(c) The cation is magnesium, Mg^{2+}, and the anion is phosphate, PO_4^{3-}. The compound is magnesium phosphate.

Example 3.14   **Writing formulas based on ions**

Write the formula of the following compounds.

(a) lithium nitrite   (b) sodium acetate   (c) calcium cyanide

Thinking Through the Problem Now we reverse the strategy. We first write the formula, using the correct charges, of both the cation and the anion. Using the principle of charge balance, we make sure that the compound has a neutral charge.

Working Through the Problem

(a) Lithium is Li^+ and nitrite is NO_2^-. The formula has one of each.

(b) Sodium is Na^+ and acetate is $C_2H_3O_2^-$. The formula has one of each.

(c) Calcium is Ca^{2+} and cyanide is CN^-. We need one Ca^{2+} and two CN^- ions.

Answer (a) $LiNO_2$ (b) $NaC_2H_3O_2$ (c) $Ca(CN)_2$

✔ **Meeting the Goals**

Ionic compounds that contain a metal with only one possible charge and a polyatomic ion are named by writing the name of the metal followed by the name of the polyatomic ion. Formulas of such compounds are written using the principle of charge balance—the total positive charge must equal the total negative charge in the compound.

How are you doing?   3.5

Fill in the missing names or formulas for the following ionic compounds.

Name of Compound	Formula of Compound
	$LiC_2H_3O_2$
Cadmium phosphate	
	$Mg(NO_3)_2$

PROBLEMS

11. Write correct formulas for the compounds formed from the following ions.

(a) Na^+ and chromate ion, CrO_4^{2-}
(b) Ba^{2+} and acetate ion, $C_2H_3O_2^-$
(c) Al^{3+} and nitrite ion, NO_2^-

12. Write formulas for the following compounds.

(a) sodium carbonate
(b) barium sulfite
(c) aluminum chlorate

13. Write formulas for the following compounds.

(a) potassium permanganate
(b) cadmium cyanide
(c) silver nitrate

14. Write formulas for the following polyatomic ions.

(a) ammonium ion
(b) chlorite ion
(c) oxalate ion

15. Discussion question: Write formulas for the following polyatomic ions.

(a) sulfate ion   (d) nitrite ion
(b) sulfite ion   (e) phosphate ion
(c) nitrate ion   (f) phosphite ion

Discuss whether or not there is a trend in the use of the suffixes -ite and -ate in polyatomic ions.

16. Name each of the following polyatomic ions. What is the same in each formula? What is different?

(a) ClO_4^-
(b) ClO_3^-
(c) ClO_2^-
(d) ClO^-

17. Name the following compounds.

(a) $K_2C_2O_4$
(b) Ag_2CrO_4
(c) $Al_2(CO_3)_3$
(d) $Mg(BrO_4)_2$

PRACTICAL B   Is calcium important only for bones?

Calcium is the fifth most abundant element in our bodies. This is not surprising when we remember that calcium is a major ingredient in our bones and teeth. Our teeth and skeletons account for about 99% of the calcium in our bodies, but the remaining 1% performs some very interesting functions. This calcium helps prevent muscle cramps, aids in the digestion of food, and, by promoting the clotting of blood, helps heal wounds. In some cases, increased calcium intake has resulted in lowered blood pressure.

Calcium forms many compounds that are very hard. A simple example is calcium phosphate, $Ca_3(PO_4)_2$, shown in Figure 3-5.

The calcium in our bones and teeth is combined with other elements to form a lattice-like structure. The major calcium compound in bones and teeth contains both phosphate and hydroxide ion, two polyatomic ions. The formula of this compound is $Ca_5(PO_4)_3OH$. Is $Ca_5(PO_4)_3OH$ electrically neutral? Use what you know about the charges of calcium, phosphate, and hydroxide to determine the answer.

Calcium and other minerals found in dietary supplements often come "packaged" with polyatomic ions. Say you are told different supplements contain calcium carbonate, calcium acetate, and calcium hydrogen carbonate. What are their formulas?

Most of the polyatomic ion names are based on the name of the nonmetal that is combined with oxygen. For example, carbonate ion is made of carbon and oxygen, borate ion is made of boron and oxygen, and phosphate ion is made of phosphorus and oxygen. What nonmetal combines with oxygen to form the polyatomic ions silicate ion and selenate ion?

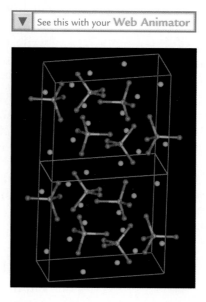

▼ | See this with your **Web Animator**

Figure 3-5 ▲ Structure of Calcium Phosphate, $Ca_3(PO_4)_2$ Calcium atoms are shown in green, oxygen atoms are shown in red, and phosphorus atoms are shown in orange.

You can determine the charge on silicate and selenate ions by examining the following balanced formulas containing these ions and by applying principles of charge balance. Calcium silicate has the formula Ca_2SiO_4. What is the charge on silicate ion, SiO_4^x? Calcium selenate has the formula $CaSeO_4$. What is the charge on selenate ion, SeO_4^x?

3.3 Oxidation Numbers and Chemical Formulas

SECTION GOALS

✔ Assign oxidation numbers to each element in a binary ionic compound.

✔ Assign oxidation numbers to each element in a compound containing a polyatomic ion.

✔ Name ionic compounds containing metals with either a fixed or a variable oxidation number.

Oxidation Numbers and Ionic Compounds

We have just seen that by understanding the expected charges on certain metal and nonmetal ions, we are able to understand and write formulas for ionic compounds. The concept of charged ions can be extended to include polyatomic ions. The charge on an ion is so central to how chemists analyze compounds that charge can be extended to all compounds by using the concept of **oxidation number.** The concept of oxidation number is easiest to understand in connection with *binary ionic compounds. There, the oxidation number is the same as the charges we have been assigning to the cation and the anion, and these charges balance to give the overall formula.*

KEY IDEA 3.7

Example 3.15 Determining oxidation numbers

What is the oxidation number of each element in the compound Li_3P?

Thinking Through the Problem Li is a Group I element; it will have a charge of $+1$ and an oxidation number of $+1$ as well. Since P is in Group V, in this binary ionic compound it has a charge of -3 and an oxidation number of -3. (Note that in writing charges in chemical notations like Mg^{2+} or N^{3-}, the charge sign *follows* the charge number.)

Answer Lithium has an oxidation number of $+1$, phosphorus -3.

Example 3.16 Determining oxidation numbers

Determine the oxidation numbers on the ions in Al_2S_3, $CdBr_2$, and $AgCl$.

Thinking Through the Problem The oxidation number on an ion in a binary ionic compound is the same as the charge on the ion. Therefore, we first think of what charges we expect for these cations and anions, then convert them to oxidation numbers.

Answer For Al_2S_3, there are Al^{3+} and S^{2-} ions. The oxidation number for Al in this compound is $+3$ and that for S is -2. For $CdBr_2$, the oxidation number on Cd is $+2$ and on Br is -1. For AgCl, the silver has an oxidation number of $+1$ and the chlorine has an oxidation number of -1.

How are you doing? 3.6

What are the oxidation numbers of the ions in the following compounds?

(a) Rb_3As (b) Ca_3N_2 (c) Li_2S

The concept of oxidation number prepares us to include metals with more than one possible charge in our understanding of ionic compounds. For example, lead forms two important compounds with oxygen, PbO and PbO_2. In both cases, we recognize that the oxygen, as the oxide ion, has a charge of -2. The oxidation number of Pb in both compounds can be obtained by applying the principle of charge balance.

For PbO: charge on Pb + charge on O = 0
                charge on Pb + (-2) = 0
                charge on Pb = $+2$

For PbO_2:   charge on Pb + $(2 \times$ charge on O$)$ = 0
                charge on Pb + $2(-2)$ = 0
                charge on Pb = $+4$

REMEMBER

Metals with single expected charges were discussed earlier in this chapter. You will recall that all metals in groups I and II, Al, Zn, Cd, and Ag are characterized by having only one possible charge. All other metals we will study can have more than one charge.

Example 3.17   **Determining oxidation numbers**

What is the oxidation number of the metal in $Ti(NO_3)_3$?

Thinking Through the Problem We recognize that Ti does not have a single possible charge, but also that the anion in this compound is NO_3^-, with a single negative charge.

Working Through the Problem

charge on Ti + $(3 \times$ charge on $NO_3^-)$ = 0
charge on Ti + $3(-1)$ = 0
charge on Ti = $+3$

Answer   The oxidation number on Ti in $Ti(NO_3)_3$ is $+3$.

Example 3.18   **Determining the charge on metals in a compound**

Determine the charge on the metal in $FeBr_3$, $FeSO_4$, and Fe_2O_3.

Thinking Through the Problem In each case, we have a recognizable anion—bromide, Br^-, sulfate, SO_4^{2-}, and oxide, O^{2-}. We can determine the charge on Fe algebraically in each case.

Working Through the Problem

$FeBr_3$:   charge on Fe + $(3 \times$ charge on $Br^-)$ = 0
                charge on Fe = $+3$. Oxidation number = $+3$

$FeSO_4$:   charge on Fe + charge on SO_4^{2-} = 0
                charge on Fe = $+2$. Oxidation number = $+2$

Fe_2O_3: $(2 \times$ charge on Fe$)$ + $(3 \times$ charge on $O^{2-})$ = 0
                $(2 \times$ charge on Fe$)$ = $+6$.
                charge on Fe = $+6/2 = +3$. Oxidation number = $+3$.

Answer   Fe has a charge of $+2$ in $FeSO_4$ and a charge of $+3$ in $FeBr_3$ and Fe_2O_3.

> **How are you doing?** 3.7
>
> Determine the oxidation number of each element in the following formulas.
>
> (a) CaC_2 (b) SiO_2

Oxidation Numbers for Molecules and Polyatomic Ions

If electrons are as important as they seem in the formation of chemical compounds, then we need a method to account for the oxidation numbers within compounds or ions that have more than one atom. For example, in the peroxide ion O_2^{2-}, there are two oxygens and two negative charges. What amount of charge resides on each oxygen? It might seem the answer is that each oxygen atom is simply equivalent to an O^- ion.

But is that *really* what is going on? It is helpful to think of the peroxide ion as being composed of two discrete ions with a single negative charge on each, but the atoms themselves are not so picky. *The charge is spread out over the whole molecule.* Thus, the only "real" charge is the -2 charge on the two oxygen atoms as a whole. To say that each atom possesses a -1 charge is a *convention,* by which we refer to the charge as the oxidation number. Oxidation numbers sometimes refer to real charges; more often, they are a handy bookkeeping method. Oxidation numbers can also be applied to molecular compounds.

Molecular compounds are formed by covalent bonding (electron sharing) and therefore do not contain ions. Nonetheless we can assign oxidation numbers by thinking of the compound as if the atoms in it had charges. Then we can calculate them according to established rules.

To assign oxidation numbers, we rely on the rules presented in Table 3-6. You must memorize these rules and apply them *in sequence*—rule 1 is applied before rule 2; rule 2 is applied before rule 3, etc. If you apply the rules in order, the exceptions will take care of themselves.

Another important point is that the oxidation number must add up to the total charge on the formula under consideration. An algebraic equation can be written and solved to determine an unknown oxidation number, given the oxidation numbers of all other elements in the formula. For example, in the compound CO, we recall that oxygen will be O^{2-}. If the molecule is to have zero total charge, then there must be an oxidation number of $+2$, on the carbon.

Algebraically, what we have done is:

$$\text{(oxidation number C)} + \text{(oxidation number O)} = \text{total charge on CO}$$
$$\text{(oxidation number C)} + (-2) = 0$$
$$\text{oxidation number C} = 0 - (-2) = +2$$

To find the oxidation number of each element in a formula, apply the Table 3-6 rules in order until all elements but one have been

Table 3-6 Rules for Assigning Oxidation Numbers

1. The oxidation number of a pure element is zero, even if that element is present in a molecule. For example, the oxidation numbers of sodium in Na, oxygen in O_2 or O_3, and phosphorus in P_4 are all zero.
2. The oxidation number of an alkali metal (Group I) in any compound is $+1$. The oxidation number of an alkaline earth metal (Group II) in a compound is $+2$.
3. The oxidation number of fluorine in a compound is -1.
4. The oxidation number of hydrogen in compounds is $+1$, except for binary compounds when only H and a metal are present. Then the oxidation number of hydrogen is -1.
5. The oxidation number of oxygen in any compound is -2. Exceptions are $OF_2(\overset{+2}{O})$, the peroxide ion O_2^{2-} (each O is $\overset{-1}{O}$), and the superoxide ion O_2^{-} (each O is $\overset{-\frac{1}{2}}{O}$).
6. Halogens are -1 in compounds, except those with oxygen and fluorine, where rules 3 and 5 take precedence.
7. Sulfur, selenium, and tellurium are -2 in compounds, except those with oxygen and halogens.

assigned an oxidation number. Find the oxidation number of the remaining element algebraically. The following example includes the logical thinking necessary in order to arrive at a correct answer.

Example 3.19 **Determining oxidation numbers**

What is the oxidation number of each element in SO_3?

Thinking Through the Problem We begin to apply the rules in order. There are no Group I or II elements; the first rule that applies is rule 5. Assign a value of -2 to oxygen. Determine the oxidation number of S algebraically.

Working Through the Problem

$$x + 3(-2) = 0$$
$$x = +6$$

Answer S has an oxidation number of $+6$; O has an oxidation number of -2. We may write this as $\overset{+6\,-2}{S\,O_3}$.

✓ Meeting the Goals

The easiest way to assign oxidation numbers to each element in a compound is to follow the rules given in Table 3-6 and to follow them in order. The order in which elements should be considered is: elements in group I ($+1$) and group II ($+2$), fluorine (-1), hydrogen ($+1$), oxygen (-2).

How are you doing? **3.8**

Determine the oxidation number of each element present in Mn_2O_7.

Polyatomic Ions and Oxidation Numbers

KEY IDEA 3.9

When deciding oxidation numbers for a compound that contains a polyatomic ion, use the rules to determine the oxidation numbers of the first and the last element. Determine the middle element algebraically. For ex-

ample, let's determine the oxidation numbers for the elements in $NaNO_3$, sodium nitrate. The oxidation number of sodium (group I) is $+1$; we can write $\overset{+1}{Na}$. Oxygen must have its preferred oxidation number of -2, so it is written $\overset{-2}{O}$. Note that, for bookkeeping purposes, we must count all *three* oxygens equally. Remember that, in a neutral compound, the sum of the total positive and the total negative charges is zero. Then, the formula is $\overset{+1}{Na}\overset{+x}{N}\overset{-2}{O_3}$ and the oxidation number of nitrogen is determined as follows. Count the number of times the element appears in the formula. Multiply the number by the oxidation number of that element. Repeat this for each element in the formula. The sum of these must equal the total charge on the formula. Neutral substances will have a total charge equal to zero. For this problem, there are one Na, one N, and three O's in the formula. Thus,

$$(1)(+1) + (1)(x) + (3)(-2) = \text{total charge on } NaNO_3 = 0$$
$$(+1) + x + (-6) = 0$$
$$x = 0 - (-6 + 1) = +5$$

The oxidation number of N is $+5$. We write $\overset{+1}{Na}\overset{+5}{N}\overset{-2}{O_3}$.

Example 3.20 **Determining oxidation numbers**

Determine the oxidation number of each element in $CaSO_4$.

Thinking Through the Problem Find the oxidation number of each element by applying the rules in order. Ca is an alkaline earth metal (group II) and is written as $\overset{+2}{Ca}$. Oxygen is written $\overset{-2}{O}$.

Working Through the Problem $\overset{+2}{Ca}\overset{x}{S}\overset{-2}{O_4}$

Determine the oxidation number of S algebraically.

$$(1)(+2) + (1)(x) + (4)(-2) = 0$$
$$x = +6; \text{ the oxidation number of S is } +6.$$

Answer $\overset{+2}{Ca}\overset{+6}{S}\overset{-2}{O_4}$

In $CaSO_4$, the sum of the positive oxidation numbers is $(2 + 6) = 8$ and the sum of the negative oxidation numbers is $(4)(-2) = -8$. This results in an overall neutral compound.

How are you doing? **3.9**

Determine the oxidation number of each element in the following compounds.

(a) $NaMnO_4$ (b) K_3BO_3 (c) $LiHCO_3$

Now we can apply what we know to polyatomic ions, keeping in mind that polyatomic ions have a nonzero overall charge. The sum of the oxidation numbers of atoms in a polyatomic ion will equal the charge on that ion, not zero.

In a formula such as sulfate, SO_4^{2-}, be careful to recognize that the superscript $2-$ describes the charge on the *entire* ion. You may

✔ **Meeting the Goals**

When a formula contains a polyatomic ion, use the rules to determine oxidation numbers for the first and last elements in the formula. Then determine the oxidation number of the middle element algebraically.

want to think of this as $(SO_4)^{2-}$ to distinguish more clearly the negative two charge of the sulfate ion from the negative two charge of a single oxygen ion, but remember that these parentheses must be omitted when you write formulas that do not call for parentheses.

Example 3.21

Determining oxidation numbers

Determine the oxidation number of N in NO_2^-.

Thinking Through the Problem Proceed as before, noting that this species is not neutral. In this case, the oxidation numbers will add up to the charge on the ion, -1, not zero.

Working Through the Problem Applying the rules for oxidation numbers, we find that rule 5 is the first rule that applies. Rule 5 tells us that the oxidation number of each oxygen is -2. The problem we face, then, can be expressed in this way: $(\overset{+x\ -2}{N\ O_2})^-$.

Algebraically this translates into:
$$1(+x) + 2(-2) = -1$$
$$x + (-4) = -1$$
$$x = +3$$

Answer   The oxidation number of N in NO_2^- is $+3$. We write $(\overset{+3\ -2}{N\ O_2})^-$.

How are you doing? 3.10

Determine the oxidation number of each element in the following ions:

(a) $Cr_2O_7^{2-}$ (b) $C_2O_4^{2-}$ (c) MnO_4^-

Naming Ionic Compounds Containing Metals with Variable Oxidation Numbers

Transition metals and certain heavier main group elements, such as tin and lead, can have more than one oxidation number, depending on the compounds they form. These elements exhibit *variable* oxidation numbers and the oxidation number of the metal must be determined algebraically. Earlier in the chapter, we compared two formulas of ionic compounds of Pb and O, namely, PbO and PbO_2. Although we can easily see the difference in these two formulas as written, the name "lead oxide" does not allow us to distinguish between these two compounds. From the last section, we can determine that the oxidation numbers of Pb in PbO and in PbO_2 are $+2$ and $+4$, respectively. We distinguish between these two compounds by including the oxidation number of lead in each name. PbO is named lead (II) oxide and PbO_2 is named lead (IV) oxide.

We extend this technique to all ionic compounds that contain metals with variable oxidation numbers. The oxidation number is explicitly written in roman numerals within parentheses immediately following the metal's name. To determine the oxidation number of the transition metal, all other atoms in the formula must be assigned their preferred oxidation numbers so there is only *one* unknown in the problem. Table 3-7 describes this process.

Table 3-7 Determining the Oxidation Number of a Metal with a Variable Oxidation Number

REMEMBER

1. Determine the charge on the anion by assigning the preferred charge to a monatomic anion or by recalling the charge on a polyatomic anion.
2. Determine algebraically the oxidation number of the metal.

Example 3.22 Determining oxidation numbers and naming compounds

What is the oxidation number of chromium in Cr_2O_3? What is the name of the compound?

Thinking Through the Problem Cr is a metal with variable oxidation number. We must determine the oxidation number of Cr algebraically, assigning oxygen a value of -2.

Working Through the Problem $\overset{x}{Cr_2}\overset{-2}{O_3}$

As the compound is neutral, the total negative charge must equal the total positive charge. Two Cr must *together* contribute $+6$; one Cr then has an oxidation number of $+3$. The oxidation number of Cr appears as a roman numeral immediately following chromium in the name.

Remember that we can set up and use an algebraic equation to determine a missing oxidation number. The problem is written:

$$2x + (3)(-2) = 0$$

Solving, we have:

$$2x - 6 = 0$$
$$2x = +6$$
$$x = +3$$

The oxidation number of chromium in this formula is $+3$. This gives us the name of the compound, chromium (III) oxide. Figure 3-6 shows compounds containing CrO_4^{2-} and $Cr_2O_7^{2-}$ in solution.

Answer Cr has an oxidation number of $+3$. The name of the compound is chromium (III) oxide.

✔ Meeting the Goals

To name a compound that contains a metal with a variable oxidation number, assume the preferred oxidation number for the nonmetal and then determine the metal's oxidation number algebraically. Include the metal's oxidation number as a roman numeral within parentheses immediately following the name of the metal.

How are you doing? **3.11**

Determine the oxidation number of each element in the following formulas.

(a) V_2O_5 (b) MnO_2

Example 3.23 **Naming compounds**

Name $Co(HSO_4)_2$.

Thinking Through the Problem Determine the oxidation number of Co in this formula algebraically by looking at the oxidation number of the anion, hydrogen sulfate ion.

Working Through the Problem

$$\text{charge on cobalt} + (2 \times \text{charge on hydrogen sulfate ion}) = 0$$
$$\text{charge on cobalt} + (2 \times -1) = 0$$
$$\text{charge on cobalt} = +2$$

The oxidation number of Co is $+2$ and must be shown in roman numerals following the word *cobalt*.

Answer Cobalt (II) hydrogen sulfate

How are you doing? **3.12**

Name $CuSO_4$, which is shown in Figure 3-7.

Figure 3-6 ▲ Compounds containing chromium exhibit different colors depending on the oxidation number of chromium in the compound. The chromate ion, CrO_4^{2-}, is yellow. When acid is added to a chromate solution, the ions form dichromate ions, $Cr_2O_7^{2-}$, which are orange. (W. H. Freeman photo by Ken Karp.)

Figure 3-7 ▲ Copper (II) sulfate, $CuSO_4$, is a white powder in the absence of water. When water is added, copper (II) sulfate reacts to form blue crystals. Copper (II) sulfate has such a strong attraction for water that it is usually colored a very pale blue from reaction with the water in the air. (W. H. Freeman photo by Ken Karp.)

PRACTICAL C — Does the oxidation number of an element affect its nutritional benefit to us?

Many of the trace minerals we need for good health have more than one oxidation number in compound form. Does the oxidation number of the element make any difference in our ability to make use of it?

Determine the oxidation number of the mineral in each of the following compounds. Then name the compound. The first one has been done for you.

Mineral	Easily Absorbed Form	Oxidation State of Mineral	Name of Compound
Fe	$FeSO_4$	+2	iron (II) sulfate
Mn	$MnSO_4$		
Cu	CuO		
Sn	SnF_2		

Other forms of these elements are not so readily absorbed. Write formulas for each of the following compounds.

Compound Name	Compound Formula
iron (III) sulfate	
manganese (III) sulfate	
copper (I) oxide	
tin (IV) fluoride	

PROBLEMS

18. Concept question: The heavier main group metals on the right side of the periodic table have variable oxidation numbers. But, except for a few unusual compounds, all of these elements have just *two* oxidation numbers in their compounds. For example, Pb is usually either $+2$ or $+4$, never $+3$. In (indium) is $+1$, $+3$, never $+2$. And Bi (bismuth) is $+5$ or $+3$, never $+4$. Look at the position of these elements in the periodic table. Suggest why there might be just these two oxidation numbers for these elements.

19. Concept question: The text emphasizes that oxidation numbers are a convention, not a complete factual picture of chemical reality. Give two other examples of "conventions" that we use when a complete factual picture is too complicated.

20. Discussion question: Determine the oxidation number for each element in the following ions.

(a) NO_3^- (c) SO_3^{2-} (e) CO_3^{2-}
(b) SO_4^{2-} (d) ClO_4^-

Compare the oxidation number of the first element in each of these polyatomic ions with

the preferred charge for that ion itself. For example, compare the oxidation number of S in MgS with the oxidation number of S in SO_4^{2-}. Is there a trend here?

21. Determine the oxidation number for each element in the following compounds. Some oxidation numbers may be fractions.

(a) CsO_2 (b) NaS_5 (c) Pb_3O_4

22. Determine the oxidation number of each element in the following formulas.

(a) CH_4 (c) HF (e) Cl_2O_7
(b) NaH (d) ClF

23. Determine the oxidation number of the metal in each of the following compounds.

(a) PbI_2 (c) Cu_2O
(b) SnO_2 (d) Fe_2S_3

24. Name the compounds in Problem 23.

25. Indicate the charge and the number of valence electrons on each of the elements in the compounds in the following table. Note which elements do *not* have either zero or eight valence electrons.

Compound	Charge on Ions	Number of Valence Electrons in Ions
PbO_2	Pb^{4+} O^{2-}	Pb^{4+}: 0; O^{2-}: 8
PbO		
In_2Se_3		
InI_3		
$GaCl_2$		

26. Determine the oxidation number of the metal in each of the following compounds.

(a) $Fe(NO_3)_2$ (c) $Ca(OCl)_2$
(b) $Al(ClO_4)_3$ (d) $U(SO_4)_2$

27. Name the compounds in Problem 26.

28. Write formulas for the following binary ionic compounds.

(a) tin (IV) iodide
(b) lead (II) sulfide
(c) manganese (I) oxide

29. Write formulas for the following ionic compounds.

(a) silver nitrate
(b) plutonium (IV) sulfate
(c) lead (II) sulfate

Chapter 3 Summary and Problems

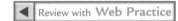 ◀ Review with **Web Practice**

VOCABULARY

ionic bond
A kind of bond that involves the very strong forces associated with charged ions. In this case, a positive charge and a negative charge are present, and the charges attract one another to form an ionic bond.

ionic compounds
Compounds that contain positive and negative ions in an extended structure.

principle of charge balance
Charge balance means that, in an ionic compound, the sum of the charges on the cations must be equal and opposite to the charges on the anions:

$$\text{charges on cations} + \text{charges on anions} = 0$$

or

$$\text{charges on anions} = -\text{charges on cations}$$

extended structures
Large structures formed by the attraction between cations and anions in ionic compounds.

closed shell
An outer, or valence, shell that contains 0 valence electrons or 8 valence electrons.

oxidation number
The oxidation number of an atom in a compound indicates the effective charge on an ion. Oxidation numbers are assigned according to the rules in Tables 3-6 and 3-7.

SECTION GOALS

Can you determine the charge and the number of electrons on each element in a formula of an ionic compound?

Metals lose all of their valence electrons to make a positive ion that has a closed shell of zero valence electrons. The resulting charge on the metal ion is positive and is determined by the number of electrons lost. Nonmetals gain valence electrons to achieve a closed shell of 8 valence electrons. The resulting charge on the nonmetal ion is negative and is determined by the number of electrons gained.

Can you write the formula for a binary ionic compound by using the preferred charge of each of its elements?

To write the formula of a common binary ionic compound, determine the group number of the elements from the periodic table, determine the preferred charge on each of the elements, and write the formula in a manner that balances the electrical charge.

Are you able to recognize and name the common polyatomic ions?

The names and formulas of the common polyatomic ions are given in Table 3-5. These should be memorized so that you can write the formula given the name and vice versa.

Are you able to write a correctly balanced formula containing a polyatomic ion and an element having a single possible charge?

Ionic compounds containing a metal with only a single possible charge and a polyatomic ion are named by writing the name of the metal followed by the name of the polyatomic ion. Formulas of such compounds are written using the principle of charge balance—the total positive charge must equal the total negative charge in the compound.

How do we name ionic compounds with polyatomic ions?

These are named as with binary ionic compounds. We give the name of the cation and then the name of the ion.

Are you able to assign oxidation numbers to each element in a binary ionic compound?

The easiest way to assign an oxidation number to each element in a binary ionic compound is to follow the rules given in Table 3-6 and to follow them in order. The order in which elements should be considered is: elements in group I ($+1$) and group II ($+2$), fluorine (-1), hydrogen ($+1$), oxygen (-2).

Are you able to assign oxidation numbers to each element in a compound containing a polyatomic ion?

When a formula contains a polyatomic ion, use the rules to determine oxidation numbers for the first and last elements. Then determine the middle element algebraically.

Are you able to name ionic compounds containing metals with either a fixed or a variable oxidation number?

To name a binary ionic compound in which the metal has only one possible oxidation number, first name the metal. Then use the root of the nonmetal name, changing the ending to -ide. To name a compound that contains a metal with a variable oxidation number, assume the preferred oxidation number for the nonmetal and then determine the metal's oxidation number algebraically. Include the metal's oxidation number as a roman numeral within parentheses immediately following the name of the metal.

PROBLEMS

30. Write a correct formula for each of the following compounds.

(a) barium hydrogen carbonate
(b) aluminum phosphate
(c) copper (I) oxide
(d) silver permanganate
(e) iron (III) chloride

31. Determine the oxidation state of each element in the following formulas.

(a) $K_2Cr_2O_7$ (c) $Co(NO_3)_3$
(b) $SnBr_2$ (d) $MnCrO_4$

32. Write the ions present in the substances in the following table. Next to each ion, write its name. Finally, write the name of the compound. The first one has been done for you.

Compound	Cation	Anion	Name of Compound
FeF_3	Fe^{3+} iron (III)	F^- fluoride	Iron (III) fluoride
$MnBr_2$			
CaS			
UF_6			
Ag_2O			

33. Write the ions present in the substances in the following table. Next to each ion, write its name. Then write the name of the compound.

Compound	Cation	Anion	Name of Compound
$Fe(NO_3)_3$	Fe^{3+} iron (III)	NO_3^- nitrate	Iron (III) nitrate
$CuSO_4$			
Li_2CO_3			
K_3N			
$Sn(OH)_2$			

34. In the following table, write the formula and name of each ion, and the formula of the corresponding compound.

Name of Compound	Cation	Anion	Compound
Lithium phosphate	Li^+ lithium	PO_4^{3-} phosphate	Li_3PO_4
Manganese (III) sulfide			
Zinc iodide			
Calcium cyanide			
Aluminum oxide			
Magnesium hydrogen carbonate			
Ammonium hydrogen sulfate			
Titanium (III) chloride			

35. Write formulas for the combinations of ions in the following table.

	Cl^-	S^{2-}	N^{3-}	F^-	NO_3^-	SO_4^{2-}	PO_4^{3-}
Na^+	NaCl						
Mg^{2+}							
Al^{3+}							
NH_4^+							
Cr^{3+}							
Cu^{2+}							

36. Indicate the number of valence electrons in the following ions. The first one has been done for you.

Ion	Valence Electrons in the Ion
Cl^+	8
Pb^{2+}	
Pb^{4+}	
Se^{2-}	
S^{6+}	
O^-	

37. Name the following compounds.

(a) $BaSO_3$ (c) $Cd(NO_2)_2$

(b) $Ca(ClO_2)_2$ (d) $Al(C_2H_3O_2)_3$

38. Determine the oxidation number of each element in the following formulas.

(a) BaO_2 (b) OF_2 (c) BrF_5 (d) IF_7

Compound	Charge on Ions	Number of Valence Electrons in Ions
Na_2O		
NaH		
SrO_2		
$BaSe$		
Al_2O_3		

39. Determine the oxidation number of each element in the following formulas.

(a) $CrCl_3$ (c) Cu_2O (e) ZnF_2

(b) MnO_2 (d) CuO

40. Determine the oxidation number of each element in the following ions.

(a) HCO_3^- (c) ClO_3^- (e) CN^-

(b) ClO_2^- (d) BrO_4^-

41. Determine the oxidation number of the metal in each of the following compounds.

(a) MnO_2 (c) Ni_2O

(b) $CoCl_3$ (d) CrF_2

42. Indicate the charge and the number of valence electrons in each of the elements in the compounds listed below. Note which elements do *not* have either zero or eight valence electrons.

43. Determine the oxidation number of the metal in each of the following compounds.

(a) Na_3PO_4 (c) $MnSO_4$

(b) $Ti_2(CO_3)_3$ (d) KNO_2

44. Write formulas for the following binary ionic compounds.

(a) iron (II) chloride

(b) copper (II) phosphide

(c) nickel (I) bromide

45. Write formulas for the following ionic compounds.

(a) calcium sulfite

(b) iron (II) carbonate

(c) aluminum hydrogen phosphate

46. Use the principle of charge balance to determine whether the following formulas are correctly written or not.

(a) SiF_3 (b) Al_2S_3 (c) Al_2F_3 (d) Al_2O_3

47. Can you think of a reason why Al_2S_3 and Al_2O_3 have similar formulas? Does the periodic law support these formulas? Predict a formula for Ga and O; Ga and S.

48. Both $ZnCl_2$ and $CdCl_2$ exist. Do you think that $HgCl_2$ exists? Why or why not?

49. Both PO_4^{3-} and PO_3^{3-} exist. Predict a formula for a polyatomic ion formed by combining As and O.

50. Suppose that Astatine formed a series of four polyatomic ions with oxygen, similar to that shown in Table 3-7. Name the following ions.

(a) AtO_3^- (b) AtO^- (c) AtO_4^- (d) AtO_2^-

51. Write correct formulas for

(a) magnesium hypoiodite

(b) manganese (III) dichromate

(c) magnesium permanganate

52. Compare the formula of manganese (II) chloride with that of manganese (III) chloride. Why is the roman numeral important in these names?

53. Determine the oxidation number of each element in these compounds.

(a) $Co(NO_3)_2$ (b) $Pt(SO_4)_2$

54. Name the following compounds.

(a) $Co(NO_3)_2$ (b) $Pt(SO_4)_2$

55. Determine the oxidation number of each element in these compounds.

(a) $Ag_2Cr_2O_7$ (b) $BaCr_2O_7$ (c) $Al_2(Cr_2O_7)_3$

56. The peroxide ion, O_2^{2-}, is a polyatomic ion made of a highly reactive form of oxygen. In chemical compounds, peroxide ion appears with a subscript of 2 and an overall charge of -2. Write the formulas for the following compounds.

(a) hydrogen peroxide
(b) sodium peroxide
(c) barium peroxide

57. The superoxide ion, O_2^-, is found in only a few compounds with very reactive metals. Write formulas for the following compounds.

(a) cesium superoxide
(b) potassium superoxide

The Mole and Chemical Equations

The ignition of a match is a multistep chemical process for which we can write several chemical equations. (Joseph P. Sinnot/ Fundamental Photographs.)

PRACTICAL CHEMISTRY The Match

Matches are so familiar to us that we rarely ask "How does that work?" But the match—a portable source of flame—is an excellent example of how chemical processing and careful control of reactions in a system allow a very dangerous thing, fire, to be created safely and reliably.

There are three components needed to start a fire: a fuel, oxygen or another "oxidizing agent," and a source of ignition. In the burning of a match, the fuel is the wood or cardboard of the match and the oxygen comes from the air. The spark that gets things going comes from the friction caused by striking the match against a rough surface.

Most of us are familiar with "safety matches" that require us to strike the match head against a specially prepared "striker" that is on the outside of the box or book of matches. This rough surface contains small amounts of red phosphorus, a chemical that easily gives rise to sparks as a result of friction. But the spark generated when the match head is rubbed against the striker is too small to ignite the wood or cardboard of the match. This spark initiates a second reaction of the chemical substances in the match head. Small amounts of potassium chlorate, an oxidizing agent, and sulfur are present in the match head. The burning (oxidation) of the small amount of sulfur by the chlorate gives a much bigger flame than the original spark, and this flame ignites the cardboard or wood of the match.

There often are other components in safety matches (such as wax used for additional fuel, or other additional oxidizing agents) but the simple chemical reactions just discussed—potassium chlorate with phosphorus, or a phosphorus compound, and potassium chlorate with sulfur—are the most important.

This chapter will focus on the following questions:

PRACTICAL A How is a match put together?

PRACTICAL B What are some of the chemical reactions that occur in a safety match?

PRACTICAL C What are the balanced chemical equations for some of the reactions that take place in matches?

4.1 Measuring and Counting Chemical Substances

SECTION GOALS

✔ Describe the meaning of the phrase "measurement by counting."

✔ Explain why chemists use a unit called the mole in counting chemical quantities.

✔ Know how the relative amounts of elements in a compound relate to the relative number of moles of elements in a compound.

✔ Know how moles of atoms or ions relate to the formulas of compounds containing them.

Measurement by Counting

The study of chemistry requires that we measure chemical substances, including the atoms that compose them. In this chapter we introduce the special way that chemists measure by counting numbers of atoms, molecules, and other substances. But first we need to discuss what we mean by measurement.

There are two important aspects of scientific measurement: quantity and quality. The first thing that a measurement tells us is "how much" or "how many" of something we have. This measurement gives us the quantity. We are all familiar enough with quantity, but quality, the second aspect of a scientific measurement, may be a less familiar idea to you. This is discussed in a later chapter. For now we focus on what is meant by quantity in a measurement.

We cannot answer the question "how much?" without knowing two things: the type (or "dimension") of measurement, and the unit

A pharmacist measures out a customer's prescription by counting pills.
(Dan McCoy/Rainbow.)

Figure 4-1 ▶ There are many ways to answer the question "What is in this bag of candy?" including information on nutrition. But when we eat them, we usually care most about the number of pieces that we get! (Erv Schowengerdt.)

of measurement. By "type" or "dimension" we mean, for example, length, area, volume, mass, and time. The most important types of measurement you will use for chemical substances are the volume and mass.

The unit of measurement also specifies a standard size for the measurement. In our everyday lives we often use English units. For example, we measure mass in units of pounds or ounces. We measure volume in units such as cups, quarts, and teaspoons. Scientists use the more universal, decimal-based units found in the metric system. Metric system units, and their equivalents in English units, are discussed in more detail in Chapter 7.

The measurement of mass, volume, and time are so important that it is easy to neglect a more fundamental type of measurement—the measurement of the *number* of things. This type of measurement is known as **measurement by counting.** For example, when we measure "three eggs" for a cake batter, we are measuring the *number* of eggs, not their mass or volume. When we eat candy, we are usually most concerned about the number of pieces, but there are many other things measured, as shown in Figure 4-1. *Measurement by counting is the key concept of this chapter.*

Counting and Chemical Formulas

In Chapter 1 you learned that a chemical formula indicates the type and number of atoms present in a substance. In this chapter you will see that, in terms of the atoms of different elements, the ratios of the atoms in molecular compounds are the same whether we have one, ten, or a billion molecules of that compound.

When we ask questions about the quantity of a chemical substance, we can mean several different things. We must look carefully at the wording of the question in order to determine exactly what is being asked. For example, suppose that we have 1,000,000 molecules of ammonia, NH_3. We can ask three different quantity questions:

- *How many molecules of ammonia are present?* There are 1,000,000 molecules of ammonia.

- *How many atoms of N are present?* There are 1,000,000 atoms of N present.

- *How many atoms of H are present?* There are 3,000,000 atoms of H present.

As another example, consider the molecular compound water, which has the chemical formula H_2O. This formula has a ratio of two hydrogen atoms for every one oxygen atom. *We can use the chemical formula of any substance to determine how many atoms of an element are in a sample of that compound.* The chemical formula H_2O gives us a ratio for the hydrogen atoms in water:

$$\frac{2 \text{ H atoms}}{1 \text{ } H_2O \text{ molecule}}$$

This ratio will be the same in any sample of water we consider. If we know the number of water molecules in a sample, then we can determine the number of H atoms by using this ratio. We can also calculate the number of O atoms in a sample of water, if we know the number of water molecules. Conversely, if we know the number of H atoms or O atoms in a sample of water, we can calculate the number of molecules of water.

This ability to move from one measurement—say, 25,000 molecules of water—to another—say, 50,000 atoms of H—is critical to the success of any study of chemical formulas. In this chapter we further explore the meanings of the ratios indicated by a chemical formula. You will learn how to count like a chemist. Anyone who works with a chemical substance should be able to discuss the different elements present and the amounts of these elements. Figure 4-2 shows chemical formulas being used to label the ingredients in a product.

✔ Meeting the Goals

The atomic nature of matter means that we can count chemical substances by (a) the number and type of atoms present, (b) the number of molecules present in a molecular substance, or (c) the number of formula units present in an extended structure.

Figure 4-2 Ingredients on a bag of fertilizer, showing chemical formulas ▶

When we use chemical substances we need to know how much of each element is present. (Erv Schowengerdt.)

Figure 4-3 A 100-mL Erlenmeyer flask ▲
Almost all laboratory glassware has a clear reference to the measurement of volume it provides. (Premium Stock/Corbis.)

In summary, whenever we want to indicate the "amount" of a chemical substance we must make a choice. We use either convenient laboratory measurements, like volume and mass, or we can count the number of atoms, molecules, or formula units, using methods described in this chapter. For example, we could easily measure the volume of a sample of water as 100 milliliters, using a common piece of laboratory glassware, which is shown in Figure 4-3. But this measurement does not give us direct information about "how many" atoms or molecules of the substance are present. **Chemical counting** involves counting the number of atoms, molecules, and formula units. We will soon see how it is possible to count the number of molecules of water in a particular volume.

Practical Counting: Moles as a Counting Unit

Chemical counting is a kind of measurement. There are some chemists, especially those who work with very small amounts of substances in biochemistry, who do count small numbers of molecules, atoms, and formula units. But this is unusual. Molecules are so small that it is more convenient to count large groups of them. Just as we measure mass in grams, volume in liters, and temperature in degrees, chemists have a preferred way to measure the number of atoms, molecules, and formula units in a sample of a substance. Chemists count these using a unit called the **mole,** abbreviated *mol.* The mole is a counting unit for a very large number of things and is used in very much the same way as the counting unit *dozen.* The word *dozen* is a widely known term for a counting of twelve items, as shown in Figure 4-4. Other such grouping units that you may have run across are the score (20), the pair (2), and the gross (144).

The mole has a value that is much bigger than the value of the dozen. Since atoms are so small, we study approximately 1,000,000,000,000,000,000,000 atoms in even the smallest sample. The mole must accommodate such large numbers. In fact, the mole is equal to **602,213,700,000,000,000,000,000** items. Using scientific notation and rounding for the sake of convenience, this number is 6.02×10^{23}, and is called **Avogadro's number.**

REMEMBER

✔ Meeting the Goals

The mole is very important in chemistry and has the numerical value of Avogadro's number. But the key concept is not the value of Avogadro's number; it is that chemists count by using mole amounts.

For baseballs:

One *dozen* baseballs is *12* baseballs.

One *mole* of baseballs is *602,213,700,000,000,000,000,000* baseballs.

For doughnuts:

One *dozen* doughnuts is *12* doughnuts.

One *mole* of doughnuts is *602,213,700,000,000,000,000,000* doughnuts.

For molecules:

One *dozen* molecules of methane is *12* molecules of methane.

One *mole* of methane is *602,213,700,000,000,000,000,000* molecules of methane.

Figure 4-4 ▲ One box of baseballs contains a particular number of individual balls. One box of doughnuts contains a certain number of doughnuts, in this case a dozen. One mole of a substance contains a certain number of individual molecules or atoms of that substance, called Avogadro's number. (a: Andy Levin/Photo Researchers; b: Erv Schowengerdt.)

Any statement we can make about the relative numbers of individual atoms, molecules, or formula units can also be made about the relative mole amounts of these things. We have to be careful to remember that we can count the whole molecule, its atoms, etc., all using the mole. It is common for students to think that the mole can be used only to measure a whole substance. But the mole can also apply to anything in that substance. For example, if we have one mole of methane, CH_4, we have one mole of carbon atoms and four moles of hydrogen atoms, or a total of five moles of atoms. The term "mole" is used in one case to refer to the whole methane molecule, and in the other cases to refer to the individual atoms. Figure 4-5 shows one molecule of water and one mole of water.

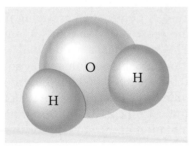

(a)

Figure 4-5 ▲ There are 6.022×10^{23} molecules of H_2O in one mole of water. But the same $2:1$ ratio of hydrogen atoms to oxygen atoms applies to both (a) one molecule of water and (b) one mole of water. (b: Terry Gleason/Visuals Unlimited.)

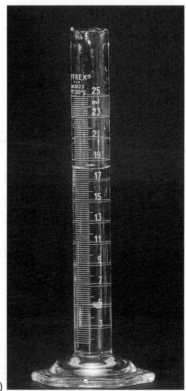

(b)

When we speak in terms of atoms and molecules, we are using the **atomic** *or* **molecular scale.** *When we speak in terms of moles of substances, we are using the* **macroscopic scale.** Chemists routinely shift between these two scales, depending on the point they wish to make.

Example 4.1 Rewriting atomic and molecular statements in mole terms

Rewrite the following statements about elements in chemical formulas as statements about mole amounts.

(a) The substance pyridine, C_5H_5N, contains five carbon atoms for every molecule of pyridine.
(b) One formula unit of silver phosphate, Ag_3PO_4, has one PO_4^{3-} ion.
(c) We find that there is one iron atom in every molecule of hemoglobin.

Thinking Through the Problem The sentences need to be rewritten with mole amounts. ➟ (4.2) The key idea here is that if we use the idea of the mole to discuss one substance, then we must use it to discuss all the substances. For example, it would be *wrong* to say, "The substance pyridine, C_5H_5N, contains five moles of carbon atoms for every pyridine molecule." Instead we say that pyridine contains five moles of carbon atoms for every mole of pyridine molecules.

Answer

(a) Pyridine, C_5H_5N, contains five moles of carbon atoms for every mole of pyridine molecules.
(b) One mole of silver phosphate, Ag_3PO_4, has one mole of PO_4^{3-} ions.
(c) We find that there is one mole of iron atoms in every mole of hemoglobin.

Example 4.2 Rewriting statements about moles in atomic and molecular terms

Rewrite the following sentences about mole amounts to reflect individual atoms, molecules, and formula units.

(a) There are two moles of oxygen atoms in a mole of oxygen gas, O_2.
(b) A mole of Al_2O_3 contains two moles of Al atoms for every three moles of O atoms.
(c) A mole of the molecular substance Vitamin B_{12}, shown in Figure 4-6, contains one mole of Co atoms.

Thinking Through the Problem This is the reverse of the previous example. Here we must be careful to know when to introduce the term "molecule" (for molecular substances) and the term "formula unit" (for extended structures).

Answer

(a) There are two oxygen atoms in a molecule of oxygen gas, O_2.
(b) A formula unit of Al_2O_3 contains two Al atoms for every three atoms of O.
(c) A molecule of vitamin B_{12} contains one atom of cobalt.

How are you doing? 4.1

Rewrite the following sentences to reflect a mole scale:

(a) One molecule of ethane contains six hydrogen atoms.

(b) Gold-containing arthritis drugs commonly contain a core of one gold atom and three sulfur atoms.

Rewrite the following sentences to reflect an atomic or molecular scale:

(c) This sample of four moles of calcium carbonate has four moles of calcium ions.

(d) I have determined that each mole of the compound contains two moles of nitrogen, three moles of carbon, and ten moles of hydrogen.

Figure 4-6 Molecular structure of B_{12} ▼
Large molecules like vitamin B_{12} often contain one essential component. Here it is cobalt, Co.

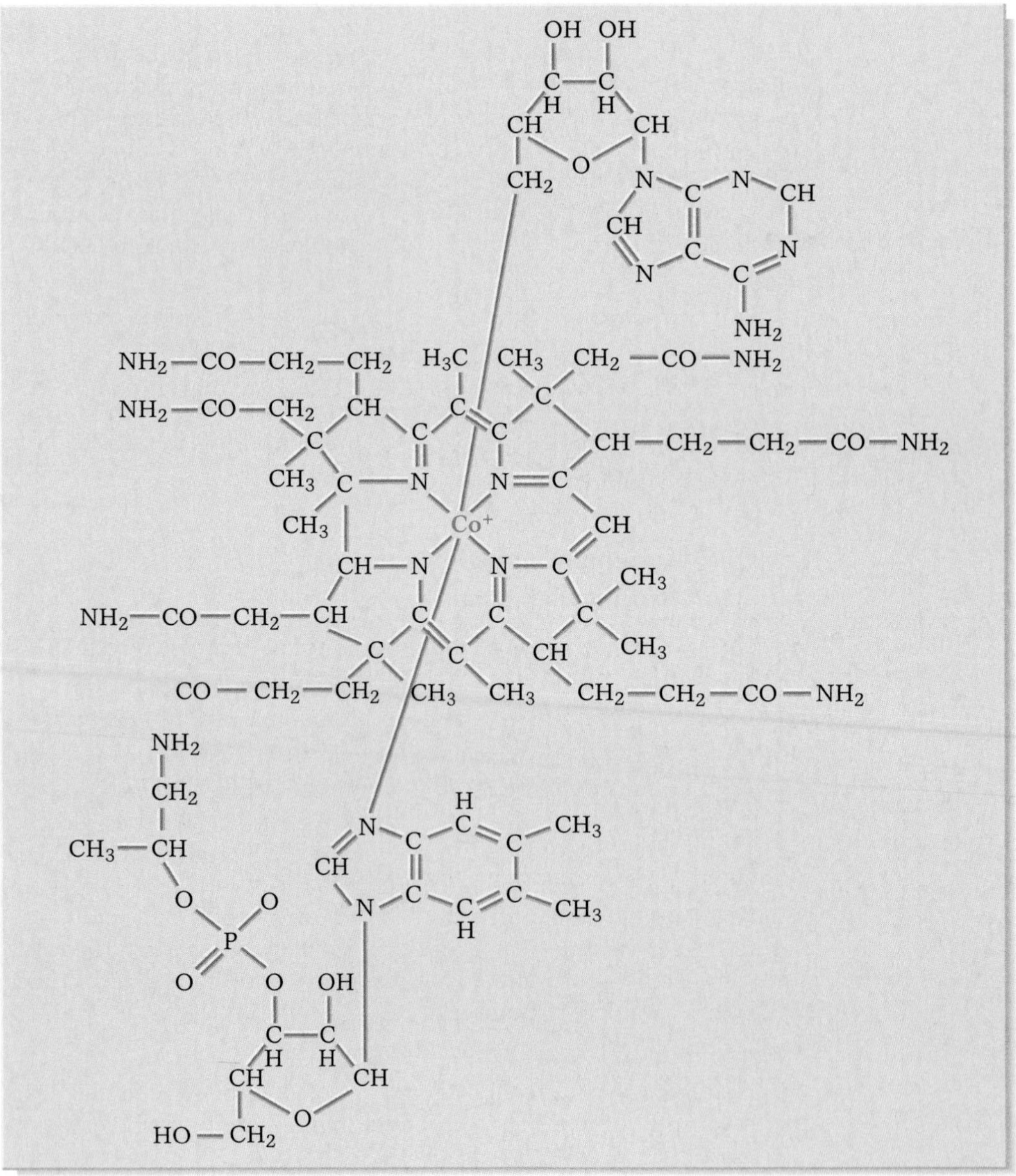

Moles, Counting Ratios, and Chemical Formulas

In the preceding section we showed how statements about relative amounts can be expressed using different counting units—individual amounts, dozens, and moles (to name a few). We express relative amounts by using ratios of different components of a substance or thing.

Using ratios to relate different mole amounts of atoms, molecules, and formula units is so important that when we refer to the ratios involved we will use the term **mole ratio.** Remember that the mole ratio has the same numerical values as the simple atomic or molecular scale ratio that we saw earlier in this section.

For example, we showed earlier that for H_2O, the ratio

$$\frac{2 \text{ H atoms}}{1 \text{ H}_2\text{O molecule}}$$

applies. We can use this same 2 : 1 ratio with different counting units:

atom/molecule ratio

$$\frac{2 \text{ H atoms}}{1 \text{ H}_2\text{O molecule}} \longrightarrow \frac{2 \text{ dozen H atoms}}{1 \text{ dozen H}_2\text{O molecules}} \longrightarrow$$

mole ratio

$$\frac{2 \text{ moles H atoms}}{1 \text{ mole H}_2\text{O molecules}}$$

All three ratios reduce to the 2 : 1 ratio shown first because the units *dozen* and *mole* stand for particular numbers that can be divided out.

Example **4.3** **Writing atom/atom and atom/molecule ratios**

Write the ratio of C to H atoms in the molecular substance C_2H_5N and then rewrite the ratio with the counting units *dozen* and *mole*. Do the same for the ratio of C atoms to molecules of C_2H_5N.

Thinking Through the Problem We need to look at the formula and convert it into a ratio with "C atoms" in the numerator and "H atoms" in the denominator. Then we apply the correct numerical values to both parts and, finally, rewrite using the units *dozen* and *mole*. The same process is then applied to a ratio with "C atoms" in the numerator and "C_2H_5N molecules" in the denominator.

Answer

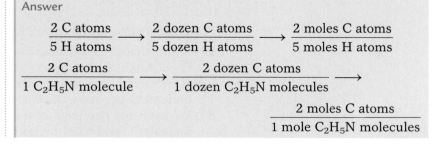

$$\frac{2 \text{ C atoms}}{5 \text{ H atoms}} \longrightarrow \frac{2 \text{ dozen C atoms}}{5 \text{ dozen H atoms}} \longrightarrow \frac{2 \text{ moles C atoms}}{5 \text{ moles H atoms}}$$

$$\frac{2 \text{ C atoms}}{1 \text{ C}_2\text{H}_5\text{N molecule}} \longrightarrow \frac{2 \text{ dozen C atoms}}{1 \text{ dozen C}_2\text{H}_5\text{N molecules}} \longrightarrow$$

$$\frac{2 \text{ moles C atoms}}{1 \text{ mole C}_2\text{H}_5\text{N molecules}}$$

How are you doing? 4.2

Write the following ratios as atom/molecule ratios, then as mole/mole ratios.

(a) The ratio of C to H in ethane, C_2H_6
(b) The ratio of atoms of O to molecules for the substance N_2O_5
(c) The ratio of atoms of C to atoms of H in C_3H_9N

We can use mole ratios to determine a chemical formula. If, for example, we determine that one mole of a molecular compound contains two moles of oxygen and one mole of sulfur, then we can write:

✔ Meeting the Goals

Mole ratios give us access to the atom/molecule ratios for a molecular formula.

$$\frac{2 \text{ moles O}}{1 \text{ mole compound}} \longrightarrow \frac{2 \text{ O atoms}}{1 \text{ molecule}} \quad \text{and}$$

$$\frac{1 \text{ mole S}}{1 \text{ mole compound}} \longrightarrow \frac{1 \text{ S atom}}{1 \text{ molecule}}$$

Because we know the number of O and S atoms in one molecule, we can write the formula of this substance: one S atom and two O atoms, or SO_2. Thus, we can determine the molecular formula of a substance by looking at all of its mole ratios.

Example 4.4

Determining a chemical formula from mole amounts

Determine the chemical formula of a compound where one mole of a molecular substance contains two moles of Cl and three moles of O.

Thinking Through the Problem We can set up a mole ratio that can be rewritten as an individual ratio. We have a molecular substance, so we determine a molecular formula.

Working Through the Problem

$$\frac{2 \text{ moles Cl}}{1 \text{ mole compound}} \longrightarrow \frac{2 \text{ Cl atoms}}{1 \text{ molecule}} \quad \text{and}$$

$$\frac{3 \text{ moles O}}{1 \text{ mole compound}} \longrightarrow \frac{3 \text{ O atoms}}{1 \text{ molecule}}$$

Answer The chemical formula is Cl_2O_3.

How are you doing? 4.3

Write the chemical formulas of the following molecular compounds.

(a) A substance that contains four moles of Cl and one mole of C per mole of compound.
(b) A substance with four moles of H, two moles of C, and one mole of O per mole of the compound.
(c) We find that one mole of a molecular substance has four moles of C and eight moles of H.

✔Meeting the Goals

Mole ratios give us access to the atom/formula ratios for the formula of an ionic compound.

Figure 4-7 ▲ These samples of (clockwise from top) MnO_2, Mn_3O_4, MnO, and Mn_2O_3, illustrate the colorful changes that occur as atom/atom ratios change. (Charles Winters/Photo Researchers.)

Ionic Compounds and Chemical Counting

We have derived the concept of the mole ratio from examples of atoms in molecules. Molecules contain fixed numbers of atoms in each individual molecule. But the same kind of reasoning can be applied in cases of ionic compounds where the formula tells us the number of the different types of elements in the compound's **formula unit** (the smallest whole number ratio of the different types of atom in the extended structure). For example, we might find that a compound of manganese (Mn) and oxygen contains three moles of Mn and four moles of O in one mole of the substance. Then we have:

$$\frac{3 \text{ moles Mn}}{1 \text{ mole compound}} \longrightarrow \frac{3 \text{ Mn atoms}}{1 \text{ formula unit}} \text{ and}$$

$$\frac{4 \text{ moles O}}{1 \text{ mole compound}} \longrightarrow \frac{4 \text{ O atoms}}{1 \text{ formula unit}}$$

The formula of the compound, then, is three Mn atoms and four O atoms per formula unit, or Mn_3O_4. Figure 4-7 shows four different ionic compounds of manganese oxide, including Mn_3O_4.

Determining the formula of an ionic compound from mole amounts

Example 4.5

Determine the chemical formula of an ionic compound containing two moles of iron and three moles of sulfate ions per mole of the compound.

Thinking Through the Problem In this problem we determine the formula unit of an ionic compound. The reasoning process is the same as in Example 4.4. In this case we have ions present, but they are treated as parts of the formula just like individual atoms.

Working Through the Problem

$$\frac{2 \text{ moles Fe}}{1 \text{ mole compound}} \longrightarrow \frac{2 \text{ Fe atoms}}{1 \text{ formula unit}} \text{ and}$$

$$\frac{3 \text{ moles sulfate}}{1 \text{ mole compound}} \longrightarrow \frac{3 \text{ sulfate ions}}{1 \text{ formula unit}}$$

Answer The chemical formula is $Fe_2(SO_4)_3$.

PROBLEMS

1. Concept question: We previously discussed why it is wrong to refer to "one half" of an ethane molecule. But is it all right to refer to "one half *of a mole*" of ethane molecules?

2. Concept question: How can it be that one mole of a substance has four moles of atoms? Include an example in your answer.

3. Rewrite the following statements to reflect dozens instead of individual amounts.

   (a) For every cake, you will need three eggs.

   (b) When you make sandwiches for six students, you will need one-half of a loaf of bread.

   (c) Building a house requires about one thousand pieces of wood.

PRACTICAL A   How is a safety match put together?

There are three essential parts to a safety match. A small amount of material that will ignite when rubbed must be present on the "striker" that is on the box. The match head must contain material that allows the initial fire to grow large enough to burn the rest of the match. And there is the main material of the match, usually wood or cardboard.

As we think about putting a match together, we can use some of what we have learned about measuring and counting chemical substances. First, take a look at Table 4-1. For which of these substances can you write formulas?

In this chapter, we focus on certain reactions that are central to the process of making a match. Table 4-1 provides a complete list of the match ingredients, including the purpose of each.

Which of the substances in the striker and in the match head do you think play an essential role in the "chemistry" of the match? In what kind of reaction do you think they are involved?

Table 4-1 Components of the Safety Match

Ingredient	Purpose
Safety Match—Match Head	
Gelatin	Glue for components
Starch	Supports structure of the match head
Sulfur	Main fuel
Potassium chlorate	Oxidizing agent
Zinc oxide	Provides location for combustion
Calcium carbonate	Neutralizer
Silica compounds	Increase friction
Other oxides	Catalyst for the combustion
Safety Match—Striker	
Gelatin	Glue for components
Red phosphorus	Fuel
Calcium carbonate	Neutralizer
Powdered glass	Friction

(Larry Stepanowicz/Visuals Unlimited.)

4. Rewrite the following statements to reflect mole amounts instead of individual amounts.

(a) There are four hydrogen atoms in one ammonium molecule.

(b) A molecule of carbonic acid contains six atoms.

(c) Three atoms of titanium and four atoms of nitrogen are in one formula unit of titanium nitride.

5. Rewrite the following statements to reflect individual amounts instead of mole amounts:

(a) We find that there are two moles of iron in one mole of iron (III) oxide.

(b) For the molecular substance sulfur hexafluoride there are six moles of fluorine for every mole of sulfur.

(c) "White" phosphorus is a molecular substance with four moles of P atoms per mole of white phosphorus.

6. Write the following mole ratios as individual ratios. Assume all substances are ionic.

(a) $\dfrac{3 \text{ moles Mg}}{1 \text{ mole compound}}$ (c) $\dfrac{3 \text{ moles nitrate}}{1 \text{ mole compound}}$

(b) $\dfrac{1 \text{ mole Pb}}{1 \text{ mole compound}}$ (d) $\dfrac{2 \text{ moles phosphate}}{1 \text{ mole compound}}$

7. Write the following mole ratios as individual ratios. Assume all substances are molecular.

(a) $\dfrac{3 \text{ moles Br}}{1 \text{ mole compound}}$ (c) $\dfrac{13 \text{ moles H}}{1 \text{ mole compound}}$

(b) $\dfrac{1 \text{ mole P}}{1 \text{ mole compound}}$ (d) $\dfrac{4 \text{ moles Si}}{1 \text{ mole compound}}$

8. Write the mole ratios and determine the molecular formula consistent with the following statements.

(a) A mole of a compound contains three moles of C, two moles of O, and six moles of H.

(b) There are five moles of Cl and one mole of P in one mole of a compound.

(c) Fifteen moles of hydrogen, one of nitrogen, and six of carbon are in one mole of a compound.

9. Discussion question: Write the mole ratios and determine the ionic formulas consistent with the following statements.

    (a) There is one mole of Pt and six moles of F in one mole of a compound.

    (b) Three moles of titanium and four moles of phosphate are in a mole of a compound.

    (c) One mole of ammonium ions, one mole of iron atoms, and two moles of sulfate ions are in one mole of a compound.

Discuss whether or not the formulas would have been any different if, instead of moles, we had presented this information in terms of the number of atoms present. How does the mole concept explain whether or not there would be a difference?

10. A sample of a compound contains 20,000 atoms of H. If we have a second sample of the same compound that is twice this size, then how many atoms of H are in the second sample?

11. "Macromolecules," such as proteins, are very large molecules containing thousands of atoms. But the mole concept applies to macromolecules also. If you are told "A molecule of this protein contains 1,200 hydrogen atoms," then how many moles of hydrogen atoms are in one mole of that protein?

12. Web question: For the illustrations in Web Animator Problem 4-12, create at least three mole ratios that describe the relative mole amounts of elements and the compounds.

4.2 Chemical Reactions and Equations

SECTION GOAL

✔ Identify the most important parts of a balanced chemical equation.

Describing Chemical Reactions

In Section 4.1 we described how chemists count using moles. We applied counting with moles to the formulas of molecular and ionic compounds. Now we will see that the same counting concepts are useful as we write equations for **chemical reactions.** A chemical reaction is the change of a substance into a new substance through the rearrangement of atoms. Recall from Chapter 1 that a substance undergoes a chemical change, that is, a chemical reaction, when it forms an entirely new substance with different properties.

We usually describe what happens in a reaction by indicating that there are substances that react (called **reactants**) and substances that result from the reaction (called **products**). An arrow indicating the direction of the change links reactants and products:

reactants $\longrightarrow$ products

One important characteristic of a reaction is the identity of the substances involved. We need to identify which substances are reactants and which are products. A second characteristic of a chemical reaction is that *the amount and kind of elements present must remain the same throughout the course of the reaction.* There must be an equal number of each element at the beginning and at the end of the reaction, as shown in Figure 4-8. This equality is why we often refer to a description of a chemical reaction as a **chemical equation.**

The chemical equation summarizes the essential components in a chemical reaction. Correct formulas and physical states of all reactants and products are included in the equation. In addition, the number and kind of each element initially present in the reactants is equal to the number and kind of each element present in the products.

Chemical equations are similar to mathematical equations in important ways. In a mathematical equation there are two distinct expressions that have the same value, each on opposite sides of an equals sign. A standard chemical equation contains two expressions also, separated by an arrow.

Each reactant and each product must be represented by a correctly written formula, which is often followed by an abbreviation for its physical state. The abbreviations for the physical states of matter are: (*s*) for solid materials, (*l*) for pure liquids, (*g*) for gases, and (*aq*) for aqueous solutions (in which substances are dissolved in water).

The equation is read in normal fashion from left to right, whether the equation is written as words or as chemical formulas. For example, the word statement for the synthesis of carbon dioxide from its elements is:

Solid carbon *reacts with* oxygen gas *to form* carbon dioxide gas.

Notice that this word statement contains all the information necessary for you to understand what is involved in the reaction, including the physical states of the reactants and products: "solid carbon," "oxygen gas," and "carbon dioxide gas." Don't be intimidated by the fact that this may be an unfamiliar chemical sentence. The operative words are the verbs, which are all you need to understand the chemistry that is occurring. *Verbs like "react," "burn," or "decompose" are the action words that describe the chemical change happening to the reactants.*

For example, the phrase:

Solid carbon *reacts with* oxygen gas

identifies the two reactants that will undergo a chemical change. In symbols, this is translated into:

Figure 4-8 ▶ Measuring the correct amount of reactants is an essential part of many chemical procedures. (Doug Martin/Photo Researchers.)

$$C\,(s) + O_2\,(g)$$

Products in a word statement are the substances that are "produced" or "formed." Often the products are separated from the reactants by verb phrases such as "result in," "to produce," "to form," and "give." In our example the product carbon dioxide gas appears in the phrase "to form *carbon dioxide gas.*" The complete word statement above is transcribed or written in chemical symbols as:

Solid carbon	reacts with	oxygen gas	to form	carbon dioxide gas.
$C\,(s)$	$+$	$O_2\,(g)$	$\longrightarrow$	$CO_2\,(g)$

Example **4.6** | **Recognizing reactants and products**

Identify the reactants and products in each of the following sentences. Note the word that describes the kind of chemical reaction.

(a) Solid sulfur burns in air to form gaseous sulfur dioxide.
(b) Solid potassium chlorate ($KClO_3$) decomposes when heated to produce oxygen gas and solid KCl.
(c) Plants synthesize glucose and oxygen from carbon dioxide and water.

Thinking Through the Problem We will be taking apart the sentences to indicate the reactants, the products, and the word or phrase that links them. We identify the reactants at the "start" of the reaction and the products in the "result" of the reaction. It is *not* always true that the sentence is written with the reactants first. ➡ (4.5) A key idea is that the word describing the chemical reaction is usually a verb that expresses a kind of change.

Answer

(a) The reaction is identified by the word *burns* and it begins with sulfur and the oxygen in air as the reactants and ends with sulfur dioxide as the product.
(b) The reaction is identified by the word *decomposes*. The compound $KClO_3$ is the reactant that gives rise to the products oxygen and KCl.
(c) In this case, note that the subject of the sentence (plants) is not a chemical substance. But the word *synthesize* identifies a chemical reaction in which the reactants carbon dioxide and water are changed into the products glucose and oxygen.

Figure 4-9 ▲ The production of gas is an important indicator that a reaction is occurring. Here, magnesium dissolves in acid, producing hydrogen gas. (Richard Megna/Fundamental Photographs.)

How are you doing? 4.4

Indicate the substances that are the reactants and products in the following word statements. Note the word that describes the reaction.

(a) Magnesium metal reacts with aqueous solutions of hydrochloric acid (HCl) to liberate hydrogen gas and to form an aqueous solution of magnesium chloride ($MgCl_2$), as shown in Figure 4-9.
(b) Hydrogen peroxide, H_2O_2, decomposes to water and oxygen gas when poured on a wound.

Picturing Chemical Reactions

In addition to showing reactants and products, a correctly written chemical equation shows the *ratio* that the amounts of these substances have to one another. For example, in the reaction of the phosphorus trichloride with chlorine to give phosphorus pentachloride, we have the chemical equation:

$$PCl_3 \ (l) + Cl_2 \ (g) \longrightarrow PCl_5 \ (s)$$

In this case we see that each side of the equation has one P atom and five Cl atoms. The ratio of substances is one PCl_3 to every one Cl_2, and one PCl_5. In addition to writing the formulas, we can draw this equation using Lewis structures, which we learned to write in Chapter 2. Here we omit lone pairs for simplicity:

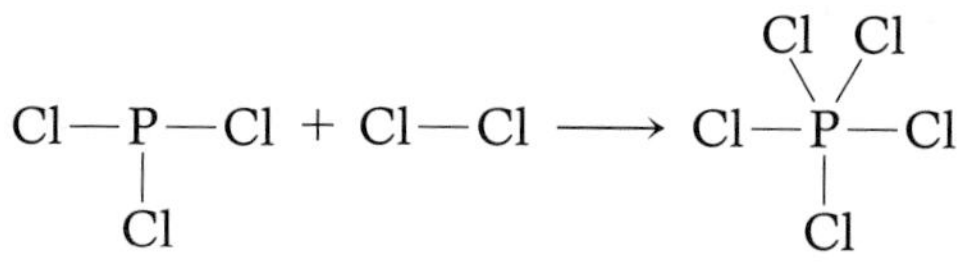

$$Cl-P-Cl + Cl-Cl \longrightarrow Cl-P-Cl$$

Notice that there is one Lewis structure drawn for each reactant and product because the equation shows one of each substance. Not all substances react in a one-to-one ratio. In fact, most do not. For example, hydrogen peroxide decomposes to give water and oxygen according to the chemical equation:

$$2 \ H_2O_2 \ (aq) \longrightarrow 2 \ H_2O \ (l) + O_2 \ (g)$$

A picture of this reaction would look like:

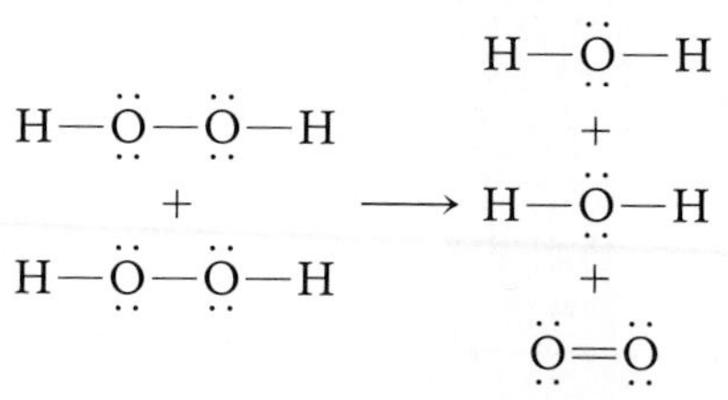

In this case, we have drawn two Lewis structures for H_2O_2, two for H_2O, and only one for O_2, in order to correspond to the numbers of these substances given in the equation. The Lewis structures allow us to check that equal numbers of H and O are present in the reactants and in the products for this reaction. Counting the atoms shows that four H atoms and four O atoms are on each side of the equation.

This is called a **balanced chemical equation.** The number of Lewis structures drawn for each substance is given by the **coefficients** in the original equation. Coefficients are the numbers written

PRACTICAL B — What are some of the chemical reactions that occur in a safety match?

To produce a safety match, the match manufacturer must carefully control chemical amounts. These amounts are determined by the chemical "recipe" of the reactions that occur when the match is struck. Some of these reactions can be depicted using Lewis structures.

The genius of the safety match is that the initial spark requires the reaction of a component (potassium chlorate) in the match head with a component (red phosphorus) on the striker, the part of the matchbook that also provides the needed friction. Red phosphorus has an extended structure, so its formula is usually written simply as "P." When red phosphorus and potassium chlorate make brief and violent contact as the match is struck, potassium chloride and

diphosphorus pentaoxide are the initial products. Write this reaction with the names and the correct formulas. Identify the reactants and the products.

Keeping a stable flame going requires the action of air on the sulfur and the wax in the match head. Elemental sulfur usually exists as a *ring* of eight S atoms, each one connected to the next by a single bond. The molecular formula of sulfur is therefore S_8. When sulfur burns in air, the oxygen in the air reacts with S_8 to form molecules of SO_2. Draw Lewis structures for this reaction. You will need to include the Lewis structures of S_8, O_2, and SO_2. Count the atoms involved. How many molecules of SO_2 will this reaction produce? How many molecules of S_8 are needed?

in front of the chemical formula representing the substance. *The coefficients in a balanced equation show how many molecules react and how many are produced in the chemical reaction.* Thus, the equation

$$2\ H_2O_2\ (aq) \longrightarrow 2\ H_2O\ (l) + O_2\ (g)$$

means "two molecules of H_2O_2 decompose to form two molecules of H_2O and one molecule of O_2."

Example 4.7 — Representing a chemical reaction using Lewis structures

Rewrite the following chemical equation using Lewis structures.

$$CH_4\ (g) + 2\ Cl_2\ (g) \longrightarrow 2\ HCl\ (g) + CH_2Cl_2\ (l)$$

Thinking Through the Problem We need to write correct Lewis structures of the four substances. We need to represent Cl_2 and HCl twice in the drawing.

Answer

Example 4.8 **Translating Lewis structures for a reaction into a chemical equation**

Determine the chemical equation for the reaction shown by the following Lewis structures.

$$H-C\equiv C-H + H-H + H-H \longrightarrow H-\underset{\overset{|}{H}}{\overset{\overset{|}{H}}{C}}-\underset{\overset{|}{H}}{\overset{\overset{|}{H}}{C}}-H$$

Thinking Through the Problem We can check that each side of the equation has the same number of carbon atoms (2) and hydrogen atoms (6). Then we write the formulas that we find, with a coefficient of 2 for the H_2 molecule.

Answer $C_2H_2 + 2\,H_2 \rightarrow C_2H_6$

PROBLEMS

13. Magnesium metal reacts rapidly with oxygen gas to form solid magnesium oxide. Identify the reactants and the products in this reaction. Write formulas for each reactant and each product.

14. Nitrogen dioxide gas is produced when nitrogen monoxide reacts with oxygen gas. Identify the reactants and the products in this reaction. Write formulas for each reactant and each product.

15. Aluminum metal reacts with fluorine gas to give solid aluminum fluoride. Identify the reactants and the products in this reaction. Write formulas for each reactant and each product.

16. Discussion question: Solid calcium carbonate decomposes into carbon dioxide gas and solid calcium oxide. Identify the reactants and the products in this reaction. Write formulas for each reactant and each product. One way to characterize this reaction is that it "releases" carbon dioxide. Discuss if that means that calcium carbonate should be viewed as a compound of CaO and discrete carbon dioxide molecules, or as a different kind of substance entirely.

17. Draw Lewis structures for the reaction $CH_4 + 2\,O_2 \rightarrow CO_2 + 2\,H_2O$. Verify that the number of C, H, and O atoms is the same in the reactants and in the products.

18. Draw Lewis structures for the reaction $H_2N - NH_2 + O_2 \rightarrow N_2 + 2\,H_2O$. Verify that the number of N, H, and O atoms is the same in the reactants and in the products.

19. Draw Lewis structures for the chemical reaction $H_2 + Cl_2 \rightarrow 2\,HCl$. Verify that the number of C, H, and Cl atoms is the same in the reactants and in the products.

4.3 Chemical Counting in Chemical Reactions

SECTION GOALS

✔ Tell what the coefficients in a balanced equation represent on an atomic and on a macroscopic scale.

✔ Balance a chemical equation.

Chemical Stoichiometry: Mole Amounts in Chemical Reactions

We saw in Section 4.2 that in an equation there are *equal* numbers of the different elements on each side of the reaction arrow. The "equality" present in a chemical equation rests on the atomic theory and the law of conservation of mass. Briefly, this states that the elements present in the reactants are present in equal chemical amounts in the products. Thus, when we count the elements in the reactants, we must find the same elements in the same amount in the products. *What differs between the reactants and the products is the arrangement of the elements, not the amounts.* Because the same atoms are present in identical amounts, the mass before the reaction is the same as the mass after the reaction.

The study of the amounts of chemical substances, especially those involved in chemical reactions, is called **chemical stoichiometry.** The word *stoichiometry* is derived from Greek words meaning "measurement of parts" (or more literally, "measurement of step"; "stoich" means "stair"). In the present context, stoichiometry is a measurement of the components of a chemical reaction.

Although chemical reactions occur between atoms and molecules, we usually think of them in terms of mole amounts. This brings us back to our discussion of the atomic/molecular scale and the macroscopic scale. A chemist uses both scales to interpret an equation, often switching back and forth between them as is necessary to answer a particular question. Thus, an equation such as

$$PCl_3\,(l) + Cl_2\,(g) \longrightarrow PCl_5\,(s)$$

means

$$1 \text{ molecule } PCl_3 + 1 \text{ molecule } Cl_2 \longrightarrow 1 \text{ molecule } PCl_5$$
(atomic/molecular scale)

as well as

$$1 \text{ mole } PCl_3 + 1 \text{ mole } Cl_2 \longrightarrow 1 \text{ mole } PCl_5 \text{ (macroscopic scale)}$$

The scale used to discuss an equation depends on the size of the sample or the point we are trying to make. Very small samples are discussed more conveniently using the atomic or molecular scale, whereas we use the macroscopic scale for the large samples we use in the lab.

In short, when we talk about the chemical reaction

$$2\,C_2H_6\,(g) + 7\,O_2\,(g) \longrightarrow 4\,CO_2\,(g) + 6\,H_2O\,(g)$$

either (or both) of the following interpretations can be used:

$$2 \text{ moles } C_2H_6\,(g) + 7 \text{ moles } O_2\,(g) \longrightarrow$$
$$4 \text{ moles } CO_2\,(g) + 6 \text{ moles } H_2O\,(g)$$
$$2 \text{ molecules } C_2H_6\,(g) + 7 \text{ molecules } O_2\,(g) \longrightarrow$$
$$4 \text{ molecules } CO_2\,(g) + 6 \text{ molecules } H_2O\,(g)$$

KEY IDEA 4.6

✔ **Meeting the Goals**

On the atomic or molecular scale, the coefficients of a balanced equation represent atoms and molecules. On the macroscopic scale, the coefficients of a balanced equation represent moles of atoms and molecules.

We must be careful not to mix the interpretations. (It would be wrong, for example, to use moles on one side of the equation and molecules on the other.)

Example 4.9   **Rewriting chemical reactions in mole terms**

Rewrite the following sentences to describe mole amounts.

(a) In the reaction of ethylene, C_2H_4, with oxygen to make acetaldehyde, C_2H_4O, we need one molecule of oxygen for every two molecules of ethylene, according to the equation $2\,C_2H_4 + O_2 \rightarrow 2\,C_2H_4O$.

(b) Sulfur dioxide gas can be "scrubbed" from factory gases, as shown in Figure 4-10, by a reaction of two formula units of calcium sulfide with each molecule of sulfur dioxide, as in the equation $SO_2 + 2\,CaS \rightarrow 2\,CaO + 3\,S$.

(c) According to the reaction $2\,NaI + H_2O_2 \rightarrow I_2 + 2\,NaOH$, we need two formula units of sodium iodide and one molecule of hydrogen peroxide, H_2O_2, if we want to make a molecule of iodine from sodium iodide.

Thinking Through the Problem As in the first part of this chapter, we have to convert the individual reference for every substance into a mole reference.

Answer

(a) In the reaction of ethylene, C_2H_4, with oxygen to make acetaldehyde, C_2H_4O, we need one mole of oxygen for every two moles of ethylene, according to the equation $2\,C_2H_4 + O_2 \rightarrow 2\,C_2H_4O$.

(b) Sulfur dioxide gas can be "scrubbed" from factory gases by a reaction of two moles of calcium sulfide with each mole of sulfur dioxide, as in the equation $SO_2 + 2\,CaS \rightarrow 2\,CaO + 3\,S$.

(c) According to the reaction $2\,NaI + H_2O_2 \rightarrow I_2 + 2\,NaOH$, if we want to make a mole of iodine from sodium iodide, we need two moles of sodium iodide and one mole of hydrogen peroxide, H_2O_2.

Figure 4-10 ▶ Removing unwanted gases by scrubbing exhausts is important in controlling pollution. Calcium sulfide in the scrubber pictured here reacts with sulfur dioxide gas, removing all sulfur-containing substances from the factory exhaust. (Betty Crowell.)

Coefficients of a balanced equation give mole amounts. We use coefficients to compare the mole amounts of two substances in the equation. This comparison is called a mole/mole ratio.

Properties of a Balanced Chemical Equation

In a balanced chemical equation, each coefficient should be a whole number in lowest ratio to all the other coefficients in the equation. For example, the balanced equation for the reaction between sodium sulfate and barium chloride is:

$$Na_2SO_4\ (aq) + BaCl_2\ (aq) \longrightarrow BaSO_4\ (s) + 2\ NaCl\ (aq) \qquad (1)$$

Compare this equation with the following three equations. How do these equations differ from equation (1)?

$$3\ Na_2SO_4\ (aq) + 3\ BaCl_2\ (aq) \longrightarrow 3\ BaSO_4\ (s) + 6\ NaCl\ (aq) \qquad (2)$$

$$1.5\ Na_2SO_4\ (aq) + 1.5\ BaCl_2\ (aq) \longrightarrow$$
$$1.5\ BaSO_4\ (s) + 3\ NaCl\ (aq) \qquad (3)$$

$$500\ Na_2SO_4\ (aq) + 500\ BaCl_2\ (aq) \longrightarrow$$
$$500\ BaSO_4\ (s) + 1000\ NaCl\ (aq) \qquad (4)$$

Equation (2) has coefficients that are three times those in equation (1). Dividing the coefficients by three will give us equation (1). Equation (3) has coefficients that are 1.5 times those in equation (1). Dividing the coefficients in equation (3) by 1.5 will give us back equation (1). Equation (4) has coefficients that are 500 times those in equation (1). Dividing the coefficients in equation (4) by 500 will give us back equation (1). Mathematically, the three equations (2) through (4) are each equivalent to equation (1). Although each equation has correct ratios between substances, it is customary to write the coefficients of a balanced equation in lowest terms, as in equation (1). Thus, only equation (1) is an acceptable final description.

Another important property of coefficients is that *they multiply the number of atoms in the formulas they precede*. For example, NH_3 contains one N atom and three H atoms. When the coefficient 2 is placed in front of the formula, as in the expression "2 NH_3," we have 2 N atoms and 6 H atoms. This is important to know when we balance a chemical equation.

KEY IDEA 4.7

Example 4.10

Reducing a chemical equation to the simplest whole numbers

Write the following equation with the smallest whole numbers possible for the coefficients.

$$4\ MgO\ (s) + 2\ C\ (s) \longrightarrow 2\ CO_2\ (g) + 4\ Mg\ (s)$$

Thinking Through the Problem We must determine whether or not the coefficients are in the lowest terms. We can copy them down, separate from the chemical substances, to make this easier. The coefficients are 4, 2, 2, and 4. All are even numbers, so they all share a common factor of 2. To solve the problem, divide each coefficient by 2.

Answer $2\ MgO\ (s) + C\ (s) \rightarrow CO_2\ (g) + 2\ Mg\ (s)$

Example **4.11**

Converting fractions in a chemical equation into whole numbers

Write the following equation with the smallest whole numbers possible for the coefficients.

$$C_2H_6 \ (g) + 7/2 \ O_2 \ (g) \longrightarrow 2 \ CO_2 \ (g) + 3 \ H_2O \ (g)$$

Thinking Through the Problem We see that the coefficient of O_2 is a fraction. To clear the fraction, we multiply it and all of the other coefficients in the equation by 2 (the denominator of 7/2).

Answer $2 \ C_2H_6 \ (g) + 7 \ O_2 \ (g) \rightarrow 4 \ CO_2 \ (g) + 6 \ H_2O \ (g)$

How are you doing? **4.5**

Rewrite each of these equations with coefficients in lowest terms.

(a) $8 \ Fe + 6 \ O_2 \rightarrow 4 \ Fe_2O_3$
(b) $6 \ Al + 9 \ Br_2 \rightarrow 3 \ Al_2Br_6$

KEY IDEA 4.8

Another important property of coefficients is that *they multiply the number of atoms in the formulas they precede.* For example, NH_3 contains one N atom and three H atoms. When the coefficient 2 is placed in front of the formula, as in the expression "2 NH_3," we have 2 N atoms and 6 H atoms. This is important to know when we balance a chemical equation.

A balanced chemical equation represents what actually happens in a chemical reaction. No atoms are lost and no atoms are gained during the course of the reaction. Instead, the atoms are rearranged into new substances.

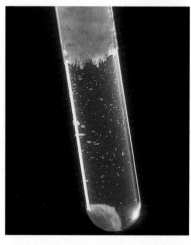

Figure 4-11 ▲ The reaction of a solution of sodium sulfate with a solution of barium chloride Mixing solutions of sodium sulfate and barium chloride results in the production of solid barium sulfate and aqueous sodium chloride. The solid formed is called a precipitate and the reaction is called a precipitation reaction. (Larry Stepanowicz/Visuals Unlimited.)

Balancing Equations

All the equations we have seen so far are balanced equations. Often, however, we are given only the formulas of the reactants and the products, and must determine the coefficients that will give us a balanced equation. Let's examine the equation-balancing process more closely. How do we balance the equation for the reaction shown in Figure 4-11? We can begin by writing down the correct chemical formulas for the reactants and products.

$$\_\_ \ Na_2SO_4 \ (aq) + \_\_ \ BaCl_2 \ (aq) \longrightarrow \_\_ \ BaSO_4 \ (s) + \_\_ \ NaCl \ (aq)$$

We know that, in a balanced equation, we must have the same number of each atom on both sides of the arrow. We begin by counting the various kinds of atoms, checking to see whether or not the number of reactant atoms is equal to the number of product atoms for each element.

Beginning with the first atom in the equation, we find that there are two atoms of Na in Na_2SO_4 and only one atom of Na in NaCl. To balance the Na, we insert a coefficient of 2 *in front of* the formula for NaCl and a 1 in front of Na_2SO_4. Keep in mind that the coefficient

2 multiplies everything in the formula for NaCl. Therefore, 2 NaCl gives us 2 Na atoms *and* 2 Cl atoms. Our equation now becomes

1 Na_2SO_4 (*aq*) + __ $BaCl_2$ (*aq*) $\longrightarrow$ __ $BaSO_4$ (*s*) + **2** NaCl (*aq*)
2 Na   =   2 Na

Now there are two sodium atoms on each side of the equation. Notice that we will have the same number of chlorine atoms on each side of the equation if we give $BaCl_2$ a coefficient of **1**. The chlorine atom count is satisfied.

1 Na_2SO_4 (*aq*) + **1** $BaCl_2$ (*aq*) $\longrightarrow$ __$BaSO_4$ (*s*) + **2** NaCl (*aq*)
2 Na   =   2 Na
                              2 Cl   =   2 Cl

Examining the other elements in the reaction we see that there will be equal numbers of S, Ba, and O atoms on each side of the equation if we give a coefficient of **1** to $BaSO_4$. The equation is now balanced:

1 Na_2SO_4 (*aq*) + **1** $BaCl_2$ (*aq*) $\longrightarrow$ **1** $BaSO_4$ (*s*) + **2** NaCl (*aq*)

Total atom count:

2 Na + 1 S + 4 O + 1 Ba + 2 Cl = 2 Na + 1 S + 4 O + 1 Ba + 2 Cl

A coefficient of 1 is implied if no other number is written for that substance. Therefore, we can drop the coefficients of 1, and the balanced equation is:

Na_2SO_4 (*aq*) + $BaCl_2$ (*aq*) $\longrightarrow$ $BaSO_4$ (*s*) + 2 NaCl (*aq*)

The conservation of atoms is very important here. Note that the "total atom count" gives us the assurance that *the same number and kinds of atoms appear in both the reactant and the product.* Only the arrangement of these atoms changes during the reaction.

Balancing a chemical equation is a matter of "counting atoms." There are several ways to do this. In Table 4-2 we present one that works reliably for all different chemical reactions. If we follow this technique correctly we will not need to "back up" to change any co-efficient we have already written.

✔ **Meeting the Goals**

To "balance" an equation, we put coefficients in front of each chemical formula in the equation in such a way that the number of atoms of each element in the reactants is equal to the number of atoms of each element in the products.

KEY IDEA 4.9

REMEMBER

Table 4-2 A Method for Balancing Chemical Equations

1. Pick an element that occurs *in only one substance on each side of the equation* and balance it. If possible, pick an element that will not give a "1" as a coefficient.
2. Pick an element that is present in only one substance that is left to be balanced, and balance that element. If possible, balance the other elements in the formula containing the element balanced in step one.
3. Pick a third element that is in only one substance that is left to be balanced, and balance that element. Then continue with a fourth element, and so on.

Here is how the technique given in Table 4-2 works: Suppose that we want to balance the equation for the reaction that forms ammonia, NH_3, from N_2 and H_2. We begin with the equation:

$$\_\ H_2 + \_\ N_2 \longrightarrow \_\ NH_3$$

Using step 1, we can pick either H or N. Let's pick H. *Notice the different subscripts* on H here. The presence of H in H_2 on the left side of the equation and in NH_3 on the right means that we must select a coefficient of 3 for H_2 and 2 for NH_3 to balance the equation, giving us a total of six H atoms on each side:

$$3\ H_2 + \_\ N_2 \longrightarrow 2\ NH_3$$

Using step 2, we note that the substance NH_3 has one element left to be balanced: nitrogen. There are two N's on the right, and these need to be balanced with the N's on the left of the equation. We can do this by using the coefficient 1 for N_2:

$$3\ H_2 + 1\ N_2 \longrightarrow 2\ NH_3$$

There are no elements remaining to be balanced. In our final answer we can leave the "1" out, since it is implied if there is no other coefficient specified for that substance.

$$3\ H_2 + N_2 \longrightarrow 2\ NH_3$$

Example 4.12 **Balancing a chemical equation**

Balance the following equation:

$$\_\ C_2H_6O\ (g) + \_\ O_2\ (g) \longrightarrow \_\ CO_2\ (g) + \_\ H_2O\ (l)$$

Thinking Through the Problem First inspect the equation and notice that oxygen appears in every formula, including O_2. Leave the coefficient for O_2 until last. Using step 1, we can pick either C or H. We will begin by balancing carbon. ➡ (4.10) The key idea is to notice that there are two carbon atoms in C_2H_6O but only one carbon atom in CO_2. Put a coefficient of 2 in front of CO_2. Also put a coefficient of 1 in front of C_2H_6O to remind yourself that something in this formula has been counted.

$$1\ C_2H_6O + \_\ O_2 \longrightarrow 2\ CO_2 + \_\ H_2O$$

The count of carbon atoms is now balanced. Next, balance hydrogen. The equation as written has 6 hydrogens in C_2H_6O on the left. No H's appear in counted substances on the right. We put a coefficient of 3 in front of H_2O to get 6 hydrogens on the right side of the equation.

$$1\ C_2H_6O + \_\ O_2 \longrightarrow 2\ CO_2 + 3\ H_2O$$

The count of hydrogen atoms is now balanced and only the oxygen count remains to be balanced. At this point there are coefficients in front of every substance except O_2. (Be very careful at this point *not to count the oxygen atoms in O_2*. That is the coefficient we are looking for! If necessary, cover the O_2 with your finger while you are counting the oxygens.) We have 1 O in a counted substance on the left, and seven in the counted substances on the right. We therefore need six more O's on the left. The six O's on the left can be obtained by placing a coefficient of 3 in front of the O_2.

$$\textbf{1 } C_2H_6O + \textbf{3 } O_2 \longrightarrow \textbf{2 } CO_2 + \textbf{3}H_2O$$

Answer $C_2H_6O + 3\,O_2 \rightarrow 2\,CO_2 + 3\,H_2O$

Example 4.13 Writing and balancing a chemical equation

Balance the reaction in which zinc metal (Zn) reacts with hydrogen chloride (HCl) to give zinc chloride and hydrogen gas.

Thinking Through the Problem We start with $\_\_ Zn + \_\_ HCl \rightarrow$ $\_\_ ZnCl_2 + \_\_ H_2$. Following step 1 from Table 4-2, we pick an element that occurs once on each side of the equation and balance this. In this case, all the elements occur only once on each side. But it is best to choose either H or Cl, and not Zn, because Zn would give a 1 as the coefficient. We begin, then, by balancing H:

$$\_\_ Zn + \textbf{2 } HCl \longrightarrow \_\_ ZnCl_2 + \textbf{1 } H_2$$

Next, we balance the second element in HCl. We have two Cl's on the left. Therefore, we need to have two Cl's in $ZnCl_2$, which we get from one $ZnCl_2$. Then:

$$\_\_ Zn + \textbf{2 } HCl \longrightarrow \textbf{1 } ZnCl_2 + \textbf{1 } H_2$$

When we finally turn to Zn, we just put a 1 in front of Zn to balance the one zinc in $ZnCl_2$.

Answer $Zn + 2\,HCl \rightarrow ZnCl_2 + H_2$

Example 4.14 Balancing a chemical equation

◀ | Try this with your **Personal Web Tutor**

Balance the equation $\_\_ Fe_2O_3 + \_\_ Cl_2 \rightarrow \_\_ FeCl_3 + \_\_ O_2$.

Thinking Through the Problem All the elements appear only once on each side of the equation, so we can choose to first balance any of the three, Fe, O, or Cl. However, balancing Fe first will give us a coefficient of 1 because there is a simple 2 : 1 ratio of Fe atoms.

Instead, we will choose O, which has a 2 : 3 ratio for the Fe_2O_3 and O_2. Balancing O, we get:

$$2\ Fe_2O_3 + \underline{\ \ }\ Cl_2 \longrightarrow \underline{\ \ }\ FeCl_3 + 3\ O_2$$

For the second step we must choose to balance Fe, because that element is left unbalanced in only one substance, $FeCl_3$. There are 4 Fe's in a balanced substance on the left. Therefore, we need to obtain 4 Fe's on the right by placing the coefficient 4 in front of $FeCl_3$.

$$2\ Fe_2O_3 + \underline{\ \ }\ Cl_2 \longrightarrow 4\ FeCl_3 + 3\ O_2$$

For the third step, only Cl remains to be balanced. We have 12 Cl's counted in a balanced substance on the right, and none yet counted on the left. We need to obtain 12 Cl's on the left, so we place the coefficient 6 in front of Cl_2.

Answer $2\ Fe_2O_3 + 6\ Cl_2 \rightarrow 4\ FeCl_3 + 3\ O_2$

Example 4.15 Balancing a chemical equation

Balance the following equation:

$$\underline{\ \ }\ Na\ (s) + \underline{\ \ }\ H_2O\ (l) \longrightarrow \underline{\ \ }\ NaOH\ (aq) + \underline{\ \ }\ H_2\ (g)$$

Thinking Through the Problem To balance this equation, we can begin with either Na or O. Both appear only once on either side of the equation. If we begin with O we note that there is 1 oxygen atom in H_2O and 1 oxygen atom in NaOH. There are coefficients of 1 before H_2O and NaOH:

$$\underline{\ \ }\ Na + 1\ H_2O \longrightarrow 1\ NaOH + \underline{\ \ }\ H_2$$

The coefficient of 1 in front of NaOH then requires a coefficient of 1 to go in front of the Na in the second step:

$$1\ Na + 1\ H_2O \longrightarrow 1\ NaOH + \underline{\ \ }\ H_2$$

Now both the oxygen and the sodium atom counts are balanced. Only hydrogen atoms remain to be counted and balanced. We have 2 H atoms in balanced substances on the left, and 1 H in a balanced substance on the right. The unbalanced substance, H_2, must provide 1 H atom. For this we need only 1/2 of H_2. So we put 1/2 in front of H_2:

$$1\ Na + 1\ H_2O \longrightarrow 1\ NaOH + 1/2\ H_2$$

The coefficients of the balanced equation are 1, 1, 1, 1/2. These coefficients represent the correct ratio for this reaction.

However, as we are required to have only whole numbers as coefficients, we must multiply each number in the ratio by the denominator of the fractional coefficient. We must multiply by 2.

Answer $2 Na + 2 H_2O \rightarrow 2 NaOH + H_2$

How are you doing? 4.6

Balance the following equations.

(a) $Pb + HCl \rightarrow PbCl_2 + H_2$

(b) $Fe(SO_4) + (NH_4)_2S \rightarrow FeS + (NH_4)_2SO_4$

PRACTICAL C What are the balanced equations for some of the reactions that take place in matches?

Potassium chlorate is the most important oxidant in the operation of safety matches. When potassium chlorate reacts with phosphorus, the products are P_2O_5 and KCl. In Practical B we described this reaction using words and formulas. Now, let us write a properly balanced chemical equation for the reaction.

Some of the potassium chlorate also reacts with sulfur in the match head (as does air, as discussed in the last section). If we write the formula of sulfur as "S_8," what is the balanced chemical equation for the reaction of sulfur with potassium chlorate to give sulfur dioxide and potassium chloride?

$$\_\_KClO_3 + \_\_S_8 \longrightarrow \_\_SO_2 + \_\_KCl$$

What will be the balanced chemical equation for this reaction if we write the formula of sulfur as "S"?

$$\_\_KClO_3 + \_\_S \longrightarrow \_\_SO_2 + \_\_KCl$$

For both equations, write the ratio of $KClO_3$ units per S atom. Is there a pattern?

Match heads often contain other oxides, which keep the combustion reaction proceeding smoothly. These, too, react with the sulfur in the match. For example, MnO_2 reacts with S to give Mn_2O_3 and SO_2. What is the balanced chemical equation for this reaction?

Because the amount of phosphorus that reacts is very small, the other products are not a concern. However, the sulfur dioxide produced in the match head is a noxious gas. It is responsible for the characteristic odor of freshly burning matches. To protect the user from the dangerous effects of sulfur dioxide gas, match manufacturers add a neutralizer, calcium carbonate ($CaCO_3$), which reacts with sulfur dioxide to give calcium sulfite, $CaSO_3$, and a gas that is harmless. Write the reaction of calcium carbonate and sulfur dioxide and suggest the formula of the harmless gas that is formed. (Hint: think about the formula of the sulfite ion and compare this to the formula of sulfur dioxide.)

Some matches, called "strike-anywhere matches," do not require a striker in order to ignite. They can be lit by rubbing them against any rough surface. The ingredients necessary to start and maintain the combustion reactions are all present in the match head, which is designed so that the substances capable of rapid reaction are not brought into contact until the match head is crushed by the rubbing action. A special coating covers the main, sulfur-containing part of the match head to prevent the match from igniting spontaneously. This coating includes a mixture of potassium chlorate and tetraphosphorus trisulfide (P_4S_3). When a strike-anywhere match is rubbed against a rough surface, the initial reaction converts both the phosphorus and the sulfur in P_4S_3 into P_2O_5 and SO_2. KCl also is produced. After this reaction begins, a reaction between sulfur and $KClO_3$ in the inner layers of the match head takes over, promoting further burning. Write the balanced chemical equation of tetraphosphorus trisulfide and potassium chlorate.

PROBLEMS

20. Rewrite the following statements in terms of mole amounts.

    (a) The displacement of two silver ions, Ag^+, from solution uses one atom of Cu metal.

    (b) Combustion of an octane molecule (C_8H_{18}) results in the formation of eight molecules of carbon dioxide.

    (c) Every two molecules of ammonia requires three molecules of hydrogen for its formation.

21. Rewrite the following statements in terms of individual atoms or molecules.

    (a) To make a mole of copper (II) nitrate from copper requires two moles of nitric acid, HNO_3.

    (b) When we get two moles of oxygen by decomposition of hydrogen peroxide we also form a mole of water.

    (c) Conversion of two moles methane to one mole of ethanol also requires one mole of oxygen.

22. Balance the following equations.

    (a) $C + O_2 \rightarrow CO$

    (b) $S + O_2 \rightarrow SO_2$

    (c) $Fe + O_2 \rightarrow Fe_2O_3$

    (d) $Al + S_8 \rightarrow Al_2S_3$

    (e) $Al + O_2 \rightarrow Al_2O_3$

    (f) $N_2 + O_2 \rightarrow NO$

23. Balance the following equations.

    (a) $HgO \rightarrow Hg + O_2$

    (b) $H_2O_2 \rightarrow H_2O + O_2$

24. Discussion question: Balance the following equations.

    (a) $C_2H_2 + O_2 \rightarrow CO_2 + H_2O$

    (b) $C_8H_{18} + O_2 \rightarrow CO_2 + H_2O$

    (c) $C_4H_8 + O_2 \rightarrow CO_2 + H_2O$

The reaction of hydrocarbons (molecules of just carbon and hydrogen) with oxygen is very important. Discuss the trends you see in the number of oxygen molecules required to react with each of these hydrocarbons.

25. Balance the following equations.

    (a) $Cu + AgNO_3 \rightarrow Ag + Cu(NO_3)_2$

    (b) $Al + Sn(NO_3)_2 \rightarrow Al(NO_3)_3 + Sn$

    (c) $Ba(C_2H_3O_2)_2 + (NH_4)_3PO_4 \rightarrow$
$$Ba_3(PO_4)_2 + NH_4C_2H_3O_2$$

Chapter 4 Summary and Problems

 Review with **Web Practice**

VOCABULARY

measurement by counting	The measurement of the number of things in a quantity.
chemical counting	The application of counting rules to the counting of atoms, molecules, and formula units.
mole	A unit that represents a very large number, 602,213,700,000,000,000,000,000. This number is often called Avogadro's number. "Mole" is abbreviated "mol."
Avogadro's number	The number 602,213,700,000,000,000,000,000.
atomic or molecular scale	A scale used to describe and count very small samples of matter; this scale is used when we describe a chemical reaction in terms of individual atoms, molecules, and formula units.
macroscopic scale	A scale used to describe and count large samples of matter; this scale is used when we describe a chemical reaction in terms of moles of reactants and products.
mole ratio	A fractional comparison of mole amounts of atoms, molecules, and formula units. Just as the number of atoms in a molecule or formula unit can be counted, so can the number of moles of atoms in a mole of molecules or formula units be counted. Thus, the formula $C_3H_6O_3$ has the

atom/molecule ratio $\dfrac{3 \text{ O atoms}}{1 \text{ molecule}}$ and the mole/mole ratio $\dfrac{3 \text{ moles O}}{1 \text{ mole compound}}$.

formula unit The smallest whole number ratio of the different types of atoms in the extended structure of an ionic compound.

chemical reaction A process whereby the atoms present in reactant substances are rearranged to form different substances called products.

reactants Elements and compounds that appear on the left side of a chemical equation; the substances that react or undergo a chemical change into other elements and compounds called products.

products The elements and compounds formed in a chemical reaction.

chemical equation A statement, in words or in symbols, that describes a chemical reaction; names (or formulas) of all reactants and products, together with their physical states of matter, are included in the equation.

balanced chemical equation A balanced chemical equation contains equal numbers of atoms of each element on the reactant side and on the product side of the equation.

coefficients Numbers written in front of chemical formulas in a balanced chemical equation; coefficients should be written as whole numbers in lowest terms.

chemical stoichiometry The measurement of the components of a chemical reaction.

SECTION GOALS

What is meant by measurement by counting?

If we measure something by referring to individual units, whether one by one or in larger groups, then we are measuring by counting.

Why do chemists use the mole in counting chemical quantities?

The mole is used because it is equal to a very large number of things, and when we count even the smallest measurable amounts of chemical quantities, we are counting very large numbers.

How do the relative amounts of elements in a compound relate to the relative number of moles of elements in a compound?

As with other counting units like the dozen, when we want to count relative amounts by moles, we just use the *mole* unit in *all* the parts of the relative amounts.

How does knowing the number of moles of atoms or ions in a compound help us to determine the formula for that compound?

The mole ratio of an atom or ion in a compound to one mole of that compound gives us the subscript number of that atom or ion in the formula for the compound.

When chemical substances react, we often use chemical formulas to describe what is happening. What are the most important parts of a proper chemical equation?

The most important parts are the group of reactants, the group of products, and an arrow to indicate the transformation of reactants to products.

What do the coefficients in a balanced equation represent on the atomic or molecular and the macroscopic scales?

The coefficients represent individual atoms, molecules, and formula units on the atomic or molecular scale; the same coefficients indicate different mole amounts on the macroscopic scale.

What does "balancing" an equation mean?

Balancing an equation means taking a description of the chemical substances in a reaction, converting to chemical formulas, and then determining the correct coefficients.

PROBLEMS

26. Rewrite the following statements to reflect mole amounts instead of individual amounts. Write the relevant mole ratios in each case.

(a) There are four hydrogen atoms in one acetic acid molecule.

(b) When methane combusts, two water molecules are formed for each molecule of methane produced.

(c) Forming one unit of silver phosphate requires three units of silver nitrate in the reaction.

(d) There are eighteen hydrogen atoms in one molecule of octane.

27. Rewrite the following statements to reflect individual amounts instead of mole amounts. Write the relevant ratio of atoms and molecules in each case.

(a) We find that there are two moles of carbon in one mole of the molecule ethane.

(b) Decomposition of germane, GeH_4, gives two moles of hydrogen gas for every mole of germane.

(c) The destruction of nitrogen dioxide in a car's catalytic converter gives one mole of nitrogen gas for two moles of nitrogen dioxide.

28. Industrial chemists work on very large scales. One notes that a sample of a rock contains 15 moles of copper atoms. How much copper is in a sample of the same rock that is ten times this size?

29. The oxygen-transport protein hemoglobin contains four atoms of iron in one unit of hemoglobin. How many moles of iron are in one mole of hemoglobin?

30. Balance the following equations.

(a) $NH_3 + O_2 \rightarrow NO + H_2O$

(b) $C_5H_{10} + O_2 \rightarrow CO_2 + H_2O$

(c) $SO_2 + O_2 \rightarrow SO_3$

31. Balance the following equations.

(a) $Fe(OH)_3 + H_2SO_4 \rightarrow Fe_2(SO_4)_3 + H_2O$

(b) $NaCl + H_2O \rightarrow Cl_2 + H_2 + NaOH$

(c) $Na_2NH + H_2O \rightarrow NH_3 + NaOH$

(d) $Al + NH_4ClO_4 \rightarrow Al_2O_3 + AlCl_3 + NO + H_2O$

(e) $SCl_2 + NaF \rightarrow SF_4 + S_2Cl_2 + NaCl$

(f) $UO_2 + HF \rightarrow UF_4 + H_2O$

(g) $N_2O_5 + H_2O \rightarrow HNO_3$

32. Balance the following equations.

(a) $CH_4 + Cl_2 \rightarrow HCl + CCl_4$

(b) $OF_2 + H_2O \rightarrow O_2 + HF$

(c) $N_2H_4 + H_2O_2 \rightarrow N_2 + H_2O$

33. Draw Lewis structures for each equation in Problem 30, showing the number of each atom before and after the reaction.

34. Balance the following equations.

(a) $C_3H_5O_9N_3 \rightarrow CO_2 + N_2 + O_2 + H_2O$

(b) $(NH_4)_2Cr_2O_7 \rightarrow Cr_2O_3 + N_2 + H_2O$

35. For each of the following reactions, write the equation using the names of the reactants and products.

(a) When potassium bromide is exposed to chlorine gas, potassium chloride and bromine are produced.

(b) A white powder, ammonium chloride, is formed when hydrogen chloride and ammonia gases are mixed.

(c) Magnesium sulfide produces magnesium sulfate and clouds of odorous hydrogen sulfide when mixed with sulfuric acid, H_2SO_4.

(d) Carbon dioxide is emitted from a solution when potassium carbonate reacts with hydrogen chloride to give water and potassium chloride.

36. For the reactions in Problem 35, write the chemical formulas for the reactants and products.

37. Write a balanced chemical equation for each reaction given in Problem 35.

38. Correct the following statements.

(a) 1 mole CH_4 reacts with 2 molecules Cl_2 to give 2 molecules HCl + 1 mole CH_2Cl_2.

(b) 2 molecules H_2O decompose to give 1 mole O_2 + 2 moles H_2.

39. Write the chemical formula for the compound that contains the following mole ratios.

(a) 5 moles C/1 mole compound, 12 moles H/1 mole compound

(b) 2 moles Ag/2 moles compound, 2 moles N/2 moles compound, 6 moles O/2 moles compound

(c) 1 mole N/0.5 mole compound, 2 moles O/0.5 mole compound

40. Determine whether the following statements are true or false. If false, suggest a correction that would make the statement true.

(a) 3 moles Cl atoms/1 mole molecules = 3 dozen Cl atoms/1 dozen molecules

(b) 12 H atoms/2 molecules = 6 moles H atoms/1 molecule

(c) 2 NO_3^- ions/6 formula units = 5 moles NO_3^- ions/6 moles formula units

41. One mole of a compound contains 3 moles Mn and 4 moles O. Determine the correct formula for this compound and the oxidation number of Mn in the compound. What is the name of the compound?

42. Three moles of a compound contains 6 moles Mn and 9 moles O. Determine the correct formula for this compound and the oxidation number of Mn in the compound. What is the name of the compound?

43. Two moles of a compound contains 2 moles Mn and 4 moles O. Determine the correct formula for this compound and the oxidation number of Mn in the compound. What is the name of the compound?

44. At what temperatures would you expect the following substances to exist?

(a) H_2O (s)
(b) O_2 (g)
(c) H_2O (g)

45. Balance $C + O{=}O \rightarrow C{\equiv}O$.

46. Elemental sulfur usually exists in a ring of eight S atoms connected to one another by single bonds. Draw the Lewis structure for the reaction $S_8 + O_2 \rightarrow SO_2$

47. Write and balance the equation for the reaction red phosphorus + potassium chlorate $\rightarrow$ potassium chloride + diphosphorus pentoxide.

48. Rewrite the following sequences of numbers as simple whole numbers in lowest terms.

(a) $2 : 1/2 : 1/3$
(b) $1 : 1/5 : 1/2$
(c) $3 : 1/2 : 1/4$

49. Rewrite the following equations so that the coefficients are simple whole numbers in lowest terms.

(a) $2\ Al + 3/2\ O_2 \rightarrow Al_2O_3$

(b) $3\ Mg(NO_3)_2 + 6\ NaCl \rightarrow 6\ NaNO_3 + 3\ MgCl_2$
(c) $2/3\ Al + 2\ HCl \rightarrow 2/3\ AlCl_3 + H_2$

50. Determine the total number of atoms of each element in one molecule of the following.

(a) $3\ Al_2Br_6$
(b) $5\ C_5H_{10}O$
(c) $4\ C_3H_5N_3O_9$
(d) $3\ S_8$

51. Determine the total number of atoms of each element in one formula unit of the following.

(a) $16\ KClO_3$
(b) $2\ Al_2(SO_4)_3$
(c) $7\ U_3(PO_3)_4$
(d) $5\ (NH_4)_2CO_3$

52. Determine the ratio of 1 formula unit of $KClO_3$ to one S atom in the balanced equation for the reaction $KClO_3 + S_8 \rightarrow SO_2 + KCl$.

53. Determine the ratio of 1 formula unit of $KClO_3$ to one S atom in the balanced equation for the reaction $KClO_3 + S_8 \rightarrow SO_2 + KCl$.

54. The dangerous effects of sulfur dioxide gas are neutralized by reaction with calcium carbonate. The reaction produces calcium sulfite and carbon dioxide gas. Write the balanced equation for this reaction.

55. Write a balanced equation for the reaction $P_4S_3 + KClO_3 \rightarrow P_2O_5 + SO_2 + KCl$.

(a) Name each of the substances in this reaction.
(b) Determine the oxidation numbers of S in P_4S_3 and in SO_2.
(c) Determine the oxidation number of Cl in KCl and in $KClO_3$.

Chemical Reactions

The reaction of magnesium in air gives off a great deal of light.
(Richard Megna/Fundamental Photographs.)

PRACTICAL CHEMISTRY What's in that Metal and How Do We Prove It?

In Chapter 1, we learned that metals are an important kind of element. Many of the metals have properties in common, but each has certain distinct properties as well. We can use chemical reactions as a way to distinguish between different metals.

We can also use chemical reactions to determine whether a metallic sample is a pure element. For example, the aluminum in cans or the copper in common electrical wire are both pure metals, aside from decorative or protective coatings. Other metallic substances are mixtures of metals, called alloys. Common examples of alloys include steel, dental fillings, and the 14K gold used in jewelry. We can identify the particular metals present in an alloy by using chemical reactions. If, for example, we wanted to determine whether there is copper present in a "silver" bracelet, we could test to see whether the bracelet reacts in ways that are specific to copper.

This chapter will give you practice in recognizing whether or not chemical reactions are taking place. It will also help you to distinguish between several different types of chemical reactions. In the Chapter Practicals we will use what we know about chemical reactions to answer these questions:

PRACTICAL A How is color used to identify metals?

PRACTICAL B How does the synthesis of oxides determine the reactivity of metals in the environment?

PRACTICAL C What can we infer when a metal reacts with a solution?

PRACTICAL D What kinds of chemical reactions are used to test for metal ions?

5.1 Chemical Properties and Changes

SECTION GOALS

✔ Differentiate between chemical and physical properties.

✔ Recognize a chemical change from a word description or from a chemical equation.

Chemical Properties

We first discussed the difference between a chemical and a physical property in Chapter 1. A chemical property describes the reactivity of a substance. It tells us how a material reacts—or doesn't react—under different conditions. A chemical property, then, is a description of a possible chemical reaction.

Chemical reactions occur when two or more reactive substances are combined, or when a single substance is decomposed. When we describe whether or not a substance reacts with another substance to form new substances, we are stating a chemical property.

For example, one chemical property of sodium is that it reacts with chlorine to produce sodium chloride. Sometimes, however, it is just as important to know that a substance does not react with some other substance. Some substances react spontaneously with oxygen, O_2, and must be kept away from oxygen for safe storage. Many of these same substances do not react with nitrogen, however. Therefore, they are often kept in a nitrogen gas environment as a safety measure. Nonreactivity (sometimes called inertness) is an important chemical property because it helps us to know how to safely store substances. *Words that help us recognize chemical properties are verbs, such as "reacts with," "forms," "gives," "results in the formation of," or "does not react."* Thus, "gold does not burn" is a statement about a chemical property of gold just as "sulfur burns readily" is a statement of a chemical property of sulfur.

The stability of a pure substance is also a chemical property of the substance. In Chapter 2, some of the "power molecules" that were discussed in the Chapter Practicals were capable of explosions if improperly handled. Thus, a chemical property of TNT is its ability to spontaneously react with itself if triggered. Other more familiar pure substances have important chemical properties of their own. Sodium hydrogen carbonate (baking soda) will react when heated to give off carbon dioxide; this is why baking soda can be applied to flames to help extinguish them: it produces a smothering coat of carbon dioxide.

Some changes of substances are not chemical properties. The most important of these are phase changes. The chemical composition of water, for example, does not change when ice melts to liquid water or when liquid water turns into water vapor.

Example 5.1   **Identifying descriptions of chemical properties**

Which of the following descriptions involves a chemical property?

(a) Potassium reacts violently with water to produce a caustic solution of potassium hydroxide.
(b) Candle wax melts when exposed to high temperatures.
(c) Candle wax burns to form carbon dioxide and water vapor.
(d) Cesium does not react with nitrogen.

Thinking Through the Problem We must know the difference between chemical and physical properties. ◀◼ (5.1) The key idea is that the important words are the verbs. The words *react* and *burns to form* indicate chemical properties. On the other hand, words that describe or predict a physical change of state—such as *melt, freeze, evaporate, condense,* or *sublime*—indicate a physical change.

Answer Descriptions (a), (c), and (d) are chemical properties; description (b) is a physical property. If these processes actually occurred, then (a) and (c) would be chemical changes and (b) would be a physical change. Description (d) does not describe a change, but is a chemical property because it describes how the substance behaves with another substance—it does not react.

✔**Meeting the Goals**

A chemical property tells us whether or not a substance reacts with another chemical substance to form new substances. A physical property tells us characteristics of a substance that do *not* involve reactions with other substances, and do *not* involve the formation of new substances.

How are you doing? 5.1

Classify the following properties as either chemical or physical properties.

(a) floats on water
(b) silvery white in color
(c) will not react with cold water
(d) burns with a white flame

Chemical Changes

Chemical reactions bring about **chemical changes.** When substances undergo a chemical change, new and different substances are formed. In a chemical reaction, as we saw in Chapter 4, all the atoms present in the reactants are present in the product, but in different arrangements. The properties of the products are different from the properties of the reactants.

For example, when sodium and chlorine react, they form a new substance, sodium chloride. Both sodium and chlorine undergo a chemical change: their original identities are lost as they enter into the composition of a new chemical substance (sodium chloride) with its own set of physical and chemical properties. We recognize a process as a chemical change when the name, formula, and properties of a substance change as a result of the process. We write the equation for the chemical reaction of sodium and chlorine as follows:

$$2 \, Na \, (s) + Cl_2 \, (g) \longrightarrow 2 \, NaCl \, (s)$$

sodium + chlorine $\longrightarrow$ *sodium chloride*

The names and formulas of the products are different from those of the reactants. We see in Table 5-1 that the properties of the products differ from the properties of the reactants as well. Both physical and chemical properties are listed to give you a better idea of the magnitude of the differences among these three substances. As shown in Figure 5-2, none of these substances could be mistaken for any of the others!

There are many ways in which these properties can be used to see if a chemical change has occurred. For example, Table 5-1 shows that sodium and sodium chloride are solids at room temperature and

Table 5-1 Properties of Sodium, Chlorine, and Sodium Chloride

	Sodium	Chlorine	Sodium Chloride
Melting point, °C	97.8	−101	801
Boiling point, °C	882	−34.6	1413
Density (g/cm³)	0.97 g/cm³	0.0032 g/cm³	2.16 g/cm³
Physical phase at 25°C	Solid	Gas	Solid
Physical appearance	Soft, bright, silvery	Greenish yellow gas	Colorless, crystalline solid
Chemical reactivity	Sodium is a very reactive metal; it is never found free in nature.	Chlorine combines directly with nearly all elements; it is never found free in nature.	Sodium chloride is ordinary table salt. It reacts as sodium ions and chloride ions.
Safety precautions	Sodium metal must be handled with care. It should be stored in an environment protected from water and air.	Chlorine gas is extremely poisonous. Concentrations greater than 1 part per million are considered fatal.	Some NaCl is needed for health, but an excessive amount may increase hypertension.

Figure 5-2 ◄ Chlorine gas and sodium metal have physical properties very different from the compound formed by sodium and chlorine, sodium chloride. (Richard Megna/ Fundamental Photographs.)

that chlorine is a gas at room temperature. The reactants have two phases, the products only one. The two solids are different, also. When we compare them we see that sodium has much lower melting and boiling temperatures than does sodium chloride. It may be difficult to think about the melting point of a gas, but gases can be converted into liquids and even solids by increasing the pressure or lowering the temperature, or both (think of what needs to be done to change steam into liquid water and then into ice). We will talk more about properties and reactions of gases in Chapter 10.

A very important chemical property of a substance is its ability to react with oxygen. Substances that combine easily with oxygen are not expected to be durable or safe in air. Sodium reacts spontaneously with oxygen, and is therefore often stored in nitrogen, with which it does not react. If, on the other hand, a substance resists reaction with oxygen, then it will be durable in air. Many substances resist reacting with oxygen, unless a spark is provided to start the reaction. For example, it is safe to handle many organic compounds, even gasoline, in the open air, as long as sparks and other ignition sources are excluded. **Burning** is the rapid combination or reaction of a substance with oxygen; it is a chemical property. Figure 5-3 shows isopropyl alcohol burning in air.

Figure 5-3 ◄ Alcohol is a colorless liquid. These are two physical properties. Alcohol reacts with air to give a flame. This is a chemical property. (Richard Megna/Fundamental Photographs.)

Example 5.2   Recognizing properties and changes

In the following examples, decide whether the descriptions given refer to a *property* (or *properties*) or to a *change* and whether this property or change is *chemical* or *physical* in nature.

(a) Nitrogen is a colorless, odorless gas at room temperature.
(b) Nitrogen gas can be converted to liquid nitrogen by lowering its temperature and increasing its external pressure.
(c) Nitrogen gas does not burn in air.
(d) In the Haber process, nitrogen gas reacts with hydrogen gas to form ammonia gas.

Thinking Through the Problem We must determine whether the statement describes a chemical change, a chemical property, a physical change, or a physical property. We carefully read the examples and try to determine whether or not the nitrogen is converted into a different chemical substance. If we detect a change in the chemical substances, or a reaction with another substance, then we know that we have a chemical property or a chemical change.

Answer

(a) This is a description of properties and of a physical state. "Colorless and odorless" refer to physical properties of nitrogen. "Gas" refers to its physical state.
(b) This is a physical change of state. Nitrogen gas is changed to liquid nitrogen, but it is still nitrogen. It is the *same* chemical substance before and after the change.
(c) This is a description of how nitrogen gas *acts;* it is a chemical property.
(d) Nitrogen reacts to form a new substance, ammonia. This is a chemical change.

How are you doing?   5.2

Classify the following as a property or a change. Then tell if it is chemical or physical in nature. (a) Fuel in a camp lantern is ignited with a spark. (b) The fuel is a red liquid.

PRACTICAL A   How is color used to identify metals?

We test a substance to find out what elements it contains by a process called *chemical analysis*. Often the elements in the substance being tested are first separated from one another by causing them to enter aqueous (water) solution. We then use the known chemical and physical properties of the elements we are looking for as guidelines to their presence. After the separation process, the solution can be tested for the presence of particular ions. A chemical reaction is used to verify the presence or absence of these specific ions. Many of the tests depend on our ability to "see" the results. For example, we know in principle that we can recognize a chemical reaction by looking for the formation of a new substance. But how can we be sure whether or not a reaction has taken place—whether a new substance is formed—when we

combine unknown substances in the laboratory? Sometimes it is obvious that a reaction has occurred, other times it is not.

Suppose we mix two clear solutions. If we see gas bubbling out of the solution, or see that a solid has formed (precipitated), then we know that a new substance has been formed and that a chemical reaction has taken place. Another sign that a chemical reaction has occurred is a color change when the original solutions are mixed. For example, under suitable conditions a light pink solution containing manganese (II) ions will turn purple when mixed with Bi_2O_5. This color change occurs as manganese (II) is oxidized to manganese (VII) in MnO_4^- according to the following equation:

$$18 \text{ H}^+(aq) + 4 \text{ Mn}^{2+}(aq) + \text{Bi}_2\text{O}_5 \, (aq) \longrightarrow 4 \text{ MnO}_4^- \, (aq) + 10 \text{ Bi}^{3+}(aq) + 9 \text{ H}_2\text{O} \, (l)$$

pale pink   purple

Using very specific procedures, reactions such as this one are used to identify metal ions in solution.

Let's look at another example of how color is used to identify a metal ion. Suppose that we want to see whether a sample of a metal is iron or cobalt. First we dissolve a small amount of the sample in

acid. After adjusting the solution to the proper conditions, thiocyanate ion (SCN^-) is added as a test reagent (chemical). The table below lists certain characteristics of the reaction of thiocyanate with iron (III) and with cobalt (II). These reactions are shown in Figure 5-1.

Metal Ion in Solution	Reacting Substance in the Testing Solution	Resulting Color When Solutions are Mixed
Fe^{3+} (pale yellow)	SCN^- (colorless)	red solution
Co^{2+} (pink)	SCN^- (colorless)	blue solution

Suppose that you add thiocyanate ion to a solution that contains one of these ions.

(a) What can you conclude about the metal if the solution remains lightly colored?

(b) What color do you expect the solution to be if cobalt (II) is present?

Sometimes color is used to distinguish between solids that are formed when the two solutions are

Figure 5-1 ▲ (a) When a solution of iron is mixed with a solution of thiocyanate the resulting solution is red. (b) When a solution of cobalt is mixed with a solution of thiocyanate the resulting solution is blue. (a: Richard Megna/Fundamental Photographs; b: Larry Stepanowicz/Visuals Unlimited.)

mixed together. Suppose your metal ion solution contains one or more of the following metal ions: lead (II), mercury (II), tin (IV) cadmium (II), or antimony (III). The following table shows the results obtained when a solution containing one of these ions is mixed with a solution containing sulfide ion.

Metal Ion in Solution	Reacting Substance in the Testing Solution	Resulting Color of Solid Formed When the Solutions are Mixed
Pb^{2+}	S^{2-}	PbS (black)
Hg^{2+}	S^{2-}	HgS (red)
Sn^{4+}	S^{2-}	SnS_2 (yellow)
Cd^{2+}	S^{2-}	CdS (yellow)
Sb^{3+}	S^{2-}	Sb_2S_3 (orange)

Use this table to answer the following questions. Assume that the only possible ions present are those listed in the table and that the proper procedure has been followed to allow for a possible chemical reaction to occur.

Which of the metal ions is present if, after mixing the solutions, the solid formed is

(a) black?
(b) red?
(c) yellow?
(d) some black and some orange?

How sure are you of your answers? Of which answers are you least sure? Why?

Example 5.3   **Evidence of a chemical change**

> When a candle burns, several things occur. We can see that the wax underneath the wick melts. If we look closely, there is a vapor rising from the surface of this liquid. Of course, the flame itself glows brightly, and a gentle stream of hot gas rises from the candle.
>
> (a) What evidence do you have that a chemical reaction occurs here?
> (b) Can you think of an experiment to determine whether the change of the solid wax to liquid wax is a physical or chemical change?
>
> **Thinking Through the Problem** In this case, the questions are answered by reflecting on the evidence given.
>
> **Answer**
>
> (a) Evidence for a reaction includes the emission of light, the release of heat, and the possible formation of a new gaseous substance.
> (b) A physical change can be reversed. If we cool the liquid and it becomes solid wax again, this is probably a physical change.

✔ **Meeting the Goals**

A chemical change is the enactment of what a chemical property predicts. The result of a chemical change is the formation of a new substance.

PROBLEMS

1. Concept question: What is the difference between a physical property and a chemical property?

2. Concept question: What is the difference between a chemical property and a chemical change?

3. Can you conclude from Table 5-1 that either sodium or chlorine reacts readily with oxygen? Explain your reasoning.

4. Classify the following as a chemical change, physical change, chemical or physical property, or none of these.

   (a) A solid is attracted by a magnet.
   (b) Grapes ferment to wine.
   (c) A substance does not burn.
   (d) A tennis ball bounces 5 times.
   (e) A substance reacts with acid to liberate a gas.
   (f) An iron nail rusts.
   (g) An opening is cut in a brick wall.
   (h) The pine freshener scent in the closet "disappears" over time.

5. List the physical properties described in Table 5-1.

6. List the chemical properties described in Table 5-1.

7. The table below lists certain properties of several substances. Use them to answer the questions below.

Certain Properties of Acetic Acid, Calcium Carbonate, Water, Carbon Dioxide, and Calcium Acetate

	Acetic Acid	Calcium Carbonate	Water	Carbon Dioxide	Calcium Acetate
Phase	Liquid	Solid	Liquid	Gas	Solid
Dissolves in water?	Yes	No	Yes	Very slightly	Yes
Color	Colorless	Colorless	Colorless	Colorless	Colorless

Consider the reaction of acetic acid and solid calcium carbonate (the major component of marble). The reaction gives calcium acetate, water, and carbon dioxide as products.

(a) Write the chemical equation and balance it.
(b) When this reaction occurs, what changes do you expect? Will any phases form or disappear? What evidence is there for chemical change?
(c) If you place a piece of calcium carbonate in a clear colorless solution and carbon dioxide forms, can you say that acid was present?

5.2 Chemical Reactions and Prediction

SECTION GOALS

✔ Name the three main types of chemical reactions.
✔ Be able to identify a given chemical reaction by reaction type.
✔ Be able to predict the outcomes of simple synthesis and decomposition reactions.

Prediction and Chemistry

"What do you think happens next?" This is one of the most basic questions that we can ask, in science and in other parts of our lives. Answering this question is actually one of the very first skills that human babies develop. We learn that nature behaves in certain ways, we figure out which sounds correspond to which things or actions, and we try to determine how people around us will act. This is usually done through a process of trial and error, and we find it unusual when a pattern doesn't repeat. Our investigations aren't flawless, and surprises (especially with the behavior of people!) will always occur. But the bottom line of our development includes being able to predict that a set of conditions will lead to the same outcome each time.

Life would be pretty limited if all we knew was based on specific cases that we had observed. But, as we learn patterns of nature, language, and behavior, we come to recognize that we can predict, sometimes with great reliability, the outcomes of situations we have not seen before. Different people use the same words for the same things. All things fall, sometimes with dramatic results. And our big sister likes being tickled as much as our big brother. To do this extra step of prediction, however, requires that we understand—or, at least, think we understand—the reasons why certain outcomes follow certain initial conditions.

KEY IDEA 5.2

This prediction in other systems is one of the things that science does particularly well. *Chemistry is based on the understanding that atoms combine in expected ways to make substances whose properties are derived from the atoms in the substance.* There are two particularly powerful applications of this. The first involves the formulas that we expect for certain substances. When we have a compound of sodium and oxygen, we expect the formula to be Na_2O because we know this is what the component Na^+ and O^{2-} ions will form. The second prediction we can make is derived from this: if we react Na and O_2, we expect the reaction to be $4\,Na + O_2 \rightarrow 2\,Na_2O$.

Another example is a reaction in which we expect a new covalent compound to form. We have considered the bonding of molecules; for example, we saw that second period elements are surrounded by eight electrons in their compounds. Thus, when we predict the reaction of carbon and fluorine, we are not surprised that CF_4 forms, whereas CF_2 does not.

The next sections will introduce you to a systematic way of making similar predictions about a host of different reactions. It will seem very daunting at first, but remember that in predicting chemical reactions, you are using one of the most powerful thinking skills you have: being able to go from a set of conditions to an outcome you *expect* based on prior experience and understanding.

Understanding the first four chapters is essential to predicting what will happen in a chemical reaction, because predicting products involves considering reactants and determining what new arrangements are possible. If we don't know what arrangements are possible, we will have trouble predicting a reaction outcome. For example, if we want to predict the product of the reaction of aluminum and chlorine, we can ask, "What compound do we expect aluminum and chlorine to make?" The answer, as you learned in Chapter 3, is $AlCl_3$.

There is no doubt that your understanding of chemistry will allow you to do the kind of prediction we discuss here. But in its full complexity, chemistry still has plenty of surprises—and predicting and controlling complicated chemical reactions remains one of the most interesting, and challenging, parts of chemical research today.

Types of Reactions

If we are asked to predict the outcome of a chemical reaction, one of the most important questions we can ask at the start is "What kind of chemical reaction may occur?" For the purpose of this text, we will see that reactions fall into three different main categories, or types. A great deal of research in chemistry and allied fields involves chemical reactions. Even the most skilled chemist cannot predict everything that will happen when two or more substances are mixed for the first time. But your understanding of chemical formulas and bonding already gives you insights into many different reactions.

In the first three chapters of this text you learned powerful ways to think about the makeup of a chemical substance. You learned to read a chemical formula and understand what it represents. You saw how the ability to form ionic bonds or covalent bonds is very important in determining how a substance holds together. You learned how to distinguish different types of substances based on the elements in them. You then learned how to recognize chemical reactions and write balanced equations describing reactions in Chapter 4.

In characterizing chemical reactions, the chemical substances define the likely outcome of the type. For example, ionic compounds usually react to form new ionic compounds; molecular compounds form new molecular compounds.

In this section, we will take your skills one step further. You will learn how to distinguish different reactions based on the substances

✔ **Meeting the Goals**

Most chemical reactions can be classified as being one or more of the following three reaction types: synthesis (or decomposition), displacement, or oxidation-reduction reactions.

in them. We will see that, in general, reactions fall into one of the following three categories, or types:

(a) Synthesis or **decomposition reactions** occur when a new substance is synthesized, usually from its elements, or when a substance breaks down into other substances. Synthesis occurs when two or more substances react to give a new compound. Synthesis reactions include cases where two elements combine to make a binary compound. A common example of a synthesis reaction is the reaction of sodium and chlorine to form sodium chloride. The opposite of a synthesis reaction is a decomposition reaction, where two or more substances are formed by the decomposition of a single reactant. The formation of sodium hydroxide and carbon dioxide when sodium hydrogen carbonate is heated, for example, is a decomposition reaction.

(b) Displacement reactions involve a compound reacting with an element or with another compound. The second compound or element is said to "displace" a component of the compound, resulting in two new substances. An important type of displacement reaction involves the exchange of ions in solution and is called an ionic reaction. Another type of displacement reaction involves the exchange of hydrogen ion (H^+), and this is an example of an acid-base reaction (discussed in Chapter 15.)

(c) Oxidation and **reduction** are chemical processes that occur when one or more atoms undergoes a change in its oxidation state. An important kind of **oxidation-reduction reaction** is a combustion reaction, where oxygen gas is used to burn, or "combust," a substance.

Recognizing Chemical Reactions

Let's now examine each of these three reaction types, and learn how to recognize them. Once you have learned to determine the reaction type, you can use your understanding of chemical bonding to make predictions of the products formed in a reaction. Descriptions and examples of each reaction type are given in Table 5-2. Study the examples given in this table until you understand each type of reaction. You should become competent in two skills:

1. Can you recognize a reaction type when given the entire equation?

2. Can you predict possible products given the reactants and the reaction type?

The formulas for the products of many different reactions can be written by following the general patterns presented in Table 5-2. All formulas must be correctly written.

Synthesis or Decomposition Reactions

Remember that a chemical compound is composed of two or more chemical elements. In many cases it is possible to combine two elements and have them react to form a compound. In these cases, we say that the compound is synthesized from its elements. The following are equations for four reactions where a compound is directly synthesized from its elements. A graphic example of a synthesis reaction is given in Figure 5-2.

Synthesis from elements:

$$2\,H_2 + O_2 \longrightarrow 2\,H_2O$$
$$4\,Fe + 3\,O_2 \longrightarrow 2\,Fe_2O_3$$
$$Ti + 2\,Cl_2 \longrightarrow TiCl_4$$
$$2\,Na + Cl_2 \longrightarrow 2\,NaCl$$

✔Meeting the Goals

A synthesis reaction combines two reactant elements into a single product. In a decomposition reaction, a substance is broken down into smaller substances.

Table 5-2 Reaction Types

Type of Reaction	Description of Reaction	Example
Synthesis or decomposition reactions		
Synthesis	Two elements or compounds combine to form a compound.	*Synthesis from elements:* $C + O_2 \longrightarrow CO_2$ *Synthesis from compounds:* $C_4H_8 + HBr \longrightarrow C_4H_9Br$
Decomposition	A single reactant is broken down into two other substances.	*Decomposition into elements:* $Cu_2S \longrightarrow S + 2\,Cu$ *Decomposition into compounds:* $CaCO_3 \longrightarrow CaO + CO_2$
Displacement reactions		
Single displacement reactions	An element reacts with a binary compound to produce a new compound and a different element.	*Displacement of elements:* $Fe_3O_4 + 4\,H_2 \longrightarrow 3\,Fe + 4\,H_2O$
Double displacement reactions	Two compounds react so that their component elements or ions are switched.	*Exchange of elements:* $CaO + 2\,HCl \longrightarrow CaCl_2 + H_2O$ *Exchange of ions:* $2\,AgNO_3 + CaCl_2 \longrightarrow 2\,AgCl + Ca(NO_3)_2$
Oxidation-reduction reactions		
Combustion reactions	A compound reacts with oxygen to give all the elements as compounds with oxygen; most common with compounds containing C and H, to give CO_2 and H_2O.	$C_3H_8 + 5\,O_2 \longrightarrow 3\,CO_2 + 4\,H_2O$
Other oxidation-reduction reactions	Two or more substances react and an element in one substance is oxidized and an element in another substance is reduced.	$Mg + Cl_2 \longrightarrow MgCl_2$ $2\,CH_4 + O_2 \longrightarrow 2\,CH_3OH$

In these reactions we see that one product is made from two reactants. Whenever we see two reactants forming one product we know that we have a synthesis reaction. Remember that the product is a new substance with properties that are different from those of the elements it contains. The properties of water, for example, are very different from those of hydrogen and oxygen, and the properties of sodium chloride (table salt), as you can see in Table 5-1, are very different from those of chlorine gas and sodium metal.

Making compounds by combinations of elements is just one example of a synthesis reaction. In other cases, substances that react in synthesis reactions are compounds.

Synthesis from compounds:

$$C_2H_4 + H_2 \longrightarrow C_2H_6$$
$$CaO + SO_3 \longrightarrow CaSO_4$$
$$Al_2O_3 + 3\ H_2O \longrightarrow 2\ Al(OH)_3$$

Some synthesis reactions are very useful. The synthesis of calcium sulfate from calcium oxide and sulfur trioxide, for example, is an excellent way to remove sulfur trioxide from industrial gases. And the reaction of aluminum oxide with water can be used to dry up surfaces—which is why Al_2O_3 is used in many antiperspirants.

If we can form a substance from other substances, it is reasonable that there are many cases where we can take a substance and decompose it into two or more different substances. All of the preceding synthesis reactions can occur in the opposite direction, as decomposition reactions. For example,

$$2\ H_2O \longrightarrow 2\ H_2 + O_2$$
$$2\ NaCl \longrightarrow 2\ Na + Cl_2$$

Other decomposition reactions include:

$$2\ HgO \longrightarrow 2\ Hg + O_2$$
$$GeH_4 \longrightarrow Ge + 2\ H_2$$
$$C_4H_{10}O \longrightarrow H_2O + C_4H_8$$

Example 5.4   **Recognizing reaction types**

For each of the following chemical reactions, indicate if it is a synthesis reaction, a decomposition reaction, or neither.

$$CaCO_3 \longrightarrow CaO + CO_2$$
$$4\ Li + O_2 \longrightarrow 2\ Li_2O$$
$$2\ C_4H_6 \longrightarrow C_8H_{12}$$
$$C_4H_8N_2O_3 + H_2O \longrightarrow 2\ C_2H_5NO_2$$

Thinking Through the Problem Synthesis reactions have two or more reactants forming one product. The second, third, and fourth reactions all have a single compound in the product. They are all synthesis reactions. Note that two molecules of C_4H_6 are the reactants in the third reaction.

Decomposition reactions have a single reactant that produces two or more products. This is what happens in the first reaction.

Answer The first reaction is decomposition; the others are synthesis reactions.

How are you doing? 5.3

Determine if the following reactions are decomposition reactions, synthesis reactions, or neither.

$$MgCl_2 \longrightarrow Mg + Cl_2$$
$$2\,Mn + 3\,S \longrightarrow Mn_2S_3$$
$$C_4H_8 + NH_3 \longrightarrow C_4H_{11}N$$

Predicting Synthesis Reactions

CONNECT TO
SECTION 3.1

It is easy to predict the product in many *synthesis reactions*. In most cases, a synthesis reaction begins with two elements as the reactants (sodium and chlorine, for example). The two elements will probably form only one stable compound. As we saw in Chapter 3, if the compound is ionic we can use the common charges of certain elements and the principle of charge balance to predict what compound will form. We have seen this with sodium and chlorine. If these were to form a compound, that compound would have to contain sodium ions, Na^+, and chloride ions, Cl^-. These would combine in the substance NaCl, and in no other substance.

Suppose, for another example, we are asked what substance will form when we combine aluminum and oxygen. We would need to recall that the aluminum ion is Al^{3+} and the oxide ion is O^{2-}. The only new substance we can predict to form is aluminum oxide, Al_2O_3.

Example 5.5

Predicting a synthesis reaction with a metal and a nonmetal

Predict the product of the synthesis reaction between aluminum metal and chlorine gas.

$$Al\ (s) + Cl_2\ (g) \longrightarrow$$

Thinking Through the Problem Again, both reactants are elements so they will combine to form a compound. In this case, however, aluminum is a metal and chlorine is a nonmetal. Therefore, the product will be an *ionic* compound of Al^{3+} and Cl^-.

Working Through the Problem First we must determine the preferred oxidation states of the two ions formed. Because aluminum has three valence electrons, a stable Al^{3+} cation will form. Chlorine has seven valence electrons, therefore a stable Cl^- anion will form. Lastly, we combine the ions so that charge neutrality is achieved.

Answer $2Al\ (s) + 3\ Cl_2\ (g) \longrightarrow 2AlCl_3\ (s)$

CONNECT TO
SECTION 2.2

When two or more elements combine to form a compound that is *not* ionic, we use a different strategy. In this case, we look to find the product that has a reasonable Lewis structure, in view of the starting materials, using the skills we developed in Chapter 2. One simple example is the reaction of oxygen and hydrogen. The simplest compound of oxygen and hydrogen has each hydrogen sharing one electron with oxygen and, therefore, two H—O bonds. The product, we expect, will have an O bonded to two H's, as in H_2O, water.

Example 5.6   **Predicting a synthesis reaction from nonmetal reactants**

Predict the product of the reaction between hydrogen gas and elemental bromine.

$$H_2\ (g) + Br_2\ (l) \longrightarrow$$

Thinking Through the Problem Because both reactants are elements, we can predict that they will combine in a *synthesis* reaction to form a compound. But will this be a covalent compound or an ionic compound? Looking at the periodic table, we see that both reactants are nonmetals. Therefore, the product will be a *covalent* compound. We must think about Lewis structures to help us write the formula. Hydrogen (one valence electron) and bromine (seven valence electrons) both can form a closed shell by gaining one electron. Thus, they form a *single bond* in the compound HBr (g).

Answer $H_2\ (g) + Br_2\ (l) \longrightarrow 2\ HBr\ (g)$

✔**Meeting the Goals**

To predict the product of a synthesis reaction, combine the two reactant elements and determine the correct ratio between them by using Lewis dot structures (for molecular compounds) or charge neutrality (for ionic compounds).

How are you doing? 5.4

Predict the product of the following synthesis reactions. Write a correctly balanced formula for this product and balance the equations that result.

1. $Mg\ (s) + P_4\ (g) \rightarrow$
2. $Si\ (s) + H_2\ (g) \rightarrow$

Predicting Decomposition Reactions

In many ways, decomposition reactions are the easiest to predict. We will examine only those in which a binary compound decomposes to the elements. In this case, the only thing to consider is what the elements will be, especially for a compound with a nonmetal. For

PRACTICAL B — How does the synthesis of oxides determine the reactivity of metals in the environment?

A very important aspect of the reactivity of chemical elements is their ability to persist in different environments. Metals, in particular, can react with oxygen in the air and form oxides. If this happens, the metal corrodes, and the corrosion of metals is a very expensive form of environmental damage. Because corrosion reactions involve an elemental metal and oxygen in a reaction to form an oxide, these are examples of synthesis reactions.

Perhaps the most common form of corrosion is rust, as shown in Figure 5-4. If oxygen is present, then the reaction of Fe and O_2 to form Fe_2O_3 occurs, although slowly. Preventing this reaction is a major task in working with iron-containing products. One way to do this is to make the iron into an alloy known as steel. But other ways to protect iron involve using other elements in creative ways.

Two different elements have played an important role in the history of corrosion prevention. One is tin. The "tin can" was a major technological innovation for two reasons. First, it made it possible to coat steel and iron to prevent the attack of oxygen. The oxide of tin (IV) does not spontaneously form in the presence of air. Hence, a coating of tin forms a protective blanket. Can you write the formula for the oxide of tin, and write a reaction for the reaction of tin and oxygen?

Another important protectant is zinc. Zinc does react with air, in fact more readily than iron. If a piece of zinc is in contact with a piece of iron, the zinc is often corroded selectively. This zinc, often applied as a *galvanizing* coat, may eventually be used up, but usually over a long time period. Nevertheless, a zinc oxide does form. What will be its formula and its reaction with oxygen?

The final case of oxide protection occurs for certain metals that may be very reactive with air. But they react in a fortunate way, creating a coating that forms a protective layer. Two excellent examples of this are aluminum and magnesium. The rapid reaction of magnesium and air shown in Figure 5-5 is a powerful indication of how favorable this reaction is. However, it never occurs without first exposing some of the magnesium to air and applying heat to start the reaction. Similarly, aluminum metal, even when spread in very thin foil sheets, is always covered by a very thin layer of clear, colorless aluminum oxide. This layer blocks oxygen, and the aluminum can persist for many years.

Write the reactions that provide protective coating for aluminum and magnesium. Name the products.

Finally, we return to zinc. Zinc metal that is not acting as a protective material for another metal lasts for a long time. As with aluminum and magnesium, it also reacts to form a protective layer. But in this case, carbon dioxide and oxygen both react to form zinc carbonate. Write these formulas, and the balanced reaction, for the protection of zinc.

Figure 5-4 ▲ Iron undergoes an oxidation reaction in air, producing rust.
(Erv Schowengerdt.)

Figure 5-5 ▲ The rapid reaction of Mg and O_2.
(Richard Megna/Fundamental Photographs.)

example, when a compound containing hydrogen decomposes, we get H_2. When a compound decomposes from a bromide, then Br_2 will result.

Example 5.7 **Determining the product of a decomposition**

When the compound CaO decomposes, what will be the likely products?

Thinking Through the Problem We know that we will get calcium and oxygen. We expect oxygen to form O_2 molecules.

Answer The likely products of the decomposition of CaO are Ca and O_2.

PROBLEMS

8. Predict the product of the reaction $Zn + Cl_2$.

9. What do you expect from the reaction of Cd and O_2?

10. What compound will form in the reaction of Ag and O_2?

11. Predict the products of the following reactions.
    (a) $Zn + P_4 \rightarrow$
    (b) $Ba + F_2 \rightarrow$
    (c) $Sr + N_2 \rightarrow$
    (d) $Al + F_2 \rightarrow$

12. Complete the following equations for decomposition reactions. Write the correct formulas for the predicted product and balance the final equation.
    (a) $Ag_2O \rightarrow$
    (b) $NaN_3 \rightarrow$
    (c) $N_2O \rightarrow$

    (d) $B_2O_3 \rightarrow$
    (e) $Fe_2O_3 \rightarrow$
    (f) $FeO \rightarrow$

13. Identify whether the following reactions are synthesis or decomposition reactions. Balance all equations.
    (a) $Bi + O_2 \rightarrow Bi_2O_3$
    (b) $Hg(OH)_2 \rightarrow HgO + H_2O$
    (c) $N_2O_3 + H_2O \rightarrow HNO_2$

14. Identify whether the following reactions are synthesis or decomposition reactions. Balance all equations.
    (a) $Sn + Cl_2 \rightarrow SnCl_4$
    (b) $C_6H_{12}O_6 \rightarrow C_2H_5OH + CO_2$
    (c) $H_2O_2 \rightarrow H_2O + O_2$

5.3 Single Displacement and Oxidation-Reduction Reactions

SECTION GOALS

✔ Predict the product of single displacement reactions.

✔ Identify oxidation-reduction reactions.

Single Displacement Reactions

CONNECT TO
SECTION 3.1

Single displacement reactions take place when an element reacts with a compound. These reactions are most common for the ionic compounds discussed in Chapter 3. The direction of the displacement reaction depends on the element that reacts. If it is a metal, then the metal of the ionic compound is displaced. Figure 5-6 shows an example of a displacement reaction. If the element that reacts is a non-metal, then it displaces the nonmetal of the ionic compound. Correct formulas must be written for the predicted products: first predict the new combinations of elements, then write correct formulas for the new combinations present in the products.

Example 5.8

✔ Meeting the Goals

To predict the products in a single displacement reaction, identify whether the reactant element is a metal or a nonmetal. Metals displace metals and nonmetals displace nonmetals from the reactant compound. This produces a new element and a different compound.

Figure 5-6 ▲ When a strip of copper metal is placed in a solution of silver nitrate, the copper displaces Ag from the solution. (Peticolas/Megna/ Fundamental Photographs.)

Predicting a single displacement reaction

Predict the products of the reaction between fluorine gas and lead (IV) chloride.

$$PbCl_4 \ (s) + F_2 \ (g) \longrightarrow$$

Thinking Through the Problem In this example, the reactant element, F, is a nonmetal. It will displace the nonmetal, Cl, in the compound. Lead is not a representative element but we can find its initial oxidation state from the reactant formula given, $PbCl_4$. In displacement reactions, we will assume that there is no change in oxidation state for the element that remains in the compound.

Working Through the Problem The predicted products are lead (IV) fluoride and elemental chlorine. Remember that chlorine, like fluorine and the other halogens, is diatomic, as we learned in Chapter 1.

Elements in the Predicted Products	New Combination of Elements	Balanced Formula
Lead and fluorine	$Pb^{4+}F^-$	PbF_4
Elemental chlorine	Chlorine is a diatomic molecule	Cl_2

The reaction can be summarized by:

Fluorine displaces chlorine

$$PbCl_4 + F_2 \longrightarrow PbF_4 + Cl_2$$

nonmetal   nonmetal

Answer $PbCl_4 + 2 \ F_2 \longrightarrow PbF_4 + 2 \ Cl_2$

Example 5.9

Predicting a single displacement reaction

Predict the products of the reaction between aluminum metal and iron (III) oxide.

$$Al \ (s) + Fe_2O_3 \ (s) \longrightarrow$$

Thinking Through the Problem Aluminum is a metal. It therefore will displace the metal Fe from the compound. The new combinations will be aluminum oxide and elemental iron.

Working Through the Problem

Predicted Compound	New Combination of Elements	Formula
Aluminum oxide	$Al^{3+}O^{2-}$	Al_2O_3
Elemental iron	Fe	Fe

Answer $2 \text{ Al } (s) + \text{Fe}_2\text{O}_3 (s) \longrightarrow 2 \text{ Fe } (s) + \text{Al}_2\text{O}_3 (s)$

How are you doing? **5.5**

Predict products for these single displacement reactions. Write correct formulas for the products, then balance the equation that results.

(a) $\text{TiCl}_4 (s) + \text{O}_2 (g) \rightarrow$
(b) $\text{NaI} (aq) + \text{Cl}_2 (g) \rightarrow$
(c) $\text{Mg} (s) + \text{SiCl}_4 (l) \rightarrow$

Single Displacement Reactions with Hydrogen and Carbon

Two of the most important elements for single displacement reactions are carbon and hydrogen. These two elements are capable of displacing metal atoms in metal oxides to make elemental metal and either water or carbon dioxide.

Many oxides of the transition elements (such as copper, manganese, and zinc) react with hydrogen to give the pure metal and water. One example is copper (I) oxide:

$$\text{Cu}_2\text{O} (s) + \text{H}_2 (g) \longrightarrow 2 \text{ Cu } (s) + \text{H}_2\text{O} (g)$$

Carbon is also used to displace metals from metal oxides, producing the pure metal and CO_2. For example, a special form of graphite called coke is used in the preparation of iron from iron ore. This reaction requires high temperatures, so some of the carbon is also burned to give heat energy.

$$2 \text{ Fe}_2\text{O}_3 (s) + 3 \text{ C } (s) \longrightarrow 4 \text{ Fe } (s) + 3 \text{ CO}_2 (g)$$

In this reaction, the carbon displaces iron. When aluminum oxide reacts with carbon to form aluminum metal, the carbon displaces the aluminum.

$$2 \text{ Al}_2\text{O}_3 (s) + 3 \text{ C } (s) \longrightarrow 4 \text{ Al } (s) + 3 \text{ CO}_2 (g)$$

Most of the cost of pure aluminum metal is for the electricity required to make this reaction happen.

Similar reactions occur when solutions containing certain chloride salts react with hydrogen. The metal is converted into the pure element as it is displaced by hydrogen. The hydrogen then forms a compound with chloride, namely HCl. An example is the reaction of palladium (II) chloride, $PdCl_2$, with hydrogen:

$$PdCl_2 \, (aq) + H_2 \, (g) \longrightarrow Pd \, (s) + 2 \, HCl \, (aq)$$

Example 5.10   **Predicting a displacement of a metal by carbon**

Predict the reaction when carbon reacts with copper (I) oxide.

Thinking Through the Problem Copper (I) oxide is Cu_2O. The carbon displaces the copper, and carbon dioxide results along with copper metal.

Answer $2 \, Cu_2O \, (s) + C \, (s) \longrightarrow 4 \, Cu \, (s) + CO_2 \, (g)$

Example 5.11   **Predicting a displacement of a metal by hydrogen**

Predict the products of the reaction between hydrogen and nickel (II) chloride. Write correct formulas for the predicted products. Finally, balance the resulting equation.

$$H_2 \, (g) + NiCl_2 \, (aq) \longrightarrow$$

Thinking Through the Problem We are told that there is a reaction, and we know that the hydrogen will displace the metal, not the chloride. Metallic nickel and hydrogen chloride form.

Answer $H_2 \, (g) + NiCl_2 \, (aq) \longrightarrow Ni \, (s) + 2 \, HCl \, (aq)$

How are you doing? 5.6

Predict the products of the reaction between carbon and lead (II) oxide. Write correct formulas for the predicted products. Finally, balance the resulting equation.

$$C \, (s) + PbO \, (s) \longrightarrow$$

Oxidation-Reduction Reactions

Most synthesis and single displacement reactions are examples of a chemical process called oxidation-reduction. Oxidation is defined as the loss of electrons. Reduction is the gain of electrons. In oxidation, a substance loses electrons and becomes more positive. In the example below, elemental iron, Fe, is oxidized to iron (II) ion, Fe^{2+} by losing two electrons. In reduction, a substance gains electrons so that its oxidation number decreases. In the following outline, iron (II) ion is reduced to elemental iron by gaining two electrons. Recall that elements have oxidation numbers of zero.

✔ **Meeting the Goals**

Oxidation and reduction occur together. When a substance is oxidized in a reaction, another substance is reduced.

Oxidation is the loss of electrons.	$Fe^0 - 2\,e^- = Fe^{2+}$	During oxidation, the oxidation number of the element increases $(0 \rightarrow +2)$.
Reduction is the gain of electrons.	$Fe^{2+} + 2\,e^- = Fe^0$	During reduction, the oxidation number of the element decreases $(+2 \rightarrow 0)$.

CONNECT TO
SECTION 3.3

The result of the oxidation of iron is apparent in Figure 5-4.

We can tell when a reaction is an oxidation-reduction reaction by assigning oxidation numbers to each element present in the balanced chemical equation for the reaction. We determine the oxidation numbers of all elements present in both the reactants and in the products by applying the rules for assigning oxidation numbers found in Table 3-6. Then we examine the oxidation numbers of each element to see if any change occurs in going from the reactant to the product side of the equation. Whenever elements show changes in oxidation number, the reaction is an oxidation-reduction reaction.

Example 5.12   **Recognizing an oxidation-reduction reaction**

Determine whether the following is an oxidation-reduction reaction by assigning oxidation numbers to each element present:
$$H_2\,(g) + Br_2\,(l) \longrightarrow 2\,HBr\,(g)$$

Thinking Through the Problem   We will apply the rules for assigning oxidation numbers found in Table 3-5.

Working Through the Problem

Step 1: Assign oxidation numbers. H_2 and Br_2 are both elements. They have oxidation numbers of zero.

$$\overset{0}{H_2}\,(g) + \overset{0}{Br_2}\,(l) \longrightarrow 2\,HBr\,(g)$$

In the compound HBr, hydrogen has an oxidation number of $+1$, bromine has an oxidation number of -1.

$$\overset{0}{H_2}\,(g) + \overset{0}{Br_2}\,(l) \longrightarrow 2\,\overset{+1\,-1}{HBr}\,(g)$$

Step 2: Determine which elements show a change in oxidation numbers: Hydrogen goes from zero to $+1$; bromine goes from zero to -1.

Answer   This is an oxidation-reduction reaction as hydrogen and bromine each undergo a change in oxidation number.

Example 5.13   **Recognizing an oxidation-reduction reaction**

Determine whether the following is an oxidation-reduction reaction by assigning oxidation numbers to each element present:
$$H_2(g) + NiCl_2(aq) \longrightarrow Ni(s) + 2\ HCl(aq)$$

Thinking Through the Problem   We will assign oxidation numbers, by using the rules of Table 3-5, and check to see if an element's oxidation number changes during the reaction.

Working Through the Problem

Step 1: Assign oxidation numbers to each element. Hydrogen and nickel are elements. Assign them values of zero.

$$\overset{0}{H_2}(g) + NiCl_2(aq) \longrightarrow \overset{0}{Ni}(s) + 2\ HCl(aq)$$

In the compound HCl, H is $+1$, Cl is -1. In the compound $NiCl_2$, Cl is -1 so Ni must be $+2$.

$$\overset{0}{H_2}(g) + \overset{+2\ -1}{NiCl_2}(aq) \longrightarrow \overset{0}{Ni}(s) + 2\ \overset{+1\ -1}{HCl}(aq)$$

Step 2: Determine which elements show a change in oxidation numbers: Hydrogen goes from zero to $+1$; nickel goes from $+2$ to zero.

Answer   This is an oxidation-reduction reaction as both elements undergo a change in oxidation number.

Example 5.14   **Recognizing the oxidized and reduced elements in an oxidation-reduction reaction**

Determine which element is oxidized and which is reduced in the following reaction.

$$2\ Al(s) + Fe_2O_3(s) \longrightarrow 2\ Fe(s) + Al_2O_3(s)$$

Thinking Through the Problem   Assign oxidation numbers as before.

Working Through the Problem

Step 1: Assign oxidation numbers to each element. Aluminum and iron are elements: assign them values of zero.

$$2\ \overset{0}{Al}(s) + Fe_2O_3(s) \longrightarrow 2\ \overset{0}{Fe}(s) + Al_2O_3(s)$$

In the compound Fe_2O_3, oxygen is -2; this makes Fe $+3$. In the compound Al_2O_3, oxygen is -2; Al is $+3$.

$$2\ \overset{0}{Al}(s) + \overset{3+\ 2-}{Fe_2O_3}(s) \longrightarrow 2\ \overset{0}{Fe}(s) + \overset{3+\ 2-}{Al_2O_3}(s)$$

Step 2: Determine which elements show a change in oxidation numbers: Aluminum goes from zero to $+3$; Fe goes from $+3$ to zero. Show which element is oxidized and which is reduced.

$$\overset{0}{Al} \longrightarrow \overset{3+}{Al}$$ This is oxidation: a loss of electrons is shown by the increase in oxidation number.

$$\overset{3+}{Fe} \longrightarrow \overset{0}{Fe}$$ This is reduction: a gain of electrons is shown by the decrease in oxidation number.

Answer This is an oxidation-reduction reaction as both elements undergo a change in oxidation number. Aluminum is oxidized and iron is reduced.

How are you doing? 5.7

Assign oxidation numbers to each element in the equations below. Show the change in oxidation number for the two elements. Show which element is oxidized and which is reduced, as in Example 5.14.

(a) $2 Al_2O_3 (s) + 3 C (s) \longrightarrow 4 Al (s) + 3 CO_2 (g)$
(b) $16 Al (s) + 3 S_8 (s) \longrightarrow 8 Al_2S_3 (s)$

Predicting Reactions Using an Activity Series

Figure 5-6 on page 156 shows the reaction of Cu metal with solution of $AgNO_3$. This is a single displacement reaction in which the copper metal in the strip displaces Ag from the solution of $AgNO_3$. If we immerse a strip of Mg or a strip of Ni into a solution of $AgNO_3$, we see similar results. However, we see no evidence of a reaction when metal strips of Pt or Au are used. Some metals appear to have a greater ability to displace other metal ions from aqueous solutions of their compounds. The ability to displace other metals in a single displacement reaction is called the activity of a metal. More active metals displace less active metals: in this example, Cu is more active than Ag, as it displaces Ag^+ from $AgNO_3$. The balanced equation is:

See this with your **Web Animator** ▶

$$Cu (s) + 2 AgNO_3 (aq) \longrightarrow Cu(NO_3)_2 (aq) + 2 Ag (s)$$

Example 5.15 **Determining the more active metal in a single displacement reaction**

You have finished a lab experiment on single displacement reactions and have obtained the following results:

(1) Mg + Zn(NO$_3$)$_2$	Reaction occurs
(2) Al + Zn(NO$_3$)$_2$	Reaction occurs
(3) Al + Mg(NO$_3$)$_2$	No reaction occurs

List the metals Mg, Zn, and Al from most to least active based on the experimental observations given.

Thinking Through the Problem We know that more active metals will displace less active metals. We also know that both Mg and Al react with (displace) Zn but Al does not displace Mg.

Working Through the Problem As Mg and Al both displace Zn, they are both more active than Zn: Zn is the least active in this series.

The third experiment shows no reaction between Al and $Mg(NO_3)_2$. Al must be less active than Mg, as it does not displace Mg in this experiment.

Answer Mg is the most active metal and Zn is the least active metal in this series. The order of activity from most active to least active is Mg > Al > Zn.

Similar experiments can be carried out for many different metal/metal ion combinations, and the order of activity determined.

Example 5.16 Determining the more active metal in a single displacement reaction

You have finished a lab experiment on single displacement reactions and have obtained the following results:

(1) $Fe + Ni(NO_3)_2 \rightarrow$ Reaction occurs
(2) $Fe + Al(NO_3)_3 \rightarrow$ No reaction occurs
(3) $Fe + Co(NO_3)_2 \rightarrow$ Reaction occurs
(4) $Ni + Co(NO_3)_2 \rightarrow$ No reaction occurs

List the metals Fe, Ni, Al, Co from most to least active based on the experimental observations given.

Thinking Through the Problem We will use the same method to determine activity as developed in the previous example.

Working Through the Problem For reactions that occur, the metal written as an element is more active than the metal contained in the compound. Reaction (1) shows Fe > Ni, reaction (2) gives Fe < Al, reaction (3) gives Fe > Co, and reaction (4) gives Ni < Co.

Starting with reaction (1), we have Fe > Ni. Adding the information from reaction (2), we have Al > Fe > Ni. Adding the information from reaction (3), we have Al > Fe > Ni and Co.

Answer Reaction (4) lets us distinguish between Ni and Co so that finally we have Al > Fe > Co > Ni.

These examples demonstrate the procedure used to determine the order of activity among common metals. Table 5-3 contains a listing of metals in order of their activities. This listing can be used to predict whether or not a particular displacement will occur.

Table 5-3 Activity Series

Li > K > Ca > Na > Mg > Al > Mn > Zn > Fe > Cd > Co > Ni > Sn > Pb > H_2 > Cu > Ag > Pt > Au

Example 5.17 Predicting products of single displacement from activity series

Predict which of the following reactions will occur and which reactions will not occur. Write balanced equations for those reactions that occur and N.R. (no reaction) for those reactions that do not occur.

(1) $Sn + Cu(NO_3)_2 \rightarrow$
(2) $Zn + KNO_3 \rightarrow$
(3) $Cd + Mn(NO_3)_2 \rightarrow$
(4) $Cd + Cu(NO_3)_2 \rightarrow$

Thinking Through the Problem More active metals are able to displace less active metals from their compounds. A reaction will occur whenever the metal written as an element is more active than the metal in compound form. A reaction will not occur if the metal written as an element is less active than the metal in the compound. We will use the listing in Table 5-3 as a guide to metal activity.

Working Through the Problem Reaction (1) will occur, as Sn is more active than Cu (Sn > Cu). Reaction (2) will not occur, as Zn < K (less active). Reaction (3) will not occur, as Cd is less active than Mn. Reaction (4) will occur, as Cd is more active than Cu.

Answer
(1) $Sn + Cu(NO_3)_2 \rightarrow Sn(NO_3)_2 + Cu$
(2) $Zn + KNO_3 \rightarrow N.R.$
(3) $Cd + Mn(NO_3)_2 \rightarrow N.R.$
(4) $Cd + Cu(NO_3)_2 \rightarrow Cu + Cd(NO_3)_2$

✔**Meeting the Goals**

The ability of a metal to displace other metals from compounds is called its activity. More active metals displace less active metals from their compounds. Thus, activity trends can be used to predict whether or not a particular reaction will occur.

PRACTICAL C   What can we infer when a metal reacts with a solution?

The chemical reactions we can write based on the activity series in Table 5-3 give us a guide to identifying unknown metals. Under the right circumstances, we can take an unknown metal and place it in solutions of metal salts. Whether there is a reaction depends on the position of the metals in the activity series.

For example, suppose we place a piece of cleaned metal in a solution of nickel (II) chloride. After waiting a while, we see that the nickel is beginning to form a blackish coating on the metal. We can tell that a reaction is occurring! The activity series indicates that the metal must be higher in the series (to the left in Table 5-3) than nickel. We still do not know if the metal is zinc, aluminum, etc. But we can be certain that it is not a metal lower (to the right) in the series.

Let us apply this reasoning in three different situations. First, we will learn about three different experiments with a piece of metal and see if we can determine logically what the metal might be. Second, we will use different metals to determine what might be in an unknown solution. And finally, we will consider the use of different metals to pull precious metals out of solution.

Imagine that you have the metal strip that we discussed a short while ago. We know that it reacts with nickel (II) chloride. If, on the other hand, we place it in a solution of iron (II) nitrate, we find that there is no reaction. If you want to further investigate what metal we have, would you then use a solution of aluminum chloride or a solution of cobalt (II) chloride? Explain your reasoning with chemical equations and predictions of what might happen. You may assume throughout that the unknown metal forms an ion in the +2 oxidation state in all reactions.

The same elimination strategy works just as well with known metals and other elements reacting with a solution. Let's say we have a solution that we know reacts with zinc metal, depositing an unknown metal in place of the zinc. We find, however, that the solution does not deposit the metal when hydrogen gas is bubbled through the solution. What does the activity series in Table 5-3 tell us about the unknown metal ion in the solution? Finally, suggest two other metals to test with that would further restrict the range of possible metals in the solution.

Our final practical application of the activity series may be familiar to anyone who has worked in photography. Most effective photographic processes rely on the absorption of light by finely divided silver halide salts. After exposure, these can be reduced to silver metal. Unexposed silver is then washed away in solution. Recovering this silver is important, for economic and environmental reasons. The easiest way to do this is to treat the solution with a metal that exchanges with the silver in solution. Use your understanding of the activity series to discuss how iron (in the form of steel wool), zinc, or copper can be used to recover silver ions from solution.

| How are you doing? | 5.8 |

Predict which of the following reactions will occur and which reactions will not occur. Write balanced equations for those reactions that occur and N.R. (no reaction) for those reactions that do not occur.

(1) $Cu + Zn(NO_3)_2 \rightarrow$
(2) $Zn + AgNO_3 \rightarrow$

PROBLEMS

15. Indicate the likely product of the following single displacement reactions. Write a balanced chemical equation in all cases.

(a) $CuCl_2 + Ag \rightarrow$
(b) $KBr + Cl_2 \rightarrow$
(c) $NiO + H_2 \rightarrow$
(d) $Cr_2O_3 + C \rightarrow$

16. For each of the reactions in Problem 15, determine the oxidation numbers of each element. Which element is oxidized? Which is reduced?

17. Predict the likely products of the following reactions. Indicate if the reaction is a synthesis, a single displacement, or a double displacement reaction.

(a) lead (II) iodide + fluorine $\rightarrow$
(b) zinc oxide + hydrogen $\rightarrow$
(c) potassium + sulfur $\rightarrow$

18. Balance the following equations and identify the type of reaction each equation represents.

(a) $H_2\ (g) + S_8\ (s) \rightarrow H_2S\ (g)$
(b) $C_5H_{10}O_5\ (s) + O_2\ (g) \rightarrow CO_2\ (g) + H_2O\ (l)$
(c) $Fe\ (s) + O_2\ (g) \rightarrow Fe_2O_3\ (s)$

19. Answer the following questions about the compound Fe_2S_3.

(a) What is the oxidation number of Fe in this compound?
(b) What is the name of the compound?
(c) Balance the equation $Fe + S_8 \rightarrow Fe_2S_3$
(d) What is oxidized and what is reduced in the reaction $Fe + S_8 \rightarrow Fe_2S_3$?
(e) Predict the products and balance the equation for the single displacement reaction $Fe_2S_3 + Al \rightarrow$.

20. How do you recognize an oxidation-reduction reaction? Which of the reactions in the previous problem are also oxidation-reduction reactions?

21. For each of the following reactions, find the oxidation numbers for each of the elements in the reaction and determine which reactant is oxidized and which is reduced.

(a) $8\ Fe + S_8 \rightarrow FeS$
(b) $2\ Al + 3\ F_2 \rightarrow 2\ AlF_3$

5.4 Double Displacement Reactions

SECTION GOAL

✔ Predict the products of double displacement reactions.

✔Meeting the Goals

In a double displacement reaction, two compounds react so that their component elements or ions are switched.

In double displacement reactions, two compounds react in such a way that the elements or ions that make them up are exchanged or "switched." One example is the production of silver chloride when solutions of silver nitrate and sodium chloride are mixed:

$$AgNO_3\ (aq) + NaCl\ (aq) \longrightarrow AgCl\ (s) + NaNO_3\ (aq)$$

Aqueous solutions of $AgNO_3$ contain silver ions and nitrate ions. Aqueous solutions of NaCl contain sodium ions and chloride ions.

Ions in Solution 1, $AgNO_3$	Ions in Solution 2, NaCl	Ions in the Mixture of Solutions 1 and 2	Possible New Combinations of Ions
$Ag^+ + NO_3^-$	$Na^+ + Cl^-$	$Ag^+ + NO_3^-$ and $Na^+ + Cl^-$	$Na^+ + NO_3^-$ and $Ag^+ + Cl^-$

When the two solutions are mixed, the possible new combinations are AgCl and $NaNO_3$. Each of these formulas is balanced for charge neutrality. If the new combination of ions results in a compound that is insoluble (does not dissolve to a significant extent) in water, then we see a solid called a precipitate, and we know that a reaction has occurred. If a mixture of two aqueous solutions does not result in an insoluble compound, then it appears that no reaction has occurred. In that case, the solution contains four ions and we write "N.R." for "no reaction."

KEY IDEA 5.3

We can determine which compounds are insoluble in water by consulting Table 5-4. For the reaction above, we see that AgCl forms an insoluble solid (trend 3); $NaNO_3$ remains in aqueous solution as sodium ions and nitrate ions (trends 1 and 2). The key to understanding double displacement reactions and to predicting their products lies in breaking up each of the reactants into its ions and then recombining the ions in two new pairs. For mixtures of aqueous solutions, a reaction occurs when an insoluble compound is formed.

Table 5-4 Solubility Trends

1. Nearly all ionic compounds of lithium, sodium, potassium, rubidium, cesium, and ammonium are soluble in water.
2. The nitrates and acetates of all metals are soluble.
3. The chlorides, bromides, and iodides of all metals except silver, lead, mercury (I), and copper (I) are soluble.
4. The sulfates of all metals except lead, barium, and calcium are soluble.
5. The hydroxides of all metals (except those given in trend 1) are insoluble.
6. Carbonates, phosphates, and sulfides are insoluble (except as given in trend 1).

Example 5.18 **Predicting a double displacement reaction**

Predict possible products for the double displacement reaction that occurs between solutions of lead (II) nitrate and potassium iodide. Write correct formulas for the predicted products. Finally, balance the chemical equation that results.

$$Pb(NO_3)_2\ (aq) + KI\ (aq) \longrightarrow$$

Thinking Through the Problem Pb^{2+} and K^+ ions will switch places, giving the new combinations of Pb^{2+} with I^- and K^+ with NO_3^-.

✔Meeting the Goals

In double displacement reactions, two compounds react in such a way that the elements or ions that make them up are exchanged or "switched." One way this occurs is when two solutions are mixed and one or more of the new ionic combinations are compounds that are insoluble in water. Solubility properties can be determined by checking Table 5-4.

Working Through the Problem Pb has an oxidation number of $+2$ in the reactant material; it is expected to have the same oxidation number in the product material. ◀▬▷ (5.3) The key idea is to consult the table of solubility trends. We see from Table 5-4 that PbI_2 is insoluble in water (trend 3). Therefore, a reaction takes place. A solid material, PbI_2 results. This is shown in Figure 5-7.

New Combination of Ions	Balanced Formula of Predicted Product	Solubility of Product	Chemical Formula and Physical State of Product
Pb^{2+} and I^-	PbI_2	Insoluble solid	$PbI_2 \ (s)$
K^+ and $(NO_3)^-$	KNO_3	Soluble; remains in solution.	$KNO_3 \ (aq)$

Answer $Pb(NO_3)_2 \ (aq) + 2 \ KI \ (aq) \longrightarrow PbI_2 \ (s) + 2 \ KNO_3 \ (aq)$

How are you doing? 5.9

Predict possible products for the following double displacement reactions. Write correct formulas for the new combinations of ions. Consult Table 5-4 to see if either of the products is insoluble. Write the balanced equation for the reaction.

(a) $CuSO_4 \ (aq) + Na_3PO_4 \ (aq) \rightarrow$
(b) $Na_2SO_4 \ (aq) + Ba(NO_3)_2 \ (aq) \rightarrow$
(c) $MgBr_2 \ (aq) + NaOH \ (aq) \rightarrow$

Figure 5-7 ▶ Combining lead (II) nitrate and potassium iodide results in a colorful precipitate. (Terry Gleason/Visuals Unlimited.)

PRACTICAL D What kinds of chemical reactions are used to test for metal ions?

In Practical A, we discussed using color changes to identify metal ions during chemical analysis. If a color change occurs as the result of a chemical reaction, the reaction is often an oxidation-reduction reaction. Color depends on electronic energy. The change in oxidation number that occurs in an oxidation-reduction reaction changes the number of valence electrons and often the color of the substance. The most highly colored substances contain transition metal ions. These are easy to recognize when a colored solution results after mixing two colorless solutions or when one of the reacting solutions is colorless and the color of the other changes after the mixing process. Show that the following reactions are oxidation-reduction reactions by finding the elements that undergo a change in oxidation number.

Figure 5-8 ▲ This photo shows the precipitation of either PbS, CdS, or Sb_2S_3. Which of these compounds has been formed here? How do you know? (Richard Megna/Fundamental Photographs.)

$$18 \ H^+ \ (aq) + Bi_2O_5 \ (s) + 4 \ Mn^{2+} \ (aq) \longrightarrow 4 \ MnO_4^- \ (aq) + 10 \ Bi^{3+} \ (aq) + 9 \ H_2O \ (l)$$
$$\text{pale pink} \qquad\qquad\qquad \text{purple}$$

(a) What color change indicates a chemical reaction has occurred?
(b) Which element is oxidized in this reaction?
(c) Which element is reduced in this reaction?

A test used to determine alcohol in breath uses an oxidation-reduction reaction involving chromium.

$$2 \ Cr_2O_7^{2-} \ (aq) + C_2H_5OH \ (l) + 16 \ H^+ \ (aq) \longrightarrow 4 \ Cr^{3+} \ (aq) + 11 \ H_2O \ (l) + 2 \ CO_2 \ (g)$$
$$\text{orange} \qquad\qquad\qquad\qquad \text{green}$$

(a) What color change indicates a chemical reaction has occurred?
(b) Is Cr being oxidized or reduced in this reaction?
(c) What gas is evolved in this reaction?

It is easier to be sure that a chemical reaction has occurred when an insoluble solid is formed from the mixing of two clear solutions. The chemical properties of the reacting substances determine whether or not a reaction occurs and a new substance is formed. But a *physical* property of the product determines whether or not the new substance is an insoluble solid. This physical property is the solubility. Figure 5-8 shows the precipitation of an insoluble solid from a solution. Predict possible products for the following reactions. Then, using the solubility rules, determine whether or not an insoluble solid is formed.

1. $H_2S \ (aq) + Pb(NO_3)_2 \ (aq) \rightarrow$
   (a) Write the formula for any solid formed.
   (b) What is the name of this compound?
   (c) What is the color of this compound? (See Practical A.)

2. $H_2S \ (aq) + Cd(NO_3)_2 \ (aq) \rightarrow$
   (a) Write the formula for any solid formed.
   (b) What is the name of this compound?
   (c) What is the color of this compound?

3. $H_2S \ (aq) + Sb(NO_3)_3 \ (aq) \rightarrow$
   (a) Write the formula for any solid formed.
   (b) What is the name of this compound?
   (c) What is the color of this compound?

4. Use solubility rules to predict a compound containing sulfide ion that is soluble in water solution.

PROBLEMS

22. Indicate the likely products of the following double displacement reactions. Write a balanced chemical equation in all cases.

(a) $CuSO_4 + Na_3PO_4 \rightarrow$
(b) $H_2SO_4 + PbCl_2 \rightarrow$
(c) $H_2SO_4 + Ca_3(PO_4)_2 \rightarrow$
(d) $K_2SO_4 + Ba(ClO_4)_2 \rightarrow$

23. For each of the reactions in Problem 22, consult the table of solubility to determine whether or not a reaction took place.

24. Predict the likely products of the following reactions.

(a) aluminum hydroxide + hydrogen chloride $\rightarrow$
(b) sodium iodide + silver nitrate $\rightarrow$

25. Use the solubility trends to predict whether or not the following double displacement reactions will occur. In each case, write the formulas for the predicted products, then determine if any product is insoluble in water solution, and write balanced equations for any reaction that occurs.

(a) $Mg(NO_3)_3 + NaOH \rightarrow$?
(b) $Fe(C_2H_3O_2)_3 + K_2SO_4 \rightarrow$?
(c) $Na_3PO_4 + Cd(NO_3)_2 \rightarrow$?

26. A solid forms when Na_2S and $Fe(NO_3)_3$ are mixed. What is the likely formula of the solid?

27. We mix solutions of zinc chloride and potassium phosphate. A solid forms. What is its formula and its name?

Chapter 5 Summary and Problems

 ◀ Review with Web Practice

VOCABULARY

chemical change	A reaction that transforms (changes) a substance into a new substance or substances with different properties.
burning	A chemical reaction involving the rapid combination of a substance with oxygen.
synthesis reaction	A reaction in which two or more substances combine to give a new compound.
decomposition reaction	A reaction in which a substance breaks down to form two or more substances.
displacement reaction	A reaction in which a component of a compound is "displaced" by a second reactant.
oxidation	A process by which an atom loses electrons, signified by an increase in the value of the atom's oxidation number.
reduction	A process by which an atom gains electrons, signified by a decrease in the value of the atom's oxidation number.
oxidation-reduction reaction	A reaction in which one substance is oxidized and a second substance is reduced, signified by changes in oxidation numbers from the reactant to the product side of an equation.

SECTION GOALS

How can we differentiate between chemical and physical properties?

A chemical property tells us how a substance acts or reacts to form new substances. A physical property tells us those characteristics of a substance that do not result in the formation of a new chemical substance.

How can we recognize a chemical change from a word description or from a chemical equation?

A chemical change is the enactment of what a chemical property predicts. The result of a chemical change is the formation of a new substance. We recognize a chemical change in a word description by the words "acts," "reacts," "forms," and "produces." We recognize a chemical change in an equation when the formulas on the product side differ from those on the reactant side of the equation.

Can you name the three main types of chemical reactions?

Most chemical reactions can be classified as being one or more of the following three reaction types: synthesis (or decomposition), displacement, or oxidation-reduction reactions.

How can you identify a given chemical reaction by reaction type?

A synthesis reaction combines two reactant elements into a single product. In a decomposition reaction, a substance is broken down into other substances. In a single displacement reaction, an element reacts with a binary compound to produce a different element and a new compound. In a double displacement reaction, two compounds react so that their component elements are switched. In an oxidation-reduction reaction, the oxidation number of two elements change during the course of the reaction. The element oxidized will show an increase in the value of its oxidation number and the element reduced will show a decrease in the value of its oxidation number.

Can you predict the products of a synthesis or displacement reaction?

To predict the product of a synthesis reaction, combine the two reactant elements and determine the correct ratio between them by using Lewis dot structures (for molecular compounds) or charge neutrality (for ionic compounds). To predict the products in a single displacement reaction, first identify whether the reactant element is a metal or a nonmetal. Metals will displace metals and nonmetals will displace nonmetals from the reactant compound. This produces a new element and a different compound. In double displacement reactions, two compounds react in such a way that the elements or ions that make them up are exchanged or "switched." One type of double displacement occurs when two solutions are mixed and a solid is produced.

Can you predict the products of single displacement reactions?

In a single displacement reaction, a pure element displaces another element from a compound. In general, metals displace metals and nonmetals displace nonmetals. Two important exceptions are carbon and hydrogen when they react with ionic compounds; typically, the carbon and the hydrogen displace the metal.

How do you identify oxidation-reduction reactions?

Oxidation-reduction reactions are based on the change in oxidation state of an element during a reaction. If we find that one or more elements change in oxidation number, then the reaction is an oxidation-reduction reaction.

Can you predict the outcome of a double displacement reaction?

Double displacement reactions are based on the exchange of anions and cations between ionic compounds. One way we can predict what will happen is to consider the solubility of the potential products.

PROBLEMS

28. Classify each of the following reactions as a synthesis, single displacement, or double displacement reaction. Use the table to indicate your answer by a check mark.

	Synthesis	Single Displacement	Double Displacement
Ba (s) + Br$_2$ (aq) $\longrightarrow$ BaBr$_2$ (aq)			
TiO$_2$ (s) + 4 HCl (g) $\longrightarrow$ TiCl$_4$ (l) + 2 H$_2$O (l)			
MnO$_2$ (s) + Si (s) $\longrightarrow$ SiO$_2$ (s) + Mn (s)			
H$_2$ (g) + NiCl$_2$ (aq) $\longrightarrow$ Ni (s) + 2 HCl (g)			
C (s) + O$_2$ (g) $\longrightarrow$ CO$_2$ (g)			
AgNO$_3$ (s) + CsCl (aq) $\longrightarrow$ CsNO$_3$ (aq) + AgCl (s)			
N$_2$ (s) + 3 H$_2$ (g) $\longrightarrow$ 2 NH$_3$ (g)			
Ba(NO$_3$)$_2$ (aq) + Na$_2$SO$_4$ (aq) $\longrightarrow$ 2 NaNO$_3$ (aq) + BaSO$_4$ (s)			
2 Ta (s) + 5 Br$_2$ (l) $\longrightarrow$ 2 TaBr$_5$ (l)			
2 Al$_2$O$_3$ (s) + 3 Si (s) $\longrightarrow$ 2 Al (s) + 3 SiO$_2$ (s)			

29. Balance the following reactions. Show that they are oxidation-reduction reactions by identifying the element that is oxidized and the element that is reduced.
 (a) CH$_4$ + Cl$_2$ $\rightarrow$ HCl + CCl$_4$
 (b) OF$_2$ + H$_2$O $\rightarrow$ O$_2$ + HF
 (c) N$_2$H$_4$ + H$_2$O$_2$ $\rightarrow$ N$_2$ + H$_2$O

30. Balance the following reactions.
 (a) Fe(OH)$_3$ + H$_2$SO$_4$ $\rightarrow$ Fe$_2$(SO$_4$)$_3$ + H$_2$O
 (b) NaCl + H$_2$O $\rightarrow$ Cl$_2$ + H$_2$ + NaOH
 (c) Na$_2$NH + H$_2$O $\rightarrow$ NH$_3$ + NaOH
 (d) Al + NH$_4$ClO$_4$ $\rightarrow$ Al$_2$O$_3$ + AlCl$_3$ + NO + H$_2$O
 (e) SCl$_2$ + NaF $\rightarrow$ SF$_4$ + S$_2$Cl$_2$ + NaCl

31. Balance the following decomposition reactions.
 (a) C$_3$H$_5$O$_9$N$_3$ $\rightarrow$ CO$_2$ + N$_2$ + O$_2$ + H$_2$O
 (b) (NH$_4$)$_2$Cr$_2$O$_7$ $\rightarrow$ Cr$_2$O$_3$ + N$_2$ + H$_2$O

32. Indicate the likely product of the following single displacement reactions. Write a balanced chemical equation in all cases.
 (a) LiCl + Na $\rightarrow$

 (b) Pd(ClO$_4$)$_2$ + H$_2$ $\rightarrow$
 (c) Rh(NO$_3$)$_3$ + Ca $\rightarrow$
 (d) PbSO$_4$ + Al $\rightarrow$

33. Indicate the likely products of the following double displacement reactions. Write a balanced chemical equation in all cases.
 (a) AgNO$_3$ + Na$_2$CO$_3$ $\rightarrow$
 (b) HCl + Al$_2$O$_3$ $\rightarrow$
 (c) CaF$_2$ + H$_2$SO$_4$ $\rightarrow$
 (d) KI + Pb(NO$_3$)$_2$ $\rightarrow$

34. Predict the likely products of the following reactions. Indicate if the reaction is a synthesis, a single displacement, or a double displacement reaction.
 (a) potassium + titanium (IV) chloride
 (b) hydrogen + mercury (I) chloride
 (c) water + boron trifluoride
 (d) hydrogen fluoride + nickel (II) oxide

35. Water contains oxygen, but oxygen supports burning. Why does water put out a fire instead of fuel it?

36. Water contains hydrogen and oxygen. Hydrogen burns; burning is the rapid combination with oxygen. Can water be used as a fuel? Support your decision with a statement.

37. Concept question: A closed, flexible vessel is filled with a pure gas. When the top of the vessel is opened, the gas inside immediately bursts into flame and a new gas is formed. What can you say about the physical and chemical properties of the *product* substance?

38. Concept question: When sugar is treated with concentrated sulfuric acid, steam is given off and a mass of pure carbon is left behind. This suggests that sugar contains carbon. Why doesn't carbon taste sweet?

39. What is the likely product of the reaction
$Sr + N_2 \rightarrow$?

40. What is the likely product of the reaction
$Cs + N_2 \rightarrow$?

41. Predict the products of the reaction $H_2 + VO \rightarrow$.

42. What kind of reaction is
$Cu_2S + H_2 \rightarrow 2Cu + H_2S$?

43. Predict the product of the synthesis reaction of postassium and phosphorus.

44. Classify the following as either chemical or physical changes.
(a) $N_2(l) \rightarrow N_2(s)$
(b) $2 NO_2(g) \rightarrow N_2O_4(g)$

45. Name three observable signs that a chemical reaction has occurred. (See Practical A.)

46. Write the correct formula and name for the products of the following reactions between metal ions and sulfide ions.

Metal ion + Sulfide ion $\rightarrow$	Formula of Product	Name of Product
$Co^{2+} + S^{2-} \longrightarrow$		
$Fe^{3+} + S^{2-} \longrightarrow$		
$Ga^{3+} + S^{2-} \longrightarrow$		
$Mn^{2+} + S^{2-} \longrightarrow$		
$Ni^{3+} + S^{2-} \longrightarrow$		

47. Define combustion. How do you recognize a combustion reaction?

48. Balance the following combustion reactions.
(a) $C_4H_8O + O_2 \rightarrow CO_2 + H_2O$
(b) $C_6H_7O + O_2 \rightarrow CO_2 + H_2O$
(c) $C_5H_{12}O_2 + O_2 \rightarrow CO_2 + H_2O$

49. Write a word equation for the following reactions. Balance the equations.
(a) $CO_2(s) \rightarrow CO_2(g)$
(b) $Na_2CO_3(s) + HCl(aq) \rightarrow$
$CO_2(g) + H_2O(l) + NaCl(aq)$

50. Indicate the likely product of the following synthesis reactions. Write a balanced chemical equation in each case.
(a) $Al + Cl_2 \rightarrow$?
(b) $Al + Se \rightarrow$?
(c) $Ba + N_2 \rightarrow$?

51. Indicate the likely product of the following single displacement reactions. Write a balanced chemical equation in each case.
(a) $AgCl + F_2 \rightarrow$?

(b) $GaBr_3 + Cl_2 \rightarrow$?
(c) $Al_2O_3 + C \rightarrow$?

52. Use the activity series of metals to predict whether or not the following single displacement reactions will occur. Write a balanced chemical equation for any reaction that occurs.
(a) $Fe(NO_3)_3 + Zn \rightarrow$?
(b) $KF + H_2 \rightarrow$?
(c) $Mn(NO_3)_2 + Cd \rightarrow$?

53. Answer the following questions about the compound $PtCl_4$.
(a) What is the oxidation number of Pt in this compound?
(b) What is the name of the compound?
(c) Balance the equation $Pt + Cl_2 \rightarrow PtCl_4$.
(d) What is oxidized and what is reduced in the reaction $Pt + Cl_2 \rightarrow PtCl_4$?
(e) Predict the products and balance the equation for the single displacement reaction $PtCl_4 + F_2 \rightarrow$.

54. Answer the following questions about the compound SnO_2.

(a) What is the oxidation number of Sn in this compound?

(b) What is the name of the compound?

(c) Balance the equation $Sn + O_2 \rightarrow SnO_2$.

(d) What is oxidized and what is reduced in the reaction $Sn + O_2 \rightarrow SnO_2$?

(e) Predict the products and balance the equation for the single displacement reaction $SnO_2 + Mg \rightarrow$.

Quantitative Properties of Matter

Chemists carefully measure the quantities of chemical substances that exist before and after experiments. Here a chemist examines a flask to determine the volume of a chemical solution. (Rosenfeld Images/Rainbow.)

PRACTICAL CHEMISTRY Interpreting Chemical Experiments

In order to understand a chemical experiment, we need information about the quantities of chemical substances that we have both before and after the experiment. In this chapter, we will look at measurements of chemical quantities that are commonly obtained and used in chemical experiments, and we will practice interpreting these measurements. As many of the fundamental relationships in chemistry involve mathematical proportions, we will use proportional reasoning to better understand how chemists think about these relationships.

This chapter will focus on the following questions:

PRACTICAL A What measurements do chemists report from the lab?

PRACTICAL B How can we measure the density of a liquid?

PRACTICAL C What ratios do we encounter in chemistry?

PRACTICAL D How does a chemical experiment give us the data for a chemical formula?

6.1 Methods of Measurement

SECTION GOALS

✔ Understand the difference between an intensive property and an extensive property.

✔ Know the common metric units used to measure length, mass, and volume.

✔ Understand how temperature influences the physical state (phase) of a pure substance.

✔ Use melting and normal boiling temperatures to predict the physical state (phase) of a substance.

Intensive and Extensive Properties

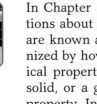

In Chapter 1, we began to work with certain characteristic observations about elements and chemical substances. These characteristics are known as physical properties. For example, metals can be recognized by how they look and feel. Appearance and hardness are physical properties. We can observe whether a substance is a liquid, a solid, or a gas. The state or phase of a substance is also a physical property. In addition to physical properties that can be seen or felt, chemists study properties that must be *measured,* such as temperature, density, and mass. Properties that have numerical measurements are called **quantitative properties.** Knowing how chemists measure quantitative properties will help you to understand and interpret numerical data in chemistry.

In this chapter, we consider two kinds of quantitative properties, *extensive and intensive properties. Some of these are shown in Table 6-1.* An **extensive property** is a property that depends on the size of the sample. Mass is an example of an extensive property. A larger sample of water, for example, has a greater mass than does a smaller sample. An **intensive property** does not depend on the size of the sample, and is the same whether the sample is large or small. Figure 6-1 shows temperature as an example of an intensive property.

✔ **Meeting the Goals**

Extensive properties depend on the sample size; intensive properties do not. There are good reasons for making intensive and extensive measurements. Both measurements are used in the laboratory to characterize substances and their behavior.

Table 6-1 Quantitative Properties That Are Measured in Chemistry

Extensive Properties *depend on size of sample*	Intensive Properties *do not depend on size of sample*
Mass	Temperature
Volume	Pressure
Distance	Ratios of extensive variables
Area	• density
Number	• percentage
Time	• concentration
	• molar mass

Figure 6-1 ▲ Temperature is an intensive property. After baking, the temperature of the small cookie is the same as that of the large cookie.

Example **6.1** **Distinguishing between intensive and extensive properties**

For each of the following situations, indicate whether an intensive or extensive property is measured.

(a) A cook adds three more eggs to a waffle batter.

(b) In the process of inspecting a used car before purchase, you ask to see a record of its gas mileage.

(c) An eighteenth-century soap maker checks to see if a lye solution is ready for use by determining whether or not an egg will float in it.

(d) In order to determine the amount of yarn needed for a sweater, a knitter makes a careful measurement of arm, chest, and neck size.

Thinking Through the Problem We can solve this problem by thinking about whether the measurement would change if the person doing the procedure had a larger sample of the object to measure. We know that extensive measurements are larger when the measured object is larger. Intensive measurements do not change when the size of the object being measured changes.

Answer

(a) Because a greater number of eggs would change the amount of the waffle batter and the number or size of the waffles produced, this must be a measurement of an *extensive* property.

(b) The gas mileage is a record of gas used per mile. It will not change if the car traveled a greater distance. Gas mileage is an *intensive* property.

(c) The solution will presumably cause an egg to float even if the volume of solution is larger. So this is a measurement of an *intensive* property.

(d) The amount of yarn needed to knit the sweater depends on the size of the measurements: more yarn is needed for a large measurement than for a small measurement. Thus, the person's size is an *extensive* property.

How are you doing? 6.1

For each of the following cases, indicate whether the measurement involves an extensive or an intensive property.

(a) You keep a notebook recording the amount of fertilizer applied in your backyard garden.

(b) A fertilizer is applied to a field with young corn plants, and a yield of corn is obtained from each acre at harvest time.

(c) An ice cube tray of water is placed in a freezer to determine the time it takes for the water to freeze solid.

Extensive Measurements and Units

Measurements in chemistry will always have units. In fact, you will soon discover that numbers without units are meaningless. A note that simply says "It is 5" is very ambiguous. Five what? Even if you know the "5" refers to a time or distance, there is a big difference between 5 minutes and 5 years, or between 5 blocks and 5 miles. If you don't clearly state what you mean when you write down a measurement, a reader or listener (or even yourself at a later point) may misinterpret that number. Imagine asking a friend to pay you five dollars for his part of a meal. If you say, "Just give me five" what might happen? Figure 6-2 illustrates the importance of units.

Chemists use a system of internationally agreed-upon units called "SI units," the *International System of Units*. SI units use the **metric system,** which is a decimal system. The most commonly used metric units are the gram (for mass), the centimeter (for length), and the second (for time). Volume has units of "centimeters cubed," cm^3, or cubic centimeters, since linear dimensions are recorded in centimeters, as shown in Figure 6-3. You may recognize the abbrevi-

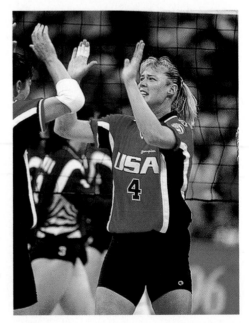

Figure 6-2 ▶ A number without units and identity is meaningless to a chemist, just as the expression "Give me five" can be misinterpreted. (Bob Daemmrich/Stock Boston.)

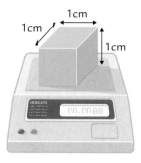

Figure 6-3 ▲ The length of each side of the cube is 1 cm. Volume has units of cm × cm × cm = cm³. The volume of this cube is 1 cm³.

✔ Meeting the Goals

KEY IDEA 6.2

In comparing particular samples, we often compare extensive quantities when we want to know "how much" or "how big." Any unit in Table 6-2 might be used to answer these questions.

ation for cubic centimeter, "cc," that is commonly used by medical personnel. Chemists also use the unit *liter* (L) for volume. One liter is 1,000 cubic centimeters. We will spend more time learning the metric system in Chapter 7.

A second requirement of a good measurement is that the *identity* of whatever is being measured be stated, unless it is obvious. In the case of five dollars, we could say "five dollars of money" but it is obvious when we say "five dollars" that the substance being measured is money. But the identity of the measured object is not always obvious. For example, "one liter" could describe the volume of almost any substance. One liter of ethyl alcohol, for instance, is not the same sample of matter as one liter of water. Ethyl alcohol and water are different substances even though they look the same and have the same volume. We must state the identity of the substance being measured. Just as we must properly label containers of chemicals, we must properly label the answer to a calculation. When we write "1 liter of ethyl alcohol" we leave no doubt as to either the amount or the identity of the chemical substance being measured.

Table 6-2 lists the extensive properties you will use in this course. This table includes some English and metric units that are commonly used to measure extensive properties.

The most common measurements you will need in chemistry experiments are the mass and the volume of a sample. Examples of volume units derived from SI units are cubic centimeters or cubic meters. Typical laboratory volume measurements tend to be in the range of 1 to 1,000 cubic centimeters (or 1 to 1,000 milliliters). Figure 6-4 shows some common devices for measuring volume.

Although most of us use the terms *mass* and *weight* interchangeably, *these words actually mean two different things*. **Mass** is a measure of the amount of material present; it is always the same for a given sample. **Weight** is a measure of the amount of material present *as it is acted upon by a gravitational force*. Thus, the *weight* of a sample varies with its location relative to the earth. A sample will weigh slightly less when it is measured at high altitudes, such as on a high mountain, because the gravitational force is less than at sea level. However, the mass of this sample will be the same at both locations, as the amount of matter present does not change. Although mass and

Table 6-2 Some Common Units of Measurement

Extensive Property	English Unit	Metric Unit
Mass	Ounce, pound, ton	Gram, kilogram
Distance	Inch, foot, mile	Centimeter, meter
Time	Second, minute, day	Second, minute, day
Chemical amount	Not applicable	Mole
Volume	Gallon, quart, pint, cup	Cubic meter, liter, cubic centimeter (These units are derived from the meter.)

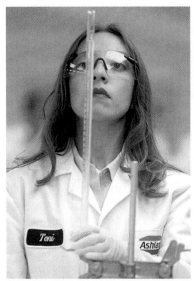

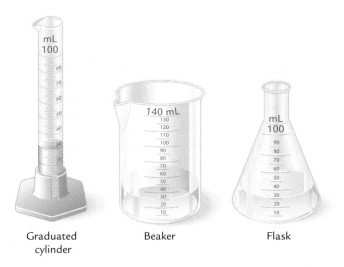

Graduated   Beaker   Flask
cylinder

Figure 6-4 ▲ Different types of glassware are used to measure volume amounts in chemistry.

weight are different quantities, the numerical difference between the two measurements in earth-bound chemistry laboratories is negligible. This is why we commonly refer to the mass of an object when in fact we record its weight.

Counting is a common way to measure objects, as we saw in Chapter 4. When we count "three eggs," "four skeins of yarn," or "150 molecules," we have measured these amounts. The unit we use to count larger numbers of molecules or atoms is the *mole,* shown in Table 6-2.

Intensive Properties: Temperature and Phase Changes

Intensive properties do not depend on size. Therefore, the numerical value of an intensive property remains the same whether the sample size is large or small. Two common intensive properties that we measure are temperature and pressure. These measurements are the same for any part of a pure sample, assuming the sample is not actively changing (for example, by being heated).

Temperature is a measure of the hotness or coldness of a substance compared to some relative temperature scale. *Temperature influences how a substance will appear to us, because a substance may undergo a change in its physical state or phase as it is heated or cooled.* You will recall that the three most important phases in chemistry are solids, liquids, and gases. Table 6-3 shows the shape, volume, and relative temperature characteristics of solids, liquids, and gases.

Temperature is related to the energy possessed by the molecules of a substance. As we increase the temperature of a substance, its atoms and molecules begin to vibrate more vigorously. As long as the substance is not constrained in a rigid container, this increased vibration will cause the system to expand. If the substance is a gas, it will

Table 6-3 Physical States and Characteristics

Physical States of Matter	Shape and Volume Characteristics
Solid	Both the volume and shape of a solid are definite.
Liquid	Liquids have a definite volume but take the shape of their container.
Gas	Both the volume and the shape of a gas are determined by its container.

✔ **Meeting the Goals**

The general trend is for substances to be solids at lower temperatures, liquids at intermediate temperatures, and gases at higher temperatures. However, not all substances have a liquid phase under normal conditions. Some substances such as dry ice (solid CO_2) sublime, changing from a solid directly into a gas.

REMEMBER

expand significantly. If it is a liquid, it will expand by a lesser degree, and if it is a solid, it will expand still less. If a sample is a solid or a liquid, an increase in temperature may bring the sample to a point where it has enough energy to change from one physical state to another. In most cases, heating a solid substance can convert it first into a liquid and then into a gas, as shown in Figure 6-5. Some substances, such as solid carbon dioxide ("dry ice"), *sublime* under heating—they pass directly from the solid to the vapor state. Phase changes were discussed in Chapter 1 and can be summarized as:

$$\text{SOLID} \underset{\text{freezing}}{\overset{\text{melting}}{\rightleftharpoons}} \text{LIQUID} \underset{\text{condensing}}{\overset{\text{evaporating}}{\rightleftharpoons}} \text{GAS}$$

$$\text{SOLID} \underset{\text{depositing}}{\overset{\text{sublimating}}{\rightleftharpoons}} \text{GAS}$$

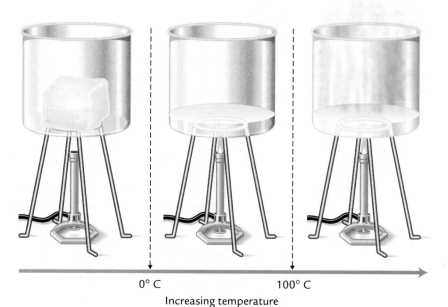

Figure 6-5 ▶ Under normal pressure conditions, water is a solid (ice) at temperatures less than 0 °C. It is a liquid between 0 and 100 °C, and it is a gas (water vapor) at temperatures above 100 °C.

0° C   100° C

Increasing temperature

PRACTICAL A   What measurements do chemists report from the lab?

Chemists usually report their research in the form of papers published in chemical journals. A critical part of many papers is the "Experimental Section," which includes a discussion of what was done in a particular experiment. Reading experimental sections can be difficult because chemists often present material with a lot of abbreviations and terms that are known to other chemists, but not to the general public. The following three reports, which have been changed only to remove some abbreviations, indicate the kind of measurements that researchers often use to report their work.

Read these three reports, paying close attention to the measurements. Then answer the questions that follow. (Notice that, to save space and make it easy to refer to compounds in an article, the authors of many papers use numerical designations for certain compounds).

1) Into a 500-mL round-bottom flask were placed **2b** (6.41 g, 0.0205 mole), dimethylformamide (250 mL), imidazole (11.2 g, 0.164 mole) and TBSCl (12.4 grams, 0.0822 mole). The reaction was heated to 50 °C for 6 hours. The mixture was cooled to room temperature and quenched with NH_4Cl (200 mL).

2) Azide binding studies were performed using freshly prepared solutions of $[Fe(III)S_2Me_2N_3(Pr,Pr)] \, PF_6$ **(2)** and NaN_3 in MeOH. Equilibrium constants were determined by monitoring electronic spectral changes at four different temperatures (-77 °C, -41 °C, -15 °C, and 0 °C). Low-temperature spectra were obtained using a custom-designed low-temperature quartz optical dewar filled with the appropriate cooling bath (acetone/CO_2 (-77 °C), acetonitrile/N_2 (-41 °C), ethlyene glycol/CO_2 (-15 °C), and ice water (0 °C)).

3) A modified cell was used for illumination, which consists of a stainless steel pressure vessel equipped with a 2-cm-thick quartz window. The electrolyte solution [3 cubic centimeters, 0.3 moles per liter tetrabutylammo-

Figure 6-6 ▲ As the temperature increases, the snowman melts into water. However, the melted snow may refreeze into ice if the temperature drops below 0 °C. (Erv Schowengerdt.)

The temperature at which a solid melts is called its **melting temperature** ("melting point"). The melting temperature of a substance is the same as its **freezing temperature** ("freezing point"), and thus marks the boundary between its solid and liquid phases. Figure 6-6 shows a familiar example of this. The substance will be solid at temperatures lower than the melting temperature, it will exist in both the solid and liquid phases at the melting temperature, and will be a liquid at temperatures above its melting point and below its boiling point.

The temperature at which a liquid boils is called its **boiling temperature** ("boiling point"). Boiling occurs when a liquid sample begins to vaporize at the same pressure as the external pressure. The gas formed by the evaporation or boiling of a liquid is often called a **vapor** (for example, water vapor). At temperatures above the boiling point, the substance is a gas; between the boiling point and the freezing point, the substance is a liquid.

Because gases are strongly affected by pressure, the boiling point temperature depends on the external pressure. To standardize boiling point temperatures, chemists often refer to the **normal boiling point,** which is the boiling temperature measured at sea level pressure.

Both the melting point and the boiling point of a pure substance are **physical constants** (under specified pressure conditions.) A physical constant measures a physical property of a substance, and because this value does not change, physical constants can be used to help identify substances. For example, the normal boiling point of

nium perchlorate (TBAP, reagent grade, Tokyo Kasei) in CH_3OH (reagent grade, Nacalai Tesque)] was placed in a glass cell liner in the stainless steel vessel. CO_2 was introduced into the pressure vessel and was allowed to equilibrate for 1 hour at the designated pressures (1–40 atmospheres). In control experiments, 1 atmosphere Ar was used.

For the three reports, answer the following questions:

- Where do the authors identify chemical substances? Which ones do you recognize? When do you think the authors may be using a trivial name?
- When the authors measure a chemical substance, what units do they use? In cases where more than one unit is used, can you suggest why the different units are important?
- When do the authors measure an intensive quantity? Are there any examples where they measure a ratio of units (look for the word "per").
- Do you think that a skilled scientist could use these descriptions to repeat the experiments? Why or why not?

1) From: "The Pinene Path to Taxols" by Paul A. Wender and coworkers, *Journal of Organic Chemistry*, **1996**, *61*, page 7662.

2) From "Reactivity of Five-Coordinate Models for the Thiolate-Ligated Fe Site of Nitrile Hydratase" by Julie A. Kovacs and coworkers, *Journal of the American Chemical Society*, **1998**, *120*, pages 5691–5700.

3) From "Photoelectrochemical Reduction of CO_2 in a High-Pressure CO_2 + Methanol Medium at *p*-Type Semiconductor Electrodes" by Akira Fujishima and coworkers, *Journal of Physical Chemistry*, **1998**, *102*, pages 9834–9843.

water is 100 °C, the normal boiling point of acetone is 56 °C, and the normal boiling point of ethanol is 78.2 °C. Suppose we had a sample of one of these liquids but didn't know its identity. We could identify the liquid by slowly heating a small sample of it and then measuring the temperature at which it boils. Because the boiling temperatures of these three chemicals are very different from one another, we can easily identify which liquid we have.

We saw in Chapter 1 that *we can use temperature to create "boundaries" to the phases of a substance if we know its melting and boiling temperatures.* This kind of reasoning allows us to predict the phases of a substance at different temperatures or to predict the limits of the temperature if we know the phase.

KEY IDEA 6.4

Example 6.2   Determining physical state from temperatures

Bromine has a melting point of −7.2 °C and a normal boiling point of 58.8 °C. Will it be a solid, liquid, or gas at −0.12 °C?

Thinking Through the Problem (6.4) The key idea is that the melting point and boiling point are the temperatures at which bromine undergoes its phase changes under normal pressure. We can look at a temperature scale to determine the phase. In this case, we note that −0.12 °C is between −7.2 °C and 58.8 °C.

Answer   Bromine will be a liquid at −0.12 °C.

Example 6.3   **Determining physical states from temperatures**

Consider the following melting point and boiling point information for the metals copper, sodium, and gold.

Metal	Melting point, °C	Normal boiling point, °C
Copper	1083	2567
Sodium	97	883
Gold	1064	2807

(a) Determine the physical state of each metal at 750 °C.
(b) Determine the physical state of each metal at 1000 °C.
(c) Determine the physical state of each metal at 1072 °C.

Thinking Through the Problem For each metal, the melting point and the normal boiling point define the temperatures for transitions among solid, liquid, and gas. One way to look at this problem is to insert the melting and boiling temperatures into the equation scheme.

Working Through the Problem

$$\text{For copper, SOLID} \xleftrightarrow{1083\ °C} \text{LIQUID} \xleftrightarrow{2567\ °C} \text{GAS}$$

$$\text{For sodium, SOLID} \xleftrightarrow{97\ °C} \text{LIQUID} \xleftrightarrow{883\ °C} \text{GAS}$$

$$\text{For gold, SOLID} \xleftrightarrow{1064\ °C} \text{LIQUID} \xleftrightarrow{2807\ °C} \text{GAS}$$

Answer (a) At 750 °C, copper and gold are solids. Sodium is liquid. (b) At 1000 °C, copper and gold are solids. Sodium is a gas. (c) At 1072 °C, copper is a solid, gold is a liquid, and sodium is a gas.

✔ Meeting the Goals

The freezing and boiling temperatures mark the boundaries between solid and liquid and between liquid and gas, respectively, for a pure substance under ordinary pressure conditions. The substance will be a solid at temperatures below the freezing temperature, a liquid at all temperatures between the freezing and the boiling temperatures, and a gas at temperatures above the boiling temperature.

How are you doing?   6.2

The normal boiling point for carbon (C) is 3652 °C; that of CO is −191.5 °C. Both of these substances contain carbon but they have very different identities. Suppose we have one of these two substances at a temperature low enough that it is a solid. We observe that it turns into a gas as the temperature warms to room temperature (25 °C). Which substance is it? What kind of a change is involved?

PROBLEMS

1. Suppose that you must measure the volumes of various substances but you have only a broomstick, a ballpoint pen (with the cap on), and a paper clip as possible measuring tools. Using these tools, estimate the length (l), width (w), and height (h) for each item in the following table. These are only estimates; the first one has been done for you. Which measuring tool do you think is the best "fit" for each case?

Substance to Be Measured	Unit of Measurement		
	A broomstick	A ballpoint pen	A paper clip
A shoe box	l = 1/6 broomstick w = 1/12 broomstick h = 1/12 broomstick	l = w = h =	l = w = h =
Your kitchen	l = w =	l = w =	l = w =

2. Units are important in measurement.

(a) When you exchange currency, it is important to know your relative units. Look in a daily paper and decide which of the following you would want: 25 US dollars, 25 Australian dollars, or 25 Canadian dollars.

(b) Equivalent volumes of different things may have very different effects when they are put to use. Which would you rather have for a snack, a cup of ice cream or a cup of molasses? If you were baking, which would you use?

3. For each of the following sentences, indicate whether an intensive or extensive property is being measured.

(a) A city council realizes that the conversion of factory buildings to apartments will mean they need new school classrooms.

(b) A farmer calculates how many kilograms of phosphorus fertilizer is needed to convert an acre of land from soybeans to corn.

(c) Fewer vehicles on the road in the summer means there are fewer accidents.

(d) The "lift" of a hot air balloon increases on days when the air pressure is higher.

4. Concept question: Measurements should always include the units and the identity of the substance measured. What would be the difference between adding 25 cubic centimeters of water to a fire, and placing 25 cubic centimeters of acetone (the major ingredient in nail polish remover) on a fire? Do you know any of the properties of acetone that would have to be considered if it were to be used around open flames?

5. Measurement of mass depends on the law of conservation of mass. Consider a reaction in which 23.2 grams of a solid compound is heated. After heating, 21.2 grams of material is left. Where did the remaining mass go? Suggest an experiment that might determine whether or not you are correct.

6. Common thermometers use a liquid to indicate the temperature. Safe operation requires that the substance be a liquid at all times. Mercury has a melting point of -38.9 °C. What effect does this have on using a mercury thermometer?

7. Sulfur is a solid at room temperature. It is often purified by melting a sample of the *impure* material (melting point 115 °C) and then spraying the liquid into a water bath. What will happen when melted sulfur hits the cold water?

8. Discussion question: An advertisement for an antacid tablet suggests that it "neutralizes its weight in stomach acid." What does weight refer to here? With what you know about substances and mixtures, suggest some questions that need to be answered before this claim has meaning from a chemist's viewpoint.

9. Consider the table on the following page.

(a) What trends do you observe in the freezing and boiling points of these compounds as you move down a column of the periodic table? Which do *not* fit a trend?

(b) What trends do you observe in the freezing and boiling points moving across a row of the periodic table?

(c) Which compounds are solids at room temperature, 30 °C? Which are liquids? Which are gases?

AlCl$_3$ boiling point: sublimes freezing point: 192.4 °C	SiCl$_4$ boiling point: 57.6 °C freezing point: −70 °C	PCl$_3$ boiling point: 76.1 °C freezing point: −93.6 °C
GaCl$_3$ boiling point: 201 °C freezing point: 77.9 °C	GeCl$_4$ boiling point: 83.1 °C freezing point: −49.5 °C	AsCl$_3$ boiling point: 130.2 °C freezing point: −16.2 °C
InCl$_3$ boiling point: sublimes freezing point: 586 °C	SnCl$_4$ boiling point: 114 °C freezing point: −33.3 °C	SbCl$_3$ boiling point: 223 °C freezing point: 73.4 °C

6.2 Density

SECTION GOALS

✔ Solve a formula for a specified variable.

✔ Explain why density is an intensive property.

✔ Solve the density formula for each of the three variables, d, m, and V.

✔ Understand properties measured in a density experiment and apply proper units.

Intensive Properties: Density

Density is an important intensive property that gives the number of grams per cubic centimeter of a substance. It is the ratio between two extensive variables, mass and volume (see Table 6-1).

The densities of some common materials are given in Table 6-4. Note that the two gases listed have a much smaller density than any of the liquids or solids listed. Density can be used to help us distinguish between a gas and a liquid or solid. A substance with a very low density is likely to be a gas at room temperature. Can we also

A balloon will rise or fall in air, depending on the density of the gas filling it. (William Thomas Cain/AP.)

Table 6-4 Densities of Some Common Substances (25 °C, normal pressure)

Substance	Density, g/cm^3	Substance	Density, g/cm^3
Water (*l*)	0.996	Sulfur (*s*)	1.96
Ethanol (*l*)	0.789	Phosphorus (*s*)	1.88
Glycerin (*l*)	1.261	Manganese (*s*)	7.47
Mercury (*l*)	13.546	Fluorine (*g*)	0.00155
Acetone (*l*)	0.790	Chlorine (*g*)	0.00290
Carbon tetrachloride (*l*)	1.59	Bromine (*l*)	3.10
Stearic acid (*s*)	0.85	Naphthalene (*s*)	1.16

use density to distinguish between liquids and solids? The answer, perhaps surprisingly, is no. There are two liquids listed in Table 6-4 that have densities greater than solid sulfur. Can you find them? Density is not a good property to use to distinguish between liquids and solids.

We can use density to predict whether a substance will float or sink when it is placed in a second substance that is fluid. If the two substances do not react or mix with each other, they will form layers based on density. When placed in a denser substance, a substance that is less dense will float, as Figure 6-7 shows. A substance that is denser will sink.

KEY IDEA 6.5

Example 6.4

Using density to separate substances

A laboratory worker has a mixture of flakes of the compound naphthalene and flakes of the compound stearic acid. They must be separated, so she mixes them well with water. Neither dissolves in the water and there is no chemical reaction. Will this procedure result in the separation of the two compounds? Why or why not?

Thinking Through the Problem ━━▷ (6.5) A key idea is that, in a mixture, a material that is less dense will float. Therefore, a mixture of nonreactive substances can be separated if the substances have different densities. In this case, we see from Table 6-3 that the density of stearic acid is less than that of water and the density of water is less than the density of naphthalene.

Answer We predict that the stearic acid will float on the water (it is less dense than water) and the naphthalene will sink to the bottom (it is more dense than water). Yes, the procedure will separate naphthalene from stearic acid because of the difference in densities.

Figure 6-7 ▲ Vinegar and oil salad dressings separate into an oil layer and a vinegar layer. Because the oil layer is less dense than the vinegar layer, it lies on top. (Coco McCoy/ Rainbow.)

How are you doing? **6.3**

Use the information in Table 6-4 to explain the following observations:

(a) The metal manganese is very dense, but it floats on a pool of mercury.
(b) If we mix water and carbon tetrachloride, we get two layers.

MAKING IT WORK WITH MATH

Algebraic Formulas and Variables

Chemists use basic mathematical models such as equations and graphs to describe real-world relationships. In this section we will examine how mathematical formulas are used to relate chemical measurements. You should expect to use many mathematical equations in chemistry. An equation provides a rule by which two or more quantities are related; thus, equations are great tools to help us define the relationship that exists between different quantities. Their usefulness will depend to some degree on your ability to *recognize what the relationship is* and to *manipulate the formula into an equivalent relationship* that focuses on the particular quantity that you are most interested in. You may be familiar with several of the equations given below.

$$d = \frac{m}{V} \qquad \text{density of sample}$$

$$P = 2l + 2w \qquad \text{perimeter of a rectangle}$$

$$A = \pi r^2 \qquad \text{area of a circle}$$

$$V = lwh \qquad \text{volume of a rectangular solid}$$

KEY IDEA 6.6

Each of these mathematical equations has been solved for one of the variables it contains. *Solving for a variable means isolating it on one side of the equation.* We see that the first equation is solved for density because d has been isolated. But there may be situations where we are more interested in knowing the mass of the sample or the volume of the sample. In those cases, we have to solve the equation for that desired variable. We can think of the equation as something that has "packaged" the variable. We must "unwrap" that package by asking the following questions:

1. **What is the last operation done to that variable?** The last operation may be an *addition or subtraction.* For example, if we were solving for w in $P = 2l + 2w$, then the last operation was the addition of $2l$ to the right side of the equation.

    If no addition or subtractions are left on the side of the equation where we have our variable, then the last operation will be *multiplication or division.* For example, if we are solving for h in $V = lwh$, then the last operation done was multiplication of h by lw.

2. **How do you "undo" that operation?** We undo an operation by reversing it. The reverse of subtraction is addition, and the reverse of multiplication is division. For example, we would begin to solve for h in $V = lwh$ by *dividing* both sides by lw:

$$V = lwh$$

$$\frac{V}{lw} = \frac{lwh}{lw}$$

$$\frac{V}{lw} = h$$

We would begin to solve $P = 2l + 2w$ for w by *subtracting* $2l$ from both sides.

$$P = 2l + 2w$$

$$P - 2l = 2l + 2w - 2l$$

$$P - 2l = 2w$$

Notice that we are not done yet.

✔ **Meeting the Goals**

Mathematical equations are solved for a given variable by "undoing" the last operation on that variable until the variable is isolated. The density formula is

$$d = \frac{m}{V}.$$

Solving this for m gives

$$m = dV;$$

solving for V gives

$$V = \frac{m}{d}.$$

3. **Repeat the process until the variable is isolated.** In some cases, we may need to do additional steps to completely solve for the variable. Here, we solve for w by dividing both sides by 2.

$$\frac{P - 2l}{2} = \frac{2w}{2}$$

$$\frac{P - 2l}{2} = w$$

Example 6.5 Solving for variables

Given $d = rt$, solve for t.

Thinking Through the Problem To solve for t we carry out steps that isolate t on one side of the equation. The variable t is being multiplied by r. To "undo" this operation we must divide by r. Always remember that when working with equations, if we perform an operation on one side of the equation, then we must perform the same operation on the other side. This keeps the two sides equal.

Working Through the Problem

$$d = rt$$

Divide by r. $\quad \dfrac{d}{r} = \dfrac{rt}{r}$

Simplify. $\quad \dfrac{d}{r} = t$

Answer $\dfrac{d}{r} = t$ A similar process could be used to solve the original equation for r. We would need to divide both sides of the equation $d = rt$ by t to isolate r.

Example 6.6 Solving for variables

Given $\dfrac{P_1 V_1}{T_1} = \dfrac{P_2 V_2}{T_2}$, solve for T_1.

Thinking Through the Problem There are three variables, P, V, and T, which represent pressure, volume, and temperature. Each variable is subscripted by a number to distinguish the first measurement (1) from the second measurement (2) of the same property. Therefore, there are actually six variables in this equation since we expect that P_1 is different from P_2, etc. The variable we are interested in is the denominator, so we multiply both sides of the equation by T_1. This establishes T_1 in the numerator. We then carry out operations to isolate the other variables from T_1.

Working Through the Problem

$$\frac{P_1 V_1}{T_1}(T_1) = \frac{P_2 V_2}{T_2}(T_1)$$

$$P_1 V_1 = \frac{P_2 V_2}{T_2} T_1$$

$$\frac{T_2}{P_2 V_2} \times P_1 V_1 = \frac{P_2 V_2}{T_2} T_1 \times \frac{T_2}{P_2 V_2}$$

$$\frac{P_1 V_1 T_2}{P_2 V_2} = T_1$$

Answer $\dfrac{P_1 V_1 T_2}{P_2 V_2} = T_1$

Example 6.7   Solving for variables

Solve $C = \dfrac{5}{9}(F - 32)$ for F.

Thinking Through the Problem We will solve this problem in a similar manner as the previous one.

Working Through the Problem

Multiply by 9. $9C = \cancel{9}\left(\dfrac{5}{\cancel{9}}\right)(F - 32)$

Simplify. $9C = 5(F - 32)$

Distribute. $9C = 5F - 160$

Add 160. $9C + 160 = 5F - 160 + 160$

Simplify. $9C + 160 = 5F$

Divide by 5. $\dfrac{9C + 160}{5} = \dfrac{\cancel{5}F}{\cancel{5}}$

Simplify. $\dfrac{9C + 160}{5} = F$

This is equivalent to: $\dfrac{9}{5}C + 32 = F$

Answer $\dfrac{9}{5}C + 32 = F$

Solve each formula for the indicated variable.

$$PV = nRT \text{ for } V \quad (\text{ideal gas law})$$

$$A = \frac{1}{2}bh \text{ for } h \quad (\text{area of triangle})$$

$$\frac{P_1 V_1}{T_1} = \frac{P_2 V_2}{T_2} \text{ for } T_2 \quad (\text{gas law})$$

$$d = \frac{m}{V} \text{ for } V \quad (\text{density})$$

Density—Making the Formula Do the Work!

Scientists often use mathematical formulas to describe physical properties. Density is a good example of this, as it is a physical property determined by dividing the mass of a substance by its volume. It is a physical property because we can measure both mass and volume without changing or destroying the identity of the substance measured. Both mass and volume are extensive properties, that is, they depend on the size of the sample. The ratio of mass to volume, however, is an intensive property. We expect, then, that a small sample of a pure substance will have the same density as a larger sample of that substance, as shown in Figure 6-8.

(a)

(b)

Figure 6-8 ▲ The density of water is a constant value no matter how large or small the sample is. Density is an intensive property. The density of fresh water is the same in (a) as it is in (b). (a: Erv Schowengerdt; b: Tom McHugh/Photo Researchers.)

Fractions such as $\dfrac{\text{mass}}{\text{volume}}$ are called **ratios.** Thus, density is the ratio of mass to volume,

$$\text{density} = \frac{\text{mass}}{\text{volume}}$$

If we replace the words with letters (variables), we get an algebraic equation that looks like:

$$d = \frac{m}{V}$$

The equation can be solved for any one of the three variables, given values for the other two. For a pure substance, the $\dfrac{\text{mass}}{\text{volume}}$ ratio gives a constant value when the temperature and pressure are specified. Density is treated just like any other constant in an algebraic equation and it will apply for any sample of that substance at the specified temperature and pressure.

The units for density depend on the units used for mass and volume. In this chapter, we will restrict these units to *grams* for mass and *cubic centimeters* for volume. Thus, the units of density will be *grams per cubic centimeter,* $\dfrac{\text{g}}{\text{cm}^3}$, or g/cm^3. Changes in the temperature and pressure (for gases) may change the value for density. For this reason, the temperature and the pressure are usually stated in a density problem. If temperature and pressure are not stated, we assume we are working at 25 °C and at standard pressure. Density is a physical constant under specified temperature and pressure and can be used, along with other information, to help identify a substance.

✔ Meeting the Goals

Note that density does what we expect a ratio to do: it expresses a property of a substance that does not change from sample to sample at the same temperature. Density is an intensive property because we measure the same value for density using any size sample when the temperature and pressure are the same.

Example 6.8   **Calculating mass from density values**

> Determine the mass of 15.00 cm^3 of ethanol.
>
> Thinking Through the Problem We recognize 15.00 cm^3 as a volume because it has the units *cubic centimeters.* We begin with the formula for density $d = m/V$ and rearrange to isolate m. We find the density of ethanol in Table 6-4. Then we substitute the known values for volume and density and solve for the unknown value, the mass. Notice that the unit, cm^3, divides out leaving grams as the final unit.
>
> Answer $m = dV = 0.789\dfrac{\text{g}}{\text{cm}^3} \times 15.00\ \text{cm}^3 = 11.8\ \text{g}$

Example 6.9   **Calculating volume from density**

> Mercury is the only metal that is a liquid under ordinary conditions of temperature and pressure. (Room temperature is about 25 °C and room pressure is close to 1 atmosphere.) What is the

Mercury, the only liquid metal, has a very high density. (Richard Megna/Fundamental Photographs.)

volume occupied by 999 grams of mercury at room temperature and pressure?

Thinking Through the Problem We recognize 999 grams as a mass quantity because of the units *grams*. We can find the density of mercury by referring to Table 6-4. We will use the formula for density and solve for the volume.

Working Through the Problem The formula for density is: $d = m/V$. In this case, we want to solve for the volume, V, and so we rearrange the equation. After substituting the known values given in the problem or in Table 6-4, we get:

$$V = \frac{m}{d} = \frac{999 \text{ g}}{13.546 \text{ g/cm}^3}$$

We can make use of the fact that division by some number n is equivalent to multiplication by the inverse of n. We can write the equation as:

$$V = 999 \text{ g} \times \frac{1 \text{ cm}^3}{13.546 \text{ g}} = 73.7 \text{ cm}^3$$

The unit gram (g) divides out, leaving final units of volume, cm^3.

Answer   73.7 cm^3

How are you doing?   6.5

(a) Find the mass of a cube of gold (density $= 19.320 \text{ g/cm}^3$). The volume of the cube is 16.1 cm^3.

(b) Sulfur hexafluoride, SF_6, is one of the densest gases that can be handled in air. It is often used as a blanket to protect electronics from sparks. Under room conditions, SF_6 has a density of 0.0598 g/cm^3. Determine the mass of SF_6 that fills a 250-cm^3 container.

Example   6.10   Calculating density

A fixed volume of a gas is weighed to measure its mass. It is found that 250 cm^3 of the gas has a mass of 0.450 g. What is the density of the gas?

Thinking Through the Problem We are given a volume and a mass. We substitute these values into the formula for density and solve for d. The final answer will have the mass and volume units of the quantities given in the problem.

PRACTICAL B   How can we measure the density of a liquid?

Suppose you are asked to determine the density of an unknown liquid. You are told to measure volumes of approximately 10 cm³, 20 cm³, 30 cm³, and 40 cm³ of the liquid and to weigh and record the mass of each sample. To do this, you use a graduated cylinder or a beaker to measure the volumes, as shown in Figure 6-9, and a balance to measure the masses. You obtain and record the measurements given in Table 6-5. Notice the organization and labeling in this table.

The temperature of the liquid and the atmospheric pressure are also recorded. For each sample listed in Table 6-5, one of the assigned volumes is measured and weighed. The volume for each sample is given first.

Table 6-5 Experiment: Density of a Liquid

	Data	Variable
Temperature of liquid (room temperature)	30.0 °C	
Barometric pressure[1]	0.998 atmospheres	
Mass of empty cylinder	62.025 g	m_0
Sample 1		
Total volume of liquid	10.10 cm³	V_1
Mass of cylinder + liquid	75.569 g	m_1
Mass of liquid	13.544 g	$m_1 - m_0$
Sample 2		
Total volume of liquid	20.05 cm³	V_2
Mass of cylinder + liquid	87.789 g	m_2
Mass of liquid	25.764 g	$m_2 - m_0$

[1]Some density tables assume atmospheric pressure and report only the temperature values. (SW Productions/Photo Disc.)

Figure 6-9 ▲ Density values are determined experimentally by measuring the mass and volume of several samples of the same substance. Density is a physical property because we can measure both mass and volume without changing the identity of the substance measured; it is also an intensive property, because we obtain the same value regardless of the size of the sample. (SW Productions/PhotoDisc.)

✔ **Meeting the Goals**

To find the density of a sample, we measure the mass and the volume of the sample. Density is the ratio of mass to volume. We expect units that are a ratio of mass units to volume units. Typical units for density are g/cm³.

Working Through the Problem

$$d = \frac{m}{V} = \frac{0.450 \text{ g}}{250 \text{ cm}^3} = \left(\frac{0.450}{250}\right)\frac{\text{g}}{\text{cm}^3} = 0.0018 \text{ g/cm}^3$$

Answer 0.0018 g/cm³

How are you doing? 6.6

Determine the density of sodium chloride if 1,000. g of sodium chloride has a volume of 462 cm³.

The mass of the liquid in each case is found by subtracting the mass of the empty cylinder from the mass of the cylinder + liquid. For sample 1, the measured volume is 10.10 cm^3. The mass of this sample is calculated as follows.

m_1	m (cylinder + liquid)	75.569 g
$-m_0$	m (cylinder)	-62.025 g
$m_1 - m_0$	m (liquid)	13.544 g

Using this data, calculate a density for each of the four samples. Record the four density values in the space provided in Table 6-6.

Table 6-5 Experiment: Density of a Liquid (Continued)

	Data	Variable
Sample 3		
Total volume of liquid	29.95 cm^3	V_3
Mass of cylinder + liquid	99.612 g	m_3
Mass of liquid	37.587 g	$m_3 - m_0$
Sample 4		
Total volume of liquid	39.90 cm^3	V_4
Mass of cylinder + liquid	113.975 g	m_4
Mass of liquid	51.950 g	$m_4 - m_0$

[1]Some density tables assume atmospheric pressure and report only the temperature values.

Table 6-6 Data Summary For the Density Experiment

Sample	Mass, g	Volume, cm^3	Density, g/cm^3
1	_____	_____	_____
2	_____	_____	_____
3	_____	_____	_____
4	_____	_____	_____

Measuring the Density of an Irregularly Shaped Object

A particularly useful procedure for measuring density involves measuring the volume of water displaced by an object. The volume of an irregularly shaped, insoluble solid can be measured as the difference in water volume before and after adding the solid to a cylinder containing a measured volume of water. See Figure 6-10. If the solid is denser than and insoluble in the water, it will displace a volume of water equal to its own volume.

Figure 6-10 ◀ The volume of an irregularly shaped object can be determined by measuring the volume of the water that it displaces. The difference in the measured volume of water after the object has been added is the volume of the object. The volume of the object pictured here is approximately 10 cm^3.

Example 6.11   **Calculating density using water displacement**

An irregularly shaped solid with a mass of 43.0 g is put into a cylinder containing 22.5 cm^3 of water. It sinks and the volume increases to 45.6 cm^3. What is the density of this solid?

Thinking Through the Problem The mass of the object is clearly stated but the volume is not. We recognize that this is a water displacement problem because of the description given.

Working Through the Problem When the object is put into a cylinder containing 22.5 cm^3, the volume of water in the cylinder increases to 45.6 cm^3. The difference in the two volumes is the volume of the solid object. Thus, volume (final) − volume (initial) = volume of solid: 45.6 cm^3 − 22.5 cm^3 = 23.1 cm^3. Now density can be calculated using the formula.

Answer $\dfrac{43.0 \text{ g}}{23.1 \text{ cm}^3} = 1.87 \text{ g/cm}^3$

PROBLEMS

10. Concept question: Density values sometimes but not always give us a clue about the physical state of a substance. Go through this chapter and tabulate all of the densities that are given as part of the text or the examples. What trends do you observe for gas vs. liquid vs. solid?

11. A laboratory procedure requires 25.0 g of carbon tetrachloride, CCl_4. What volume will this mass occupy? (See Table 6-4 for density information.)

12. Concept question: The densities of substances are the same under any gravitational conditions. Does that mean that manganese will float on mercury in outer space?

13. The average density of air is 0.00129 g/cm^3. Balloons filled with a gas that is less dense than air will rise. Those filled with a gas denser than air will remain on the ground. Which, if any, of the following balloons will rise? (Assume the mass of the balloon itself does not matter.)

(a) The density of gas inside balloon 1 is 0.00145 g/cm^3.

(b) The density of gas inside balloon 2 is 0.00115 g/cm^3.

(c) The density of gas inside balloon 3 is 0.00122 g/cm^3.

(d) The density of gas inside balloon 4 is 0.00086 g/cm^3.

14. You are asked to weigh out 155 g of ethyl alcohol. Which of the following containers is the *smallest* container you can use without spillage?

Container A will hold exactly 155 cm^3.
Container B will hold exactly 190 cm^3.
Container C will hold exactly 200 cm^3.
Container D will hold exactly 125 cm^3.

15. Density measurements are made for a substance at temperature T_1. The measurements are repeated at temperature T_2. Choose a mass value and make a point on the axis that measures mass. Now draw a line through your point and parallel to the axis that measures volume so that the line cuts through both lines that are graphed. Now you have a way to compare the volumes at T_1 and at T_2. Did the volume of the substance expand or contract when measured at T_2?

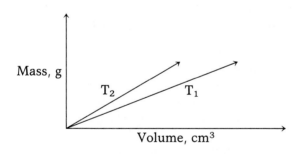

16. Discussion question: The density of a gas is a property that helps determine whether the gas will rise or fall. Gas will collect in your basement if the gas is denser than air, or in upper floors if the gas is less dense than air. Will carbon monoxide gas (d = 0.00113 g/cm^3) rise or sink in air? What about propane (d = 0.00178 g/cm^3)? What safety issues are associated with these facts?

17. Concept question: The mass and the volume of a substance are measured at 25 °C and the following graph is made. On the same graph, sketch in the line that would result if the same substance were measured at 100 °C. Does the new line fall above or below the existing line on the graph?

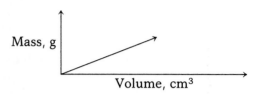

18. Suppose that you could "layer" liquids one on top of another so that they did not mix with one another. Which liquid would be found in the top layer of a container of water (d = 0.995 g/cm^3) and toluene (d = 0.946 g/cm^3)?

19. Castor oil has a density of 0.985 g/cm^3 at a temperature of 20 °C and at room pressure. What is the volume of 55 g of castor oil under these same conditions of temperature and pressure?

20. An irregularly shaped piece of metal having a mass of 14.25 grams is placed into a graduated cylinder containing 21.5 cm^3 of water. The metal sinks to the bottom of the cylinder resulting in a final water volume of 22.6 cm^3. Calculate the density of the metal.

21. Determine the density (in g/cm^3) of 10.0 cm^3 of silicon dioxide that has a mass of 21.6 g.

22. Gold has a density of 19.30 g/cm^3. What volume of gold has a mass of 20.0 g?

23. A 60.0-g ceramic figurine sinks to the bottom of a graduated cylinder originally containing 40.0 cm^3 water. The water level rises to 73.0 cm^3. What is the density of the ceramic figurine?

24. An experiment is conducted to determine the density of a compound. A graduated cylinder with a mass of 23.99 g is used. When 25 cm^3 of the compound is added to the cylinder, then the combined mass of the cylinder and the compound is 41.23 g. What is the density of the compound?

6.3 Proportions in Chemical Measurement

SECTION GOALS

✔ Understand that extensive measurements can be used to determine an intensive measurement, using a ratio called a proportionality.

✔ Recognize a constant of proportionality.

✔ Find and write a proportionality using a relationship among extensive units given in a problem.

✔ Write and use ratios that involve percentage, parts per million, and parts per billion.

MAKING IT WORK WITH MATH

✔ **Meeting the Goals**

Whenever two variables are directly proportional, their ratio gives a constant value called a constant of proportionality. For example, density is the constant of proportionality between mass and volume.

Constants of Proportionality

The ratio of two extensive variables is called a **proportionality.** If this ratio is the same whenever we compare these variables, the constant is called the **constant of proportionality.** When we use proportionalities to help us solve problems, the process is called **proportional reasoning,** which is a very important skill in chemistry and other sciences. For example, you can measure the distance from your house to a bus stop and the amount of time it takes to jog this distance. These are both simple, extensive measurements. The ratio of distance to time gives an intensive measurement—the rate—which tells us how quickly or slowly we cover a particular distance. See Figure 6-11.

Mass and volume are both extensive variables. When we relate these two extensive variables, we get a proportionality called the density. Proportionalities like density free us from measuring every extensive property in every sample: if we know the proportionality, we need to measure only one extensive variable. The other variable can be calculated.

Constants of proportionality are found whenever one measurement changes with respect to another by direct variation. In mathe-

Figure 6-11 ▲ Although both distance and time are extensive properties, their ratio (the rate) is an intensive property. (Larry Brownstein/Rainbow.)

Figure 6-12 ▲ A direct relationship can exist between many different substances. If we restrict one person to a maximum of two bags of groceries, we can easily determine the number of bags that ten people could carry. We could also determine how many people would be needed to carry one hundred bags of groceries. (Dana White/PhotoEdit.)

KEY IDEA 6.7

matical terms, direct variation requires that a variable y "varies directly with" or "is directly proportional to" the variable x. Direct variations therefore have the form

variable y = constant of proportionality × variable x

In symbols, we have $y = kx$ where k is the constant of proportionality. For a given system, the value of k does not change from sample to sample. We can determine k from any sample, and the same value should be found for all other samples regardless of size. In other words, the constant of proportionality k is *fixed* for all samples. So, if we determine k for one sample, then it applies to all samples of that substance.

We use proportional reasoning to solve problems in which a relationship is given between two quantities. Sometimes the relationship is explicitly stated in the problem. For example, the problem may state that a factory manufactures 250 radios a day. The relationship here is between the number of radios and a time period, one day. Another simple example of a relationship between two quantities is shown in Figure 6-12. Some relationships, such as density, are measured experimentally. Still other relationships are defined, such as the conversion of inches to feet: 12 inches = 1 foot. In this section, we will explore how proportional reasoning is used to solve problems that involve proportionalities.

How do we recognize a proportional relationship in a problem that is new to us? *To find a proportional relationship, we look for the amount and the identity of the two quantities in the problem.* For example, suppose we are told that a titanium mine is capable of producing 0.100 tons of titanium for every 7 tons of rock. We can write an appropriate ratio containing these two quantities, and then evaluate the ratio to find the constant that relates them.

This sentence gives us the amount and the identity of two substances, titanium and rock. The two related quantities are italicized in the following sentence.

"A titanium mine is capable of producing *0.100 tons of titanium* for every *7 tons of rock*."

We can write the relationship by forming a ratio in which one quantity appears as the numerator and the other appears as the denominator. We normally choose to set up the ratio using the order given in the problem. In these problems,

1. We assume the relationship is constant.
2. We write a ratio having the first quantity in the numerator and the second quantity in the denominator.
3. If the quantities vary directly with one another, then we know the ratio is equal to some constant value k. This gives us an equation with two variables and one constant.

Thus, for this example, the equation is:

$$k = \frac{0.100 \text{ ton Ti}}{7 \text{ tons rock}}$$

We can now solve for k:

$$\frac{0.100 \text{ ton Ti}}{7 \text{ tons rock}} = k = 0.0143 \frac{\text{tons Ti}}{\text{tons rock}}$$

The answer is read "0.0143 tons of Ti *per* ton of rock."

Example 6.12 Solving problems with proportional reasoning

An oil well produces 84,000 barrels of oil during a one-week period. What is the ratio for the production of oil by this well?

Thinking Through the Problem We will emphasize the two quantities related in this sentence and write an appropriate ratio containing these two quantities. Then we can solve the ratio to find the constant of proportionality that relates the two quantities.

Working Through the Problem The information given in the problem indicates that we get *84,000 barrels of oil per week*. The variables are *barrels of oil* and *week*. They have values of 84,000 and 1, respectively. Note that the word *per* implies "one" here. We can write two ratios with these variables, depending on what we choose to be the numerator in the ratio. One ratio is:

$$\frac{84,000 \text{ barrels of oil}}{1 \text{ week}}$$

The second ratio is the inverse of this:

$$\frac{1 \text{ week}}{84,000 \text{ barrels of oil}}$$

Both ratios are correct. Each ratio can be divided out but the resulting answer will have different units. The first ratio relates barrels of oil to weeks; the second ratio relates weeks to barrels of oil.

Answer The ratio $\dfrac{84,000 \text{ barrels of oil}}{1 \text{ week}} = 84,000$ barrels of oil/week is a constant of proportionality for the well. So is the inverse, $\dfrac{1 \text{ week}}{84,000 \text{ barrels of oil}} = 0.000012$ weeks/barrel of oil.

How are you doing? 6.7

The town of Boron, California, is named for a massive boron mine nearby, capable of producing 5,500 tons of mineral ore per day. This amount of ore can be refined to yield 500 tons of boron. Write a ratio for the production of boron from the mineral.

Predictions Using Proportional Reasoning

A proportion is an equation in which two fractions, or ratios, are set equal to one another. At first glance you may wonder how we make variations such as density fit the definition of a proportion. Remember that for a direct variation $y = kx$. We find the constant of proportionality by isolating k.

$$\frac{y}{x} = k$$

The power of proportional reasoning is that we can calculate the value k for any $\frac{y}{x}$ ratio and then use this k to solve other $\frac{y}{x}$ ratios for the same system.

Suppose, for example, that we know that the density of some substance equals one gram per two cubic centimeters. Then, $d = \frac{m}{V} = \frac{1\ g}{2\ cm^3}$. Now we can use the value of d, $\frac{1\ g}{2\ cm^3}$, to calculate the mass for any other volume—or the volume for any other mass of this substance. Suppose the volume changes to 4 cm^3. We calculate the new mass in the following manner.

$$\frac{1\ g}{2\ cm^3} = \frac{m}{4\ cm^3}$$
$$2m = 4\ g$$
$$m = 2\ g$$

Because density is a constant value, doubling the denominator from 2 to 4 forces the numerator also to double from 1 to 2. The resulting proportion, $\frac{2}{4} = \frac{1}{2} = d$ demonstrates why density is called the constant of proportionality between the mass and the volume for a pure substance. Any two mass-volume ratios for the same substance must be equal as they both equal the density. We show this mathematically for any general case in the following manner.

If we have two sets of corresponding data, we can say:

$$\frac{y_1}{x_1} = k \qquad \text{and} \qquad \frac{y_2}{x_2} = k$$

Then, because the value of k is constant, it follows that

$$\frac{y_1}{x_1} = \frac{y_2}{x_2}$$

✔ **Meeting the Goals**

The constant of proportionality density is a simple ratio. We will encounter such ratios throughout the course.

Using this method, we eliminate the need to find the constant, k. Chemists often use proportions in solving physical problems. Here are just a few examples.

- We want to know how much of a substance is in a mixture. If the mixture has constant proportions, the *percentage* ratio is often used to report how much of the substance is in any sample of the mixture.

- We want to know the volume of a large chunk of metal. If the metal is pure or if it is an alloy with a constant composition, we can measure its mass and use the density ratio to calculate its volume.

- We need to know how many atoms of an element are in a sample of a molecular substance. We use the chemical formula of the molecular substance to get an *atom-to-atom* or *mole-to-mole* ratio for that substance, as we saw in Chapter 4.

- We want to know how many molecules of reactant or product we have in a particular reaction. We use the balanced chemical equation to get a *mole-to-mole ratio of substances in the reaction*. We have already seen the mole ratio in Chapter 5.

- We need to find the amount of a substance that is dissolved in a particular solution. If we know the *concentration* of the solution, which can be expressed in terms of how much of the substance is dissolved in one liter of the solution, we can calculate the amount of the substance in any volume of the solution.

The key to working with proportions is to set them up correctly. The proportion will contain two ratios set equal to each other. One ratio (which we call the *known* ratio) gives some known relationship between two quantities. The known relationship can come from a previously learned definition (such as 12 inches = 1 foot) or it may be stated in the problem. The known ratio is the same as the constant of proportionality we discussed earlier. The second ratio contains the same quantities in the numerator and denominator as does the known ratio but in this case, one of the values is not known. For this reason, we refer to this ratio as the *unknown* ratio.

Proportional reasoning is the foundation of many different kinds of calculations you will encounter in this course and beyond. Problem-solving methods such as "unit factors," "dimensional analysis," "unit conversion," and "conversion factor method" are all based on the mathematics of proportional reasoning. Let's go back to the titanium problem and see how proportional reasoning handles a new question.

Example 6.13   Solving by using proportional reasoning

A titanium mine produces 0.100 ton titanium for every 7 tons of rock processed. How many tons of titanium will be produced if the plant processes 250 tons of rock?

Thinking Through the Problem The first sentence gives the relationship between tons of titanium and tons of rock processed. This gives us the quantities for our known ratio, $\dfrac{0.100 \text{ ton Ti}}{7 \text{ tons rock}}$.

✔ Meeting the Goals

These problems contain a sentence that relates "so many of this" to "so many of that." The ratio is written with one of these quantities in the numerator and the other in the denominator. Both the numbers and the units should be included in the ratio. It does not matter which quantity goes into which position (numerator or denominator) for the known ratio. The quantities in the unknown ratio must be in the same positions as those in the known ratio.

We will now write a second ratio that has the same units in the numerator and denominator as our known ratio.

Working Through the Problem We are *looking for* tons of titanium and are *given* 250 tons of rock. The unknown ratio is $\dfrac{x \text{ tons Ti}}{250 \text{ tons rock}}$.

We set up a proportion by making the two ratios equal and solving for x.

$$\text{known ratio} = \text{unknown ratio}$$

$$\frac{0.100 \text{ ton Ti}}{7 \text{ tons rock}} = \frac{x \text{ tons Ti}}{125 \text{ tons rock}}$$

$$125 \text{ tons rock} \times \frac{0.100 \text{ ton Ti}}{7 \text{ tons rock}} = \frac{x \text{ tons Ti}}{125 \text{ tons rock}} \times 125 \text{ tons rock}$$

$$125 \times \frac{0.100 \text{ ton Ti}}{7} = x \text{ tons Ti}$$

$$1.79 \text{ tons Ti} = x \text{ tons Ti}$$

Answer   1.79 tons Ti

How are you doing?   6.8

For the boron mine discussed in "How are you doing?" exercise 6.7, determine how many tons of mineral must be processed to obtain 220,000 tons of boron.

When setting up proportions, the two ratios must have the same units in the numerator and in the denominator. "Tons Ti" does not have to be in the numerator. We could set up a ratio in which "tons rock" is in the numerator and "tons Ti" is in the denominator. *It doesn't make any difference to the problem.* What is critical is that the units in the *second* ratio are set up in an identical manner.

Many people find it easier to set up the unknown ratio first by reading it directly from the problem. In the example above, the question is "How many tons of titanium will be produced if the plant processes 250 tons of rock?" This makes the unknown ratio easy to write. "How many tons Ti" goes in the numerator as the unknown and the "250 tons rock" becomes the denominator:

$$\frac{x \text{ tons Ti}}{250 \text{ tons rock}}$$

The known ratio is set up identically with "tons Ti" in the numerator and "tons rock" in the denominator, and the proportion is solved.

Example 6.14 **Solving problems using proportional reasoning**

The resistance of a solution of copper sulfate is measured at 0.41 ohms when two electrodes are 3.00 cm apart. If resistance is directly proportional to the distance between electrodes, then what is the resistance when the electrodes are 10.0 cm apart?

Thinking Through the Problem The relationship in the problem is between distance and resistance. The first sentence gives us the numbers and the units we need for our known ratio, $\dfrac{0.41 \text{ ohm}}{3.00 \text{ cm}}$.

We set up an unknown ratio with corresponding units and solve for the unknown quantity.

Working Through the Problem

$$\frac{0.41 \text{ ohm}}{3.00 \text{ cm}} = \frac{x}{10.0 \text{ cm}}$$

$$10.0 \text{ cm} \times \frac{0.41 \text{ ohm}}{3.00 \text{ cm}} = \frac{x}{10.0 \text{ cm}} \times 10.0 \text{ cm}$$

$$1.4 \text{ ohms} = x$$

Answer 1.4 ohms

How are you doing? **6.9**

From the previous example, what is the distance between the electrodes if the resistance is 10.0 ohms?

Proportional Reasoning and Percentage

Many statements signal the use of proportional reasoning with the word *per*. When we say that there are "25 grams per liter," we relate 25 grams to one liter. The "one" is usually omitted from the sentence but, when no other number is explicitly stated, it is always implied by the word *per*.

$$25 \text{ grams per liter} = \frac{25 \text{ g}}{1 \text{ L}} = 25 \text{ g/L}$$

In some cases, we use a specific number with the word *per*. For example, in a study of a community, one might read, "There are 32 birdhouses per one thousand homes." In this case, the number of homes is specified and is used in the denominator of the proportionality between birdhouses and homes, $\dfrac{32 \text{ birdhouses}}{1,000 \text{ homes}}$. In the last example, the ratio $\dfrac{0.41 \text{ ohms}}{3 \text{ cm}}$ is read "0.41 ohms *per* 3 cm." The word *per* indicates a ratio; the quantity following the word *per* is the denominator of the ratio. The number preceding the word is meant to be the numerator of the ratio.

Because "per" problems are proportionalities, we can use proportional reasoning to solve them. There are other uses of "per" in chemistry. Percentage is probably the most familiar of the "per" statements. When we refer to 25 percent of a sample, we are talking about 25 out of every 100 parts of the sample. The word *parts* is a general term that refers to a unit such as grams or liters that is specified in the problem. Percentage, as well as other "per" statements we will discuss in this section, are proportionality constants for some specified system. We often use the symbol % to mean percent.

$$k = x\% = \frac{x \text{ parts}}{100 \text{ parts}}$$

✔ **Meeting the Goals**

The word *per* alone means "per one." Fill in the value of 1 in the denominator in all cases. "Percent" means per 100 or x parts/100 parts. "Per million" means x parts/1,000,000 parts. "Per billion" means x parts/1,000,000,000 parts.

Some other common "per" amounts are given special abbreviations. Examples include "parts per million" (ppm), "parts per billion" (ppb), and "parts per thousand" (ppt). These are encountered in discussions of very small amounts of something in a larger sample of something else.

$$x \text{ ppt} = \frac{x \text{ parts}}{1000 \text{ parts}}; \quad x \text{ ppm} = \frac{x \text{ parts}}{1,000,000 \text{ parts}}$$

and

$$x \text{ ppb} = \frac{x \text{ parts}}{1,000,000,000 \text{ parts}}$$

The word *parts* in these general ratios is replaced by a unit such as grams or liters that is specified in the problem.

Example 6.15 Writing ratios based on "per" relationships

Each of the following statements includes a "per" relationship. Write each ratio that is defined in the sentence; include numbers and units in both the numerator and the denominator of the ratio.

(a) Carbon dioxide contains 27% carbon by mass.
(b) Years after it was banned, the pesticide DDT is still present in some soils at a level of 25 ppm by mass.
(c) Seven out of every 1,000 pregnancies result in twins.

Thinking Through the Problem Each statement contains a ratio—a quantity in the numerator divided by a quantity in the denominator.

Working Through the Problem Part (a) relates carbon to carbon dioxide. The reference is to mass, so the units will be mass units such as grams. The numerator contains grams of C, the denominator contains grams of CO_2. The values are 27 in the numerator and, because this is a percentage problem, 100 in the denominator. Therefore, 27% by mass means there are 27 grams of C for every

100 grams of CO_2. Be aware that any mass units (kilograms, pounds, tons) could be used here as long as the same unit is used in both the numerator and denominator of the ratio.

Part (b) uses ppm, parts per million. Again the units will be mass units so we will use grams. 25 ppm by mass means there are 25 grams of DDT per million (1,000,000) grams of pesticide. These two numbers make up the numerator and denominator of our ratio.

Part (c) directly gives the relationship between twins and births. There are 7 sets of twins born per 1,000 pregnancies. We will use these quantities in our ratio.

Answer (a) $\dfrac{27 \text{ g C}}{100 \text{ g CO}_2}$ (b) $\dfrac{25 \text{ g DDT}}{1,000,000 \text{ g soil}}$ (c) $\dfrac{7 \text{ births with twins}}{1,000 \text{ total births}}$

How are you doing? 6.10

Determine the ratio that can be derived from the following statements.

(a) The production of nitrogen is 190% of the production of ammonia by mass.

(b) There are fractional amounts of many elements in the body, including 3 ppm selenium by mass.

(c) Online courses currently account for 2% of all course enrollments at the University of Utah.

Now that we have practiced recognizing and writing ratios based on powers of ten such as percentage, we can use proportional reasoning to solve problems using these ratios.

Example 6.16   **Proportional reasoning and "per" statements**

A factory finds 2.5 g of impurities present per 100 g of chemical dye produced. How many grams of impurities do you expect to be present in a shipment of 25 grams of dye?

Thinking Through the Problem The first sentence gives the relationship between grams of impurities and grams of dye produced. We will use this sentence to help us write our known ratio. We will write an unknown ratio, set it equal to the known ratio, and solve.

Working Through the Problem

$$\text{known ratio} = \text{unknown ratio}$$

$$\frac{2.5 \text{ g impurities}}{100.0 \text{ g dye}} = \frac{x \text{ g impurities}}{25 \text{ g dye}}$$

$$25 \text{ g dye} \times \frac{2.5 \text{ g impurities}}{100.0 \text{ g dye}} = \frac{x \text{ g impurities}}{25 \text{ g dye}} \times 25 \text{ g dye}$$

$$25 \times \frac{2.5 \text{ g impurities}}{100.0} = x \text{ g impurities}$$

$$0.62 \text{ g impurities} = x \text{ g impurities}$$

Answer 0.62 g impurities

So far, we have set up proportions in which we have written the known ratio on the left. But we can write either ratio first. All that is important is that the ratios are written with the same units in the numerator and in the denominator. For instance, Example 6.16 asks how many grams of impurities are expected in 25 grams of dye. Then you might have written 25 grams dye in the first numerator and x grams impurities in the first denominator. That's fine. But now units in the second ratio must match. Your problem would look like this:

$$\text{unknown ratio} = \text{known ratio}$$
$$\frac{25 \text{ g dye}}{x \text{ g impurities}} = \frac{100.0 \text{ g dye}}{2.5 \text{ g impurities}}$$

The solution of this proportion will give the same answer as in Example 6.16:

$$\text{unknown ratio} = \text{known ratio}$$
$$\frac{25 \text{ g dye}}{x \text{ g impurities}} = \frac{100.0 \text{ g dye}}{2.5 \text{ g impurities}}$$
$$25 \text{ g dye} \times 2.5 \text{ g impurities} = x \text{ g impurities} \times 100.0 \text{ g dye}$$
$$x \text{ g impurities} = \frac{25 \text{ g dye} \times 2.5 \text{ g impurities}}{100.0 \text{ g dye}}$$
$$= 0.62 \text{ g impurities}$$

If you compare this procedure with the one in Example 6.16, you will notice that we did get the same answer, but with more steps.

Example 6.17 **Proportional reasoning and "per" relationships**

A shampoo is listed as having 0.22% selenium sulfide (SeS_3) by mass. What mass of selenium sulfide is present in a 250-g bottle of the shampoo?

Thinking Through the Problem The unit system—grams—is specified for us when we are told "250-g bottle." The percentage indicates how many *grams* of SeS_3 are in 100 grams of shampoo.

Working Through the Problem

$$\text{unknown ratio} = \text{known ratio}$$
$$\frac{x \text{ g SeS}_3}{250 \text{ g shampoo}} = \frac{0.22 \text{ g SeS}_3}{100 \text{ g shampoo}}$$
$$250 \text{ g shampoo} \times \frac{x \text{ g SeS}_3}{250 \text{ g shampoo}} = $$
$$\frac{0.22 \text{ g SeS}_3}{100 \text{ g shampoo}} \times 250 \text{ g shampoo}$$
$$x \text{ g SeS}_3 = \frac{0.22 \times 250}{100} \text{ g SeS}_3$$
$$x \text{ g SeS}_3 = 0.55 \text{ g SeS}_3$$

Answer We expect to find 0.55 g SeS_3 in 250 g shampoo.

(Felicia Martinez/PhotoEdit.)

Long ago you may have learned to do percentage problems another way. If 0.22% of the shampoo was selenium sulfide then you could find the amount of selenium sulfide by multiplying the percent (after converting it to a decimal) by the amount of shampoo. This is exactly the math shown in the third equation in Example 6.17. Percent is converted to a decimal form by dividing by 100. Then it is multiplied by 250 g. Although you may find some percent problems straightforward, for others the use of proportional reasoning is very helpful.

Example 6.18 **Proportional reasoning and percent**

A company produces cars and trucks. If 21.2% of the production of vehicles is cars, and there are 192,222 cars produced in a year, how many vehicles were produced that year?

Thinking Through the Problem The relationship is between cars and vehicles. 21.2% of the production is cars. This means there are 21.2 cars for every 100 vehicles. Now we can set up our known and unknown ratio and solve the problem.

Working Through the Problem

$$\text{unknown ratio} = \text{known ratio}$$

$$\frac{x \text{ vehicles}}{192{,}222 \text{ cars}} = \frac{100 \text{ vehicles}}{21.2 \text{ cars}}$$

$$192{,}222 \text{ cars} \times \frac{x \text{ vehicles}}{192{,}222 \text{ cars}} = \frac{100 \text{ vehicles}}{21.2 \text{ cars}} \times 192{,}222 \text{ cars}$$

$$x \text{ vehicles} = \frac{100 \times 192{,}222}{21.2} \text{ vehicles}$$

$$x \text{ vehicles} = 906{,}707 \text{ vehicles}$$

Answer 906,707 vehicles

How are you doing? 6.11

We are told that 85% of all hydrogen is made from coal. If 2.50 million tons of hydrogen were produced last year, what mass of hydrogen was produced from coal?

Example 6.19 **Solving a parts per million problem**

A sample of river water contains 34.5 ppm Cd. How many grams of Cd will there be in 250.0 g of this water?

Thinking Through the Problem The wording of this question helps us to set up one of the ratios. "How many grams of Cd will there be in 250.0 g of water?" becomes $\dfrac{x \text{ g Cd}}{250.0 \text{ g water}}$. This ratio contains the unknown. The known ratio must come from the first sentence but there seems to be only one number there. We look back in the text and find that ppm means "parts per million" so we put 1,000,000 in the denominator of the known ratio. Now we can write the known ratio, $\dfrac{34.5 \text{ g Cd}}{1,000,000 \text{ g water}}$, set the two ratios equal, and solve for the unknown. Here we will cross-multiply.

Working Through the Problem

$$\frac{x \text{ g Cd}}{250.0 \text{ g water}} = \frac{34.5 \text{ g Cd}}{1,000,000 \text{ g water}}$$

$$(x \text{ g Cd}) \times (1,000,000 \text{ g water}) = (250.0 \text{ g water}) \times (34.5 \text{ g Cd})$$

$$x \text{ g Cd} = \frac{(250.0 \ \cancel{\text{g water}}) \times (34.5 \text{ g Cd})}{(1,000,000 \ \cancel{\text{g water}})}$$

$$= 0.00863 \text{ g Cd}$$

Answer 0.00863 g Cd

Many problems in chemistry require more than a one-step solution. The density and percent relationships you have studied in this chapter often appear in the same problem. We may sequence two or more proportional reasoning steps to solve the problem.

Example 6.20   **Extended calculations using proportional reasoning**

A solution is 12.5% Na_3PO_4 by mass. How many grams of Na_3PO_4 are there in a 35.0-mL solution of Na_3PO_4 if the density of the solution is 1.10 g/mL?

Thinking Through the Problem This problem links several concepts together. We need to look closely at each detail given to find the key ideas of density and proportionality. Begin with the question "How many grams of Na_3PO_4 are there in a 35.0-mL solution of Na_3PO_4?" We know that g and mL are related by the density and we see that a density value is given. Before we can continue with this problem, we must convert the volume of solution to grams of solution using density.

Step 1. The density given is the density of the solution. We have the volume of the solution. Density will allow us to calculate grams of solution. It is important to notice that all these values refer to the solution.

Step 2. After finding grams of solution we must find grams of Na_3PO_4 in the solution. Percent by mass will let us calculate the mass of Na_3PO_4 in the solution.

Working Through the Problem

Step 1:
$$\frac{1.10 \text{ g}}{1 \text{ mL}} = \frac{x}{35.0 \text{ mL}}$$

$$35.0 \text{ mL} \times \frac{1.10 \text{ g}}{1 \text{ mL}} = x$$

$$x = 38.5 \text{ g of solution}$$

Step 2: Now that we know that we have 38.5 grams of solution, we will use percent to find out the mass of Na_3PO_4 in that amount of solution.

$$\frac{12.5 \text{ g } Na_3PO_4}{100 \text{ g solution}} = \frac{x}{38.5 \text{ g solution}}$$

$$38.5 \text{ g solution} \times \frac{12.5 \text{ g } Na_3PO_4}{100 \text{ g solution}} = x$$

$$x = 4.81 \text{ g } Na_3PO_4$$

Answer There are 4.81 g Na_3PO_4 in 35.0 mL of this solution.

When you learn to recognize proportional relationships, you will be able to write a single calculator setup for an entire problem (discussed in Chapters 7–9). Stringing both steps together, we get:

| Volume → grams solution | | g solution → g Na₃PO₄ |

$$35.0 \text{ mL solution} \times \frac{1.10 \text{ g solution}}{1 \text{ mL solution}} \times \frac{12.5 \text{ g } Na_3PO_4}{100 \text{ g solution}} = x$$

$$x = 4.81 \text{ g } Na_3PO_4$$

PROBLEMS

25. For each of the following ratios, suggest the two extensive variables that are involved in forming the ratio.

(a) miles per gallon
(b) tons of chlorine per ton of NaOH
(c) atoms of C per methane molecule
(d) molecules of ammonia per molecule of hydrogen

26. Concept question: Proportional reasoning is used in many fields, such as population trends. Can you find an example of proportional reasoning that was cited in the news this week?

27. The tension on a spring varies in direct proportion to the distance it is stretched. If the tension is 300 lb when the spring is stretched 9 in, what is the tension when the distance stretched is 1 ft?

28. It is found that 32% of the entering class at a medical school intends to major in family practice. If the entering class has 234 students, how many intend to major in family practice?

29. A chemical company employs 65 technicians, who comprise 12.2% of its workforce. How many people work for the company?

30. It is found that 0.42% of all Chicago residents move each month. If there are 1.8 million people living in Chicago, how many will move each year?

31. An alloy contains 3% copper, 23% silver, 28% mercury, and the rest cobalt. What mass of each metal is present in 0.253 g of the alloy?

32. A sample of drinking water contains 688 ppm lead. How many grams of Pb are in 100.0 grams of this water?

PRACTICAL C   What ratios do we encounter when we talk about chemistry?

Practical A introduced you to the language that chemists use when reporting their research in official journals. Here we introduce some conversational language that you might hear in the lab. In reading it, you will be able to note where important ratios are identified and then used. This is an exercise in recognition.

The first column of the table below contains a quotation that you might hear from a coworker or a supervisor in the laboratory. In the next column, the ratio is identified, in the form of the two measurements that are specified. The third column shows the ratio in mathematical form, both in terms of the reported numbers and on a "per" basis. Finally, a question in the last column prompts us to use the ratio in a calculation.

The first row has been completed for you as an example. See whether you can fill in the remainder of the table. If you have difficulty, review Section 6.3.

Comment	Identifying the Ratio	Writing the Ratio	Using the Ratio
"The solution should be prepared by mixing three parts of acid to eight parts of water."	The ratio involves *parts of acid* and *parts of water*	$\dfrac{3 \text{ parts acid}}{8 \text{ parts water}}$, $\dfrac{0.375 \text{ parts acid}}{1 \text{ part water}}$	What should we do if we have ten parts of water already measured? *We adjust the amount of acid accordingly:* $\dfrac{x \text{ parts acid}}{10 \text{ parts water}} = \dfrac{0.375 \text{ parts acid}}{1 \text{ part water}}$ $x \text{ parts acid} = 3.75 \text{ parts acid}$
"We found that 25 grams of soil contained 2.5 grams of organic matter."	The ratio involves . . .		What is the percentage of the soil that is organic matter?
"The reaction consumed 0.250 moles of compound in 15 minutes."	The ratio involves . . .		How long will it take for 1.00 mole of the compound to be consumed?
"Hydrogen peroxide solution, 30% water by weight, is all the storeroom has."	The ratio involves . . .		What mass of the peroxide solution should we use to get 6.0 grams of hydrogen peroxide?

33. A sample of iron ore contains 45% iron. How many grams of Fe are in 500.0 grams of the ore?

34. The compound $(NH_4)_2CO_3$ contains 8.40% H. How many grams of $(NH_4)_2CO_3$ will you need if you must have 150.0 g H?

35. A solution is 4.82% calcium by mass. How many grams of calcium are in 500.0 mL of this solution if the solution density is 1.25 g/mL?

36. A solution is 23.7% glucose by mass. How many grams of glucose are in 3000.0 mL of this solution if the solution density is 1.14 g/mL?

37. An experiment produces 15.0 grams H_2 for every 71.0 grams O_2 used. How many grams of O_2 will you need to produce 6.0 g H_2?

38. Discussion question: MTBE is the abbreviation for the organic chemical methyl-*tert*-butyl ether. This is a major, and controversial, gasoline additive that promotes cleaner burning of fuel. It is, however, easily dissolved in water and therefore can contaminate water supplies over a wide area.

(a) 20 million metric tons of MTBE were produced in a recent year. This is the equivalent of 20,000,000,000,000 grams. If the U.S. population is currently 265,000,000, how much MTBE is this per person?

(b) MTBE can be injected directly into the gall bladder to dissolve away gallstones, which are mostly pure cholesterol deposits. If MTBE has a density of 740 g per liter, then what is the mass of a 0.125-L injection of MTBE?

(c) A recent EPA document suggests that MTBE should make up 11.0 percent by volume of gasoline. If one gallon of gasoline has a volume of 3.8 L, then what volume, in liters, of MTBE is in one gallon of gasoline?

(d) Assuming that the MTBE in part (c) has a wholesale price of 16¢ a liter, what is the cost of the MTBE in one gallon of gasoline?

(e) In the winter of 1998, the officials from the state of Maine found a town well contaminated with 3,500 parts per billion MTBE, about 100 times the level the state considers unhealthy. Determine the mass of MTBE in one glass (about 250 g) of water.

Discuss which of the proportions you used were defined. Which required an experimental number? Which were stated by a person or agency?

39. In a solution used for disinfecting a large mixing unit at a food factory, we are told to prepare a bleach solution that is 10.0% bleach by volume. We have 25 liters of bleach. How much cleaning solution can we make?

40. Inspectors at a factory check 3.5% of the daily production. If the factory produces 2,500 articles, how many of them were inspected?

41. A photographer wishes to enlarge a print that is 13 cm long and 7 cm wide.

(a) How long will the enlarged print be if its width is 28 cm?

(b) If each of the original dimensions of a rectangular photo is doubled, how is the area of the photograph changed?

42. A pump can raise the water level in a rectangular aquarium 10 inches in one half hour. How long will it take to raise the water level to 4 ft?

43. In wildlife management, ecologists estimated that there was a population of 100,000 alligators in the Everglades National Park. In 1958 a researcher counted and tagged 300 alligators in a small section of the park, and then let them mix with the others. Later, in 1966, the researcher resampled 1,000 alligators, counting and tagging a portion of them. Assume that the ratio of the tagged alligators in the population is the same as the ratio of the tagged alligators in the resample, and estimate the number of alligators that were tagged in the resample.

6.4 Proportional Reasoning and Chemical Substances

SECTION GOALS

✓ Use atom/atom and atom/molecule ratios to solve for unknown numbers of atoms or molecules.

✓ Write mole ratios for atom/atom and atom/molecule relationships.

Proportions in Chemical Substances

CONNECT TO
SECTION 2.2

We can apply what we have learned about proportional reasoning to proportions within chemical substances. This means we must look closely at chemical formulas, as in Chapter 4, and at counting atoms in a molecule, as in Chapter 2. We can then write the atom/molecule ratios for both simple and complicated molecules. For example, we can write the ratio of C atoms to the molecules CO_2, C_2H_6, and $C_6H_{12}O_6$ quite easily. They are, respectively:

$$\frac{1 \text{ atom C}}{1 \text{ molecule } CO_2}, \frac{2 \text{ atoms C}}{1 \text{ molecule } C_2H_6}, \frac{6 \text{ atoms C}}{1 \text{ molecule } C_6H_{12}O_6},$$

The ratio we use depends on which quantities we want to relate. Consider the compound C_7H_8O. We can write several different ratios based on this chemical formula. We can write atom/atom ratios for any two of the three elements in the formula or we can write atom/molecule ratios for any one of the elements in the formula. For example:

Atom/atom ratios for C_7H_8O	$\dfrac{1 \text{ atom O}}{8 \text{ atoms H}}, \dfrac{8 \text{ atoms H}}{7 \text{ atoms C}}$
Atom/molecule ratios for C_7H_8O	$\dfrac{7 \text{ atoms C}}{1 \text{ molecule } C_7H_8O}, \dfrac{8 \text{ atoms H}}{1 \text{ molecule } C_7H_8O}$

Many other ratios can be written as well.

Example 6.21   Writing atom/atom and atom/molecule ratios

Write three different atom/atom ratios and two different atom/molecule ratios for the compound methyl red, $C_{15}H_{15}N_3O_2$.

Thinking Through the Problem We can choose any pairs of elements to write atom/atom ratios and then choose any two elements to make the atom/molecule ratios. It doesn't matter which atom we put in the numerator and which we put in the denominator of an atom/atom ratio. But we should put "molecules of $C_{15}H_{15}N_3O_2$" in the denominator of the atom/molecule ratio, because it is easier to think about atoms per molecule than about molecules per atom.

Answer   Atom/atom ratios: $\dfrac{15 \text{ atoms C}}{15 \text{ atoms H}}, \dfrac{15 \text{ atoms C}}{3 \text{ atoms N}}, \dfrac{3 \text{ atoms N}}{15 \text{ atoms H}}$

Atom/molecule ratios: $\dfrac{15 \text{ atoms C}}{1 \text{ molecule } C_{15}H_{15}N_3O_2},$

$\dfrac{2 \text{ atoms O}}{1 \text{ molecule } C_{15}H_{15}N_3O_2}$

Other ratios are also valid answers to this problem.

1. Write four atom/molecule ratios for the dye indigo, $C_{10}H_{12}N_2O_2$.

2. Write at least four different atom/atom ratios for indigo.

Using Atom/Atom and Atom/Molecule Ratios

CONNECT TO
SECTION 4.1

We saw in Chapter 4 that the ratios of the atoms in compounds are the same whether we have one, ten, or a billion molecules of a compound. Now we will use these ratios in calculations like any other constants of proportionality. This means that we can use the chemical formula to determine how many atoms of an element are in any sample of a compound. The ratio for the carbon atoms in glucose, $C_6H_{12}O_6$, is:

$$\frac{6 \text{ C atoms}}{1 \text{ } C_6H_{12}O_6 \text{ molecule}}$$

This 6:1 ratio is the same for *any* sample of glucose we consider. If we know the number of glucose molecules, we can determine the number of C atoms by setting up a proportion and solving for C atoms. Suppose we want to know the number of C atoms in 132 molecules of $C_6H_{12}O_6$.

✔ Meeting the Goals

Atom/atom or atom/molecule ratios are *known* whenever we have the formula for a substance. Once we know the atom/atom ratio for a substance, we can calculate any number of one of the atoms, given a number of the other atom. The unknown ratio contains an *x* (or some other variable) for the quantity we must calculate. The two ratios are set equal to one another and solved for the unknown.

$$\text{unknown ratio} = \text{known ratio}$$

$$\frac{x \text{ C atoms}}{132 \text{ } C_6H_{12}O_6 \text{ molecules}} = \frac{6 \text{ C atoms}}{1 \text{ } C_6H_{12}O_6 \text{ molecule}}$$

$$\cancel{132 \text{ } C_6H_{12}O_6 \text{ molecules}} \times \frac{x \text{ C atoms}}{\cancel{132 \text{ } C_6H_{12}O_6 \text{ molecules}}} =$$

$$\frac{6 \text{ C atoms}}{1 \cancel{C_6H_{12}O_6 \text{ molecules}}} \times 132 \cancel{C_6H_{12}O_6 \text{ molecules}}$$

$$x \text{ C atoms} = \frac{6 \times 132}{1} \text{ C atoms}$$

$$x \text{ C atoms} = 792 \text{ C atoms}$$

Of course, this is the same as counting 6 C atoms in one molecule and then multiplying by 132 molecules. For short problems such as this one, it is usually sufficient to multiply 6 C atoms/molecule by 132 molecules. The use of proportions becomes more critical when we get to extended calculations in which we must use several relationships to reach a final answer. Your ability to recognize and use the relationships developed in this chapter will help you solve the more complicated problems in the next several chapters.

Example 6.22

Using proportions to relate atoms and molecules

Sulfur tetrafluoride has the formula SF_4. How many F atoms are in a sample of 123,000,000 molecules of SF_4?

Thinking Through the Problem We count 4 F atoms in one molecule SF_4. We know there will be 123,000,000 $\times$ 4 F atoms in the sample. This is a straightforward multiplication problem you should have little difficulty with. To illustrate how proportion is used to solve such a problem, we will use a proportional method. Then we can use the proportional method in other problems that are less clear to us.

Working Through the Problem

$$\text{unknown ratio} = \text{known ratio}$$

$$\frac{x \text{ F atoms}}{123{,}000{,}000 \text{ } SF_4 \text{ molecules}} = \frac{4 \text{ F atoms}}{1 \text{ } SF_4 \text{ molecule}}$$

$$\cancel{123{,}000{,}000 \text{ } SF_4 \text{ molecules}} \times \frac{x \text{ F atoms}}{\cancel{123{,}000{,}000 \text{ } SF_4 \text{ molecules}}} =$$

$$\frac{4 \text{ F atoms}}{1 \text{ } \cancel{SF_4 \text{ molecule}}} \times 123{,}000{,}000 \text{ } \cancel{SF_4 \text{ molecules}}$$

$$x \text{ F atoms} = \frac{4 \times 123{,}000{,}000}{1} \text{ F atoms}$$

$$x \text{ F atoms} = 496{,}000{,}000 \text{ F atoms}$$

Answer 496,000,000 F atoms

Example 6.23 **Using proportions to relate atoms and molecules**

How many molecules of SO_3. are needed to supply 24 atoms of oxygen?

Thinking Through the Problem In this problem, we are given atoms of oxygen, so it is the opposite of the previous problem. Here it may be easier to think about the solution in terms of proportion. The known ratio will have molecules of SO_3 and atoms of O. We are looking for molecules of SO_3, so we write a ratio with molecules of SO_3 in the numerator and atoms of oxygen in the denominator.

Working Through the Problem

$$\text{unknown ratio} = \text{known ratio}$$

$$\frac{x \text{ } SO_3 \text{ molecules}}{24 \text{ atoms O}} = \frac{1 \text{ } SO_3 \text{ molecule}}{3 \text{ atoms O}}$$

$$\cancel{24 \text{ atoms O}} \times \frac{x \text{ } SO_3 \text{ molecules}}{\cancel{24 \text{ atoms O}}} = \frac{1 \text{ } SO_3 \text{ molecule}}{3 \text{ } \cancel{\text{atoms O}}} \times 24 \text{ } \cancel{\text{atoms O}}$$

$$x \text{ } SO_3 \text{ molecules} = \frac{1 \times 24}{3} \text{ } SO_3 \text{ molecules}$$

$$x \text{ } SO_3 \text{ molecules} = 8 \text{ } SO_3 \text{ molecules}$$

Answer 8 SO_3 molecules

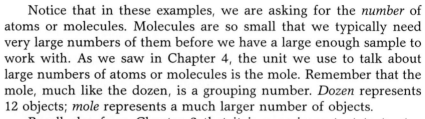

How are you doing? 6.13

(a) Determine the number of O atoms in a sample containing 45 molecules of carbon dioxide, CO_2.

(b) Determine the number of molecules of boron trifluoride needed to supply 457,500 atoms of F.

(c) How many atoms of carbon are there in 16 molecules of C_3H_7COOH?

CONNECT TO
SECTION 4.1

✔ Meeting the Goals

The ratio of moles to moles is numerically identical to the simple atom/atom or atom/ molecule ratio.

Notice that in these examples, we are asking for the *number* of atoms or molecules. Molecules are so small that we typically need very large numbers of them before we have a large enough sample to work with. As we saw in Chapter 4, the unit we use to talk about large numbers of atoms or molecules is the mole. Remember that the mole, much like the dozen, is a grouping number. *Dozen* represents 12 objects; *mole* represents a much larger number of objects.

Recall also from Chapter 2 that it is very important to treat a chemical formula as a whole entity. It is not OK to say "half of a molecule of C_2H_6" because that implies we can physically get one-half of a molecule of C_2H_6. But it is OK to say "half a dozen" or "half a mole" because these are equal to a whole number of molecules of C_2H_6.

Because *dozen* and *mole* both represent numbers, a dozen/dozen or mole/mole ratio of atoms is numerically the same as a simple atom/atom ratio.

For example, we know that the reduced C to H ratio for C_2H_6 is $\dfrac{1 \text{ C atom}}{3 \text{ H atoms}}$. We will have this same 1:3 ratio with group counting numbers (dozen and mole):

$$\frac{1 \text{ C atom}}{3 \text{ H atoms}} = \frac{1 \text{ dozen C atoms}}{3 \text{ dozen H atoms}} = \frac{1 \text{ mol C atoms}}{3 \text{ mol H atoms}}$$

$$\frac{1 \text{ C atom}}{3 \text{ H atoms}} = \frac{(1 \times \cancel{12}) \text{ C atoms}}{(3 \times \cancel{12}) \text{ H atoms}} =$$

$$\frac{(1 \times \cancel{602,213,700,000,000,000,000,000}) \text{ C atoms}}{(3 \times \cancel{602,213,700,000,000,000,000,000}) \text{ H atoms}} = \frac{1 \text{ C}}{3 \text{ H}}$$

As numbers, both *dozen* and *mole* divide out in these ratios, leaving the initial 1:3 ratio between C and H atoms.

The following two examples use the quantity *dozen* to count atoms and molecules.

Example 6.24   Using proportions with dozen as a grouping number

Determine how many dozen H atoms are in a sample containing 25 dozen C_2H_6 molecules.

Thinking Through the Problem The ratio we need relates H atoms to C_2H_6 molecules. We know how to write this ratio. Dozen is just a multiplier (by twelve) of the basic relationship of 6 to 1. To solve the problem, then, we will write the known and unknown ratios and use proportional reasoning.

Working Through the Problem

$$\frac{x \text{ dozen H atoms}}{25 \text{ dozen } C_2H_6 \text{ molecules}} = \frac{6 \text{ dozen H atoms}}{1 \text{ dozen } C_2H_6 \text{ molecules}}$$

$$x \text{ dozen H atoms} =$$

$$\frac{6 \text{ dozen H atoms}}{1 \text{ dozen } \cancel{C_2H_6 \text{ molecules}}} \times 25 \text{ dozen } \cancel{C_2H_6 \text{ molecules}}$$

$$= 150 \text{ dozen H atoms}$$

Answer 150 dozen H atoms

Example 6.25 Using proportions with dozen as a grouping number

Determine how many dozen H atoms are in a sample of C_2H_6 that contains 11 dozen C atoms.

Thinking Through the Problem This example is similar to the last one but it requires a different viewpoint that may not be as simple to "see." It requires an atom/atom ratio between atoms of C and of H in one molecule of C_2H_6. The basic relationship is between C and H. There are 2 C atoms for every 6 H atoms. *Dozen* is a constant multiplier that can be divided out in the known ratio, which is what we did in the last example. We can also leave *dozen* in the known ratio. It still divides out during the calculation but it may be easier to set up the two corresponding ratios with the word *dozen* left in. We will leave *dozen* in the known ratio for this example.

Working Through the Problem

$$\frac{x \text{ dozen H atoms}}{11 \text{ dozen C atoms}} = \frac{6 \text{ dozen H atoms}}{2 \text{ dozen C atoms}}$$

$$11 \text{ dozen } \cancel{C \text{ atoms}} \times \frac{x \text{ dozen H atoms}}{11 \text{ dozen } \cancel{C \text{ atoms}}} =$$

$$\frac{6 \text{ dozen H atoms}}{2 \text{ dozen } \cancel{C \text{ atoms}}} \times 11 \text{ dozen } \cancel{C \text{ atoms}}$$

$$x \text{ dozen H atoms} = \frac{(6 \times 11 \text{ dozen}) \text{ H atoms}}{2}$$

$$x \text{ dozen H atoms} = 33 \text{ dozen H atoms}$$

Answer 33 dozen H atoms

We can apply the same reasoning process that we used with *dozen* to problems that use the unit *mole*. Using proportional reasoning to relate different mole amounts of atoms, molecules, and formula units is so important that the ratios we use have the "shorthand" term *mole ratio* to indicate a mole/mole ratio.

First, consider some of the mole ratios we can write for C_7H_8O:

Atom/atom ratios for C_7H_8O	$\dfrac{1 \text{ mole O}}{8 \text{ moles H}}, \dfrac{8 \text{ moles H}}{7 \text{ moles C}}$
Atom/molecule ratios for C_7H_8O	$\dfrac{7 \text{ moles C}}{1 \text{ mole } C_7H_8O}, \dfrac{1 \text{ mole } C_7H_8O}{8 \text{ moles H}}$

Example 6.26 **Writing mole/mole ratios**

Write at least four mole ratios for the molecule nicotine, $C_{10}H_{14}N_2$. Include ratios that relate moles of atoms to one another and that relate moles of atoms to moles of the molecule.

Thinking Through the Problem We start with the chemical formula. The ratios of atoms can have any element in the numerator or the denominator. But as before, we chose to write the ratio with molecules in the denominator.

Answer $\dfrac{10 \text{ moles C}}{14 \text{ moles H}}, \dfrac{10 \text{ moles C}}{2 \text{ moles N}}, \dfrac{2 \text{ moles N}}{14 \text{ moles H}}, \dfrac{10 \text{ moles C}}{1 \text{ mole } C_{10}H_{14}N_2},$

$\dfrac{2 \text{ moles N}}{1 \text{ mole } C_{10}H_{14}N_2}$

How are you doing? 6.14

Write at least three mole ratios for the molecule DDT, $C_{13}H_5Cl_5$.

Of course, one convenient aspect of the mole is that we may correctly talk about fractional amounts, as the next example shows.

Example 6.27 **Solving mole/mole ratio problems**

Determine the number of moles of C_5H_5N that contains 0.0280 moles of C.

Thinking Through the Problem We can find the ratio of C to C_5H_5N by counting the C's in the formula. This gives us a known ratio of $\dfrac{5 \text{ moles C}}{1 \text{ mole } C_5H_5N}$. We then use the number of moles of C, given in the problem, to set up the unknown ratio. The number of moles of C_5H_5N is what we are solving for; so we write the known ratio with "moles of C_5H_5N" in the numerator. Note that we have written the known ratio in order to accommodate the correct substance in the numerator of the unknown ratio.

Working Through the Problem

$$\text{unknown ratio} = \text{known ratio}$$

$$\frac{x \text{ moles } C_5H_5N}{0.0280 \text{ mole C}} = \frac{1 \text{ mole } C_5H_5N}{5 \text{ moles C}}$$

$$\cancel{0.0280 \text{ mole C}} \times \frac{x \text{ moles } C_5H_5N}{\cancel{0.0280 \text{ mole C}}} =$$

$$\frac{1 \text{ mole } C_5H_5N}{5 \cancel{\text{ moles C}}} \times 0.0280 \cancel{\text{ mole C}}$$

$$x \text{ moles } C_5H_5N = \frac{0.0280}{5} \text{ mole } C_5H_5N$$

$$x \text{ moles } C_5H_5N = 0.00560 \text{ mole } C_5H_5N$$

Answer 0.00560 mole C_5H_5N

How are you doing? **6.15**

Determine the number of moles of glucose, $C_6H_{12}O_6$ that contain 0.52 mole of H atoms. How many moles of H atoms are in 3.25 moles of NH_3?

Proportional Reasoning and Atoms in Extended Structures

CONNECT TO
SECTION 3.1

Our examples so far have involved only molecular substances. But proportional reasoning also applies to nonmolecular substances. Recall from Chapter 3 that sodium chloride consists of a huge number of NaCl *units* connected in an extended structure. To refer to one NaCl unit we use the term *formula unit* instead of *molecule*, because we do not want to suggest that NaCl molecules are present in solid sodium chloride. The same reasoning is used for other extended substances, including pure elements.

Example 6.28 **Using proportion to relate atoms and formula units**

How many atoms of Na are in 250 formula units of Na_2SO_4?

Thinking Through the Problem We are solving for the number of Na atoms per formula unit, so we will start with an atom/formula unit ratio that has "atoms Na" in the numerator and "formula units of Na_2SO_4" in the denominator. Once we have the correct starting ratios, we are ready to follow the proportional reasoning route to the answer.

Working Through the Problem

$$\text{unknown ratio} = \text{known ratio}$$

$$\frac{x \text{ atoms Na}}{250 \text{ formula units Na}_2\text{SO}_4} = \frac{2 \text{ atoms Na}}{1 \text{ formula unit Na}_2\text{SO}_4}$$

$$\cancel{250 \text{ formula units Na}_2\text{SO}_4} \times \frac{x \text{ atoms Na}}{\cancel{250 \text{ formula units Na}_2\text{SO}_4}} =$$

$$\frac{2 \text{ atoms Na}}{1 \cancel{\text{ formula unit Na}_2\text{SO}_4}} \times 250 \cancel{\text{ formula units Na}_2\text{SO}_4}$$

$$x \text{ atoms Na} = \frac{2 \times 250}{1} \text{ atoms Na}$$

$$x \text{ atoms Na} = 500 \text{ atoms Na}$$

Answer 500 atoms Na

Example 6.29 Solving mole/mole problems

How many moles of Na atoms are in 2.0 moles of Na_2SO_4?

Thinking Through the Problem We proceed as before using a mole ratio that relates moles Na atoms to moles Na_2SO_4.

Working Through the Problem

$$\text{unknown ratio} = \text{known ratio}$$

$$\frac{x \text{ moles Na}}{2.0 \text{ moles Na}_2\text{SO}_4} = \frac{2 \text{ moles Na}}{1 \text{ mole Na}_2\text{SO}_4}$$

$$\cancel{2.0 \text{ moles Na}_2\text{SO}_4} \times \frac{x \text{ moles Na}}{\cancel{2.0 \text{ moles Na}_2\text{SO}_4}} =$$

$$\frac{2 \text{ moles Na}}{1 \cancel{\text{ mole Na}_2\text{SO}_4}} \times 2.0 \cancel{\text{ moles Na}_2\text{SO}_4}$$

$$x \text{ moles Na} = 2 \times 2.0 \text{ moles Na}$$

$$x \text{ moles Na} = 4.0 \text{ moles Na}$$

Answer 4.0 moles Na

How are you doing? 6.16

Determine the number of moles of Cl atoms in 5.21 moles of the compound $AlCl_3$.

PROBLEMS

44. Review molar ratios in formulas with the compound ammonium oxalate monohydrate, $(NH_4)_2C_2O_4 \cdot H_2O$.

Complete the following table of ratios by determining x for each. The first is completed for you.

$\dfrac{2 \text{ moles C atoms}}{1 \text{ mole } (NH_4)_2C_2O_4 \cdot H_2O}$	$\dfrac{x \text{ moles H atoms}}{1 \text{ mole } (NH_4)_2C_2O_4 \cdot H_2O}$	$\dfrac{x \text{ moles N atoms}}{1 \text{ mole } (NH_4)_2C_2O_4 \cdot H_2O}$
$\dfrac{x \text{ moles O atoms}}{1 \text{ mole } (NH_4)_2C_2O_4 \cdot H_2O}$	$\dfrac{x \text{ moles N atoms}}{5 \text{ moles O atoms}}$	$\dfrac{x \text{ moles H atoms}}{2 \text{ moles C atoms}}$

45. How many atoms of oxygen are there in 275 formula units of $Al_2(CrO_4)_3$?

46. **Discussion question:** Write the mole/mole ratio for H in each of the following substances.

   (a) $Fe(C_9H_6NO)_3$
   (b) $Na_2C_{10}H_{14}N_2O_8 \cdot 2 H_2O$
   (c) $Pb(C_2H_3O_2)_4$
   (d) $Fe(NH_4)_2 (SO_4)_2 \cdot 4 H_2O$

Discuss how parentheses and the raised dot $\cdot$ are interpreted to get your answers.

47. How many moles of Cl atoms are there in 0.20 mole of Cl_2 gas?

48. A sample of H_2S is found to contain 22,000 atoms of S. How many atoms of H are present?

49. How many moles of S are contained in 1,432 moles of $Na_2S_2O_4$?

50. How many moles of hydrogen gas contain 0.20 mole of hydrogen atoms?

51. A sample of $TiCl_4$ is found to contain 22,000 atoms of Ti. How many atoms of Cl are present?

52. How many moles of H atoms are contained in 2.0 moles of hydrogen gas?

53. Determine the number of moles of S in 12,300 moles of Al_2S_3.

PRACTICAL D   **How does a chemical experiment give us the data for a chemical formula?**

Experiments that lead to the chemical formula of a substance are very important in chemistry. Such experiments provide information about the number of moles of different elements in a substance. In the next several chapters we will describe measurements of volume, mass, and pressure that involve numbers of moles. Here we will look at some experiments that can be interpreted in the form of a mole ratio:

(a) A sample of a compound of barium and oxygen is found to have 0.137 mole of barium and 0.274 mole of oxygen. What is the ratio of moles of oxygen to moles of barium?

(b) When 0.0215 mole of a gaseous compound of nitrogen and hydrogen are decomposed, 0.0215 mole of nitrogen and 0.0430 mole of hydrogen are formed. Use the ratio of N to H to find the formula of the compound.

(c) A sample of 0.00112 mole of a compound of sodium and sulfate contains 0.00112 mole of sodium, 0.00112 mole of sulfate ions, and 0.00112 mole of hydrogen. What is the formula of this ionic compound?

(d) We find that a 25.0 moles of an oil with the formula $C_{18}H_{32}O_2$ reacts with 75.0 moles of hydrogen gas. A waxy substance with 18 carbon atoms forms. What is the formula of the new compound?

(e) A solution containing 0.00542 mole of a palladium compound reacts with 0.00271 mole of carbon monoxide to make a new compound. What is the mole ratio of CO to the palladium compound?

Chapter 6 Summary and Problems

VOCABULARY

quantitative property	A property that has a numerical measurement.
extensive property	A property that depends on the size of the sample; the measured value of an extensive property will change as the size of the sample changes.
intensive property	A property that is intrinsic to a sample of material; the measured value of an intensive property does *not* depend on the size of the sample.
metric system	The numerical system of measurement used by scientists; it is based on powers of ten.
mass	The measure of the amount of matter present.
weight	A measure of the amount of matter present in a sample as it is acted upon by gravity.
temperature	A measure of the hotness or coldness of a substance compared to some temperature scale.
melting/freezing temperature	The temperature at which a substance melts/freezes, i.e., converts between the solid and liquid phases.
boiling temperature	The temperature at which a substance boils/condenses, i.e., converts between the liquid and gaseous phases.
vapor	The gas formed when a liquid evaporates.
normal boiling point	The boiling point temperature measured at sea level.
physical constant	A measurement of a physical property of a substance that does not change when measured at the same conditions of temperature and pressure.
density	A physical property determined by dividing the mass of a substance by its volume; density is an intensive property that can be used (with other criteria) to identify a substance.
proportionality	A proportionality occurs when we form the ratio of two extensive variables.
constant of proportionality	The k in the equation $y = kx$, where the variable y is directly proportional to the variable x.
proportional reasoning	A problem-solving technique based on the use of proportionality. One ratio in the proportion gives some known relationship; the second ratio contains the same quantities in the numerator and denominator as does the known ratio but in this case, one of the values is not known.

SECTION GOALS

What is the difference between an intensive and an extensive property?

Extensive properties depend on the size of the sample; intensive properties do not. These properties are used in the laboratory to characterize how substances exist or act; they are also used to make predictions.

What common metric units are used to measure length, mass, and volume?

Common scientific units that measure length, mass, and volume are centimeter, gram, and cubic centimeter, respectively.

How does temperature influence the physical state of a pure substance?

The general trend is for substances to be solids at lower temperature, liquids at intermediate temperatures, and gases at higher temperature. However, not all substances have a liquid phase under normal conditions.

Some substances such as dry ice (solid CO_2) sublime, changing from a solid directly into a gas.

What is the relationship between freezing and boiling temperatures and the physical state of matter for a substance?

The freezing temperature marks the boundary between solid and liquid for a pure substance under ordinary pressure conditions. Substances will be solids at temperatures below the freezing temperature, liquids at all temperatures between the freezing and the boiling temperatures, and gases at temperatures above the boiling temperature.

Can you solve a formula for a specified variable? In particular, can you solve the density formula for m and for V?

Mathematical equations are solved for a given variable by "undoing" the last operation on that variable until the variable is isolated. The density formula is $d = \dfrac{m}{V}$. Solving this for m gives $m = dV$; solving for V gives $V = \dfrac{m}{d}$.

Can you explain why density is an intensive property?

Density is an intensive property because we get the same density value for any sample size of a pure substance. Note that density does what we expect a ratio to do: it expresses a property of a substance that does not change from sample to sample when the temperature and pressure are held constant.

What two properties are measured in a density experiment? What kind of units do we expect for a density value?

To find an experimental value of density, we measure the mass and the volume of the same sample. Density is the ratio of mass to volume. We expect units that are a ratio of mass units to volume units. Typical units for density are grams/cm^3; other mass and volume units will be introduced in the next chapter.

What is a constant of proportionality? What is the constant of proportionality in the equation $d = \dfrac{m}{V}$?

Whenever two quantities are directly proportional to one another, their ratio gives a constant value called a constant of proportionality. For the density relationship, mass and volume are directly proportional to one another. The ratio of mass to volume for the same sample of a pure substance gives a constant value called the density. Density is the constant of proportionality between mass and volume.

Can you write a proportionality using a known and unknown ratio that is given in a problem?

All these problems contain a sentence that relates "so many of this" to "so many of that." The ratio is written with one of these quantities in the numerator and the other in the denominator. Both the number and the identity should be included in the ratio. It does not matter which quantity goes into which position (numerator or denominator) for the known ratio. The quantities in the unknown ratio, however, must be in the same positions as those in the known ratio.

How do you write and use ratios that involve percentage, parts per million, or parts per billion?

The word *per* alone means "per one." "Percent" means "per 100," or $\dfrac{x \text{ parts}}{100 \text{ parts}}$. "Per million" means "per 1,000,000," or $\dfrac{x \text{ parts}}{1,000,000 \text{ parts}}$. "Per billion" means "per 1,000,000,000," or $\dfrac{x \text{ parts}}{1,000,000,000 \text{ parts}}$. The word *parts* usually stands for some unit, such as gram or liter, that is specified in the problem.

Can you use atom/atom and atom/molecule ratios to solve for unknown numbers of atoms or molecules?

Atom/atom or atom/molecule ratios are *known* whenever we have the molecular formula for that substance. Once the known ratio is formed, the unknown ratio is easily written. Both ratios have the same quantity in the numerator and in the denominator. The unknown ratio contains an x (or some other variable) for the quantity we must calculate. The two ratios are set equal to one another and solved for the unknown.

Can you write mole ratios for atom/atom and atom/molecule relationships?

A mole ratio (i.e., a mole/mole ratio) is a fraction that relates moles of atoms X (in the numerator) to moles of atoms Y (in the denominator). Mole ratios for atoms have the same value as the simple atom/atom ratios because *mole* is a constant numerical value that can be divided out. Thus, mole ratios are written exactly the same as the simple atom/atom and atom/molecule ratios, but include the number represented by *mole*.

$$\frac{1 \text{ C atom}}{3 \text{ H atoms}} = \frac{(1 \times 602{,}213{,}700{,}000{,}000{,}000{,}000{,}000) \text{ C atoms}}{(3 \times 602{,}213{,}700{,}000{,}000{,}000{,}000{,}000) \text{ H atoms}} = \frac{1 \text{ mole C atoms}}{3 \text{ moles H atoms}}$$

PROBLEMS

54. An experiment is conducted to examine the reaction of copper metal with acid. When different amounts of copper metal react, the amount of hydrogen changes, too, as shown in the chart below.

Mass of Copper (g)	Volume of Hydrogen (cm^3)
1.59	0.671
3.50	1.48
5.59	2.36

(a) Determine the volume of hydrogen you expect from the reaction of 2.0 g of copper.
(b) If we want 1,000. cm^3 of hydrogen, what mass of copper should we react?

55. An alloy contains 6% copper, 27% silver, 26% mercury, and the rest is cobalt. What mass of each metal is present in 0.376 g of the alloy?

56. What is the volume of an 8.2-gram sample of a pure substance if a 20.8-gram sample of the substance has a measured volume of 26.2 cm^3?

57. Silver has a density of 8,920 g per liter. What mass of silver has volume of 0.020 L?

58. Find the mass of a perfect cube of gold ($d = 19.320$ g/cm^3) given that each edge of the cube measures 1.5 centimeters.

59. Disilane, Si_2H_6, is a gas with a density of 0.00278 g/cm^3. Determine the mass of 95.0 cm^3 of disilane.

60. Determine the mass of mercury in a typical mercury thermometer. The volume of this mercury is about 2.5 cm^3 ($d = 13.546$ g/cm^3).

61. A sample of a liquid has a mass of 0.532 g and a volume of 0.653 cm^3. What is the density of the liquid?

62. The initial volume of 24.5 cm^3 of water has a mass of 23.954 g. When a metal is added, the total volume increases to 39.8 cm^3 and the mass increases to 108.435 g. What is the density of the metal?

63. (a) What is the volume of a 17.2-gram sample of a pure substance if 21.4 grams of the substance has a measured volume of 27.6 cm^3?
(b) Using the information in part (a) identify the unknown substance from the following list.

Substance	Density, g/cm^3
A	0.642
B	0.780
C	0.841
D	0.1242

(c) Suppose that substance C in part (b) is powdered and then sprinkled on the liquids shown in the following table. Decide whether substance C will float or sink in these liquids. Record your choice by putting an X in the appropriate column.

Substance	Density, g/cm^3	Float?	Sink?
Water	1.001		
Ethyl alcohol	0.791		
Benzene	0.877		

64. How many atoms of Na are in 345 formula units of Na_3PO_4?

65. A solution is 34.6% NaCl by mass. How many grams of NaCl are in 5000.0 cm^3 of this solution if the solution density is 1.22 g/cm^3?

66. A copper ore contains 14.8% Cu by mass. How many grams of copper are in 50.0 g of this ore?

67. A solution contains 8.6% sodium, 10.5% glucose, and 4.8% potassium by mass. How many grams of each are in 125.0 g of this solution?

68. A polluted stream contains 6.9 parts per billion of a particular bacteria. How many grams of this bacteria are in 1 quart of this water? (1 quart is 946.3 cm^3) The density of the river water is 1.050 g/cm^3.

69. A 50.0-g sample of shampoo contains 0.000068 gram of coloring agent. How many ppm by mass of coloring agent is in this shampoo?

70. A 2.00-liter volume of air contains 0.0050% CO. How many parts per million CO is this?

71. A sample contains 6 parts per million Pb. How many grams of Pb will there be in 4000.0 grams of this sample?

72. A solution of bleach contains 5.0% NaClO by mass. How many grams of NaClO will there be in a 20,000-cm^3 container of this solution if the density of the solution is 1.080 g/cm^3?

73. You are given a pure sample of a metal, and the following data is measured. What is the density of this metal sample?

Temperature	26.7 °C
Mass of empty cylinder	53.23 g
Initial volume of water added to cylinder	25.05 cm^3
Mass of cylinder + water	78.30 g
Volume of water after adding metal	28.30 cm^3
Mass of cylinder + water + metal	83.18 g

Counting and Measurement in Chemical Experiments

The number of grains of sand on a beach is too great to count, but we can make measurements that lead to the answer.
(Erv Schowengerdt.)

PRACTICAL CHEMISTRY The Measure and Pleasure of Sand

What can you say about the sand on your favorite beach? Most people think of a light brown, grainy material when we ask this question. But this is a physical description of only one kind of sand. There is also black sand, white sand, red sand, and green sand. Grains of sand can range from very coarse to extremely fine.

Geologically speaking, sand consists of small particles of rocks and minerals that have been broken down by erosion. At the beach or in riverbeds, the different constituents become separated by the action of water currents, tides, or waves. Less dense particles are carried away, while denser particles are left behind. Air currents move sand particles as well—as in arid climates such as deserts, where winds move the sand. The lighter the particles, the farther they travel.

Qualitative descriptions like these are important in chemistry, but it is also important to be able to answer *quantitative* questions—questions that ask "how much?" When it comes to sand, however (and other materials, for that matter), it is almost impossible to say "how much" in an accurate way. So we use measurements—and measurements have some uncertainty associated with them. Being able to make quantitative statements and understanding issues of uncertainty in measurement are important parts of experimental science. So is understanding how different scales of measurement are used in different situations.

In this chapter, we will discuss the use of mathematical exponents in scientific measurements and in counting. You will see that, by using scientific measurement and notation, we can characterize samples of sand ranging in quantity from a mere cupful to the entire Sahara desert.

This chapter will focus on the following questions:

PRACTICAL **A** How is sand measured?

PRACTICAL **B** How much sand must be purchased to build a patio?

PRACTICAL **C** How many moles of sand are on your favorite beach?

7.1 Recording Measurements

SECTION GOALS

✔ Understand the difference between accuracy and precision.

✔ Understand why the last digit in a measured number is uncertain.

✔ Understand the concept of significant figures and why we count them in a measured number.

✔ Know when zeroes are counted as significant figures in a measured number.

✔ Know how many significant figures to include in a calculated answer when measured numbers are used.

Measured Numbers: How Good Are They?

Chemistry is an experimental science. This means that much of what we know in chemistry is derived from carefully performed experiments. In the first part of this book, we used the results of experiments to describe many different chemical systems, from molecules to reactions. In Chapter 6, we discussed the kind of measurements that chemists and other scientists use in studying nature. A measurement, you will recall, includes both a numerical value (which we usually just call a "number") and a unit.

The precision of a measurement depends on the quality of the measuring tool—in this case, a tape measure. (Corbis.)

It is easy to think that all numbers we encounter have the same quality, or reliability. If we run a business, for example, we expect that an accountant can compute a financial statement "to the penny." But such accounting cannot be done when we use measured amounts. Scientists distinguish between two kinds of numbers: exact numbers and estimated numbers. We get an **exact number** when *we count* every member of a set (as in "there are four tires on a standard automobile") or when a unit is *defined exactly* (such as 1 dozen is 12 objects). In many other cases we do not have the ability to get an exact number. When we *measure* something we get an **estimated number.** This does not mean that we are making a wild guess about the value; we are merely acknowledging that measurements typically contain some uncertainty.

We judge estimated values using two criteria: accuracy and precision. **Accuracy** indicates how closely the experimentally measured value agrees with some known value. For example, if you know that your weight is 125 pounds, then a scale that gives this weight gives an accurate value. Accuracy is one criterion used to judge the usefulness of a measurement.

Precision is the second criterion used to judge measurements. Precision indicates how many digits can be reported about an experimentally measured value. We determine the precision of a measurement by repeating it and finding out whether or not the newly measured values agree with the first value. For example, suppose that you weigh yourself five times in a row. Figure 7-1 shows two different sets of measurements. Values of 125.2, 125.0, 124.8, 125.3, and 125.4 pounds agree with each other more closely, and are therefore more precise, than val-

Figure 7-1 ▶ Accurate measurements are close in value to the true value. Precise measurements are close in value to each other.

ues of 124.0, 128.5, 126.4, 123.6, and 122.0 pounds. In the first case, the values range from 124.8 to 125.4, which is a narrow range. In the second case, the values range from 122.0 to 128.5, a much broader range that is less precise than the first group of measurements.

Measurements can be very precise (great lab technique!) but not very accurate (the instrument gives consistently incorrect readings). Precise measurements agree closely with one another but lie far from the accepted value for that measurement (accuracy). Ideally, we want values that are both accurate and precise.

Although we want to obtain **measured numbers** with the highest accuracy and precision possible, *there is always error present.* We can distinguish between two types of error: random error and systematic error. **Random error** is due to normal fluctuations that occur when we repeat the same task a number of times. It is difficult to perform every single step of an experiment *exactly* the same way, and instruments don't always give identical results from trial to trial. Random error can be minimized by improving one's lab technique, but it is not possible to completely eliminate random error. Random error generally affects the precision of the experimental results.

Systematic error is due to some error in the experimental setup. We might have chemicals that are impure, lab equipment that doesn't work correctly, or a technique that involves a repeated error. Every measurement made under these circumstances will be inaccurate in a consistent way. We can have high precision in this case but be wildly "off" from the correct numerical value.

Reporting and Writing Measured Numbers

How do we report the precision of experimental results? The precision of the instruments we use in the experiment determines the precision of our measurements. *Whenever we report measured numbers we should include one estimated digit—no more and no fewer.* But how do we know when to estimate a number? The answer is to use a good measuring device with well-marked divisions. Then we estimate the number that lies in between these divisions.

First, we determine what value is associated with the smallest markings on the measuring device. For example, does each line stand for ten units? For one unit? For one-half unit? With proper laboratory equipment we can be certain that all the digits represented by a line or mark are reliable.

Second, we estimate the value for one additional digit in the next smallest decimal place value *not* indicated by a marking on the device. An estimated digit that is not represented by an actual line on the measuring device is called "uncertain." Measured numbers contain all of the "certain" digits plus one "uncertain," estimated digit.

Consider the graduated cylinders shown in Figure 7-2. (The illustration shows only part of each cylinder: between 40 and 60 mL on the left and between 40 and 55 mL on the right.) The cylinder on the left has lines or markings for every 10 milliliters. The cylinder on the right has smaller divisions—one line for every 1 mL. The arrows point to the levels at the top of the liquid we are measuring. What volume will we record in the first case?

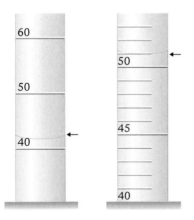

Figure 7-2 ▲ The instrument used affects the precision of the measurement.

Because the left cylinder is marked off in ten-unit divisions, we can read all the values in the tens position, and expect to *estimate* in the ones position. The liquid in this cylinder has a volume that falls between 40 and 50 mL. A good estimated report of this first volume measurement is 42 mL. Reporting 42 mL is more precise than reporting 40 mL in this case. It is also better to report 42 mL than to report 42.5 mL, which contains *two* uncertain, estimated digits. Remember that when we record measured numbers we should include no more and no fewer than *one* estimated digit. Recording anything other than 42 mL in this example decreases the quality of the reported value.

Since the cylinder on the right in Figure 7-2 has markings for every 1 mL, we expect to estimate in the *tenths* place. The liquid in this cylinder has a volume very close to 51 mL and, in fact, appears to exactly touch the line that marks 51 mL. Therefore, a good reported value of the volume in the second cylinder is 51.0 mL. There is one estimated digit in this number. If we thought that the liquid was slightly above the line that marks 51 mL, we could report the volume as 51.1 mL. The reported value contains 3 digits, with the last one being estimated.

To get some additional practice in the proper reporting of measured numbers, consider how different rulers determine the length of the rectangular box in Figure 7-3. (Only a portion of each ruler is shown.)

The top ruler in Figure 7-3 is marked only at 100 cm; we expect to estimate in the tens place. The rectangular box appears to be a little more than half the length of the measuring stick, so we report that the length is around 60 centimeters. The middle measuring stick is marked off in 10-centimeter divisions; we expect to estimate in the ones place. We see that the box being measured is a little longer than 60 cm, so we estimate it to be 61 cm. The bottom measuring stick has markings every 1 cm. We expect to report a value in the tenths place. This measurement clearly lies between 60 and 61 cm, and so we estimate this measurement to be 60.5 cm.

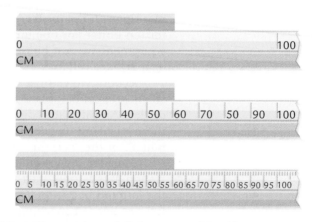

Figure 7-3 ▲ The number of digits reported in this measurement is determined by the precision of the ruler.

What Are Significant Figures?

We have just introduced a very important concept—the precision of a reported number. We must remember to include no more and no fewer than one estimated digit whenever we report a measured number. All of the certain numbers, plus the first estimated number, are called **significant figures.** Our reported value for the length of the rectangular box in Figure 7-3 was 60.5 cm, using the bottom measuring stick. This number has one estimated digit (5), so this number has three significant figures. *Significant figures (s.f.) are used only with estimated numbers because the estimation process incorporates a degree of uncertainty.* Because exact numbers have *no uncertainty* associated with them, they are excluded from discussions of significant figures. See Figure 7-4.

Let's look again at the measurements we recorded for the box in Figure 7-3. How many significant figures are there in each of those measurements? We estimated the top measurement as 60 cm. The zero at the end of this number is called a **trailing zero** because of its position at the end of the number. Are trailing zeroes counted as significant figures? Sometimes they are, and sometimes they are not. Trailing zeroes are counted as significant figures when there is an *expressed decimal* in the number. When there is no expressed decimal point in the number, trailing zeroes are considered to be the result of "rounding off" and are *not* counted as significant figures. Our number, 60, has no expressed decimal point. Therefore, the zero is not counted as a significant figure, and the number 60 has only one significant figure.

In contrast, the increased precision of the middle and bottom rulers resulted in measurements with more significant figures—61 cm has 2 significant figures and 60.5 cm has 3 significant figures.

Zeroes that appear as placeholders at the beginning of a number that is between $+1$ and -1 are called **leading zeroes.** For example, the first zero or zeroes in the numbers 0.0103, 0.112, or -0.0099 are leading zeroes. Leading zeroes are never counted as significant figures. Table 7-1 contains a summary of the rules to use in counting significant figures in a measured number. Table 7-2 provides some examples of the number of significant figures in a particular number.

KEY IDEA 7.3

✔**Meeting the Goals**

All of the digits in a measured number (except for leading zeroes or trailing zeroes where there is no expressed decimal) are considered significant and include the certain digits and the final uncertain (estimated) digit.

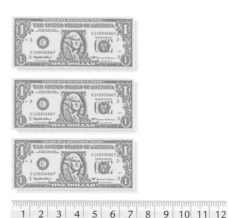

Figure 7-4 ▶ Individual objects such as dollar bills can be counted exactly. There are exactly three bills shown in this figure. Measurements such as the length of a one-dollar bill always involve some uncertainty. The length of a bill is about 6.9 inches. This number contains two significant figures.

Table 7-1 Rules for Counting Significant Figures in a Measured Number

REMEMBER

1. *All nonzero digits and zeroes included between nonzero digits are significant.* The number <u>1.0089</u> has five significant figures (5 s.f.).
2. *Leading zeroes are never significant.* The numbers 0.00<u>25</u>, 0.<u>25</u>, 0.000000<u>25</u> each contain two significant figures (2 s.f., underlined).
3. *Trailing zeroes in numbers without a decimal point are usually not significant.* The number <u>34</u>00 does not have an expressed decimal point. It has two significant figures (2 s.f.).
4. *Trailing zeroes in numbers with a decimal point are always significant.* The number 3400. has an expressed decimal point; it contains 4 significant figures (4 s.f.). The number 0.0<u>30</u> has two significant figures (2 s.f.). It also contains two leading zeroes, which are never significant!

Table 7-2 Significant Figures In Decimal Notation

Measured Number	Number of Significant Figures
4,250	3
6,000,000	1
0.00010	2
6,000,000.	7
0.0030	2
6.00	3
250.0	4
0.0025	2
2,050	3

Example **7.1**

Determining the number of significant figures in decimal numbers

Determine the number of significant figures in each of the following measured numbers.

(a) 104.50   (b) 0.0010   (c) 300   (d) 10004.0

Thinking Through the Problem All of the nonzero digits and zeroes included between nonzero digits are significant figures. It is usually the leading or trailing zeroes that cause us problems in counting significant figures.

Working Through the Problem Remember that leading zeroes and trailing zeroes in numbers without an expressed decimal point are never counted as significant figures. In numbers with an expressed decimal point, only trailing zeroes are counted as significant figures. Part (a) has a trailing zero that is significant, since this number has an expressed decimal. Part (b) has three leading zeroes that are not significant, but the trailing zero is significant. Part (c) has two trailing zeroes that are not significant, since there is no expressed decimal. Part (d) has three zeroes between the 1 and the 4 and one trailing zero after an expressed decimal.

Answer

(a) 5 s.f.   (b) 2 s.f.   (c) 1 s.f.   (d) 6 s.f.

How are you doing? 7.1

Determine the number of significant figures in these measured numbers: (a) 0.001   (b) 200.0   (c) 2000

Significant Figures in Calculations

When we carry out calculations with measured numbers we are very likely to get numbers that appear too precise. That is, *the calculated number may have more digits than the measured numbers, and reporting all of these would imply that we have a greater number of significant figures than actually provided by our measurements.* We need to correct the answer by rounding off the number to the appropriate number of significant figures. But before we discuss how many significant figures to keep, we need to discuss rounding.

When we round off a number we first determine the number of digits that should be in our answer. If the number following the last digit that we want to retain is less than 5, then we drop that number and all the digits following it. For example, if we want three significant figures in our answer, then we will round off the number 23.4276 to 23.4. If the number following the last retained digit is greater than 5, then we round off the number by increasing the last retained digit by one. For example, the number 23.4776 rounded off to 3 significant figures, is 23.5. Finally, if the number following the last retained digit is exactly 5 with nothing or with only zeroes following it, the final answer should be written so that it is an even number. For example, the number 23.45000 rounded to three significant figures, is made even by rounding *down* to 23.4. The number 23.35000 is made even by rounding *up* to 23.4.

Rule 1: Multiplication and division (Count significant figures in each number.)

$1.4551 \times 2.04 = 2.9684 \approx 2.97 \leftarrow$ The answer is rounded to 3 s.f.

5 s.f.   3 s.f.

Rule 2: Addition and subtraction (Look at position after the decimal point!)

   1.4551
$+2.0 \leftarrow$ This measurement limits the answer to **one decimal place.**
   $3.4551 \approx 3.5 \leftarrow$ The answer, after rounding off, contains 2 s.f.

Figure 7-5 ▶ Examples of the application of significant figure rules.

When measured numbers are used in calculations, how do we decide how many significant figures to include in the final answer? Each measured number in a calculation contributes uncertainty to the calculated answer. We use the rules in Table 7-3 to decide how many digits to report in a calculation. An application of each rule is shown in Figure 7-5.

Table 7-3 Rules for Determining Significant Figures in a Measured Number

REMEMBER

1. *For multiplication or division calculations* the answer must contain only the *fewest number of significant figures* of any number used in the calculation.
    (a) Find the smallest number of significant figures of the numbers given in the problem.
    (b) Round off the calculated answer to the correct number of significant figures.
2. *For addition or subtraction operations* the calculated answer must contain only the *fewest number of places after the decimal point* of any number used in the calculation.
    (a) Find the number with the fewest digits following the decimal point.
    (b) Round off the calculated answer to this number of digits after the decimal point.

Example 7.2

Significant figures in a calculated number: Multiplication and division

Calculate the answer to the problem, $\dfrac{(34.0) \times (0.003)}{(1.090) \times (20.0)}$.

Assume that these are measured numbers, and express your answer with the correct number of significant figures.

Thinking Through the Problem This is a multiplication/division problem involving measured numbers. According to our rules for determining significant figures in calculations involving measured numbers (see Table 7-3) we must find the smallest number of significant figures in the problem and then report the final answer with this number of significant figures.

Working Through the Problem In the numerator, 34.0 has 3 s.f. and 0.003 has 1 s.f. In the denominator, 1.090 has 4 s.f. and 20.0 has 3 s.f. The smallest number of significant figures, then, is 1. The calculator answer must be rounded off to 1 significant figure.

$$\frac{(34.0) \times (0.003)}{(1.090) \times (20.0)} = 0.004679 \text{ (calculator answer)}$$

$$\approx 0.005 \text{ (answer rounded to 1 significant figure)}$$

Answer 0.005 (1 s.f.)

Example 7.3

Significant figures in a calculated number: Addition and subtraction

Solve this problem involving measured numbers: 12.098 + 0.27 − 11.6. Report the answer with the correct number of significant figures.

Thinking Through the Problem In this case we must follow rule 2 in Table 7-3. The final calculated answer must be rounded to the fewest number of digits after the decimal point of any number in the calculation. The first number, 12.098, has 3 digits following the decimal point, the second number has 2 digits following the decimal point, and the third number has 1 digit following the decimal point. The smallest number of digits following the decimal point is 1 digit. The final calculated answer must therefore be rounded to 1 digit following the decimal point.

Working Through the Problem

$$12.098 + 0.27 - 11.6 = 0.768 \text{ (calculator answer)}$$
$$\approx 0.8 \text{ (correct answer rounded to one place after the decimal point)}$$

Answer 0.8 (1 s.f.)

Example 7.4

Significant figures in a calculated number: Addition and subtraction

Compute the answer to this addition problem involving measured numbers: 12.098 + 0.27 + 11.6.

Thinking Through the Problem This is similar to example 7.3, but here the problem involves addition only, rather than addition and subtraction. Again we follow rule 2 in Table 7-3.

Working Through the Problem The smallest number of digits following the decimal point is found in the third number (1 digit). The final calculated answer must be rounded to 1 digit following the decimal point.

$12.098 + 0.27 + 11.6 = 23.968 \approx 24.0$. Here the final answer has been rounded to one decimal place (according to the rules for addition/subtraction). This number contains 3 significant figures.

Answer 24.0.

✔ Meeting the Goals

For multiplication or division problems, the final answer cannot contain any more significant figures than the smallest number of significant figures appearing in the problem. For addition and subtraction problems, the final answer cannot contain any more places after the decimal point than the smallest number of decimal places appearing in the problem.

Example 7.5

Significant figures in calculations: Answers much greater than 1

Perform the following calculation involving measured numbers and give the answer with the correct number of significant figures.

$$\frac{(6.023) \times (802,000)}{(0.0001986)} =$$

Thinking Through the Problem This is a straightforward calculation using your calculator. It is best to enter the entire calculation as it appears and not to do a series of partial calculations. You would enter "6.023 × 802,000 ÷ 0.0001986 = ". The number of significant figures for this answer is determined by the least number of significant figures in the problem. 802,000 has 3 significant figures; the other two numbers have four significant figures.

Working Through the Problem

$$\frac{(6.023) \times (802,000)}{(0.0001986)} = 24{,}322{,}487{,}411.9 \text{ (calculator answer)}$$

Answer 24,300,000,000 (3 s.f.)

Example 7.6 Significant figures in calculations: Answers less than 1

Compute the answer to this calculation involving measured numbers and report the correct number of significant figures.

$$\frac{(12.898 - 13.73)}{(2{,}090{,}000)} =$$

Thinking Through the Problem This problem is a combination of subtraction and division. We will need the rules for both operations. We first perform the subtraction in the numerator. Then we do the division.

Working Through the Problem The numerator simplifies to 12.898 − 13.73 = −0.832 (calculator answer). But, according to the rules for subtraction, the answer can contain no more than two digits after the decimal (13.73 has only 2 digits following the decimal). The correct answer to the subtraction, if we stopped here, would contain 2 significant figures, the 8 and the 3. However, we are not finished with the problem. We need to know the number of significant figures in the numerator *before* we finish the problem *even though we will carry all of the digits* along for the division process.

$$\frac{(-0.832)}{(2{,}090{,}000)} = -0.0000003981 \text{ (calculator answer)}$$

KEY IDEA 7.5

Notice that the entire calculator answer is carried along in the numerator; *rounding off is done at the end of the problem.* However, in order to determine the number of significant figures in the final answer, we should remember the number of significant figures resulting from the subtraction in the numerator. The final answer will contain 2 significant figures because the subtraction yielded an answer with a precision of 2 significant figures. Remember to do the addition/subtraction part of the calculation first, mentally determine the number of significant figures that result, and then go on to the rest of the problem.

Answer 0.00000040

PRACTICAL A   How is sand measured?

Making quality measurements involves trying to eliminate any unnecessary errors, especially systematic error. To do this we should think about the possible effects of any assumptions we make. For example, when you think about measuring sand, you might make several assumptions. You might assume that the sand is dry, that it is pure, and that there is no difficulty in transferring any sand into or out of the measuring device.

What kinds of error could be involved in making a measurement on a simple load of sand? What if the sand is wet or contains contaminants? A container of wet sand is heavier than the same container filled with dry sand because of the added water. If we weigh the wet sand thinking it to be dry, our masses will be *systematically* wrong. How do you think this systematic error can be corrected? Tell how the error can be detected for this example.

When we transfer substances there is always the danger of some spillage. What kind of error is there if we always spill some of a sample before the measurement? What if we consistently spill some after the measurement? What if we do a bad job of cleaning the container that we weigh the substance in?

Will these errors be systematic or random? Can they be prevented?

Finally, there are errors associated with whether the sample is actually what we think it to be. Suppose that you assume the sand is pure, but it is actually contaminated with other materials throughout. What kind of error will exist in our mass measurements of this mixture, random or systematic error?

You should also be aware of the importance of significant figures in determining the precision of a mass, volume, or density value. For example, a glass graduated cylinder has a mass of 75.2 g. Water is added until the cylinder reads 0.0503 L. The mass of the cylinder and water is measured as 124.3 g. Next, sand (dried at 120 °C) is added until the cylinder reads 0.0759 L. The mass of the cylinder and mixture is now 190.2 g.

What is the mass of the sand?
What is the volume of the sand?
What is the density of the sand?
What if the same measurements were made with wet sand? Would the density of the wet sand be the same as the density we calculated for the dry sand? Why or why not?

(Jeff Greenberg/Visuals Unlimited.)

How are you doing? 7.2

Perform the following calculation and report the answer to the correct number of significant figures:

$$\frac{(7.982) \times (0.0236)}{(22.6 + 0.0356 - 22.34)} =$$

PROBLEMS

1. Concept question: Most people strongly resist reporting a "guessed" or "estimated" number. They want the data they report to be 100% accurate. As a result they report *fewer* numbers than they should, and their answers are less precise than necessary. Discuss the tension you may feel between reporting "guessed" numbers vs. reporting imprecise numbers.

2. What is the difference between precision and accuracy?

3. Why do we count significant figures in a measured number?

4. What is the difference between a systematic error and a random error?

5. What is the cause of uncertainty in a measured number?

6. When are zeroes considered significant in a measured number?

7. How do you determine the number of significant figures allowed in an answer to a multiplication or division calculation?

8. How do you determine the number of significant figures allowed in the answer to an addition or subtraction problem?

9. Indicate the number of significant figures in each of these measured numbers:

(a) 84.3   (d) 0.021

(b) 3.210   (e) 9,900

(c) 0.00008

10. Perform the following operations. Report answers to the correct number of significant figures.

(a) 0.123×0.02342

(b) 23450×232.0

(c) $0.00002345 + 0.003230$

(d) $\dfrac{(2.335 + 21.09)}{(13.75)}$

(e) $\dfrac{(1.305 + 2.97 - 5.0)}{(5.21) \times (11.01)}$

7.2 Units of Measure

SECTION GOALS

✔ Know how to use exponents in calculations.

✔ Understand scientific notation.

✔ Estimate the magnitude of a product by using the rules of exponents.

✔ Know which metric units are used for mass, volume, and length.

✔ Understand how metric prefixes correspond with the relative size of the measured object.

✔ Know the abbreviations for the metric units and metric prefixes used in this chapter.

MAKING IT WORK WITH MATH

Working with Exponents

Exponents are used as a type of mathematical shorthand for multiplication. If you have the factored form of a number such as $81 = 3 \cdot 3 \cdot 3 \cdot 3$ it is easier to write the expression in exponential form, $81 = 3^4$. In this case 3 is the base and 4 is the exponent. *The exponent indicates the number of times the base appears as a factor.*

Remember that an exponent operates only on the factor *immediately below and to the left* of the exponent. *If the exponent appears beside a parentheses, it operates on everything inside the parentheses.* Thus $(-2)^3 = -2 \times -2 \times -2 = -8$.

KEY IDEA 7.6

Example 7.7 **Simplifying numbers with exponents**

Simplify the following numbers.

(a) 2^3 (b) $(-5)^2$ (c) -4^2

Thinking Through the Problem ➡ (7.6) We know that the exponent "works" only on the factor immediately preceding it.

Working Through the Problem Part (a) is straightforward as it has only one digit with the exponent: $2^3 = 2 \cdot 2 \cdot 2$. Part (b) shows the exponent above and to the right of a parenthesis. The exponent acts on everything *inside* the parenthesis: $(-5)^2 = (-5) \times (-5)$. In part (c), the exponent works only on the 4: $-4^2 = -(4 \times 4)$. This expression should read "the negative of 4 squared." Notice that part (c) is different from part (b) where the negative sign is included within the parentheses.

Answer

(a) $2^3 = 2 \cdot 2 \cdot 2 = 8$
(b) $(-5)^2 = (-5)(-5) = 25$
(c) $-4^2 = -(4)(4) = -16$

There are several properties mathematicians associate with exponents. These are listed in Table 7-4. The properties in this table are also important in chemistry.

Table 7-4 Properties of Exponents

REMEMBER

Property	Mathematical Notation	Example from Powers of Ten
When multiplying two monomials with the same base, add the exponents.	$x^a \cdot x^b = x^{a+b}$	$10^2 \cdot 10^5 = 10^7$
When a monomial involves a power and is raised to a power, multiply the exponents.	$(x^a)^b = x^{ab}$	$(10^3)^2 = 10^6$
When dividing monomials with like bases, subtract the exponents.	$\dfrac{x^a}{x^b} = x^{a-b}$	$\dfrac{10^7}{10^4} = 10^3;\ \dfrac{10^8}{10^{10}} = 10^{-2}$
Any nonzero expression raised to the zero power is equal to 1.	$x^0 = 1$ $(a+b)^0 = 1$	$10^0 = 1$ $(10^3 + 10^6)^0 = 1$
Raising an expression to a negative power is the same as raising its reciprocal to the positive power. We call this the negative exponent rule.	$x^{-1} = \dfrac{1}{x}$ $x^{-a} = \dfrac{1}{x^a}$	$10^{-1} = \dfrac{1}{10}$ $10^{-6} = \dfrac{1}{10^6}$

Example 7.8   **Multiplying expressions with exponents**

Multiply $x^4 \cdot x^8$.

Thinking Through the Problem We know that if we multiply two monomials with the same base (x), we add the exponents. In factored form: $(x \cdot x \cdot x \cdot x)(x \cdot x \cdot x \cdot x \cdot x \cdot x \cdot x \cdot x)$. In exponential form this yields: $x^4 \cdot x^8$.

Answer $x^4 \cdot x^8 = x^{12}$

We use these laws to rearrange the factors in multiplication problems, including cases where there is more than a single variable. Exponent rules apply even when there are several variables. We add the exponents of factors with like bases.

$$(x^a y^b)(x^c y^d) = x^{a+c} y^{b+d}$$

Coefficients (numerical constants) are treated in the same way as variables, as shown in the following example.

Example 7.9   **Multiplying values with different bases for the exponents**

Multiply $(5x^5 y^2)(8x^3 y^4)$.

Thinking Through the Problem We will add the exponents of the same variable; we will multiply the coefficients together. Using commutative laws we obtain $5 \cdot 8 \cdot x^5 \cdot x^3 \cdot y^2 \cdot y^4$.

Answer Simplifying we obtain: $40x^8 y^6$.

Example 7.10   **Applying exponents in sequence**

Evaluate $(5x^2 y^4)^3$.

Thinking Through the Problem The exponent 3 works on the entire parentheses. In factored form: $5 \cdot 5 \cdot 5 \cdot x^2 \cdot x^2 \cdot x^2 \cdot y^4 \cdot y^4 \cdot y^4$.

Answer Simplifying: $125x^6 y^{12}$

Example 7.11   **Exponents in fractions**

✔ Meeting the Goals

To multiply two monomials with the same base, add the exponents: $x^a x^b = x^{a+b}$

To raise a monomial with a power to a power, multiply the exponents: $(x^a)^b = x^{ab}$

To divide monomials with the same base, subtract the exponents: $\dfrac{x^a}{x^b} = x^{a-b}$

Simplify $\dfrac{x^5}{x^3}$.

Thinking Through the Problem Exponents are subtracted when variables are divided.

Working Through the Problem $\dfrac{x^5}{x^3} = x^{5-3} = x^2$

Answer $\dfrac{x^5}{x^3} = x^2$

There are situations, especially in science, when we encounter zero and negative exponents. Suppose that we want to simplify the expression, $\dfrac{x^5}{x^5}$.

$$\frac{x^5}{x^5} = 1$$

since any nonzero number divided by itself is 1.

Now let us apply a rule we just learned to the same problem. We learned that $\dfrac{x^a}{x^b} = x^{a-b}$.

Applied to this problem we get $\dfrac{x^5}{x^5} = x^{5-5} = x^0$.

Therefore, $x^0 = 1$ since both are equivalent to $\dfrac{x^5}{x^5}$.

This is a very important idea. Any quantity that is raised to the zero power is equal to 1. This is called the **zero exponent rule.**

What about reporting a negative exponent in an answer? Many algebra texts require answers with only positive exponents, and so you might simplify the following expression by subtracting the smaller exponent from the larger exponent and expressing the result like this:

$$\frac{x^2}{x^3} = \frac{1}{x^1} \text{ or } \frac{1}{x}$$

✔ **Meeting the Goals**

Any nonzero expression raised to the zero power is equal to 1: $x^0 = 1$. An expression raised to a negative power is the same as the reciprocal of the expression raised to the positive power: $x^{-a} = \dfrac{1}{x^a}$.

If we did not have this restriction, however, we could simply subtract the exponent in the denominator from the exponent in the numerator. Now we have

$$\frac{x^2}{x^3} = x^{2-3} = x^{-1}$$

Both this result and the previous result are equivalent, and we can write:

$$x^{-1} = \frac{1}{x}.$$

This is called the **negative exponent rule.**

A negative exponent of some quantity can be changed to a positive exponent by taking the reciprocal of that quantity. Negative exponents also apply to units. Density, as we saw in Chapter 6, is an example of this. Suppose that the units of density in some problem are grams per liter, $\dfrac{g}{L}$. Textbooks commonly avoid such ratios because they take up too much space on the page. Instead, $\dfrac{g}{L}$ is rewritten as $g\ L^{-1}$; both are read "grams per liter." With practice, you will become adept at both representations.

CONNECT TO
SECTION 6.2

Example 7.12   **Writing units with negative exponents**

Write the units *miles per hour* both as a fraction and as a product using the negative exponent rule.

Thinking Through the Problem The position of the word *per* identifies the word preceding it (miles) as the numerator and the word following it (hour) as the denominator. We begin by writing the fraction. We can then rewrite the fraction as a product of the numerator and denominator by using the negative exponent rule.

Answer $\dfrac{\text{miles}}{\text{hour}} = \text{miles hour}^{-1}$

How are you doing?   **7.3**

(a) Simplify: $(15y^2)^0 \, (3y^3)^{-2}$

(b) Rewrite this ratio as a product, using the negative exponent rule: $\dfrac{(\text{liter})^2}{(\text{moles})^2 \, (\text{second})}$

Scientific Notation

Exponential or scientific notation allows us to write both very large and very small numbers more compactly. As an example, consider the distance from the earth to the sun, approximately 150,000,000 kilometers. Or consider the diameter of a living cell, approximately 0.000010 meters. Figures 7-6(a) and 7-6(b) illustrate these large and small distances. The string of zeroes in these numbers makes them tedious to write. To manage this problem, scientists use a notation called **scientific notation** to represent numbers that contain many zeroes at the end (very large numbers) or at the beginning (very small numbers).

    A number written in scientific notation is written as the product of a number, n, between 1 and 10 ($1 \leq n < 10$), *and a power of 10.* It can also be a negative number, comprised of a number between -10 and -1 ($-10 < n \leq -1$) multiplied by a power of 10. In scientific notation, the distance to the sun is written as 1.5×10^{11} kilometers. The size of a living cell might be written as 1.0×10^{-5} meter. The US deficit in recent years has been, for example, -1.8×10^{10} dollars.

    To convert a number to scientific notation:

1. Move the decimal point to the position after the first nonzero digit, so that the number is between 1 and 10;

2. Determine the appropriate power of ten by counting the number of decimal places that you need to move the decimal point.
   (a) The exponent of 10 will be negative when the original number is between -1 and 1;
   (b) The exponent of 10 will be positive when the original number is greater than 10 or less than -10.

The number written in scientific notation must have the same value as the original number. The number has not changed, only its representation has changed.

✔**Meeting the Goals**

A positive number written in scientific notation consists of a number *n* between 1 and 10 ($1 \leq n < 10$), multiplied by ten raised to some power. The exponent of 10 equals the number of places the decimal in the original number was moved in order to give an expression in scientific notation. A negative exponent indicates that the decimal number was less than 1. A positive exponent indicates that the decimal number was greater than 10 or less than -10.

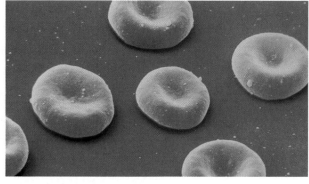

(a)   (b)

Figure 7-6 ▲ Scientific notation is useful for recording measurements that are very large or very small. (a) The distance from the earth to the sun is 1.5×10^8 km. (b) The diameter of one of these blood cells is 1×10^{-5} meters. (a: NASA/Science Source/Photo Researchers; b: Stanley Flegler/Visuals Unlimited.)

Example 7.13 · **Converting a number greater than 1 to scientific notation**

Write 950,000 in scientific notation.

Thinking Through the Problem We must rewrite the number with only one digit before the decimal. Then we must multiply this number by ten raised to a power such that the two expressions are equal.

Working Through the Problem The number in decimal notation is a large number; we expect a positive exponent for 10. The understood decimal point follows the final zero in this number. It must be moved 5 places to the left. The exponent of 10 is 5.

Answer $950,000 = 9.5 \times 10^5$

Example 7.14 · **Converting a number less than 1 to scientific notation**

Write 0.0257 in scientific notation.

Thinking Through the Problem The number in decimal notation is less than 1. We expect a negative exponent for 10. The decimal point must be moved two places to the right; the exponent of 10 is -2.

Answer $0.0257 = 2.57 \times 10^{-2}$

Example 7.15 · **Converting from scientific notation with a positive exponent to decimal notation**

Convert 5.97×10^3 to decimal form.

Thinking Through the Problem This is a multiplication problem; $10^3 = 1,000$. Then, $5.97 \times 10^3 = 5.97 \times 1,000 = 5,970$. We see that the effect of the positive exponent of 10 was to move the decimal point three places to the right.

Answer 5,970

Example 7.16

Converting from scientific notation with a negative exponent to decimal notation

Convert 7.9×10^{-4} to decimal form.

Thinking Through the Problem The exponent of 10 is negative. We know from the negative exponent rule that $10^{-4} = \dfrac{1}{10^4}$. We also know that $10^4 = 10,000$. The problem is rewritten as $\dfrac{7.9}{10^4} = \dfrac{7.9}{10,000} = 0.00079$. We see that the effect of the negative exponent of 10 was to move the decimal point four places to the left.

Answer 0.00079 is the decimal form of this number.

How are you doing? 7.4

Convert to decimal form: (a) 5.93×10^5 (b) 1237×10^{-2}
Convert to scientific notation: (c) 0.00890   (d) 125.6

Counting significant figures is much easier when a number is written in scientific notation. For one thing, scientific notation eliminates all leading zeroes. For instance, it would have been much easier to determine the number of significant figures to include in the answers to Examples 7.5 and 7.6 if the problem had been written in scientific notation. Then, Example 7.5 would be $\dfrac{(6.023) \times (8.02 \times 10^5)}{(1.986 \times 10^{-4})} = 2.43 \times 10^{10}$ and Example 7.6 would be $\dfrac{(-8.32 \times 10^{-1})}{(2.09 \times 10^6)} = -3.980 \times 10^{-7} \approx -4.0 \times 10^{-7}$.

Whether a measured number is written as a decimal or in scientific notation has no bearing upon the number of significant figures that it contains. Table 7-5 shows five measurements. Column one shows each measurement written in its decimal form, column two shows the number of significant figures for each measurement, and column three shows each measurement written in scientific notation. Verify for yourself that the measurements in column three have the same number of significant figures as the measurements in column one.

Table 7-5 Measurements in Decimal and Scientific Notation Contain the Same Number of Significant Figures

Decimal Form of Measurement	Significant Figures in Decimal Form	Measurement in Scientific Notation
93,000,000	2	9.3×10^7
951,000	3	9.51×10^5
0.0207	3	2.07×10^{-2}
−5970	3	-5.97×10^3
−0.00079	2	-7.9×10^{-4}

MAKING IT WORK WITH MATH

Numerical Estimation of the Magnitude of a Product

Scientific notation makes it easy to estimate the magnitude of many calculated answers and thus to increase our assurance that we have done our calculation correctly. The **magnitude** of a number is the size or extent of the sample being measured and is correlated with the exponent of 10 when the number is written in scientific notation.

A number written in scientific notation has two parts that are multiplied together, a decimal number and an exponential (power of ten). Multiplication and/or division problems can be rewritten as the product of the decimal numbers and the exponential parts. For example, consider the following calculation:

$$\frac{(2.1 \times 10^3)\,(3.2 \times 10^{-2})^2}{(6.0 \times 10^{-5})\,(9.4 \times 10^{22})}$$

Suppose that you got 1.1×10^{-29} as your calculator answer. How can you know whether or not this is the correct answer?

You can verify this answer by doing a rough estimation. First regroup as follows:

$$\frac{(2.1) \times (3.2)^2}{(6.0) \times (9.4)} \times \frac{(10^3)\,(10^{-2})^2}{(10^{-5})\,(10^{22})}$$

Then you can *approximate* the decimal part as whole numbers and mentally solve:

$$\frac{2 \times 9}{6 \times 9} = \frac{1}{3} \approx 0.33$$

The exponential part is simplified by following the rules of exponents:

$$\frac{(10^3)\,(10^{-2})^2}{(10^{-5})\,(10^{22})} = \frac{(10^3)\,(10^{-4})}{(10^{17})} = \frac{(10^{-1})}{(10^{17})} = 10^{-18}$$

The answer should have a magnitude close to 10^{-18}. The "answer" 1.1×10^{-29} is incorrect.

This method of regrouping is especially useful in problems where the exponent is larger than 99. Some calculators will not show exponents containing more than two digits. Instead, they will show an error message when the final answer is of a magnitude larger than 10^{99} or smaller than 10^{-99}. Regrouping as above allows you to use a calculator for the decimal part of the problem and your personal calculator (your brain!) to determine the exponent of ten in the answer!

✔ **Meeting the Goals**

To estimate the magnitude of the answer to a multiplication/division calculation, use the properties of exponents to find the exponent of 10 in the calculated answer. Often this estimation is enough to allow you to check the accuracy of your original answer.

Example 7.17   **Managing calculations with very large powers of ten**

Compute the answer to the following and report it with the correct number of significant figures:

$$\frac{(3.8 \times 10^{61})(4.01 \times 10^5)^3}{(2.90 \times 10^{-45})(3.197 \times 10^{-23})^2}$$

Thinking Through the Problem This appears to be a straightforward calculation. Start by entering each number into a calculator. If you get an error message, you must approach this problem in parts.

Working Through the Problem ➡ (7.7) First regroup the problem into a decimal part and an exponential part:

$$\frac{(3.8 \times 4.01^3)}{(2.90 \times 3.197^2)} \times \frac{10^{61} \times (10^5)^3}{10^{-45} \times (10^{-23})^2}$$

Use a calculator to solve the decimal portion:

$$\frac{(3.8 \times 4.01^3)}{(2.90 \times 3.197^2)} = 8.2667$$

Use the exponent rules to find the correct power of ten for this result:

$$\frac{10^{61} \times (10^5)^3}{10^{-45} \times (10^{-23})^2} = \frac{10^{61} \times 10^{15}}{10^{-45} \times 10^{-46}} = \frac{10^{76}}{10^{-91}} = 10^{167}$$

Combining the decimal and exponential parts of the calculation, obtain a calculator answer, 8.2667×10^{167}. The correct answer must be rounded to two significant figures because the number 3.8 limits the answer to two significant figures.

Answer 8.3×10^{167}

✔ **Meeting the Goals**

The basic units of the metric system we will use in this chapter are meter (length), gram (mass), liter (volume) and mole (chemical amount). The metric unit for temperature (Kelvin) will be used in later applications.

✔ **Meeting the Goals**

Metric prefixes correlate with an exponent of 10 when a number is written in scientific notation. As the exponent of 10 tells where the decimal belongs when the number is written in decimal notation, the metric prefix also indicates the relative size of the measured object. A positive exponent indicates a number greater than 1, whereas a negative exponent indicates a number less than 1.

The Metric System

The **metric system of units** is used in most countries of the world and in all areas of science. In 1960, the General Conference of Weights and Measures convened in France to set up an internationally agreed-upon system of units called the International System of Units, commonly referred to as *SI units* (for the abbreviation of their name in French). The SI units are based on the metric system. The most common units you will encounter in this course are those of length, mass, and volume. These, together with other common units, are listed in Table 7-6, which also includes an important non-SI unit, the milliliter, often used to measure volumes in chemistry.

The metric system is a decimal system, so the smaller and larger parts of a unit are related to one another by powers of ten. A prefix is used with the basic unit to express the power of ten needed. Each **metric prefix,** then, stands for a power of ten. The prefixes may

Table 7-6 Some Units of Measurement

Quantity	Base Unit	Common Unit	Abbreviations	English Equivalent
Length	meter	centimeter	m, cm	1 m = 1.0936 yd
				2.540 cm = 1 in
Mass	gram	kilogram	g, kg	1 kg = 2.2 lb
				453.6 g = 1 lb
Temperature	Kelvin	Celsius	K, °C	$°F = 32 + \dfrac{5}{9}(°C)$
Time	second	second	s	s
Chemical amount	mole	mole	mol	mol
Volume	cubic meter	cubic centimeter	m^3, cm^3	1,057 qt = 1 m^3
				3,785 cm^3 = 1 gal
	liter	milliliter	L, mL	0.9463 L = 1 qt

either be written out explicitly or abbreviated as shown in Table 7-7. This table lists the most commonly used metric prefixes, and should be committed to memory.

The most important columns in Table 7-7 are the first, second, and fourth. You should be able to associate each prefix with the appropriate power of ten that it represents. For example, any unit with the prefix *giga-* signifies that the measurement is some number times 10^9. This represents a very large number, one that is multiplied by 1,000,000,000. Learn to associate the prefixes with the power of

Table 7-7 Some Metric Prefixes

Prefix	Symbol	Numerical Value	Exponential Form	For a Meter
giga-	G-	1,000,000,000	1×10^9	1 Gm $\longleftrightarrow$ 1×10^9 m
mega-	M-	1,000,000	1×10^6	1 Mm $\longleftrightarrow$ 1×10^6 m
kilo-	k-	1,000	1×10^3	1 km $\longleftrightarrow$ 1×10^3 m
deca-	da-	10	1×10^1	1 dam $\longleftrightarrow$ 1×10^1 m
no prefix		1	1×10^0	1 m $\longleftrightarrow$ 1 m
deci-	d-	0.1	1×10^{-1}	1 dm $\longleftrightarrow$ 1×10^{-1} m
centi-	c-	0.01	1×10^{-2}	1 cm $\longleftrightarrow$ 1×10^{-2} m
milli-	m-	0.001	1×10^{-3}	1 mm $\longleftrightarrow$ 1×10^{-3} m
micro-	μ-	0.000001	1×10^{-6}	1 μm $\longleftrightarrow$ 1×10^{-6} m
nano-	n-	0.000000001	1×10^{-9}	1 nm $\longleftrightarrow$ 1×10^{-9} m
pico-	p-	0.000000000001	1×10^{-12}	1 pm $\longleftrightarrow$ 1×10^{-12} m

ten each signifies. Positive exponents of ten represent large numbers, while negative exponents of ten represent small numbers. Notice that the only prefix abbreviations beginning with an uppercase letter are the very large designations, *giga-* (G-) and *mega-* (M-).

Metric prefixes are used with basic metric units such as gram, meter, and liter to represent larger and smaller quantities than 1. You must learn to *see both the prefix and the unit.* You also should be aware of what the unit is measuring. For example, 23 gigameters is a measurement of length; giga- represents a number much larger than 1; 1 gigameter is 1×10^9 meters.

Example 7.18 **Metric measurements**

Fill in appropriate responses for each of the following measurements:

Metric Unit	Measurement of Length, Mass, or Volume?	Measurement Larger or Smaller than 1?	Exponential Form?
1 centigram			
1 microliter			
1 kilometer			

Thinking Through the Problem We try to see both "parts" of each unit by using italics for the prefix.

Working Through the Problem The prefixes *centi-* and *micro-* indicate values less than 1, whereas the prefix *kilo-* indicates a value greater than 1. The base units gram, meter, and liter indicate measurements of mass, length, and volume, respectively. Thus, *centi*gram is a small unit that measures mass, *micro*liter is a very small measurement of volume, and *kilo*meter measures length or distance.

Answer

Metric Unit	Measurement of Length, Mass, or Volume?	Measurement Larger or Smaller than 1?	Exponential Form?
1 centigram	Mass	Smaller than 1	1×10^{-2} grams
1 microliter	Volume	Smaller than 1	1×10^{-6} liters
1 kilometer	Length	Larger than 1	1×10^3 meters

How are you doing? 7.5

Perform the same exercise as above with the following measurements: 1 megaliter, 1 nanogram, 1 millimeter.

Use of Abbreviations in Units of Measurement

The most common units are those of length (meter), mass (gram), and volume (liter). *We use the metric system and scientific notation together to describe many different measurements.* Any of the units in Table 7-6 can be combined with any of the prefixes in Table 7-7 to get a unit that is a multiple or a fraction of the base unit. For example, the standard unit of length in the metric system is the meter (1.0936 yd). The abbreviation for meter is m. The metric prefixes are used with meter to designate larger and smaller parts of a meter, and can be used to replace the exponential form:

<table>
<tr><td>**giga**meter</td><td>6 **giga**meters</td></tr>
<tr><td>1×10^9 meters</td><td>6×10^9 meters</td></tr>
</table>

In this case, giga- is replaced by 10^9, which means that the measurement is multiplied by 10^9. Of course, this also works in reverse! A measurement such as 2.57×10^9 grams can be rewritten as 2.57 gigagrams (2.57 Gg).

$$2.57 \times 10^9 \text{ grams}$$

is

$$2.57 \textbf{ giga}\text{grams}$$

You should memorize the equivalency expressions in which the prefixes are associated with their corresponding powers of 10. For example, the equivalency expression that you should learn for gigaliter is: 1 GL $\leftrightarrow 1 \times 10^9$ L. Remember that the exponential term 1×10^9 is written with the base unit, as 1×10^9 L. The prefix G stands alone with the unit, as 1 GL. This practice will help you learn how to convert between metric units in Section 7.3.

✔ **Meeting the Goals**

The metric prefixes in Table 7-7 can be abbreviated. Generally, the abbreviation is the first letter of the prefix. There are two exceptions: *deca-* is abbreviated *da-* to distinguish it from *deci-* (*d-*); and *micro-* is written by using the Greek lowercase letter mu, μ, to distinguish it from *milli-* (*m-*) and from *mega-* (*M-*).

Example 7.19   **Writing metric prefixes**

Rewrite these expressions using metric prefixes to replace the exponential term.

(a) 3×10^{-3} meters $\longleftrightarrow$ _____ meter
(b) 4.6×10^1 grams $\longleftrightarrow$ _____ gram
(c) 2.5×10^{-12} liters $\longleftrightarrow$ _____ liter

Thinking Through the Problem The decimal part of the number remains the same, but each exponential term has a metric prefix that replaces it. 10^{-3} is replaced by *milli*, 10^1 is replaced by *deca*, and 10^{-12} is replaced by *pico*.

Answer

(a) 3×10^{-3} meters $\longleftrightarrow$ 3 millimeters
(b) 4.6×10^1 grams $\longleftrightarrow$ 4.6 decagrams
(c) 2.5×10^{-12} liters $\longleftrightarrow$ 2.5 picoliters

PRACTICAL B   How much sand do you need to build a patio?

If you visit a construction site, such as the one shown in Figure 7-7, you will probably see huge amounts of sand in use. In the United States, sand is usually sold by the cubic yard. Of course, in the metric system we use the cubic meter instead. This is a *lot* of sand, but it takes a lot to build even simple things, as we will discover here. We can use what we've learned about the metric system and significant figures to calculate the amount of sand needed to build a patio.

According to the plans for a townhouse community, each of 180 townhouses is to have a patio, made either of concrete or paving bricks. Here are the specifications given to the building contractor:

Plan A: A concrete patio

The depth of the concrete patio is to be 0.015 m, the length is to be 4.5 m, and the width is to be 3.5 m. Calculate the volume of concrete needed to fill this space. How many significant figures did you include in your answer?

Concrete is a mixture of cement, water, sand, and crushed stone. If the concrete mixture is 22% sand by volume, calculate the volume of sand needed to give 90 townhouses a concrete patio. How many significant figures are in your answer? Why?

Plan B: A brick patio

A second option for the prospective home-owners is a patio made with paving bricks. Sand is still needed as it provides the "bed" for the bricks. A

Figure 7-7 ▲ Measurement is important at a construction site. (Geoff Tompkinson/Science Photo Library/Photo Researchers.)

patio of the same length and width as the concrete patio is to be constructed. But instead of concrete, it needs a bed of sand that is 5.0 cm deep.

Convert 5.0 cm to m. Next, determine the volume of sand needed for one brick patio. How many significant figures are in your answer? Why?

How much sand should be ordered to provide 90 townhouses with brick patios? Give your answer in cubic meters and in kilograms, assuming that the density of sand is 2.2×10^6 g m^{-3}.

Example 7.20   Converting from metric prefixes

Rewrite the following measurements by replacing the prefix with its exponential equivalent.

(a) 7.8 kilograms   (b) 3.9 deciliters   (c) 2.6 centimeters

Thinking Through the Problem Following the reverse of the procedure used in the previous example, we replace each metric prefix with the corresponding exponential term.

Answer

(a) 7.8×10^3 grams
(b) 3.9×10^{-1} liters
(c) 2.6×10^{-2} meters

How are you doing?   7.6

Complete the following equations.

(a) 4.1×10^{-9} meters ↔ _____ meters
(b) 1.90 microliters ↔ _____ liters

PROBLEMS

11. Concept question: Exponent rules apply only to expressions involving multiplication and division rather than addition or subtraction. Choose some number for a and test to see if it is true or not that $a^2 + a^2 = (a^2)^2 = a^4$. Also verify that $a^2 + a^2 = 2a^2$.

12. What are the rules for treating exponents in calculations?

13. What is meant by scientific notation?

14. What metric units are used for mass? Volume? Length? Chemical amount?

15. Fill in the metric prefix for the following:

(a) 1×10^9 g $\leftrightarrow$ _____ g
(b) 1×10^{-1} g $\leftrightarrow$ _____ g
(c) 1×10^3 g $\leftrightarrow$ _____ g
(d) 1×10^{-3} g $\leftrightarrow$ _____ g
(e) 1×10^{-6} liters $\leftrightarrow$ _____ L

16. Fill in the exponent of 10 (give the value of x) for the following metric measurements.

(a) 1 megameter $= 1 \times 10^x$ meter
(b) 1 nanoliter $= 1 \times 10^x$ liter
(c) 1 decagram $= 1 \times 10^x$ gram
(d) 1 centimeter $= 1 \times 10^x$ meter
(e) 1 picogram $= 1 \times 10^x$ gram

17. How many significant figures are there in each of the following measured numbers?

(a) 1,000   (b) 0.00020   (c) 200.00

18. How many significant figures should be kept when these expressions are calculated out?

(a) $\dfrac{(34.0) \times (0.003)}{(0.090) \times (20.0)} =$

(b) $\dfrac{(3.06 \times 10^6) \times (2.000 \times 10^{-8})}{(1.0 \times 10^{17}) \times (2.00 \times 10^{42})} =$

19. Simplify the following.

(a) -1^3 (b) $(-1)^3$

20. Simplify: $(8y^{-5})(2y^{-3})$

21. Simplify: $\dfrac{3x^{-2}}{4x^3}$

22. Simplify the following expressions completely.

(a) $\dfrac{(6a^2b)^3}{(4a^3b)^2}$ (c) $\dfrac{(5x^2y^3)^2\,(8xy)^6}{15x^3(2x^3)^3}$

(b) $(x^2)^4\,(x^2y^3)^2\,(y^4)^3$ (d) $\left(\dfrac{3y^3}{4z^2}\right)^2 \left(\dfrac{2z}{6y^2}\right)^4$

23. Simplify the following expression, writing answers with only positive exponents. Repeat with negative exponents.

$\dfrac{5x^{-2}y^{-3}}{3z^{-1}}$

24. Simplify: $\dfrac{(6.94 \times 10^{41})}{(1.02 \times 10^{-12})} \cdot \dfrac{(8.15) \times 10^{27})}{(6.52 \times 10^{-35})}$

25. Rewrite using scientific notation:

(a) 46,500
(b) 0.0003
(c) 3,610,000
(d) 41.7089
(e) 2,198

26. Suppose that each of the values in Problem 25 refers to a mass in grams. Express this mass in the most convenient metric unit from Table 7-7. For example, 0.0023 g $\leftrightarrow$ 2.3×10^{-3} g $\leftrightarrow$ 3.2 mg, while 0.00001125 g $\leftrightarrow$ 11.25×10^{-6} g $\leftrightarrow$ 11.25 μg.

27. Write the following numbers in decimal form.

(a) 1.63×10^{-4}
(b) 2.317×10^5
(c) 7.3×10^6

28. Imagine that each of the values in Problem 27 refers to a volume in liters. Express this volume in the most convenient metric unit from Table 7-7.

29. For each measurement given in the table below, check the column designating the type of measurement (length, volume, mass). Finally, determine the number of significant figures (s.f.) contained in each measurement.

Measurement	Length, Volume, or Mass?	Number of s.f.
17.25 cm		
2.83 μg		
1.0004×10^2 mL		
32.400 kg		
0.00068 pm		
1.48×10^{-2} dL		
5.230×10^2 cm^3		

7.3 Using Units of Measure

SECTION GOALS

✔ Use proportional reasoning to convert from one metric unit to another.

✔ Use the conversion factor method to convert from one metric unit to another.

✔ Define Avogadro's number and understand how to use it.

Unit Conversions Using Proportional Reasoning

Your progress through chemistry will be greatly aided by the ability to convert between metric units. In this section, we use two methods—proportional reasoning and the conversion factor method—for solving unit conversion problems. When we use proportional reasoning to solve a unit conversion problem we set an "unknown ratio" equal to a certain "known ratio." The unknown ratio comes from the problem; the known ratio comes from definitions such as those contained in Table 7-7. Proportions can be used for conversions between any two types of units as long as we know the relationship between the units.

Example 7.21

Converting a unit with a metric prefix to a base unit using proportional reasoning

Convert 6.2 mm to meters.

Thinking Through the Problem We know the relationship:

$$1 \times 10^{-3}\ m \longleftrightarrow 1\ mm.$$

We can use this equivalency to write the "known ratio" between m and mm: $\dfrac{1 \times 10^{-3}\ m}{1\ mm}$. The "unknown" ratio comes from the statement of the question, and is the ratio of an unknown number of meters to the given number of millimeters: $\dfrac{y\ m}{6.2\ mm}$.

Working Through the Problem We solve the problem by making the known ratio equal to the unknown ratio and solving for the variable y.

$$\frac{1 \times 10^{-3}\ m}{1\ mm} = \frac{y\ m}{6.2\ mm}$$

$$y\ m = \frac{1 \times 10^{-3}\ m}{1\ \cancel{mm}} \times 6.2\ \cancel{mm}$$

Answer $6.2\ mm = 6.2 \times 10^{-3}\ m$

Example 7.22

Converting a unit with a metric prefix to a base unit using proportional reasoning

Convert 2.57 cm to kilometers.

Thinking Through the Problem We know the relationship between cm and m; we also know the relationship between m and km. This will be a two-step problem. First we will use a proportion to convert centimeters to meters, then we will use a second proportion to convert from meters to kilometers.

Working Through the Problem We have 2.57 cm in our problem. Our first proportion is:

$$\frac{x \text{ m}}{2.57 \text{ cm}} = \frac{1 \times 10^{-2} \text{ m}}{1 \text{ cm}}$$

$$x \text{ m} = \frac{1 \times 10^{-2} \text{ m}}{1 \text{ cm}} \times 2.57 \text{ cm}$$

We solve:

$$x \text{ m} = 2.57 \times 10^{-2} \text{ m}$$

We now must convert 2.57×10^{-2} meters to kilometers. For the known ratio we use the definition: $1 \text{ km} = 1 \times 10^{3}$ m. Our proportion is:

$$\frac{2.57 \times 10^{-2} \text{ m}}{x \text{ km}} = \frac{1 \times 10^{3} \text{ m}}{1 \text{ km}}$$

Solving for x, we get:

$$x = 2.57 \times 10^{-5} \text{ km}$$

Answer 2.57 cm is equivalent to 2.57×10^{-5} km

There are two things we could do to streamline calculations such as those shown in Example 7.22.

1. We could recognize that 2.57 cm is equivalent to 2.57×10^{-2} m.

2. We could use conversion factors.

Unit Conversions Using Conversion Factors

Unit conversions using proportional reasoning, in one or more steps, will be the heart of many of the calculations we do in this and later chapters. Now that you are familiar with proportional reasoning we will show how it can be streamlined into the **conversion factor method** (also known as the unit factor method, or dimensional analysis).

To convert a measurement from one metric unit to another, begin with the measurement given in the problem and multiply it by the ratio relating the two units. The ratio must be written so that the desired unit is in the numerator and the unit from which it is being converted is in the denominator. In this way, the original unit divides out and the desired unit remains. This technique is called "the conversion factor method."

Recall that the "known ratios" we use in proportional reasoning problems are the same as the constant of proportionality k in the equation $y = kx$. This means that, if we have the right form of k, our known ratio, we can solve for y by multiplying x by k. In example 7.21 we had to determine the value y in meters. To do this, we multiply the "given" value for x (6.2 mm) by the appropriate ratio (in this case 1×10^{-3} m/1 mm):

$$y = \qquad k \qquad \times \quad x$$
$$= \frac{1 \times 10^{-3} \text{ m}}{1 \text{ mm}} \times 6.2 \text{ mm}$$

The key idea behind the conversion factor method is that we convert x into y by the appropriate ratio. This is mathematically identical to the known/unknown ratio method we have used thus far. But the conversion factor method is much faster. From now on, we will use proportional reasoning in the first of each set of examples, and the conversion factor method thereafter.

We have a choice of writing a ratio in either of two forms when we use our conversion factor. Which ratio is correct to use? *In the conversion factor method, we seek to determine the ratios that convert one unit into another.* This is also known as the *dimensional analysis* method because we analyze the types or dimensions of the units. It has the following general form:

$$\text{starting unit} \times \frac{\text{desired unit}}{\text{starting unit}} = \text{desired unit}$$

KEY IDEA 7.9

Example 7.23

Converting a unit with a metric prefix to a base unit using the conversion factor method

Convert 9.25 Mg to g.

Thinking Through the Problem M is the abbreviation for mega-; Table 7-7 tells us that mega means 10^6.

Working Through the Problem Here the y is in g and the x is in Mg. For the conversion factor k we use the ratio from the definition given in Table 7-7:

$$y \text{ g} = 9.25 \text{ Mg} \times \frac{1 \times 10^6 \text{ g}}{1 \text{ Mg}}$$

Answer $9.25 \text{ Mg} = 9.25 \times 10^6 \text{ g}$

The conversion factor method allows us to do multistep problems in an efficient manner. We string together the conversion factors to form a sequence of calculations. In the case of metric conversions, this sequence usually involves converting from the first metric prefix unit to the base unit and then to the second metric prefix unit:

$$\text{metric prefix 1} \longrightarrow \text{base unit} \longrightarrow \text{metric prefix unit 2}$$

Example 7.24   **Converting among units with a metric prefix**

Convert 0.0925 μg to ng.

Thinking Through the Problem We do not have a direct conversion between μg and ng. Instead, we carry out two steps: Convert μg to g and then g to ng. "μ" is the abbreviation for *micro-*; Table 7-7 tells us that this means 10^{-6}. On the other hand, n is the abbreviation for *nano-*, which means 10^{-9}.

Working Through the Problem ➠ (7.9) We will need to arrange the conversion factors so that the first factor converts 0.0925 μg to g and then the second converts g to ng:

$$0.0925 \ \mu\text{g} \times \frac{1 \times 10^{-6} \ \text{g}}{1 \ \mu\text{g}} \times \frac{1 \ \text{ng}}{1 \times 10^{-9} \ \text{g}} = 9.25 \times 10^1 \ \text{ng}$$

Answer 0.0925 μg $= 9.25 \times 10^1$ ng

Example 7.25   **Converting among units with a metric prefix**

Use the conversion factor method to convert 2.57 centimeters to kilometers.

Thinking Through the Problem This problem was solved in two steps using proportional reasoning in Example 7.22. Here we will use the conversion factor method to help us set up a single calculator entry. As before, we start with the given number and unit. We note that this is a metric-metric conversion; only the prefixes are changed. We expect to use two ratios, one to relate cm to m and the second to relate m to km. We also expect the final answer to be 2.57×10^x, where the exponent x signifies the position of the decimal point for the final unit.

Working Through the Problem

$$2.57 \ \text{cm} \times \frac{1 \times 10^{-2} \ \text{m}}{1 \ \text{cm}} \times \frac{1 \ \text{km}}{1 \times 10^3 \ \text{m}}$$
$$= 2.57 \times 10^{-2-(3)} \ \text{km} = 2.57 \times 10^{-5} \ \text{km}$$

Answer 2.57 cm $= 2.57 \times 10^{-5}$ km

Example 7.26   **Converting among units with a metric prefix**

Convert 1.63×10^{-7} millimeters to picometers.

Thinking Through the Problem We will approach this problem in a similar manner as in the previous example. We need a ratio to convert mm to m and a second ratio to convert from m to pm.

✔Meeting the Goals

Proportional reasoning can be used to convert from one metric unit to another. To do so, use the definitions of each metric prefix in Table 7-7 to write a known ratio or conversion factor.

Working Through the Problem

$$1.63 \times 10^{-7} \text{ mm} \times \frac{1 \times 10^{-3} \text{ m}}{1 \text{ mm}} \times \frac{1 \text{ picometer}}{1 \times 10^{-12} \text{ m}}$$
$$= 1.63 \times 10^{-7+(-3)-(-12)} \text{ pm} = 1.63 \times 10^2 \text{ pm}.$$

We could also think of this process as a *sequence of conversions.* The "unit route" is: mm → m → pm.

Answer 1.63×10^{-7} mm = 1.63×10^2 pm

How are you doing? 7.7

Convert (a) 7.94 ng to grams, (b) 1.8 m to km, and (c) 3.45 mg to μg.

Avogadro's Number and Numerical Calculations

KEY IDEA 7.10

✔Meeting the Goals

Avogadro's number, 6.022×10^{23}, is the number of particles in 1 mole of a substance. The substance whose particles are counted by Avogadro's number must be the same substance as that whose mole quantity is given. We see this in the two ratios,

$$\frac{6.022 \times 10^{23} \text{ molecules H}_2\text{O}}{1 \text{ mole H}_2\text{O}}$$

and $\dfrac{6.022 \times 10^{23} \text{ atoms S}}{1 \text{ mole S}}$. In the first ratio, H_2O molecules are being counted, and in the second ratio, S atoms are being counted.

In Chapter 4, we encountered the concept of the mole for the first time. The key idea of the mole concept is that *we can convert mole statements about substances into statements about individual atoms and substances.* This links the atomic or molecular level to the macroscopic level.

The numerical link between these two scales is, as we said, the ratio that relates the number of particles in 1 mole. In fact, the number in scientific notation is 6.022×10^{23}. We call this **Avogadro's number,** which is sometimes symbolized as N_0. This means that:

(a) 1 mole of C contains 6.022×10^{23} C atoms
(b) 1 mole of CO_2 contains 6.022×10^{23} molecules CO_2

Notice that the number 6.022×10^{23} counts particles of the *same substance.* Statements (a) and (b) each describe a single substance. For an element such as C, the basic unit is the atom. We can only count *atoms* of C. For compounds such as CO_2, the basic unit is the molecule so this is what we count. If we replace the word *contains* with the phrase *is equivalent to* in these sentences, we have two equivalencies that can be used as ratios in unit conversion problems.

In such relationships we are dealing with the known ratios that are the basis of proportional reasoning and conversion factor calculations. In this case, the ratios are (in general form):

$$\frac{6.022 \times 10^{23} \text{ units}}{1 \text{ mole}} \qquad \text{or} \qquad \frac{1 \text{ mole}}{6.022 \times 10^{23} \text{ units}}$$

Converting moles to atoms using proportional reasoning

Example 7.27

How many atoms of Cu are there in 7.04×10^2 moles of Cu?

Thinking Through the Problem The unknown amount in this case is atoms of Cu and we are given moles of Cu. ◄━━ (7.10) We can solve for the unknown by using the appropriate known ratio, in this case Avogadro's number, for Cu atoms and moles of Cu.

Working Through the Problem

$$\text{unknown ratio} = \text{known ratio}$$

$$\frac{y \text{ atoms Cu}}{7.04 \times 10^2 \text{ moles Cu}} = \frac{6.022 \times 10^{23} \text{ atoms Cu}}{1 \text{ mole Cu}}$$

$$7.04 \times 10^2 \text{ moles Cu} \times \frac{y \text{ atoms Cu}}{7.04 \times 10^2 \text{ moles Cu}}$$

$$= \frac{6.022 \times 10^{23} \text{ atoms Cu}}{1 \text{ mole Cu}} \times 7.04 \times 10^2 \text{ moles Cu}$$

$$y \text{ atoms Cu} = 4.24 \times 10^{26} \text{ atoms Cu}$$

Answer 4.24×10^{26} atoms Cu (3 s.f. because of 7.04)

Example 7.28

Converting molecules to moles using the conversion factor method

How many moles of CO are in a sample containing 8.26×10^{26} molecules of CO?

Thinking Through the Problem We are converting from a number of molecules to moles of CO. This is a job for Avogadro's number! The number, 6.022×10^{23}, relates molecules CO to moles of CO. We can use it as a conversion factor. We begin with the number, the unit, and the identity of the given substance.

Working Through the Problem

$$8.26 \times 10^{26} \text{ molecules CO} \times \frac{1 \text{ mole CO}}{6.022 \times 10^{23} \text{ molecules CO}}$$

$$= 1.3716 \times 10^3 \text{ moles CO (calculator answer)}$$

Answer 1.37×10^3 moles of CO (3 s.f. because of 8.26)

Example 7.29

Converting moles to molecules

How many molecules of H_2O are there in 0.0245 mole H_2O?

Thinking Through the Problem We are given moles H_2O and asked for molecules of the same substance. We will use Avogadro's number in an appropriate ratio, set up a calculation sequence, and evaluate the expression.

Working Through the Problem

$$0.0245 \text{ mole } H_2O \times \frac{6.022 \times 10^{23} \text{ molecules } H_2O}{1 \text{ mole } H_2O}$$

$$= 1.4753 \times 10^{22} \text{ molecules } H_2O \text{ (calculator answer)}.$$

Notice that the original unit "mole H_2O" divides out, leaving the desired unit "molecules H_2O."

Answer 1.48×10^{22} molecules H_2O

How are you doing? **7.8**

How many molecules of HNO_3 are contained in a 2.88×10^2-mole sample of HNO_3?

Extended Unit Conversion Problems with Avogadro's Number

The strategy developed in earlier sections can be generalized and used to solve any problem involving conversions between atoms, molecules, and moles.

unit route: moles $\longleftrightarrow$ molecules $\longleftrightarrow$ atoms

conversion factor: N_0 *numerical count*

Avogadro's number (N_0) is the conversion factor between moles and molecules. The relationship between atoms and a molecule containing that atom is found by counting the number of times that element appears in the formula of the compound.

The conversion from moles to molecules always includes Avogadro's number. The conversion from molecules to atoms always involves counting. The only thing that changes in these problems is the chemical substance involved. Suppose that you must convert from moles of a compound to atoms of an element in that compound. The given number has units of moles. You know that you first must convert from moles to molecules or formula units, and so the first ratio must have moles in the denominator (so that it will divide out) and molecules or formula units in the numerator. In the conversion between moles and molecules or formula units, the number 6.022×10^{23} always goes with the unit molecule or formula unit. Because molecules or formula units were in the numerator of the first ratio, they must be in the denominator of the next ratio (so that it also will divide out). This means that atoms of the element must be in the numerator of this second ratio.

KEY IDEA 7.11

Example 7.30

CONNECT TO SECTION 4.1

Converting molecules to atoms using the conversion factor method

How many atoms of O are there in 18 molecules of O_2?

Thinking Through the Problem We will repeat the reasoning we first used in Chapter 4. ➥ (7.11) The key idea is to use a conversion factor method involving a numerical count. It may be tempting to always use Avogadro's number when atoms and molecules are involved, but Avogadro's number counts the number of things in a *mole*. It is *never used in a ratio where the unit mole is not also used*.

Working Through the Problem This problem does *not* involve moles. We merely have to count the number of O atoms in each formula of O_2 and multiply by 18.

$$18 \text{ molecules } O_2 \times \frac{2 \text{ atoms O}}{1 \text{ molecule } O_2} = 36 \text{ atoms O}$$

Answer 36 atoms of O (All the digits are used because both of these numbers are counted numbers. We take account of significant figures only when we have measured numbers.)

Example 7.31   **Converting moles to atoms**

How many atoms of O are there in 42.1 moles of $KMnO_4$?

Thinking Through the Problem In this problem, we have to look at *both* the unit change (moles to atoms) *and* the substance change ($KMnO_4$ to O). This will be a two-step problem. We can use conversion factors to help us solve this problem if we can find the two ratios we need. Avogadro's number relates formula units of $KMnO_4$ to moles of the same substance, $KMnO_4$. O is related to $KMnO_4$ because there are four atoms of O in every formula unit of $KMnO_4$. The unit route here is an important one to note:

$$\text{moles } KMnO_4 \longrightarrow \text{formula units } KMnO_4 \longrightarrow \text{atoms O}$$

Working Through the Problem The first conversion is done using Avogadro's number; the second conversion is done by counting the O's in the formula, $KMnO_4$.

$$42.1 \text{ moles } KMnO_4 \times \frac{6.022 \times 10^{23} \text{ formula units } KMnO_4}{1 \text{ mole } KMnO_4}$$

$$\times \frac{4 \text{ atoms O}}{1 \text{ formula unit } KMnO_4}$$

$$= 1.014 \times 10^{26} \text{ atoms O (calculator answer)}$$

Answer 1.01×10^{26} atoms of O (Both moles and Avogadro's number are measured numbers so the rules for counting significant figures applies here. The answer has 3 s.f.)

Example 7.32   **Converting atoms to moles**

How many moles of H_2SO_4 contain 7.08×10^{22} atoms of O?

Thinking Through the Problem Using the unit route from the previous example, we see that we must convert from atoms of O to molecules of H_2SO_4 by counting the number of O's in the formula. Then we must convert from molecules of H_2SO_4 to moles of H_2SO_4 by using Avogadro's number.

PRACTICAL C   How many moles of sand are on your favorite beach?

We generally consider a particle of sand to be a *grain* of sand. Let's assume that each grain is a spherical particle with a diameter of 1.00 mm. Calculate the volume of a grain of sand in cm^3 ($V = 4/3 \pi r^3$).

Now calculate the mass of this grain of sand, assuming it is pure SiO_2 (the density of which is 2.64 g cm^{-3}). Be careful of units!

If we assume that the sand at your local beach is 3 m deep, what volume of sand is in a 0.50-km stretch of beach that is 100 m wide?

Use the volume of one grain of sand that you calculated above to determine how many grains of sand are on this beach. ($1 \times 10^{-6} m^3 \leftrightarrow 1 cm^3$)

How many moles of sand (SiO_2) are on the beach?

If the area of the Sahara Desert is $8.6 \times 10^{12} m^2$ and we assume that the desert sand has a depth of 20 m, how many grains of sand are in this desert?

Working Through the Problem

$$7.08 \times 10^{22} \, \cancel{atoms \, O} \times \frac{1 \, molecule \, \cancel{H_2SO_4}}{4 \, \cancel{atoms \, O}}$$

$$\times \frac{1 \, mole \, H_2SO_4}{6.022 \times 10^{23} \, \cancel{molecules \, H_2SO_4}} = 0.02939 \, \text{mole of } H_2SO_4$$

(calculator answer)

Answer 2.94×10^{-2} mole of H^2SO^4

How are you doing? 7.9

How many moles of C_3H_6 contain 2.94×10^{20} atoms of C?

PROBLEMS

30. Concept question: Most of us are accustomed to American (or English) units of measurement—inches, feet, pounds, etc. So it is easy for us to see that the statement "6 feet equals 1/2 foot" is wrong. The correct statement is "6 *inches* equals 1/2 foot." But without much experience in the metric system, it is easy to make a mistake with some proportionalities, such as "2 meters equals 0.02 cm." The correct statement is "2 meters equals 200 cm." Suggest a way for you and other students to develop an intuitive grasp of metric sizes.

31. The definition of density is the ratio of mass to volume. In metric units, the density is given as the mass in grams per 1 mL. If the density of a liquid is 1.50 g/mL, what is the mass of 25 mL of the liquid?

32. The speed of light is 3.00×10^{10} cm/sec. The moon is about 384,000 km from earth. How many seconds would it take a radio signal from

earth traveling at the speed of light to reach the moon?

33. The volume of blood flowing through an adult liver is 25.0 mL sec^{-1}. How many liters per minute is this?

34. Simplify: $\dfrac{(2.94 \times 10^{23})}{(8.02 \times 10^{-52})} \cdot \dfrac{(7.15 \times 10^{21})}{(4.52 \times 10^{-65})}$

35. Simplify: $\dfrac{(3.64 \times 10^{81})}{(1.75 \times 10^{-72})} \cdot \dfrac{(8.85 \times 10^{-7})}{(6.94 \times 10^{135})}$

36. Simplify: $\dfrac{(5.12 \times 10^{-91})}{(1.84 \times 10^{27})} \cdot \dfrac{(4.75 \times 10^{23})}{(9.24 \times 10^{75})}$

37. Convert 1.07 Mm to km.

38. Convert 3.44×10^{-2} dm to mm.

39. Convert 7.53 GL to μL.

40. Convert $6.83 \times 10^{-4} m^3$ to cm^3.

41. Convert 2.03×10^{-4} mm to pm.

42. Convert 2.65 mL to μL.

43. Convert 1.89×10^{-3} megaliter to deciliters.

44. Convert 3.72 nanoliters to milliliters.

45. Perform the following conversions. Report answers with the correct number of significant figures.

(a) 10.52 mL to L
(b) 1.50 L to mL
(c) $18.3\ cm^3$ to mL
(d) $125.0\ cm^3$ to L
(e) 0.1095 g to mg
(f) 65.2 kg to g
(g) 5837 g to kg

46. What is the mass in grams of a 40.0-mL sample of mercury? Mercury has a density of 13.55 $g(mL)^{-1}$.

47. How many moles of CO_2 contain 6.022×10^{23} atoms of O?

48. How many atoms of H are contained in 3.44×10^{-2} mole of NH_4HSO_3?

49. What is the volume of a 34.5-g sample of cork? Cork has a density of $0.740\ g\ cm^{-3}$. (Volume will have units of cm^3 in this problem.)

Chapter 7 Summary and Problems

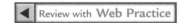 ◀ Review with **Web Practice**

VOCABULARY

exact number	An exact number results when a unit is defined without any uncertainty, that is, either by complete counting or by definition.
estimated number	We get an estimated number when we measure something and thus need to estimate part of our answer.
accuracy	The accuracy of an experimental value tells us how closely it agrees with some known value.
precision	Precision means that the experimental value is roughly the same every time that a measurement is done. We report numbers with as many digits as are consistent with the precision of the instrument.
measured numbers	The numbers that result when a measurement is taken.
random error	Random error is due to normal fluctuations that occur when the same task is repeated a number of times.
systematic error	Systematic error may be due to some fault in the lab equipment, such as a poorly calibrated thermometer, or contamination in a standard solution.
significant figures	The significant figures in a measured number include all digits you can read with certainty plus one uncertain estimated digit.
trailing zeroes	Zeroes that appear at the end of a number.
leading zeroes	Zeroes that appear as placeholders at the beginning of a number that is between $+1$ and -1.
exponent	A superscript number that indicates the number of times the base appears as a factor.
zero exponent rule	Any nonzero expression raised to the zero power is equal to 1.
negative exponent rule	Raising an expression to a negative power is the same as raising its reciprocal to the positive power.
scientific notation	A number written as the product of a number, n, between 1 and 10 (i.e., $1 \leq n < 10$) and a power of 10.
magnitude	The size or extent of the sample being measured, correlated with the exponent of 10 when the number is written in scientific notation.

metric system of units	An international system of units used in most countries of the world and in all areas of science. The metric unit system is called the International System of Units and the units are commonly referred to as SI units.
metric prefix	A metric prefix is used with basic units such as gram, meter, and liter to indicate the magnitude of the measurement. (See Table 7-7.)
conversion factor method	A method of converting one unit to another using a ratio containing the two units in the correct proportion.
Avogadro's number	Avogadro's number (sometimes symbolized as N_0) is equal to 6.022×10^{23} objects. N_0 is the number of particles in one mole. Thus, there are 6.022×10^{23} atoms of C in one mole of C and 6.022×10^{23} molecules of CO_2 in one mole of CO_2.

SECTION GOALS

What is the difference between accuracy and precision?	An accurate measurement is one that is close to the correct number for that measurement. A precise measurement is one that agrees closely with other measurements of the same quantity.
Why is the last digit in a measured number estimated?	The last digit in a measured number should be an estimate made by judging its value between the gradations on our instrument.
What are significant figures? Why do we count them in a measured number?	All the digits in a measured number are considered significant, both the certain digits and the final uncertain digit. We count significant figures to get an idea of how precise the measurement is. Measurements with more digits following the decimal point are more precise than those with fewer digits after the decimal point.
When are zeroes counted as significant figures in a measured number?	The concept of significant figures applies only to measured numbers. In a measured number, the following rules apply. 1) All nonzero digits in a measured number are significant. 2) Leading zeroes are never significant. 3) Trailing zeroes are significant only when there is an expressed decimal in the number. 4) Trailing zeroes in numbers without an expressed decimal point are never counted as significant figures.
How many significant figures can you include in a calculated answer when measured numbers are used?	For multiplication or division problems, the final answer cannot contain any more significant figures than the smallest number of significant figures appearing in the problem. For addition and subtraction problems, the final answer cannot contain any more places after the decimal point than the smallest number of decimal places appearing in the problem.
What properties of exponents do we need to know for our calculations?	To multiply two monomials with the same base, add the exponents: $x^a x^b = x^{a+b}$
	To raise a monomial with a power to a power, multiply the exponents: $(x^a)^b = x^{ab}$
	To divide monomials with the same base, subtract the exponents: $\dfrac{x^a}{x^b} = x^{a-b}$
	Any nonzero expression raised to the zero power is equal to 1: $x^0 = 1$.
	An expression raised to a negative power is the same as its reciprocal raised to the positive power: $x^{-a} = \dfrac{1}{x^a}$
What is scientific notation?	A number written in scientific notation consists of a number n between -10 and -1 or between 1 and 10 multiplied by ten raised to some power. The exponent of 10 equals the number of places the decimal in the

original number was moved in order to have scientific notation. A negative exponent indicates that the decimal number was less than 1 and greater than -1. A positive exponent indicates that the decimal number was greater than 10 or less than -10.

How can you estimate the magnitude of a product by using the rules of exponents?

To estimate the magnitude of the answer to a multiplication/division calculation, use the properties of exponents to find the exponent of 10 in the calculated answer. Often this is enough of an estimation.

What metric units are used for mass, volume, and length?

The basic units of the metric system are meter (length), gram (mass), liter (volume), and mole (chemical amount). The metric unit for temperature (kelvin degrees) will be used in later applications.

How do metric prefixes correspond with the relative size of the measured object?

Metric prefixes correlate with an exponent of 10 when a number is written in scientific notation. As the exponent of 10 tells where the decimal point belongs when the number is written in decimal notation, the metric prefix also indicates the relative size of the measured object.

What are the abbreviations for the metric units and metric prefixes used in this chapter?

The metric prefixes in Table 7-7 can be abbreviated. Generally the abbreviation is the first letter of the prefix. There are two exceptions: *deca-* is abbreviated *da-* to distinguish it from *deci-* (*d-*); and *micro-* is written with the Greek lowercase mu, μ, to distinguish it from *milli-* (*m-*) and from *mega-* (*M-*).

How is proportional reasoning used to convert from one metric unit to another?

Proportional reasoning techniques can be used to convert from one metric unit to another. To do so, use the definitions of each metric unit (Table 7-7) to write a "known ratio." Write the corresponding "unknown ratio," using the measurement and question in the problem. Finally, set the two ratios equal and solve for the unknown quantity.

How is the conversion factor method used to convert from one metric unit to another?

To convert a measurement from one metric unit to another, begin with the measurement given in the problem and multiply it by the ratio relating the two units. The ratio must be written so that the desired unit is in the numerator and the unit that is being converted from is in the denominator. In this way, the original unit divides out and the desired unit remains. This technique is sometimes called *dimensional analysis*.

What is Avogadro's number? How is it used?

Avogadro's number, 6.022×10^{23}, expresses the number of atoms in 1 mole of an element or the number of molecules in 1 mole of a compound. The substance whose particles are being counted by Avogadro's number must be the same substance as that measured by the mole. We see this in the two ratios, $\dfrac{6.022 \times 10^{23} \text{ molecules } H_2O}{1 \text{ mole } H_2O}$ and $\dfrac{6.022 \times 10^{23} \text{ atoms } S}{1 \text{ mole } S}$.

In the first ratio, H_2O is the substance whose particles are being counted. S is the substance in the second ratio.

PROBLEMS

50. The following numbers (grams) were reported as the result of an experimental measurement in which a precisely manufactured machine part was weighed: 22.1, 22.4, 21.8, 22.3, 22.1, 22.3, 21.9, 22.2.

(a) Find the average value of these measurements by summing them and dividing by the number of measurements.

(b) What is the lowest value given? the highest value? (The lowest and highest values together give what is called the range of values.)

(c) How precise (close together) are these values?

(d) The manufacturer tells us that the "true" value for this measurement is exactly 22.3 g. Comment on the accuracy of the values listed in this problem.

(e) The error in your measurement is the difference between the "true" value given by the manufacturer and the average of the experimental values. What is the error for these measurements?

51. Each member of the chemistry class measured the volume of the same irregularly shaped object. These volumes (in liters) are as follows: 0.01235, 0.01242, 0.01233, 0.01250, 0.01225, 0.01238, 0.01230.

(a) What is the average value of the volume determined by this class?

(b) What is the range of values?

(c) What is the mass in grams of the object if the average density reported by the class is 125 g per liter?

52. The diameter of an atom is approximately 1 Å, where 1 Å is 1×10^{-10} m. How many atoms can be lined up side by side to make a line that measures 1.00 mm long?

53. Simplify: $7x^{-4} \cdot 6x^{-3}$

54. Simplify: $(8x^{-2}y^3)^2 \cdot (2x^4y^{-5})^{-3}$

55. A 15.0-g sample of a pure substance has a volume of 22.4 mL. What volume would a 50.0-gram sample have?

56. Perform this calculation and report the answer with the correct number of significant figures. (You may have to regroup and figure out the exponent of 10 using the rules of exponents.)

$$\frac{(4.00 \times 10^{48})\,(0.0050)}{(2.08 \times 10^{52})\,(6.022 \times 10^{23})}$$

57. A sample of tap water contains 26.80 ppm Pb. How many atoms of Pb are there in a glass of tap water (120.0 g)?

58. How many molecules of $HClO_4$ can be made from 2.8×10^4 atoms of O?

59. How many moles of $KHC_8H_4O_4$ are there in a sample that contains 6.022×10^{22} molecules of $KHC_8H_4O_4$?

60. A sample of water contains 50 ppb of arsenic. How many atoms of As are there in 225 grams of this water?

61. Indicate the number of significant figures in each of these measured numbers:

(a) 10.00   (d) 0.100
(b) 0.001320 (e) 10035
(c) 14.00674 (f) 0.010

62. Perform the following operations. Report answers to the correct number of significant figures.

(a) $0.00002345 \times 0.003230$

(b) $0.123 + 0.02342$

(c) $\dfrac{(0.123 + 24.095)}{(1.084)^3}$

(d) $\dfrac{(1.305 + 2.97 - 5.0)}{(5.21) \times (11.01)}$

63. Rewrite using scientific notation:

(a) 0.000000873
(b) 6008.35
(c) 5,260,000,000
(d) 0.00000914
(e) 8.904

64. Write the following numbers in decimal form.

(a) 3.12×10^{-1}
(b) 9.64×10^{-7}
(c) 5.35×10^2

65. Perform the following conversions. Report answers with the correct number of significant figures.

(a) 250.0 mL to L
(b) 15.0 L to mL
(c) 25.95 cm^3 to mL
(d) 9.87 cm^3 to L
(e) 1.68×10^5 μg to g
(f) 4.975 g to μg
(g) 4.027×10^{19} mg to kg

Measurement of Chemical Substances

Medicines and vitamin supplements contain carefully measured amounts of chemical substances. (Kim Fennema/Visuals Unlimited.)

PRACTICAL CHEMISTRY Medicinal Drugs—Molecules with Special Uses

The pharmaceutical industry supplies Americans with numerous medicinal drugs and supplemental nutrients each year. In fact, most of us would hardly recognize a city or town in which there was no drugstore. Medicinal drugs are chemicals, both naturally-occurring and synthetic organic molecules. Many drugs belong to "families" of molecules that have similar structures and formulas.

This chapter contains additional information about measuring chemical substances and understanding chemical formulas. The chapter Practicals apply what you know about measurement and chemical formulas to some special molecules that are most likely found in your medicine cabinet. This chapter will focus on the following questions:

PRACTICAL A What are B vitamins and why are they important?

PRACTICAL B Have you had your ascorbic acid today?

PRACTICAL C What common medicines and drugs come from plants?

PRACTICAL D Drug therapy—where do the drugs come from?

8.1 Chemical Formula and Molar Mass

SECTION GOALS

✔ Define atomic mass, formula mass, and molar mass.

✔ Relate mass measurements to the number of moles of a substance.

✔ Calculate a molar mass to the correct precision, using the precision of atomic masses as a guide.

Atomic Mass

In the preceding four chapters, we introduced some very important ideas in chemistry. In Chapters 4 and 5, we discussed chemical reactions and introduced the mole as a quantity that measures chemical amounts. In Chapter 6, our discussion of measuring matter covered other extensive properties, including mass and volume. We saw how mass and volume are proportional to one another, and how mole amounts give another set of ratios for proportional reasoning. Finally, we discussed in Chapter 7 the practical issues of measurement in the laboratory, where significant figures and different measurement scales are important.

In this chapter, we connect the way we measure chemical substances in the laboratory—using mass especially—and chemical counting. Although we measure mass in the laboratory, it is more important for chemists to be able to count moles. Fortunately, mass and moles have a proportional relationship, so we can use proportional reasoning techniques to relate mass and mole amounts for a given substance.

Before we begin to use the mass/mole ratio, we need to take a step back and ask, "How do we know the masses of the chemical elements?" Each different element has a characteristic mass. We call this mass the **atomic mass.**

Chemists have a standard for measuring atomic mass. *It is the atom of carbon with six protons, six neutrons, and six electrons. Because this atom has a mass number of 12, we call it "carbon-12."* We define the **atomic mass scale** as follows: the mass of an atom of carbon-12 has a mass of exactly 12. All atomic masses are relative to the mass of carbon-12. Note that we don't have a unit here, because the atomic mass scale is a relative scale that will work in *any* mass measurement system. (Soon we will link the atomic mass scale to the mass unit chemists use most—the gram.)

The atomic masses of the element are given to us on the periodic table or on a separate atomic mass table. If you examine a complete table of atomic masses, like the one inside the back cover, you will see that different atomic masses have different precisions. This is because of variations in the amounts of different isotopes in naturally occurring elements. Some elements, such as F, have only one isotope in nature. These elements have very precise atomic masses. Other

Figure 8-1 ▲ Mass spectrometers are instruments that can measure the masses of individual molecules or atoms. (Dan McCoy/Rainbow.)

elements, such as Pb, have many isotopes in different amounts in different samples. These elements have the least precise atomic masses. How the amounts and masses of isotopes affect atomic mass will be discussed further in Chapter 12.

At the level of measuring individual atoms, we can define a unit called the **"atomic mass unit"** or amu. The atomic mass unit is defined as exactly 1/12 the mass of a carbon-12 atom. A single carbon-12 atom has a mass of exactly 12 amu.

The amu is useful in cases where we study the mass of individual atoms and molecules. Chemists can measure the masses of individual atoms and molecules using an instrument called a mass spectrometer, which is shown in Figure 8-1.

Example 8.1 — **The precision of atomic masses**

Locate the masses given for the following elements in the periodic table and in the table of atomic masses inside the cover of the book. Which element has a mass that is known with the greatest precision? Which element has a mass that is known with the least precision?

(a) Pb   (b) C   (c) H   (d) Cl   (e) O   (f) N   (g) Fe

Thinking Through the Problem We examine the table of atomic masses and the periodic table, then note the precision of each atomic mass by looking at the number of places after the decimal point.

✔Meeting the Goals

The atomic mass is related to the number of protons, neutrons, and electrons in an atom. An atom of carbon-12 has an atomic mass of exactly 12.

Working Through the Problem From the table of atomic masses:
(a) 1 (b) 4 (c) 5 (d) 3 (e) 4 (f) 4 (g) 3 (number of decimal places)
From the periodic table:
(a) 1 (b) 3 (c) 4 (d) 3 (e) 3 (f) 3 (g) 3 (number of decimal places)
Answer Pb is the least precise; H is the most precise.

Formula Mass

KEY IDEA 8.2

When atoms combine to make a substance, we can assume by the law of conservation of mass that the mass of the substance will be the sum of the masses of all the atoms in the substance. This sum is known as the **formula mass** and is also the sum of the atomic masses of all the elements present in the formula unit. A formula mass of a substance is determined by multiplying the atomic mass of each element present (read from a periodic table or table of atomic masses) by the number of times its symbol appears in the formula. Like atomic masses, formula masses have no units.

Example 8.2 Calculating formula mass

What is the formula mass of HCl?

Thinking Through the Problem We are given the chemical formula. We then use the atomic masses of H and Cl from the periodic table. This problem is a simple addition of the atomic mass of H and the atomic mass of Cl.

Working Through the Problem The formula mass of HCl is then:

$$
\begin{array}{ll}
1\ \text{H} & 1.0079 \\
1\ \text{Cl} & \underline{35.453} \\
& 36.4609
\end{array}
$$

Answer The formula mass of HCl without rounding off is 36.4609.

Example 8.3 Calculating formula mass

What is the formula mass of ammonia, NH_3?

Thinking Through the Problem We proceed as in the last example, but we must multiply the atomic mass of H by 3 because there are three H atoms in one molecule of NH_3.

Working Through the Problem

$$
\begin{array}{lllllll}
1\ \text{nitrogen} & 1 & \times & 14.007 & = & 14.007 \\
3\ \text{hydrogen} & 3 & \times & 1.0079 & = & \underline{3.0237} \\
& & & & & 17.0307
\end{array}
$$

Answer The formula mass of NH_3 without rounding off is 17.0307.

✔Meeting the Goals

A formula mass is the sum of the atomic masses of all elements present in the formula unit. Look at the periodic table for the atomic masses or a table of atomic masses of each element present.

CONNECT TO
SECTION 7.1

Notice that the atomic masses in Example 8.3 are both significant to the same number of decimal places. But what happens when this is not the case? Figure 8-2 illustrates a macroscopic analogy of this situation. When we combine two masses that are known to different degrees of precision, we use the rules of addition and significant figures that we learned in Chapter 7. For example, let's determine the formula mass of carbon tetrafluoride, CF_4. Carbon has an atomic mass that is precise to four places after the decimal point (12.0107). The atomic mass of fluorine is precise to *seven* places after the decimal point (18.9984032). When we combine these atomic masses, here is what happens:

1 carbon	1	$\times$	12.0107	=	12.0107
4 fluorine	4	$\times$	18.9984032	=	75.9936128

88.0043128 (calculator answer)

88.0043 (correct answer)

We have written "calculator answer" because calculators are not programmed to determine uncertainty. Because the atomic mass of C is precise only to the fourth place after the decimal, the value for the formula mass of CF_4 is also precise only to the fourth place after the decimal. Rather than report insignificant figures, our best answer is rounded off to the fourth position after the decimal, 88.0043.

Example 8.4

Finding precision of formula masses

Determine the formula masses of the following substances to the correct precision.

(a) SF_4 (b) N_2O_5 (c) HI

Thinking Through the Problem We do additions of the atomic masses in each case. The final result is rounded to the number of digits in the least precise atomic mass.

Answer

(a) SF_4: 108.059 (S least precise)
(b) N_2O_5: 108.0104 (N and O equally precise)
(c) HI: 127.91241 (H and I equally precise)

How are you doing? 8.1

Determine the formula masses of the following substances to the correct precision.

(a) IF_7 (b) N_2O_3 (c) HBr

Figure 8-2 ▲ There are many practical situations where the precision of our measurements can be very different.
(Tom Pantages.)

Molar Mass

Atomic and formula masses can be used to measure compounds with any mass unit. But in chemistry the preferred mass unit is the *gram* and the preferred counting unit is the *mole*. The **molar mass** of a substance is the mass, in grams, of one mole of the substance.

Figure 8-3 ▶ The molar masses of different substances can vary greatly. Here we see one mole each of copper, salt, and water. (Richard Megna/ Fundamental Photographs.)

$$\text{mass of substance} = k \times \text{moles of substance}$$

$$k = \frac{\text{mass of substance}}{\text{moles of substance}} = \text{molar mass}$$

The molar masses of different substances can vary greatly. For example, each of the samples in Figure 8-3 is one mole, but the masses are quite different.

Molar mass is a constant value for a specific substance. It is the constant of proportionality between the mass of a substance and the corresponding number of moles of the substance. We get the molar mass from the formula mass of the substance by adding the units g mol^{-1}. For example:

$$\text{formula mass of HCl} = 36.461;$$

$$\text{molar mass of HCl} = \frac{36.461 \text{ g HCl}}{1 \text{ mol HCl}} = 36.461 \text{ g mol}^{-1}$$

$$\text{formula mass of NH}_3 = 17.0305;$$

$$\text{molar mass of NH}_3 = \frac{17.0305 \text{ g NH}_3}{1 \text{ mol NH}_3} = 17.0305 \text{ g mol}^{-1}$$

KEY IDEA 8.3

✔ **Meeting the Goals**

Note that a ratio of grams/ moles expresses the same quantity as g mol^{-1}.

The units of molar mass are important. In general chemistry classes, the units will always be *grams* of the substance *per mole* of the substance.

Although many atomic masses are known to six or seven decimal places, as shown in the alphabetical listing in the back of the book, most of the numerical answers we calculate will include at least one measured number with only 3 or 4 significant figures. Therefore, the precision of the atomic masses listed in the periodic table is sufficient for all further calculations.

Example 8.5 **Finding molar mass**

What is the molar mass of acetic acid, CH_3COOH?

Thinking Through the Problem We first determine the formula mass of acetic acid by adding the atomic masses of the atoms that are in the molecule. The formula mass is then converted into the molar mass by indicating the appropriate units, $g \, mol^{-1}$.

Working Through the Problem

2 carbon atoms	2	×	12.011	= 24.022
2 oxygen atoms	2	×	15.999	= 31.998
4 hydrogen atoms	4	×	1.0079	= 4.0316

60.0516 (calculator answer)

Formula mass: 60.052 (correct answer)

In this example, the number of places after the decimal point is determined by the mass of carbon, which has three places after the decimal point.

$$\text{Molar mass:} \quad \frac{60.052 \text{ g } CH_3COOH}{1 \text{ mol } CH_3COOH} \quad \text{or} \quad 60.052 \text{ g mol}^{-1}$$

Answer The molar mass of CH_3COOH is $60.052 \text{ g mol}^{-1}$.

Example 8.6 **Finding molar mass**

What is the molar mass of the amino acid glycine, $C_2H_5NO_2$?

Thinking Through the Problem We set up this problem in the same way that we did example 8.5.

Working Through the Problem

✔Meeting the Goals

The molar mass is the formula mass of a substance expressed in units of $g \, mol^{-1}$.

2 carbon atoms	2	×	12.011	= 24.022
2 oxygen atoms	2	×	15.999	= 31.998
1 nitrogen atom	1	×	14.007	= 14.007
5 hydrogen atoms	5	×	1.0079	= 5.0395

75.06650 (calculator answer)

Formula mass: 75.067 (correct answer)

$$\text{Molar mass:} \quad \frac{75.067 \text{ g } C_2H_5NO_2}{1 \text{ mol } C_2H_5NO_2} \quad \text{or} \quad 75.067 \text{ g mol}^{-1}$$

Answer The molar mass of $C_2H_5NO_2$ is $75.067 \text{ g mol}^{-1}$.

> **How are you doing?** **8.2**
>
> (a) What is the molar mass of dioxin, $C_{12}H_2O_2Cl_4$?
> (b) What is the molar mass of cholesterol, $C_{26}H_{45}O$?
> (c) What is the molar mass of hydrazine, N_2H_4?

The Molar Mass Is a Ratio for Converting Between Moles and Mass

The molar mass of a chemical substance is one of its most important properties. Why? Because we cannot directly measure the number of formula units or molecules in a sample; there are no "mole-meters" in the laboratory. However, in the lab we can measure the *mass* of a sample quite easily. Because molar mass is a ratio between grams and moles of a substance, we can use it to calculate either grams or moles when we know one of these amounts, as shown in Figure 8-4. Molar mass is a known ratio that can be used as a conversion factor in mass to mole calculations. For example, if we know that we need 3.000 moles of HCl for an experiment, we first calculate the molar mass (36.4606 g mol^{-1}) and then multiply this by the number of moles desired:

$$3.000 \text{ mol HCl} \times \frac{36.4609 \text{ g HCl}}{1 \text{ mol HCl}} \approx 109.4 \text{ g HCl}$$

We need 109.4 grams of HCl in order to have 3.000 moles of HCl.

Example **8.7**

Calculating grams from moles using molar mass (proportional reasoning)

We need to measure out 3.9970 moles of iron in order to do a particular experiment. What is the mass of iron we should measure?

Thinking Through the Problem ➧ (8.3) The molar mass ratio relates moles of iron to grams of iron. This molar mass is easy to obtain because it has a value equal to the atomic mass of Fe, 55.847 g mol^{-1}. Because we know that grams Fe and moles Fe are proportional, we can calculate grams Fe using the molar mass ratio, $\dfrac{55.847 \text{ grams of Fe}}{1 \text{ mole of Fe}}$, as our known ratio.

Working Through the Problem

$$\text{unknown ratio} = \text{known ratio}$$

$$\frac{x \text{ g Fe}}{3.9970 \text{ mol Fe}} = \frac{55.845 \text{ g Fe}}{1 \text{ mol Fe}}$$

$$3.9970 \text{ mol Fe} \times \frac{x \text{ g Fe}}{3.9970 \text{ mol Fe}} = \frac{55.845 \text{ g Fe}}{1 \text{ mol Fe}} \times 3.9970 \text{ mol Fe}$$

$$x \text{ g Fe} = \frac{55.845 \times 3.9970}{1} \text{ g Fe}$$

$$x \text{ g Fe} = 223.21 \text{ g Fe}$$

Answer 223.21 g Fe are needed to give 3.9970 mol Fe.

✔ **Meeting the Goals**

Grams and moles of a substance are related by the molar mass of that substance. Don't forget that the molar mass is different for each substance. When solving a problem be sure first to note which substance is being described by a particular molar mass.

Figure 8-4 ▶ Molar mass is the constant of proportionality between grams and moles.

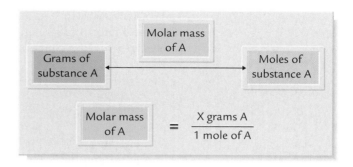

Calculating grams from moles using molar mass (proportional reasoning)

Example 8.8

An experiment calls for 0.200 mole of $CaCl_2$. What mass of $CaCl_2$ is needed to obtain this amount?

Thinking Through the Problem We start by calculating the molar mass of $CaCl_2$. The molar mass ratio is used to convert moles to grams of $CaCl_2$.

Working Through the Problem First, calculate the formula mass of $CaCl_2$.

1 calcium atom	1	×	40.078	=	40.078
2 chlorine atoms	2	×	35.453	=	70.906

$$110.984$$
(calculator answer)

$$110.984$$
(correct answer)

The molar mass of $CaCl_2$ is $110.984 \text{ g mol}^{-1}$. Next, convert moles of $CaCl_2$ to grams of $CaCl_2$. We need to find the number of grams of $CaCl_2$, so we will write the molar mass as a simple ratio, with grams of $CaCl_2$ in the numerator and moles of $CaCl_2$ in the denominator.

$$\text{unknown ratio} = \text{known ratio}$$

$$\frac{x \text{ g CaCl}_2}{0.200 \text{ mol CaCl}_2} = \frac{110.984 \text{ g CaCl}_2}{1 \text{ mol CaCl}_2}$$

$$x \text{ g CaCl}_2 = \frac{(0.200 \text{ mol CaCl}_2) \times (110.984 \text{ g CaCl}_2)}{1 \text{ mol CaCl}_2}$$

$$x \text{ g CaCl}_2 = \frac{110.984 \times 0.200}{1} \text{ g CaCl}_2$$

$$x \text{ g CaCl}_2 = 22.2 \text{ g CaCl}_2$$

Answer We need to weigh out 22.2 g $CaCl_2$.

We solved Examples 8.7 and 8.8 by using the unknown/known ratio method of proportional reasoning. We showed that molar mass is "just another ratio" and that we can solve problems with it as with any other ratio. Now, in the following examples, we will use the conversion factor method to carry out more grams-from-moles and moles-from-grams conversions.

Example 8.9   **Calculating grams from moles using molar mass**

Determine the mass of argon, Ar, that is equivalent to 22.94 moles of Ar.

Thinking Through the Problem We are seeking a mass amount from a mole amount. We will use the molar mass of Ar and solve by using a conversion factor. The molar mass of Ar is 39.948 g mol^{-1}. To convert from moles to grams, we multiply by the conversion factor $\dfrac{39.948 \text{ g Ar}}{1 \text{ mol Ar}}$.

Working Through the Problem

$$22.94 \text{ mol Ar} \times \frac{39.948 \text{ g Ar}}{1 \text{ mol Ar}} = 916.4 \text{ g Ar}$$

Answer   916.4 g Ar

Example 8.10   **Calculating moles from grams using molar mass**

There are 0.250 g of iron (II) sulfate, $FeSO_4$, in a standard iron supplement. How many moles of $FeSO_4$ is this?

Thinking Through the Problem We start with the number of grams that we are given. We must also calculate the molar mass of $FeSO_4$, so we start with the correct ratio of grams of $FeSO_4$ per moles of $FeSO_4$. We need to find the number of moles of $FeSO_4$. We use the molar mass of $FeSO_4$ as a conversion factor, writing it as its inverse, with moles of $FeSO_4$ in the numerator and grams of $FeSO_4$ in the denominator.

Working Through the Problem

1 iron atom	1	×	55.847	=	55.847
4 oxygen atoms	4	×	15.9994	=	63.996
1 sulfur atom	1	×	32.066	=	32.066
					151.909
					(calculator answer)
					151.909
					(correct answer)

The molar mass of $FeSO_4$ is 151.909 g mol^{-1}.

$$0.250 \text{ g FeSO}_4 \times \frac{1 \text{ mol FeSO}_4}{151.909 \text{ g FeSO}_4} = 0.001646 \text{ mol FeSO}_4$$

Answer   0.001646 or 1.646×10^{-3} mol $FeSO_4$

PRACTICAL A   What are B vitamins and why are they important?

Vitamins are nutrients essential to our bodies' health. In the best of all possible worlds, we would get our full recommended daily requirement of vitamins from the foods we eat, but there are times when we may require a supplement, either from specially fortified foods, or from vitamin "pills." The B vitamins are a group of nutrients that include thiamin (B_1), riboflavin (B_2), and niacin (B_6). Niacin, which lowers the risk of heart disease, is found in fortified breads and cereals. Riboflavin and thiamin also are found in fortified pastas, cereals, and breads. They are essential B vitamins, necessary for various chemical processes in the body. The names and formulas of these three B vitamins are given below, along with the masses of these nutrients contained in common vitamin pills. Determine the molar mass for each of these vitamins. Then, using the mass contained in each pill, calculate the number of moles of each vitamin per pill.

See this with your Web Animator ▼

Vitamin	Name	Chemical Formula	Molar Mass	Mass/pill	Moles/pill
B_1	Thiamin	$C_{12}H_{18}Cl_2N_4OS$		2.5 mg	
B_2	Riboflavin	$C_{17}H_{20}N_4O_6$		2.5 mg	
B_6	Niacin	$C_8H_{11}NO_3$		20 mg	

Suppose that a vitamin pill also contains other B vitamins such as folic acid, biotin, and pantothenic acid. Determine the molar mass and the moles per pill for each of these.

Name	Chemical Formula	Molar Mass	Mass/pill	Moles/pill
Folic acid	$C_{19}H_{19}N_7O_6$		200 μg	
Biotin	$C_{10}H_{16}N_2O_3S$		20 μg	
Pantothenic acid	$C_{18}H_{34}NO_{10}$		5 μg	

How are you doing? 8.3

Determine the number of moles of substance in each of the following samples.

(a) 25.00 g $PbCl_2$
(b) 100. g $C_3H_8O_3$

Determine the mass of substance in each of the following samples.

(c) 0.0211 mol $FeCl_3$
(d) 18.23 mol $Fe(NO_3)_3$

PROBLEMS

1. Complete the following table by supplying the missing quantities. The first has been done for you as an example.

Conversion	Ratio Needed in Calculation
Grams $H_3PO_4 \longrightarrow$ moles of H_3PO_4	$\dfrac{1 \text{ mol } H_3PO_4}{97.994 \text{ g}}$
Number of H atoms $\longrightarrow$ molecules of H_3PO_4	
Molecules $H_3PO_4 \longrightarrow$ number of O atoms	
Number of H atoms $\longrightarrow$ moles of $C_6H_{12}O_6$ atoms	
Moles of P atoms $\longrightarrow$ grams of P	

2. For each of the following substances, determine the molar mass in grams, give the symbol and name of the element whose atomic mass has the *fewest* places after the decimal, and determine the number of significant figures in the final answer.
 (a) H_2O_2 (c) HIO_4
 (b) H_2SO_3 (d) $C_2F_4H_2$

3. Calculate the number of moles of $(NH_4)_3PO_3$ in 266 grams of $(NH_4)_3PO_3$.

4. Calculate the number of grams of $Ba(C_2H_3O_2)_2$ contained in 8.022 moles of $Ba(C_2H_3O_2)_2$.

5. Calculate the number of moles of NO_2 in 0.0123 grams of NO_2.

6. Calculate the number of grams of Cl_2O_3 that contain 12.5 moles of Cl_2O_3.

7. Calculate the mass of 3.25 moles of potassium nitrate, KNO_3.

8. Calculate the number of moles of NO in 34.6 grams of NO.

9. Calculate the number of grams of N_2O_5 in 4.95 moles of N_2O_5.

10. Calculate the number of moles of compound in each of the following samples.
 (a) 125 g $SnCl_4$
 (b) 91.2 g $FeCO_3$

11. Calculate the number of grams of each compound in each of the following samples.
 (a) 0.0010 mol SF_4
 (b) 23.4 mol $CaCO_3$

12. Discussion question: Calculate the molar masses of the alkali metal carbonates. Suppose that you have a "mystery" compound containing carbonate and an alkali metal. The molar mass is between 90 and 100 g per mole. Which compound do you think this is? Discuss your answer to the same question if you are told that the compound contains carbonate and an alkaline earth element.

8.2 Multiple Conversions and Chemical Amounts

SECTION GOALS

✔ Set up multiple-step calculations that involve masses and moles.

✔ Use density to determine the number of moles of a substance in a measured volume of that substance.

Mass and Mole Conversions

By now, you have worked on several problems using a known proportionality to determine an unknown quantity in some other situation. This process is often thought of as "converting" from one quantity to another. The word *converting* deserves some comment, however. When we convert, for example, 120.11 grams of C (carbon) to 10.000 moles of C, *we are not changing the C in any way. Instead, we are altering how we describe the carbon,* as Figure 8-5 shows.

As in Chapters 6 and 7, we use proportional reasoning as a way to move from one measurement system, such as grams, to another, such as moles. *Molar mass is the way we convert between mass and mole amounts.*

Converting between mass amounts and mole amounts for the same substance requires only a single calculation, as we saw in Section 8.1. For example, we calculate the moles of CH_4 in a 12.1-g sample of CH_4 by solving the proportion

$$\frac{12.1 \text{ g } CH_4}{x \text{ mol } CH_4} = \frac{16.0426 \text{ g } CH_4}{1 \text{ mol } CH_4}$$

$$x \text{ mol } CH_4 = 12.1 \text{ g } CH_4 \times \frac{1 \text{ mol } CH_4}{16.0426 \text{ g } CH_4}$$

$$= 0.754 \text{ mol } CH_4$$

In this problem, all numerical values refer to the same substance, CH_4. The molar mass ratio relates grams and moles of the *same* substance.

Suppose instead that we have a sample of 12.1 g CH_4 and wish to know how many moles of H are in this sample. This problem requires more than one step. We must convert grams to moles of CH_4 (step 1) and then we must convert moles CH_4 to moles H (step 2).

Given: 12.1 g CH_4 Step 1: Molar mass will convert grams of CH_4 to moles of CH_4 (the same substance).

Find: moles H   Step 2: Count 4 moles of H per 1 mole of CH_4.

So we can carry out our conversion in two steps. We can set up this two-step problem as follows:

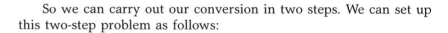

$$\text{g of } CH_4 \xrightarrow[\text{Step 1}]{\frac{1 \text{ mol } CH_4}{16.0426 \text{ g } CH_4}} \text{mol of } CH_4 \xrightarrow[\text{Step 2}]{\frac{4 \text{ mol } H}{1 \text{ mol } CH_4}} \text{mol of } H$$

The problem can be solved using proportional reasoning:

Step 1: g CH_4 to mol CH_4

$$\frac{x \text{ mol } CH_4}{12.1 \text{ g } CH_4} = \frac{1 \text{ mol } CH_4}{16.0426 \text{ g } CH_4}$$

$$12.1 \text{ g } CH_4 \times \frac{x \text{ mol } CH_4}{12.1 \text{ g } CH_4} = \frac{1 \text{ mol } CH_4}{16.0426 \text{ g } CH_4} \times 12.1 \text{ g } CH_4$$

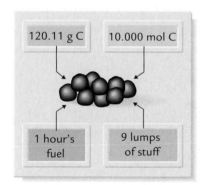

Figure 8-5 ▲ There are many ways to describe a chemical sample such as these lumps of carbon. The best way depends on the viewpoint of the observer.

✔ Meeting the Goals

With so many ratios available, careful planning of a calculation is a required first step. Do not just plunge in with the first number you can think of.

$$x \text{ mol CH}_4 = \frac{1 \times 12.1}{16.0426} \text{ mol CH}_4$$

$$x \text{ mol CH}_4 = 0.7542 \text{ mol CH}_4$$

Step 2: mol CH$_4$ to mol H

$$\frac{x \text{ mol H}}{0.7542 \text{ mol CH}_4} = \frac{4 \text{ mol H}}{1 \text{ mol CH}_4}$$

$$0.7542 \text{ mol CH}_4 \times \frac{x \text{ mol H}}{0.7542 \text{ mol CH}_4} = \frac{4 \text{ mol H}}{1 \text{ mol CH}_4} \times 0.7542 \text{ mol CH}_4$$

$$x \text{ mol H} = \frac{4 \times 0.7542}{1} \text{ mol H}$$

$$x \text{ mol H} = 3.016 \approx 3.02 \text{ mol H}$$

The conversion factor method allows us to solve this problem more easily. In this problem we begin with 12.1 grams of CH$_4$. We can convert grams of CH$_4$ to moles of CH$_4$ using the molar mass ratio and then apply the second ratio to convert moles of CH$_4$ to moles of H in a second multiplication:

$$12.1 \text{ g CH}_4 \times \frac{1 \text{ mol CH}_4}{16.0426 \text{ g CH}_4} \times \frac{4 \text{ mol H}}{1 \text{ mol CH}_4} = 3.02 \text{ moles of H}$$

The mole/mole ratio is written so that mol CH$_4$ divides out. The final result is 3.02 moles of H.

Example 8.11

Calculating moles of an element in a mass of a compound

Determine the number of moles of Mg in 95.0 grams of Mg$_3$N$_2$.

Thinking Through the Problem Look carefully at what is given and what you must find.

Given 95.0 grams of Mg$_3$N$_2$, find moles of Mg.

🔑 (8.5) The key idea is to use molar mass to convert between grams and moles. This requires a conversion from grams to moles of Mg$_3$N$_2$ and a second conversion from moles of Mg$_3$N$_2$ to moles of Mg. This is a two-step calculation. The chemical formula allows us to determine both the molar mass (100.929 g mol^{-1}) and the mole/mole ratio that we need for the calculation (3 mol Mg per mol Mg$_3$N$_2$). The problem setup is:

$$\text{grams of Mg}_3\text{N}_2 \xrightarrow[\text{Step 1}]{\frac{1 \text{ mol Mg}_3\text{N}_2}{100.929 \text{ g Mg}_3\text{N}_2}} \text{moles of Mg}_3\text{N}_2$$

$$\text{moles of Mg}_3\text{N}_2 \xrightarrow[\text{Step 2}]{\frac{3 \text{ mol Mg}}{1 \text{ mol Mg}_3\text{N}_2}} \text{moles of Mg}$$

Working Through the Problem

$$95.0 \text{ g } \cancel{\text{Mg}_3\text{N}_2} \times \frac{1 \text{ mol } \cancel{\text{Mg}_3\text{N}_2}}{100.929 \text{ g } \cancel{\text{Mg}_3\text{N}_2}} \times \frac{3 \text{ mol Mg}}{1 \text{ mol } \cancel{\text{Mg}_3\text{N}_2}} =$$

$$2.823 \approx 2.82 \text{ mol Mg}$$

Answer 2.82 mol Mg

Example 8.12

Calculating moles of a compound from grams of an element

Determine the number of moles of Au_2O_3 that contains 0.012 gram of Au.

Thinking Through the Problem Compare what is given and what you must find.

Given 0.012 gram of Au, find moles Au_2O_3.

This is the reverse of the last problem. Convert grams Au to moles Au, then use the mole/mole ratio to find moles of Au_2O_3 from this number of moles of Au. The problem setup is:

$$\text{grams of Au} \xrightarrow[\text{Step 1}]{\dfrac{1 \text{ mol Au}}{196.97 \text{ g Au}}} \text{moles of Au} \xrightarrow[\text{Step 2}]{\dfrac{1 \text{ mol Au}_2\text{O}_3}{2 \text{ mol Au}}} \text{moles of Au}_2\text{O}_3$$

Working Through the Problem

$$0.012 \text{ g } \cancel{\text{Au}} \times \frac{1 \text{ mol } \cancel{\text{Au}}}{196.97 \text{ g } \cancel{\text{Au}}} \times \frac{1 \text{ mol Au}_2\text{O}_3}{2 \text{ mol } \cancel{\text{Au}}} \approx$$

$$3.0 \times 10^{-5} \text{ mol Au}_2\text{O}_3$$

Answer 0.000030 or 3.0×10^{-5} mol Au_2O_3

How are you doing? 8.4

Determine the number of moles of calcium in 25.0 grams of $CaCl_2$ and in 25.0 grams of $CaCO_3$. Which has more Ca?

REMEMBER

The most complex of these kinds of calculations involves relating mass amounts of two different substances. This requires *three* steps, summarized by the diagram shown in Figure 8-6. We relate mole amounts using mole/mole ratios. We relate mole amounts and masses using molar masses.

Example 8.13

Calculating grams of a compound from grams of an element

Determine the number of grams of U_3O_8 that contain 250.0 grams of U.

Figure 8-6 ▶ Multiple conversions in a chemical substance.

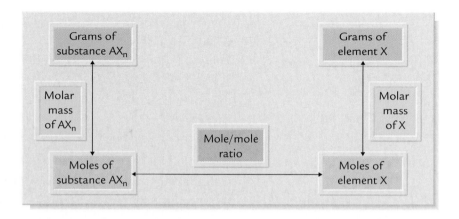

Thinking Through the Problem This problem is aimed at getting a mass amount from another mass amount. Figure 8-6 guides us from grams of U to moles of U (molar mass of U 238.03 g mol^{-1}), then moles of U to moles of U_3O_8 (mole/mole ratio from the formula), and finally moles of U_3O_8 to grams of U_3O_8 (molar mass of U_3O_8 842.09 g mol^{-1}). The problem setup is:

$$\text{g U} \xrightarrow[\text{Step 1}]{\dfrac{1 \text{ mol U}}{238.03 \text{ g U}}} \text{mol U} \xrightarrow[\text{Step 2}]{\dfrac{1 \text{ mol U}_3O_8}{3 \text{ mol U}}} \text{mol U}_3O_8$$

$$\xrightarrow[\text{Step 3}]{\dfrac{842.082 \text{ g U}_3O_8}{1 \text{ mol U}_3O_8}} \text{g U}_3O_8$$

Working Through the Problem

$$250.0 \text{ g U} \times \frac{1 \text{ mol U}}{238.03 \text{ g U}} \times \frac{1 \text{ mol U}_3O_8}{3 \text{ mol U}} \times \frac{842.082 \text{ g U}_3O_8}{1 \text{ mol U}_3O_8} =$$
$$294.8 \text{ g U}_3O_8$$

Answer   294.8 g U_3O_8

How are you doing?   8.5

Determine the number of grams of iron in 1.00 gram of $FeSO_4$.

Volume, Mass, and Mole Calculations

CONNECT TO
SECTION 6.2

Multiple steps are also useful when we use density and molar mass to determine the number of moles in a sample measured by volume. Recall from Chapter 6 that the density of a substance is the ratio of its mass to its volume. In this case, our calculations will follow the diagram in Figure 8-7.

The density ratio relates mass to volume of the same substance; the molar mass ratio relates mass to moles of the same substance.

Example 8.14

Calculating moles from volume using density and molar mass

The volume of Hg in a simple mercury thermometer is 0.245 cm³. How many moles of Hg are present in this volume? The density of Hg is 13.6 g cm⁻³.

Thinking Through the Problem The volume and the density are given. Compare what is given with what we must find.

Given 0.245 cm³ of Hg, find moles of Hg.

We calculate grams of Hg from cm³ of Hg by using the density ratio. We can calculate moles of Hg from grams of Hg by using the molar mass ratio. We need the molar mass of Hg from the periodic table.

We can plan our work using Figure 8-7. We must carry out two steps in this problem. The problem setup is:

Step 1: $\text{cm}^3 \text{ Hg} \longrightarrow \text{g Hg}$ *use density of Hg*

Step 2: $\text{g Hg} \longrightarrow \text{mol Hg}$ *use molar mass of Hg*

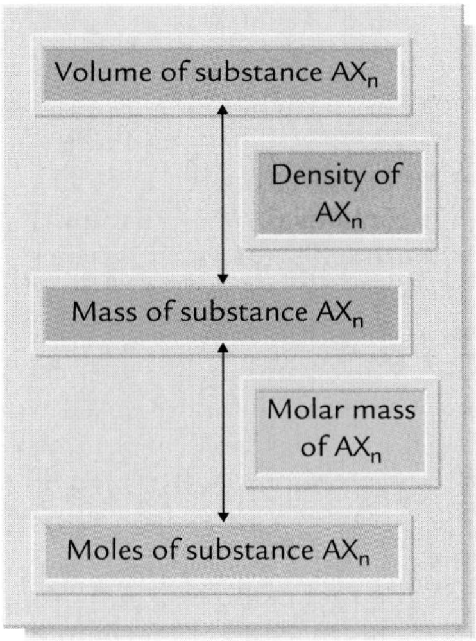

Working Through the Problem

Step 1: cm³ Hg to g Hg (given d = 13.6 g/cm³)

$$\frac{x \text{ g Hg}}{0.245 \text{ cm}^3 \text{ Hg}} = \frac{13.6 \text{ g Hg}}{1 \text{ cm}^3 \text{ Hg}}$$

$$\frac{x \text{ g Hg}}{0.245 \text{ cm}^3 \text{ Hg}} \times 0.245 \text{ cm}^3 \text{ Hg} = 0.245 \text{ cm}^3 \text{ Hg} \times \frac{13.6 \text{ g Hg}}{1 \text{ cm}^3 \text{ Hg}}$$

$$x \text{ g Hg} = 0.245 \times 13.6 \text{ g Hg}$$

$$x \text{ g Hg} = 3.332 \text{ g Hg}$$

Step 2: 333.2 g Hg to mol Hg

$$\frac{x \text{ mol Hg}}{3.332 \text{ g Hg}} = \frac{1 \text{ mol Hg}}{200.59 \text{ g Hg}}$$

$$3.332 \text{ g Hg} \times \frac{x \text{ mol Hg}}{3.332 \text{ g Hg}} = \frac{1 \text{ mol Hg}}{200.59 \text{ g Hg}} \times 3.332 \text{ g Hg}$$

$$x \text{ mol Hg} = \frac{1 \times 3.332}{200.59} \text{ mol Hg}$$

$$x \text{ mol Hg} = 0.0166 \text{ mol Hg}$$

This problem can also be solved by using conversion factor methods. In this case, the calculator setup is:

$$0.245 \text{ cm}^3 \text{ Hg} \times \frac{13.6 \text{ g Hg}}{1 \text{ cm}^3 \text{ Hg}} \times \frac{1 \text{ mol Hg}}{200.59 \text{ g Hg}} = 0.0166 \text{ mol Hg}$$

Answer 0.0166 mol Hg

Figure 8-7 ▲ Conversions from volume to moles of a chemical substance.

✔ **Meeting the Goals**

For a given substance, density gives the relationship between grams and volume, and molar mass gives the relationship between grams and moles. Because both density and molar mass deal with mass amounts, they can be used together to relate volume amounts to mole amounts, as shown in Figure 8-7.

PRACTICAL B   Have you had your ascorbic acid today?

Ascorbic acid, or vitamin C, is present in or routinely added to many of the foods we eat. Orange juice is perhaps the most popular and well-known source of vitamin C. Vitamin C is reputed to possess healing properties and might play a role in the treatment of conditions as varied as cardiovascular disease and allergies. Yet its formula, $C_6O_6H_8$, is relatively simple. Determine the molar mass of ascorbic acid, $C_6O_6H_8$.

The recommended daily value (DV) for vitamin C is 60 mg (100 mg for cigarette smokers). The nutrition panel on an orange juice carton states that each serving (1 glass) supplies 120% DV of vitamin C. Can you calculate how many molecules of vitamin C are in one glass of orange juice?

This is an extended calculation that uses Avogadro's number! One glass of orange juice supplies 120% of the daily requirement of vitamin C. The daily requirement of vitamin C is 60 mg. One glass will give us 72 mg vitamin C.

See this with your **Web Animator** ▼

$$\frac{120 \text{ mg vitamin C}}{100 \text{ mg vitamin C}} \times 60 \text{ mg} = 72 \text{ mg vitamin C}$$

We can convert mg vitamin C to molecules if we first convert mg to g, then g to moles, and finally moles to molecules. Write in the correct numbers for the molar mass of $C_6O_6H_8$ and for Avogadro's number in the following calculator setup to calculate molecules of vitamin C. Then calculate the answer.

$$72 \text{ mg } C_6O_6H_8 \times \frac{1 \text{ g}}{1000 \text{ mg}} \times \frac{1 \text{ mol } C_6O_6H_8}{\_\_\_ \text{ g } C_6O_6H_8}$$

$$\times \frac{\_\_\_ \text{ molecules } C_6O_6H_8}{1 \text{ mol } C_6O_6H_8} =$$

How many moles of $C_6O_6H_8$ are there in one serving? How many molecules?

Suppose that a different brand of orange juice has a density of 1.06 g/mL. It contains 0.050% vitamin C by mass. Find the volume of this orange juice that contains the DV for vitamin C (60 mg).

Vitamin K helps blood to clot. It is present in green leafy vegetables, such as broccoli, and in liver. Suppose that a solution of natural vitamin K has 1.6×10^{-6} moles of vitamin K in 1.00 mL. What volume of this solution is needed to supply the DV (60 µg) of vitamin K ($C_{31}H_{46}O_2$)?

Determine the volume, in mL, that contains 60 µg of $C_{31}H_{46}O_2$. Determine the mole amounts and the number of molecules present in a 1.00-gram sample of each of the vitamins in the following table.

Name	Chemical Formula	Molar Mass	Moles in 1.00 g	Molecules in 1.00 g
Vitamin E	$C_{28}H_{48}O_2$			
β-carotin (provitamin A)	$C_{40}H_{56}$			
Vitamin K	$C_{31}H_{46}O_2$			
Vitamin D_2	$C_{28}H_{44}O$			

How are you doing? 8.6

Determine the volume of 100.0 mol of water, which has a density of 0.995 g cm^{-3}.

PROBLEMS

13. Calculate the following quantities.

(a) How many moles of C are in 24.4 grams of $Ba(C_2H_3O_2)_2$?

(b) How many moles of H are in 264.2 moles of $(NH_4)_3PO_3$?

14. Calculate the number of moles of each element in 100. grams of each of the following substances.

(a) $SnCl_4$ (c) $Ca(ClO_4)_2$

(b) $FeCO_3$ (d) $(NH_4)_2HPO_4$

15. Calculate the mass of metal in one mole of the following compounds.

   (a) KBr

   (b) Mn_2O_3

   (c) Fe_3O_4

16. How many moles of $Ca(ClO_4)_2$ are in a sample of 1.12 grams of $Ca(ClO_4)_2$?

17. Calculate the number of grams of Fe and Cl present in 0.4318 gram of $FeCl_2$.

18. What is the mass of a sample of $MnCl_2$ that contains 1.22 moles of Cl?

19. Calculate the number of moles of liquid water in 0.025 liter of water, if water has a density of 995 g L^{-1}.

20. Calculate the volume of liquid water in 25.0 moles of water, if water has a density of 995 g L^{-1}.

21. Discussion question: Calculate the number of moles of each of the following gases, liquids, or metals in 0.0010 liter of the substance. The density of each is given. You will need to determine the molar mass.

GASES:

   (a) H_2 (hydrogen), 0.084 g L^{-1}

   (b) Cl_2 (chlorine), 2.95 g L^{-1}

   (c) CH_4 (methane), 0.666 g L^{-1}

   (d) Xe (xenon), 5.59 g L^{-1}

LIQUIDS:

   (e) $C_3H_8O_3$ (glycerol), 1,260 g L^{-1}

   (f) H_2O, 995.0 g L^{-1}

   (g) C_2H_5Br (ethyl bromide), 1,460 g L^{-1}

   (h) $C_{10}H_{22}$ (decane), 730 g L^{-1}

   (i) C_5H_{12} (pentane), 630 g L^{-1}

METALS:

   (j) Pb (lead), 11,400 g L^{-1}

   (k) Fe (iron), 7,860 g L^{-1}

   (l) Al (aluminum), 2,700 g L^{-1}

   (m) Pu (plutonium), 19,800 g L^{-1}

Using these data, discuss any trends (or lack of trends) when the densities of different gases are compared. Discuss any trends (or lack of trends) when the numbers of moles of each gas in 0.0010 liter of that gas are compared. Do the same for the liquids and for the metals.

8.3   Mass Composition of Compounds

SECTION GOALS

✔ Know how to calculate percentage, by mass, of a compound.

✔ Understand how the percent by mass ratio is used.

Determining the Percentage Composition of a Compound

In Section 8.1, we saw how to calculate the formula mass of a compound from the elements in the compound. We used the number of elements in the compound (from the chemical formula) and the atomic mass of those elements. In Section 8.2, we developed a general procedure (summarized in Figure 8-6) to calculate the mass amounts of the elements in a sample of a compound. There is a simple way to streamline this procedure, for cases where we need to repeatedly calculate the mass of the elements in many different samples of a compound. This is to use the mass percentage or mass percent.

We calculate the **mass percentage** for each element in the compound by using a mass/mass ratio based on the chemical formula. *To do this, we divide the mass of each element in one mole of the compound by the mass of the whole compound and multiply by 100%.* For example, the mass/mass ratio for any element A in a compound A_2X_3 is:

$$\frac{2 \times \text{ mass A}}{\text{mass } A_2X_3}$$

The mass percentage of A in A_2X_3 is:

$$\frac{2 \times \text{molar mass A}}{\text{molar mass } A_2X_3} \times 100\%$$

For example, the mass of one mole of carbon monoxide is 28.010 g. One mole of carbon atoms, with a mass of 12.011 g, is present in one mole of CO. Therefore, the mass percentage of carbon in carbon monoxide is:

$$\frac{12.011 \text{ g C}}{28.010 \text{ g CO}} \times 100\% = 42.881\% \text{ C}$$

The mass percentage of oxygen can be found in the same manner. One mole of oxygen atoms (15.999 grams) is present in one mole of CO.

$$\frac{15.999 \text{ g O}}{28.010 \text{ g CO}} \times 100\% = 57.118\% \text{ O}$$

Note that the sum of the mass percentages of all the elements in a compound is 100%.

Both of these calculations involve measured numbers with five significant figures. For the purposes of mass percentage calculations in this book, we will save only *four* significant figures in our answer. This is considered sufficient for most practical problems. Note that the mole amounts from the chemical formula (one mole C and one mole O) are counted. These do not affect significant figures.

The calculation of mass percentages gives us another property of a compound, one that relates directly to the measurement by mass. Figure 8-8 shows mass percentages being determined. Consider the values shown in Table 8-1. These are the percentages of oxygen and carbon in the various compounds formed by those two elements. These, *and only these*, percentages are found. There are no known compounds with percentages of C and O that differ from those in this table.

Figure 8-8 ▲ Chemists commonly use elemental analysis by mass percentage to determine the identity and purity of a sample. (Geoff Tompkinson/Science Photo Library/Photo Researchers.)

Table 8-1 Percentage Composition of Carbon-Oxygen Compounds

Compound	Carbon	Oxygen
A	42.88%	57.12%
B	27.29%	72.71%
C	52.96%	47.04%
D	65.24%	34.76%
E	72.43%	27.57%

This is a very important restriction on the mass composition of chemical compounds. In the next section, we will discuss how we can use mass data like this to determine the formula of a compound. For now, let's determine the mass percentage of an element in a compound from its formula.

Example 8.15 — Calculating mass percentage

Determine the mass percentage of carbon and hydrogen in ethane, C_2H_6.

Thinking Through the Problem ➔ (8.6) The key idea is that mass percentage is a mass/mass ratio. We need to find the mass of one mole of C_2H_6 and then the mass contribution of C and H.

Working Through the Problem First, we find the mass of one mole of C_2H_6.

$$\text{Mass of one mole of } C_2H_6 = 30.0694 \text{ g } C_2H_6$$

Mass of C in one mole C_2H_6:
$$2 \text{ mol C} \times \frac{12.011 \text{ g C}}{1 \text{ mol C}} = 24.022 \text{ g C}$$

Mass of H in one mole C_2H_6:
$$6 \text{ mol H} \times \frac{1.0079 \text{ g H}}{1 \text{ mol H}} = 6.0474 \text{ g H}$$

Secondly we find the mass percentage of C and H.

$$\text{Mass \% C: } \frac{24.022 \text{ g C}}{30.0694 \text{ g } C_2H_6} \times 100\% = 79.89\% \text{ C}$$

$$\text{Mass \% H: } \frac{6.0474 \text{ g H}}{30.0694 \text{ g } C_2H_6} \times 100\% = 20.11\% \text{ H}$$

Answer The compound C_2H_6 contains 79.89% by mass of C and 20.11% by mass of H.

Example 8.16 — Calculating mass percentage

Determine the mass percentage of bromine, carbon, and hydrogen in C_3H_7Br.

Thinking Through the Problem We find the mass of one mole of C_3H_7Br, then we calculate the mass percentage of each element.

Working Through the Problem

$$\text{Mass of one mole of } C_3H_7Br = 122.9926 \text{ g } C_3H_7Br$$

Mass of C in one mole of C_3H_7Br:
$$3 \text{ mol C} \times \frac{12.011 \text{ g C}}{1 \text{ mol C}} = 36.033 \text{ g C}$$

Mass of H in
mole of C_3H_7Br: $7 \; \text{mol H} \times \dfrac{1.0079 \text{ g H}}{1 \text{ mol H}}$ $=$ 7.0553 g H

Mass of Br in
mole of C_3H_7Br: $1 \; \text{mol Br} \times \dfrac{79.904 \text{ g Br}}{1 \text{ mol Br}}$ $=$ 79.904 g Br

Mass % C: $\dfrac{36.033 \text{ g C}}{122.9926 \text{ g } C_3H_5Br} \times 100\%$ $=$ 29.30% C

Mass % H: $\dfrac{7.0553 \text{ g H}}{122.9926 \text{ g } C_3H_5Br} \times 100\%$ $=$ 5.736% H

Mass % Br: $\dfrac{79.904 \text{ g Br}}{122.9926 \text{ g } C_3H_5Br} \times 100\%$ $=$ 64.97% Br

Answer The compound C_3H_7Br contains 29.30% C, 5.736% H, and 64.97% Br by mass.

How are you doing? 8.7

Determine the mass percentage of the elements in the compound $(NH_4)_2HPO_4$.

Use of Percentage Composition

In Chapter 6, we used percentage calculations in many problems. Knowing the percentage composition of a compound and the mass of an actual sample gives us a convenient way to determine how much of an element is in the sample. This is a very practical application of percentage composition. *Percentage composition gives us a mass/mass ratio that is constant for all samples of the substance.* For example, we can determine the percentage of H in NH_3 as

$$\dfrac{3.0237 \text{ g H}}{17.0307 \text{ g } NH_3} \times 100\% = 17.75\% \text{ H}$$

Now, whenever we need to know the grams of H in any mass of NH_3, we can simply calculate 17.75% of whatever mass of NH_3 is given. Suppose that we want to know the grams of H in 500.0 g NH_3. Then, using the % hydrogen ratio for H we get:

$$\dfrac{17.75 \text{ g H}}{100 \text{ g } NH_3} \times 500.0 \text{ g } NH_3 = 88.77 \text{ g H}$$

It is important to understand that this mass/mass ratio is a physical property. For example, the percentage of hydrogen in the compound water does not change, regardless of the size of a sample of water. H_2O always contains 11.19% H and 88.81% O. Because the

✔ **Meeting the Goals**

To calculate the percent composition, by mass, of a compound: (a) determine the molar mass of the compound, (b) evaluate the mass/mass ratio formed by dividing the *total* mass contribution of each element by the molar mass of the compound, and (c) convert to percent by multiplying by 100.

mass/mass ratio expressed as percent is constant, we can use it to determine the mass of an element in a compound given any known mass of that compound.

unknown ratio = known ratio

$$\frac{x \text{ g A}}{y \text{ g compound}} = \frac{\text{mass A}}{100 \text{ g compound}}$$

If y is known, then we can solve this proportion for the x grams of A in the sample.

Being able to calculate the percentage composition of an element in a compound can be very useful. For example, by calculating the percentage composition of chromium in two different chromium compounds used in mineral supplements, we can determine which supplement contains the higher dosage of chromium. Figure 8-9 shows two different mineral supplements.

Example **8.17**

Using mass percentage to find the mass of an element from the mass of a compound

Calcium sulfate is a very common source of calcium. What mass, in grams, of calcium is present in a 225-gram sample of pure $CaSo_4$?

Thinking Through the Problem ➤ (8.7) The mass/mass ratio used in percentage composition gives us a constant value. We need the ratio of $\dfrac{\text{mass Ca}}{\text{mass of one mole of } CaSo_4}$. The mass of one mole of $CaSo_4$ is 100.087 g. There is one mole of Ca per mole of $CaSo_4$. The mass of Ca in one mole of $CaSo_4$ is therefore 40.078 grams. The mass percentage is:

$$\frac{40.078 \text{ g Ca}}{136.14 \text{ g } CaSo_4} \times 100\% = 29.439\%$$

All mass percentages of Ca to $CaSo_4$ will have this value. Therefore, we set this ratio equal to a ratio containing the variables from the problem. In the statement of the problem we are asked for the mass of calcium, so this is the unknown in the mass percentage. We are given the mass amount of calcium sulfate as 225 grams.

Working Through the Problem

Using Proportional Reasoning:

unknown ratio = known ratio

$$\frac{x \text{ g Ca}}{225 \text{ g } CaSo_4} = \frac{29.439 \text{ g Ca}}{100 \text{ g } CaSo_4}$$

$$x \text{ g Ca} = 29.439 \times 225 \div 100 \text{ g Ca}$$
$$x \text{ g Ca} = 66.237 \approx 66.2 \text{ g Ca}$$

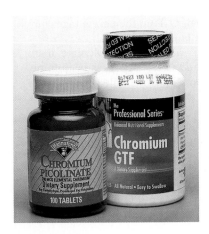

Figure 8-9 ▲ Mineral supplements for the same element may contain different compounds of that element, which affects the amount of the element we get in a single dose. (Tom Pantages.)

✔ **Meeting the Goals**

When we know the percentage by mass of an element A in a compound (%A), we can calculate the mass of that element in any mass of that compound using the percent. Percentage of A gives us the ratio:

$$\frac{\text{mass A}}{100 \text{ g of compound containing A}}$$

Using Conversion Factor Method:

$$225 \text{ g CaSO}_4 \times \frac{29.439 \text{ g Ca}}{100 \text{ g CaSO}_4} = 66.2 \text{ g Ca}$$

Answer There are 90.1 grams of calcium in a 225-gram sample of calcium sulfate.

PRACTICAL C   What common medicines and drugs come from plants?

Many illnesses in the past were treated with remedies discovered by chance. For example, two hundred years ago, sailors and ship passengers who were away from land for long periods of time often suffered from scurvy, a disease characterized by bleeding gums, muscle and joint pains, and, sometimes, dementia. Ships with a cargo that included oranges and grapefruit were found to have a much lower incidence of scurvy among their passengers and crew. Soon, sailors learned that eating oranges could prevent scurvy. Today we know that the beneficial ingredient of citrus fruits, and other fruits and vegetables, is vitamin C. A deficiency of vitamin C in the diet caused scurvy in sailors deprived of fresh fruits and vegetables during long voyages. To obtain a sufficient amount of vitamin C in your diet, you should eat five to nine servings of fruits and vegetables every day.

Another illness that was common in the past among travelers in hot, equatorial climates was malaria. It was found that people who chewed on the bark of the cinchona tree had fewer and milder symptoms of malaria. The bark of this tree contains quinine, which is still used to treat malaria today.

Practicing chemists use percentage composition data from the laboratory to help them determine chemical formulas of new or unknown substances. Conversely, we can use molecular formulas of known substances to determine percentage compositions. For example, the simplest formula for ascorbic acid (vitamin C) is $C_3H_4O_3$. What is the percentage of each element in ascorbic acid?

Now calculate the percentage compositions of the elements in $C_6H_8O_6$ and $C_{12}H_{16}O_{12}$ and compare them to the values you obtained for ascorbic acid.

What is happening here? Can you predict the %C in $C_9H_{12}O_9$?

How does this work? Each of these formulas can be reduced to $C_3H_4O_3$, the simplest formula. The *mole ratio of elements* is the *same* in all of these formulas. This leads to the identical percentages that you just calculated for C, H, and O in three very different substances.

The most obvious advantage of percentage values is that they can be used with *any* mass amount. The simplest formula for quinine, a substance found in the bark of the cinchona tree, is $C_{10}H_{12}NO$, with an element ratio of 10 moles of C, 12 moles of H, 1 mole of N, and 1 mole of O. Find the percentage composition of the elements in quinine.

These percentages are used to calculate mass amounts of C, H, N, and O for any mass of quinine. Once mass amounts are known, mole amounts can be determined, using the molar mass. Find the mass amounts and the mole amounts of each element in a 325-gram sample of quinine.

How do these mole amounts compare to the mole amounts shown by the subscript numbers in the simplest formula, $C_{10}H_{12}NO$? If the true formula for quinine has a molar mass of about 325 g mol^{-1}, what is the true formula for quinine?

PROBLEMS

22. Concept question: Chemists sometimes use mole ratios in analyzing systems. Explain why the mole ratio and the mass ratio differ by calculating each for CH_4.

23. Determine the mass percentage of carbon in the following compounds.

(a) Acetic acid: $C_2H_4O_2$
(b) Glucose: $C_6H_{12}O_6$

(c) Ribose: $C_5H_{10}O_5$
(d) Cyclopropane: C_3H_6
(e) Hexane: C_6H_{12}
(f) Ethylene: C_2H_4

Are there any trends or consistencies among your answers? Why?

24. Determine the mass percentage of sodium in each of the following compounds. Which would be the best to use as a source of sodium?

(a) Sodium sulfate: Na_2SO_4
(b) Sodium carbonate: Na_2CO_3
(c) Sodium chloride: $NaCl$
(d) Sodium hydroxide: $NaOH$

25. Determine the mass percentages for all of the elements in the following formulas.

(a) KOH (c) $LaCl_3$
(b) $RbBr$ (d) $LuCl_3$

26. Determine the mass percentages for all of the elements in the following formulas.

(a) CCl_4 (c) CCl_2H_2 (e) CH_4
(b) CCl_3H (d) $CClH_3$

27. The following ores are available for the production of copper. Which ore will give the most copper per 100 grams of the ore?

(a) Chalcocite: Cu_2S
(b) Malachite: $Cu_2CO_3(OH)_2$
(c) Chalcopyrite: $CuFeS_2$

28. Vanadium is a metal used in special steels. It is often found in complex minerals. Determine the percentage of vanadium in the ores lead chloride vanadate, $PbCl_2(Pb_3(VO_4)_2)_3$, and potassium uranyl vanadate, $KUO_2(VO_4)H_2O$.

29. Determine the mass percentage of each of the elements in the following compounds.

(a) K_2SO_4
(b) $Fe(C_5H_5)_2$
(c) $Ba(C_2H_3O_2)_2$

30. Which compound has the greatest percentage by mass of N: $(NH_4)_3PO_4$, $Al(CN)_3$, or NH_3?

31. Determine the percentage by mass of the metal in the following compounds:

(a) $NaCl$ (b) $PbCl_2$ (c) UCl_4

32. Which one in the following pairs of samples contains more *grams* of metal?

(a) 0.250 gram of $NaCl$ or 0.250 mole of $NaCl$
(b) 0.350 gram of $PbCl_2$ or 0.350 mole of $PbCl_2$
(c) 0.450 gram of UCl_4 or 0.450 mole of UCl_4

33. Discussion question: An alkali metal carbonate of the formula M_2CO_3 contains 18.79% M, 16.25% C, and the remainder O. What is the element M? Compare your work here with the "mystery" compound in Problem 12. Are mass percentages as useful as molar masses for determining the identity of an unknown metal?

34. Determine the mass percentage of each of the elements in the following compounds:

(a) $CoC_6H_{24}Cl_3N_6$
(b) $CoC_4H_{16}Cl_3N_4$

What is the mass of Co in 0.083 g of each compound? What is the number of moles of Co?

8.4 Empirical Formulas

SECTION GOALS

✔ Recognize whether a chemical formula is an empirical formula.

✔ Know how an empirical formula is calculated.

✔ Understand the relationship between empirical and molecular formulas.

✔ Calculate a molecular formula from mass data and the molar mass.

Empirical Formulas

When we know a chemical formula, it is a straightforward process to determine the mass contribution of each element given in that formula. But we can also tackle the problem in reverse. Suppose that we have an unknown substance for which we do not know the

CONNECT TO
SECTION 4.1

✔ Meeting the Goals

An empirical formula of a
compound gives the smallest
whole number mole ratios of
the elements in the compound.

formula. If we know the mass percentage of each element present in
the compound, we can find the chemical formula of the compound.
The chemical formula we obtain is the simplest, or empirical, formula.

The subscript numbers in a chemical formula are mole numbers
that tell us the number of moles of each element contained in one
mole of the compound. When the subscript numbers in the formula
are in lowest terms, we have an empirical formula. An **empirical
formula** is a formula that gives the smallest whole number mole ratio of elements in a compound. Sometimes the empirical formula is
called the *simplest* formula because it contains only the simplest information: the mole ratios of the elements present. "Smallest" means
the subscript numbers do not have a common divisor, i.e., they cannot be further reduced by division. "Whole number" has its usual
meaning—no decimal values or fractions allowed!

For example, C_2H_2 is *not* an empirical formula, because we can
reduce the subscripts by dividing them each by 2. C_2H_2 is a molecular formula and the empirical formula is CH. $C_6H_{12}O_6$ (glucose) is
not an empirical formula because the subscript numbers are all divisible by 6. CH_2O is the correct empirical formula for glucose. The
formula $C_{0.5}HO_{0.5}$ is *not* a correctly written empirical formula because the subscript numbers are not whole numbers. We can multiply the subscripts by 2 to obtain the correct empirical formula,
CH_2O.

Example 8.18 Identifying an empirical formula

Which of the following formulas can be classified as correctly written empirical formulas?

(a) $K_2Cr_2O_7$ (b) H_2O_2 (c) $C_{0.3}H_{0.6}$

Thinking Through the Problem We know that an empirical formula
has whole numbers as subscripts; these must be in lowest terms.
The first formula, $K_2Cr_2O_7$ follows the rules for empirical formulas. Since the second formula, H_2O_2, is not written in lowest terms,
it is not an empirical formula. The third formula, $C_{0.3}H_{0.6}$, does
not have whole number subscripts. It is not an empirical formula.

Answer $K_2Cr_2O_7$ is the only formula in this grouping that is an
empirical formula. The correctly written empirical formula for (b)
is HO. The correctly written empirical formula for (c) is CH_2.

How are you doing? 8.8

Which of the following formulas are correct empirical formulas?

(a) C_4H_8O (b) $Na_{0.25}O_{0.25}$ (c) B_5H_{11}

Because an empirical formula indicates the smallest whole number mole ratios, we must convert ratios that are *decimals* into whole
number ratios. We do this by assuming the number of moles of com-

pound is equal to the smallest mole amount of an element in that compound. We then divide all mole amounts by the number of moles of compound.

Example 8.19

Finding empirical formulas from moles of elements

Find the empirical formula for a compound containing 1.40 mol N and 2.80 mol H.

Thinking Through the Problem We begin with the mole information given and look for a small, whole number ratio of moles that will give us the number of each element. To convert these decimal numbers into whole numbers, we divide each one by the smaller of the two values given, 1.40 mol.

Working Through the Problem

$$\text{N:} \frac{1.40 \text{ mol N}}{1.40 \text{ mol compound}} = \frac{1 \text{ mol N}}{1 \text{ mol compound}}$$

$$\text{H:} \frac{2.80 \text{ mol H}}{1.40 \text{ mol compound}} = \frac{2 \text{ mol H}}{1 \text{ mol compound}}$$

Answer The empirical formula is NH_2.

Example 8.20

Finding empirical formulas from atoms of the elements

Find the empirical formula for the compound that contains 5.06×10^{23} atoms C and 1.012×10^{24} atoms H.

Thinking Through the Problem We convert numbers of atoms to mole amounts using Avogadro's number. Then we determine a chemical formula by finding the smallest whole number ratio of the resulting mole amounts.

Working Through the Problem
Calculate mole amounts:

$$5.06 \times 10^{23} \text{ atoms C} \times \frac{1 \text{ mol C}}{6.022 \times 10^{23} \text{ atoms C}} \approx 0.840 \text{ mol C}$$

$$1.012 \times 10^{24} \text{ atoms H} \times \frac{1 \text{ mol H}}{6.022 \times 10^{23} \text{ atoms H}} \approx 1.680 \text{ mol H}$$

Determine relative mole amounts:

$$\frac{0.840 \text{ mol C}}{0.840 \text{ mol compound}} = \frac{1 \text{ mol C}}{1 \text{ mol compound}}$$

$$\frac{1.680 \text{ mol H}}{0.840 \text{ mol compound}} = \frac{2 \text{ mol H}}{1 \text{ mol compound}}$$

Answer The empirical formula is CH_2.

Determining Empirical Formulas from Mass Composition

In Chapter 4, we saw how the mole/mole ratio of elements in a compound lets us determine a chemical formula. Now we will use the *relative* mass amounts of elements present in the substance to determine the empirical formula of that substance.

You can already use a compound's formula to determine its composition in terms of percentage by mass. Now you will do the reverse process. You will determine the formula of a compound from information about its composition by mass.

In practice, most chemical formulas are experimentally determined from some form of mass data—either percentage composition data or a listing of the masses of the elements present in a given mass of sample. In either case, we use the same procedure to determine the formula. Mass data is converted into moles of each element. As we saw before, *these mole amounts allow us to determine the small whole number ratios of the elements to the compound.*

The problem-solving steps to determine the empirical formula of a compound are given in Table 8-2.

Table 8-2 Steps for Calculating an Empirical Formula from Mass Data

1. *Convert mass data to moles.* Use molar mass as the conversion factor.
2. *Find the smallest whole number ratio of moles* present. Divide through by the smallest number of moles calculated in step 1 and adjust as needed to obtain the smallest whole number ratio.
3. *Write the formula using the mole ratios,* determined in step 2 as the subscript numbers of the elements in the formula.

Example 8.21   **Finding empirical formulas from mass**

A 25.000-gram sample of a compound containing only calcium and fluorine was analyzed and found to contain 12.825 grams of calcium and 12.175 grams of fluorine. What is the empirical formula of this compound?

Thinking Through the Problem It is a good idea to always list the elements present in the problem. Next to each element write the mass given in the problem. This will help you check if any information is missing from the problem.

$$\text{Ca: } 12.825 \text{ g} \qquad\qquad \text{F: } 12.175 \text{ g}$$

(8.8) The key idea is to convert each mass to moles by dividing by the molar mass of that element.

Working Through the Problem

$$\text{Ca: } 12.825 \text{ g Ca} \times \frac{1 \text{ mol Ca}}{40.078 \text{ g Ca}} = 0.32000 \text{ mol Ca}$$

$$\text{F:} \quad 12.175 \ \cancel{g \ F} \times \frac{1 \ \text{mol F}}{18.9984 \ \cancel{g \ F}} = 0.64084 \ \text{mol F}$$

Next, convert to simple whole numbers by dividing each mole amount by the smaller number of moles, 0.32 mol.

$$\text{Ca:} \quad \frac{0.32000 \ \text{mol Ca}}{0.32000 \ \text{mol compound}} \approx \frac{1 \ \text{mol Ca}}{1 \ \text{mol compound}}$$

$$\text{F:} \quad \frac{0.64084 \ \text{mol F}}{0.32000 \ \text{mol compound}} \approx \frac{2 \ \text{mol F}}{1 \ \text{mol compound}}$$

This indicates that for every one mole of compound there is one mole of Ca and there are two moles of F.

Answer The empirical formula for this compound is CaF_2.

Example 8.22 Finding empirical formulas from mass

A sample weighing 0.254 g is found to contain 0.192 g arsenic (As) and the rest oxygen. What is the empirical formula of this compound?

Thinking Through the Problem List the elements and their masses as given in the problem. We notice that we are not given a mass of O, but we can get this value by subtracting the mass of As from the mass of the whole sample.

Working Through the Problem

$$\text{As: } 0.192 \ \text{g} \qquad \text{O: } 0.254 \ \text{g} - 0.192 \ \text{g} = 0.062 \ \text{g}$$

Convert the mass of each element into moles:

$$\text{As: } 0.192 \ \cancel{g \ As} \times \frac{1 \ \text{mol}}{74.922 \ \cancel{g \ As}} = 2.56 \times 10^{-3} \ \text{mol As}$$

$$\text{O: } 0.062 \ \cancel{g \ O} \times \frac{1 \ \text{mol}}{15.9994 \ \cancel{g \ O}} = 3.88 \times 10^{-3} \ \text{mol O}$$

✔**Meeting the Goals**

To calculate an empirical formula from mass data, (1) convert each mass to moles by dividing by the molar mass of that element, (2) find the smallest ratio of the mole numbers obtained by dividing each by the smallest mole number, and (3) convert the resulting mole ratios to whole numbers (if necessary) and use them as subscripts to the appropriate element in the empirical formula.

Determine the smallest whole number ratio of moles by dividing the moles of each element by the smaller number of moles. ($2.56 \times 10^{-3} \ \text{mol} < 3.86 \times 10^{-3} \ \text{mol}$)

$$\text{As: } \frac{2.56 \times 10^{-3} \ \text{mol As}}{2.56 \times 10^{-3} \ \text{mol compound}} = \frac{1 \ \text{mol As}}{1 \ \text{mol compound}}$$

$$\text{O: } \frac{3.86 \times 10^{-3} \ \text{mol O}}{2.56 \times 10^{-3} \ \text{mol compound}} = \frac{1.5 \ \text{mol O}}{1 \ \text{mol compound}}$$

This molar ratio is a correct ratio and we could write the formula as $AsO_{1.5}$, but *1.5 is not a whole number.* To convert this ratio into a whole number ratio, recognize that 1.5 is the decimal equivalent of the fraction 3/2. Multiply both parts of the ratio by the denominator, 2. The whole number ratio is then:

As: $1 \times 2 = 2$ O: $1.5 \times 2 = 3$

Answer The empirical formula for the compound is As_2O_3.

How are you doing? **8.9**

Suppose that you have a 100.0-g sample of a compound known to be composed of 40.0 g carbon, 6.7 g hydrogen, and the rest oxygen. Calculate the mass of oxygen in the sample, then calculate the empirical formula of this compound.

What is the relationship between *percent* and *grams* of each element in the sample when the sample mass is 100.0 grams?

Empirical Formulas from Mass Percentage

KEY IDEA 8.9

Mass percentage is a special kind of mass ratio, where the sum of all percentages is 100%. How do we find the empirical formula of a compound when given a breakdown of the compound's composition in terms of mass percentage? *If we choose a sample size of 100 grams, then the mass percentage of each element is equal to its mass in grams.* We then convert these mass amounts to mole amounts, and the ratio of mole amounts gives us the formula, as we have seen.

Example 8.23   Finding empirical formulas from mass percentage

A compound containing only magnesium and oxygen is analyzed to be 60.3% Mg and 39.7% O by mass. What is the empirical formula of this compound?

Thinking Through the Problem First of all, there are two elements; their percentage by mass values are given in the problem. ◁━▷ (8.9) The key idea is to assume a 100-g sample. This gives us 60.3 g Mg and 39.7 g O.

Working Through the Problem

Mg: 60.3 g   O: 39.7 g

We then convert each mass to moles. The molar mass of Mg is 24.305 g mol^{-1} and the molar mass of O is 15.999 g mol^{-1}.

$$\text{Mg: } 60.3 \text{ g Mg} \times \frac{1 \text{ mol Mg}}{24.305 \text{ g Mg}} = 2.48 \text{ mol Mg}$$

$$\text{O: } 39.7 \text{ g O} \times \frac{1 \text{ mol}}{15.9994 \text{ g O}} = 2.48 \text{ mol O}$$

Divide both moles found in the last step by the smaller number of moles. Here the mole numbers are the same and it is easy to see that the molar ratio is 1:1.

$$\text{Mg:} \quad \frac{2.48 \text{ mol Mg}}{2.48 \text{ mol compound}} = \frac{1 \text{ mol Mg}}{1 \text{ mol compound}}$$

$$\text{O:} \quad \frac{2.48 \text{ mol O}}{2.48 \text{ mol compound}} = \frac{1 \text{ mol O}}{1 \text{ mol compound}}$$

Answer The empirical formula of this compound is MgO.

Example 8.24   **Finding empirical formulas from mass percentage**

✔Meeting the Goals

To calculate empirical formula from mass percentage data, assume a sample mass of 100 grams. This allows us to use the number given by mass percentage as grams and proceed with the problem.

Upon chemical analysis, a compound is found to contain 32.37% sodium, 22.53% sulfur, and 45.05% oxygen. What is the empirical formula of this compound?

Thinking Through the Problem Given mass percentage values and assuming a 100.0-gram sample, we have 32.37 g Na, 22.53 g S, and 45.05 g O. List the elements and their masses as given in the problem. Check that all elements are included.

Na: 32.37 g   S: 22.53 g   O: 45.05 g

Convert each mass to moles, then write as a ratio to the smallest number of moles of any element.

Working Through the Problem

(a) Convert grams to moles.
(b) Divide through by smallest mole amount from (a).

$$\text{Na: } 32.37 \text{ g Na} \times \frac{1 \text{ mol}}{22.990 \text{ g Na}} = 1.41 \text{ mol Na} \qquad \frac{1.41 \text{ mol}}{0.70 \text{ mol}} \approx 2$$

$$\text{S: } \quad 22.53 \text{ g S} \times \frac{1 \text{ mol}}{32.066 \text{ g S}} = 0.70 \text{ mol S} \qquad \frac{0.70 \text{ mol}}{0.70 \text{ mol}} \approx 1$$

$$\text{O: } 45.05 \text{ g O} \times \frac{1 \text{ mol}}{15.9994 \text{ g O}} = 2.82 \text{ mol O} \qquad \frac{2.82 \text{ mol}}{0.70 \text{ mol}} \approx 4$$

Answer There are 2 moles Na, 1 mole S, and 4 moles O in the formula. The empirical formula of the compound is Na_2SO_4, sodium sulfate.

How are you doing? 8.10

A compound is analyzed and found to contain 62.0% C, 10.4% H, and 27.5% O. Determine the empirical formula of this compound.

Table 8-3 The Relationship Between Empirical and
            Molecular Formulas

Compound	Empirical Formula	Molecular Formula	Multiples
Water	H_2O	H_2O	1
Hydrogen peroxide	HO	H_2O_2	2
Sodium peroxide	NaO	Na_2O_2	2
Benzene	CH	C_6H_6	6
Acetylene	CH	C_2H_2	2
Ethylene	CH_2	C_2H_4	2

Molecular Formulas

✔Meeting the Goals

The molecular formula is the true formula of a substance; its formula mass is the same as the molar mass of the substance. The molecular formula is the same as, or some multiple of, the empirical formula.

The **molecular formula** for a substance may be the *same* as its empirical formula or it may be some whole number *multiple* of the empirical formula. H_2O is both the empirical *and* the molecular formula of water. Hydrogen peroxide, however, has an empirical formula of HO and a molecular formula that is twice the empirical formula, H_2O_2. Some examples of the relationship between empirical and molecular formulas are shown in Table 8-3.

The last column in Table 8-3 is labeled "Multiples." The numbers listed in this column give the "multiplier" used with each subscript number in the empirical formula to give the subscript numbers in the corresponding molecular formula. For example, benzene has a multiplier of 6. Both subscript numbers in its empirical formula (CH) are ones. We get the molecular formula by multiplying both subscript numbers by 6 to get C_6H_6. Likewise, the molecular formula for ethylene is obtained by multiplying the subscript numbers in its empirical formula (CH_2) by 2 to get C_2H_4. How are these multiples determined? It is very simple.

$$\frac{\text{molar mass of substance}}{\text{molar mass of empirical formula}} = \text{multiples}$$

Let's verify this relationship with the two substances already mentioned, benzene and ethylene.

For benzene, $\dfrac{\text{molar mass of } C_6H_6}{\text{molar mass of CH}} = \dfrac{78.113 \text{ g mol}^{-1}}{13.0189 \text{ g mol}^{-1}} = 5.999 \approx 6.$

The molecular formula for benzene, then, is $(CH)_6 = C_6H_6$. (It is interesting to note here that rounded mass numbers give the same result, $\dfrac{78 \text{ g}}{13 \text{ g}} = 6$ because we are looking for a whole number multiple.)

For ethylene, $\dfrac{\text{molar mass of } C_2H_4}{\text{molar mass of } CH_2} = \dfrac{28 \text{ g mol}^{-1}}{14 \text{ g mol}^{-1}} = 2$. The molecular formula for ethylene, then, is $(CH_2)_2 = C_2H_4$.

Figure 8-10 ▲ These three substances have the same empirical formula but different molecular formulas and very different properties. (a [vinegar, $C_2H_4O_2$]: Richard Megna/Fundamental Photographs; b [formaldehyde, CH_2O]: Kjell Sandved/Visuals Unlimited; c [glucose, $C_6H_{12}O_6$]: Randy Faris/Corbis.)

In most problems, *you* must determine how many multiples of the empirical formula make up the molecular formula. It is important to be able to distinguish between these two types of formulas. Compounds with the same empirical formula often have very different properties because they have different molecular formulas. Examples of this are shown in Figure 8-10. How do you know if the molecular formula is the same as the empirical formula or if the molecular formula is some multiple of the empirical formula? You cannot tell by just looking at the empirical formula. You must have some *additional* information. You need to know the molar mass of the compound, because *the molar mass of the compound is the mass of the molecular formula.*

To determine if the molecular formula is the same as the empirical formula, ask if the molar mass of the substance is equal to the empirical molar mass. There are two possibilities:

- If the molar mass is *equal* to the empirical formula mass, then the molecular formula is the same as the empirical formula.

- If the molar mass is *not equal* to the empirical formula mass, use the empirical formula and the molar mass to determine the number of multiples needed to make the molecular formula.

Example **8.25** **Finding molecular formulas from mass data**

The molar mass of borane is approximately 27.6 g mol^{-1}. A 5.00-gram sample of borane is found to contain 3.91 grams of B and the rest H. What is the molecular formula of borane?

Thinking Through the Problem ◄ (8.10) The key idea is that the molar mass is the mass of the molecular formula. To find the molecular formula, we need both the empirical formula and the molar mass of borane. We are given the molar mass and must use the masses of B and H given to find the empirical formula. Then we can use the empirical formula and the molar mass to find the molecular formula.

Working Through the Problem We find the mass of H in the compound by subtracting the mass of B from the mass of the sample: 5.00 g sample − 3.91 g B = 1.09 g H.

$$\text{B: } 3.91 \text{ g} \div \frac{1 \text{ mol B}}{10.811 \text{ g B}} = 0.3616 \text{ mol B}$$

$$\text{H: } 1.09 \text{ g} \div \frac{1 \text{ mol H}}{1.0079 \text{ g H}} = 1.081 \text{ mol H}$$

Dividing through by the smaller number of moles, we get:

$$\frac{0.3616 \text{ mole B}}{0.3616 \text{ mole}} = 1 \text{ mole B}$$

$$\frac{1.081 \text{ moles H}}{0.3616 \text{ mole}} \approx 2.99 \approx 3 \text{ moles H}$$

The empirical formula for borane is BH_3.

Now we calculate the number of multiples of BH_3:

$$\frac{\text{molar mass}}{\text{mass of BH}_3} = \frac{27.7 \text{ g}}{13.8 \text{ g}} \approx \frac{28}{14} = 2.$$

Answer The molecular formula of borane is B_2H_6.

Molecular formulas can always be determined from an empirical formula and a molar mass. However, it is also possible to find the molecular formula *directly*—in a one-step calculation—when the *percentage composition* of the substance and its *molar mass* are given. Because the molar mass is the mass of the molecular formula, we can use mass percentage data and the molar mass of the compound to calculate *the mass of each element in the molecular formula.* We can then convert to moles. *The resulting mole numbers are the subscript numbers in the molecular formula. Do not reduce them!* Reducing subscript numbers in a molecular formula will give the empirical formula.

Example 8.26 Calculating molecular formulas

A compound is composed of 40.0% carbon, 6.72% hydrogen and 53.3% oxygen. Determine its empirical formula. Determine the molecular formula if the molar mass is 180.15 g mol^{-1}.

Thinking Through the Problem We must find the molecular formula of a compound, given percentage by mass composition and the mass of one mole of the compound. We will use this information to calculate the *mass of each element present in the molecular for-*

mula. Then we will convert mass amounts to mole amounts to get the subscript numbers for the molecular formula. To find the empirical formula, reduce the subscript numbers in the molecular formula to lowest terms.

Working Through the Problem

$$C: \frac{40.0 \text{ g C}}{100 \text{ g compound}} \times 180.15 \text{ g compound} = 72.06 \text{ g C}$$

$$72.06 \text{ g C} \times \frac{1 \text{ mol C}}{12.011 \text{ g C}} = 5.999 \approx 6 \text{ mol C}$$

$$H: \frac{6.72 \text{ g H}}{100 \text{ g compound}} \times 180.15 \text{ g compound} = 12.106 \text{ g H}$$

$$12.106 \text{ g H} \times \frac{1 \text{ mol H}}{1.0079 \text{ g H}} = 12.01 \approx 12 \text{ mol H}$$

$$O: \frac{53.3 \text{ g O}}{100 \text{ g compound}} \times 180.15 \text{ g compound} = 96.019 \text{ g O}$$

$$96.019 \text{ g O} \times \frac{1 \text{ mol O}}{15.999 \text{ g O}} = 6.001 \approx 6 \text{ mol O}$$

Answer The molecular formula is $C_6H_{12}O_6$. The empirical formula is CH_2O.

Example 8.27 Calculating formulas

✔ Meeting the Goals

To calculate a molecular formula of a compound, use percentage data and molar mass to calculate the grams of each element present in the molecular formula. Convert grams of each element to moles. These mole numbers are the subscript numbers of the elements in the molecular formula. To find the empirical formula from the molecular formula, reduce subscript numbers to lowest terms.

A compound containing only aluminum and oxygen contains 52.93% aluminum by mass. What is the formula for this compound if the molar mass is determined to be 101.96 g mol^{-1}?

Thinking Through the Problem We find the percentage of oxygen by subtracting the percentage of aluminum in the compound from 100%: 100.00 g sample − 52.93 g Al = 47.07 g O. Here we combine into a single calculation the two steps shown in the previous example.

Working Through the Problem

$$Al: \frac{52.93 \text{ g Al}}{100 \text{ g compound}} \times \frac{101.96 \text{ g compound}}{1 \text{ mol compound}} \times \frac{1 \text{ mol Al}}{26.982 \text{ g Al}}$$

$$= \frac{2.00 \text{ mol Al}}{1 \text{ mol compound}}$$

$$O: \frac{47.07 \text{ g O}}{100 \text{ g compound}} \times \frac{101.96 \text{ g compound}}{1 \text{ mol compound}} \times \frac{1 \text{ mol O}}{15.9994 \text{ g O}}$$

$$= \frac{2.999 \approx 3 \text{ mol O}}{1 \text{ mol compound}}$$

Answer The formula for this compound is Al_2O_3.

8.11

A substance is composed of 24.8% C, 2.08% H, and 73.1% Cl. What is the molecular formula of this compound if its molar mass is 290 g mol^{-1}?

PRACTICAL D   Drug therapy—where do the drugs come from?

With some forms of drug therapy, chemicals are used that function either to increase the concentration of a particular substance required for healthy bodily functioning, or to eliminate a disease-causing bacteria or virus from the body. Therapeutic drugs may be naturally-occurring substances or may be synthetically manufactured in a laboratory. "Experimental" drugs are chemicals that are being developed for possible therapeutic use. The process of drug development is long and arduous because drugs must be tested extensively before they are made available to the public. Any new drug must be shown to be both safe and effective in properly conducted clinical trials.

The development of a new drug begins with a study of its chemical structure because similar chemical structures are associated with similar chemical behavior and function. Consider dopamine, which is a naturally-occurring chemical substance important for proper brain functioning. A loss of dopamine is associated with Parkinson's disease and some forms of dementia. Chemists have developed a way to manufacture the drug L-dopa, which has a structure similar to that of dopamine. The chemical behavior and physiological function of L-dopa also are similar to those of dopamine, which makes L-dopa helpful in treating Parkinson's disease. A structural drawing would show almost identical substances. The molecular formulas of dopamine and L-dopa are given

Substance	Molecular Formula	Empirical Formula
Dopamine	$C_8H_{11}NO_2$	$C_8H_{11}NO_2$
L-dopa	$C_9H_{10}NO_4$	$C_9H_{10}NO_4$

here. Both are also empirical formulas; the subscript numbers in each formula are present in the lowest whole number ratio.

Some other chemical substances found in the brain, such as histamine $(C_5H_9N_3)$ and serotonin $(C_{10}H_{12}N_2O)$, have molecular formulas that are the same as their empirical formulas. When you look at chemical formulas, notice the elements present as well as their mole ratios.

In both cases, these formulas contain one subscript number that prevents numerical reduction to lower values. The subscript 5 in the molecular formula of histamine makes this formula impossible to reduce; the subscript 1 (understood for oxygen) makes the molecular formula of serotonin impossible to reduce. Thus, these molecular formulas are also empirical formulas.

Acetylsalicylic acid (aspirin, $C_9H_8O_4$) is derived from salicylic acid $(C_7H_6O_3)$. Compare the molecular formulas of these two compounds and, in each case, *circle the subscript number* that causes the molecular formula to be the empirical formula as well.

PROBLEMS

35. Concept question: Look up the word *empirical*. How is this used in the phrase "empirical formula"?

36. Which of the following formulas are empirical formulas?

    (a) H_2O

    (b) H_2O_2

    (c) $Mg(OH)_2$

    (d) $C_{10}H_{14}O$

    (e) $Zn_2P_2O_7$

    (f) $K_3Fe(CN)_6$

    (g) $H_2C_2O_4 \cdot 2\ H_2O$

37. Find the empirical formula for a compound that contains 0.25 mol C and 0.625 mol H.

38. Find the empirical formula for the compound that contains 2.505×10^{23} atoms of N and 6.262×10^{23} atoms of O.

39. Calculate the empirical formula of a compound that contains 28.03% Na, 39.0% O, 3.69% H, and 29.3% C.

40. An 8.75-gram sample of a compound containing only bromine and fluorine was found to contain 3.64 grams of fluorine. What is the empirical formula of this compound?

41. Determine the empirical formula of a compound that contains 35.9% Al and 64.1% S.

42. Upon analysis a compound is found to contain 26.58% K, 35.35% Cr, and 38.07% O. What is the empirical formula for this compound?

43. Determine the empirical formula for the compound with the following composition: 79.96% carbon, 9.39% hydrogen, and 10.65% oxygen.

44. A compound is found to have the following composition: 27.74% magnesium (Mg), 48.69% oxygen (O), and 23.57% phosphorus (P). What is its empirical formula?

45. A substance with a molar mass of 28 g mol^{-1} has a simplest formula of CH_2. What is its molecular formula?

46. A substance with an empirical formula of $CuCl_2 \cdot 6 H_2O$ has a molar mass of 242.5 g mol^{-1}. What is the molecular formula for this compound?

47. **Discussion question:** A compound containing only hydrogen and oxygen is found to contain 94.07% oxygen. Its molar mass is 34.0 g mol^{-1}. What is the molecular formula for this compound? Use both methods you have seen, as shown in Examples 8.25 and 8.26. Discuss which of the two methods you prefer and why.

48. A compound is composed of 79.8% carbon and the rest hydrogen. The molar mass of this compound is 30.07 g mol^{-1}. What is the molecular formula of this compound?

49. A 5.000-gram sample of a compound containing only phosphorus and oxygen is analyzed and found to contain 2.182 grams of phosphorus and 2.818 grams of oxygen. Determine its molecular formula if its molar mass is 283.89 g mol^{-1}.

50. A compound contains 24.27% carbon, 4.07% hydrogen, and 71.66% chlorine. Its molar mass is 98.96 g mol^{-1}. What is its molecular formula?

51. A compound contains 13.88% Li, 23.57% C, and the rest oxygen. It has a molar mass of about 101 grams. Calculate the empirical and molecular formulas for this compound.

52. The analysis of nickel ore shows the following percent composition: 21.15% Ni, 3.54% Co, 3.35% Fe, 71.96% As. Calculate the empirical formula of the ore.

53. 5.3 grams of tin reacts with fluorine to form 8.7 grams of product. Calculate the empirical formula of the metal fluoride.

54. The empirical formula for a compound is C_4H_9, and its molar mass is 171 g mol^{-1}. What is its molecular formula?

Chapter 8 Summary and Problems

◀ Review with **Web Practice**

VOCABULARY

atomic mass	The mass of an atom expressed with reference to the standard atom, carbon-12.
atomic mass scale	A mass scale in which the standard of mass is an atom of carbon-12.
atomic mass unit (amu)	A defined unit in the atomic or molecular scale of measurement; the atomic mass unit is defined as exactly 1/12 the mass of a carbon-12 atom.
formula mass	The sum of the masses of all elements in the formula of the substance.
molar mass	The mass, in grams, of one mole of a substance; it is numerically the same as the formula mass but has units of grams per mole (g mol^{-1}).
mass percentage	The percentage of an element in a compound containing that element, based on the mass/mass ratio of element to compound.
empirical formula	The empirical formula of a substance gives the smallest whole number mole ratios of the elements present.
molecular formula	The molecular formula of a substance has a mass equal to the molar mass of that substance; it is either the empirical formula or some whole number multiple of the empirical formula.

SECTION GOALS

How do we define the terms "atomic mass," "formula mass," and "molar mass"?

An atomic mass is the mass of an atom, defined relative to the mass of carbon-12. A formula mass is the sum of the atomic masses of all the elements in the formula of a substance. A molar mass is the mass, in grams, of one mole of a substance. In practice, molar mass is a formula mass with units of grams per mole (g mol^{-1}).

How do we use mass measurements to determine the number of moles of a substance?

We relate measurements of mass to the number of moles through the molar mass, which tells us the mass of one mole of a substance.

How do we calculate a molar mass and how do we use atomic masses as a guide?

The periodic table and tables of atomic masses give us the mass of atoms (amu). When we add the atomic masses of all the elements in a formula, we get the formula mass. This is converted to a molar mass by relating the value of the formula mass to the value for the ratio "grams per mole."

How do we structure multiple-step calculations that involve masses and moles?

Calculations involving mass and mole amounts of different species, for example an element in a compound, are done using molar masses and the mole ratio. Both of these can be calculated if we know the formula of the substance; the sum of the atomic masses of the elements gives us the value of the molar mass of the substance, and the chemical formula allows us to directly read the number of moles of each element in the substance.

How can we use density to determine the number of moles of a substance in a measured volume of that substance?

We can convert from the volume of a substance to the moles of that substance by using two ratios, the density and the molar mass. Because the density of a substance is the ratio of mass to volume, we can use mass and volume to calculate the mass of the substance: $d = m/V$ so $m = dV$. Molar mass is the ratio of grams of a substance to moles of the same substance: molar mass = grams/mole so moles = grams/(molar mass).

How is percentage by mass of a compound calculated?

To find the percentage composition by mass of a compound: (a) determine the molar mass of the compound, (b) evaluate the mass/mass ratio formed by dividing the total mass contribution of each element by the molar mass of the compound, and (c) convert to a percentage by multiplying by 100.

How is the percentage by mass ratio used?

When we know the percentage by mass of an element A in a compound (%A), we can calculate the mass of that element in any mass of compound. The %A is used to write a known ratio. This is made equal to a similar ratio containing the specified quantities given by the problem. Notice that we can calculate x given y or vice versa.

$$\frac{x \text{ g A}}{y \text{ g compound}} = \frac{\text{mass A}}{100 \text{ g compound}} = \%A$$

If y is known, we can solve this proportion for the x grams of A in the sample. If x is known, we can solve this proportion for the y grams of compound.

What does the term "empirical formula" mean?

An empirical formula gives the symbols of the elements and their smallest whole number mole ratio to each other in a particular compound.

How is an empirical formula calculated?

To calculate an empirical formula from mass data, (a) convert each mass to moles by dividing by the molar mass of the element, (b) find the smallest ratio of the mole numbers obtained by dividing each by the smallest mole number, and (c) convert the resulting mole ratios to whole numbers (if necessary) and use them as subscripts to the appropriate element in the empirical formula.

What is the relationship between empirical and molecular formulas?

The molecular formula is the true formula of a substance. The molecular formula is the same as or is some multiple of the empirical formula.

How do you calculate the molecular formula when given mass data and the molar mass of the substance?

To calculate a molecular formula of a compound, use percentage data and molar mass to calculate the grams of each element present in the molecular formula. Convert grams of each element to moles. These mole numbers are the subscript numbers of the elements in the molecular formula.

PROBLEMS

55. Determine the molar mass of the following compounds:

(a) PBr_3
(b) YCl_3
(c) $Ba(C_2H_3O_2)_2$
(d) $(NH_4)_3PO_3$
(e) $NaHSO_4$

56. Determine the following quantities.

(a) How many moles of $FeSO_4$ are in a 0.250-g tablet of $FeSO_4$?
(b) What mass of K_3PO_4 is in a sample of 25.00 moles of K_3PO_4?

57. Determine the number of moles of bromine in a sample of 250 cm^3 of bromine. Bromine is a liquid with a density of 3.119 g cm^{-3}.

58. Determine which of these minerals has the highest percentage of vanadium (V) by mass: vanadinite $(Pb_3(VO_4)_3Cl)$ or descloizite $(ZnPbVO_4(OH))$.

59. Determine the mass of the metal in 1.00 gram of the following compounds.

(a) OsO_4
(b) $Ca(ClO_4)_2$

60. Determine which has more moles of potassium:

(a) 0.225 gram of KCl
(b) 0.448 gram of K_2SO_4

61. Determine the mass of zinc present in 2.33 moles of $ZnCl_2 \cdot (NH_3)_4$.

62. Determine the number of moles of sodium in 32.5 grams of sodium phosphate, Na_3PO_4.

63. Correct the following formulas so that they follow the rules for empirical formulas.

(a) $C_2H_4Cl_2$
(b) $Hg_2(C_2H_3O_2)_2$
(c) $N_6H_{24}S_6$

64. Find the empirical formula for a compound that contains 2.5 mol N and 7.5 mol H.

65. Find the empirical formula for the compound that contains 1.40×10^{-2} mol $CaSO_4$ and 2.80×10^{-2} mol H_2O.

66. Determine the percentage composition by mass of PCl_3.

67. A compound is found to contain only phosphorus and chlorine. Determine its empirical formula after chemical analysis reveals that it contains 85.13% Cl and 14.87% P.

68. A compound containing only lead and oxygen is analyzed and found to be 90.66% lead. What is the empirical formula for this compound?

69. The empirical formula of a substance is NaO. Its molar mass is 78 g mol^{-1}. What is the molecular formula of this substance?

70. A certain hydrocarbon has the empirical formula CH_2. Determine its true molecular formula if its molar mass is 70.134 g mol^{-1}. Determine the molecular formula of this compound if its molar mass is 112.214 g mol^{-1}.

71. Find the molecular formula of a compound that contains 30.45% nitrogen and 69.55% oxygen. Its molar mass is 92.01 g mol^{-1}.

72. Calculate the number of moles of compound in each of the following samples.

(a) 0.213 g $Ca(ClO_4)_2$ (b) 25,000 g $(NH_4)_2HPO_4$

73. Calculate the number of grams of each compound in each of the following samples.

(a) 4.00 mol $Ca(ClO_3)_2$ (b) 2100 mol $NH_4H_2PO_4$

74. Determine the molar mass of the following compounds:

(a) CF_4 (c) BeF_2
(b) NH_4Br (d) Na_2O_2

75. Correct the following formulas so that they follow the rules for empirical formulas.

(a) $NaS_{0.5}O_2$ (b) $(NH_4)_2S_2O_8$

Chemical Stoichiometry

Stoichiometry can be used to study mass and energy relationships in living things, such as these trees, which harness energy from the sun to build complex molecules out of carbon dioxide and water. (Erv Schowengerdt.)

PRACTICAL CHEMISTRY Metabolism and the Chemical Budget of Life

Life as we know it depends on the use of chemical substances as fuel and as the building blocks of organisms. We can learn a lot about living things by following the flow of chemical substances between organisms and their nonliving surroundings and from one organism to another. Metabolism is the special area of biochemistry that studies the way organisms extract energy and substances from the environment and put them to use in the manufacture of important cell parts. The reactions of metabolism show that living things follow a chemical "budget" for their existence.

Organisms can be classified by their primary energy source. Some organisms harness energy from the sun and use it to build complex molecules out of simpler ones, such as carbon dioxide and water. Such organisms are called phototrophs. Green plants are a prime example of phototrophs. Other organisms use the chemical substances made by phototrophs or other organisms as their sources of energy and of the complex chemical compounds that are the building blocks of their cells. Because their energy comes primarily from organic substances, these are called organotrophs. Humans are organotrophs. Our bodies obtain energy and building blocks by eating and digesting plants and animals.

In this chapter, we will see that the amounts of chemicals made by phototrophs and consumed by organotrophs are linked by chemical reactions. We will also see how the chemical reactions of metabolism let us discover how much energy is "stored" in certain molecules. To do this, we will use stoichiometry, which is the study of mole and mass relationships in chemical reactions.

This chapter will focus on the following questions:

PRACTICAL A How can we use stoichiometry to describe the metabolism of sugars and other food molecules?

PRACTICAL B What is "fixed" carbon, and how much carbon is fixed in a living plant?

PRACTICAL C What is the role of nitrogen fertilizer in controlling the growth of organisms?

9.1 Mole Amounts and Chemical Equations

SECTION GOAL

✓ Apply proportional reasoning and conversion factors to chemical equations when using mole ratios that relate different amounts of substances in a chemical equation.

Chemical Stoichiometry

In Chapter 4, we discussed what happens during a chemical reaction—how the atoms of the reactants recombine into new substances. We saw that the number and the types of the atoms do not change in a reaction. Only the arrangement of atoms changes. Because the number of atoms does not change at the atomic or molecular level, the mole amounts of the atoms do not change at the macroscopic level, either. The conservation of the number and identity of the elements in a reaction is the basis for writing balanced chemical equations.

Because the mole is a group counting unit that can be used to count large numbers of atoms and molecules, the chemical equation says something about the number of moles present. For example, in Chapter 4, we discussed the chemical equation:

$$PCl_3\ (l) + Cl_2\ (g) \longrightarrow PCl_5\ (s)$$

We can read this equation as "one molecule of PCl_3 and one molecule of Cl_2 give one molecule of PCl_5." But we can also read it as "one mole of PCl_3 and one mole of Cl_2 give one mole of PCl_5." These two sentences represent two scales—the atomic or molecular scale and the macroscopic scale. The same chemical equation can be discussed using either scale. The scale is determined by the size of the sample. Very small samples are discussed more conveniently using the atomic or molecular scale, whereas we use the macroscopic scale for the large samples we use in the lab.

For example, the equation: $C(s) + O_2(g) \rightarrow CO_2(g)$ can be interpreted in terms of single atoms and molecules or in terms of moles of substances. Thus, we can read:

1 atom of C + 1 molecule of O_2 $\longrightarrow$ 1 molecule of CO_2
(atomic or molecular scale)

or

1 mole C + 1 mole O_2 $\longrightarrow$ 1 mole CO_2 (macroscopic scale)

✔Meeting the Goals

On the atomic or molecular scale, the coefficients of a balanced equation represent atoms and molecules. On the macroscopic scale, the coefficients of a balanced equation represent moles of atoms and molecules.

Each atom of carbon that reacts is present in the product and each atom of oxygen that reacts is present in the product. All atoms are accounted for; none are lost and no extra atoms appear in the products. The result is a "balanced" equation.

Because the coefficients in a balanced equation represent the number of moles of each substance in the reaction, we can use the coefficients to write mole/mole ratios. Once the equation is balanced, all mole/mole ratios will remain constant. Let's examine this statement for the reaction:

$$2\ C_2H_6(g) + 7\ O_2(g) \longrightarrow 4\ CO_2(g) + 6\ H_2O(g)$$

This equation tells us that we need 2 moles C_2H_6 for every 7 moles O_2. In ratio form, we could write the ratio:

$$\frac{7 \text{ moles } O_2}{2 \text{ moles } C_2H_6}$$

We can write similar ratios for any two substances—either reactant or product—in the equation.

Similarly, the equation:

$$1\ C_2H_6O + 3\ O_2 \longrightarrow 2\ CO_2 + 3\ H_2O$$

can be read as: "1 molecule C_2H_6O + 3 molecules O_2 gives 2 molecules CO_2 + 3 molecules H_2O." This equation can also be read as: "1 mole C_2H_6O + 3 moles O_2 gives 2 moles CO_2 + 3 moles H_2O." The coefficients in front of each formula in a balanced equation represent the number of moles of that substance. We can use the coefficients of a balanced equation to form a ratio between the moles of any two reactants and products in the equation. For example, the mole ratios of CO_2 to each of the other substances in the reaction are:

$$\frac{2 \text{ mol CO}_2}{1 \text{ mol C}_2\text{H}_6\text{O}}, \qquad \frac{2 \text{ mol CO}_2}{3 \text{ mol O}_2}, \quad \text{and} \quad \frac{2 \text{ mol CO}_2}{3 \text{ mol H}_2\text{O}}$$

Notice that the numbers in these ratios are the coefficients of the balanced equation. We say that CO_2 and C_2H_6O are present in a 2-to-1 mole ratio (written 2:1). CO_2 and O_2 are present in a 2:3 mole ratio. CO_2 and H_2O are in a 2:3 mole ratio.

So far we have been writing **mole ratios for a chemical reaction.** Consider the following balanced equation: 1 mol $C_5H_{10}O_5 \rightarrow$ 5 mol C + 5 mol H_2O. We can write mole ratios relating both products to the reactants. Thus, we relate the moles of C to the moles of $C_5H_{10}O_5$ with the ratio $\dfrac{5 \text{ mol C}}{1 \text{ mol C}_5\text{H}_{10}\text{O}_5}$ and moles of H_2O to moles of $C_5H_{10}O_5$ with the ratio $\dfrac{5 \text{ mol H}_2\text{O}}{1 \text{ mol C}_5\text{H}_{10}\text{O}_5}$.

This is not the only kind of mole ratio we see in chemistry. Mole ratios can also apply to a formula for a chemical substance. In Chapter 4, we saw a **mole ratio for a chemical substance,** in which we write a ratio that relates the mole amount of one element to the overall formula, or one that relates the mole amount of one element to the mole amount of another element in the same formula. For example, for the formula Ag_2S we can write the mole ratio $\dfrac{1 \text{ mol S}}{1 \text{ mol Ag}_2\text{S}}$ to relate the element S to the formula of the compound Ag_2S. Or, we can write $\dfrac{2 \text{ mol Ag}}{1 \text{ mol S}}$ to relate the element S to the element Ag.

Example 9.1 Interpreting mole ratios

Using the following balanced chemical equations, indicate whether the given mole ratios apply to a chemical formula or to a chemical equation. The precipitation of calcium phosphate from two solutions is shown in Figure 9-1.

$$2 \text{ K}_3\text{PO}_4 + 3 \text{ CaSO}_4 \longrightarrow \text{Ca}_3(\text{PO}_4)_2 + 3 \text{ K}_2\text{SO}_4$$
$$2 \text{ C}_5\text{H}_5\text{N} + 13 \text{ O}_2 \longrightarrow \text{N}_2\text{O} + 10 \text{ CO}_2 + 5 \text{ H}_2\text{O}$$

(a) $\dfrac{3 \text{ mol K}}{1 \text{ mol K}_3\text{PO}_4}$ (c) $\dfrac{1 \text{ mol Ca}_3(\text{PO}_4)_2}{2 \text{ mol K}_3\text{PO}_4}$

(b) $\dfrac{2 \text{ mol C}_5\text{H}_5\text{N}}{1 \text{ mol N}_2\text{O}}$ (d) $\dfrac{2 \text{ mol N}}{1 \text{ mol N}_2\text{O}}$

Thinking Through the Problem We can recognize mole ratios for chemical equations because both the numerator and the denominator refer to reactants or products in the equation. If this is not the case, then we look to find the mole ratio from a chemical formula.

Figure 9-1 ▲ The amount of solids formed depends on the mole ratio in the chemical reaction. (Michael Dalton/ Fundamental Photographs.)

Answer Parts (a) and (d) relate a single element (K and N, respectively) to a formula containing that element. They are ratios for a chemical formula. Parts (b) and (c) refer to reactants and products in an equation so they are mole ratios for chemical equations.

Example 9.2   Choosing mole ratios

Using the balanced equations below, write the relevant mole ratios that would be needed to answer the questions that follow. Do not answer the questions here; simply write the relevant mole ratios.

$$2 \, K_3PO_4 + 3 \, CaSO_4 \longrightarrow Ca_3(PO_4)_2 + 3 \, K_2SO_4$$
$$2 \, C_5H_5N + 13 \, O_2 \longrightarrow N_2O + 10 \, CO_2 + 5 \, H_2O$$

(a) Determine how many moles of calcium phosphate are made by reacting 25 moles of calcium sulfate.
(b) Determine how many moles of CO_2 are made by burning 10.0 moles of C_5H_5N.
(c) Determine how many moles of Ca are in 12.2 moles of calcium phosphate.

Thinking Through the Problem We can look at each sentence and underline the two substances we must relate with mole amounts. These give rise to the known ratio. ◀◉ (9.1) A key idea is that the coefficients of these substances in the balanced equation are the numbers in the ratio. We can work this by underlining the mole statements in the questions as a guide to the ratios.

Working Through the Problem

(a) Determine *how many moles of* calcium phosphate are made by reacting *25 moles* of calcium sulfate.
Relate: $Ca_3(PO_4)_2$ to $CaSO_4$

Source of mole ratio:
Chemical reaction

(b) How many *moles of CO₂* are made by burning *10.0 moles of C₅H₅N?*
Relate: CO_2 to C_5H_5N

Chemical reaction

(c) How many *moles of Ca* are in *12.2 moles of calcium phosphate?*
Relate: Ca to $Ca_3(PO_4)_2$

Chemical formula

Answer

(a) $\dfrac{1 \text{ mol } Ca_3(PO_4)_2}{3 \text{ mol } CaSO_4}$

(b) $\dfrac{10 \text{ mol } CO_2}{2 \text{ mol } C_5H_5N}$

(c) $\dfrac{3 \text{ mol Ca}}{1 \text{ mol } Ca_3(PO_4)_2}$

Propane gas burns in air to form carbon dioxide and water vapor according to the equation:

$$C_3H_8\,(g) + 5\,O_2\,(g) \longrightarrow 3\,CO_2\,(g) + 4\,H_2O\,(g)$$

Write mole ratios taken from the balanced equation that compare:

(a) moles CO_2 to moles O_2
(b) moles C_3H_8 to moles CO_2
(c) moles O_2 to moles C_3H_8

A stoichiometric calculation allows us to determine the quantity of one substance in a chemical experiment or procedure when given the quantity of some other substance also in the experiment or procedure. The two substances may be one reactant and one product, two reactants, or two products. *The essential requirements of a stoichiometric calculation are*:

- *The mole amount of the initial substance*
- *A mole ratio that relates moles of the initial substance to moles of the second substance*

We use these known mole ratios and given amount to determine the actual amount of the second substance. Figure 9-2 maps these essential steps in a stoichiometry calculation.

Mole/Mole Calculations in Chemical Equations

The mole ratios for the substances in chemical equations relate the different amounts of substances involved in a reaction. We use these known mole ratios to solve for an unknown quantity using either a proportional reasoning method or a conversion factor method.

When we carry out a stoichiometric calculation, the stoichiometric coefficients in the balanced chemical equation are counted numbers. They do not influence the number of significant figures in the answer.

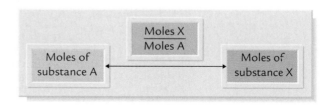

Figure 9-2 ▲ Strategy map for mole/mole stoichiometric calculations.

Example 9.3   **Mole/mole calculations using proportional reasoning**

Determine the number of moles of CO_2 gas produced by the combustion of 2.5 moles of C_2H_6O. The balanced chemical equation for this reaction is $C_2H_6O + 3\ O_2 \rightarrow 2\ CO_2 + 3\ H_2O$.

Thinking Through the Problem ◀━◀▶ (9.2) The key idea in this problem is to start with the given mole amount of the initial substance, 2.5 moles of C_2H_6O. We then relate moles of CO_2 and moles of C_2H_6O. From the chemical equation we know that these two substances have a stoichiometric ratio of 2 moles of CO_2 to one mole of C_2H_6O:

$$\frac{2\ \text{mol } CO_2}{1\ \text{mol } C_2H_6O}$$

Working Through the Problem In this example, we use proportional reasoning to calculate the number of moles of CO_2 produced by 2.5 moles of C_2H_6O.

$$\text{unknown ratio} = \text{known ratio}$$

$$\frac{x\ \text{mol } CO_2}{2.5\ \text{mol } C_2H_6O} = \frac{2\ \text{mol } CO_2}{1\ \text{mol } C_2H_6O}$$

$$2.5\ \cancel{\text{mol } C_2H_6O} \times \frac{x\ \text{mol } CO_2}{2.5\ \cancel{\text{mol } C_2H_6O}} =$$

$$\frac{2\ \text{mol } CO_2}{1\ \cancel{\text{mol } C_2H_6O}} \times 2.5\ \cancel{\text{mol } C_2H_6O}$$

$$x\ \text{mol } CO_2 = \frac{2 \times 2.5}{1}\ \text{mol } CO_2$$

Answer $\qquad x\ \text{mol } CO_2 = 5.0\ \text{mol } CO_2$ (2 s.f.)

Example 9.4   **Mole/mole calculations using the conversion factor method**

Magnesium metal reacts with an aqueous solution of hydrochloric acid, according to the balanced equation:

$$Mg + 2\ HCl \longrightarrow MgCl_2 + H_2$$

This reaction is shown in Figure 9-3. How many moles of HCl are required to react completely with 3.25 moles of Mg?

Thinking Through the Problem We are asked for the number of moles of HCl required. This is a reactant. We are given 3.25 moles of Mg, another reactant. We can compare mole amounts for two reactants just as we compared mole amounts of reactants and products in Example 9.3. ◀━◀▶ (9.2) The key idea is to use the given mole amount, 3.25 moles of Mg, and the mole ratio that relates moles of Mg to moles of the substance we are looking for, HCl.

Figure 9-3 ▲ Magnesium in a solution of hydrochloric acid.
(Richard Megna/Fundamental Photographs.)

Again, we can use either proportional reasoning, or the conversion factor method. Here we will use the ratio $\dfrac{2 \text{ mol HCl}}{1 \text{ mol Mg}}$ as a conversion factor.

Working Through the Problem

$$3.25 \text{ mol Mg} \times \frac{2 \text{ mol HCl}}{1 \text{ mol Mg}} = 6.50 \text{ mol HCl}$$

Answer 6.50 mol HCl

How are you doing? **9.2**

Using the balanced equation: $C_3H_8\ (g) + 5\ O_2\ (g) \rightarrow 3\ CO_2\ (g) + 4\ H_2O\ (g)$, answer the following questions.

(a) Determine how many moles of O_2 are needed to make 25.0 moles of CO_2.
(b) If 2.00 moles of C_3H_8 react, how many moles of CO_2 should be formed?
(c) If an experiment determines that 12.0 moles of CO_2 are formed, how many moles of C_3H_8 are used?

Multiple Conversions with Mole Ratios

Some problems may require that you use more than a single type of mole ratio in order to answer the question. For example, you may need to determine the number of moles of a certain element that appears in the formula for a substance in the reaction. In this case, you may first need to use the mole ratio arising from the balanced equation to relate two substances in the chemical reaction and then use the mole ratio that relates the formula of a substance to the element you are looking for. The solution of this problem includes an additional step, as we see in the following example.

Example 9.5 Mole calculations involving multiple conversions

Ethane gas burns in air to produce gaseous carbon dioxide and water according to the equation:

$$2\ C_2H_6 + 7\ O_2 \longrightarrow 4\ CO_2 + 6\ H_2O$$

How many moles of C atoms in the C_2H_6 are needed to produce 1.70 moles of CO_2 according to this equation?

Thinking Through the Problem Look carefully at the information given and asked for in this problem. We need to produce 1.70 moles of CO_2. We can calculate the number of moles of C_2H_6 required for this. Then we calculate the number of moles of C in that mole amount of C_2H_6.

Working Through the Problem The mole ratio of C_2H_6 to CO_2 is given by the balanced equation; 2 mol C_2H_6/4 mol CO_2. We will use this ratio with the given amount of CO_2 to find the mol C needed.

$$1.70 \; \cancel{mol \; CO_2} \times \frac{2 \; mol \; \cancel{C_2H_6}}{4 \; \cancel{mol \; CO_2}} = 1.70 \; mol \; C \times \frac{2 \; mol \; C}{1 \; \cancel{mol \; C_2H_6}}$$

Answer 1.70 mol C in the C_2H_6 is needed to produce 1.70 mol CO_2.

Example 9.6   **Mole/mole calculations involving multiple conversions**

Oxygen can be prepared in the lab by the decomposition of metal chlorates such as potassium chlorate.

$$2 \, KClO_3 \, (s) \longrightarrow 2 \, KCl \, (s) + 3 \, O_2 \, (g)$$

How many moles of $KClO_3$ are required to produce 2.25 moles of oxygen atoms in the oxygen gas?

Thinking Through the Problem We need to find the number of moles of $KClO_3$ that will produce 2.25 moles of oxygen atoms. A mole O is related to moles O_2 by the formula. Moles of O_2 are related to moles of $KClO_3$ by the stoichiometric ratio. The problem is mapped as: mol O → mol O_2 → mol $KClO_3$

$$2.25 \; \cancel{mol \; O} \times \frac{1 \; \cancel{mol \; O_2}}{2 \; \cancel{mol \; O}} \times \frac{2 \; mol \; KClO_3}{3 \; \cancel{mol \; O_2}} = 0.75 \; mol \; KClO_3$$

Answer 0.75 mole of $KClO_3$ will give 2.25 moles of O in the product O_2.

How are you doing?   **9.3**

1. Elemental arsenic may be produced by reacting As_2O_3 with carbon according to the equation:

$$As_2O_3 \, (s) + 3 \, C \, (s) \longrightarrow 2 \, As \, (s) + 3 \, CO \, (g)$$

How many moles of arsenic are produced when 150,000 moles of As_2O_3 are used?

2. An important use for ammonia is in the first step in the commercial production of nitric acid. Ammonia is first oxidized to NO according to the equation:

$$4 \, NH_3 \, (g) + 5 \, O_2 \, (g) \longrightarrow 4 \, NO \, (g) + 6 \, H_2O \, (g)$$

How many moles of NH_3 are required to produce 1.25 moles of NO?

3. How many moles of oxygen are used in the preceding reaction?

PRACTICAL A — How can we use stoichiometry to describe the metabolism of sugars and other food molecules?

We were introduced at the beginning of this chapter to the term *organotrophs*. Organotrophs eat plants or other animals to obtain molecules for energy and for their basic building material. For example, glucose is a molecule contained in some foods. Our bodies get energy from glucose by breaking it down into smaller molecules. A simple decomposition reaction of glucose takes place in our muscles to make a molecule called lactic acid:

$$C_6H_{12}O_6 \longrightarrow 2\,C_3H_6O_3$$

glucose   lactic acid

Energy is released in this reaction. But a much larger release of energy from the glucose molecule occurs when oxygen is involved in the reaction. Oxygen reacts with the two molecules of lactic acid produced by the simple splitting of glucose. This reaction produces carbon dioxide and water, and releases energy:

$$2\,C_3H_6O_3 + 6\,O_2 \longrightarrow 6\,CO_2 + 6\,H_2O$$

The overall reaction of glucose is therefore:

$$C_6H_{12}O_6 + 6\,O_2 \longrightarrow 6\,CO_2 + 6\,H_2O \qquad (1)$$

Use this equation to answer the following questions.

- How many moles of carbon dioxide are produced by the complete reaction of 0.052 mole of glucose?
- How many moles of glucose are required to produce 3.02 moles of carbon dioxide?
- How many moles of oxygen are required to react completely with 22.5 moles of glucose?

Just as our bodies break down complex molecules from phototrophs (plants) to obtain energy, our factories, machines, and automobiles use such molecules as well! The complex molecules in the decayed remains of plants and animals have become the coal, oil, and natural gas used today. When burned, these fossil fuels produce large amounts of energy and carbon dioxide. If we take the molecule octane, C_8H_{18}, as a "typical" fossil fuel, we can see that it reacts with oxygen as follows:

$$2\,C_8H_{18} + 25\,O_2 \longrightarrow 16\,CO_2 + 18\,H_2O \qquad (2)$$

In the case of both octane and glucose, we can ask important questions about how much energy is produced. The key to the analysis is not in how much carbon dioxide is produced but in determining how much oxygen is used in a reaction that produces a particular amount of carbon dioxide. To do this, we will also have to account for the O_2 needed to make the H_2O byproduct.

Use equations (1) and (2) to calculate answers for the following questions:

- How many moles of oxygen are needed in a reaction that produces 100 moles of carbon dioxide from octane?

- How many moles of oxygen are needed in a reaction that produces 100 moles of carbon dioxide from glucose?

- Which molecule, octane or glucose, uses more oxygen?

- If the "energy content" of a substance is directly related to the amount of oxygen needed to produce a given amount of carbon dioxide, then which molecule has greater energy content?

Carry out the same analysis for the fatty acid $C_{18}H_{34}O_2$, known as oleic acid, and the amino acid $C_3H_7NO_2$, known as alanine. Amino acids are the building blocks of proteins. You will need to write a balanced chemical equation first, then determine how many moles of O_2 are needed to get 100 moles of CO_2 from each substance.

Use your data to predict which foods have the highest energy content in terms of food calories: carbohydrates such as glucose, fats such as oleic acid, or amino acids such as alanine.

PROBLEMS

1. Concept question: The use of proportional reasoning in analyzing chemical reactions requires that the number of atoms does not change in going from reactants to products. Is this statement true? Why or why not?

2. Consider the reaction:

$$2\,C_2H_6 + 7\,O_2 \longrightarrow 4\,CO_2 + 6\,H_2O$$

(a) How many molecules of O_2 are needed for

the complete reaction of 136 molecules of C_2H_6?

(b) How many molecules of H_2O are produced from the complete reaction of 42 molecules of O_2?

(c) How many molecules of CO_2 are produced in this reaction when 312 molecules of H_2O are produced?

(d) How many atoms of H are required in order to produce 68 molecules of CO_2?

3. Butane gas (C_4H_{10}) burns in air to form a gaseous mixture of carbon dioxide and water according to this equation:

$$2\,C_4H_{10} + 13\,O_2 \longrightarrow 8\,CO_2 + 10\,H_2O$$

(a) How many molecules of CO_2 will be produced by the complete reaction of 26 molecules of C_4H_{10}?

(b) How many molecules of CO_2 will be produced by the complete reaction of 26 molecules of O_2?

4. When heated, solid calcium carbonate, $CaCO_3$, forms solid calcium oxide and gaseous carbon dioxide according to this equation:

$$CaCO_3 \longrightarrow CaO + CO_2$$

(a) How many atoms of C are contained in 35 molecules of CO_2?

(b) As you recall from Chapter 3, we refer to formulas of substances with extended structures as formula units. How many formula units of $CaCO_3$ are required to produce 125 formula units of CaO?

5. Aluminum metal oxidizes in air to form Al_2O_3 according to this equation:

$$4\,Al + 3\,O_2 \longrightarrow 2\,Al_2O_3$$

(a) How many formula units of Al_2O_3 can be produced by the complete reaction of 126 atoms of Al?

(b) How many formula units of Al_2O_3 can be produced by the complete reaction of 126 atoms of O?

(c) What is the maximum number of formula units that can be produced by a mixture containing 126 atoms of Al and 126 atoms of O? (Consider your answers from parts (a) and (b) above.)

(d) How many molecules of O_2 are needed to produce 28 formula units of Al_2O_3?

6. Repeat the calculations in Problem 5 with all atomic or molecular amounts converted to mole amounts.

7. Solid potassium metal reacts with water to form an aqueous solution of potassium hydroxide, KOH, and hydrogen gas according to this equation:

$$2\,K + 2\,H_2O \longrightarrow 2\,KOH + H_2$$

(a) How many moles of H_2 are produced by the complete reaction of 314 moles of K?

(b) How many moles of H are required in order to produce 54 moles of KOH?

(c) How many moles of H_2O are required in order to produce 150 moles of H_2?

(d) How many moles of H are produced in the complete reaction of 82 moles of K?

8. Repeat the calculations in Problem 7 using atomic and molecular units.

9. How many moles of calcium oxide can be produced from the decomposition of 4,520 moles of calcium carbonate according to the balanced equation:

$$CaCO_3\,(s) \longrightarrow CaO\,(s) + CO_2\,(g)$$

10. In the process of making steel, Fe_3O_4 is reduced by carbon monoxide to iron (II) oxide and carbon dioxide according to this equation:

$$Fe_3O_4\,(s) + 4\,CO\,(g) \longrightarrow 3\,FeO\,(s) + 4\,CO_2\,(g)$$

(a) How many moles of $Fe_3O_4\,(s)$ are required to produce 152 mol $FeO\,(s)$?

(b) How many moles of $CO\,(g)$ are required to reduce 3 moles of $Fe_3O_4\,(s)$?

11. Determine the number of moles of each reactant, Na_2CO_3 and Fe_3Br_8, needed to produce 2,450 moles of NaBr according to the reaction:

$$4\,Na_2CO_3 + Fe_3Br_8 \longrightarrow 8\,NaBr + 4\,CO_2 + Fe_3O_4$$

12. Discussion question: Chromium metal reacts with molecular sulfur (S_8) to form Cr_2S_3 according to this equation:

$$16\,Cr + 3\,S_8 \longrightarrow 8\,Cr_2S_3$$

(a) How many moles of S_8 are needed to use 16 moles Cr?

(b) How many moles of Cr are needed to completely react with 16 moles of S_8?

Suppose that you have a mixture of 16 moles of Cr and 16 moles of S_8. Discuss what will be left when the reaction is complete.

13. Lithium metal combines with nitrogen gas to form lithium nitride according to this equation:

$$3\,Li\,(s) + N_2\,(g) \longrightarrow Li_3N\,(s)$$

(a) How many moles of $Li_3N\,(s)$ can be formed from the complete reaction of 0.304 mole of Li (s)?

(b) How many moles of $N_2\,(g)$ are required for the complete reaction of 3.04×10^2 moles of Li (s)?

14. In Problem 10 we considered the reduction of Fe_3O_4 by CO:

$$Fe_3O_4\,(s) + CO\,(g) \longrightarrow 3\,FeO\,(s) + CO_2\,(g)$$

(a) How many moles of FeO are produced from 152 moles of Fe_3O_4?

(b) How many moles of CO_2 are produced from 3 moles of Fe_3O_4?

15. Aluminum sulfide can be produced by the reaction of aluminum with sulfur according to this equation:

$$2\,Al + 3\,S \longrightarrow Al_2S_3$$

(a) How many moles of Al_2S_3 can be produced by the complete reaction of 25 moles of Al?

(b) How many moles of Al_2S_3 will result from the complete reaction of 15.5 moles of S?

9.2 Mass Amounts and Chemical Equations

SECTION GOALS

✔ Connect mole ratios from chemical equations with mass measurements.

✔ Calculate relative amounts when several different conversions are involved.

Mass/Mole Problems

CONNECT TO
SECTION 8.1

The coefficients in a balanced equation are the numbers used in the stoichiometric ratio for calculations of mole amounts in that chemical reaction. But direct measurement of moles is usually not possible. It is easy to measure gram amounts directly, then determine moles using the molar mass of that substance to convert from grams to moles. In this section, we combine what we have learned about mole/mole calculations with our knowledge of molar mass from Chapter 8.

Suppose we want to calculate the mass of $PbSO_4$ that will form when solutions of lead nitrate and sodium sulfate are mixed. The balanced equation for this reaction is:

$$Na_2SO_4\,(aq) + Pb(NO_3)_2\,(aq) \longrightarrow PbSO_4\,(s) + 2\,NaNO_3\,(aq)$$

How many grams of $PbSO_4$ will be produced by the complete reaction of 4.5 moles of Na_2SO_4?

First of all, remember that the coefficients of a balanced equation relate moles of substances. The coefficients do *not* describe the mass relationships. The balanced equation on the previous page indicates that one mole of Na_2SO_4 will produce one mole of $PbSO_4$:

$$1 \text{ mol } Na_2SO_4 \longrightarrow 1 \text{ mol } PbSO_4$$

Next, we use the molar mass of $PbSO_4$ (302.3 g mol^{-1}) to convert moles of $PbSO_4$ into grams of $PbSO_4$.

The strategy map needed to solve this problem is:

$$\text{Moles of } Na_2SO_4 \longrightarrow \text{moles of } PbSO_4 \longrightarrow \text{grams of } PbSO_4$$

              use coefficients of equation   *use molar mass*

The calculation for this problem, using the appropriate ratios in order, is:

$$4.5 \text{ mol } Na_2SO_4 \times \frac{1 \text{ mol } PbSO_4}{1 \text{ mol } Na_2SO_4} \times \frac{302.3 \text{ g } PbSO_4}{1 \text{ mol } PbSO_4} = 1400 \text{ g } PbSO_4$$

                coefficients of equation   *molar mass*

The stoichiometric ratio is central to this problem. Figure 9-2 showed the strategy map for converting moles of A to moles of X. *When mass amounts are introduced into the problem, the appropriate molar mass is needed for the gram-to-mole conversion.* Figure 9-4 shows the strategy map for mass/mole stoichiometric calculations.

KEY IDEA 9.3

For any general equation (2 A + 3 X → A_2X_3), moles of A or moles of X are converted to moles of A_2X_3 by multiplication with the appropriate stoichiometric ratio. When grams of A are given, the molar mass of A is used to convert from grams of A to moles of A. Then the stoichiometric ratio is used to convert to moles of A_2X_3:

$$\text{grams } A \times \frac{1 \text{ mol } A}{\text{grams } A} \times \frac{1 \text{ mol } A_2X_3}{2 \text{ mol } A} = \text{moles of } A_2X_3$$

             molar mass

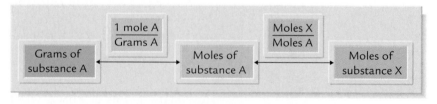

Figure 9-4 ▲ Strategy map for mass/mole stoichiometric calculations.

When we are looking for grams of A_2X_3, we first use the mole/mole strategy shown in Figure 9-2. Then we use the molar mass of A_2X_3 to convert moles of A_2X_3 to grams of A_2X_3:

$$\text{mols A} \times \frac{1 \text{ mol } A_2X_3}{2 \text{ mol A}} \times \underbrace{\frac{\text{grams } A_2X_3}{1 \text{ mol } A_2X_3}}_{molar\ mass} = \text{grams of } A_2X_3$$

✔ **Meeting the Goals**

We are using a sequence of ratios here. We should carefully plan and map our calculation before we put the numbers down on paper.

A mass/mole problem either begins with mole information and asks us to calculate mass information or vice versa. We look for answers to such questions as: "What mass should I begin with?" and "What mass of product do I expect to get from this reaction?"

Example 9.7   **Moles-to-mass stoichiometry**

Xenon can be fluorinated by the following process:

$$Xe\,(g) + F_2\,(g) \longrightarrow XeF_2\,(g)$$

How many grams of $XeF_2\,(g)$ will be produced by 2.30 moles of $F_2\,(g)$?

Thinking Through the Problem ◀▬ (9.3) The key idea is to use both a mole ratio and a molar mass. We are given mole information for F_2 and need to find mass information for XeF_2. Think how to map this problem. We must start with 2.30 moles of F_2 and then use the strategy shown in Figure 9-4 to complete the mapping. The strategy map for this problem is: moles $F_2 \rightarrow$ moles $XeF_2 \rightarrow$ grams XeF_2.

Working Through the Problem This is a two-step conversion, moving from moles $F_2 \rightarrow$ moles XeF_2 using the stoichiometric ratio and from moles $XeF_2 \rightarrow$ grams XeF_2 using the molar mass.

$$2.30 \text{ mol } F_2 \times \frac{1 \text{ mol } XeF_2}{1 \text{ mol } F_2} \times \frac{169.29 \text{ g } XeF_2}{1 \text{ mol } XeF_2} \approx 389 \text{ g } XeF_2$$

Example 9.8   **Mass-to-moles stoichiometry**

Chlorine is in the same family of elements as fluorine and has properties similar to those of fluorine. The process of "adding" chlorine to a substance is called chlorination. Consider the chlorination of silicon according to this equation:

$$Si\,(s) + 2\,Cl_2\,(g) \longrightarrow SiCl_4\,(l)$$

How many moles of $SiCl_4$ will be produced from 15.2 grams of silicon?

Thinking Through the Problem We are given 15.2 grams of Si and need to find moles of $SiCl_4$. We will use the strategy shown in Figure 9-4, beginning with grams of Si. The strategy map for this problem is: grams Si → moles Si → moles $SiCl_4$.

Working Through the Problem

$$15.2\,\text{g Si} \times \frac{1\,\text{mol Si}}{28.086\,\text{g Si}} \times \frac{1\,\text{mol SiCl}_4}{1\,\text{mol Si}} \approx 0.541\,\text{mol SiCl}_4$$

Answer 0.541 mol $SiCl_4$ (3 s.f.)

Example 9.9   **Mass/mole calculation**

Chlorine reacts with hydrogen to produce hydrogen chloride gas according to the following equation.

$$Cl_2(g) + H_2(g) \longrightarrow 2\,HCl(g)$$

How many moles of Cl_2 gas are needed to produce 15.50 grams of HCl?

Thinking Through the Problem We are given 15.50 grams of HCl and need to find moles of Cl_2. We use the strategy map in Figure 9-4: grams of HCl → moles of HCl → moles of Cl_2.

Working Through the Problem

$$15.50\,\text{g HCl} \times \frac{1\,\text{mol HCl}}{36.461\,\text{g HCl}} \times \frac{1\,\text{mol Cl}_2}{2\,\text{mol HCl}} \approx 0.2126\,\text{mol Cl}_2$$

Answer 0.2126 mol Cl_2 (4 s.f.)

How are you doing? 9.4

Liquid dinitrogen trioxide decomposes into the two gases nitrogen dioxide and nitrogen monoxide according to the equation $N_2O_3(l) \rightarrow NO_2(g) + NO(g)$. How many moles of NO will be produced if you start with 20.5 grams of N_2O_3?

Mass/Mass Problems

KEY IDEA 9.4

A mass/mass problem in stoichiometry is a problem in which both the quantity given and the quantity sought have mass units. *Mass/mass problems are logical extensions of the work we have been doing to this point and require no new conversion factors.* It is now a matter of reading the problem, mapping the problem strategy, and solving. The problem-solving strategy for mass/mass problems is shown in Figure 9-5.

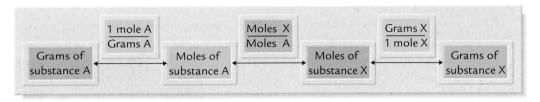

Figure 9-5 ▲ Strategy map for mass/mass stoichiometric calculations.

Example 9.10 **Mass/mass problem**

One of the steps in the recovery of metals from their sulfide ores is to roast them in air to form the metal oxide. The oxide is then reduced to the metal. In the recovery of zinc, the following reaction is used:

$$2\,ZnS\,(s) + 3\,O_2\,(g) \longrightarrow 2\,ZnO\,(s) + 2\,SO_2\,(g)$$

How many grams of zinc oxide will result when 25.0 grams of zinc sulfide is roasted in air?

Thinking Through the Problem ➡️ (9.4) The key idea is to carry out the mass → mole → mass conversion using known ratios. We are given grams of ZnS and we need to find grams of ZnO. To get started, we need the stoichiometry ratios given by the balanced equation, and the molar masses of ZnO and ZnS. We will use the strategy shown in Figure 9-5 to map the problem: 25.0 grams of ZnS → moles of ZnS → moles of ZnO → grams of ZnO.

Working Through the Problem

$$25.0\,g\,ZnS \times \frac{1\,mol\,ZnS}{97.46\,g\,ZnS} \times \frac{2\,mol\,ZnO}{2\,mol\,ZnS} \times \frac{81.39\,g\,ZnO}{1\,mol\,ZnO}$$

$$\approx 20.9\,g\,ZnO$$

Answer 20.9 g ZnO (3 s.f.)

Example 9.11 **Mass/mass problem**

Uranium ores are converted into fluorides by reaction with hydrogen fluoride according to the following balanced equation:

$$UO_2\,(s) + 4\,HF\,(aq) \longrightarrow UF_4\,(s) + 2\,H_2O\,(l)$$

How many grams of UF_4 can be produced from reaction of 125.0 grams HF with excess ore? Figure 9-8 on page 319 shows a uranium processing plant.

Thinking Through the Problem We are given grams of HF and need to find grams of UF_4. Following the strategy shown in Figure 9-5, we map the problem: 125.0 grams of HF → moles of HF → moles of UF_6 → grams of UF_4.

PRACTICAL B   How much carbon is "fixed" in a living plant?

We learned earlier that phototrophs (green plants) use simple molecules like CO_2 and H_2O as the building blocks of their cells. But where do these molecules come from? Plants absorb carbon dioxide from the atmosphere. The carbon is converted into plant material, and oxygen is released. Some plants absorb carbon from the soil as well. Converting carbon into plant material is known as "fixing" carbon.

Plants also need water. Most plants obtain water from the soil, but certain plants absorb most of the water they need directly from the air around them. An excellent example of such a plant is the orchid, which grows among rocks in the wild, and can be grown in a pot with little moisture or soil. An orchid, shown in Figure 9-6, grows by fixing carbon and absorbing water from the air. Let's assume that an orchid has a "dry weight" (the weight of just the solid, carbon-containing material after all free water is removed) of 10.0 grams. The majority of this weight is the carbohydrate known as starch, which has an empirical formula of $C_6H_{10}O_5$. We can calculate that 10.0 grams of starch is equal to 0.0618 mole of starch, and contains 0.370 mole of carbon. Mole/mass calculations allow us to determine how many moles of carbon dioxide and water are used to make a particular amount of plant material.

We start by writing a balanced chemical equation for a simplified chemical reaction for the formation of starch:

Figure 9-6 ▲ Orchids are plants that obtain all their components, including water, from the air alone. (David Stone/Rainbow.)

$$6\,CO_2 + 5\,H_2O \longrightarrow C_6H_{10}O_5 + 6\,O_2 \qquad (1)$$

We can use this equation to calculate the number of grams of carbon dioxide and water needed to make the 10.0 grams of carbohydrate we assume to be in an orchid. If we find that the air in the room where the orchid grows contains 0.025 gram of carbon dioxide per liter of air, then how many liters of this air are needed to make the orchid?

Working Through the Problem

$$125.0\,\cancel{\text{g HF}} \times \frac{1\,\cancel{\text{mol HF}}}{20.006\,\cancel{\text{g HF}}} \times \frac{1\,\cancel{\text{mol UF}_4^-}}{4\,\cancel{\text{mol HF}}} \times \frac{314.02\,\text{g UF}_4}{1\,\cancel{\text{mol UF}_4^-}}$$

$$= 490.5\,\text{g UF}_4$$

Answer   490.5 g UF$_4$ (4 s.f.)

How are you doing?   **9.5**

1. Given the equation $Mg(NO_3)_2 + 2\,NaCl \rightarrow MgCl_2 + 2\,NaNO_3$, calculate the mass of $MgCl_2$ that will be produced when 24.52 grams of NaCl are used.

2. Given the equation $NaCl + AgNO_3 \rightarrow NaNO_3 + AgCl$, calculate the mass in grams of NaCl that is needed in order to produce 5.22×10^{-2} mole of AgCl. What is the mass of this amount of AgCl?

Certain plants, especially those that make bulbs or other root storage structures, require even more carbon. An example is the potato, shown in Figure 9-7, which (except for moisture) is basically pure starch. A 120-gram potato contains 22 grams of starch. Determine the amount of carbon dioxide and water needed to make this much starch.

The ability of plants to absorb carbon dioxide has important implications. Increasing levels of carbon dioxide in the atmosphere caused by the burning of fossil fuels may trap heat in the atmosphere through the so-called greenhouse effect, contributing to global warming. One possible solution to this problem may be the planting of new forests to take up the carbon dioxide produced each year.

Consider the following problem. What would you need to plant in order to absorb the carbon dioxide emitted by your car during a 30-mile trip? The combustion of gasoline can be approximated as:

$$2\,C_8H_{18} + 25\,O_2 \longrightarrow 16\,CO_2 + 18\,H_2O \qquad (2)$$

If your car gets 15 miles per gallon, it will use 2 gallons of gasoline for a 30-mile trip. Two gallons is equal to about 7000 grams of gasoline (C_8H_{18}). Calculate the number of grams of carbon dioxide produced by the combustion of 7000 grams of C_8H_{18}. Use equation (1) to calculate the grams of starch that

Figure 9-7 ▲ The leaves on a potato plant are responsible for a tremendous amount of carbon fixing in the form of starch. (Scott Bauer/Agricultural Research Service, USDA.)

can be made by this much CO_2. Would this be equivalent to an orchid? A potato? Something larger, such as a tree?

Figure 9-8 ▲ Careful control of chemical reactions is important in processing uranium. (AP Photo/APTV.)

PROBLEMS

Questions 16–20 refer to this balanced equation:

$$2\,KMnO_4\,(aq) + 16\,HCl\,(aq) \longrightarrow$$
$$2\,KCl\,(aq) + 2\,MnCl_2\,(aq) + 5\,Cl_2\,(g) + 8\,H_2O\,(l)$$

16. How many moles of $MnCl_2$ can be produced by the complete reaction of 1.05×10^3 moles of HCl?

17. How many molecules of HCl are needed for the complete reaction of 1.05×10^3 moles of $KMnO_4$?

18. How many grams of chlorine are produced by the complete reaction of 125.0 grams of $KMnO_4$?

19. How many grams of water are produced as a product along with 216.5 grams of chlorine?

20. Discussion question: How many grams of HCl are necessary for the complete reaction of 2.55 kilograms of $KMnO_4$? What about 2.55 grams of $KMnO_4$? Or 2.55 milligrams?

Compare your answers and discuss where the unit (gram, kilogram, or milligram) of the starting material mattered in your answer. Could you, without additional calculation, determine the mass of HCl from 2.55 megagrams of starting material?

Questions 21–28 refer to this balanced equation:

$$4\,HNCO\,(g) + 6\,NO\,(g) \longrightarrow$$
$$5\,N_2\,(g) + 2\,H_2O\,(g) + 4\,CO_2\,(g)$$

21. How many moles of CO_2 will result from the complete reaction of 1.55 moles of HNCO?

22. If 7.05 moles of H_2O are produced, how many moles of N_2 are also produced?

23. How many moles of NO are needed to react completely with 11.3 moles of HNCO?

24. How many grams of H_2O are produced when 251 moles of CO_2 are generated?

25. How many molecules of N_2 are produced when 525 grams of HNCO react?

26. How many grams of NO are required to completely react with 12.5×10^{-2} mole of HNCO?

27. How many moles of CO_2 will result from the complete reaction of 2.05 grams of HNCO?

28. How many grams of NO are required to completely react with 101 grams of HNCO?

9.3 Limiting Reactants and Yields

SECTION GOALS

✔ Know how a limiting reactant effects a reaction.
✔ Calculate a theoretical yield.
✔ Calculate a percentage yield.

Stoichiometry and the Law of Conservation of Mass

Atoms are conserved in ordinary chemical reactions. The atoms initially present in the reactants will be present in the products when the reaction ends. The balanced equation for the reaction shows there is an equal count of each element in the reactants and in the products. Let's see how this works in the oxidation of aluminum to aluminum oxide. The equation for this reaction is:

$$4\,Al + 3\,O_2 \longrightarrow 2\,Al_2O_3$$

The coefficients are the number of moles of each reactant and each product.

$$4\,Al + 3\,O_2 \longrightarrow 2\,Al_2O_3$$
$$4\,mol\,Al + 3\,mol\,O_2 \longrightarrow 2\,mol\,Al_2O_3$$

Mole quantities are converted to mass quantities using the molar mass of each substance.

$$4\,Al + 3\,O_2 \longrightarrow 2\,Al_2O_3$$

$$4\;mol\,Al\;\frac{26.982\,g\,Al}{1\,mol\,Al} + 3\,mol\,O_2\;\frac{31.998\,g\,O_2}{1\,mol\,O_2}$$

$$\longrightarrow 2\,mol\;\frac{101.961\,g\,Al_2O_3}{1\,mol\,Al_2O_3}$$

$$107.928\,g\,Al + 95.994\,g\,O_2 \longrightarrow 203.922\,g\,Al_2O_3$$

$$203.922\;grams\;of\;reactants = 203.922\;grams\;of\;product$$

A 4:3 ratio of aluminum to oxygen indicates the **stoichiometric amounts** of these substances. In other words, exactly 4 moles of aluminum (107.928 g Al) and 3 moles of oxygen (95.994 g O_2) are reactants expressed in stoichiometric amounts. In a theoretical or "perfect" world, the stoichiometric amount of aluminum oxide will be formed with nothing left over. You will notice that, when exact stoichiometric amounts of reactants are used, the sum of the reactant masses is equal to the mass of the product. This is the law of conservation of mass in action! *When you have the exact stoichiometric amounts of reactants, all of the reactants are consumed in the reaction.* All reactants "disappear," and only the product remains. This is the only case in which you can add the masses of the reactants to find the mass of product formed.

This last statement requires a bit of explanation. The law of conservation of mass always holds true, but we must remember its exact meaning, which states that we do not "lose" matter during an ordinary chemical reaction. The law specifies neither that *all* the reactants will be used up nor that *only* products will remain. If we start with ten grams of reactants in a sealed reaction vessel, we expect to have ten grams of matter when the reaction is completed.

In many cases we do not have exact stoichiometric amounts of all the reactants. We may have more of some of the reactants than we need. While it is true that the total mass of all of the substances in a closed reaction vessel will remain constant when measured before and after the reaction, some of the reacting substances may be left over (in excess). The interesting question when conducting an experiment with **excess reactants** is "How much product will result from this reaction?" To answer this, we turn to the concept of limiting reactants.

The small number of seats for this ride is the limiting factor for the people waiting on line.
(Frank Siteman/Stock Boston/ PictureQuest.)

Limiting Reactants

The notion of a limit imposed by a small quantity of things is one that we face nearly every day. For example, suppose you are in line for a ride at an amusement park. The ride can accommodate 32 people at a time. There are 50 people waiting in line. The number of available seats is the limiting factor in this case.

Or suppose that you promised to bake brownies for the class picnic and you have 10 boxes of brownie mix and 6 eggs at your disposal. If each batch of brownies requires one egg, how many batches of brownies can you bake with your available supplies?

Materials Available	Restraints	Product Formed
10 boxes brownie mix	Requires 1 box of mix/batch	10 batches
6 eggs	Requires 1 egg/batch	6 batches

One way to answer this question is to determine the stoichiometric amount of product that will result, using each of the given quantities one at a time. Figure 9-9 shows the batches of brownies that can be baked in the brownie example.

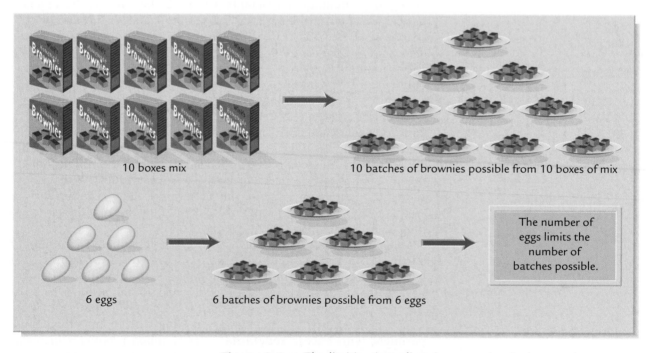

10 boxes mix

10 batches of brownies possible from 10 boxes of mix

6 eggs

6 batches of brownies possible from 6 eggs

The number of eggs limits the number of batches possible.

Figure 9-9 ▲ The limiting ingredient in any recipe is the one that runs out. This limits the amount of product formed.

In this example, the number of eggs limits the batches of brownies we can bake. The eggs are the limiting factor. Knowing the number of boxes of brownie mix does not help us determine the actual number of batches of brownies we can make because we have an excess of brownie mix. Once we know there can be only six batches of brownies, we can easily determine that we will have an excess of four boxes of mix. *When we know the limiting ingredient in a process, we can find the amount of product possible and the amount of ingredient that is left over.*

Just as there was a limiting ingredient in the brownie recipe as there would be for a recipe "from scratch" (Figure 9-10), there can be a limiting reactant in a chemical reaction. In practice, we almost never use exact stoichiometric amounts of reactants (the "ingredients" in an experiment). Other factors, including time and cost considerations, enter the planning of our experiment. How do we know how much product we will get if we don't use exact stoichiometric amounts of reactants? We use a strategy similar to that used in the brownie example to identify the limiting reactant, summarized as follows:

Figure 9-10 ▲ Cooks must have all the ingredients, in the correct amounts, as given in the recipe. (Erv Schowengerdt.)

- Determine the amount of product formed by each of the reactants.
- The smallest amount of product formed is the actual amount of product possible. The reactant that gives the smallest amount of product is the **limiting reactant.**
- All reactants other than the limiting reactant are excess reactants.

✔ Meeting the Goals

When the limiting reactant is consumed, the reaction stops. The limiting reactant literally "limits" the amount of product that is formed in the reaction. When given mass amounts for two reactants, the reactant that produces the smaller amount of a specified product is the limiting reactant for that reaction. Any other reactant will be present in excess and some small amount will remain when the reaction stops.

Limiting reactants work the same way for chemical reactions as for brownies. Suppose that we start the aluminum oxidation reaction discussed earlier with nonstoichiometric amounts—say, 100.0 grams of each reactant. Before the reaction begins we have a total of 200.0 grams of reactants.

$$4\,\text{Al} \quad + \quad 3\,\text{O}_2 \quad \longrightarrow \quad 2\,\text{Al}_2\text{O}_3$$
$$\textit{Initially:} \quad 100.0\,\text{g} \quad 100.0\,\text{g} \quad\quad\quad 0.0\,\text{g}$$

As Al reacts with O_2, the product Al_2O_3 begins to form. Both initial masses decrease during the reaction until one reactant is used up. Then the reaction stops. We use the strategy developed in the brownie example (Figure 9-9) and stoichiometry, to calculate the amount of Al_2O_3 that would be produced from each reactant if that reactant were completely used up.

$$100.0\,\text{g Al} \times \frac{1\,\text{mol Al}}{26.982\,\text{g Al}} \times \frac{2\,\text{mol Al}_2\text{O}_3}{4\,\text{mol Al}} \times \frac{101.961\,\text{g Al}_2\text{O}_3}{1\,\text{mol Al}_2\text{O}_3}$$
$$\approx 188.9\,\text{g Al}_2\text{O}_3$$

$$100.0\,\text{g O}_2 \times \frac{1\,\text{mol O}_2}{31.998\,\text{g O}_2} \times \frac{2\,\text{mol Al}_2\text{O}_3}{3\,\text{mol O}_2} \times \frac{101.961\,\text{g Al}_2\text{O}_3}{1\,\text{mol Al}_2\text{O}_3}$$
$$\approx 212.4\,\text{g Al}_2\text{O}_3$$

The reactant that produces the smaller amount of product is the limiting reactant. In this case, Al is the limiting reactant because it produces the smaller amount of product. The reaction will stop when the limiting reactant is used up, and therefore the actual amount of Al_2O_3 formed in this reaction will be 188.9 g. What has happened to our original 200.0 grams of reactants in this process? Let's see.

$$4\,Al \;+\; 3\,O_2 \longrightarrow 2\,Al_2O_3$$

At end: 0.0 g   ? g   188.9 g

The limiting reactant, Al, has a final mass of 0.0 grams because it was used up. The original 100.0 grams of O_2 is reduced by the amount needed to react with Al and form 188.9 g Al_2O_3. We can calculate the mass of O_2 that remains by first calculating the mass of O_2 that is *used* to react with 100.0 g Al, the limiting reactant.

$$100.0\,\text{g}\,\cancel{Al} \times \frac{1\,\text{mol}\,\cancel{Al}}{26.982\,\text{g}\,\cancel{Al}} \times \frac{3\,\text{mol}\,\cancel{O_2}}{4\,\text{mol}\,\cancel{Al}} \times \frac{31.998\,\text{g}\,O_2}{1\,\text{mol}\,\cancel{O_2}} = 88.9\,\text{g}\,O_2\ \text{used}$$

The mass of O_2 that remains at the end of the reaction, then, is: $100.0 - 88.9 = 11.1$ g O_2.

$$4\,Al \;+\; 3\,O_2 \longrightarrow 2\,Al_2O_3$$

At end: 0.0 g   11.1 g   188.9 g

Note that the total mass at the end is still 200.0 g. Because there are only three substances in this reaction, we could have used the law of conservation of mass to find the mass of O_2 left at the end of the reaction: $200.0\ \text{g} - 188.9 = 11.1$ g O_2. Figure 9-11 summarizes the steps in a limiting reactant calculation.

The concept of a limiting reactant is the same no matter what the number of products or reactants. Once the limiting reactant is determined, it is used to find mass amounts of any other reactant or product. The same procedure could be followed to determine mole amounts of each substance.

Figure 9-11 ▶ Only the limiting reactant can be used in further calculations because we know the exact amount of the limiting reactant used in the reaction. This is not true for substances that are in excess.

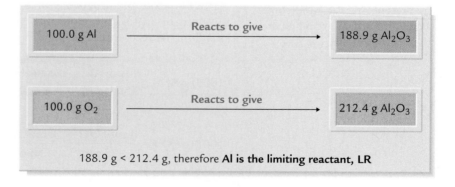

188.9 g < 212.4 g, therefore **Al is the limiting reactant, LR**

Example 9.12   **Limiting reactant calculation**

A reaction mixture of 10.0 grams of Al_2O_3 and 2.00 grams of C reacts completely to form products. Calculate the mass of each product formed and any reactant that remains in the reaction vessel after the reaction stops. Assume that the reaction vessel is sealed and that no chemicals enter or leave the vessel during the reaction.

$$2\,Al_2O_3 + 3\,C \longrightarrow 4\,Al + 3\,CO_2$$

Thinking Through the Problem ◀═━◗ (9.5) The key idea is to determine the amount of product formed by each of the reactants, and to determine both the limiting reactant and the reactant that is in excess. We can then calculate the mass of every substance in the reaction vessel after the reaction stops. It is a good idea to list the substances and masses present at the beginning, to give us the total mass of substances present in the container. We expect the total mass to be the same at the end of the reaction. We know that both products will be formed, and that one of the reactants may be left over. The sum of the masses of all substances present at the end will equal the sum of the reactant masses. We list what we know and what we must calculate.

Before the reaction:
10.0 g Al_2O_3
2.00 g C

After the reaction stops:
___ g Al_2O_3
___ g C
___ g Al
___ g CO_2

Working Through the Problem We first must determine the limiting reactant. Following a procedure like the one mapped out in Figure 9-11, we calculate the mass of Al produced from each of the reactant masses.

$$10.0\,\text{g}\,\cancel{Al_2O_3} \times \frac{1\,\cancel{\text{mol}\,Al_2O_3}}{101.961\,\text{g}\,\cancel{Al_2O_3}} \times \frac{4\,\cancel{\text{mol}\,Al}}{2\,\cancel{\text{mol}\,Al_2O_3}} \times \frac{26.982\,\text{g}\,Al}{1\,\cancel{\text{mol}\,Al}}$$

$$\approx 5.29\,\text{g}\,Al \quad \text{(limiting case)}$$

$$2.0\,\text{g}\,\cancel{C} \times \frac{1\,\cancel{\text{mol}\,C}}{12.001\,\text{g}\,\cancel{C}} \times \frac{4\,\cancel{\text{mol}\,Al}}{3\,\cancel{\text{mol}\,C}} \times \frac{26.982\,\text{g}\,Al}{1\,\cancel{\text{mol}\,Al}} \approx 5.99\,\text{g}\,Al$$

The reaction stops when 5.29 g Al is produced. At this point, the Al_2O_3 is used up. Al_2O_3 is the limiting reactant. We now know the following information about this reaction system:

Before the reaction:
10.0 g Al_2O_3

2.00 g C

After the reaction stops:
0 g Al_2O_3 (The L.R. is used up; zero grams remain.)
___ g C
5.29 g Al (3 s.f. because of 10.0 in calculation)
___ g CO_2 (Calculate next using L.R.)

We must use the limiting reactant in all further calculations on this system. Suppose we now calculate the mass of CO_2 produced.

$$10.0\,g\,Al_2O_3 \times \frac{1\,mol\,Al_2O_3}{101.961\,g\,Al_2O_3} \times \frac{3\,mol\,CO_2}{2\,mol\,Al_2O_3} \times \frac{44.009\,g\,CO_2}{1\,mol\,CO_2}$$

$$\approx 6.47\,g\,CO_2$$

Before the reaction:
10.00 g Al_2O_3
2.00 g C
12.0 g total

After the reaction stops:
0 g Al_2O_3
___ g C (Calculate next, using L.R.)
5.29 g Al
6.47 g CO_2 (3 s.f. because of 10.0 in calculation)

We now calculate the mass of C that reacts with the limiting reactant.

$$10.0\,g\,Al_2O_3 \times \frac{1\,mol\,Al_2O_3}{101.961\,g\,Al_2O_3} \times \frac{3\,mol\,C}{2\,mol\,Al_2O_3} \times \frac{12.011\,g\,C}{1\,mol\,C}$$

$$\approx 1.76\,g\,C\,used$$

We find the mass of C remaining by subtracting 1.76 g from the initial mass of C, 2.00 g. The mass of C that remains is: 2.00 g C at beginning − 1.76 g C used = 0.24 g C remaining.

Answer   0 g Al_2O_3, 0.24 g C, 5.29 g Al, and 6.47 g CO_2

Notice that we can verify that 0.24 g of C remains by using the law of conservation of mass. To find the mass of C remaining, subtract the product masses from 12.0: 12.0 − 5.29 − 6.47 = 0.24 g C.

Example 9.13   **Limiting reactant calculations**

A reaction mixture of 8.00 g FeO and 2.00 g Si is put into a sealed reaction vessel and the reaction is allowed to proceed until it stops. Calculate the mass of Fe formed and the mass of reactant left when the reaction ends. Assume the reaction occurs according to the following equation:

$$2\,FeO + Si \longrightarrow 2\,Fe + SiO_2$$

Thinking Through the Problem We must determine the limiting reactant and use it to calculate the mass of Fe formed and the mass of reactant that remains when the reaction stops.

Working Through the Problem We start by determining the limiting reactant.

$$8.00\,g\,FeO \times \frac{1\,mol\,FeO}{71.884\,g\,FeO} \times \frac{2\,mol\,Fe}{2\,mol\,FeO} \times \frac{55.845\,g\,Fe}{1\,mol\,Fe} \approx 6.22\,g\,Fe$$

(limiting case)

$$2.00 \, \text{g Si} \times \frac{1 \, \text{mol Si}}{28.086 \, \text{g Si}} \times \frac{2 \, \text{mol Fe}}{1 \, \text{mol Si}} \times \frac{55.845 \, \text{g Fe}}{1 \, \text{mol Fe}} \approx 7.95 \, \text{g Fe}$$

The reaction stops when 6.22 g Fe is formed. FeO is the limiting reactant and will be used to calculate the mass of Si used in the reaction.

$$8.00 \, \text{g FeO} \times \frac{1 \, \text{mol FeO}}{71.844 \, \text{g FeO}} \times \frac{1 \, \text{mol Si}}{2 \, \text{mol FeO}} \times \frac{28.086 \, \text{g Si}}{1 \, \text{mol Si}}$$
$$\approx 1.56 \, \text{g Si used}$$

Mass of Si remaining, then, is: $2.00 - 1.56 = 0.44$ g Si remaining.

Answer At the end of the reaction, we have 0 g FeO (limiting reactant), 0.44 g Si (remaining from initial mixture), 6.22 g Fe and some mass of SiO_2 that is produced but not calculated in this problem. The total mass of the vessel contents at the beginning and at the end of the reaction is 10.0 g.

How are you doing? **9.6**

Suppose that a mixture of 20.0 grams of Al and 50.0 grams of HCl in water is put into a sealed reaction vessel and that the reaction is allowed to proceed according to this equation:

$$2 \, \text{Al} \, (s) + 6 \, \text{HCl} \, (aq) \longrightarrow 2 \, \text{AlCl}_3 \, (s) + 3 \, \text{H}_2 \, (g)$$

Calculate the mass of each substance that is left in the reaction vessel when the reaction stops.

Theoretical and Actual Reaction Yields

The amount of product or products formed in a reaction is called the **reaction yield** of the reaction. Chemists do not always obtain all of the reaction yield that is predicted by the stoichiometry of a reaction. Let us consider how to handle cases where the amount of product isolated in a reaction is less than that amount predicted by the stoichiometry.

The amounts of products actually obtained in reactions often vary from the theoretical amounts predicted by stoichiometry. Why? There are several reasons, including:

- Some of the product may be lost in handling.
- A reaction may not be complete at the time that the product is isolated.
- The chemical substances may engage in side reactions that we did not predict or that we must tolerate to get some of the reaction we want.
- Some reactants may be impure.

The amount of product actually isolated or obtained from a reaction is called the **actual yield** *of the reaction. The amount of product predicted by stoichiometry is called the* **theoretical yield.** The actual yield

KEY IDEA 9.6

never should be more than the theoretical yield; it is usually less—much less in some cases. (Poor lab techniques or procedures occasionally result in an actual yield that is more than the theoretical yield, but this is not the usual case.) The ratio of the actual yield to the percentage yield is called the **percentage yield,** where:

$$\text{percentage yield} = \frac{\text{actual yield}}{\text{theoretical yield}} \times 100$$

For example, the synthesis of silver chloride involves mixing known amounts of a chloride salt—such as potassium chloride—and silver nitrate, a soluble silver salt:

$$KCl\ (aq) + AgNO_3\ (aq) \longrightarrow AgCl\ (s) + NO_3\ (aq)$$

According to the stoichiometry of this balanced equation, if we mix one mole of KCl and one mole of $AgNO_3$, then we get a theoretical yield of one mole of AgCl, which has a mass of 143.32 grams. In a typical reaction, however, perhaps 125 grams of AgCl are isolated. This amount is the actual yield. The percentage yield is then:

$$\frac{125\ \text{g AgCl}}{143.32\ \text{g AgCl}} \times 100 = 87.22\%$$

The percentage yield can be used to predict the actual amounts of product formed by other, identical, reactions. For example, suppose that we again react potassium chloride and silver nitrate, but this time use smaller amounts of reactants so that the theoretical yield is 15.2 g of AgCl. If the percentage yield is again 87.2%, what will be the actual yield? We can answer this question by rearranging the equation for percentage yield and then substituting the two known values:

$$\text{actual yield} = \frac{\text{percentage yield}}{100\%} \times \text{theoretical yield}$$

$$\text{actual yield} = \frac{87.2\%}{100\%} \times 15.2\ \text{g AgCl} = 13.3\ \text{AgCl}$$

We also can use the percentage yield and an actual yield to find the theoretical yield. Finding a theoretical yield from an actual yield is very useful when we want to obtain a certain actual yield. In such a case we need to know how much theoretical yield we should plan for, to be sure that we get the desired actual yield.

For example, let's say that we want to obtain 250. g of AgCl. What theoretical yield should we plan for, if the percentage yield for the reaction is expected to be 87.2%? We rearrange the percentage yield equation, and then insert the known values:

$$\text{theoretical yield} = \frac{\text{actual yield}}{\text{percentage yield}} \times 100$$

$$= \frac{250.\ \text{g AgCl}}{87.2\%} \times 100 = 287\ \text{g AgCl (3 s.f.)}$$

✔**Meeting the Goals**

Percentage yield is the ratio between what is actually produced in an experiment and what is theoretically possible.

percentage yield
$$= \frac{\text{actual yield}}{\text{theoretical yield}} \times 100$$

✔**Meeting the Goals**

The theoretical yield is calculated from initial mass amounts of a substance using the stoichiometry given by the equation. Theoretical yield can also be calculated from the actual yield and the percentage yield by rearranging the formula given for percentage yield. Thus,

theoretical yield
$$= \frac{\text{actual yield}}{\text{percentage yield}} \times 100$$

Example 9.14 | Finding percentage yield

In the synthesis of ammonia from nitrogen and hydrogen, large quantities are used. If 2.56×10^6 grams of N_2 and excess H_2 are used, what is the theoretical yield? If only 2.4×10^6 grams of NH_3 are obtained, what is the percentage yield? The equation is $N_2 (g) + 3\ H_2 (g) \rightarrow 2\ NH_3 (g)$.

Thinking Through the Problem ◄═◄▷ (9.6) The key idea is to use stoichiometry to obtain the theoretical yield. We must determine the percentage yield for this reaction. To find the percentage yield, we need both the theoretical and the actual yields. We are given the actual yield (2.4×10^6 g NH_3) but not the theoretical yield. We must begin by calculating the theoretical yield.

Working Through the Problem

$$2.56 \times 10^6 \ \cancel{\text{g N}_2} \times \frac{1 \ \cancel{\text{mol N}_2}}{28.014 \ \cancel{\text{g N}_2}} \times \frac{2 \ \cancel{\text{mol NH}_3}}{1 \ \cancel{\text{mol N}_2}} \times \frac{17.031 \ \text{g NH}_3}{1 \ \cancel{\text{mol NH}_3}}$$
$$= 3.112 \times 10^6 \approx 3.11 \times 10^6 \ \text{g NH}_3$$

Now we can calculate percentage yield:

$$\frac{\text{actual yield}}{\text{theoretical yield}} \times 100 = \text{percentage yield}$$

$$\frac{2.4 \times 10^6}{3.11 \times 10^6} \times 100\% = 77.17 \approx 77\%$$

Answer 77% yield

Example 9.15 | Finding theoretical yield from percentage yield and actual yield

The heating of sodium hydrogen carbonate can be used to make CO_2 in small amounts in the laboratory, according to the stoichiometry of this equation:

$$2\ NaHCO_3 \ (s) \longrightarrow Na_2CO_3 \ (s) + H_2O \ (g) + CO_2 \ (g)$$

If we want to make 2.5×10^{-3} moles of CO_2 gas, and we typically obtain a percentage yield of 76% in this reaction, what mass of sodium hydrogen carbonate should be reacted?

Thinking Through the Problem We must calculate the mass of $NaHCO_3$ needed in order to produce 2.5×10^{-3} moles of CO_2. We are told that this amount of CO_2 represents a 76% yield. In this case, we are given the actual yield and the percentage yield of the CO_2 gas. We must use these to get the theoretical yield of CO_2 gas, which can then lead us to the mass of $NaHCO_3$ required. First, the theoretical yield:

PRACTICAL C   How are fertilizers the limiting reactants in plant growth?

We have discussed how plant growth depends on the simple synthesis of starch and similar substances from carbon dioxide and water. Both substances are abundant on earth, although water can be very scarce in certain areas or at certain times. But even given a steady source of air and water, most plants will need to obtain small amounts of other elements for their existence. Plant growth is stunted—or limited—when certain elements are missing. Potassium, for example, is important in maintaining the salt balance in a plant; nitrogen is needed to make the proteins involved in the biological activity of the plant; and plants need phosphorus for the manufacture of new DNA, which is the hereditary material in the plant's cells.

Thus, while carbon dioxide and water are the chemical substances needed in the greatest amounts for plant growth, small amounts of other substances are critical as well. In many cases, these substances are applied in the form of natural and artificial fertilizers.

Nitrogen is the most abundant component of the atmosphere. For every carbon dioxide molecule there are about 1,000 N_2 molecules in the air. You may wonder, then, why farmers and gardeners find it necessary to apply nitrogen as fertilizers. The answer is that nitrogen in the air is in a chemical form that cannot be used by most plants. Instead, plants (and almost all other organisms) must use nitrogen in chemical compounds, such as in derivatives of ammonia and salts containing nitrates. Nitrogen fertilizers can be a major expense, constituting over 10% of the operating cost of growing corn.

Consider the formation of the simple amino acid alanine, an essential component of plants and animals. It has the chemical formula $C_3H_7NO_2$. Write a balanced chemical equation for the formation of alanine from carbon dioxide, water, and ammonia (NH_3). Assume that the other product of the reaction is oxygen.

Assume that a plant needs to make 1.23 grams of alanine. How much carbon dioxide, water, and ammonia are needed? If the ammonia is provided by the salt ammonium sulfate, then how much ammonium sulfate is needed?

$$\text{theoretical yield} = \frac{100\%}{\text{percentage yield}} \times \text{actual yield}$$

$$= \frac{100\%}{76\%} \times 2.5 \times 10^{-3} \text{ mol } CO_2$$

$$= 3.3 \times 10^{-3} \text{ mol } CO_2$$

To get the mass of $NaHCO_3$, we next carry out a moles-to-mass stoichiometric calculation:

$$3.3 \times 10^{-3} \text{ mol } CO_2 \times \frac{2 \text{ mol } NaHCO_3}{1 \text{ mol } CO_2} \times \frac{84.006 \text{ g } NaHCO_3}{1 \text{ mol } NaHCO_3}$$

$$\approx 0.55 \text{ g } NaHCO_3$$

Answer The mass of $NaHCO_3$ needed to provide an actual yield of 2.5×10^{-3} mole of CO_2 is 0.55 gram of $NaHCO_3$.

How are you doing?   9.7

In a reaction of 45.6 grams of WCl_6 with NaOH, WCl_4O is the most important product:

$$WCl_6 \; (s) + 2 \, NaOH \; (aq) \longrightarrow WCl_4O \; (s) + H_2O \; (l) + 2 \, NaCl \; (aq)$$

Figure 9-12 ▲ Workers in a North Dakota field where fast-growing alfalfa is used to clean up a spill from a railroad car. (Bruce Fritz/Agricultural Research Service, USDA.)

When we consider agriculture, of course, we do not consider individual plants and particular amino acids. Instead, we can look at the mass percentages in the plant. An ear of corn contains about 5% nitrogen in its dry weight. If the ear of corn has a mass of 190 grams, calculate the mass of nitrogen in the corn, and determine the number of moles of ammonium sulfate needed to provide that nitrogen.

Sometimes there is too much nitrogen in soil. This was the case following a nitrogen fertilizer spill in North Dakota in 1989. The spill continued to pollute the soil years later. To solve the problem, a special alfalfa crop was planted, much like the one shown in Figure 9-12. Recent data from the U.S. Agricultural Research Service showed that the alfalfa, when planted and harvested three times a year, removed 170 kg of nitrogen (as N) per acre. This is an example of cleanup through biology—known as bioremediation. Calculate this amount of nitrogen in terms of moles of nitrogen and mass of ammonium nitrate.

If a percentage yield of 89.2% $WOCl_4$ is obtained, what is the actual yield?

PROBLEMS

29. Concept question: Think of an example from your everyday life that uses the limiting reactant idea.

30. Consider the following balanced equation:
$C_5H_{12} (g) + 8\ O_2 (g) \rightarrow 5\ CO_2 (g) + 6\ H_2O (l)$

(a) How many moles of CO_2 can be produced from a reaction mixture composed of 10.0 moles of C_5H_{12} and 10.0 moles of O_2?

(b) How many moles of CO_2 can be produced from a reaction mixture composed of 3.0 moles of C_5H_{12} and 12.0 moles of O_2?

(c) How many moles of CO_2 can be produced from a reaction mixture composed of 12.0 moles of C_5H_{12} and 3.0 moles of O_2?

(d) If 5.0 moles of C_5H_{12} and 15.0 moles of O_2 are placed in a closed container and are allowed to react as completely as possible according to the balanced equation, what compounds are present in the container after the reaction and how many moles of each compound are present?

31. Coke (elemental carbon) will reduce calcium oxide to form calcium carbide and carbon monoxide according to the following balanced equation:

$CaO (s) + 3\ C (s) \longrightarrow CaC_2 (s) + CO (g)$

(a) If 50.0 grams of CaO and 34.0 grams of C are put into a sealed vessel and allowed to react according to the balanced equation, what is the limiting reactant in this problem?

(b) Calculate the mass of CaC_2 that results from the reaction mixture in part (a).

(c) What is the total mass in the reaction vessel when the reaction ends?

(d) Give the mass of each substance in the reaction vessel at the end of the reaction.

32. Powdered aluminum displaces manganese from manganese (II) oxide when a mixture of the two substances is heated according to the equation:

$2\ Al (s) + 3\ MnO (s) \longrightarrow 3\ Mn (s) + Al_2O_3 (s)$

(a) Calculate the number of moles of manganese that could be obtained in this reaction if 14.2 grams of MnO is heated with 3.31 grams of Al.

(b) Calculate the number of moles of excess reactant remaining at the end of this reaction.

33. Ammonia is burned to form nitrogen monoxide and water in the first step in the commercial production of nitric acid. What is the maximum amount of NO that can be produced by mixing 0.85 gram of NH_3 and 1.28 grams of O_2 in a closed vessel if they react according to the equation:

$$4 NH_3 (g) + 5 O_2 (g) \longrightarrow 4 NO (g) + 6 H_2O (g)$$

34. What is the maximum mass, in grams, of $Ca_3(PO_4)_2$ that can be prepared from a reaction mixture containing 7.4 grams of $Ca(OH)_2$ and 9.8 grams of H_3PO_4 in the reaction:

$$3 Ca(OH)_2 + 2 H_3PO_4 \longrightarrow Ca_3 (PO_4)_2 + 6 H_2O$$

35. Consider the following balanced equation:

$$2 C_2H_6 + 7 O_2 \longrightarrow 4 CO_2 + 6 H_2O$$

(a) How many grams of CO_2 can be produced from a reaction mixture composed of 22.0 grams of C_2H_6 and 16.0 grams of O_2? What is the limiting reactant for this reaction?

(b) How many grams of H_2O can be produced from 22.0 grams of C_2H_6 and 16 g of O_2? What is the limiting reactant?

(c) Compare your answers from (a) and (b). Discuss whether or not the two answers agree as to which compound is the limiting reactant.

36. Nitrogen monoxide gas burns in oxygen gas to produce gaseous nitrogen dioxide.

(a) Write the balanced equation for this reaction.

(b) 25.5 grams of nitrogen monoxide and 28.4 grams of oxygen are placed in a sealed reaction vessel and react as far as possible according to the balanced equation in part (a). What is the mass of the contents of the container after the reaction ceases?

37. Determine the percentage yield if an actual yield of 1.62 grams of $PbCrO_4$ is obtained from the reaction of lead (II) nitrate and sodium chromate to give lead chromate and sodium nitrate. The limiting reactant is 1.85 grams of lead (II) nitrate.

38. What is the actual yield expected if a reaction has a percentage yield of 95.2% and a theoretical yield of 2.98 grams?

Chapter 9 Summary and Problems

◄ Review with **Web Practice**

VOCABULARY

mole ratio for a chemical substance	The ratio of the number of moles of an element to the number of moles of the chemical substance containing the element, or to the number of moles of a different element in the substance.
mole ratio for a chemical reaction	The ratio of the number of moles of two substances in a chemical reaction; the values of these ratios are taken from the stoichiometric coefficients of the balanced chemical equation.
stoichiometric amount	An amount of a substance that is exactly equal to the amount predicted by the stoichiometry of the reaction.
limiting reactant	The reactant that produces the smallest amount of product in a chemical reaction.
excess reactant	The reactant or reactants that are left over when all of the limiting reactant has been consumed.
reaction yield	The amount of product formed in a reaction.
actual yield	The amount of product actually isolated from a reaction.
theoretical yield	The amount of product predicted by the stoichiometry of the balanced equation for the reaction.
percentage yield	The ratio, in percent terms, of the actual yield to the theoretical yield.

SECTION GOALS

How are proportional reasoning and conversion factors used in mole/mole stoichiometry problems?

The mole ratios given by the stoichiometric coefficients of a balanced chemical equation are the known ratios or conversion factors used to relate given mole amounts to unknown mole amounts.

What steps must be taken to solve a mole/mass stoichiometry calculation?

The coefficients in the balanced equation are mole numbers. In stoichiometry, we use these mole numbers to relate moles of reactants and products to each other. When given grams of a substance, we must first convert grams to moles by dividing the given mass by the molar mass of that substance. We can use mole/mole ratios derived from the equation only when we have the mole amount of a substance.

How do we calculate relative amounts when several different conversions are involved?

We carry out multiple conversion problems by multiplying a sequence of conversion factors. In stoichiometry, these are usually molar masses and mole ratios.

What steps must be taken to solve a mass/mass stoichiometry calculation?

To solve a mass-mass stoichiometric problem, first convert the given mass to moles of that substance; then use the mole/mole ratio to find moles of a second substance (for which mass information is sought); and finally, convert moles of the second substance to grams.

What effect does a limiting reactant have on a reaction?

A limiting reactant stops the reaction when it is used up. In a very real sense, the limiting reactant "limits" the amount of product that is formed in the reaction. When given mass amounts for two reactants, the reactant that produces the smaller amount of a specified product is the limiting reactant for that reaction.

What is a theoretical yield and how is it calculated?

The theoretical yield is calculated from initial mass amounts of a substance using the stoichiometry given by the equation. A theoretical yield can also be calculated from the actual yield and the percentage yield by rearranging the formula given for percentage yield. Thus,

$$\text{theoretical yield} = \frac{\text{actual yield}}{\text{percentage yield}} \times 100$$

What is percentage yield and how is it calculated?

Percentage yield gives the ratio between what is actually produced in an experiment and what is theoretically possible.

$$\text{percentage yield} = \frac{\text{actual yield}}{\text{theoretical yield}} \times 100$$

PROBLEMS

39. Determine the number of moles of O_2 that can be obtained if 25 grams of H_2O_2 reacts according to the equation:

$$2\ H_2O_2 \longrightarrow 2\ H_2O + O_2$$

40. The following reaction forms copper (II) phosphate from a solution of copper (II) nitrate and potassium phosphate:

$$3\ Cu(NO_3)_2\ (aq) + 2\ K_3PO_4\ (aq) \longrightarrow$$
$$Cu_3(PO_4)_2\ (s) + 6\ KNO_3\ (aq)$$

What mass of copper (II) phosphate should form if the reaction uses 0.191 mole of $Cu(NO_3)_2$?

41. The reaction of calcium and hydrochloric acid proceeds as follows:

$$Ca + 2\ HCl \longrightarrow CaCl_2 + H_2$$

In a typical reaction, 0.25 gram of calcium is used. How many moles of H_2 will be formed?

42. Scientists studying the mud along the coast of Germany have found an organism that metabolizes iron as a food. The essential part of the stoichiometry of the reaction for the growth of the organism is:

$$4 FeCO_3 (s) + 7 H_2O (l) \longrightarrow$$
$$CH_2O (s) + 4 Fe(OH)_3 (s) + 3 CO_2 (g)$$

The formula CH_2O is a "shorthand" notation indicating the carbohydrates formed by the organism. Determine the mass of $FeCO_3$ consumed in making 0.325 gram of the CH_2O.

43. Chlorine dioxide, ClO_2, is a gas that can be used as a replacement for liquid chlorine bleaches. It can be made by the following reaction:

$$2 HClO_3 + H_2C_2O_4 \longrightarrow$$
$$2 ClO_2 + 2 CO_2 + 2 H_2O$$

Determine the mass of ClO_2 that can be formed by the complete reaction of 23.5 grams of $H_2C_2O_4$ with 15.8 grams of $HClO_3$ in this reaction.

44. Bromine is made by the treatment of seawater with chlorine gas. This reaction can be represented as follows:

$$2 NaBr + Cl_2 \longrightarrow Br_2 + 2 NaCl$$

(a) If this reaction produces 12.3 moles of Br_2, how many moles of NaBr are required?
(b) Determine how many grams of Br_2 are produced if 12.2 moles of Cl_2 are consumed.

45. Pentane gas burns in air according to this equation:

$$C_5H_{12} + 8 O_2 \longrightarrow 5 CO_2 + 6 H_2O$$

(a) How many moles of CO_2 will be produced from a reaction mixture of 0.176 mole C_5H_{12} and 0.352 mole of O_2?
(b) Calculate the mole amount of the reactant that is in excess.

46. Consider this reaction for the synthesis of indium sulfide:

$$2 In + 3 S \longrightarrow In_2S_3$$

What mass of S is required to completely react with 1.000 gram of In?

47. Determine the mass of NaOH is needed to react with 25.0 grams of $Sr(NO_3)_2$ according to the following chemical equation:

$$Sr(NO_3)_2 + 2 NaOH \longrightarrow Sr(OH)_2 + 2 NaNO_3$$

48. The reaction to make K_2SnF_6 is given by this equation:

$$2 KF (s) + SnF_4 (s) \longrightarrow K_2SnF_6 (s)$$

What mass of K_2SnF_6 is formed from a reaction mixture of 20.0 grams of KF and 12.0 grams of SnF_4?

49. Determine the mass of $FeCO_3$ that will form in the following reaction, starting from 0.100 gram of $C_6H_{12}O_6$ and excess CO_2 and $Fe(OH)_3$:

$$18 CO_2 + C_6H_{12}O_6 + 24 Fe(OH)_3 \longrightarrow$$
$$24 FeCO_3 + 42 H_2O$$

50. Ammonia is used in the first step in the production of nitric acid. In this reaction, ammonia is oxidized to water and nitrogen monoxide. The equation for this reaction is:

$$4 NH_3 (g) + 5 O_2 (g) \longrightarrow 6 H_2O (g) + 4 NO (g)$$

Calculate the mass of NO produced when this reaction begins with a mixture of 1.20 grams of NH_3 and 4.88 grams of O_2. Which is the limiting reactant in this problem?

51. Nitrogen monoxide can be oxidized to nitrogen dioxide, as shown in this equation:

$$2 NO (g) + O_2 (g) \longrightarrow 2 NO_2 (g)$$

Calculate the maximum amount of NO_2 produced from a reaction mixture of 10.0 grams of NO and 11.0 grams of O_2.

52. In the following reaction, what mass of C_2H_5OH should result from the complete reaction of 1.25 grams of $C_6H_{10}O_4$, assuming excess NaOH is present?

$$C_6H_{10}O_4 (s) + 2 NaOH (aq) \longrightarrow$$
$$Na_2C_2O_4 (s) + 2 C_2H_5OH (aq)$$

Discovering the Gas Laws

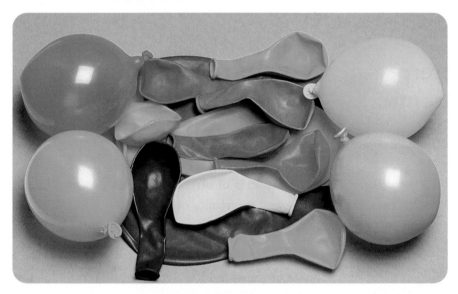

The behavior of a gas, such as the air in some of these balloons, can be described using mathematical models. (Tom Pantages.)

PRACTICAL CHEMISTRY Mathematical Models in Chemistry

In the Practicals for this chapter we take a slight departure from the "real world" questions that we usually ask. As the chapter title suggests, we can discover important relationships in nature by applying our understanding of chemistry and the gas laws. To do so, we use math to characterize the way that nature behaves in different circumstances. This is the way that scientific laws—mathematical statements of relationships we find in nature—are developed. Just as storytellers and movie producers sometimes use visual concepts to get across some point, chemists rely on mathematical laws and graphs to describe chemical behavior. This allows us to make certain conclusions about natural behaviors and to develop theories that explain those behaviors. These laws, theories, and concepts are used to develop "models" of chemical behavior.

In this chapter we will use mathematical models to interpret the physical relationships that exist between the volume, pressure, temperature, and mole amount of gas samples. Mathematical models describe the phenomena of nature in two ways: through mathematical expressions and through graphical presentations of data.

This chapter will focus on the following questions:

PRACTICAL **A** What makes a good graph?
PRACTICAL **B** How is an equation related to its graph?
PRACTICAL **C** What do balloons have to do with chemistry?
PRACTICAL **D** Where are the gas molecules when a balloon is flattened?

10.1 Interpreting Relationships Among Chemical Phenomena

SECTION GOALS

✔ Understand the relationship of the Cartesian coordinate system to mathematical descriptions of a relationship of two variables.

✔ Determine the slope and intercepts of a line.

✔ Relate independent and dependent variables in chemical experiments and graphical presentations of data in the Cartesian system.

✔ Determine the units of the slope for a line given the units for the chemical properties that are graphed.

MAKING IT WORK WITH MATH

Graphing in the Cartesian Coordinate System

Because scientific data are often presented graphically, it is important to review graphing. Most graphing is based on the **Cartesian coordinate system.** The Cartesian coordinate system, named for the French mathematician René Descartes, is a rectangular coordinate system formed by the intersection of two perpendicular lines, as shown in Figure 10-1. *The intersecting lines divide the plane containing them into four equal parts called quadrants.* The lines are drawn as arrows to indicate that these lines continue without end. The point of intersection of the two lines is the *origin*. The origin has coordinates (0, 0). All points in the plane are referenced from the origin.

The horizontal axis is the *x*-axis and the vertical axis is the *y*-axis. Each point on the graph is given as an ordered pair of the form (x, y). In the ordered pair (x, y), *x* measures the horizontal distance from the origin and *y* measures the vertical distance from the origin. The positive horizontal motion is to the right and the negative horizontal motion is to the left—the same as on the number line. By the same token, the positive vertical motion is up and the negative vertical motion is down. For example, to find the point $(-2, 5)$ we start at the origin and move 2 units to the left and 5 units up. Based on these descriptions, we can locate any point in the plane. Given the coordinates, we can determine the location of a point. Figure 10-2 shows the coordinate system imposed on Figure 10-1 with four points, including $(-2, 5)$, identified. Note that in this case the markings on the axes represent one unit, but this will not always be the case.

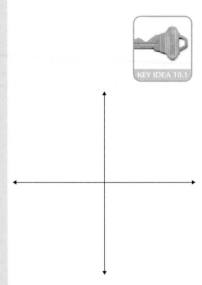

Figure 10-1 ▲ A coordinate axis system.

Figure 10-2 ▶ A coordinate axis system with marked axes and four ordered pairs.

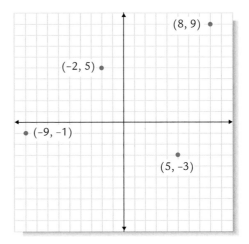

Example 10.1 Locating a point on a Cartesian coordinate graph

Locate the point $(-3, -8)$ on a Cartesian coordinate graph.

Thinking Through the Problem The point $(-3, -8)$ corresponds to the ordered pair (x, y). We know that the x-coordinate is -3 and the y-coordinate is -8. We must start at the origin, $(0, 0)$. The negative direction for x is left; the negative direction for y is down.

Working Through the Problem Starting at $(0, 0)$, move three spaces to the left $(x = -3)$ and then eight spaces down $(y = -8)$.

Answer The point $(-3, -8)$ is located three spaces to the left and eight spaces down from the origin.

How are you doing? 10.1

Locate the point $(2, 5)$ on a Cartesian coordinate graph.

MAKING IT WORK WITH MATH

Linear Relationships

We will start our examination of mathematical relationships among variables by looking at linear relationships. This implies that the relationships can be presented as a single line. Lines in the Cartesian system can be presented in the form of an equation $y = mx + b$, where m and b are constants. If the data fit the relationship in the equation, then we can "plug in" the values of x and y in an ordered pair and find that the equation is true. If an ordered pair (x, y) satisfies the equation, we can say that they "fit" the equation.

Example 10.2 Determining whether or not a point is on a line

(a) Is the point $(1, 3)$ on the line $y = 6 - 3x$?
(b) Is the point $(4, -3)$ on this line?

Thinking Through the Problem $(1, 3)$ defines an (x, y) data point. If $(1, 3)$ is on the line $y = 6 - 3x$, we can substitute $x = 1$ and $y = 3$ into the equation and obtain a true statement. For (b) we do the same thing for $x = 4$ and $y = -3$.

Working Through the Problem

$$y = 6 - 3x$$

(a)
$$3 = 6 - 3(1)$$
$$3 = 6 - 3$$
$$3 = 3$$

(b)
$$y = 6 - 3x$$
$$-3 = 6 - 3(4)$$
$$-3 = 6 - 12$$
$$-3 = -6$$

Answer The point (1, 3) is on the line, the point (4, −3) is not.

How are you doing? **10.2**

Determine whether or not the points (6, 10) and (1, −5) are on the line $y = 5x - 10$.

✔ **Meeting the Goals**

To find the x-intercept, let $y = 0$ and solve for x. To find the y-intercept, let $x = 0$ and solve for y.

The advantage of an equation for a linear relationship of two variables is that, in principle, we can use the equation to determine the value of x that corresponds to a given value of y, and vice versa. This is something we first saw in Chapter 6, when we discussed how to solve an equation for a variable. Then, of course, we were working with proportions, but here we are learning about equations in connection with graphing in the Cartesian system. An important case is the value of one variable when the other variable has a value of zero. These are called the intercepts of x or y. The **x-intercept** is the place where the line crosses the x-axis, that is, where the value of y is zero. The **y-intercept** is the place where the line crosses the y-axis, that is, where the value of x is zero. A general case is shown in Figure 10-3. To find an intercept, we let the other variable equal zero and solve. Thus, to find the y-intercept, we let the value of $x = 0$ and solve for y. To find the x-intercept, let the value of $y = 0$ and solve for x.

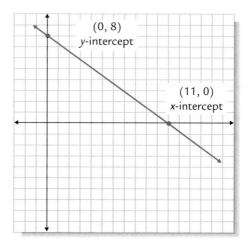

Figure 10-3 ▶ A line on a coordinate system with the x- and y-intercepts marked.

Example **10.3**   **Finding the intercepts of a line**

Find the x- and the y-intercept for the equation $y = 12x + 3$.

Thinking Through the Problem We want to find the x-intercept and the y-intercept. To find the y-intercept, let $x = 0$ and solve for y. To find the x-intercept, let $y = 0$ and solve for x.

Working Through the Problem

$$y = 12x + 3 \qquad\qquad y = 12x + 3$$
$$y = 12(0) + 3 \qquad\qquad 0 = 12x + 3$$
$$y = 3 \qquad\qquad\qquad -3 = 12x$$
$$-\frac{3}{12} = x$$
$$-\frac{1}{4} = x$$

Answer The y-intercept is the point $(0, 3)$ and the x-intercept is the point $(-1/4, 0)$.

Let's now consider the slant of a line. The slant of a line is called the **slope of the line**. We use m to indicate slope.

$$m = \frac{\text{vertical change}}{\text{horizontal change}} = \frac{\text{difference in } y \text{ values}}{\text{difference in } x \text{ values}} = \frac{\Delta y}{\Delta x}$$

For any two points (x_1, y_1) and (x_2, y_2) that are on the line

$$m = \frac{\Delta y}{\Delta x} = \frac{y_2 - y_1}{x_2 - x_1}$$

The symbol "Δ" indicates we take the *difference* between two values. If the line is steep, the vertical change Δy will be greater than the horizontal change Δx, and this ratio will be greater than 1. If the vertical change Δy is less than the horizontal change Δx, the ratio is less than 1 because the numerator is smaller than the denominator.

✔ **Meeting the Goals**

The slope of a linear plot is found by evaluating the ratio

$$m = \frac{y_2 - y_1}{x_2 - x_1}$$ for any two points (x_1, y_1) and (x_2, y_2) that are on the line.

Let's apply this formula to the two points, $(2, 1)$ and $(5, 9)$. The slope of the line connecting them is $m = \frac{9 - 1}{5 - 2} = \frac{8}{3}$. If you plot the x and y coordinates of these two points in the Cartesian coordinate system, you should notice that the line which contains them rises from left to right. The slope of this line is a positive number; lines with positive slopes rise from left to right. Remembering this fact may help you to avoid arithmetic errors. If your calculation of slope yields a positive value but your line falls from left to right, there is an error.

Example 10.4   **Finding a slope using two points**

Find the slope of the line containing the points $(-3, 8)$ and $(2, 1)$.

Thinking Through the Problem Find the slope of the line given two points. Begin with the formula for slope: $m = \dfrac{y_2 - y_1}{x_2 - x_1}$.

Working Through the Problem $m = \dfrac{8 - 1}{-3 - 2} = \dfrac{7}{-5} = -\dfrac{7}{5}$

Answer $m = -\dfrac{7}{5}$

How are you doing? 10.3

Find the slope of the line containing the points $(5, 2)$ and $(1, 10)$.

Note that the slope in example 10.4 is negative. If you plot the given points and draw the line containing them, the line will decrease from left to right. *This is true of any line with a negative slope.* Note in the last example which ordered pair was used as (x_1, y_1). Which point is "first" does not matter. If we interchange the points, we get:

$$m = \frac{1 - 8}{2 - (-3)} = \frac{-7}{5} = -\frac{7}{5}$$

Note that the result is the same.

There are many relationships where the line describing two variables is y = mx (or, as we have seen repeatedly, y = kx). This is the case in which there is a simple proportion relating y and x. In that case, the only time that y is 0 is when x is 0. The x- and the y-intercept both lie at the point (0, 0).

Independent and Dependent Variables in Chemical Experiments

In chemistry we often use the Cartesian system to show how different variables, such as temperature, volume, and amount of product, are related to each other. This requires that we think of one variable as changing in response to another. When we do an experiment, we often can change one variable at will; for example, we can change the temperature of a substance by turning on a laboratory burner or by packing a beaker in ice. A change that we can make at will is said to occur independently, so this variable is called the **independent variable.** We then see what effect this change has on another variable—called the **dependent variable.**

When we make measurements of an independent variable and a dependent variable we create an ordered pair for a graph. We usually make the independent variable the x value and the dependent

In most experiments, we see changes in a dependent variable when we make changes in an independent variable. The values of these measurements become (x, y) pairs for graphs.

variable the y value. Thus, we graph data in the (x, y) system of Cartesian coordinates using pairs in the form (independent variable, dependent variable). We are then able to read the graph and say how a change in the independent variable (seen as changes in the x values) relates to a change in the dependent variable (seen as changes in the y value).

The language we use to describe an experiment usually indicates what we mean by the independent and dependent variables in an experiment. For example, suppose that an experiment is described as follows: "We measured the way the volume changed as we added different masses of rock to the liquid." We can tell from this statement that *first* a mass of rock was added to a volume of liquid and *then* the volume was measured. In this case, the "mass of rock" is the independent variable and the "volume" is the dependent variable. If we graphed the data, we would put "mass of rock" on the x-axis and "volume" on the y-axis. There may be some additional variables that do *not* change during the experiment. For example, the temperature does not change during the "rock" experiment. A variable that is deliberately held constant during an experiment is known as a **controlled variable.**

Example 10.5 Determining dependent and independent variables

For the following experiments, indicate which is the independent variable and which is the dependent variable. If a controlled variable is present, indicate what it is.

(a) We mix different amounts of water into a fixed amount of isopropyl alcohol. The final volume of the solution is measured in each case.

(b) Different masses of Ca are dissolved in a consistent amount of acid. The temperature change is measured in each case.

(c) The volume of a sample of mercury is measured at different temperatures.

Thinking Through the Problem The independent variable is something that we deliberately change during the course of the experiment. The variable that changes as a result of the experiment is the dependent variable.

Answer

(a) Because the experiment starts with different amounts of water, the volume of water is the independent variable. The volume of the solution after mixing is measured at the end of the experiment, so it is the dependent variable. The volume of isopropyl alcohol is "fixed" and does not change, so it is neither independent nor dependent in this experiment. It is a controlled variable.

(b) Because we deliberately use different masses of Ca at the beginning of the experiment, the mass of Ca is the independent variable. When the experiment is done we measure a temperature change; this is the dependent variable. The "consistent amount of acid" is a controlled variable.

(c) The phrase "at different temperatures" implies that the experiment involves selecting a temperature and then measuring the volume of mercury. The temperature is the independent variable and the volume is the dependent variable.

How are you doing? **10.4**

For the following experiments, determine the independent variable and the dependent variable.

(a) When a fixed amount of Ca is dissolved in acid, a measured volume of gas forms.

(b) Different masses of $AgNO_3$ react with excess NaCl, and the mass of the product AgCl is determined.

Graphing and Interpreting Chemical Data

When we graph the data from experiments we often see trends in the data. If these mathematical trends seem to apply to many different experiments, then we may be able to assess the relationship between two variables.

REMEMBER

Not all graphs of chemical systems are linear, but linear relationships are the easiest to understand and to use. For linear relationships, the **slope-intercept equation** gives both the slope m and the y-intercept b immediately. The slope-intercept equation is $y = mx + b$, where m is the slope and b is the y-intercept. Some important chemical properties can be evaluated using the slope.

In Table 10-1 we see data obtained from an experiment involving gases. Data such as this can be represented by a graph, as shown in Figure 10-4. This graph shows the relationship between the measured mass and the volume of a particular gas.

Table 10-1 Data for a Density Experiment on a Gas

Reading	Volume (mL)	Mass (g)
A	2.00	0.120
B	5.10	0.305
C	10.20	0.610
D	16.40	0.970

Reading the graph from left to right we see that, as we increase the volume size of the sample, the mass also increases. In Figure 10-4(a) we show just the data. In Figure 10-4(b) we have drawn a line that best fits the data.

Figure 10-4 ▶ The graph of mass versus volume of a substance from the data in Table 10-1.

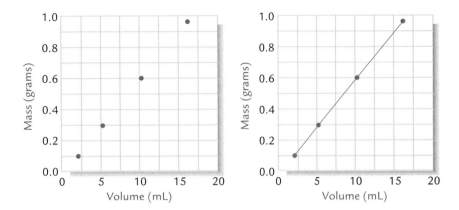

We can tell that the graph in Figure 10-4 is linear and represents the equation:

$$mass = slope \times volume + 0$$

as the *y*-intercept appears to pass through the origin. Simple proportional relationships between two variables will give a zero *y*-intercept. For the experiment in Figure 10-4 there will be a zero volume when there is no substance, just as we expect.

We evaluate the slope from the graph by finding the (*x*, *y*) coordinates of two points on the line. If we take readings A and C from Table 10-1, for example, we would have the two points (2.00, 0.120) and (10.20, 0.610). From this we get the slope:

$$slope = \frac{(0.610 - 0.120)}{(10.20 - 2.00)} = \frac{0.490}{8.20} = 0.0598$$

We also can use the linear regression (or "line-fitting") program available on most scientific calculators and in spreadsheet programs for computers.

If we were doing mathematical studies only, we might think we were done. But in this case we need to consider not just the *value* of the slope (0.0598), but also the *units*. The slope is determined as $\frac{y_2 - y_1}{x_2 - x_1}$ or $\frac{\Delta y}{\Delta x}$. Therefore, *the units of the slope will be the units of the quantity graphed on the y-axis divided by the units of the quantity graphed on the x-axis.* In this case, since the *y* measurement is in grams and the *x* measurement is in mL, the units of the slope is grams/mL, or g mL^{-1}. The proper way to report the slope is, therefore, 0.0598 g mL^{-1}.

KEY IDEA 10.3

Example 10.6   **Fitting chemical data to a slope-intercept equation**

An experiment is done to examine the volume of a gas formed when different amounts of calcium carbonate react with excess aqueous HCl (hydrochloric acid). The following data are obtained.

Reading	Mass (g)	Volume (mL)
A	0.125	30.9
B	0.211	52.0
C	0.284	70.2
D	0.333	83.1

(a) What is the independent variable and what is the dependent variable?

(b) Graph the data in an appropriate manner.

(c) Is there a linear relationship between the independent and the dependent variables? If so, what is the slope of the line, and what are its units?

Thinking Through the Problem There are three parts to this problem, but we can "think" our way to the answer of the first part. We do the experiment by adding measured masses of calcium carbonate to a solution, then observing the volume of gas. The independent variable is the mass of calcium carbonate; the dependent variable is the volume of the gas. We then assign the values of mass of calcium carbonate to the x-axis and the volume of gas to the y-axis.

Working Through the Problem The graph of the results is shown here, with a line drawn through the different points. We can see that there is a linear relationship.

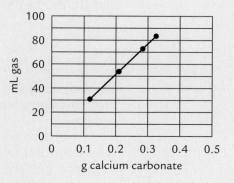

To get the slope of the line, we can select any two points that are on the line. If we use D and C, we get:

$$\text{slope} = \frac{(83.1 - 52.0)}{(0.333 - 0.211)} = \frac{31.1}{0.122} \approx 255$$

The units of the slope are the units of the y-axis divided by the units of the x-axis. This is mL/grams, or mL g^{-1}.

Answer The data for the independent variable, grams of calcium carbonate, are graphed on the x-axis, and the data for the dependent variable, mL gas, are graphed on the y-axis. The data have a linear relationship, with a slope of 255 mL g^{-1}.

For an experiment involving limiting reactants, students find that the mass of PbI_2 produced in a reaction varies depending on the mass of $Pb(NO_3)_2$ that is reacted with excess KI. The data are shown below.

(a) What is the independent variable and what is the dependent variable in this case?
(b) Graph the data.
(c) Determine the slope of the line relating the data. Include the units.

Reading	Mass $Pb(NO_3)_2$ (g)	Mass PbI_2 (g)
A	0.25	0.35
B	0.56	0.78
C	0.77	1.08
D	0.98	1.37

There are many other cases in chemical experiments where there is not a linear relationship or where the data include a nonzero y-intercept. We will see examples of these in Practical A and in Section 10.2, where we work with the gas laws.

PROBLEMS

1. Concept question: Mathematicians have a very precise definition of a line: it is straight. But many other scientists—including chemists—use the word "line" to describe things that aren't straight, like a circle. Suggest an alternative word to describe objects that are not straight.

2. Find the x-intercept and the y-intercept for the equation, $7x + y = -6$.

3. Find the x-intercept and the y-intercept for the equation, $x - 14y = -2$.

4. Find the x-intercept and the y-intercept for the equation, $5x + 6y = 10$.

5. Use the slope-intercept model to write the equation of a line containing the points $(C_1, K_1) = (0, 273)$ and $(C_2, K_2) = (300, 573)$. The resulting equation describes the relationship between the Celsius (°C) and the Kelvin (K) temperature scales. (Hint: What is the x-coordinate for the y-intercept?)

6. Use the equation derived in the previous question to convert the boiling temperature of water (100 °C) to Kelvin degrees.

7. Find the units of k when a plot of Π against C is made for the relationship, $\Pi = kC$. Π has units of atmospheres and C has units of mol L^{-1}.

8. A reaction is done to examine the reaction of zinc metal with acid. When different amounts of zinc metal react, the amount of hydrogen changes, too, as shown in the table below.

Moles of Zinc	Volume of Hydrogen (L)
0.025	0.671
0.055	1.48
0.088	2.36

(a) Identify the independent variable and the dependent variable.
(b) Graph the data, including a line connecting the points, on ruled paper.
(c) What is the slope of the line?
(d) What are the units of the slope?
(e) Write in slope-intercept form the equation of the line generated from the data.

PRACTICAL A   What makes a good graph?

Graphing chemical data is done in the same way that we graph ordinary x, y data in math. A key difference is that in math, the x and y values are specified and we merely have to determine where to put the point on the graph paper. In working with chemical data, several other steps must be followed. We must determine which variable to put on which axis, lay out the graph to make good use of the available space, and correctly label the axes.

A common experiment in general chemistry laboratories involves the measurement of the amount of light absorbed by a solution of a compound. We vary the number of moles of the compound per liter (with units of mol L^{-1}) and then see how much light is absorbed with an instrument that reports in units of absorbance, abbreviated A.

The following table shows data from such a report. Different amounts of the dye carmine indigo were dissolved in water and then the absorbance was measured at an appropriate wavelength of light.

Concentration (mol L^{-1})	Amount of light (A)
1.72×10^{-5}	0.204
3.43×10^{-5}	0.421
5.75×10^{-5}	0.652
6.86×10^{-5}	0.890
8.59×10^{-5}	1.12

Draw a graph of this data and include the following:

- a *title* indicating the nature of the graph, the system studied and, when known, the origin of the data.
- *axis labels* that indicate the variables and include units.

The numbers used on the axis should be appropriate to scale the data. A common error in graphing is the use of inappropriate scales for the axes. Two issues are important: showing the origin point and proper use of available space. In this case we expect that there will be no light absorbed ($A = 0$) when the concentration is zero. Take a moment to look at the available graph paper. What is the largest value you have for a variable? What is the smallest value (it may not always be zero). How does the range of values relate to the number of major grid points on the paper? Try to label and scale the axes so they fit the grid neatly while using as much of the paper as possible. The grid lines do *not* have to be integers, or even decimals, but they must be spaced uniformly.

We are now ready to see whether we can interpret the graph in a meaningful way. Does the dependent variable increase or decrease as the independent variable increases? Is it possible to fit this data to a slope-intercept model? If so, then what is the value of the slope and what are its units?

Let's look back at the scientific report we saw in Chapter 6 Practical D. There we read of an experiment where "Equilibrium constants were determined by monitoring electronic spectral changes at four different temperatures (-77 °C, -41 °C, -15 °C, and 0 °C)." What is the independent variable and what is the dependent variable in this study? The data reported have the values shown in the following table.

Temperature (°C)	K (L mol^{-1})
-77	1200
-41	200
-15	66
0	23

Graph the data in an appropriate manner. Does the dependent variable increase or decrease as the independent variable increases? Is it possible to fit this data to a slope-intercept model? If you can, what is the value and what are the units of the slope?

9. Discussion question: Liquids exert a pressure, known as the vapor pressure, on the atmosphere above them. This pressure varies when substances are mixed. When the substances benzene and toluene are mixed, pressure measurements such as those in the table below result. The amount of toluene is kept constant.

Moles of Benzene	Pressure (atm)
0.25	0.937
0.55	1.29
0.88	1.47

(a) Identify the independent and the dependent variables.
(b) Graph the data, including a line connecting the points, on ruled paper.
(c) What is the slope of the line?
(d) What are the units of slope?

(e) Write in slope-intercept form the equation of the line generated from the data.
(f) How can your answers be used to predict the pressure when 0.40 mole of benzene is mixed with this amount of toluene?

10. Often, to graph an equation we first must algebraically rearrange the equation so that it is in the slope-intercept form. Given the equation $PV = nRT$, graph P against T.

(a) Correctly label the axes of this graph.

(b) Rearrange the equation, $PV = nRT$, into the slope-intercept model represented by the graph shown in part (a).
(c) What variable or variables from this rearranged equation represent the slope of the line resulting from a plot of P against T?

10.2 Properties of Gases

SECTION GOALS

✔ Understand which variables must be specified when discussing a gas sample.

✔ Use the common pressure units.

✔ Apply Boyle's Law to gas problems.

Gas Properties and Variables

CONNECT TO
SECTION 1.4

We have already learned quite a bit about gases. In Chapter 1 we saw that substances in the gaseous state require containment on all sides, and that gases form homogeneous mixtures, with air being the most common example. Gases play an important role in many of the physical and biochemical processes that are the basis of our health and well-being. They also are important in many industrial processes.

Gases have the same physical properties we have seen for other substances, including properties of density, color, and odor. We also can use the extensive measurements of volume (V), mass, and chemical amount (n, moles) to tell us "how much" of a gas is present. But there are two other measurements that are especially important when we deal with gases: pressure and temperature, which are intensive measurements. When we study a gas, the four variables *pressure, temperature, mole amount*, and *volume* are all important.

Pressure is usually described using the variable P. Pressure is created by a gas as the molecules of the gas strike a surface, exerting a force on that surface. It describes the amount of force on a given area. The temperature, T, is a property related to molecular motion and energy. The temperature of a gas sample must be specified because it has an effect on both the pressure and the volume of a gas.

Previously, we studied the relationships between two variables at a time. For example, the molar mass relates the variables of mass and chemical amount. Density relates the variables of mass and volume. But because gases have *four* different variables that are important, we have to look more carefully at the relationships among them. We can study the relationships between pressure, volume, moles, and temperature, by following two rules:

1. We *keep the values of two properties constant* in order to study the relationship between the remaining two properties. For example, to discover the relationship between P and V, we keep T and n constant.

2. Noting the original values of the two variables under study, *we record what happens to one variable when we change the other*. Suppose that we are studying the relationship between pressure and volume. We can increase (or decrease) the pressure and measure the change in the volume. We ask ourselves, "Does volume increase or decrease when we increase (or decrease) the pressure?"

✔ **Meeting the Goals**

Important gas properties are pressure (P), temperature (T), volume (V), and chemical amount in moles (n).

In this section, we will study the relationships between *pairs* of gas properties and learn how to write mathematical equations that describe these relationships. The symbols P for pressure, V for volume, T for temperature, and n for moles are the variables that we use in these mathematical equations.

See this with your **Web Animator** ▶

Measurement of Gas Pressure

The pressure of a gas sample is a measurement of the force exerted on or by the gas molecules in the sample on a given area. Gas pressure is measured in various ways. Sometimes we need to know the pressure of the air surrounding a sample of gas. **Atmospheric pressure** is the pressure or force of all the gases in the atmosphere on a particular system. Atmospheric pressure is measured with a **barometer.** A simple barometer can be made by inverting a glass tube filled with mercury into an open dish of mercury. The level of the mercury inside the glass tube falls until the pressures on the mercury inside and outside of the tube are exactly equal. The pressure inside is due to the extra weight of the mercury column; the pressure outside is due to the atmospheric gas pressure. Figure 10-5 illustrates a simple barometer. The units of this pressure measurement are *mm Hg* or *cm Hg*, depending on the linear scale used to measure the height of the mercury column.

Figure 10-5 ▶ In a simple barometer, the column of mercury is balanced between internal and external pressures.

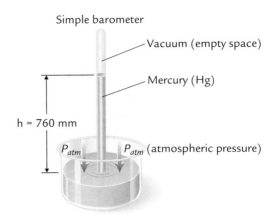

Simple barometer

Vacuum (empty space)

Mercury (Hg)

h = 760 mm

P_{atm} P_{atm} (atmospheric pressure)

As with many other variables, chemists use several different units to measure pressure. The most common pressure unit in chemistry is the *atmosphere,* usually abbreviated as *atm.* Because pressure is force divided by area, another pressure unit is *newtons per square meter.* A newton m^{-2} is equivalent to one *pascal* of pressure. When barometers are used to measure pressure, the unit of pressure is *millimeters of mercury,* or *mm Hg.* The relationship between these units is given in Table 10.2.

Standard pressure is approximately the pressure of Earth's atmospheric coating that we find at sea level at 0 °C. If we measure standard pressure with a barometer, the column of mercury is raised to a level that is 760 mm high at 0 °C.

$$standard\ pressure = 1\ atm = 760\ mm\ Hg\ at\ 0\ °C$$

✔ Meeting the Goals

The most common pressure units for gases are the mm Hg and the atmosphere (atm). Standard pressure is 760 mm Hg at 0 °C or 1 atmosphere of pressure.

There is, however, one problem with mm Hg as a pressure unit, because the density of mercury varies with temperature. To avoid this problem we sometimes use a unit that is similar in magnitude to mm Hg but does not change with temperature. This unit is the *torr,* named after the Italian scientist Torricelli who pioneered the careful study of gases.

Table 10-2 Some Units of Pressure

Unit	Equivalent
1 atm	760 mm Hg at 0 °C
760 mm Hg at 0 °C	760 torr
1 torr	$\dfrac{1}{760}$ atm
1 atmosphere (atm)	101,325 pascals

Example 10.7   **Converting units of atmosphere to millimeters of mercury**

Convert 4.93 atm to mm Hg at 0 °C.

Thinking Through the Problem We see that this is a unit conversion problem. Table 10-2 gives us the relationship between atmospheres and mm Hg: 1 atm ↔ 760 mm Hg. We solve this problem with the conversion factor method. We want to obtain mm Hg, so we multiply the starting value (4.93 atm) by the ratio of 760 mm Hg/1 atm.

Working Through the Problem

$$4.93 \ \cancel{atm} \times \frac{760 \ mm \ Hg}{1 \ \cancel{atm}} = 3.746 \times 10^3 \ mm \ Hg$$

$$\approx 3.75 \times 10^3 \ mm \ Hg$$

Answer 3.75×10^3 mm Hg

How are you doing?   **10.6**

Convert 864.3 mm Hg to units of atmospheres.

Boyle's Law

Let's consider the relationship between volume and pressure. According to the rules stated at the beginning of this section, studying these two gas variables requires that we hold constant the other two variables—the number of moles of the gas and the temperature of the gas. This can be done with a device called a "J-tube," shown in Figure 10-6. A J-tube is so named because it is shaped like the letter *J*. The bottom of the tube is filled with a liquid—mercury is the most common—and the short arm of the J is closed and contains a trapped gas. The long arm of the tube is open to the surrounding atmosphere.

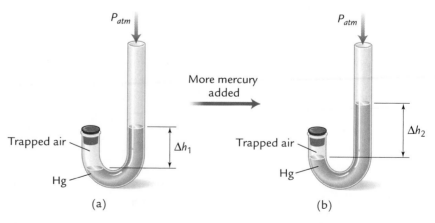

Figure 10-6 ▲ The volume of air trapped in a J-tube decreases when pressure increases.

Robert Boyle (1627–1691) was an Irish chemist who performed very early studies of the relationship between a volume of trapped air and the pressure exerted on it. He poured some mercury into a J-tube open at one end, trapping air in the closed part of the tube. Because of the trapped air, the levels of mercury in the arms of the J-tube were different, as shown in Figure 10-6(a).

Boyle found that, as he poured more mercury into the J-tube, the volume of trapped air continued to decrease. The addition of mercury increased the pressure on the trapped air. This is shown in Figure 10-6(b).

What Boyle observed is that the volume decreased as the pressure increased. This kind of a relationship is called an **inverse variation,** or inverse proportion. *In inverse variation, as the magnitude of one measurement increases, the magnitude of a second measurement decreases, and vice versa.* According to Boyle's experiments, measurements of the volume and the pressure varied inversely when the temperature and mole amount were held constant. We can write this as a mathematical relationship using the variables for volume and pressure: $V \propto \dfrac{1}{P}$. (The symbol $\propto$ means "is proportional to.")

Note that this is different from the relationship for mass and volume that is described in density, where volume varies *directly* with mass, $V \propto m$.

Boyle's experiments indicate that pressure and volume are inversely related. Therefore, we can make predictions about either P or V if we know that the other variable is increasing or decreasing.

✔Meeting the Goals

Boyle's Law states that at constant temperature, the volume occupied by a fixed quantity of gas is inversely proportional to the applied pressure.

Example **10.8**   Reasoning about volume and pressure relationships

The pressure of 12.5 L of gas is 0.82 atm. If the pressure changes to 1.32 atm, will the final volume be larger or smaller than the initial volume?

Thinking Through the Problem ➥ (10.5) The key idea is that pressure and volume are inversely proportional to each other. To answer the question we need to know whether the pressure increases or decreases in this problem. A pressure change from 0.82 atm to 1.32 atm is an increase in pressure. If pressure increases, then volume must decrease.

Answer   The final volume is smaller than the initial volume.

How are you doing?   **10.7**

A gas sample has a measured pressure of 1.05 atm. The temperature and the amount of gas are kept the same but the pressure is changed to 0.95 atm. The volume of the gas at 0.95 atm is 2.90 L. Was the initial volume of gas larger or smaller than 2.90 L?

We can restate the relationship of pressure and volume in terms of an equation where volume varies inversely with pressure. This is known as **Boyle's Law.**

$$V = k\frac{1}{P} = \frac{k}{P}$$
$$PV = k$$

We see that the product of the volume and the pressure for a specified amount of gas at constant temperature is a constant value. For the same amount of gas and at the same temperature, all pressure-volume products must equal the *same* constant value. If we change P, then V will change so that the product PV = constant = k. Usually, chemists are interested in what happens to a gas sample when one of the variables is changed. We will label initial measurements with a subscript number 1 and final measurements with a subscript number 2. Thus, the initial volume of a gas is V_1 and the initial pressure of the gas is P_1. After the change is made, there is a different volume (V_2) and a different pressure (P_2), but the constant, k, remains the same. This means that:

$$P_1V_1 = k$$
$$P_2V_2 = k$$
$$P_1V_1 = P_2V_2$$

Boyle's Law gives us a very simple way to calculate a final pressure or volume of a specified amount of gas given the initial conditions and the change made to P or V. Mathematically speaking, given three of the four variables in $P_1V_1 = P_2V_2$, we can solve for the missing variable.

Knowing this relationship, we can solve pressure-volume problems.

Example 10.9   Calculating pressure and volume changes

For the problem discussed in example 10.8, present a numerical calculation of the final volume.

Thinking Through the Problem This problem involves calculating a final volume V_2. We are given V_1, P_1, and P_2.

Working Through the Problem We will have to rearrange an equation with four different variables. It is good practice to do the rearrangement *before* we put in the numerical values.

$$P_1V_1 = P_2V_2$$
$$V_2 = \frac{P_1V_1}{P_2} = \frac{12.5 \text{ L} \times 0.82 \text{ atm}}{1.32 \text{ atm}}$$
$$V_2 = 7.8 \text{ L}$$

Does this answer make sense? The answer is "yes," because we expect that an increase in pressure will result in a decrease in volume.

Answer 7.8 L. Note that the two significant figures in "0.82 atm" restrict the answer to two significant figures also.

Determine the numerical answer to the situation discussed in "How are you doing?" 10.7. Check the answer to see if the direction of the change is reasonable based on Boyle's Law.

Example 10.10   **Boyle's Law calculations**

A 150.0-mL sample of carbon dioxide has a pressure of 35.3 torr. What pressure will this gas exert if the volume is decreased to 50.0 mL while the temperature is held constant?

Thinking Through the Problem Because of the inverse relationship between pressure and volume, we expect a larger pressure when the volume is decreased. We rearrange $P_1V_1 = P_2V_2$ to solve for the variable P_2.

Working Through the Problem

$$P_2 = \frac{P_1V_1}{V_2} = \frac{(35.3 \text{ torr})(150.0 \text{ mL})}{(50.0 \text{ mL})} = 105.9 \approx 106 \text{ torr}$$

Note that this answer makes sense because the pressure increases, as it should when the volume decreases.

Answer 106 torr (3 significant figures)

A 250.0-mL sample of ammonia gas has a pressure of 125 torr at a temperature of 30 °C. What will be its pressure if the volume is decreased to 200.0 mL? (Assume temperature is held constant.)

PROBLEMS

11. Compute the following pressure conversions. Assume 0 °C for the temperature.

(a) 0.89 atm to mm Hg
(b) 2.43 atm to cm Hg
(c) 1.9×10^5 pascals to atm

12. Carry out these conversions.

(a) 745 torr to atm
(b) 0.745 atm to torr

13. State Boyle's Law in words and as an equation.

14. A sample of neon gas has a volume of 346 mL and a pressure of 0.84 atm at 25 °C. What will the gas pressure be if the gas sample is transferred to a cylinder with a volume of

500 mL while the temperature is kept constant?

15. A 2.00-liter sample of nitrogen gas is maintained at a constant temperature of −20.00 °C. Calculate the change in its volume if its pressure changes from 0.98 atm to 1.20 atm.

16. A 30.0-liter sample of gas has a pressure of 2.10 atm. Will the pressure increase or decrease when the gas is put into a smaller container?

17. A sample of gas is contained in a cylinder with a volume of 8.0 liters. In order to have the same amount of gas but at a higher pressure, should we put it into a larger or a smaller cylinder?

PRACTICAL B   How is an equation related to its graph?

One way to examine the relationships between two variables is to graph the data. For example, Robert Boyle did a very careful experiment in which he recorded the volume of a gas sample at various pressure readings. The graph of some pressure-volume data similar to that obtained by Boyle is shown in Figure 10-7.

When a chemist looks at a graph, the first task is usually to see if there are any relationships that may be important. We start by making verbal statements about what Figure 10-7 "tells us." We then move on to interpreting the graph numerically. Finally, we can see if the graphed data is consistent with an equation.

In interpreting Figure 10-7, there are several sets of questions we can ask. The first set concerns the data presented by the graph and the conclusions that can be drawn from it regarding the relationship between V and P.

- Where does volume appear on the graph? What about pressure?

- Can you draw a smooth shape connecting the data points in Figure 10-7? Do so if you can and tell what the shape looks like.

- What does this graph tell us about the relationship between V and P? That is, what does it tell us about what happens to V when P increases? Just as we read from left to right, the graph is also read from left to right.

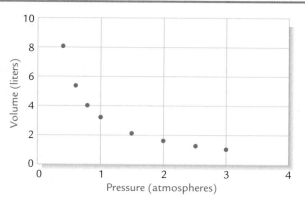

Figure 10-7 ▲ Data for Boyle's experiment.

- The shape of the line you have drawn is a *curve*. Can you recognize this as a shape you have seen in mathematics?

- Does this graph show a direct or inverse relationship between V and P?

- Is this graph consistent or inconsistent with Boyle's Law as it was used in Example 10.10?

The second set of questions we can ask about Figure 10-7 concerns using a graph as a means to tell us what might happen at values of pressure and volume that we did *not* measure. This assumes that the gas behaves in a predictable way.

18. A sample of helium gas having a pressure of 860 mm Hg is put into a different container such that its new volume and pressure are 25.0 liters and 770 mm Hg, respectively. What was the initial volume of the gas?

19. A sample of nitrogen gas is transferred from a 5.0-liter to a 6.5-liter cylinder. This move changed its pressure to a reading of 790 mm Hg. What was the pressure of the nitrogen in its original cylinder?

20. A 255-mL sample of gas has a pressure of 452 mm Hg. What is the volume of this gas when the pressure is 0.90 atm?

21. Discussion question: A sample of nitrogen gas has a volume of 0.50 L and pressure of 0.88 atm. Use Boyle's Law to determine the pressure of this gas sample when its volume is changed to 385 mL. Assume the temperature is constant. To do this problem, you needed additional steps to make all units the same. What order did you choose for these steps?

- What is the approximate value of V when P is 0.60 atm?
- What is the approximate value of V when P is 1.1 atm?
- What is the approximate value of V when P is 3.6 atm?
- Did the values of V increase or decrease in value as P values increased from 0.60 atm to 3.6 atm?

Finally, a third set of questions concerns the graph and its connection to a particular behavior presented as a mathematical equation.

- State Boyle's Law in a way that isolates the variable V on the left side of the equation. What does it suggest about the mathematical relationship of V and P?
- How does the form of the equation tell us that the relationship between V and P is inverse?
- Does the graph agree with the predictions of Boyle's Law?

If we rearrange the equation so that both V and P are separate terms, we get:

$$V = \frac{k}{P}$$

$$V = k\frac{1}{p}$$

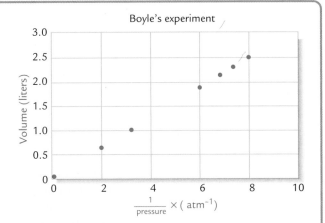

Boyle's experiment

Figure 10-8 ▲ Boyle's Law experimental data graphed as V vs. 1/P.

which has the format of

$$y = mx + b$$

where $b = 0$ and the slope is given by k. Thus, we expect that a graph of V versus $1/P$ would be linear and would intersect the y-axis at the origin. This is shown in Figure 10-8.

When these data points are connected with a smooth curve, a straight line results. It will have the units of liters divided by atm^{-1}, or L atm.

10.3 Temperature and Linearity in Chemistry

SECTION GOALS

✔ Convert between the Celsius and the Kelvin temperature scales.

✔ Recognize and use the relationship between volume and temperature.

The Kelvin or Absolute Temperature Scale

Data of volume and temperature relationships can be obtained for any substance, as long as the substance remains a gas at the temperatures studied. Figure 10-9 shows a graph of volume versus temperature for samples of helium gas; each line represents a different mole

Figure 10-9 ▶ Volume vs. temperature graphs for different mole amounts of helium using the Celsius scale.

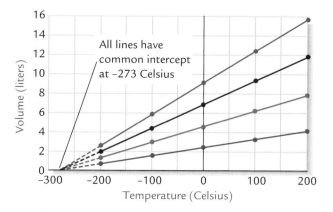

amount of gas. The graphs are linear and each sample graphed has a different line. The fact that volume versus temperature of a gas gives a linear graph was first pointed out by the French scientists Jacques Charles (1746–1823) and Joseph Gay-Lussac (1778–1850). The reliable relationship between the volume of a gas and its temperature became known as the Law of Charles and Gay-Lussac.

When we look at Figure 10-9, it is apparent that the relationship of volume and temperature in Celsius is not a simple proportion. There is a nonzero y-intercept in each case. At a temperature of 0 °C, the sample does not have a zero volume. Each line has a common x-intercept, however, at −273 °C, as indicated by the extended dotted line. The fact that all the lines converge to a zero volume at a particular temperature suggests that a temperature scale that is "offset" from the Celsius scale will give a simple proportional relationship between temperature and pressure. This became the new temperature scale called the Kelvin scale. The unit is abbreviated K. The **Kelvin temperature scale** is related to the Celsius temperature scale by the following:

$$\text{Kelvin temperature} = \text{temperature in } °C + 273.15$$

The same data, graphed using the Kelvin scale, has an intercept at 0 K (Figure 10-10).

The magnitude or size of a degree in the Kelvin scale is the same as the magnitude of a degree in the Celsius scale. Because of this equality, temperatures recorded in °C are easily converted to the Kelvin scale by adding 273.15. The degree symbol (°) is not used with

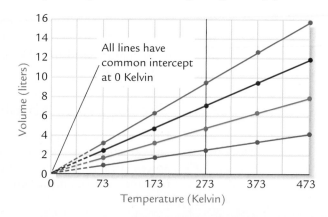

Figure 10-10 ▶ Volume vs. temperature graphs for different mole amounts of helium using the Kelvin scale.

Kelvin degrees; for example, 100 K and 273 K are temperatures on the Kelvin temperature scale. Generally, the relationship between the two temperature scales is written as:

$$T = t + 273.15$$
$$t = T - 273.15$$

where t = temperature in Celsius, T = Kelvin temperature. The Kelvin temperature must be used for all gas law calculations.

Sometimes the Kelvin scale is referred to as the *absolute temperature scale*. Because negative volume is impossible, the temperature value corresponding to a volume of zero is the lowest temperature possible. This temperature, -273.15 °C, is equal to zero on the Kelvin scale and is called **absolute zero.**

Example 10.11 — Converting Celsius and Kelvin temperatures

Convert 1000.00 °C to its equivalent Kelvin temperature.

Thinking Through the Problem We must convert a temperature measurement from the Celsius temperature scale to the Kelvin temperature scale. We know the relationship between the two temperature scales and the value of the temperature in Celsius.

Working Through the Problem

$$T = t + 273.15 = 1000.00 + 273.15 = 1273.15 \text{ K}$$

Answer 1273.15 K

✔ **Meeting the Goals**

To convert a temperature from the Celsius to the Kelvin scale, add 273.15 to the Celsius temperature. To convert from the Kelvin scale to the Celsius scale, subtract 273.15 from the Kelvin temperature.

How are you doing? 10.10

Convert the following temperature measurements as indicated.

100.00 °C = _____ K	_____ °C = 8.25 K
-174.35 °C = _____ K	_____ °C = 0.00 K
0.00 °C = _____ K	_____ °C = 100.00 K

Standard temperature for gases is defined as 0 °C or 273.15 K. Standard temperature and pressure is often abbreviated as **STP.** STP conditions for gases are 273.15 K and 1 atmosphere pressure.

Charles's Law

Let us return to the data in Figure 10-9. Once we have converted the data into the Kelvin scale, we get a simple proportion. In this form, Charles's experiments showed that the volume of a gas varies directly with the absolute temperature:

$$V \propto T$$

Introducing a constant of proportionality, k, this becomes

Balloonists. (Garry McMichael/ Photo Researchers.)

$$V = kT \quad \text{or} \quad \frac{V}{T} = \text{a constant} = k$$

This is a mathematical statement of Charles's Law.

To calculate the result of changing either the temperature or the volume of a specified amount of gas at constant pressure, then, we use an extended version of Charles's Law:

$$\frac{V_1}{T_1} = k$$

$$\frac{V_2}{T_2} = k$$

$$\frac{V_1}{T_1} = \frac{V_2}{T_2}$$

REMEMBER

In this formulation, Charles's Law allows us to calculate any one of the four variables, V_1, V_2, T_1, and T_2, when the values for the other three are given in the problem.

Example 10.12   **Reasoning about temperature and volume changes**

A sample of gas has a volume of 1.5 L when the temperature is 25 °C. If the pressure remains constant but the volume changes to 1.2 L, will the temperature increase, decrease, or remain the same?

Thinking Through the Problem Because only the volume and temperature are changing, this is an application of Charles's Law. We know that volume and temperature vary directly when the pressure and mole amount are constant.

Working Through the Problem The volume decreases from 1.5 L to 1.2 L. Because V and T vary directly with one another, the temperature must also decrease.

Answer The temperature will decrease.

Example 10.13   **Volume and temperature calculations**

A 50.0-mL sample of gas at room temperature, 26.4 °C, is slowly heated at constant pressure until its volume is 62.4 mL. What is the final temperature of the gas?

Thinking Through the Problem The relationship between the volume and the temperature is given by Charles's Law, when the pressure and the amount of gas is kept constant. In this problem, we expect the temperature to increase because the volume is increasing. We should convert the temperature to the Kelvin scale, list all the variables given in the problem, and then solve for T_2 using the extended equation for Charles's Law.

Working Through the Problem

$$T_1 = 26.4 \ °C + 273.15 = 299.55 \ K$$

Then, $\dfrac{V_1}{T_1} = \dfrac{V_2}{T_2}$ so $T_2 = \dfrac{V_2 T_1}{V_1}$ and

PRACTICAL C   What do balloons have to do with chemistry?

Part of the reason that Charles and Gay-Lussac were interested in the volume and temperature relationships of gases was that they were balloonists, with a very natural interest in what was keeping their balloons aloft! Heating the gases increased the size of their balloons and buoyed them up. This fact led Charles and Gay-Lussac to theorize about the relationship between temperature and the expansion of gases.

Charles studied the volume changes that occurred as he changed the temperature of a gas sample. Data similar to his, given in Table 10-3, were used to make the plot of volume versus temperature (at constant pressure), shown in Figure 10-11. Charles's data related the volume to the temperature when a fixed amount of gas is kept at constant pressure. We see from the graph that volume increases as temperature increases, which is what happens, for example, when a balloon is heated. Let's see what effect an increase in volume has on the density of a gas.

Assume that you have a sample of one mole of pure nitrogen gas. At 0 °C and 1 atmosphere of pressure, this sample has a volume of 22.4 liters. Determine the density, in grams per liter, of this sample of gas. (Hint: Convert one mole to the mass in grams, and then relate this mass to the volume.)

If this gas is heated while it is kept at constant pressure, the number of moles of gas will be constant but the volume will increase. Calculate the density of this gas at 25.0 °C, 50.0 °C, and 100.0 °C.

We also know that balloons float in the air when they are filled with a gas that has a density much less than that of normal air. Helium is the most common example today of such a gas, but in the 1700s helium had not yet been discovered. Instead, early balloonists used hydrogen because it could be easily generated. One mole of hydrogen gas occupies

Table 10-3   Volume of 1.009 g Neon Gas at 1.00 atm Pressure

Temperature (°C)	Volume (L)
−100.00	0.71
−50.00	0.91
0.00	1.12
50.00	1.33
100.00	1.53
200.00	1.94

22.4 liters at 0 °C and 1.0 atmosphere of pressure. Determine the density of hydrogen gas, in grams per liter, at this temperature.

Graph the data in Table 10-3 and show whether it is consistent with Figure 10-11.

The "lift" of a balloon is determined by the difference between its density and the density of the gas around it.

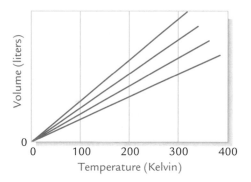

Figure 10-11 ▲ A volume vs. temperature plot of gases in Kelvin.

$$T_2 = \frac{V_2 T_1}{V_1} = \frac{62.4 \text{ mL} \times 299.55 \text{ K}}{50.0 \text{ mL}} \approx 374 \text{ K}$$

Answer   374 K ($T_2 > T_1$ as we reasoned it would be.)

How are you doing?   **10.11**

A 1.00-liter sample of gas at 0.0 °C is cooled under constant pressure. What is the volume of the gas at −100.0 °C?

✔**Meeting the Goals**

For a specified amount of gas at constant pressure, gas volume varies directly with its temperature. To calculate the value of one of the variables when given values for all of the others, use the relationship

$$\frac{V_1}{T_1} = \frac{V_2}{T_2}.$$

Temperature conversions should have as many places after the decimal point as the actual temperature being converted, and no more. In example 10.13, 26.4 °C has only one place after the decimal point. The corresponding temperature on the Kelvin scale should also have only one place after the decimal point. It is always better to keep *all* the digits until the final answer and then do any rounding off that is necessary. However, in Example 10.13 we need not worry about the significant figures in the temperature measurement because the number of significant figures in the final answer is determined by the volume measurements.

PROBLEMS

22. Each of the equations below is true only if certain variables are held constant. Complete column three in the following table by writing the variables that are held constant for each equation.

Relationship	Proportionality	Variables Held Constant
Boyle's Law	$P \propto \dfrac{1}{V}$	
Charles's Law	$V \propto T$	

23. Concept question: Volume is a quantity that must be positive. Show how this physical reality means we cannot have temperatures below absolute zero.

24. Express the proportionality, $V \propto T$ in words.

25. Using conversion factors given in this chapter, convert:

(a) 10.5 °C to K
(b) −10.5 °C to K
(c) 10.5 K to °C
(d) −182.5 °C to K
(e) 0.25 K to °C

26. Discussion question: Volume (V) varies directly with moles (n) and absolute temperature (T), or $V = knT$.

(a) How does the volume of one mole of gas change when the temperature is doubled?
(b) How must the temperature of one mole of gas change when the volume is halved?
(c) What will happen if you double the volume and double the temperature?

27. Find the missing variable in each of the cases listed in the following table. The confined gas is one mole of argon at constant pressure.

V_1	T_1	V_2	T_2
220.0 L	−23.5 °C		35.0 °C
	254.53 K	5.0 L	303.18 K
3.50 L		7.32 L	300.5 °C
25.0 L	285 K	0.500 L	
25.30 L	50.7 °C		−15.0 °C
	28.35 K	0.275 L	207.64 K

10.4 Extension and Interpretation of the Gas Laws

SECTION GOALS

✔ Know the units of the variables used in the ideal gas law equation.

✔ Carry out stoichiometric calculations involving gases.

✔ Understand the main ideas of the kinetic molecular theory.

Variation and the Ideal Gas Law

Whenever a mathematical model is used to describe a real system, a certain amount of approximation may occur. Boyle's and Charles's Laws are easy for us to use and they give approximately correct answers to problems. They apply to so-called ideal gases. At low pressures, real gases in fact approach the behavior of ideal gases. The gas laws can be combined into a single equation called the *ideal gas equation* or the **ideal gas law.**

We construct the ideal gas equation from the relationships we have studied by examining how the volume is affected by the other three variables—pressure, temperature, and mole amount. We have discussed the relationship between volume and pressure (Boyle's Law) and the relationship between volume and temperature (Charles's Law). There is also a relationship between volume and moles (Avogadro's Law).

Boyle's Law indicates that volume is inversely proportional to pressure:

$$V \propto \frac{1}{P}$$

Charles's Law indicates that volume is directly proportional to temperature:

$$V \propto T$$

where T is absolute temperature. The Kelvin scale is the proper scale to use in all gas law calculations.

The dependence of volume on the amount of a gas was originally formulated as a hypothesis by the Italian chemist Avogadro. *Avogadro's Law* states that *volume is directly proportional to the number of moles*:

$$V \propto n$$

Boyle's Law, Charles's Law, and Avogadro's Law taken together indicate that volume is directly proportional to the temperature and to the number of moles of a gas, and inversely dependent on the pressure. In equation form,

$$V = R\frac{nT}{P}$$

where R is the combined proportionality constant and is called the *universal gas constant*. We can find the numerical value for R by noting that one mole of any ideal gas at exactly 273.15 K and 1 atm pressure has a volume of 22.414 L. Thus,

$$R = \frac{PV}{nT} = \frac{(1 \text{ atm})(22.414 \text{ L})}{(1 \text{ mol})(273.15 \text{ K})} = 0.082057\frac{\text{L atm}}{\text{mol K}}$$

This value for R, 0.082057 L atm mol^{-1} K^{-1}, is the value we will use most often for ideal gas law problems. *The units for the volume, pressure, and temperature are determined by the units of R.* For this value of R, volume must have units of liters, pressure must have units of atmospheres, and temperature must have units of Kelvin.

The ideal gas law is commonly written:

$$PV = nRT$$

REMEMBER

This is very useful in gas law problems that specify a mass or mole amount of a gas.

Example 10.14   Ideal gas law calculation

15.0 grams of argon gas is transferred to a 1.50-liter container. At what temperature is the pressure 1.0263 atm?

Thinking Through the Problem This is an application of the ideal gas law because we are given grams (from which we can determine moles) and because the initial conditions are not changed in any way. Both Boyle's and Charles's Laws deal with *changes* made to a gas system.

Working Through the Problem

$$\text{amount of gas} = 15.0 \text{ g} \times \frac{1 \text{ mol Ar}}{39.948 \text{ g}} = 0.3754 \text{ mol Ar}$$

To avoid round-off error, this first answer includes one additional digit. Now solve for T and substitute into the equation

$$T = \frac{PV}{nR} = \frac{(1.0263 \text{ atm})(1.50 \text{ L})}{(0.3754 \text{ mol})(0.082057 \text{ L atm mol}^{-1} \text{ K}^{-1})} \approx 50.0 \text{ K}$$

Answer $T = 50.0$ K

A sample of 0.3055 gram of helium gas is contained in a 0.500-L glass bulb at 25.00 °C. What is the pressure of the helium gas?

Proportionality and the Gas Laws

A proportion is an equation stating that two fractions, or ratios, are equal. The ideal gas law includes such a proportion: the ratio of P, V, n, and T for any sample of gas is equal to R, the universal gas constant. Suppose that we have two samples, distinguished by subscripts 1 and 2:

$$R = \frac{P_1 V_1}{n_1 T_1} \text{ and } R = \frac{P_2 V_2}{n_2 T_2}$$

REMEMBER

We can rewrite these as the proportionality:

$$\frac{P_1 V_1}{n_1 T_1} = \frac{P_2 V_2}{n_2 T_2}$$

This is sometimes called the **combined gas law.** It is useful whenever a sample of gas changes one or more of its variables. Depending on the problem, some of these variables may be constant. Variables in the combined gas law equation that remain constant will divide out, leaving only those variables undergoing some change. This leads to the simpler, more specialized laws of Boyle and Charles.

Keep n and T constant and allow P and V to vary:

$$P_1 V_1 = P_2 V_2 \quad \text{(Boyle's Law)}$$

Keep n and P constant and allow T and V to vary:

$$\frac{V_1}{T_1} = \frac{V_2}{T_2} \quad \text{(Charles's Law)}$$

Example 10.15   Combined gas law calculation

A 0.0500-L sample of gas has a pressure of 745 mm Hg at room temperature, 26.4 °C (299.55 K). The temperature is changed until the final pressure and volume of the gas are 1.06 atm and 0.0624 L, respectively. What is the final temperature of the gas?

Thinking Through the Problem This is an application of the combined gas law equation:

$$\frac{P_1 V_1}{n_1 T_1} = \frac{P_2 V_2}{n_2 T_2}$$

To use this equation, we must solve for T_2 and substitute the given values for the variables in the equation. This equation does not contain the universal gas constant, R, so the units of P and V can be any pressure and volume unit, as long as the unit is consistent within the problem. The units of T must always be Kelvin.

Working Through the Problem First list the variables and check that the units of each variable "match."

$V_1 = 0.0500$ L $\qquad$ $V_2 = 0.0624$ L $\quad$ Both units are the same, L.

$P_1 = 745$ mm Hg $\quad$ $P_2 = 1.06$ atm $\quad$ The units are not the same.

$n_1 = n_2$ $\qquad\qquad\qquad\qquad\qquad$ The mole amounts are the same.

We must convert the pressure units so that both are mm Hg or both are atm.

$$745 \text{ mm Hg} \times \frac{1 \text{ atm}}{760 \text{ mm Hg}} = 0.980 \text{ atm}$$

Now,

$P_1 = 0.980$ atm $\qquad$ $P_2 = 1.06$ atm $\qquad$ The units are the same.

$T_1 = 299.55$ K $\qquad$ $T_2 = ?$

$$T_2 = \frac{P_2 V_2 T_1}{P_1 V_1} = \frac{(1.06 \text{ atm})(0.0624 \text{ L})(299.55 \text{ K})}{(0.980 \text{ atm})(0.0500 \text{ L})} \approx 404.4 \text{ K}$$

Answer The final temperature is 404 (3 s.f.).

How are you doing?   10.13

A 4.36-liter sample of gas at 230.00 °C has a pressure of 854 mm Hg. The temperature of the sample is changed to 95.10 °C, giving a volume measurement of 4.06 L. What is the final pressure of this system?

Gas Stoichiometry

Chemists count by using moles and measure by using other units, such as grams or liters. In Chapter 9, we focused on measurement of chemical reactants and products by mass. But now we have seen that the number of moles of a gas can be related to the volume, temperature, and pressure of the gas. This gives us a new relationship between moles and the more easily measured quantities, volume, temperature, and pressure. We can now do calculations where we relate the number of moles to the pressure, volume, and temperature of a gas involved in a chemical reaction. Such calculations are known as gas stoichiometry.

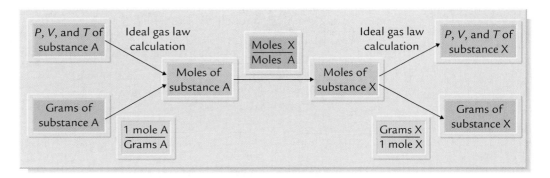

Figure 10-12 ▲ A strategy map for gas stoichiometry calculations.

CONNECT TO
SECTION 9.2

In Chapter 9, we saw that the relationship between mass and moles is a simple proportionality, with the molar mass of the substance acting as a conversion factor between mass of a substance and moles of a substance. But with gases the relationship of moles to any one other quantity (like volume, for example) is usually not a simple proportionality, because other quantities (pressure and temperature in this case) can vary. However, as long as we indicate all of the relevant gas variables, we can do gas stoichiometry problems.

In Chapter 9, Figure 9-5, we saw that we can start and end stoichiometric calculations using the masses of the compounds. We can extend that strategy to include gas calculations, as shown in Figure 10-12.

Gas stoichiometry problems include two types of calculations, stoichiometry and ideal gas law calculations. The two calculations are connected by the moles of gas.

Example 10.16   **Pressure of gas produced in a reaction**

Sodium hydrogen carbonate, $NaHCO_3$, produces carbon dioxide when heated:

$$NaHCO_3\,(s) \longrightarrow NaOH\,(s) + CO_2\,(g)$$

If we decompose 0.245 mole of $NaHCO_3$ according to this reaction, and the product CO_2 has a pressure of 1.0 atm at 453 K, what volume of CO_2 will be produced?

Thinking Through the Problem We are looking for the volume of CO_2. We can calculate this volume using the ideal gas law, $V = nRT/P$. We are given pressure (1.0 atm) and temperature (453 K), and we know the gas constant R. We need to determine the number of moles n. Because the reaction has a 1:1 relationship of $NaHCO_3$ to CO_2, we expect to get 0.245 mole of CO_2.

Working Through the Problem

$$V = \frac{nRT}{P} = \frac{(0.245 \text{ mol})(0.082057 \text{ L atm mol}^{-1} \text{ K}^{-1})(453 \text{ K})}{1.0 \text{ atm}} \approx 9.1 \text{ L}$$

Answer 9.1 L of CO_2

How are you doing? **10.14**

The reaction of zinc and hydrochloric acid generates hydrogen gas, according to the reaction

$$\text{Zn } (s) + 2 \text{ HCl } (aq) \longrightarrow \text{ ZnCl}_2 \text{ } (aq) + \text{H}_2 \text{ } (g)$$

If we react 0.100 mole of zinc, what will be the pressure of the hydrogen gas if it has a volume of 2.00 L at 315 K?

See this with your **Web Animator** ▶

Example 10.16 involved a simple stoichiometric ratio, with one mole of gas for one mole of the reactant we were given. In other cases we may have other stoichiometric ratios. These need to be included in the calculation to get the number of moles of gas.

Example 10.17 **Gas stoichiometry calculation**

The combustion of pentane gives carbon dioxide and water as products, according to the reaction:

$$\text{C}_5\text{H}_{12} \text{ } (l) + 8 \text{ O}_2 \text{ } (g) \longrightarrow 5 \text{ CO}_2 \text{ } (g) + 6 \text{ H}_2\text{O} \text{ } (l)$$

✔ **Meeting the Goals**

Gas stoichiometry calculations are a natural extension of the stoichiometric calculations we performed in Chapter 9. We use the ideal gas law to link the variables P, V, and T with the number of moles of a gas n. We then use conversion factor methods to relate the number of moles to the mass of other substances.

If we react 1.00 gram of pentane, what volume of CO_2 do we expect at a pressure of 0.050 atm and a temperature of 342 K?

Thinking Through the Problem Our strategy requires that we convert the grams of pentane to moles of pentane, then moles of pentane to moles of CO_2. We are then able to apply the ideal gas law for the calculation.

Working Through the Problem To get moles of CO_2, we note that the molar mass of pentane is 72.150 g mol^{-1} and that the stoichiometric ratio of CO_2 and pentane is 5 mol CO_2/1 mol C_5H_{12}. We then have:

$$1.00 \text{ g } C_5H_{12} \times \frac{1 \text{ mol } C_5H_{12}}{72.150 \text{ g } C_5H_{12}} \times \frac{5 \text{ mol } CO_2}{1 \text{ mol } C_5H_{12}} \approx 0.0693 \text{ mol } CO_2$$

This is then used to get the volume of the CO_2:

$$V = \frac{nRT}{P}$$

$$= \frac{(0.0693 \text{ mol})(0.082057 \text{ L atm mol}^{-1} \text{ K}^{-1})(342 \text{ K})}{0.050 \text{ atm}} = 39 \text{ L}$$

Answer   39 L of CO_2

How are you doing?   **10.15**

For the reaction described in Example 10.17, what volume of oxygen gas at a temperature of 285 K and a pressure of 0.200 atm is needed for 1.00 gram of pentane?

Note that in Example 10.17, and in most other gas stoichiometry calculations, we cannot use the conversion factor method to go all the way from the mass of one substance to the gas variables of another substance, or vice versa. This is because there is no simple proportionality between any one gas variable (P, V, or T) and the number of moles of another substance in the reaction.

We can also begin our gas stoichiometry calculations with information about the gas, and then get an answer that is a mass of another substance in the reaction.

Example   10.18   **Gas stoichiometry calculation**

For the reaction 2 Al *(s)* + 6 HCl *(aq)* → 2 $AlCl_3$ *(aq)* + 3 H_2 *(g)* we observe the production of 25.0 mL of hydrogen gas at a pressure of 0.982 atm and a temperature of 291 K. What mass of Al was consumed?

Thinking Through the Problem In this case we need to get a mass for the answer. We will need to take the pressure, volume, and temperature information about the hydrogen and calculate the moles of hydrogen used. This value can then be used in calculating the number of moles and the mass of Al consumed, using the conversion factor method.

Note that the volume of gas, 25.0 mL, must be converted to liters, which is the appropriate unit for gas law calculations: 25.0 mL is equivalent to 0.0250 liter.

Working Through the Problem We do the gas law calculation first to get the number of moles of H_2.

$$n = \frac{PV}{RT} = \frac{(0.982 \text{ atm})(0.0250 \text{ L})}{(0.082057 \text{ L atm mol}^{-1} \text{ K}^{-1})(291 \text{ K})} \approx 0.001028 \text{ mol } H_2$$

We can now use the conversion factor method to find the mass of aluminum that is consumed:

$$0.001028 \ \cancel{\text{mol } H_2} \times \frac{2 \ \cancel{\text{mol Al}}}{3 \ \cancel{\text{mol } H_2}} \times \frac{26.982 \ \text{g Al}}{1 \ \cancel{\text{mol Al}}} \approx 0.0185 \ \text{g Al}$$

Answer 0.0185 g Al

How are you doing? **10.16**

We can determine how much "fuel" an organism consumes by monitoring its uptake of oxygen. Determine the mass of glucose, $C_6H_{12}O_6$ that must be consumed if an experiment shows the consumption of 32.1 mL of oxygen gas at 0.231 atm and a temperature of 302 K. Write the chemical equation first.

Motion in Gaseous Systems: The Kinetic Molecular Theory

So far, we have looked at the use of mathematical equations as models of chemical systems, and the importance of a linear relationship in helping us to determine the value of hard-to-measure properties. We have examined some of the physical properties of gases discovered through experimentation. The ideal gas law is based on consistent observations of these physical properties of gases. But can experimentation and observation of the behavior of gases on the macroscopic level tell us anything about the microscopic level of gaseous atoms and molecules?

The *kinetic molecular theory* of gases is an attempt to do just that. A theory is a set of statements that explain the observations. In the case of gases, the theory must explain the laws we have discussed. The main ideas of the kinetic molecular theory are the following:

For an ideal gas,

KEY IDEA 10.6

1. Gases consist of molecules that are extremely small and very widely separated from one another.

2. Gas molecules are always moving in random directions.

3. Gas molecules travel in straight lines until they collide with either the sides of their container or some other gas molecule.

4. Gas molecules do not lose any energy in these collisions. They just hit and bounce away. Such collisions are called elastic collisions.

5. The kinetic energy of a gas is directly proportional to its Kelvin temperature. An increase in temperature causes an increase in kinetic energy.

How does the kinetic molecular theory explain the physical properties we have observed in gases? The first three statements provide an explanation for Boyle's Law, which models the inverse relationship between the exertion of external pressure and the volume of the

Figure 10-13 ▶ Snapshot from animation of gas phase atoms.

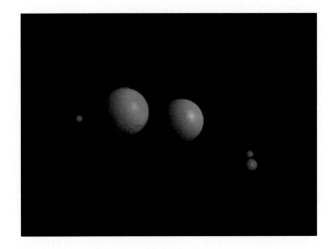

▶ See this with your **Web Animator**

✔ Meeting the Goals

For an ideal gas, the molecules are very small and widely separated. These molecules move in straight lines in random directions, until they come in contact with the sides of the container or with another gas molecule in an elastic collision. The kinetic energy of gas molecules is directly proportional to the Kelvin temperature.

gas. If gas molecules are widely separated, as shown in Figure 10-13, it is possible to compress the gas into a smaller volume by exerting some external pressure. The kinetic molecular theory of gases also lends support to the observation that gas pressure varies directly with the temperature. As the temperature increases, the molecules of gas gain more kinetic energy (energy of motion). This increase in energy increases the force with which the molecules collide with the container walls and we observe that the gas pressure has increased. When the temperature is decreased, the molecules lose kinetic energy and the gas pressure decreases.

The kinetic molecular theory allows us to discuss the relative sizes of molecules in a gas. This relates to Avogadro's Law. Statement 1 above indicates that large and small molecules are equally well separated. Gas properties, therefore, depend only on the chemical amount of gas, not the formula.

You can perform your own experiment to shed some light on the question of molecular size if you live in a cold climate. Suppose that, for your roommate's surprise birthday party, you buy some helium-filled balloons and some bottles of soda. Trudging home through snow-filled streets, you start to worry because your balloons are getting smaller. Did you buy defective balloons? What shape will they have by the time you get to the party? Inside your darkened apartment, as you wait with others for your roommate to make an appearance, the balloons regain their former size. The low temperatures outside apparently had an effect on the gas in your balloons. This effect was reversed when the balloons became warm in the heated apartment.

Example 10.19   **Reasoning about temperature and molecular motion**

Use the kinetic molecular theory to suggest what effect low temperatures have on the motion of the gas molecules in a balloon. (Hint: Look at the listing of ideas in the kinetic molecular theory.)

Thinking Through the Problem We must relate temperature to the motion of gas molecules. Consider the ideas in the kinetic molecular theory.

Working Through the Problem The idea listed as number 4 states that "the kinetic energy or energy of motion of gas molecules is directly proportional to their Kelvin temperature." For lower temperatures, then, the gas molecules move more slowly.

Answer The lower, outside temperature caused the gas molecules to have less energy and to move more slowly. The slower gas molecules did not strike the inside of the balloon with as much force as they did when they had more energy and were moving more quickly. Therefore, the balloon size decreased.

In contrast to the gas in the balloons, the soda in the bottles that you brought seemed to remain at the same volume, whether you were outside or inside the apartment. What happened to the gas that didn't affect the liquid in a noticeable way? In a gas at room temperature, the molecules are far apart; in a liquid or solid the molecules are actually in contact with one another. We know that a molecule has certain set chemical bonds, which give rise to a particular molecular structure and size. There are very small changes in the volume of a liquid or solid when it is heated or cooled, because there are only small motions possible before molecules bump into one another or into the container walls. This is why we observed very little change in the volume of the soda in the bottles that we carried through the cold to the party.

Figure 10-14 represents water in the gas, liquid, and solid phases. Note how the molecular motion and the space between molecules increase as water moves from the solid phase to the liquid phase and from the liquid phase to the gas phase.

▼ See this with your Web Animator

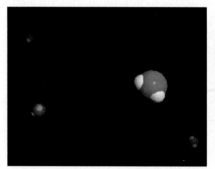  

Figure 10-14 ▲ The relative amount of motion and the space between molecules varies with gases, liquids, and solids, as shown in these snapshot images from simulations of the structure of water in all three phases.

PRACTICAL D Where are the molecules when a balloon is flattened?

Imagine that you fill a balloon by blowing into it, and then tie it shut. You immerse the balloon in liquid nitrogen ($t = -178$ °C) and watch it shrink until ultimately it appears flat.

If the balloon remains intact and is not punctured, the air you blew into it must still be present inside the balloon. Use the kinetic molecular theory of gases to explain why the balloon shrinks so much when the temperature is very low.

Suppose that the inflated balloon is a sphere with a diameter of 20.0 cm. Calculate the volume of the inflated balloon: $V = \frac{4}{3}\pi r^3$. Express this in units of cm^3, mL, and L.

Assume that all of the molecules inside the balloon are CO_2 molecules. Calculate the number of moles of CO_2 inside the balloon if the total mass of CO_2 is 8.00 g. Calculate the number of molecules of CO_2 inside the balloon.

How small are these molecules if they are still present inside the shrunken balloon? Let's see if we can find out. Under STP conditions, the density of CO_2 gas is about 1.96 g/L. Using density and molar mass calculations, calculate how many molecules of CO_2 there are in one liter of CO_2 gas. What is the volume available to one molecule of gas under these conditions?

Now let's consider what happens when the balloon collapses. The CO_2 gas becomes solid CO_2 (also known as dry ice). The density of solid CO_2 at -78 °C is 1.35 g cm^{-3}. Repeat the calculation you just did and determine the volume available to one molecule of CO_2 in the solid.

If you compare your answers to these questions, you will find that the volume available to each CO_2 molecule increases dramatically when the solid becomes a gas. Why do you think this is so? Use kinetic molecular theory to decide whether this increase in volume is because the CO_2 molecules are moving around much more in the gas than in the solid, or because the molecules increase in size.

PROBLEMS

28. Concept question: We have discussed how Boyle's Law, Charles's Law, and Avogadro's Law make up the ideal gas law. Our focus has been on how V varies as other quantities vary. Generate a new set of proportional statements for the effect of T, V, and n on pressure, P.

29. A 15.0-g sample of argon gas is transferred to a 1.50-liter container. At what temperature must this container remain if the pressure is kept at 1.05 atm?

30. A 0.3055-g sample of helium gas is contained in a 0.500-liter glass bulb at 25 °C. What is the pressure of the helium gas?

31. One mole of ammonia gas is contained in a 2.50-liter gas cylinder at room temperature. The pressure of the confined gas is 1.217 atm. What will the pressure read if one mole of ammonia

gas is contained in a 4.00-liter cylinder at the same temperature?

32. One mole of argon gas (25 °C) is confined to a volume of 4.5 liters. It has a pressure of 1.033 atm. What effect will there be on the pressure reading if there is only one-half mole of argon in the same volume and at the same temperature?

33. How does the volume of one mole of a gas change when the pressure on the gas is doubled? Assume constant temperature.

34. How does the pressure of one mole of a gas change when the container size (volume) is tripled? Assume constant temperature and amount (moles) of gas.

35. Fill in the missing information in the following table. Note that in some cases a quantity may not vary.

P_1	P_2	V_1	V_2	T_1	T_2	n_1	n_2
1.00 atm	1.00 atm	0.250 L	5.00 L	273 K		2.00 mol	2.00 mol
1.00 atm		0.250 L	5.00 L	273 K	273 K	2.00 mol	2.00 mol
1.00 atm	0.050 atm	0.250 L	5.00 L	273 K		2.00 mol	2.00 mol
	2.00 atm	0.250 L	5.00 L	273 K	273 K	2.00 mol	2.00 mol
1.00 atm	1.00 atm	0.250 L	5.00 L	273 K	273 K	2.00 mol	
3.00 atm	1.00 atm	0.250 L	5.00 L	353 K	273 K	2.00 mol	
3.00 atm	1.00 atm	0.250 L		353 K	273 K	2.00 mol	2.00 mol

36. Write a single proportionality equation to show that the volume varies jointly with the temperature and the number of moles (n) but inversely with the pressure.

37. Aluminum and oxygen react to produce aluminum oxide as shown by the equation

$$4\,Al\,(s) + 3\,O_2\,(g) \longrightarrow 2\,Al_2O_3\,(s)$$

What volume, in liters, of O_2 will completely react with 75.0 grams of Al? Assume the oxygen is at 281 K and 3.00 atm.

38. Lead (IV) chloride can be produced from its constituent elements as shown by the equation

$$Pb\,(s) + 2\,Cl_2\,(g) \longrightarrow PbCl_4\,(s)$$

How many grams of $PbCl_4$ can be produced from the reaction of 5.00 L of Cl_2 at 1.33 atm pressure and a temperature of 7.05 °C with an excess of Pb?

39. Discussion question: Suppose that 50.0 grams of CH_3OH is completely reacted according to the equation

$$CH_3OH\,(l) \longrightarrow 2\,H_2\,(g) + CO\,(g)$$

(a) Calculate the pressure of the H_2 that will result from the complete reaction of this amount of CH_3OH. Assume the product gas is kept in a 5.00-L container with a temperature of 35.20 °C.

(b) Calculate the pressure of the CO that will result from the complete reaction of this amount of CH_3OH. Assume the product gas is kept in a 5.00-L container with a temperature of 35.20 °C.

(c) If this reaction is run in a manner such that both product gases are trapped in the *same* 5.00-L container with a temperature of 35.20 °C, what will the total pressure be? Comment on why the total pressure, the pressure of the H_2, and the pressure of the CO are all different.

Chapter 10 Summary and Problems

◀ Review with **Web Practice**

VOCABULARY

Cartesian coordinate system	A rectangular coordinate system formed by the intersection of two perpendicular lines.
x-intercept	The point at which the line crosses the x-axis; to find the x-intercept, let the value of $y = 0$ and solve for x.
y-intercept	The point at which the line crosses the y-axis; to find the y-intercept, let the value of $x = 0$ and solve for y.

slope of a line	The slant of a line, indicated by m.

$$m = \frac{\text{rise}}{\text{run}} = \frac{\text{vertical change}}{\text{horizontal change}} = \frac{\text{difference in } y \text{ values}}{\text{difference in } x \text{ values}} = \frac{\Delta y}{\Delta x}$$

independent variable	In an experiment, we vary a quantity and follow the change in another variable; the variable that is varied under our control is the independent variable.
dependent variable	In an experiment we follow the effect of changes in an independent variable through changes in the dependent variable.
controlled variable	A variable that is deliberately held constant during an experiment.
slope-intercept equation	The slope-intercept equation has the form $y = mx + b$, where the slope of the line is given by m and the y-intercept is given by b.
pressure	A measurement of the forces exerted on or by the gas molecules in a gas sample.
atmospheric pressure	The pressure or force of all the gases in the atmosphere on a particular system of interest.
barometer	An instrument used to measure the height of a column of mercury that the atmosphere is able to support; the units of this pressure measurement are mm Hg or cm Hg, depending on the scale used to measure the mercury column.
atmosphere	A common unit of pressure, equal to 101,325 Pascals.
standard pressure	The pressure of the atmosphere exerted on a column of mercury when measured at sea level. 1 atm = 760 mm Hg = 760 torr = standard pressure.
Boyle's Law	Boyle's Law states that at constant temperature, the volume occupied by a specified quantity of gas is inversely proportional to the applied pressure.
Kelvin temperature scale	A scale based on the expansion properties of gases; it is related to the Celsius temperature scale by $T = t + 273.15$, where T = temperature in Kelvin and t = temperature in degrees Celsius. A Kelvin degree is the same size as a degree Celsius.
absolute zero	The lowest possible temperature is called absolute zero; on absolute temperature scales like the Kelvin scale, absolute zero has a value of 0.
STP	The abbreviation for "standard temperature and pressure"; STP for gases is 273.15 K and 1 atmosphere pressure.
ideal gas law	The ideal gas law equation is $PV = nRT$. This equation holds strictly only for hypothetical or ideal gases but real gases can approach ideal conditions at low pressures.
combined gas law	The combined gas law is $\frac{P_1 V_1}{n_1 T_1} = \frac{P_2 V_2}{n_2 T_2}$. This equation is useful whenever there is a change of one or more of variables for a gas sample.

SECTION GOALS

How is the Cartesian coordinate system used to find mathematical descriptions between variables?

The Cartesian coordinate system allows us to plot ordered pairs of points with the position (x, y). We typically take variables in an experiment and designate one as dependent and one as independent. We measure the value of the dependent variable as we change the value of the independent variable. The dependent variable is the value for y and the independent variable is the value for x. Thus, we take our variables and plot them as ordered pairs in the position (independent variable, dependent variable). By examining the points from several measurements we can look for a mathematical relationship between the two variables.

What is meant by the slope of a line?

The slope of a line is found by evaluating the ratio $m = \dfrac{y_2 - y_1}{x_2 - x_1}$ for any two points (x_1, y_1) and (x_2, y_2) that are on the line.

Can you interpret a linear equation and determine both the value and the units of the slope?

The slope-intercept form of a linear equation is $y = mx + b$. We obtain the slope-intercept form of any linear equation by solving it for the dependent variable. The slope m will be the proportionality between x and y, and it will have the units "y-unit/x-unit."

What variables must be specified when discussing a gas sample?

Important gas properties are pressure (P), temperature (T), volume (V), and chemical amount or moles (n). Pressure, P, is the force exerted when the molecules of the gas strike a surface. Volume, V, is the amount of space a gas occupies. The chemical amount (moles), n, is related to the number of molecules contained in the gas sample. The temperature, T, is a property related to molecular motion and energy. The temperature of a gas sample must be specified as it has an effect on both the pressure and the volume of a gas.

What units are commonly used to specify standard pressure?

The most common pressure units for gases are mm Hg, atmospheres (atm), and torr.

How do we recognize and apply Boyle's Law to gas problems about pressure and volume?

Boyle's Law relates pressure and volume (mole amount and temperature are held constant). A mathematical statement of Boyle's Law is $PV = k$, but it is often used to determine final conditions in a gaseous system after some change has been made. Then the formulation, $P_1V_1 = P_2V_2$, is more useful.

How do we convert between the Celsius temperature scale and the Kelvin temperature scale?

To convert a temperature from the Celsius scale to the Kelvin scale, add 273.15 to the Celsius temperature. To convert from the Kelvin scale to the Celsius scale, subtract 273.15 from the Kelvin temperature.

How do we recognize and use the relationship between volume and temperature developed by Charles?

For a specified amount of gas at constant pressure, gas volume varies directly with its temperature. To calculate the final temperature or volume condition from the change made to some initial conditions, use the relationship, $\dfrac{V_1}{T_1} = \dfrac{V_2}{T_2}$

Can we identify all of the variables in the ideal gas law equation? How do we know what units to use?

The ideal gas law is $PV = nRT$, where P, V, n, and T are the pressure, volume, mole amount, and absolute temperature of the gas. R is the universal gas constant; the units for the variables in the ideal gas law equation are determined by the units of R.

How do we extend stoichiometric calculations to cover gases?

The ideal gas law allows us to relate the number of moles of a gas to its volume, pressure, and temperature. Because the number of moles of a gas consumed or generated in a reaction depends on the amounts of other substances and on the balanced chemical equation, we have a way to connect gas law calculations with stoichiometry. We use the ideal gas law to relate n with P, V, and T. We use mole ratios to relate the mole amounts of reactants and products. And we use molar mass to relate n to mass.

What are the main ideas of the kinetic molecular theory?

The main ideas of the kinetic molecular theory are that, for an ideal gas,

1. Gases consist of molecules that are extremely small and very widely separated from one another.

2. Gas molecules are always moving in random directions.

3. Gas molecules travel in straight lines until they collide with either the sides of their container or some other gas molecule.

4. Gas molecules do not lose any energy in these collisions. They just hit and bounce away.

5. The kinetic energy of a gas is directly proportional to its Kelvin temperature.

PROBLEMS

40. When a plot is made of the initial reaction rate against initial concentration, a linear graph is obtained. What are the units of the slope of the line graphed if the units of rate are mol $L^{-1}s^{-1}$ and the units of concentration are mol L^{-1}?

41. The data in the following table are for an experiment in argon solubility, where the amount of argon gas (in moles) dissolved in one liter of water is determined by varying the pressure (in atm) of Ar above the solution.

Pressure (atm)	Moles Dissolved
5.0	0.0128
10.0	0.0256
20.0	0.0512
30.0	0.0767

(a) Identify the independent and the dependent variables.
(b) Graph the data, including a line connecting the points, on graph paper.
(c) What is the slope of the line?
(d) What are the units of slope?
(e) Write in slope-intercept form the equation of the line generated from the data.

42. The data below are for an experiment in osmosis, where the pressure (in pounds per square inch, or psi) exerted by a solution is determined by varying the amount of sugar (in moles) dissolved in one liter of water.

Pressure (psi)	Moles Dissolved
8.99	0.025
12.6	0.035
21.6	0.060
34.1	0.095

(a) Identify the independent and the dependent variables.
(b) Graph the data, including a line connecting the points, on graph paper.
(c) What is the slope of the line?
(d) What are the units of slope?
(e) Write in slope intercept form the equation of the line generated from the data.

43. A famous drawing by Leonardo daVinci, shown in Figure 10-15, suggests a relationship between height and arm-span. Use a ruler to do this measurement on the members of a group of students. Graph the data. Is there a relationship? What is it? Can you determine a mathematical equation that corresponds to this graph?

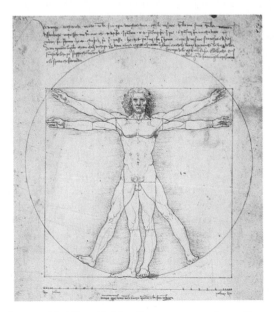

Figure 10-15 ▲ DaVinci's representation of a man includes a suggested proportionality. (Scala/Art Resource.)

44. Find y, given the following information: $(1, y)$ and $(-2, 4)$ are points on a line with slope $-\dfrac{4}{3}$.

45. Find the x-intercept and the y-intercept for the equation, $3x - 4y = 12$.

46. It is said that temperature can be determined by the number of times a cricket chirps. If a cricket chirps 40 times per minute when the temperature is 50 °F and 80 times per minute when the temperature is 60 °F, what is the temperature when the cricket stops chirping? Assume linear behavior.

47. Suppose that you consider a graph plotted by another student. He has used experimental data measured for V and T in the equation $PV = nRT$. The units of V and T are liters and K, respectively.

    (a) Rearrange the equation in the slope-intercept model so that it represents the variables as they appear on the graph.

    (b) What variable or variables can be evaluated as the slope of the line?

    (c) What is the value of the y-intercept in this graph? (Could you tell from the rearranged equation?)

    (d) What are the units of the slope?

48. An ideal gas occupies a volume of 23.86 mL at a pressure of 0.890 atm and 299 K. What volume will this gas occupy at STP?

49. A sample of gas occupies 53.91 mL at 372 K and 1.25 atm. What is the pressure of this gas when the volume is expanded to 63.81 mL at a temperature of 398 K?

50. 125.4 grams of N_2O is contained in a 6.83-L cylinder at room temperature (298 K). The entire sample of gas is transferred to a 10.00-L cylinder at the same temperature. What is the pressure of the gas in the second container?

51. When 128.90 grams of cyclopropane, C_3H_8, is put into a 8.00-L cylinder at 298 K, the pressure is 1.24 atm. When the gas is transferred into a 12.00-L cylinder at 0.88 atm pressure, what will its temperature be?

52. If $PV = nRT$, what is the value of $\dfrac{PV}{nRT}$?

(What happens when you divide a number by itself?)

53. What happens to the value of $\dfrac{PV}{nRT}$ as only the value of P changes? Will the value of the ratio change?

54. Solve $\left(P + a\dfrac{n^2}{V_2}\right)\left(V - nb\right) = nRT$ for P.

55. A sample of N_2O is contained in a 24.00-L cylinder at a pressure of 1.06 atm and a temperature of 53.12 °C. What volume cylinder is needed if the gas is to be maintained at a pressure of 0.98 atm and a temperature of 62.94 °C?

56. 174.6 grams of N_2 is contained in a gas cylinder at room temperature (298 K). The pressure reading on the cylinder is 1.63 atm. A second gas cylinder contains C_3H_8. The second cylinder has the identical volume, temperature and pressure readings as the first cylinder. How many grams of C_3H_8 are there in the second cylinder?

57. Hydrochloric acid reacts with sodium carbonate according to the following balanced equation:

$$HCl\ (aq) + Na_2CO_3\ (aq) \longrightarrow$$
$$CO_2\ (g) + H_2O\ (g) + NaCl\ (aq)$$

Calculate the mass of HCl required in order to produce 64 liters of CO_2 gas. The experiment is done at a temperature of 264 K and a pressure of 726.0 mm Hg.

58. The building material gypsum is made of $CaSO_4$. It cannot be used in very high-temperature applications because it can decompose to CaO and a gas.

    (a) What is the gas?

    (b) How many liters of this gas will be formed by the reaction of 1.00 kilogram of gypsum at 2.00 atm pressure and a temperature of 1850 °C?

59. Nitric acid can be made in a process in which nitrogen dioxide reacts with water according to this equation:

$$3\ NO_2\ (g) + H_2O\ (l) \longrightarrow 2\ HNO_3\ (aq) + NO\ (g)$$

Calculate the mass of HNO_3 produced when 386 liters of NO_2 gas completely reacts. The experiment is done at a temperature of 395 K and a pressure of 850.0 mm Hg.

60. What conditions does STP represent?

61. A sample of gas at STP is put into a 50.0-liter container and heated to 85.0 °C. Calculate the final pressure of this gas.

62. In the equation $\dfrac{P_1V_1}{n_1T_1} = \dfrac{P_2V_2}{n_2T_2}$, which variables must be held constant when we study the effect of temperature on the pressure of a gas?

63. The ideal gas equation, $PV = nRT$, tells us how P, T, V, and n vary with one another for a gas sample.

(a) In this equation, which of the variables varies directly with P?

(b) In this equation, which of the variables varies inversely with P?

(c) How does R vary as P, T, V, and n change?

64. When ammonium chloride is heated, it decomposes into ammonia and hydrogen chloride gases according to the equation

$$NH_4Cl \ (s) \leftrightarrow NH_3 \ (g) + HCl \ (g)$$

(a) Calculate the moles of NH_3 and the moles of HCl produced when 20.0 grams of NH_4Cl decomposes completely.

(b) What is the total volume of these two gases at 25.0 °C and 1 atm pressure?

65. A gas at STP is heated to 58.4 °C. Calculate the pressure of the gas after it is heated.

66. What volume of CO_2 gas can be produced by the complete metabolization of 12.0 grams of glucose, $C_6H_{12}O_6$? Assume a pressure of 748 mm Hg and 42.5 °C.

$$C_6H_{12}O_6 \ (s) + 6 \ O_2 \ (g) \rightarrow 6 \ CO_2 \ (g) + 6 \ H_2O \ (g)$$

67. A helium-filled balloon has a volume of 50.0×10^3 cm^3 when the temperature is 82.0 °C and the pressure is 754 mm Hg. How will the volume change when the balloon ascends to a height where the pressure is 695 mm Hg and the temperature is 35.0 °C?

68. A 10.0-liter evacuated container is filled with 2.46 moles nitrogen gas and 0.68 mole oxygen gas at 284 K. Assume that each gas occupies the space independently of the other.

(a) Calculate the pressure of N_2 under these conditions.

(b) Calculate the pressure of O_2 under these conditions.

(c) What is the total pressure that these gases exert on the container?

69. (a) Calculate the moles of N_2 in a 250.0-mL sample at 0.0 °C and 755 mm Hg.

(b) How many grams of N_2 are there in this sample?

(c) What is the density of N_2 in this sample?

CHAPTER 11

Chemical Systems and Heat

The energy that drives these sports cars is released when gasoline reacts with oxygen inside the engine. (Jan Halaska/Photo Researchers.)

PRACTICAL CHEMISTRY Portable Fuel

Archaeologists consider the transport of objects from one place to another to be one of the earliest marks of human civilization. Stones that were good for butchering, for example, might only be found a great distance from the place where animals were slaughtered. This gives evidence that someone was able to plan how to move the stones to where they were needed. Portable tools were very important in allowing people to live in new environments.

In more recent times a similar breakthrough occurred when humans learned to transport energy. Sources of energy tend to be very localized. This is why communities often developed near power sources such as waterfalls and fast-moving streams. But energy is often needed in places far from an energy source. Modern civilization depends on energy that can be transported. It is even important at times that we carry energy with us. We carry stored energy when we drive an automobile, use a battery-operated flashlight or radio, or simply carry a book of matches. In many cases, energy is stored in chemical form and then converted to a more useful form through a chemical reaction. In this chapter, we will discuss how energy can be stored in the form of heat and how chemical reactions that release heat are a key to portable energy.

This chapter will focus on the following questions:

PRACTICAL A How can energy from the sun be stored and used for heating?

PRACTICAL B How does the oxygen content of a fuel affect the amount of energy the fuel can provide?

11.1 Temperature and Heat

SECTION GOALS

✔ Explain what is meant by an isolated system.

✔ Explain how temperature change and molecular motion are related, and understand the relation of these concepts to heat.

✔ Describe how heat flows.

✔ Know what is meant by the terms endothermic and exothermic.

✔ Explain the relationship between heat and phase changes.

Energy and Heat

The word *energy* is so familiar to us that its scientific meaning is sometimes forgotten. **Energy** can be defined as the ability to do work or produce heat. In Table 11-1 we list six common types of energy that exist in scientific applications today. All of them, with the exception of nuclear energy, are a part of our everyday lives.

A full study of energy is outside the scope of this textbook. But we can learn a lot about chemical systems by studying the relationship between energy and heat changes. **Heat** is associated with one kind of energy, thermal energy. The release of heat by chemical reactions is a very important application of chemistry in our daily lives.

Most people are familiar with certain things that they themselves find easy to understand but difficult to explain in words to other people. Take the term *sweet*. If someone gives you a hard, shiny piece of food and says, "This is really sweet," then you know what to expect when you put it on your tongue. You may even be able to agree with that person that some things are sweeter than others. But how would you define "sweet" to someone who has never tasted a sweet thing?

Similarly, the word *hot* is easy to understand but difficult to define. We have a feel for what "hot" means, but, to study heat and the flow of heat scientifically, we must carefully describe these words in a quantitative way. Experimentally we know that *when two substances of different temperatures are placed in physical contact with each other, the warmer object will become cooler and the cooler object will become warmer, until both objects attain the same temperature.* We describe these two objects as having thermal contact. We know from

The energy released in this chemical reaction is evidenced by a spectacular color display and the evolution of heat.
(W. H. Freeman photo by Ken Karp.)

KEY IDEA 11.1

Table 11-1 Types of Energy and Examples of Their Use

Energy Type	Example of Where this Energy Is Found	An Example of this Energy	An Example of Using this Energy
Mechanical	Energy of the motion and displacement of objects	Raising a weight into the air.	The energy in a waterfall is used to turn a turbine.
Electrical	Energy of the motion of electrical charges in an electrical field	A battery is switched "on" to give electrical current, and chemical energy is released in the flow of electrons between substances.	Electrical energy is converted to light and heat energy when a lamp is turned on.
Electromagnetic	Energy of light	A "light stick" releases, in the form of light, chemical energy from a chemical reaction.	A temperature rise in the surroundings is caused by the absorption of light.
Chemical	Energy available in chemical reactions	Photosynthesis in plants transforms light energy from the sun into chemical energy as chemical reactions occur.	A solid fuel booster on a rocketship involves the reaction of chemicals to form hot gases that push the rocket.
Nuclear	Energy stored in the atomic nucleus	High-speed collision and fusion of atomic nuclei in very hot stars converts mechanical energy into nuclear energy.	Splitting (fission) of atoms in a nuclear reactor gives off heat to surroundings.
Thermal (heat)	Energy caused by the vibration and motion of atoms and molecules	A fuel burns in the presence of oxygen, and the chemical energy in the fuel-oxygen mixture is released as heat.	As air in a balloon is heated, it expands, and the balloon and its cargo rise.

experiment that *heat always flows from the hotter object to the colder object;* thus, heat transfer is always from the hotter to the cooler substance. The quantitative study of heat transfer gives us important information about objects in thermal contact.

In Table 11-2 we list some common statements about heat, along with the scientifically valid corresponding statements. In each case we see that the scientific statement is more careful and better explains what happens physically than does the common statement.

Table 11-2 Common and Scientific Statements Related to Heat

Common Statement	Chemist's Restatement	Change in Meaning	Heat Transfer
"Put a sweater on to warm yourself."	"Put a sweater on to keep yourself warm."	The sweater does not provide the warmth. It keeps the body from losing heat.	Body heat that would be transferred to the surroundings is kept close to our bodies by the sweater.
"If you want to keep the food from spoiling, pack it with lots of ice, to keep it cold."	"Ice requires heat to melt, so we can use ice to keep the temperature of the food low."	Ice does not provide cooling. Instead, ice absorbs heat so that temperatures don't rise.	The heat from the surroundings that would warm the food in the absence of ice is transferred into the constant temperature melting of ice.
"Propane is dangerous because it can give a very hot flame."	"Propane combusts in air to yield a dangerous amount of heat."	Propane itself does not produce heat. Instead, the combustion reaction of propane with air gives off heat.	The chemical energy in propane and air is transferred to the surroundings as heat energy.

In this chapter, we present the scientific view of heat that is commonly used by chemists to characterize what happens in chemical systems. Our simple experience of heat, which involves temperature change, will expand to include the heat changes that take place when chemical systems react.

Systems and Surroundings in Heat Changes

Chemists begin all studies of energy, including heat transfer and temperature change, by carefully defining two parts to any experiment: the **system** *and its* **surroundings.** As an example, think about the sweater mentioned in Table 11-2. You can consider your body and the sweater to be the system and the air to be the surroundings. In this case, the sweater helps to slow down the transfer of heat from your body to the surroundings. The systems and surroundings must be specified at the beginning of any study.

In general, we study how heat flows into or out of a system or how it flows between the parts of a system. At a constant external pressure, the total heat change for a system and its surroundings must add up to zero. Thus, as the system loses heat, the surroundings gain an equal amount of heat, and vice versa.

Example **11.1** **Heat flow between a system and surroundings**

In each of the following situations, which are shown in Figure 11-1, indicate how heat flows between a system and its surroundings.

(a) Hot water is left to stand in a pot. Both the hot water and pot are at the same temperature at the start. They both cool down.

Figure 11-1 ▲ Heat transfer is important in many different situations. (a: Jeff J. Daly/Visuals Unlimited; b: G and C Merker/ Visuals Unlimited; c: Tom Pantages.)

(b) A snake slithers out from under some cool rocks to bask on a large boulder that has been warmed by sunshine.

(c) When you place your hand on a frosty window on a cold morning, frost melts, leaving behind a ghost print of your hand.

Thinking Through the Problem ◄━━► (11.2) We need to identify something we can label as the system and something we can label as the surroundings. Then we indicate which gains heat and which loses heat.

Answer

(a) In this case, the hot water and the pot together can be considered the system, and the air around the water-filled pot comprises the surroundings. The system loses heat and the surroundings gain heat.

(b) The snake is the system here, and the warm boulder represents the surroundings. The snake is cool when it first comes out from under the rocks, but as it lies on the boulder, heat flows from the warm boulder to the snake, warming the snake. The surroundings lose heat and the system gains heat.

(c) In some cases, the definition of surroundings and system is ambiguous. Here we can say that the frosty window is the system and the hand is the surroundings. The hand gives off heat that is taken up by the window, melting the frost. Or we can say that the hand is the system, which loses heat to the surrounding window, melting the frost.

How are you doing? 11.1

In each of the following descriptions, indicate how heat flows between a system and its surroundings.

(a) Water in an ice cube tray freezes in a household freezer.

(b) You stand next to a building after sunset, and feel warmth coming from the wall next to you.

✔ **Meeting the Goals**

We define two parts to any experiment, the system and the surroundings. An isolated system is a system that does not exchange anything with the surroundings.

A special kind of a system, called an **isolated system,** exchanges nothing, not even heat, with its surroundings. In an isolated system, the heat lost by one part of the system must be gained by the other part of the system. These are special cases of the law of conservation of energy.

Heat, Temperature Change, and Molecular Motion

Heat is an important concept because it allows us to explore how energy and matter relate to each other in chemical systems. The most obvious way that we experience heat is through temperature changes; an increase in temperature is commonly called "heating" and a decrease in the temperature is called "cooling." It is easy to think of heat and cold as substances. They are not. They are actually the experience of energy changes that are associated with the thermal energy of a system. When we say *heating* we actually mean "increase in thermal energy" and when we say *cooling* we actually mean "decrease in thermal energy."

Recall from Chapter 10 that heat can cause the pressure of a gas to increase, because a change in temperature, T, causes a proportional change in pressure, P, when we have a fixed number of moles and a fixed volume. If we lower the temperature of the gas, then the pressure will decrease. What happens to the molecules in the gas as we make these changes?

Kinetic molecular theory indicates that heating increases the average speed of the gas molecules. At higher temperatures, the molecules of a gas will, on average, move faster and strike the walls of the container more often and with greater force than they will at lower temperatures. This is why gas molecules exert greater pressure on the walls of their container at higher temperatures.

The effect of temperature on a gas is an example of the following general principle, illustrated in Figure 11-2: *when a physical system warms from a lower to a higher temperature, there is increased motion of the atoms and molecules that make up the system. When a system cools, the motion of the atoms and molecules decreases.*

Atoms and molecules of solids and liquids also have a corresponding increase in motion with an increase in temperature. However, because the molecules or atoms in solids and liquids are in close contact, they move only short distances before a collision.

Sources of Heat

How do we measure heat? In chemistry, we use the abbreviation q as a measure of heat transferred. *For the simple temperature change of a system, we say that if q is positive, then the system gains heat; if q is negative, then the system loses heat.* Let's consider what happens when two objects are put in contact with each other. We know from experiment that if the objects start at different temperatures, the object at the higher temperature will become cooler and the object at the lower temperature will become warmer. We know that the change will tend to bring the two objects to the same temperature, somewhere between the two starting temperatures.

Increasing the temperature of a physical system increases the motions of the atoms and molecules that make up that system. When a gas is heated, the increased motion of the gas molecules results in increased pressure, which can be measured. The same argument, applied in reverse, explains why gas pressure decreases when a gas cools.

Figure 11-2 ▲ Heating a gas (left to right) does not change the size or the number of gas molecules. It does increase their average speed, as represented by the arrows.

We describe the transfer of heat with the variable q. If q is greater than zero, then heat flows into a system; if q is less than zero, then heat flows out of the system.

Suppose that we add some cool water to a thermos of hot coffee and then close the thermos. The cool water starts out at a lower temperature than the coffee. The initial temperature T_1 of the cool water is lower than the initial temperature T_2 of the coffee. Because the two liquids remain in contact, the final temperature of the cool water and the coffee will be some value T_{final} that is between T_1 and T_2. The heat gained by the water (q_1) is equal and opposite to the heat lost by the coffee (q_2), so we say that q is positive for the water and negative for the coffee. This is shown schematically in Figure 11-3. (Note that the final value for the temperature will *not* necessarily be the simple average of the starting temperatures. Why this is so will be considered in the next section.)

Figure 11-3 ▶ In an isolated system of two objects (1 and 2) at different starting temperatures, one object gains heat while the other loses heat.

Figure 11-4 ▲ Light energy from the sun can brew tea. (Tony Freeman/PhotoEdit.)

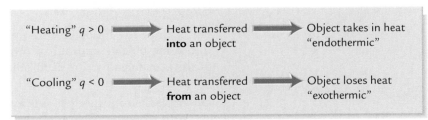

| "Heating" $q > 0$ ➡ | Heat transferred **into** an object ➡ | Object takes in heat "endothermic" |
| "Cooling" $q < 0$ ➡ | Heat transferred **from** an object ➡ | Object loses heat "exothermic" |

Figure 11-5 ▲ There are several different ways to describe heat transfer.

There are many ways to add heat to a system. For example, if we leave a bottle of water outside on a sunny day, the temperature of the water will change. Here, q for the water is positive. The transferred heat originated in light energy from the sun, as illustrated in Figure 11-4. We also use heating elements such as stovetops and ovens to heat and cook food. In this case, the q for the stovetop is negative, and the q for the food is positive.

There are two important words used to describe heat transfer. When heat is transferred into a system, q is greater than 0 and we say that the process is **endothermic** ("endo" literally means "inside"). When heat is transferred out of a system, q is less than 0 and we call the process **exothermic** ("exo" means "outside"). These definitions are summarized in Figure 11-5.

REMEMBER

Example 11.2   **Heat transfer**

Describe the heat transfer that takes place in each of the following situations. Indicate whether q is positive or negative for each object, then answer the question.

(a) A teapot at room temperature is filled with boiling water. The water stops boiling and the teapot becomes uncomfortable to touch. Why did the water stop boiling?

(b) Fresh seafood is placed on a bed of ice in a non-refrigerated display container. Why is a large supply of ice needed?

Thinking Through the Problem We need to indicate whether q is positive or negative for each object described. We also need to answer some other questions, starting with whether the object is absorbing heat and becoming warmer, or giving off heat and becoming cooler.

Answer

(a) Heat is transferred out of the boiling water and into the teapot, so $q < 0$ for the water and $q > 0$ for the teapot. The pot became uncomfortable to touch as its temperature increased. The water stopped boiling because its temperature decreased to below the boiling point.

(b) Heat is transferred from the seafood to the ice. For the seafood, $q < 0$, and for the ice, $q > 0$. As the seafood becomes cooler, some of the ice warms to its melting point and melts. A large supply of ice is needed in order to keep the temperature of the system constant for a long time.

✔ **Meeting the Goals**

The word *endothermic* indicates that heat moves into a system and describes systems for which $q > 0$. The word *exothermic* indicates that heat moves out of a system and describes systems for which $q < 0$.

For each of the following examples, indicate the heat transfer into some objects and out of others. Indicate whether q is positive or negative for each object, then answer the question for each situation.

(a) A blacksmith plunges white hot molten (liquid) iron into a bath of water. There is a hiss of steam, and a dark, hard piece of metal emerges. Why did steam form?

(b) A mold is placed in snow and allowed to stand for fifteen minutes. Melted sugar is poured into the mold. A hard sugar candy is formed in the shape of the mold. What happened to the sugar?

Phase Changes and Heat

In Chapter 1, we studied how a pure substance or a mixture can exist in more than one phase. The three simplest phases are solid, liquid, and gas. These can be as important as the identity of the chemical substance itself. For example, you do not expect to quench your thirst with either water vapor or ice without first converting these to liquid water. In Chapter 4, we discussed how the temperature of a substance is an intensive measurement that can be used to indicate the expected phase for that substance.

The change of a substance from one phase to another is accompanied by a transfer of heat between the substance and the surroundings. We know this from experience. Think of how we heat ice in order to melt it and we heat liquid water in order to make it evaporate. It is easy to think that the ice must "warm up," that is, undergo a temperature change, in order to melt. If the ice is below 0 °C, then its temperature must be raised to 0 °C before melting can occur. If you place a thermometer in a well-insulated and well-stirred container of water and ice already at 0 °C, you will probably detect no significant temperature change until all the ice is melted. Similarly, boiling water will remain at a steady temperature of 100 °C under normal atmospheric pressure.

These cases offer evidence that a system can absorb or give off heat *without a temperature change*. Such systems undergo a *change in phase*, rather than a change in temperature. The heat required to vaporize a substance is called the **heat of vaporization,** and the heat needed to melt a substance is called the **heat of fusion.** We can figure out the sign of the heat change, q, associated with a change in phase and what is known about molecular motion in solids, liquids, and gases. This is shown in Table 11-3.

✔ **Meeting the Goals**

The change from one physical state to another requires a heat transfer that increases molecular motion (endothermic process) or decreases molecular motion (exothermic process). When heated, solids can be converted to liquids or gases, and liquids can be converted to gases. Thus, the phase changes of melting, sublimation, and evaporation are endothermic processes. The opposite phase changes—freezing, deposition, and condensation—are exothermic processes.

Example 11.3

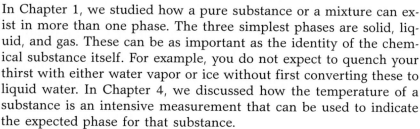

Heat transfer

When margarine in a pot is heated on a stove, it melts. If this liquid is then put into a cold mixing bowl, it solidifies. Analyze the transfer of heat, including q, for the margarine from start to finish.

Thinking Through the Problem We observe two phase changes in this example. We need to account for q as the phase changes occur.

Remember that phase changes are related to changes in the motion of molecules.

Answer In the pot, the margarine acquires heat from the stove, so $q > 0$ for the margarine. Eventually the margarine reaches its melting point, and we continue to heat it as it melts. Here, too, $q > 0$. When the liquid margarine is poured into the cold bowl, heat flows from the margarine to the bowl. Because the margarine loses heat, $q < 0$ for the margarine. The margarine solidifies and then cools further.

Table 11-3 Heat Transfer and Phase Changes

Phase Change	Heat Transfer	Type of Change
Solid → liquid	More motion → heat must be transferred into the object $q > 0$	Endothermic
Solid → gas	More motion → heat must be transferred into the object $q > 0$	Endothermic
Liquid → gas	More motion → heat must be transferred into the object $q > 0$	Endothermic
Gas → liquid	Less motion → heat must be transferred from the object $q < 0$	Exothermic
Gas → solid	Less motion → heat must be transferred from the object $q < 0$	Exothermic
Liquid → solid	Less motion → heat must be transferred from the object $q < 0$	Exothermic

Example **11.4** **Heat and phase changes**

Rubbing alcohol is made of a solution of the organic compound 2-propanol (also known as *isopropyl alcohol*) in water. When it is placed on the skin, a cooling sensation results and a sweet odor is detected. What is happening?

Thinking Through the Problem We need to account for the sweet odor and the cooling. We start by considering what happens when the 2-propanol is applied to the skin. The odor indicates that a gas is being produced, presumably by the evaporation of 2-propanol, because water vapor is odorless.

Answer The 2-propanol is evaporating on the skin. This evaporation occurs as heat is transferred into the 2-propanol ($q > 0$) from the skin ($q < 0$). We feel the lower skin temperature at the place where evaporation occurs. The odor is due to the 2-propanol vapor.

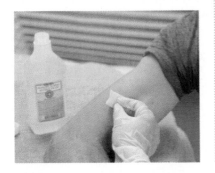

Rubbing alcohol eases inflammations and tired muscles because it evaporates when it is rubbed on the skin. (Mary Kate Denny/PhotoEdit.)

How are you doing? **11.3**

When liquid water is placed in a freezer, it cools down slowly until it reaches the freezing point of water. But it does *not* immediately freeze. Why? What has to happen to make the water freeze?

The connection between heat and phase changes can be explained in terms of the motion of molecules and atoms, just as we explained the connection between heat and temperature change. When a solid melts (becomes a liquid), the atoms or molecules move around much more freely—they are no longer held in a rigid structure and, while they are still in close contact with one another, they no longer stay in one place. Thus, a phase change is accompanied by greater motion and increased freedom of the molecules and atoms. Similarly, when a liquid or solid vaporizes (becomes a gas), there is an increase in the motion and freedom of the atoms and molecules.

We see that heat can cause either a temperature change or a phase change. In both cases, heat causes the molecules and atoms to move more rapidly. But when heat causes a phase change, the ability of the molecules and atoms to remain in fixed positions also changes radically. Their motion does increase, even though their temperature does not.

PROBLEMS

1. The following examples all involve the transfer of heat from one object to another. Indicate which object loses heat and which object gains heat.

(a) After a campfire dies down, you douse it with water, and steam forms.

(b) Hot water at its boiling point is poured into a cool teapot. The water stops boiling immediately.

(c) Steam flows through coils inside the combustion chamber of an electrical power plant. The steam becomes pressurized and is used to turn a turbine.

2. Concept question: When two objects are put into contact, one warms up and the other cools down. Is it appropriate to talk about the transfer of "cool" in the same way that we talk about the transfer of "heat"?

3. The following examples all involve the conversion of some kind of energy into thermal energy of water. Indicate the kind of energy that is being used to create the heat.

(a) A plastic bag of water is left hanging from a tree outside on a sunny day. The temperature of the water increases.

(b) Cold water is placed in a microwave oven and after 30 seconds of irradiation the water is close to boiling.

(c) Water flowing over a waterfall is warmer at the bottom than at the top.

(d) The water flowing out of a nuclear power plant is 3 °C higher than the water flowing into the plant.

4. Determine the direction of heat flow and the sign of q for both the system and the surroundings (as indicated) in the following situations:

Situation	System	Surroundings	Direction of Heat Transfer	Sign of q for the System
Cup of hot chocolate	The chocolate drink	The ceramic cup		
Testing whether candy is "done" by adding a few drops of hot candy to cold water.	The candy mixture	The water		
Holding a boiled egg under cool running water to stop the cooking process	The egg	The water used to stop the cooking process		

PRACTICAL A How is energy from the sun stored for heating?

With some exceptions, most of the thermal energy that we use every day comes from plants that have stored energy from the sun in molecules made from water and carbon dioxide. This energy may have been stored very recently, or many millions of years ago. When we obtain energy for our bodies by eating corn, for example, that energy might have been stored just last summer as the corn grew in a farmer's field. When we burn wood in a campfire to get energy for cooking, we are probably using energy stored by trees that grew during the last century or so. The peat that is used as fuel in some parts of Ireland is a compacted soil that is very rich in decomposing plant material laid down over thousands of years. The fuel that we use to power our cars and heat our homes is found deep in the earth as the fossil fuels petroleum, coal, and natural gas. The energy in these fuels was obtained from the sun perhaps millions of years ago. Some of these examples are shown in Figure 11-6.

The fuels we mention here differ in several ways. Consider their physical properties:

Fuel	Physical State	Density	Characteristics of Combustion
Natural gas	Gas	0.7 g L^{-1} (can be pressurized to $> 5 \text{ g L}^{-1}$)	Easy to burn in open flame or closed cylinder
Gasoline	Liquid	800 g L^{-1}	Dangerous to burn in open system; easy to burn in a closed cylinder
Coal	Solid	1400 g L^{-1}	Burns slowly in an open system
Wood	Solid	900 g L^{-1}	Burns slowly; a lot of moisture present
Ethanol (from corn)	Liquid	900 g L^{-1}	Burns well in open or closed system; flame is cool

Prepare a table that characterizes each of the fuels listed above according to the following characteristics:

1. Portability **2.** Safety **3.** Age and renewability **4.** Relationship to the capture of the sun's energy at some point in the past

Figure 11-6 ▲ Plants store energy from the sun. This energy is released when the plants are burned, sometimes after the passage of many thousands or millions of years. (a: Charlie Waite/Tony Stone; b: Frank Siteman/Rainbow; c: Hank Morgan/Rainbow.)

5. List three examples of processes that take up (absorb) heat. One process should be mechanical (involving objects in motion), one electrical, and one chemical.

6. List three examples of processes that produce heat. One process should be mechanical (involving objects in motion), one electrical, and one chemical.

7. Discussion question: The following sentence describes a physical process that changes the temperature of a substance.

   A rock lying in the sun warms up.

   (a) Does the rock absorb or release heat?
   (b) What is the sign of q for the rock?
   (c) Where does the heat go or come from?
   (d) What do you think is the ultimate source of the energy that became the thermal energy in the warmed rock?

8. Heat is not the same as temperature. Answer the following questions to see why this is so.

   (a) A pot of water is placed above a flame. The water reaches a temperature close to boiling, and then cold water is added. Will the temperature of the water in the pot stay the same, decrease, or increase?
   (b) A pot of water is heated until it reaches its boiling point. Then the temperature of the water remains constant at the boiling point while the water gradually boils off. The vaporization of the water requires heat, but the temperature remains constant during this process. Why?

9. When heat is transferred into an object, the process is called_____(endothermic or exothermic).

10. When heat is transferred from an object, the process is called_____(endothermic or exothermic).

11. When a substance is heated, its atoms and molecules vibrate_____(more or less).

12. A phase change from solid to liquid is accompanied by_____(an increase or a decrease) in molecular motion.

13. The phase change from solid to liquid is_____(endothermic or exothermic).

14. The phase change from liquid to solid is_____(endothermic or exothermic).

15. A phase change from gas to liquid is accompanied by_____(an increase or a decrease) in molecular motion.

16. The phase change from gas to liquid is_____(endothermic or exothermic).

17. The phase change from liquid to gas is_____(endothermic or exothermic).

11.2 The Quantitative Study of Heat Transfer

SECTION GOALS

✔ Recognize the units of heat.
✔ Define specific heat capacity, and understand its relationship to temperature change and heat transfer.
✔ Calculate q for heat transfers that do not involve phase changes.
✔ Quantify the relationship between heat and phase change.
✔ Quantify the relationship between heat and chemical reactions.

Quantifying the Relationship Between Heat and Temperature Change

In the preceding section, we discussed the concepts that are important in thinking about heat and its interactions with other kinds of energy and with materials. Now we will study these same examples of heat transfer quantitatively, and how heat is related to chemical reactions.

✔ **Meeting the Goals**

The unit of heat is the joule, J, which is used in all energy calculations.

First we must settle on a unit of heat to use in our calculations. Heat is a form of energy transfer, and therefore the units of heat (q) should be the same as the units of energy. In chemical systems, this energy unit is known as the **joule**, abbreviated J. Other units of energy that may be familiar to you, but which are not used in this book, are the calorie, the kilocalorie or Calorie (1,000 calories), and the erg.

One joule is equivalent to twice the energy contained in a one-kilogram object that is moving at a rate of one meter per second. This is an unfamiliar picture to us, but we can visualize a joule in some other more common situations. One joule is the amount of energy needed to light a dim 10-watt bulb for one tenth of a second, or to lift a simple 5-lb bag of flour a few inches into the air. In terms of heat, one joule of energy will warm one ounce of water (about 30 mL) by less than one hundredth of a degree Celsius. As all of these examples show, one joule is a very small amount of energy!

Because the joule is so small, we often carry out calculations with the kilojoule (kJ), which equals 1,000 joules. One kilojoule of energy warms an ounce of water by 8 degrees Celsius. One kilojoule is the amount of energy needed to light a 100-watt bulb for 10 seconds, or to lift 100 kilograms by a distance of one meter. A healthy human adult needs about 9,000 kilojoules of energy each day.

To simplify our study of heat transfer, we begin with an isolated system. Recall that an isolated system is a system that exchanges no energy with the surroundings. This means that any heat transfers that we observe are *among the components of the system. Because no heat is exchanged with the surroundings, if one part of the system gives off heat ($q < 0$), then another part or parts of the system must take in the heat ($q > 0$). In such a system the sum of the heat transfers is equal to zero:*

$$\Sigma q = 0$$

We use the Greek letter Σ (sigma, related to a capital S in English lettering) to say that we "find the sum." In the case of a heat transfer, this equation says that the sum of the individual heat changes must equal zero.

Let's take the simplest example—a system where a hot object, labeled 1, is placed in contact with a cooler one, labeled 2. If no heat is exchanged with the surroundings, then these objects exchange heat with each other only. Our summation is then:

$$q_1 + q_2 = 0 \qquad \text{or} \qquad q_1 = -q_2$$

If object 1 starts at a higher temperature than object 2 then $q_1 < 0$ and $q_2 > 0$. The heat lost when object 1 cools down appears as an equal amount of heat gained by object 2 as it warms up.

We know that when heat transfer occurs, there is a corresponding temperature change. But are heat change (q) and temperature change the same thing? Suppose that the temperature change for object 1 is equal to -5.2 K. Will the temperature change of object 2 be equal to $+5.2$ K? In most cases it will not! Two factors affect the temperature change for two objects in thermal contact. One concerns the masses of the different objects, the other concerns their compositions.

Let's say one object has a greater amount of material than the other. For example, when a few mL of boiling water (373 K) is added to a liter of water at a lower temperature (say, 303 K), both samples of water will soon attain the same temperature. But the larger mass of the lower temperature water ensures that the final temperature is not a simple average of 373 K and 303 K. Instead, the final temperature is much nearer to 303 K than to 373 K.

Just as the mass of an object affects temperature change during heat transfer, the composition of the object is important as well. Different substances respond differently to heat transfers. The same heat transfer q may cause equal masses of two different substances to change temperature by very different amounts. The response of an object to heat is known as the **heat capacity** of the object.

An object with a large heat capacity changes temperature less than does an object of equal mass that has a smaller heat capacity, given the same intake or output of heat. Different substances have different heat capacities. For example, if we put the same amount of heat into equal masses of water and of ethanol, we will discover that the temperature of the ethanol rises faster than that of the water. Ethanol has a smaller heat capacity than does an equal mass of water.

The heat capacity of an object is an extensive property, so it depends on the mass of the object, as we discussed in Chapter 4. A smaller mass has a smaller heat capacity than does a larger mass of the same substance. For example, if we put an equal amount of heat into two *unequal* amounts of water, the smaller volume of water will undergo the greater change in temperature. A smaller volume of water requires less heating to reach its boiling point than does a larger volume. This is illustrated in Figure 11-7.

We generally do not report extensive properties of substances. We work instead with intensive properties, because they can be applied to different samples of the same substance. We can put heat capacity into an intensive form that does not depend on the sample

Figure 11-7 ▶ A small pot of water will boil sooner than a large pot of water given the same amount of heat, because the temperature of the smaller sample will rise more quickly than the temperature of the larger sample. The heat capacity of the smaller sample is less than the heat capacity of the large sample. (Courtesy of Kathleen Civetta.)

size by calculating the heat capacity *per gram* of a substance. The heat capacity per gram of a substance is known as the **specific heat capacity,** and is abbreviated c_s. The specific heat capacity is an intensive property and is often referred to as simply the specific heat.

The specific heat capacity varies from substance to substance and can be determined from experiments. The specific heat capacities of some important substances are given in Table 11-4.

Using specific heat capacity, we are now able to quantify the relationship between heat and temperature change. We account for the temperature change that accompanies heat transfer by using a change in Kelvin temperature. The specific heat capacity is written as c_s, and the amount of heat transferred is q. We can calculate the amount of heat transfer needed to change the temperature of an object with mass m by using the following equation.

$$q = c_s \, m \, \Delta T$$

We can rewrite this in words:

$$\text{heat} = \text{specific heat capacity} \times \text{mass} \times \text{temperature change}$$

We determine a specific heat capacity by measuring the change in temperature ΔT when a quantity of heat q transfers into or out of an object that has a mass m:

$$c_s = \frac{q}{m \, \Delta T}$$

$$\text{specific heat capacity} = \frac{\text{heat}}{\text{mass} \times \text{temperature change}}$$

Table 11-4 Selected Specific Heat Capacities (298 K, 1 atm)

Substance	Specific Heat ($J \, K^{-1} \, g^{-1}$)
Water (liquid)	4.18
Acetic acid ($C_2H_4O_2$)	2.05
Copper	0.39
Gold	0.129
Oxygen	1.00
Argon	0.518
Ethanol C_2H_6O	2.42
Toluene (C_7H_8)	1.69
Mercury	0.138
Aluminum	0.900
Nitrogen	1.08
Helium	4.96

Finally, if we know the quantity of heat being transferred, the mass, and the specific heat of the object, then we can calculate the expected temperature change:

$$\Delta T = \frac{q}{c_s\, m}$$

✔ **Meeting the Goals**

We calculate the quantity of heat transferred into or out of a substance by obtaining the product of the substance mass, its specific heat capacity, and the change in temperature: $q = mc_s\Delta T$, where $\Delta T = T_{final} - T_{initial}$.

Note that in this equation we see our "common sense" notion of temperature and heat transfer. The temperature change varies directly with q; a larger value of q means ΔT is greater. The temperature change is inversely dependent on m and c_s; a larger sample or a sample with a greater specific heat both undergo a smaller ΔT than do samples with smaller values for m and c_s. More heat means more temperature change, but remember that temperature change is affected by mass. A higher mass or a higher heat capacity means a smaller change in temperature.

Example 11.5   Relating q to specific heat

Determine the amount of heat required to raise 1.00 g each of aluminum, copper, and gold by 10.0 K. How does the trend in the heat parallel the specific heat?

Thinking Through the Problem We want to calculate q for each of these substances. We have the mass and the temperature change, and can get the specific heat capacities from Table 11-4. Apply the equation $q = c_s m\ \Delta T$, where the mass is 1.00 g and the temperature change is $+$ 10.0 K.

Answer For aluminum: $q = \dfrac{0.900\ J}{1\ g\ K} \times 1.00\ g \times 10.0\ K = 9.00\ J$

For copper: $q = \dfrac{0.39\ J}{1\ g\ K} \times 1.00\ g \times 10.0\ K = 3.9\ J$

For gold: $q = \dfrac{0.129\ J}{1\ g\ K} \times 1.00\ g \times 10.0\ K = 1.29\ J$

We see that the greater the specific heat, the greater the heat required for the 10.0-K rise in temperature.

Example 11.6   Calculating a mass for a heat and temperature change

A teacher wants to show students the temperature change associated with 1.00 J of heat. He decides to use some warm water, then lets the students "feel" 1.00 J of heat as it flows into their hands while the water cools from 45.0 °C to 37.0 °C. What mass of water is needed?

Thinking Through the Problem We want to calculate the mass m. We need to look up the specific heat of water. The temperature change in this case is *negative*, as $\Delta T = T_{final} - T_{initial} = -8.0$ °C. Because the size of a degree Celsius is the same as the size of a Kelvin degree, the change in temperature in Kelvin degrees is -8.0 K. The water loses heat ($q = -1.00$ J < 0). We rearrange the equation for heat to $m = q/c_s \, \Delta T$.

Answer

$$m = \frac{-1.00 \text{ J}}{4.18 \, \dfrac{\text{J}}{\text{g K}} \times (-8.0 \text{ K})} = 0.030 \text{ g}$$

This is about one drop of water!

How are you doing? 11.4

(a) Determine the heat absorbed by a piece of gold metal with a mass of 0.250 gram as it is warmed from 24.0 °C to 37.3 °C.

(b) You want to cool a reaction that will give off 25,000 J of heat. What mass of water is needed if you want the maximum temperature change to be 10.0 K?

Heat Changes and Phase Transitions

We first discussed heat in connection with temperature change. For example, if 1.0 kJ of heat energy is put into a 100.0-g sample of liquid water, we observe that the temperature increases by 2.4 K. If we put the same amount of heat into a cup of water and ice, some ice melts, but there is no temperature change. Where is the added heat? To answer this question, remember that $q > 0$ means that heat enters a system while $q < 0$ means that heat leaves the system. In a phase transition that absorbs heat, $q > 0$; in a phase transition that releases heat, $q < 0$. For the melting of the ice, $q > 0$. Some of the heat that might have warmed the water is used instead to melt the ice.

REMEMBER

How much heat is required to cause a phase change? This must be determined experimentally for each substance. We will discuss heat flows at constant pressure only, which are easy to track. We can tabulate these, as in Table 11-5. Recall that the heat needed to melt a solid to a liquid is known as the *heat of fusion*, and the heat needed to

Table 11-5 Selected Heats of Fusion and Vaporization

Substance	Freezing Temperature (°C)	q_{fus} (kJ mol^{-1})	Boiling Temperature (°C)	q_{vap} (kJ mol^{-1})
Water	0.00	6.01	100.00	40.66
Acetic acid	16.6	12.09	118.2	24.39
Ammonia	-77.8	5.653	-33.43	23.351
Oxygen	-218.75	0.444	-182.97	6.82
Nitrogen	-210.0	0.720	-195.8	5.777

Note: The q values are given for the normal freezing and boiling temperatures.

boil a liquid to give a vapor is known as the *heat of vaporization*. Both are usually indicated in units of kJ per mole, and the values shown are valid only at the temperature at which that phase change occurs under normal pressures (1 atm).

If *n* moles of a substance melt under normal conditions, we can calculate the heat change *for the substance* from the equation $q = nq_{fus}$. This equation tells us that to accomplish the transition of the substance from a solid to a liquid we must put heat into the substance. This heat must be supplied by another part of the system or by the surroundings.

Example 11.7   Determining the direction of heat flow in melting ice

A 30.0-g ice cube requires 10.0 kJ of heat to melt. What is the direction of heat flow for the ice in this system? Does heat flow into or out of ice when it melts?

Thinking Through the Problem We need to say something about the change *q* in ice as it melts. Because $q > 0$, heat is put into the ice.

Answer Heat flows into the ice when it melts.

Example 11.8   Calculating temperature change for heated water

If another 10.0 kJ of heat is put into the water resulting from the melted ice cube in the previous example, the water will increase in temperature. Determine the temperature change you expect if 10.0 kJ of heat is put into 30.0 grams of water.

Thinking Through the Problem This problem requires that we calculate ΔT. We use the equation for relating heat, temperature change, mass, and specific heat: $q = c_s\, m\, \Delta T$.

We look up c_s from Table 11-4 and use *m* and *q* given in the problem. Next, rearrange the equation to solve for the temperature change: $\Delta T = \dfrac{q}{c_s \times m}$. We also convert kJ to J.

Working Through the Problem We have $q = 10.0$ kJ or 10.0×10^3 J and 30.0 grams of ice. We find:

$$\Delta T = \frac{10.0 \times 10^3\ \cancel{J}}{4.18\ \dfrac{\cancel{J}}{g\ K} \times 30.0\ \cancel{g}} = 79.7\ K$$

Answer The temperature change is 79.7 K.

✔ Meeting the Goals

When heat is transferred into a substance, the motion of the atoms of that substance increases. When this motion increases sufficiently, the substance changes from a solid to a liquid or from a liquid to a gas. When heat is transferred out of a substance, motion decreases. Sufficient decrease in motion converts a gas into a liquid, and a liquid into a solid.

Ice must absorb heat from its surroundings in order to melt. (J&M Studios/Liaison.)

The last two examples indicate that heat can be absorbed by a mixture of water and ice in two ways: by the melting of the ice and by the warming of the liquid water. The results of these experiments are graphed in Figure 11-8.

Figure 11-8 ▶ For the first 1,000 J of heat the T jumps to 0 °C, then it stays at 0 °C with the next 10,000 J while the ice melts. A third input of 10,000 J will warm the water to 79.7 °C, as calculated in the text.

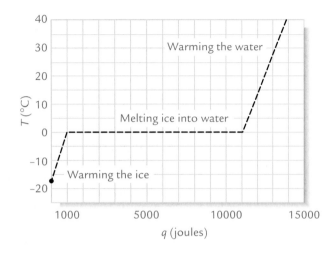

Example 11.9   Heat of vaporization

It is convenient to cool things by placing them in contact with a substance that vaporizes. Two excellent substances for this are liquid nitrogen and liquid ammonia. Determine the heat change for the vaporization of 100.0 grams of liquid nitrogen and of 100.0 grams of liquid ammonia.

Thinking Through the Problem We first need to calculate the heat change when 100.0 g of liquid nitrogen evaporates. Because this is a phase change, we need the heat of vaporization from Table 11-5. Since the units of q_{vap} are kJ/mol, we will first need to convert the gram amount given to a mole amount. We calculate heat change for the substance evaporating with the equation:

$$q = nq_{vap}.$$

Working Through the Problem

$$100.0 \ \text{g} \ \text{N}_2 \times \frac{1 \ \text{mol} \ \text{N}_2}{28.014 \ \text{g} \ \text{N}_2} = 3.570 \ \text{mol} \ \text{N}_2$$

$$q = nq_{vap} = 3.570 \ \text{mol} \ \text{N}_2 \times 5.777 \ \frac{\text{kJ}}{\text{mol}} = 20.62 \ \text{kJ}$$

We can perform the same calculation for 100.0 grams of liquid ammonia. This gives 5.872 moles and a q of 137.1 kJ.

Answer 20.62 kJ for nitrogen and 137.1 kJ for ammonia.

✔**Meeting the Goals**

The heat of fusion (q_{fus}) is the heat necessary to increase motion to the point where the substance changes phase from solid to liquid. The heat of vaporization (q_{vap}) is the heat necessary to increase motion to the point where the substance changes phase from liquid to gas. Because both q_{fus} and q_{vap} have units of J mol^{-1} or kJ mol^{-1}, q is calculated as: $q = nq_{fus}$ or $q = nq_{vap}$, where n is moles.

How are you doing? 11.5

Calculate the amount of heat needed to warm 100.0 grams of acetic acid from 25.0 °C to the boiling point. Compare this to the amount of heat needed to vaporize the same sample of acetic acid.

Heat and Chemical Change

So far in this section we have seen how heat is related quantitatively to temperature change and to phase change. Now we will explore how heat is related to chemical change (chemical reactions). Excellent examples of this relationship are found in the portable "cold packs" and "hot packs" often used in sports medicine. Simply squeezing or bending one of these packs results in the release or absorption of a great deal of heat.

We know that the heat released from a hot pack or absorbed by a cold pack must come from or go to *somewhere*. The "somewhere" is a chemical reaction. For example, a common heat pack uses the oxidation of iron to produce heat.

$$4\ Fe + 3\ O_2 \longrightarrow 2\ Fe_2O_3$$

When this reaction occurs, we observe the release of heat. If all of this heat is used to warm a sample of water, then the water will rise in temperature and $q_{water} > 0$. Since we know $q_{water} > 0$, then $q_{reaction} < 0$. We conclude that the reactants 4 Fe + 3 O$_2$ lose heat in going to the products 2 Fe$_2$O$_3$. This reaction is exothermic.

There are also chemical reactions that are endothermic. A portable cold pack relies on a dissolution reaction using NH_4NO_3. This spontaneous reaction absorbs heat from the surroundings, and we detect this absorption as cooling.

✔ **Meeting the Goals**

When heat is transferred into a chemical reaction that absorbs heat, the molecular motion of the reactants may or may not change. But the types of chemical substances *do* change during the course of the reaction. The reactants turn into the products.

Example 11.10   **Exothermic and endothermic reactions**

Determine whether the following chemical reactions are endothermic or exothermic, and indicate where the heat comes from or goes to in each case.

(a) A cigarette lighter is switched on, creating a small, hot flame.
(b) When a solution of sodium chloride and CoCl$_2$ is heated in a beaker of hot sand, a color change from blue to red occurs.

Thinking Through the Problem We need to determine whether the reactions are endothermic or exothermic, and describe the heat flow in each case. In both instances a chemical reaction occurs. We should consider whether the surrounding substance is heated or cooled as a result of each chemical reaction. If $q > 0$ for the surroundings, then $q < 0$ for the reaction, which is exothermic. If $q < 0$ for the surroundings, then $q > 0$ for the reaction, which is endothermic.

Answer

(a) The flame is a heat source, which warms the surrounding air. The reaction is exothermic.
(b) Because we observe a color change as the solution is heated, the reaction requires heat and is endothermic. There is probably a chemical reaction between the NaCl and the CoCl$_2$. The heat comes from the surrounding sand.

11.6

Describe the heat transfer that occurs when sulfuric acid and sugar are mixed, creating a cloud of steam and a mass of brittle, hot, and graphite-like solid.

We apply the same line of reasoning used for phase changes to quantitatively describe heat changes that occur during chemical reactions. For example, if we add a small amount of a catalyst to a common 3% solution of hydrogen peroxide, we observe bubbling and an increase in the temperature of the solution from the decomposition reaction:

$$2\ H_2O_2 \longrightarrow 2\ H_2O + O_2$$

KEY IDEA 11.7

The solution warms because heat has been transferred into the water. Where did this heat come from? *The heat came from the chemical reaction. We say that the decomposition of hydrogen peroxide releases heat as it proceeds. It is an exothermic reaction.*

We can use the heat associated with a reaction in a variety of calculations. In this book, we will look at how much heat is given off or taken up per mole of one of the reactants: $q_{substance}$. Thus, the units are J mol^{-1}, or kJ mol^{-1} of the substance. This value then lets us calculate $q_{reaction}$ if we know how many moles of a reactant are used up:

$$q_{reaction} = \text{moles of substance} \times q_{substance}$$

The value of $q_{substance}$ for the decomposition of hydrogen peroxide can be determined experimentally to be -191.17 kJ per mole of hydrogen peroxide. Knowing this, we can calculate how much heat is given off by any amount of hydrogen peroxide. For example, a one-pint container of hydrogen peroxide contains 470.0 mL of solution made of 15.0 g of H_2O_2 and approximately 480.0 g of water. To determine $q_{reaction}$ for this amount of hydrogen peroxide, we first convert grams to moles of H_2O_2 and then calculate the value of $q_{reaction}$:

$$\text{mol } H_2O_2 = 15.0\ \text{g } H_2O_2 \times \frac{1\ \text{mol } H_2O_2}{34.0138\ \text{g } H_2O_2} \approx 0.4409\ \text{mol } H_2O_2$$

$$\begin{aligned}
q_{reaction} &= n\ H_2O_2 \times q\ H_2O_2 \\
&= 0.4409\ \text{mol } H_2O_2 \times -191.17\ \text{kJ mol}^{-1} \\
&\approx -84.3\ \text{kJ}
\end{aligned}$$

The 84.3 kilojoules of heat released by the reaction enters the water. Thus, q for the water is $+84.3$ kJ. This allows us to calculate the *temperature* change for the water:

$$q_{water} = c_s \times m \times \Delta T$$

$$\Delta T = \frac{q}{c_s \times m}$$

Because we have $q = +84,300$ J and 480.0 g of water, we find:

$$\Delta T = \frac{84,300 \text{ J}}{4.184 \dfrac{\text{J}}{\text{g K}} \times 480.0 \text{ g}} = 42.0 \text{ K}$$

Even this simple reaction can yield a large temperature change!

Example 11.11   Calculating heat of reaction

As we have seen, a hand-warmer can be made from powdered iron. When the iron is exposed to the air, it is converted to Fe_2O_3 and gives off heat. The value of q per mole of iron is -411 kJ mol^{-1}. Determine the heat given off by the reaction of 25.0 grams of iron.

Thinking Through the Problem The statement of the problem gives us the amount of heat produced when one mole of iron reacts. We must determine the heat released by the reaction of 25.0 grams of iron. We calculate that 25.0 grams of iron is equal to 0.447 mole of Fe.

Answer We use the mole amount of Fe to calculate the heat for this reaction:

$$q_{reaction} = n_{Fe} \times q_{Fe}$$
$$= 0.447 \text{ mol Fe} \times -411 \text{ kJ mol}^{-1}$$
$$= -184 \text{ kJ}$$

PROBLEMS

18. Convert 25,500 J to kJ.

19. Convert 15.6 kJ to J.

20. The specific heat capacity of mercury is 0.139 J g^{-1} K^{-1}. Determine the amount of heat required to warm 0.050 gram of mercury in a common thermometer from 22 °C to 39 °C. Compare this to the heat required to warm the same amount of water by the same number of degrees.

21. The specific heat capacity of ice (2.09 J g^{-1} K^{-1}) is not the same as the specific heat capacity of liquid water (4.184 J g^{-1} K^{-1}). If we put 100 J of heat into liquid water and 100 J of heat into ice, which will undergo the larger temperature change?

22. Discussion question: Specific heat capacities and melting points of common metals are listed in the table at right.

(a) If you have 1.00 kg of each metal in solid form right at its melting point, determine the heat given off as it cools from that temperature

Metal	Specific Heat $\left(\dfrac{\text{J}}{\text{g} - \text{K}}\right)$	T_{melting} (K)
Cr	0.483	1900
Fe	0.449	1535
Mo	0.256	1620

to the boiling point of water (373 K). Assume that the specific heat does not vary with temperature.

(b) If the heat released in part (a) is absorbed in water at 100 °C by allowing the water to boil, how many moles and how many grams of water are boiled in the process of cooling the metal?

(c) Compare the answers you obtained for the three metals. Is there any trend that you can see related to the column or row of the transition series?

23. Heat capacity can be expressed in terms of joules per gram per degree Kelvin (specific heat capacity) or Joules per mole per degree Kelvin (molar heat capacity). Comparing the molar heat capacity of compounds can be very informative. Complete the following table, then answer Problem 24.

Substance	Molar Mass (g mol^{-1})	Specific Heat Capacity (J g^{-1} K^{-1})	Molar Heat Capacity (J mol^{-1} K^{-1})
C (diamond)		2.53	
Si		0.89	
Ge		0.36	
Sn		0.22	
Pb		0.13	

24. Not all substances follow the trend in Problem 23. Compare the molar heat capacities of the substances in the following table. Suggest a difference between these substances and those discussed in the previous problem.

Substance	Lewis Structure	Molar Mass (g mol^{-1})	Specific Heat (J g^{-1} K^{-1})	Molar Heat Capacity (J mol^{-1} K^{-1})
CH_4 (g)			2.2	
C_2H_6 (g)			1.76	
C_3H_8 (g)			1.67	
C_4H_{10} (g)			1.68	

25. The heats of fusion of common metals are listed at right.

(a) If you have 1.00 kilogram of each metal in liquid form right at its melting point, determine the heat given off as it solidifies.

(b) If the heat released in part (a) is transferred to water at 100 °C by allowing the water to boil, how many moles and how many grams of water are boiled in the process of cooling the metal?

Metal	q_{fus} (kJ mol^{-1})
Cr	21
Mo	28
Fe	13.8

26. Argon is often used in window glass to enhance its ability to insulate thermally. Determine whether this enhancement is due to argon's specific heat capacity. Determine the temperature change if

PRACTICAL B

How does the oxygen content of a fuel affect the energy it can provide?

A few winters ago, an experienced driver ran out of gas on the highway as he drove home from work in his beloved 1968 Bonneville. The old car was in very good condition, but its fuel gauge did not work. Still, the driver used this car only to drive to and from work, and he had determined that he could always drive ten days back and forth before he needed to refuel. He always used the same gasoline from the same station. Yet now, only nine days after filling the tank, the engine sputtered as the gas ran out!

What had gone wrong? Simply put, the driver unknowingly had filled his car not with "regular" gasoline, which is a mixture of hydrocarbons, but with "oxygenated" gasoline required in his area to reduce the formation of photochemical smog in the winter. The new gasoline, though it cost a little more, actually provided less energy when burned. Thus, the gas mileage dropped about 10% and the car ran out of fuel early.

We can get an idea of the difference in energy content in different fuels by comparing the heat given off when different substances are burned. This heat is called the *heat of combustion*. Gasoline, whether oxygenated or not, is a very complex mixture of substances. Representative hydrocarbons and common oxygen enhancers are given below.

For each of these fuels, calculate the heat given off by the combustion of one gram and one liter of the substance. What differences do you see?

It is clear at this point that the oxygenated fuel additives will give less energy per liter. Why is this so? Consider the amount of heat released when one liter, one kilogram, and one mole of each fuel is burned with oxygen to give carbon dioxide and water (you will need to balance the equations for the reactions in each case). The additional cost of oxygenated gas is still worthwhile. Why might this be so?

Substance (all liquid)	Density (g L^{-1})	Heat of Combustion (kJ mol^{-1})
Toluene, C_7H_8	866	−3909
Heptane, C_7H_{16}	684	−4817
Ethanol, C_2H_6O	789	−1366
MTBE, $C_5H_{12}O$	740	−3182

one *mole* of argon and one *mole* of nitrogen are warmed with the input of 500 joules of heat.

27. If you want to cause a material to undergo a large temperature change when it is placed in a preheated chamber, should the material have large or a small specific heat capacity? Do some sample calculations to explain your answer.

28. (a) If 2,300 joules of heat is released by a 25.00-g sample of marble (specific heat capacity 0.82 J g^{-1} K^{-1}), then determine the temperature change of the marble.

(b) Sand has a smaller specific heat capacity than does marble. If the same amount of heat (2,300 joules) is released by a 25.00-g sample of sand, will the temperature change be the same, greater than, or less than the value for part (a)? Explain your answer.

29. The heats of reaction ($q_{reaction}$) to form certain oxides from pure metal and oxygen are given below.

Metal	q (kJ mol^{-1})
Cr_2O_3	−1139.7
MoO_3	−754.5
Fe_2O_3	−822

(a) Which of these oxides can be formed in an endothermic reaction? An exothermic reaction?

(b) For those oxides formed in an exothermic reaction, calculate the heat given off by the formation of 100 grams of the oxide. Note that iron oxide is formed in a portable hand-warmer reaction.

(c) Compare the heat given off in forming 100 grams of Fe_2O_3 with that released in forming the other oxides. Comment on why iron may be the preferred hand-warmer material.

30. In the table for Problem 29, we see that the heat associated with the formation of one mole of Fe_2O_3 is -822 kJ. But in Example 11.11 we are told that this reaction has a heat effect of -411 kJ

per mole of iron metal. Comment on the reason why these numbers are different.

31. Food is a complex mixture of substances, but we can get some understanding of the heat content of different types of food by comparing certain substances that are good examples of carbohydrates, fats, and proteins.

(a) For the combustion reaction of each of the substances in the following table, determine the heat released per gram of that substance.

Food Type	Representative	Chemical Formula	$q_{reaction}$ (kJ mol^{-1})
Carbohydrate	glucose	$C_6H_{12}O_6$	-2753
Protein	alanine	$C_3H_7NO_2$	-1377
Fat/oil	methyl oleate	$C_{19}H_{36}O_2$	-10860

(b) The heat released in a combustion reaction is related to the Calorie content of the food. One food Calorie is equal to 4.184 kJ. Determine how many Calories are available by the combustion of 100. grams of methyl oleate, alanine, and glucose.

Chapter 11 Summary and Problems

◀ Review with **Web Practice**

VOCABULARY

energy	The ability to do work or produce heat. Energy comes in many different forms, and energy conversion involves changing energy from one form to another.
heat	Heat is associated with thermal energy; it is usually connected to temperature changes.
system	The part of the universe that is under study; it is often isolated from the surroundings.
surroundings	The part of the universe that is not under direct study.
isolated system	A special system that exchanges nothing, not even heat, with its surroundings; the heat lost by one part of the isolated system must be gained by the other part of the system.
endothermic	A process is endothermic when it absorbs heat from the surroundings such that the heat change q for the system is greater than zero.
exothermic	A process is exothermic when the components lose heat in going to the products, such that the heat change q for the system is less than zero.
heat of vaporization	The value of the heat needed to vaporize a liquid substance to a gas; it is usually expressed in units of energy per mole.

heat of fusion	The quantity of heat needed to melt a substance to a liquid; it is usually expressed in units of energy per mole.
joule	A basic unit of energy, including thermal energy and heat.
heat capacity	The ratio of heat supplied to an object divided by the actual temperature change.
specific heat capacity	The heat capacity for a substance per one gram of a substance; it is an intensive property of that substance.

SECTION GOALS

What does it mean when we call a system isolated?	A system is said to be isolated when it does not exchange anything with other systems or with the surroundings.
How are temperature changes and molecular motion related? What do these have to do with heat?	If we increase the temperature of a system, the molecules and atoms move faster, on average. Therefore, adding heat to a system increases temperature and motion, if no chemical or phase change occurs.
How do we describe heat flows?	We describe heat flows with the value q. If q is greater than zero, heat flows into a system; if q is less than zero, heat flows out of the system.
What are relationships of heat flows to the terms "endothermic" and "exothermic"?	The word *endothermic* indicates that heat moves into a system and describes systems for which $q > 0$. The word *exothermic* indicates that heat moves out of a system and describes systems for which $q < 0$.
How is heat related to phase changes?	The change from one physical state to another requires a heat transfer that either increases motion (endothermic process) or decreases motion (exothermic process). Solids are converted to liquids, liquids are converted to gases, and solids are converted to gases when heat is transferred to them. Thus, melting, evaporation, and sublimation are endothermic processes. The opposite processes—freezing, condensation, and deposition—are exothermic processes.
What are the units of heat?	The units of heat are joules (J). They are the units used in all energy calculations.
Different materials undergo a different temperature change with the same input of heat. How does the physical property of specific heat capacity characterize the relationship between a material, a temperature change, and a heat transfer?	The specific heat capacity, which has units of joules per gram per K, indicates how the temperature of a substance is affected by heat. Since it is an intensive property, the specific heat capacity is independent of the mass of the object.
How is heat calculated for a substance that undergoes a temperature change without changing phase?	We calculate the quantity of heat transferred into or out of a substance by obtaining the product of its mass, its specific heat capacity, and the change in temperature: $q = c_s m\ \Delta T$, where $\Delta T = T_{\text{final}} - T_{\text{initial}}$.
What is the relationship between heat and phase changes?	When heat is transferred into a substance, the atoms and molecules of that substance vibrate more. When this motion increases sufficiently, the substance changes from a solid to a liquid or from a liquid to a gas. When heat is transferred out of a substance, motion decreases. Sufficient decrease in molecular motion converts a gas into a liquid, a liquid into a solid.
What is the relationship between heat and chemical reactions?	When heat is transferred into a chemical reaction that absorbs heat, the atomic motion may or may not change. But the types of chemical substances *do* change during the course of the reaction.

PROBLEMS

32. Indicate whether the following processes are endothermic or exothermic.

(a) A flame burns.
(b) Acetone (the major ingredient in nail polish remover) evaporates from a bench-top spill.
(c) A lake slowly freezes.
(d) An electrical drill is used to drill a hole through hard wood. In the process the wood is scorched.

33. Analyze the following processes by indicating where heat is created and where it flows.

(a) Slow decomposition of leaves under a thick blanket of snow creates a liquid layer very close to the ground.
(b) The rapid crystallization of sodium acetate from a "supersaturated" solution is the basis of one kind of portable heat pack.

34. Indicate whether the following chemical transformations are endothermic or exothermic.

(a) When heated to a high temperature, a substance like cyclohexane, C_6H_{12}, loses hydrogen gas to give cyclohexene, C_6H_{10}.
(b) During the synthesis of ammonia from nitrogen and hydrogen gas, it is necessary to cool the reaction chamber with water to keep the temperature from climbing too high.
(c) A mixture of sugar and water, stable for many hours, becomes a solution when it is placed over a gentle flame.
(d) When a solution of the salt $Ce_2(SO_4)_3$ is warmed, solids $Ce_2(SO_4)_3$ begin to appear.
(e) On cooling a solution of sodium acetate, $NaC_2H_3O_2$, solids suddenly appear and the temperature climbs rapidly.

35. A small portable lighter consists of butane trapped inside a small chamber. The lighter can release a great deal of heat when it is turned on. Yet the butane is safe to transport as long as its container remains intact. Why won't the butane suddenly give off heat while it sits in its container?

36. A plastic beaker that has a mass of 75 grams is at room temperature. We place 2,500 joules of heat into the beaker by some means. The temperature of the beaker increases by 5.0 K.

(a) If we have a beaker of a different material with twice the specific heat capacity of the original beaker, how much heat is needed to cause a 5.0-K change in the temperature?
(b) If we fill the original beaker with some water, will we need more or less heat to cause a 5.0-K rise in temperature?

(c) The original beaker is broken up by cracking off the top several inches. We now have a beaker that has a mass of only 50 grams. We place 2,500 joules of heat into the beaker. What will be the temperature change of the beaker now? (Hint: Try to do this without calculating the specific heat capacity.)

37. Many chemical reactions are carried out in a device known as a calorimeter, where we measure the temperature change and then calculate the heat change in the reaction.

When the reaction $HCl + NaOH \rightarrow NaCl + H_2O$ is carried out in 50.0 grams of water, the reaction of 0.36 gram of HCl (0.010 mol) with 0.40 gram of NaOH (also 0.010 mol) causes a temperature increase of 2.7 K in the water.

(a) What is the heat change for the water?
(b) What is the heat change for the reaction?
(c) What is the heat change we expect if we had 0.020 mole of NaOH and 0.020 mole of HCl?
(d) What would be the temperature change for the water if 0.020 mole of NaOH and 0.020 mole of HCl is reacted in 50.0 grams of water?

38. When substances burn, the heat released often warms up the surroundings. The combustion of methane has a q of -801 kJ per mole of methane:

$$CH_4\ (g) + 2\ O_2\ (g) \longrightarrow CO_2\ (g) + 2\ H_2O\ (g)$$

If *all* of the heat released in burning 1.00 gram of methane is transferred to heating 1.000 kilogram of water (specific heat capacity $c_s = 4.184\ J^{-1}\ g^{-1}\ K^{-1}$), by how many degrees will the water heat up?

39. Calculate ΔT for a 25.00-g sample of Ar in a process where $q = -5.00$ J.

40. Calculate ΔT for a 5.00-g sample of Cu in a process where $q = 10.0$ J.

41. Calculate q for a 5.00-g sample of toluene that is heated from 10.0 °C to 25.0 °C.

42. Calculate the heat transferred, q, for a 50.0-g sample of Al that is heated from 28.0 °C to 38.0 °C.

(a) How many joules of heat are transferred for 100.0 grams of Al for the same temperature change?
(b) How many joules of heat are transferred for 25.0 grams of Al, assuming the same temperature change?

43. Define heat of vaporization and heat of fusion.

44. Calculate the heat transferred, q, when 25.0 grams of acetic acid is converted from a liquid to a gas.

45. Calculate q when 10.0 grams of acetic acid is converted from a solid to a liquid.

46. The following sentence describes a physical process that changes the temperature of a substance. For each process, answer the given questions.

Liquid water in a refrigerator cools down.

(a) Does the water absorb or release heat?
(b) What is the sign of q for the water?
(c) Where does the heat go to or come from?

47. Specific heat capacities and melting points of common metals are listed below.

Metal	Specific Heat $\left(\dfrac{J}{g - K}\right)$	$T_{melting}$ (K)
Ni	0.439	1455
Ag	0.225	961
Au	0.130	1064

(a) If you have 1.00 kg of each metal in solid form right at its melting point, determine the heat given off as it cools from that temperature to the boiling point of water (373 K). Assume that the specific heat does not vary with temperature.
(b) If the heat released in part (a) is absorbed in water at 100 °C by allowing the water to boil, how many moles and how many grams of water are boiled in the process of cooling the metal?

48. The heats of fusion of common metals are listed below.

Metal	q_{fus} (kJ mol^{-1})
Ag	11.1
Ni	17.2
Au	12.8

(a) If you have 1.00 kg of each metal in liquid form right at its melting point, determine the heat given off as it solidifies. Assume that the specific heat does not vary with temperature.
(b) If the heat released in part (a) is transferred to water at 100 °C by allowing the water to boil, how many moles and how many grams of water are boiled in the process of cooling the metal?

49. The heats of the reaction to form certain oxides from pure metal and oxygen are given below.

Metal	q (kJ mol^{-1})
Ag_2O	−30.6
NiO	−239.7
Au_2O_3	+80.7

(a) Which of these oxides can be formed in an endothermic reaction? An exothermic reaction?
(b) For those oxides with an exothermic reaction, calculate the heat given off by the formation of 100. g of the oxide. Note that iron oxide is formed in a portable hand-warmer reaction.
(c) Compare the heat given off in forming 100. g of NiO with that released in forming the other oxides. Comment on whether nickel may be the preferred hand-warmer material.

50. Calculate the heat required to bring 25.0 grams of ethanol to its boiling point (78.4 °C) from 12.0 °C.

51. A 10.0-gram sample of water at 0.0 °C was heated to its normal boiling point.

(a) Calculate the heat required to heat this sample to its normal boiling point.
(b) Calculate the amount of heat required to convert this sample into steam at its boiling point.
(c) What is the total amount of heat required for this process?

52. How is heat capacity related to the specific heat of a substance?

53. What is the specific heat of benzene (C_6H_6) if its heat capacity per mole is 136 J (K mol)$^{-1}$?

54. What is the specific heat of carbon tetrachloride if its heat capacity per mole is 133 J (K mol)$^{-1}$?

55. The formation of C_3H_8 gives off 214 kJ of heat per mole. How much heat is given off when 15.0 g C_3H_8 is formed?

56. The formation of P_4O_{10} is highly exothermic, giving off 3036 kJ of heat for every mole of P_4O_{10} formed.

$$P_4\,(g) + 5\,O_2\,(g) \longrightarrow P_4O_{10}\,(g)$$

(a) How much heat is given off when 100.0 g P_4O_{10} are formed?

(b) How many moles of O_2 are required when 100.0 g P_4O_{10} are formed?

(c) The heat calculated in part (a) resulted from the moles of O_2 calculated in part (b). Calculate the heat per mole of O_2 for this reaction.

57. You want to cool a reaction that gives off 15,000 joules of heat. What mass of water must you use if the change in temperature cannot exceed 8.50 K?

58. The value of $q_{\text{substance}}$ for the formation of $H_2C_2O_4$ is -827.0 kJ/mol.

(a) In this an exothermic or endothermic reaction?

(b) How much heat is evolved in the formation of 10.0 grams $H_2C_2O_4$?

59. The value of $q_{\text{substance}}$ for the formation of liquid CS_2 is 89.7 kJ/mol.

(a) Is this an exothermic or endothermic reaction?

(b) How much heat is needed in the formation of 10.0 grams CS_2?

CHAPTER

12

The Atomic Nucleus: Isotopes and Radioactivity

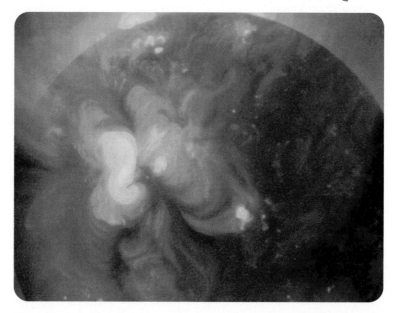

The vast amounts of heat energy coming from the sun are produced by nuclear fusion reactions, in which hydrogen atoms combine to form helium. (Solar and Astrophysics Laboratory/Lockheed-Martin Advanced Technology Center.)

12.1 The Isotopic Composition of Elements

12.2 Radioactivity: A Nuclear Reaction

12.3 The Biological Implications of Radiation

PRACTICAL CHEMISTRY What Happens When the Identities of Atoms Change?

So far in our study of chemistry, we have looked at the structures and reactions of stable atoms, seeking to develop a deeper understanding of the role of atoms in chemical substances. We have seen that the identity of an atom does not change during a chemical reaction—a carbon atom never changes into a nitrogen atom, for example. In a chemical reaction, only the *arrangement* of atoms changes. The chemical behavior we have studied depends on the interaction of valence electrons.

Now that we have developed an understanding of the role of electronic interactions in chemical behavior, we will step away from this familiar setting to learn about reactions, called *nuclear reactions*, in which the *nucleus* of the atom does change. In this chapter we will study the atomic nucleus, its makeup, its stability, and some common nuclear reactions.

408

This chapter will focus on the following questions:

PRACTICAL **A** What are some uses of isotopes in medicine?

PRACTICAL **B** The alchemists tried to make gold from lead. They couldn't do it. Can we?

PRACTICAL **C** Fission and fusion—is the difference only in the spelling?

12.1 The Isotopic Composition of Elements

SECTION GOALS

✔ Define an isotope.

✔ Write the symbol and the name for any isotope.

✔ Approximate the atomic mass of an element from the values describing the mass and natural abundance of its isotopes.

✔ Calculate the atomic mass of an element from the values of the mass and natural abundance of its isotopes.

Isotopes and Subatomic Particles

The composition of chemical elements and the nature of the atoms that make them up were important questions that chemists debated for decades. How, they wondered, do the elements relate to one another? Can one element be converted into another? Their conclusions were summarized in the early nineteenth century as part of John Dalton's atomic theory. A principle idea of Dalton's atomic theory is that the identity of an atom remains the same during a chemical reaction, but the *arrangement* of the atoms changes. A key concept of the theory is that the atoms of different elements are different from one another, and that all atoms of the same element are identical.

Is Dalton's theory correct? Well, almost. A device called the mass spectrometer, invented around 1910, enabled scientists to distinguish atoms according to their masses. This soon yielded the startling discovery that, while all the atoms of a given element show the same chemical behavior, elements are often a mixture of atoms of different masses.

Figure 12-1 shows the schematics of a mass spectrometer. The output, called a *mass spectrum*, is a graph of the masses of the atoms present in a sample. All atoms of the same element do, of course, still have the same number of protons (atomic number); this is their identity number. However, atoms of the same element do not necessarily have the same number of neutrons. In fact, most random samplings of the atoms of an element exhibit a range of mass values for the individual atoms. This difference in mass is due to a difference in the number of neutrons in the atom's nucleus.

Atoms of the same element that have different masses are called **isotopes**, derived from the Greek word meaning "same place." Isotopes are atoms with the same number of protons but with a

✔ **Meeting the Goals**

An isotope is an atom of an element which contains a different number of neutrons than do other atoms of that element.

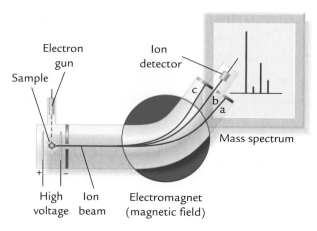

Figure 12-1 ▲ A mass spectrometer is used to measure the masses of atoms. The mass spectrum that results gives the masses of any isotopes present in a sample of a pure element.

CONNECT TO
SECTION 1.3

different number of neutrons. Neither the nuclear charge nor the number of electrons varies among the isotopes of an element. Therefore, isotopes of an element have identical chemical properties. Isotopes appear in the same place in the periodic table because the periodic table organizes elements by chemical property.

In Chapter 1, we learned that the mass of an atom is concentrated in its nucleus because the nucleus contains all of the atom's protons and neutrons. The mass number, which is the sum of the number of an atom's protons and neutrons, is used to distinguish between isotopes. A simple example of using mass number to distinguish between isotopes occurs with hydrogen, which has three isotopes. The nucleus of the most common isotope of hydrogen does not contain any neutrons at all; the mass number of this isotope is 1 and its atomic symbol is ^{1}H. Most hydrogen atoms exist as this isotope, but a sampling of many hydrogen atoms in nature shows that there are a few atoms that contain one neutron. This isotope of hydrogen is called **deuterium** (with a mass number of two). It is also possible to create hydrogen atoms with two neutrons as in **tritium** atoms (with a mass number of three). The atomic symbols for the three isotopes of hydrogen are given in Table 12-1.

Table 12-1 Isotopes of Hydrogen

Isotopes of Hydrogen	Nucleus of the Isotope protons = ● neutrons = ○	Mass Number	Charge on Nucleus (total proton charge)
^{1}H	●	1	+1
^{2}H (deuterium)	●○	2	+1
^{3}H (tritium)	●○○	3	+1

The atomic number (identifying the atom as hydrogen) is 1; it is the same for all three isotopes. The charge on the nucleus of each isotope of hydrogen is +1 because each nucleus contains only one positively charged proton. The isotopes of hydrogen differ only in the number of neutrons and, therefore, in their mass numbers.

When we talk about a particular isotope of an element, we refer to it by its mass number. *The isotopes of most elements are named by giving the element's symbol or name and the mass number.* For example, the isotope ^{12}C is called carbon-12, the isotope ^{13}C is called carbon-13, and so on. Table 12-2 lists the major isotopes of carbon, which are represented in Figure 12-2.

KEY IDEA 12.1

✔ Meeting the Goals

Isotopes are designated by mass number. Thus, an isotope of potassium with mass number 40 is designated either as ^{40}K or potassium-40.

Table 12-2 Isotopes of Carbon

Isotopes of Carbon	Name of the Isotope
^{12}C	Carbon-12
^{13}C	Carbon-13
^{14}C	Carbon-14

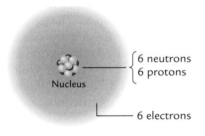
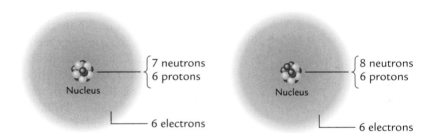

Figure 12-2 ▲ The nuclei of different isotopes contain the same number of protons but different numbers of neutrons. Here we see ^{12}C, ^{13}C, and ^{14}C.

Isotopes and Atomic Mass

The mass given in the periodic table for an element is actually a weighted average of the masses of the isotopes of that element. For example, consider the entry for hydrogen on the periodic table.

What you see	What it means
1	Atomic number = 1
H	Symbol for hydrogen = H
1.0079	Average atomic mass of the isotopes of H

The mass of hydrogen shown is 1.0079. All of the isotopes of hydrogen contribute to its atomic mass, but they don't contribute equally. As most of the hydrogen atoms in a sample are ^{1}H, we expect the average atomic mass to be very close to this value. But the atomic mass of hydrogen is *slightly* greater than this value because of the mass contributions of *small* amounts of ^2H. This example leads us to a better understanding of what is meant by the term *weighted average*. A weighted average is not a simple average, as it must consider the number of isotopes with particular isotopic masses in the mixture. The relative number of the atoms of an isotope—its **natural abundance**—in a sample of the element is given by a percentage (or decimal equivalent).

We call the relative mass of a particular isotope its **isotopic mass.** Like atomic mass, isotopic mass is defined relative to carbon-12, ^{12}C, which has an isotopic mass defined as 12. All other isotopic masses are measured relative to this. As we saw with atomic mass earlier, isotopic masses are reported independent of any particular mass unit. We impose relevant mass and counting units when necessary. For example, if we were working with mole amounts, we would use the gram for the mass unit; working on the atomic or molecular scale, we would use a mass unit known as the atomic mass unit, or amu.

We can obtain both numbers—the isotopic masses and the relative number of atoms with each isotopic mass—by means of a mass spectrometer. Figure 12-3 shows the mass spectrum of neon. The number of peaks in a mass spectrum tells us the number of different isotopes present in the element. We see three peaks in the mass spectrum of neon. Neon therefore has three isotopes. *The relative height of each peak in the spectrum gives the proportion of each isotope in the sample.* The highest peak shown in Figure 12-3 corresponds to the isotope with a mass number of 20. The other two peaks, which correspond to isotopes with mass numbers of 21 and 22, are by comparison much smaller. Thus, atoms of ^{20}Ne are significantly more abundant than atoms of ^{21}Ne and ^{22}Ne.

Table 12-3 shows the masses and natural abundances of the isotopes of some selected elements, and it illustrates two important points. First of all, the natural abundance values for each element add to 100 percent. (Their decimal equivalents add to 1.) Secondly, isotopic masses round to give us the mass number of each isotope. For example, the nitrogen isotope with a mass of 14.0031 is nitrogen-14, ^{14}N, because 14.0031 rounds to 14. The chlorine isotope with a mass of 34.9689 is chlorine-35, ^{35}Cl, because 34.9689 rounds to 35.

Example `12.1`   Writing atomic symbols

What is the symbol for the isotope of oxygen that has a mass of 17.9992?

Thinking Through the Problem The symbol for the isotope will contain the element symbol, O, and its mass number as a superscript. The mass number is the rounded off isotopic mass. 17.9992 rounds to 18.

Answer ^{18}O is the symbol for the isotope of oxygen with mass 17.9992.

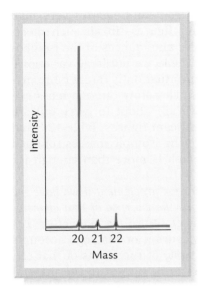

Figure 12-3 ▲ The mass spectrum of neon shows three isotopes; neon-20, neon-21, and neon-22. The highest peak in the graph corresponds to neon-20, showing that neon contains mostly neon-20.

Table 12-3 Relative Abundance Values and Masses for Naturally Occurring Isotopes of Some Common Elements

Isotopes	Natural Abundance (%)	Isotopic Mass
^{12}C	98.89	12
^{13}C	1.11	13.0033
^{14}N	99.63	14.0031
^{15}N	0.37	15.0001
^{16}O	99.759	15.9949
^{17}O	0.037	16.9991
^{18}O	0.204	17.9992
^{32}Cl	75.53	34.9689
^{37}Cl	24.47	36.9659
^{32}S	95.0	31.9721
^{33}S	0.76	32.9715
^{34}S	4.22	33.9679
^{36}S	0.014	35.9671
^{20}Ne	90.51	19.992
^{21}Ne	0.27	20.994
^{22}Ne	9.22	21.991
^{24}Mg	78.70	23.985
^{25}Mg	10.13	24.986
^{26}Mg	11.17	25.983

Example 12.2 **Finding the mass of an isotope**

Use the information in Table 12-3 to find the isotopic mass of an atom of ^{36}S.

Thinking Through the Problem This is just the opposite of example 12.1. Look for the mass that rounds to 36.

Answer The mass of ^{36}S given in Table 12-3 is 35.9671.

Precision and Atomic Masses

As we prepare to do more precise calculations on isotopic and atomic masses, it is important to note some things about the precision of these measurements. Isotopic masses are experimentally determined, but the precision of modern mass spectrometers ensures very high precision—up to 10 or more significant figures—in all such measurements. On the other hand, the natural abundances of the isotopes vary from place to place on the earth. These are not large variations, but they do mean that the abundance information is rarely more precise than five or six significant figures. Therefore, atomic mass calculations and the other calculations we are about to carry out are usually only reliable to four or five significant figures. Table 12-3 and other data for isotopic masses will have the isotopic masses rounded off to five or six significant figures, which is more than enough for our calculations.

KEY IDEA 12.3

We can often estimate an atomic mass for an element if we have the masses and the natural abundance data for the isotopes of that element. The atomic mass of the element is usually closest to the mass of the isotope that is in greatest abundance. This occurs if one isotope accounts for a large proportion, say, more than 90%, of the atoms of that element. Then we expect the atomic mass to be closest to the mass of that isotope. Example 12.3 shows how this is done.

Example 12.3 Estimating the atomic mass of an element

Estimate the atomic mass of uranium if a sample of uranium contains 0.0057% of an isotope with a mass of 234.0409, 0.72% of an isotope with a mass of 235.0439, and 99.27% of an isotope with a mass of 238.0508.

Thinking Through the Problem ⟶◗ (12.3) The key idea is that the atomic mass of an element is closest to the mass of the isotope in greatest abundance. From the percentages given, we see that more than 99% of the sample is represented by the isotope with a mass of 238.0508, ^{238}U.

Answer We expect the weighted average of the three isotopes to be very close to 238.

Example 12.4 Estimating the atomic mass of an element

Consider the information given below. Will the atomic mass of this element be closest to 23, 24, 25, or 26 grams?

Abundance	Mass
93.70%	23.985
4.13%	24.986
2.17%	25.983

Thinking Through the Problem ⟶◗ (12.3) Again, the key idea is that the isotope in greatest abundance will contribute most to the atomic mass of this element.

✔Meeting the Goals

The atomic mass of an element can be estimated when one isotope is present in more than about 90% of the abundance of that element. We expect, in these cases, that the atomic mass of the element will be close to the mass of the isotope that is in greatest abundance.

Working Through the Problem The greatest abundance is shown for the isotope with a mass of 23.985. This value is closest to 24. Looking at the other isotope masses, we note that they are each greater than 24. This will make the atomic mass of the element, which is a weighted average, slightly greater than 24.

Answer   The atomic mass of this element will be close to, but slightly greater than, 24.

How are you doing?   **12.1**

Lithium has two naturally occurring isotopes, ^{6}Li (7.42%, with a mass of 6.0151) and ^{7}Li (92.58%, with a mass of 7.0160). Predict the *approximate* atomic mass of lithium.

The Calculation of Atomic Mass from Isotope Information

According to Table 12-3, carbon consists of 99.89% of carbon-12, an isotope with an isotopic mass of exactly 12, and 1.11% of carbon-13, whose mass is 13.0033. To obtain the atomic mass of carbon, we calculate the weighted average of the masses of the two isotopes. The procedure can be outlined as follows:

(a) abundance of carbon-12 × mass of an atom of carbon-12 = contribution of carbon-12 to weighted average

(b) abundance of carbon-13 × mass of an atom of carbon-13 = contribution of carbon-13 to weighted average

(c) weighted average = (contribution of carbon-12) + (contribution of carbon-13)

The calculation is carried out in this way:

(a) $\dfrac{98.89}{100} \times 12 = 11.8668$ (contribution of ^{12}C)

(b) $\dfrac{1.11}{100} \times 13.0033 \approx 0.1443$ (contribution of ^{13}C)

(c) $11.8668 + 0.1443 \approx 12.011$ (atomic mass of C)

Generally, we express the abundance as a decimal fraction (e.g., 99.89% = 0.9989). This practice allows us to write the following formula to calculate the atomic mass of an element containing several isotopic masses:

$$M = f_1 M_1 + f_2 M_2 + f_3 M_3 + \dots$$

In this equation, the atomic mass of an element M is equal to the sum of each isotopic mass M_x multiplied by its abundance f_x, where the subscript x refers to a particular isotope. The product $f_x M_x$ determines the contribution of each isotope to the weighted average mass of the isotopes, i.e., to the atomic mass of the element.

There are several variables in this equation and you may be asked to solve for any one of them, not simply the atomic mass of an element. To do so, you must be given values for all the other variables or enough information that you can determine the values of the other variables.

Example 12.5   **Calculating the atomic mass of an element**

The element antimony has two naturally occurring isotopes. ^{121}Sb is present in 57.25% abundance and has an isotopic mass of 120.9038. ^{123}Sb is present in 42.75% abundance and has a mass of 122.9041. What is the atomic mass of antimony?

Thinking Through the Problem We start with the isotopic mass and abundance information given for antimony. ◀▷ (12.4) The key idea is that we begin by laying out our variables according to the equation for the atomic mass of an element containing several isotopes.

Working Through the Problem

$$^{121}\text{Sb} : f_1 = 0.5725; M_1 = 120.9038$$
$$^{123}\text{Sb} : f_2 = 0.4275; M_2 = 122.9041$$
$$M = f_1 M_1 + f_2 M_2$$
$$= (0.5725 \times 120.9038) + (0.4275 \times 122.9041)$$
$$= 69.22 + 52.54 = 121.76$$

Answer Based on this data, the atomic mass of this element is 121.76.

Example 12.6   **Calculating the mass of an isotope**

Boron consists of 19.78% ^{10}B, with an isotopic mass of 10.01029, and 80.22% ^{11}B. The atomic mass of boron is 10.81. What is the isotopic mass of ^{11}B?

Thinking Through the Problem We start with the mass and abundance of the other isotope of boron and with its average atomic mass. In this problem we lack information about one of the isotopic masses M_2, but we have everything else: $M = 10.81$; $f_1 = 0.1978$; $M_1 = 10.01029$; $f_2 = 0.8022$.

Working Through the Problem Before proceeding, we rearrange the equation to solve for M_2:

$$f_1 M_1 + f_2 M_2 = M$$
$$f_2 M_2 = M - f_1 M_1$$
$$M_2 = \frac{M - f_1 M_1}{f_2}$$
$$= \frac{10.81 - (0.1978 \times 10.01029)}{0.8022}$$
$$= \frac{8.829}{0.8022} \approx 11.01$$

Answer The calculated isotopic mass of ^{11}B is 11.01.

✔ **Meeting the Goals**

The atomic mass of an element is the weighted average of the masses of its isotopes. A formula that calculates the atomic mass of an element is $M = f_1 M_1 + f_2 M_2 + f_3 M_3 + \ldots$, where M is the atomic mass of the element, f_x is the abundance of the xth isotope, and M_x is the isotopic mass of the xth isotope.

Thinking about the natural abundance and isotopic mass data of an element gives important information about the atomic mass of an element. *The atomic mass must lie within the range of isotopic masses given. It cannot be less than the smallest isotopic mass nor larger than the greatest isotopic mass.* We would also expect the atomic mass of an element to be closest to the isotopic mass that is in greatest abundance. Using this information, always check to see that your calculated answer is reasonable.

Example 12.7   **Calculation of percentage abundance values**

The atomic mass of copper is 63.546. Find the percentage abundance of its two isotopes, given that their isotopic masses are 62.9298 and 64.9278. Report your answer to two decimal places.

Thinking Through the Problem There are two unknowns in this problem, but if we use two different variables, we do not have enough information to solve the problem. We will use a single variable, A, for the abundance of one of the isotopes—it does not matter which one. Because the two abundance values must add to 1, we will use $(1 - A)$ for the abundance of the second isotope. Now we must substitute these values into the general formula for atomic mass and solve for A. Remember to be careful to correctly expand the term $(1 - A)$.

Working Through the Problem

$$63.546 = 62.9298A + 64.9278 \times (1 - A)$$
$$63.546 = 62.9298A + 64.9278 - 64.9278A$$
$$63.546 = -1.9980A + 64.9278$$
$$-1.3818 = -1.9980A$$
$$0.69159 = A$$

Answer $A = 0.69159$; $(1 - A) = 0.30841$. We used the variable A with the isotopic mass of 62.9298. This means that our final answer is: 69.16% of isotope with mass 62.9298 and 30.84% of isotope with mass 64.9278.

How are you doing?   12.2

Suppose that a newly discovered element has an atomic mass of 134.0920. Calculate the fractional abundance of its two isotopes if their isotopic masses are 133.0160 and 136.0998.

PROBLEMS

1. What is an isotope? Define in terms of numbers of protons and neutrons.

2. Write the complete symbols for the neutral atoms with the following characteristics:

(a) 24 electrons and 27 neutrons

(b) a zinc atom with 36 neutrons

(c) a gallium atom with 30 neutrons

3. Which pair of the following atomic symbols represent a pair of isotopes?

(a) $^{16}_{8}X$ (b) $^{16}_{7}X$ (c) $^{17}_{7}X$ (d) $^{16}_{8}X$ (e) $^{8}_{16}X$

PRACTICAL A   What are some uses of isotopes in medicine?

Most of the isotopes we considered in the first section of this chapter are all naturally occurring. However, scientists can also make isotopes of an element in the laboratory. These are known as artificial isotopes. Many of these, and a few natural isotopes, are unstable.

Unstable isotopes emit radiation and are said to be *radioactive*. Radioactive isotopes (radioisotopes) can be absorbed by the human body and concentrate in a particular organ or system. This is often very dangerous, but radioisotopes are used in medicine to help diagnose and treat disease. For example, when the radioisotope iodine-123 is administered to a patient, the radiation given off by this isotope can be monitored and, as the iodine enters the liver, a picture of this organ can be generated, as shown in Figure 12-4. Pictures of diseased organs can be compared to pictures of healthy organs to check for abnormalities.

Iron-59, an artificially produced isotope of iron, is used in this way to study red blood cells. Ordinary samples of iron do not contain iron-59. How do we know this? Samples of naturally-occurring iron contain the four isotopes listed below. Determine the mass numbers for these isotopes to see if iron-59 is among them.

Abundance	Isotopic Mass
91.75%	55.9349
2.20%	56.9354
0.28%	57.9333
5.84%	53.9396

Thallium-201 is another radioisotope used in medicine. When thallium-201 enters the blood stream, its progress is monitored through the body's arterial system by observing the radiation it gives off. The resulting picture allows doctors to find narrowed or blocked arteries. Calculate the atomic mass of thallium from the data at right. Is thallium-201 a naturally-occurring isotope of thallium?

4. There are two naturally-occurring isotopes of bromine: $^{79}_{35}$Br has an isotopic mass of 78.9183 and an abundance of 50.54%; $^{81}_{35}$Br, has an isotopic mass of 80.9163 and an abundance of 49.46%. Determine the atomic mass of bromine.

5. There are two naturally-occurring isotopes of copper: $^{63}_{29}$Cu has a natural abundance of 69.09% and an isotopic mass of 62.9298; $^{65}_{29}$Cu has a natural abundance of 30.91%. The atomic mass of copper is 63.546. Determine the isotopic mass of $^{65}_{29}$Cu.

6. There are two naturally-occurring isotopes of silver: $^{107}_{47}$Ag has an isotopic mass of 106.90509; $^{109}_{47}$Ag has an isotopic mass of 108.9047. The atomic mass of silver is 107.868. Determine the natural abundances of the two isotopes of silver.

7. Consider the symbol $^{42}_{18}$Ar.

   (a) How many neutrons are in this atom?
   (b) Which of the following masses is closest to the mass of this atom: 41.998, 42.854, or 41.105?

8. Write the symbol for the isotope of iodine with 76 neutrons, indicating its atomic number.

9. An element has three isotopes with masses of 104.26, 103.91, and 104.98. Which of the following values cannot possibly be the atomic mass of this element?

   (a) 104.02
   (b) 103.82
   (c) 105.10

10. Discussion question: What is the symbol for each of the following?

    (a) a strontium atom with mass of 89.907
    (b) a xenon atom with a mass of 152.989
    (c) a cobalt atom with a mass of 60.992

    Look up the atomic masses of these elements. Can you estimate if the isotope you have just listed is a dominant one for this element?

11. What is the symbol for each of the following isotopes?

    (a) thorium-230
    (b) radium-224
    (c) iodine-131
    (d) zinc-72

12. Find the atomic mass of an element that has four isotopes: 0.56% of mass 83.9134, 9.86% of mass 85.9094, 7.02% of mass 86.9089, and 82.56% of mass 87.9056.

13. Identify the element in the previous problem from its atomic mass and a listing or a periodic table of elements.

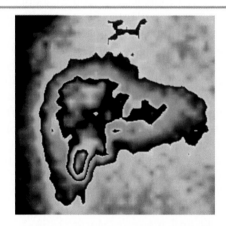

Figure 12-4 ▲ Iodine-123 is a radioactive isotope used to diagnose liver abnormalities. (CNRI/Phototake NYC.)

Abundance	Isotopic Mass
29.5%	202.9723
70.5%	204.9745

You don't need to go to a hospital to have a radioisotope in your body. Natural sources of potassium contain a very small percentage of the radioactive isotope potassium-40. Examine the following data in order to find that percentage. The mass of the last isotope of potassium is not listed. Calculate the missing isotopic mass if the atomic mass of potassium is 39.098. What is its mass number?

Abundance	Isotopic Mass
93.26%	38.9637
0.0118%	39.974
6.737%	

It is interesting to calculate how much potassium-40 (^{40}K) we ingest each day. If a typical person ingests 2.0 g of potassium, K, each day, what mass of ^{40}K is ingested? How many moles of ^{40}K is this? How many atoms of ^{40}K is this?

14. How many isotopes are shown in the mass spectrum of this element?

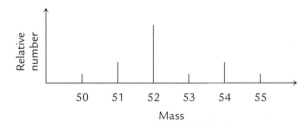

15. Which isotope in the previous problem contributes most to the atomic mass of this element?

16. Concept question: An atomic mass is calculated for the element whose isotopes are graphed in Problem 14. Based on the information in this mass spectrum, analyze the following student estimates of the atomic mass of the element.

(a) atomic mass = 55
(b) atomic mass = 49
(c) atomic mass = 50
(d) atomic mass = 44.75

12.2 Radioactivity: A Nuclear Reaction

SECTION GOALS

✔ Define radioactivity.

✔ Define and write symbols for an alpha particle, a beta particle, and gamma radiation.

✔ Write a balanced equation for alpha emission and for beta emission.

✔ Complete an equation for a nuclear reaction by writing the symbol of a missing particle.

The Discovery of Radioactivity

When you think about the word *radioactivity*, you probably think of nuclear explosions or nuclear power plants. You might be surprised to learn that radioactivity was discovered quietly, and quite by accident. Henri Becquerel (1852–1908) was studying minerals that glowed in the dark after being exposed to sunlight or ultraviolet light. The glow of these minerals was enough to expose photographic film, leaving a dark image where the light from the glowing mineral struck the film. One day, Becquerel left a uranium ore sample on top of some photographic plates that were wrapped in brown paper to protect them from light. When he later found that the plates were exposed as if they had been unwrapped, he concluded that the sample of uranium ore emitted invisible rays of radiation that could penetrate the paper. Figure 12-5 shows one of the original photographic plates used by Becquerel.

Pierre Curie (1859–1906) and his wife Marie (1859–1934) were fascinated with the work of Becquerel. They studied similar ores and discovered two new elements in them, polonium and radium, which were the major source of Becquerel's rays. Polonium emits 400 times more radiation than uranium, and radium emits almost one million times more radiation than uranium. The Curies first used the term *radioactivity* for this type of radiation.

Radioactivity is the spontaneous emission of particles and energy from the nucleus of an atom. It occurs because the radioactive nucleus is unstable. Elements with the most stable nuclei are intermediate in mass, with molar masses around 50 g mol^{-1}. There are

✔ Meeting the Goals

Radioactivity is the spontaneous emission of particles and energy from the nucleus of an atom.

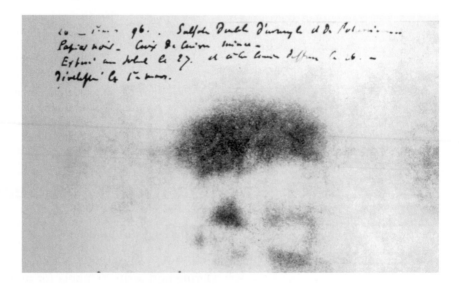

Figure 12-5 ▲ Henri Becquerel found that unexposed photographic plates became fogged when uranium ore was near. The mysterious radiation that caused the fogging led to the discovery of radioactivity. (The Granger Collection, New York.)

no stable isotopes for elements heavier than bismuth (atomic number 83); all isotopes of elements with atomic numbers larger than 83 are radioactive. Many of them occur naturally.

What factors make a nucleus stable? The atomic nucleus contains both protons and neutrons, which are bound together by nuclear forces. Without neutrons, the nucleus would fall apart because of the great repulsion between the positively charged protons. The smallest elements have stable nuclei with almost equal numbers of neutrons and protons. For larger elements, more and more neutrons are required in order to overcome the proton-proton repulsion and to bind the nucleus together. The nuclei of these elements contain more neutrons than protons.

Figure 12-6 shows the neutron-to-proton ratio for known isotopes, with a solid line representing a 1:1 ratio. We see in this figure that the elements with atomic numbers less than 20 lie very close to this solid line (N = Z). Elements with atomic numbers larger than 20, however, are clustered in a band to the left of the solid line. This band contains known stable and radioactive isotopes. A small island at Z > 100 is a theoretical prediction of stability.

Common Types of Radiation

Unstable nuclei attain a more stable neutron-to-proton ratio by emitting radiation in the form of small particles and energy. In 1919, Ernest Rutherford did an experiment that demonstrated the

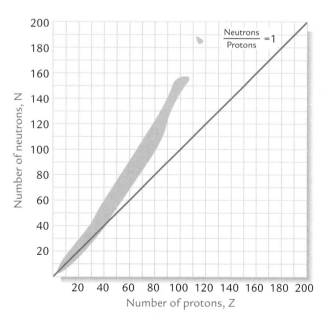

Figure 12-6 ▲ Most naturally occurring elements have nuclei that contain more neutrons than protons. These are found to the left of the line, N = Z, in a band that contains all stable nuclei and some unstable nuclei.

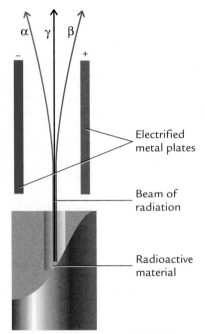

Figure 12-7 ▲
Rutherford used an electric field to study unclear radiation. He determined that alpha particles are positively charged, beta particles are negatively charged, and gamma rays are neutral.

✔ Meeting the Goals

An alpha particle is a helium nucleus; its symbol is $^4_2$He. A beta particle is an electron; its symbol is $^0_{-1}\beta$. A gamma ray is a photon of high energy; its symbol is γ.

three common types of radioactive emissions from unstable nuclei. He was very clever in his experimental design, using a radioactive material and focusing its radiation so that it passed between two electrified metal plates. One of the plates was positively charged and the other was negatively charged. A detector indicated what happened to the radiation when it passed through the electric field between the plates. Three types of events were recorded. The positive plate attracted one type of radiation, suggesting that this type must be negatively charged. The negative plate attracted a second type of radiation, suggesting that this type of radiation must be positively charged. The third type of radiation was unaffected by the charged plates, suggesting that this type of radiation was uncharged. Scientists named these radioactive emissions after the first three letters of the Greek alphabet: alpha, beta, and gamma. The three most common types of radiation are **alpha particle** emission, **beta particle** emission, and **gamma radiation.** Figure 12-7 shows a diagram of this experiment.

Alpha and beta radiation are mass-containing particles; gamma radiation consists of photons of high energy. An alpha particle is a high-energy helium nucleus that contains 2 protons and 2 neutrons but no electrons. Therefore, an alpha particle has an atomic number of 2, a mass number of 4, and a charge of +2, and is symbolized by $^4_2$He. A beta particle is an electron; it is symbolized by $^0_{-1}\beta$, having a mass number of zero and a charge of −1. A gamma ray is a photon of high energy; it has no mass and no charge. These properties of alpha, beta, and gamma radiation are listed in Table 12-4.

Equations For Nuclear Reactions

Unstable nuclei disintegrate to form more stable nuclei by emitting small, high-energy particles. As a result, the number of protons and neutrons in the nucleus is changed. Reactions in which the nucleus is changed are called nuclear reactions, which are fundamentally different from the chemical reactions you have studied so far. Nuclear reactions typically involve changes in the numbers of protons and/or neutrons in the nucleus, whereas chemical reactions involve electron interactions.

Table 12-4 Symbols Used for Alpha, Beta, and Gamma Radiation

Type of Radiation	Description of Type	Symbol for Radiation Type	Mass Number	Charge
Alpha (α)	Helium nucleus	$^4_2$He	4	2+
Beta (β)	Electron	$^0_{-1}\beta$	0	−1
Gamma (γ)	Photon	γ	0	0

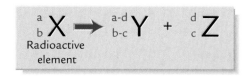

Figure 12-8 ▲ In a balanced nuclear equation, both the mass numbers (superscript numbers) and the charges (subscript numbers) must balance.

In this text, we will focus on the changes in the nucleus that accompany nuclear reactions. There are other changes and particles that may be involved, but these do not affect the nature of the new element formed after radioactive decay and are outside the scope of this text. You will not be surprised to learn that the equation for a nuclear reaction is also very different from the equation for a chemical reaction. Nuclear equations focus on what is happening to the nucleus; they keep count of the changing numbers of protons and neutrons in a given nucleus. Just as there were rules to govern how to balance chemical equations, so, too, there are rules for balancing nuclear equations. In a nuclear equation, mass numbers and charges must be conserved. The most usual change is in the atomic number.

Figure 12-8 shows the general process for balancing nuclear equations, the rules for which are as follows:

- The sum of mass numbers of the reactants must equal the sum of the mass numbers of the products. The mass numbers are the superscript numbers in the symbols of isotopes and radioactive particles.
- The sum of the charges of the reactants must equal the sum of the charges of the products. These are indicated by the subscript numbers in the symbols of isotopes and radioactive particles.

Now we can apply the procedure for balancing nuclear equations to equations for alpha and beta emission.

Equations for Alpha Emission

Alpha decay, or alpha emission, is the process in which a nucleus emits an **alpha particle,** *leaving the nucleus with two fewer protons and two fewer neutrons than it had before the emission.* Figure 12-9 shows the emission of an alpha particle from an unstable nucleus.

An example of alpha emission is the decay of ^{238}U to ^{234}Th. The equation for this nuclear reaction is:

$$^{238}_{92}\text{U} \longrightarrow \ ^{4}_{2}\text{He} + \ ^{234}_{90}\text{Th}$$

Does this equation satisfy the rules for writing nuclear equations? To check this, add the superscript numbers on the right of the

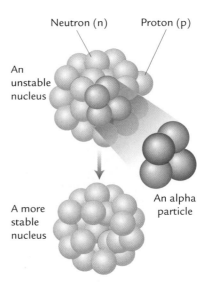

Figure 12-9 ▲ During alpha emission, a radioactive nucleus emits an alpha particle. The nucleus after emission has two fewer protons and two fewer neutrons.

equation. Do they equal 238? Does the sum of the subscript numbers on the right equal 92? If so, the equation is correctly written. This simple mathematical relationship also allows you to predict the identity of a missing symbol in a nuclear equation.

Example 12.8   **Writing nuclear equations for alpha emission**

Complete this nuclear equation for alpha emission by supplying the missing element's symbol.

$$^{230}_{90}\text{Th} \longrightarrow {}^{4}_{2}\text{He} + ?$$

Thinking Through the Problem The superscript numbers on the right have to add to 230. For the equation $4 + x = 230$, the missing number x is 226. The subscript numbers on the right must add to 90, the subscript number on the left. For the equation $2 + y = 90$, the missing number y is 88. The element with $Z = 88$ is radium.

Answer   The missing symbol is that of radium-226, $^{226}_{88}\text{Ra}$.

Example 12.9   **Writing nuclear equations for alpha emission**

Polonium-214 disintegrates by alpha emission. Write the balanced equation for this nuclear reaction.

Thinking Through the Problem Polonium has atomic number 84; polonium-214 has the symbol $^{214}_{84}\text{Po}$. This is the reactant. One of the products is an alpha particle, $^{4}_{2}\text{He}$. We can find the other product by applying the rules as above.

In alpha emission, an unstable nucleus gives off an alpha particle, $_2^4\text{He}$, which reduces the size of its nucleus by 2 protons and 2 neutrons. This means that the nucleus formed after alpha emission will have a mass number that is 4 less and an atomic number that is 2 less than those in the initial nucleus.

Working Through the Problem We start by putting in the symbols of the particles we know:

$$_{84}^{214}\text{Po} \longrightarrow {}_2^4\text{He} + ?$$

To balance this reaction, the missing particle must be $_{84-2}^{214-4}\text{X} = {}_{82}^{214}\text{X}$. The identity of element X is found by consulting a periodic table.

Answer The equation for this disintegration is $_{84}^{214}\text{Po} \rightarrow {}_2^4\text{He} + {}_{82}^{210}\text{Pb}$.

How are you doing? **12.3**

Astatine-212, ^{212}At, disintegrates by alpha emission. Write the balanced equation for this nuclear reaction.

Equations for Beta Emission

A **beta particle** is an electron that is emitted or given off when a neutron changes to yield a proton and an electron. Figure 12-10 shows an example of beta emission.

The equations for beta emission (also known as beta decay) are written according to the same rules as for alpha emission but, in this case, the particle emitted has the symbol $_{-1}^0\beta$. For example, thorium-234 decays by beta emission. The equation for this process is:

$$_{90}^{234}\text{Th} \longrightarrow {}_{-1}^0\beta + {}_{91}^{234}\text{Pa}$$

In beta emission, an unstable nucleus gives off a beta particle, $_{-1}^0\beta$, resulting in a nucleus with the same mass number but with an atomic number increased by one.

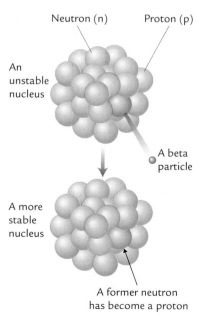

Neutron (n)   Proton (p)

An unstable nucleus

A beta particle

A more stable nucleus

A former neutron has become a proton

Figure 12-10 ▶ When a beta particle is given off by an unstable nucleus, the resulting nucleus will have the same mass number as before, but its atomic number will increase by 1.

Is the equation balanced? Do the superscript numbers on the right equal 234? Yes, since $0 + 234 = 234$. Do the subscript numbers on the right equal 90? Yes, since $-1 + 91 = 90$. The equation is balanced.

Example 12.10   Writing nuclear equations for beta emission

Sodium-25 decays by beta emission. Write the balanced equation for this nuclear reaction.

Thinking Through the Problem Sodium-25 has atomic number 11; its symbol is $^{25}_{11}\text{Na}$. This is the reactant. Since a beta particle is emitted, $^{0}_{-1}\beta$ is one of the products. We must find the second product by applying the rules for balancing nuclear equations.

Working Through the Problem Begin by writing the symbols you know:

$$^{25}_{11}\text{Na} \longrightarrow \, ^{0}_{-1}\beta + ?$$

Then we find the missing superscript and subscript numbers by subtraction; $^{25-0}_{11-(-1)}X$. Lastly, we identify X from its atomic number, $11 - (-1) = 12$; this is Mg.

Answer $^{25}_{11}\text{Na} \rightarrow \, ^{0}_{-1}\beta + \, ^{25}_{12}\text{Mg}$

Example 12.11   Writing nuclear equations for beta emission

A radioactive nucleus undergoes beta emission, producing $^{210}_{84}\text{Po}$. Write the balanced equation for this nuclear reaction.

Thinking Through the Problem This problem gives the two products, a beta particle and $^{210}_{84}\text{Po}$. We can find the reactant nucleus by adding the superscript numbers $(210 + 0 = 210)$ and the subscript numbers $(84 + (-1) = 83)$ of the two products. The identity of the reactant is found from its atomic number, 83.

Answer $^{210}_{83}\text{Bi} \rightarrow \, ^{210}_{84}\text{Po} + \, ^{0}_{-1}\beta$

✔ **Meeting the Goals**

Because the mass numbers and charges are conserved in balanced nuclear equations, we can find the identity of a missing particle. The sum of the mass numbers of the reactants must equal the sum of the mass numbers of the products. The sum of the charges of the reactants must equal the sum of the charges of the products.

How are you doing?   12.4

Rhenium-190 decays by beta emission. Write the balanced equation for the disintegration of $^{190}_{75}\text{Re}$ by beta emission.

When unstable nuclei emit small particles and energy, a more stable nucleus generally results. For elements such as uranium-238, however, alpha emission results in a nucleus that is still radioactive, thorium-234. This nucleus also decays, but the result is still an unstable nucleus. The process continues until a stable nucleus, lead-206, is formed. The natural radioactive series of unstable nuclei formed as uranium-238 that decays to lead-206 is called the *uranium series*. It is shown in Figure 12-11. Uranium-235 and thorium-232 also have a decay series called the *actinium series* and the *thorium series*, respectively.

Figure 12-11 ▶ Uranium-238 decays through successive alpha and beta emissions to the stable nucleus lead-206. The half-lives of the isotopes in this series are shown.

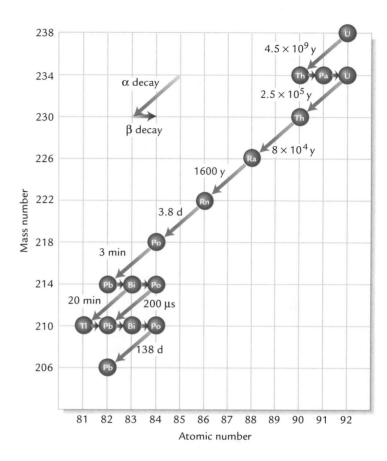

PROBLEMS

17. Write the atomic symbols for an alpha particle, a beta particle, and a gamma ray.

18. Write the isotope symbol for lead-214, uranium-134, and oxygen-18.

19. Complete the equation: $^{39}_{17}Cl \rightarrow \, ^{0}_{-1}\beta + ?$

20. Complete the equation: $^{19}_{10}Ne \rightarrow \, ^{0}_{-1}\beta + ?$

21. Complete the equation: $^{257}_{102}No \rightarrow \, ^{253}_{100}Fm + ?$

22. Complete the equation: $^{239}_{94}Pu \rightarrow \, ^{4}_{2}He + ?$

23. Complete the equation: $^{14}_{7}N \rightarrow \, ^{14}_{8}O + ?$

24. Write the names of ^{70}As, ^{106}Ag, and ^{127}Sn.

25. Bismuth-210 decays by alpha emission. Write the balanced nuclear equation for this reaction.

26. Bismuth-210 decays by beta emission. Write the balanced nuclear equation for this reaction.

27. Bromine-88 decays by beta emission. Write the balanced nuclear equation for this reaction.

28. Complete the equation $^{194}_{84}Po \rightarrow \, ^{190}_{82}O + ?$ Is it alpha or beta decay?

In the equations in Problems 29–32, the symbol $^{0}_{+1}\beta$ stands for a particle called a positron.

29. Complete: $? \rightarrow \, ^{0}_{+1}\beta + \, ^{79}_{35}Br$

30. Complete: $? \rightarrow \, ^{0}_{+1}\beta + \, ^{58}_{26}Fe$

31. Complete: $^{77}_{36}Kr \rightarrow \, ^{0}_{+1}\beta + ?$

32. Complete: $^{70}_{33}As \rightarrow \, ^{0}_{+1}\beta + ?$

33. Discussion question: Write balanced nuclear equations for the eight steps in the decay of uranium-238 to lead-214 (shown in Figure 12-11). This is a complicated path, known to occur in nature. Is it possible for it to occur in one step? What other particle would form if this happened?

PRACTICAL B

The alchemists tried to make gold from lead. They couldn't do it. Can we?

Even before the development of modern chemistry just over two hundred years ago, some pure elements were known. These were mostly metals—gold and lead were prominent examples. Workers at the time wondered if it would be possible to change (transmute) one of these into the other. In particular, they wanted to change the abundant but inexpensive element lead into the precious element gold. However, the means at their disposal, primarily chemical reactions, would never accomplish this. So the alchemists failed. But modern science, with its understandings of the true difference between the elements, can do otherwise.

In certain cases, scientists are able to change (or transmute) atoms of one element into atoms of another in a process called *bombardment*, in which very small particles are greatly accelerated so that they are traveling at tremendous speeds. When these high-energy particles collide with a nucleus, they can cause a change in the number of protons and neutrons in the target nucleus. If the number of protons is changed, the identity of the element is changed. The transuranium elements (elements beyond uranium) are all examples discovered through bombardment experiments.

Figure 12-12 shows an aerial photograph of the Fermi National Accelerator Laboratory in Batavia, Illinois.

Common types of "bullets" used in bombardment experiments are small nuclei such as one of the isotopes of hydrogen, an alpha particle, or a neutron. We have seen how to write the atomic symbols for the isotopes of hydrogen and for helium, but we can also write symbols for protons and neutrons to use in nuclear equations. A proton, symbolized by $_1^1p$, has a mass number of 1 and a charge of 1. A neutron, symbolized by $_0^1n$, has a mass number of 1 and a charge of zero.

Ernest Rutherford (1919) is believed to be the first person to successfully change one element into a different element by bombardment. He focused high-energy alpha particles onto a sample containing nitrogen and produced an isotope of oxygen, ^{17}O, and a proton, $_1^1p$. The equation for this nuclear reaction is:

$$_7^{14}N + _2^4He \longrightarrow _8^{17}O + _1^1p$$

In this case, the collision between $_7^{14}N$ and $_2^4He$ was successful; a nucleus more massive than the original nitrogen was formed and the ejected proton carried off the excess energy.

Boron can be bombarded to form an isotope of nitrogen that is radioactive. Can you determine the identity of the small particle used in this bombardment process? The partial equation is:

$$_5^{11}B + ? \longrightarrow _7^{15}N$$

And now, what about gold? Gold can be made by bombarding platinum-196 with deuterium (^2H) in a two-step process. The first step forms platinum-197 and another small particle. Complete the following reaction by writing the symbol for the second product:

$$_{78}^{196}Pt + _1^2H \longrightarrow _{78}^{197}Pt + ?$$

In the second step of this reaction, platinum-197 decays by beta emission. Write the equation for this nuclear reaction to see whether gold is really formed!

$$_{78}^{197}Pt \longrightarrow$$

The bombardment of platinum to form gold is extremely expensive, costing more than the product is worth. Still, the alchemists' idea was plausible. They simply lacked the technology to make it happen!

Figure 12-12 ▶ The Tevatron at Fermilab is the world's highest energy particle accelerator. Particles complete the four-mile course 50,000 times per second. (Fermilab.)

12.3 The Biological Implications of Radiation

SECTION GOALS

✔ Explain the concept of half-life.

✔ Define the term "radioisotope" and name some beneficial uses of radioisotopes.

The Rate of Radioactive Decay

Devices such as Geiger counters detect the high-energy emissions that accompany nuclear decay. The rate of emission by nuclear decay may be expressed in terms of seconds, minutes, days, or even years. The number of nuclei that decay per unit of time is defined as the **activity** of a sample. More active isotopes decay on a shorter time scale than less active isotopes. The time required for half of the unstable nuclei in a sample to decay has a special name, the **half-life.** The half-lives, symbolized by $t_{1/2}$, of some selected isotopes are listed in Table 12-5.

Table 12-5 Some Isotopic Half-lives

Isotope	Half-life, $t_{1/2}$
^{131}I	8.1 days
^{59}Fe	45.1 days
^{32}P	14.3 days
^{14}C	5730 years
^{238}U	4.5×10^9 years

The concept of half-life can be applied to any radioactive isotope and to any sample size. Suppose we begin with a 1.00-gram sample of ^{32}P. At the end of one half-life, the amount of ^{32}P remaining is half the initial amount, 0.500 gram. At the end of two half-lives, the amount remaining is half of 0.500 gram, or 0.250 gram. The amount of sample is reduced to half with each additional half-life. The same would be true for any of the isotopes listed in Table 12-5. The relationship between half-life and remaining sample is shown in Table 12-6.

Figure 12-13 is a graph showing the number of radioactive nuclei remaining as a function of half-life. The number of nuclei decreases by half with the passing of each successive half-life.

✔ **Meeting the Goals**

A half-life is the time it takes for half a sample to decay. It is a property of each isotope.

Table 12-6   Relationship Between Half-life and Amount of Sample Remaining

Initial Amount	Amount Remaining after 1 Half-life	Amount Remaining after 2 Half-lives	Amount Remaining after 3 Half-lives
1.00 gram ^{131}I	0.500 gram	0.250 gram	0.125 gram
1.00 gram ^{59}Fe	0.500 gram	0.250 gram	0.125 gram
1.00 gram ^{32}P	0.500 gram	0.250 gram	0.125 gram

The amount of time required for the sample of radioactive isotope to decrease by half varies from isotope to isotope. The relationship between sample size and elapsed time is shown in Table 12-7. ^{121}I requires only 8.1 days compared to 4.5 billion years for ^{238}U.

When we consider the different half-lives of radioisotopes we can understand why some are found in nature while others must be made artificially. For example, uranium-238 has such a long half-life that most of the element present when the earth formed is still around. On the other hand, hydrogen-3 (tritium) has a half-life of just a few

Figure 12-13 ▶ The number of radioactive nuclei remaining after one half-life is one-half the number initially present.

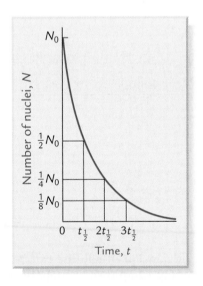

Table 12-7   Time Required for 1.00 Gram of Radioactive Sample to Decrease by Half

Initial Amount	Time Required for First Half of Sample to Decay	Time Elapsed after Decay of Next Half of Sample	Time Elapsed after Decay of Next Half of Sample
1.00 gram ^{131}I	$t_{1/2} = 8.1$ days	$2t_{1/2} = 16.2$ days	$3t_{1/2} = 24.3$ days
1.00 gram ^{14}C	$t_{1/2} = 5,730$ years	$2t_{1/2} = 11,460$ years	$3t_{1/2} = 17,190$ years
1.00 gram ^{238}U	$t_{1/2} = 4.5 \times 10^9$ years	$2t_{1/2} = 9.0 \times 10^9$ years	$3t_{1/2} = 1.35 \times 10^{10}$ years

years. Any tritium present now will decay to almost nothing in the span of a few decades. This is the difference between "short-lived" and "long-lived" nuclear waste.

The Biological Impact of Alpha, Beta, and Gamma Radiation

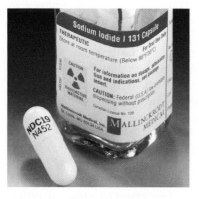

Iodide-131 is a radioactive isotope used to diagnose thyroid abnormalities. (Richard Megna/ Fundamental Photographs.)

All forms of radiation cause biological damage by interfering with molecular function or destroying living tissues. Alpha, beta, and gamma radiation have different effects on matter because they have different levels of penetration and different ionizing abilities. For example, alpha particles can penetrate a thin sheet of paper but are stopped by any material that is thicker or denser than this. Beta particles can penetrate a sheet of aluminum that is 1 cm thick. Gamma rays are able to penetrate a sheet of lead or a concrete slab that is 5 cm thick. Figure 12-14 illustrates these penetration differences.

It is not the ability to penetrate matter but the ability to *ionize* matter that makes radiation harmful. For example, alpha particles penetrate only the top layers of human tissue but rapidly ionize the molecules with which they collide before they are stopped. Tissue damage is more severe if a person is exposed to alpha radiation. Beta particles and gamma rays are able to penetrate more deeply than alpha particles but they pass through tissue so easily that less ionization takes place. Note that penetrating ability does matter when we worry about shielding ourselves. Alpha radiation is easy to shield, while gamma radiation is harder to block. Table 12-8 shows the relative biological impact of alpha, beta, and gamma radiation. All three can be extremely dangerous.

Figure 12-14 ▶ Alpha, beta, and gamma radiation have different energies and abilities to penetrate matter. While alpha particles can penetrate a thin sheet of paper, gamma rays can penetrate 5 cm of lead.

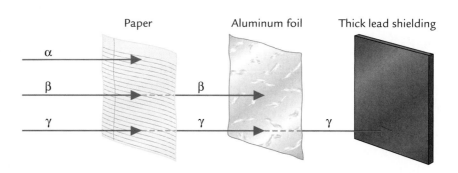

Table 12-8 Relative Ionizing Ability of Alpha, Beta, and Gamma Radiation

Radiation	Degree of Penetration	Relative Ionizing Ability	Biological Impact on Living Tissue
Alpha particles	Thin sheet of paper	Great ionizing ability	Little penetration but damaging
Beta particles	Thin sheet of aluminum foil	Medium ionizing ability	Mildly penetrating and damaging
Gamma rays	Sheet of lead (5 cm thick)	Little ionizing ability	Very penetrating but less damaging

The Effect of Radiation on Living Tissue

When radiation strikes a tissue, some of the energy of radiation is passed on to the molecules of that tissue. For example, one type of radiation with which we are all familiar is the radiation from the sun, *solar radiation*. You might think about the energy transfer from the sun's radiation to your skin when it is a sunny day. As you begin to feel warm, you are experiencing the transfer of thermal energy to your skin. Particles emitted in nuclear decay can ionize molecules. This is often called *ionizing radiation*.

The amount of damage that ionizing radiation can inflict on living tissue depends on the type of radiation, the strength of the radiation source, and the time of exposure. All these factors are usually combined in terms of an absorbed radiation *dose*. A radiation absorbed dose (1 *rad*) is equivalent to one joule of ionizing energy deposited in one kilogram of tissue. A factor called the *relative biological effectiveness* (RBE) relates biological damage resulting from alpha, beta, and gamma radiation sources. Alpha radiation has an RBE value about 10 times the RBE values for beta and gamma radiation, which are generally given an RBE of 1. The effective dose of radiation received by humans is called a *rem*.

Over the course of a year, an average person receives an accumulated dosage less than 0.100 rem of radiation from common sources such as cosmic radiation, smoke detectors, television screens, and luminous watches. Naturally occuring radioactivity (such as potassium-40) is also important, as are radiation sources like X-rays that are produced nonradioactively. Table 12-9 correlates the limits of radiation exposure (in rems) with biological damage.

We can control our exposure to ultraviolet (UV) radiation by protecting ourselves from the damaging rays of the sun. (Grant L. Gursky/AP.)

Table 12-9 Biological Damage and Exposure to Radiation

Radiation Exposure (rems)	Biological Effect of Exposure
0–25	No detectable damage
25–50	Temporary decrease in white blood cell count
50–200	More acute decrease in white blood cell count
More than 500	Death probable within short period of time

Beneficial Uses of Radiation

Although it is wise to limit human exposure to radiation, there are positive things to note, too. Controlled uses of radioactivity contribute benefits to society, most notably in medicine. *Radiation is used in medicine both as a diagnostic tool and in therapy or treatment.* Radioactive substances that produce beta particles or gamma rays are used in

A radioisotope is an isotope that is radioactive. There are many beneficial uses of radioisotopes, including tumor treatment and the use of tracers to help in the diagnosis of organ abnormalities.

medicine as diagnostic tools. The high-energy radiation released in the body by these substances is detected and projected into a pattern or image by a computer, enabling doctors to diagnose some illnesses by comparing images of healthy and diseased tissue. **Radioisotopes,** isotopes that are radioactive, are given to patients and used as "tracers" to study the function of particular organs. Iodine-131 is quickly taken up by the thyroid and so is used in thyroid therapy; technetium-99 is particularly useful to pinpoint problems in the heart; and strontium-87 is used to diagnose bone disease.

There are also many nonmedical uses of radiation. Carbon-14, a radioisotope of carbon that decays very slowly, is used to approximate the age of various materials in a process called **radiocarbon dating.** Because both the natural abundance and the rate of decay of carbon-14 are known, the age of a material that contains carbon can be calculated by measuring the amount of carbon-14 that remains in the material.

PRACTICAL C   Fission and fusion—is the difference only in the spelling?

With the advent of the nuclear age, scientists set about trying to create new isotopes. In particular, they sought to create elements with atomic numbers greater than that of uranium. It seemed that one way to do so would be to bombard uranium samples with neutrons. If some of the neutrons "stuck" in a uranium nucleus, there was the possibility that some neutrons would convert to protons, resulting in an element with an atomic number greater than 92. It didn't work. The uranium sample remained uranium.

What did happen, however, was that the uranium sample after bombardment was found to contain some barium atoms. Several scientists in Germany (including Otto Hahn, Lise Meitner, and Fritz Strassman) made this discovery in 1938, as the world teetered on the brink of war. Their conclusion? The uranium nucleus must have *split* into smaller nuclei! This was the birth of controlled nuclear fission. Nuclear fission is a process in which a larger nucleus splits or is broken into smaller nuclei, often releasing a great deal of energy.

One possible splitting of uranium-235 is:

$$^{235}_{92}\text{U} + ^{1}_{0}n \longrightarrow ^{142}_{54}\text{Xe} + ^{90}_{38}\text{Sr} + 4^{1}_{0}n$$

This is the first nuclear equation you have seen with a coefficient other than 1. We check to see that the equation is balanced as before:

Superscripts $235 + 1 = 142 + 90 + 4(1) = 236$; this is balanced.

Subscripts $92 + 0 = 54 + 38 + 4(0) = 92$; this is balanced.

There are other products possible from the splitting of a uranium-235 nucleus. Examine the following equations and determine how many neutrons are given off in the fission process. Indicate the coefficient for the product neutrons.

$$^{235}_{92}\text{U} + ^{1}_{0}n \longrightarrow ^{139}_{56}\text{Ba} + ^{94}_{36}\text{Kr} + ?^{1}_{0}n$$
$$^{235}_{92}\text{U} + ^{1}_{0}n \longrightarrow ^{144}_{55}\text{Cs} + ^{90}_{37}\text{Rb} + ?^{1}_{0}n$$

Nuclear fusion is quite a different process from nuclear fission. In a fusion reaction, two lighter nuclei are *combined* to make a larger nucleus, also releasing tremendous amounts of energy. The energy of the sun comes from the fusion of hydrogen nuclei into helium nuclei. A fusion reaction might look like this:

$$^{2}_{1}\text{H} + ^{2}_{1}\text{H} \longrightarrow ^{3}_{2}\text{He} + ^{1}_{0}n$$

Because both fission and fusion processes usually release great amounts of energy, each has been considered as a possible energy source. But, while today's nuclear power plants rely on fission, energy production from fusion remains only a remote possibility.

PROBLEMS

34. What is meant by the "activity" of a radioactive nucleus?

35. What is meant by the term "half-life"?

36. Which isotope listed in Table 12-5 has the greatest activity? Which has the least activity?

37. How long will it take for half of a sample of ^{32}P to decay?

38. Discussion question: The radioactivity in a sample of ^{131}I was measured at the beginning of an experiment. What fraction of the ^{131}I sample do you expect to find at the end of 32.4 days?

Radioactive iodine is a potential component of "dirty bombs" that spread radiation using conventional explosives. A therapy for this is to take potassium iodide containing naturally-occurring iodine-127. Discuss how long someone would need to take iodine-127 to protect against 90% of any ingested iodine-131.

39. List several factors that allow radiation to damage living tissue.

◀ Review with **Web Practice**

Chapter 12 Summary and Problems

VOCABULARY

isotope	An atom of an element that has a different mass number from other atoms of the same element because of a difference in the number of neutrons in its nucleus.
deuterium	An isotope of hydrogen that has a mass number of 2; its symbol is ^2H.
tritium	An isotope of hydrogen that has a mass number of 3; its symbol is ^3H.
natural abundance	The natural abundance of an isotope of an element is the relative number of its atoms in a sample of the element.
isotopic mass	Isotopic mass is the mass of one of the isotopes of an element.
radioactivity	The spontaneous emission of particles and energy from a nucleus.
alpha particle	A helium nucleus, ^4_2He, given off by an unstable nucleus in a process called alpha emission.
beta particle	An electron, $^0_{-1}\beta$, given off by an unstable nucleus in a process called beta emission.
gamma radiation	Gamma radiation consists of high-energy photons given off by radioactive decay processes.
activity	The activity of an isotope is the number of nuclei that decay per unit of time.
half-life	The time required for half of the unstable nuclei in a sample to decay.
radioisotope	A radioactive isotope; radioisotopes have many uses in medicine.
radiocarbon dating	A technique that correlates the amount of carbon-14 remaining in a material with its age.

SECTION GOALS

What is an isotope?	An isotope is an atom that contains the same number of protons but a different number of neutrons when compared to another atom of the same element.

How do you write the atomic symbol and the name of an isotope?	Isotopes are represented either by an atomic symbol that includes the mass number, or by the name of the element along with its mass number. For example, an isotope of potassium with mass number 40 is written as "^{40}K" or "potassium-40."
When can you approximate the atomic mass of an element from the mass and natural abundance of its isotopes?	The atomic mass of an element can be estimated when one isotope contributes more than about 90% to the abundance of that element. We expect, in these cases, that the atomic mass of the element will be close to the mass of that isotope in greatest abundance.
How can you calculate the atomic mass of an element from the mass and the natural abundance of its isotopes?	The atomic mass of an element is the weighted average of its isotopes. To calculate the atomic mass of an element, add the mass-contributions of each isotope. The mass-contribution of each isotope is found as the product of its fractional abundance and its mass. A formula that calculates the atomic mass of an element is $M = f_1 M_1 + f_2 M_2 + f_3 M_3 + ...$, where M is the atomic mass of the element, f_x is the fractional abundance of the xth isotope, and M_x is the isotope mass of the xth isotope.
What is radioactivity?	Radioactivity is the spontaneous emission of particles and energy from the nucleus of an atom.
How do you define and write symbols for an alpha particle, a beta particle, and gamma radiation?	An alpha particle is a helium nucleus; its symbol is $^4_2$He. A beta particle is an electron; its symbol is $^0_{-1}\beta$. Gamma radiation consists of photons of high energy; the symbol for a photon is γ.
How do you write a balanced equation for alpha emission?	In alpha emission, an unstable nucleus gives off an alpha particle, $^4_2$He, which reduces the size of its nucleus by 2 protons and 2 neutrons. This means that the nucleus formed after alpha emission will have a mass number that is 4 less and an atomic number that is 2 less than those of the initial unstable nucleus.
How do you write a balanced equation for beta emission?	In beta emission, an unstable nucleus gives off a beta particle, $^0_{-1}\beta$, resulting in a nucleus with an unchanged mass number and an atomic number increased by one.
How can you find the symbol of a missing particle in the equation for a nuclear reaction?	Because the mass numbers and charges are conserved in balanced nuclear equations, we can calculate the identity of a missing particle. The mass numbers of the reactants must equal the mass numbers of the products. The sum of the subscript numbers of the reactants must equal the sum of the subscript numbers of the products.
What is meant by the "half-life" of a sample?	The half-life is the time it takes for half a sample to decay. It is a fixed time for each radioisotope.
What is a radioisotope? What are some beneficial uses of radioisotopes?	A radioisotope is an isotope that is radioactive. Beneficial uses of radioisotopes include tumor treatment, and the use of tracers to help in the diagnosis of organ abnormalities or disease.

PROBLEMS

40. What is the symbol for the isotope of potassium that has a mass number of 38?

41. What is the symbol for the isotope of chlorine that has a mass number of 38?

42. Naturally-occurring lithium, Li, has two isotopes. One, lithium-6, has an abundance of 7.59% and an isotopic mass of 6.0151214. The other, lithium-71, has an isotopic mass of 7.0160030 and an abundance of 92.41%. What is the atomic mass of Li?

43. Naturally-occurring gallium, Ga, has two isotopes. One, gallium-69, has an abundance of 60.108% and an isotopic mass of 68.925580. The other, gallium-71, has an isotopic mass of 70.9247005 and an abundance of 39.892%. What is the atomic mass of Ga?

44. Describe each of the following statements as true or false. If false, rewrite the statement so that it is true.

  (a) An atom that contains the same number of protons and neutrons is neutral.

  (b) All atoms have the same number of protons and neutrons in their nucleus.

  (c) Isotopes always have the same number of neutrons.

  (d) An electrically charged atom is called an isotope.

  (e) A negative ion is an atom that has lost one or more electrons.

  (f) An atom that contains more electrons than protons has a positive charge.

  (g) Isotopes of the same element differ only in the number of neutrons they contain.

45. Iron consists of four isotopes with the abundances indicated below. What is the atomic mass of iron?

Isotopic Mass	Natural Abundance (%)
53.9396	5.82
55.9349	91.66
56.9354	2.19
57.9333	0.33

46. Boron is found as two isotopes. Boron-10 has an isotopic mass of 10.01. Boron-11 has a mass of 11.01. If the abundance of boron-10 is 20.0%, what is:

  (a) the abundance of boron-11?

  (b) the atomic mass of boron?

47. Decide whether each of the following equations is an example of alpha or beta emission.

  (a) $^{214}_{84}\text{Po} \rightarrow {}^{4}_{2}\text{He} + {}^{210}_{82}\text{Pb}$

  (b) $^{117}_{47}\text{Ag} \rightarrow {}^{0}_{-1}\beta + {}^{117}_{48}\text{Cd}$

48. Correct the following equations for beta emission so that they are properly balanced.

  (a) $^{21}_{11}\text{Na} \rightarrow {}^{0}_{-1}\beta + {}^{21}_{10}\text{Ne}$

  (b) $^{14}_{6}\text{C} \rightarrow {}^{0}_{-1}\beta + {}^{14}_{5}\text{B}$

49. Consult a listing of the elements and write the symbols for the following:

  (a) lead-214

  (b) ruthenium-103

  (c) thorium-230

  (d) protactinium-234

50. Protactinium-234 decays by beta emission. Write the balanced equation for this reaction.

51. Thorium-230 decays by alpha emission. Write the balanced equation for this reaction.

52. Ruthenium-103 decays by alpha emission. Write the balanced equation for this reaction.

53. Lead-214 decays by beta emission. Write the balanced equation for this reaction.

54. Thallium-206 decays to give lead-206. What particle is emitted?

55. Bismuth-210 decays to give thallium-206. What particle is emitted?

Electrons and Chemical Bonding

The yellow glow of sodium vapor lamps lights these Chicago streets. The absorption or emission of light by a substance tells us something about the behavior of its electrons. (Kevin O. Mooney/ Odyssey/Chicago.)

PRACTICAL CHEMISTRY Electrons and the Properties of Molecules and Atoms

In this chapter, we will more deeply examine some aspects of electron structure in relation to chemical substances. In doing so, we will see that the arrangement of the electrons around atoms and within molecules can affect the properties of those atoms and molecules. For example, the color of a substance results from the interaction of light with electrons in the atoms and molecules of that substance. Light is a form of energy, as we saw in Chapter 11. Light energy is spread over many wavelengths, only some of which we see as visible light. When visible light of a certain energy interacts with a substance, some of that light is absorbed. We experience the reflection of the remaining light as a color. We also perceive color when light is emitted from a material. Certain patterns in the absorption or emission of light give us important insight into the patterns of electron behavior in substances. The way the electrons are structured in an atom also affects many atomic properties, including such simple properties as size.

This chapter will focus on the following questions:

PRACTICAL A How does the three-dimensional structure of a molecule affect its color?

PRACTICAL B What does the length of a molecule tell us about the energy of the electrons in the molecule?

PRACTICAL C How do the "core" electrons of an atom affect its size?

13.1 Molecular Shapes

SECTION GOALS

✔ Examine a Lewis structure and relate it to the three-dimensional structure of that molecule or polyatomic ion.

✔ Use the concept of electron domains to predict the shape of a molecule or polyatomic ion.

The Electron Domain

An important consequence of our picture of a molecule or formula unit is the presence of electrons in the area around nuclei. In Chapter 2 we learned how the valence electrons of an atom can be considered to build up the structure of a molecule by forming bonding pairs and lone pairs. These electrons, whether in bonds or in lone pairs, occupy space. A Lewis structure shows only the connections and electron arrangements among atoms. The arrangement of these in three dimensions gives rise to the molecule's full structure. In this section we will examine only simple molecular structures. We describe these structures with three-dimensional geometric shapes based on the arrangement of bonds and lone pairs in the molecule.

When we consider the arrangement of atoms in molecular compounds and polyatomic ions, we must account for the space occupied by electrons. We say that each lone pair or each bond (whether it is a single, double, or triple bond) needs its own **electron domain.** Usually we are most concerned with the domains that exist around the central atom in a molecule. For example, the molecule carbon dioxide, shown in Figure 13-1, has two electron domains around the central atom.

Each domain occupies a region or space around the atom that is as far away as possible from the other domains. *Molecular shapes in three dimensions separate electron domains as completely as possible.* The

Figure 13-1 ▶ The electron domains of carbon dioxide.

shape of a molecule or ion depends on the number of electron domains.

Notice in Figure 13-1 that there are two domains around the central carbon in carbon dioxide. These domains are the double bonds to each of the two oxygens. How can we keep these as far apart as possible? The answer is to position them on opposite sides of the carbon atom. This means that an angle of 180° is formed around the carbon by the domains.

When we discuss the structure around an atom, it is useful to use words, not just degrees. In the case of carbon dioxide, the 180° angle means that the three atoms lie in a line. We call this arrangement a linear shape, and we call carbon dioxide a linear molecule.

Example 13.1 **Predicting the shape of molecules**

Predict the structure around the central atom in the molecule CS_2.

Thinking Through the Problem We want to predict the shape of a molecule. ◀━▶ (13.1) The key idea is to consider the electron domains around the central atom.

Working Through the Problem The Lewis structure of CS_2 is the same as that for CO_2, with two electron domains around the central C. We expect a linear molecule.

Answer

domain
1   2

:S=C=S:
180°

Next, what happens if we have three domains? This occurs in formaldehyde, where there are two single bonds and one double bond to the central carbon. That carbon has three electron domains. Another example, shown in Figure 13-2, is carbonate, which has a central carbon atom with four bonds: two on one oxygen and one on each of the other two oxygens (we can ignore the other resonance structures in this case; they will give the same result).

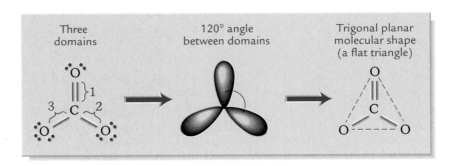

Figure 13-2 ▶ The structure of carbonate, with three electron domains.

The three domains are separated best if they have 120° angles between them. This positioning gives the ion the shape of a flat triangle, so the shape is described as trigonal planar.

Example 13.2 **Predicting the shape of molecules**

Predict the structure of the electron domains about the central S in SO_2.

Thinking Through the Problem To predict the positions of electron domains about the S in SO_2, we first draw its Lewis structure.

Working Through the Problem The Lewis dot structure of SO_2 has three electron domains about the central S. We expect the electron domains to be arranged in a triangle around the S. The arrangement of two O atoms relative to the S will not be triangular, however. Although there are three electron domains present in SO_2, only two of them are bonding domains. *The molecular shape of SO_2 is determined by the atoms actually bonded to the S.* The SO_2 molecule has only two oxygen atoms bonded to the central S. This makes the actual shape of the molecule bent.

Answer

KEY IDEA 13.2

2 resonance structures   120° angle between domains   molecular shape: bent

How are you doing? 13.1

Predict the structure of the electron domains about the central atom in N_2O, which has an N—N—O arrangement of atoms. Predict the molecular shape by ignoring any lone pairs and considering the arrangement of the atoms alone.

Three-Dimensional Shapes around Atoms with Four Electron Domains

So far, we've predicted shapes of molecules and formula units that contain two or three electron domains. The linear and triangular shapes of these molecules and polyatomic ions exist in one or two dimensions only. That is, they can be drawn on a flat sheet of paper. But what happens when we consider molecules or ions that contain four or more electron domains? The shapes of these structures are three-dimensional. This means that we need a method of drawing three-dimensional objects on paper.

Chemists adopt two conventions to show whether an object is lying above or below the plane of a piece of paper. We use a solid wedge ◄, where the wide end is understood to be above the paper, to show an electron domain positioned above the plane of the paper.

✔ **Meeting the Goals**

Electron domains consist of electrons present in bonds and in lone pairs around an atom. Electron domains are positioned as far away from each other as possible. Two electron domains result in a linear molecule. When there are three electron domains, these are best separated by 120° angles. When all three electron domains are bonding domains, the molecular shape is triangular. When only two of the three electron domains involve bonded atoms (while the third is a lone pair), the molecular shape is angular or bent.

We use a "hashed" line, ⌐⌐⌐, where the wide end is understood to be below the paper, to show an electron domain positioned below the plane of the paper. We can then draw a three-dimensional object, such as a square viewed from one corner, as shown in Figure 13-3. In Figure 13-3 we start with a flat figure lying in the plane of the paper. This is rotated so the top of the figure (the "C" position) moves back and the bottom of the figure (the "A" position) moves forward. We draw wedges from B and D to A and hashed lines from B and D to C.

Figure 13-3 ▶ The relationship of wedges and hashed lines to three-dimensional structure is shown here.

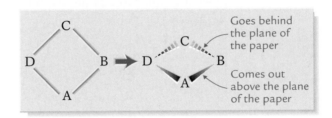

The wedge and the hashed lines also are used to show a three-dimensional arrangement of atoms in molecules. Let's consider the molecule methane (CH_4). This is one of the simplest molecules with four domains. Do these four domains point at the corners of a square? If all molecules had to be flat, then they might. But a flat structure would have angles of 90° between domains. Wider angles can be achieved if we draw the molecule with the four domains pointed at the corners of a three-dimensional tetrahedron, as shown in Figure 13-4.

See this with your **Web Animator** ▶

Figure 13-4 ▶ The electron domains of methane, a tetrahedral molecule.

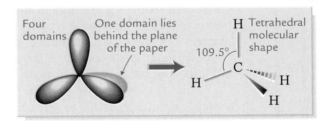

The corners of the tetrahedron form angles of 109.5° about the central atom. This is a significant increase from the 90° angles in the square.

Molecules with four electron domains do not have to have four bonded atoms. Some of the electron domains can be occupied by one or two lone pairs around the central atom. One well-known example of this case is water, which is shown in Figure 13-5.

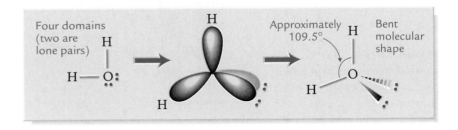

Figure 13-5 ▶ The molecular structure of water.

Example 13.3   **Predicting the structure of molecules**

Predict the structure of ammonia, NH_3.

Thinking Through the Problem To predict the structure of ammonia, we start with the Lewis structure.

Working Through the Problem There are four electron domains in ammonia, three bonding domains and one lone pair domain. When we arrange these in three dimensions, we observe that the N atom lies atop the three H atoms, forming a pyramidal structure.

Answer

$$
\text{H}\!-\!\overset{\displaystyle \cdot\cdot}{\text{N}}\!-\!\text{H} \longrightarrow \quad \overset{\displaystyle \cdot\cdot}{\underset{\text{H} \quad \text{H}}{\text{N}}}\!\diagdown\!\text{H}
$$

$$
\underset{\text{H}}{\big|}
$$

pyramid

How are you doing?   **13.2**

Predict the structure of sulfite, SO_3^{2-}.

So far, our discussion has shown how simple principles of electron domain and electron-electron repulsion explain many different structures of molecules and molecular ions. But we have not considered how different electron domains can have different sizes associated with them. For example, a single bond on an iodine atom pushes aside adjacent electron domains because the iodine atom is very large. And lone pair domains, which have no "second atom" to anchor them, also occupy a larger space than does a single bond. That is why the $\text{H}-\text{O}-\text{H}$ angle in water is slightly *less* than 109.5°.

Three-Dimensional Shapes in Atoms with More than Four Electron Domains

How do we draw the structure of a molecule containing six electron domains? We again want to arrange the six domains as far apart as possible. This will require an arrangement with the domains pointing at the corners of an octahedron, as shown in Figure 13-6.

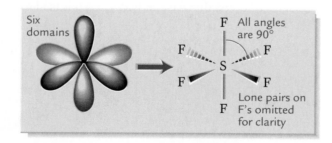

Figure 13-6 ▶ The three-dimensional structure of SF_6, a molecule with six electron domains.

F
F،،،، ،،،، F
Br
F؟ ¦ F

F،،،، ¦؟ F
Xe
F؟ ¦؟ F

If one of the electron domains is a lone pair, it will occupy one of the corners of the octahedron, as in BrF_5. If we draw the molecule with the bromine's lone pair on the bottom of the molecule, we can see that this structure should have a square pyramidal arrangement for the atoms.

When *two* lone pairs are present on the central atom, then the second lone pair goes to a position as far away from the first one as possible—that is, to the other side of the molecule. So in the molecule XeF_4 we expect the two lone pairs to be 180° apart, resulting in a *square planar* structure.

The cases we have studied previously all have equivalent positions for the electron domains—every corner of a triangle, a tetrahedron, or an octahedron is the same as any other corner. No such structure exists for a five-cornered object in three dimensions. But such molecules almost always have a trigonal bipyramidal structure, where the top and bottom positions (also called the apical sites) are 90° away from a triangle of three equatorial sites, each 120° away from one another, as shown in Figure 13-7. Note that the positions in the equatorial plane are less crowded than the positions in the apical sites, because they are 90° away from *two* positions. The 120° angle is much larger than a 90° angle.

Figure 13-7 ▶ The three-dimensional structure of PF_5, a molecule with five electron domains.

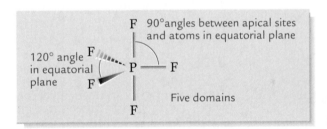

Larger Molecules and Three-Dimensional Structure

CONNECT TO
SECTION 2.3

In Chapter 2, we expanded our discussion of Lewis structures to include molecules with more than one central atom. We can use the principles developed there and in this chapter to discuss how three-dimensional structures of such molecules can be predicted, atom by atom.

Consider the molecules of ethane (C_2H_6), ethylene (C_2H_4), and acetylene (C_2H_2). Figure 13-8 shows these molecules from the perspective of electron domains. Because neutral carbon has 4 valence electrons, it will always form 4 bonds. There are four ways to do this—four single bonds, two single and one double bond, one single and one triple bond, or two double bonds. Figure 13-8 shows only three structural patterns are possible for neutral C.

We can expand this reasoning to include the different possibilities for the electron domains of O and N also, as shown in Table 13-1.

Considering electron domains allows us to see how structures develop for many different carbon-containing molecules. This discussion

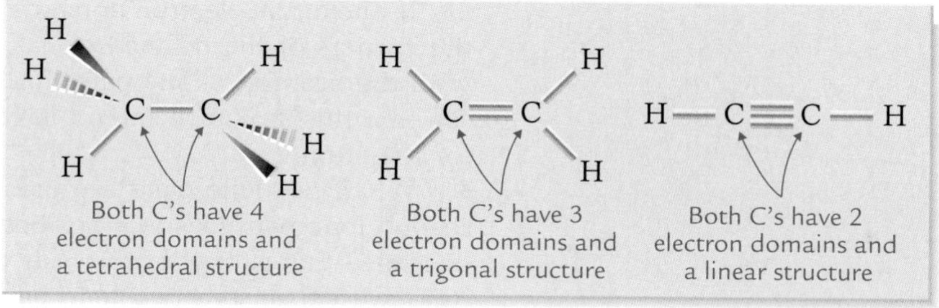

Figure 13-8 ▶ Electron domains in hydrocarbon molecules with two carbon atoms.

Both C's have 4 electron domains and a tetrahedral structure

Both C's have 3 electron domains and a trigonal structure

Both C's have 2 electron domains and a linear structure

leads us to the threshold of organic chemistry—the exploration of the very rich chemistry of carbon-containing molecules. Although it is not the purpose of this chapter to present this branch of chemistry in detail, you already have the tools needed to discuss the arrangement of C, O, and N atoms in many molecules. Consider the structure of nicotine in Figure 13-9.

We can see that nicotine is made of two rings. One ring, on the left in Figure 13-9(a), is composed of five carbon atoms and one nitrogen atom. Each of the six atoms in the ring has three electron domains. Each C is located at the center of a triangle, and the C—N—C angle is bent. The other ring has four carbons and one nitrogen, but in this case all of the atoms have four electron domains. The structure around the carbon is tetrahedral and the structure around the nitrogen is a pyramid.

Example 13.4   **Predicting the structure of molecules**

Indicate the structure around the C atom in methanol, CH_3OH, and formaldehyde, CH_2O.

Thinking Through the Problem We first need to present the Lewis structure of these molecules. We then look at the bonds around

Table 13-1 Structural Patterns for Central C, N, and O Atoms

Element	Lone Pairs	Number of Bonds	Single Bonds	Double Bonds	Triple Bonds	Electron Domains	Structure Around Atom
C	0	4	4	0	0	4	Tetrahedral
			2	1	0	3	Triangular planar
			0	2	0	2	Linear
			1	0	1	2	Linear
N	1	3	3	0	0	4	Pyramidal
			1	1	0	3	Bent
			0	0	1	2	At end of molecule
O	2	2	2	0	0	4	Bent
			0	1	0	3	At end of molecule

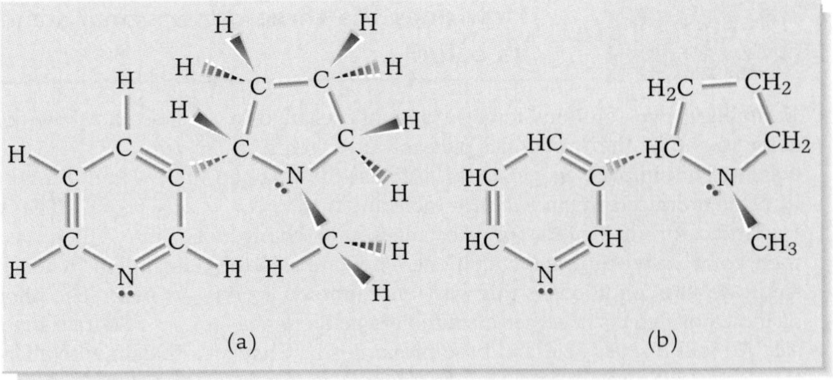

Figure 13-9 ▶ Full structure (a) and compressed structure (b) of nicotine.

(a)

(b)

the C atoms to determine the number of electron domains and the structure around the C.

Working Through the Problem The Lewis structures indicate that the C in methanol has four single bonds, and therefore four electron domains. The C in formaldehyde has two single bonds and one double bond, giving rise to three electron domains.

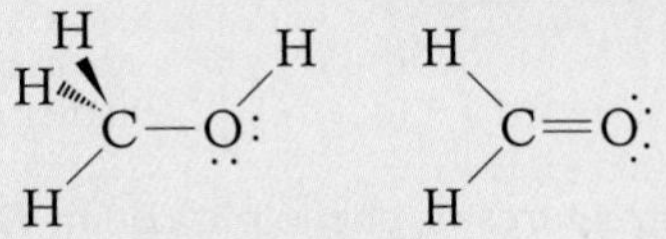

Answer The C in methanol has a tetrahedral structure around the C. The C in formaldehyde has a triangular structure.

How are you doing? **13.3**

Determine the structure around the C in HCN and in $H_2C=NH$.

A final point we can make about these structures concerns the simplicity of bonding in carbon. The vast majority of carbon atoms occurring in nature have four bonds. We don't always have to draw every one of these bonds. In some cases, we can omit the C bonds to H, and just write an abbreviated formula for each C in the structure, listing the number of hydrogens. We call structures that do not show the three-dimensional geometry of the bonds to hydrogen compressed structures. Figure 13-9(b) is a compressed structure for nicotine. Instead of drawing two C—H bonds on each of the carbons in the right ring, we simply write CH_2; and rather than write one C—H bond on each of the carbons in the left ring, we write CH.

PRACTICAL A — How does the three-dimensional structure of a molecule affect its color?

Combining two solutions to get a dramatic change in color is one of the traditional parts of any chemistry display. The simplest way to do this is to use the change in color produced by an *indicator* molecule.

Indicator molecules are substances that change their color under different conditions. They are said to "indicate" the conditions in this way. One common type of indicator detects whether a solution is acidic or basic. We will discuss acid and base properties in Chapter 15, but we already have enough knowledge to con-

sider how the color of an indicator relates to its structure.

One common indicator is phenolphthalein, $C_{20}H_{14}O_4$. The solid substance in its acidic form is white; when it is dissolved in water, it gives a colorless solution. When present in alkaline solutions, its color is pink. (See photo below.)

Redraw the following two structures of phenolphthalein with all lone pairs present. Indicate the arrangement of electron domains and atoms about each C atom.

(Chip Clark.)

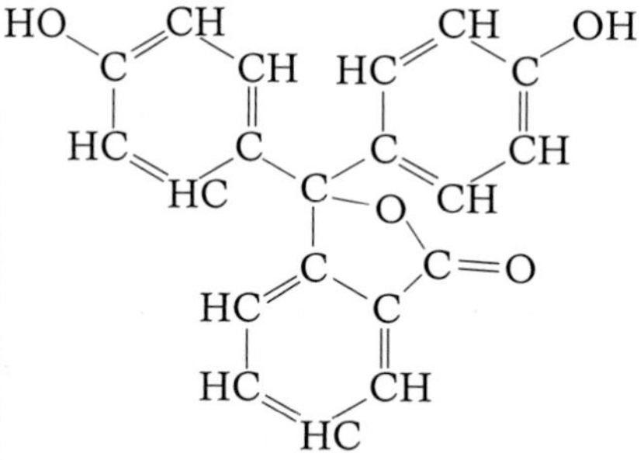

colorless—absorbs
higher energy light

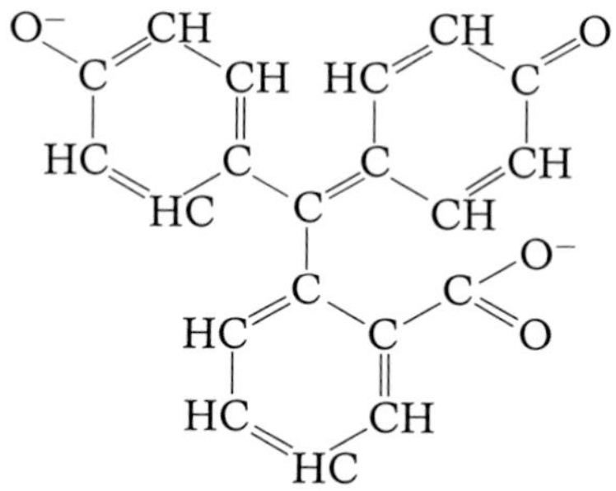

pink to red—absorbs
lower energy light

Examine these structures of phenolphthalein. Determine how the difference in structure might account for the difference in color. (Hint: Look at the electron domain structure of the central carbon and of all other carbons.)

Similar transformations occur with other indicators. For the pairs of indicators on the next page, suggest which form will absorb light at a lower energy and

which will absorb light at a higher energy. Use the information given for phenolphthalein to explain your answers.

If your instructor has covered resonance structures, you may notice that for all these indicators there are other equivalent resonance structures. Can you see how the central carbon is affected by resonance in one of the forms of each?

Example 13.5 — Drawing compressed structures

Draw compressed structures for aspirin and acetominophen, shown here as full structures.

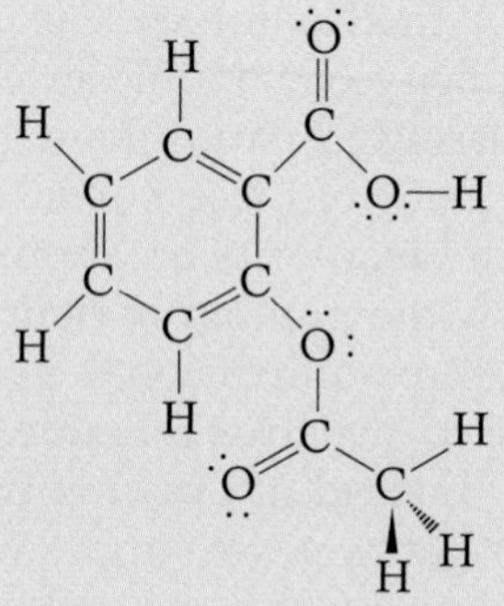

Aspirin

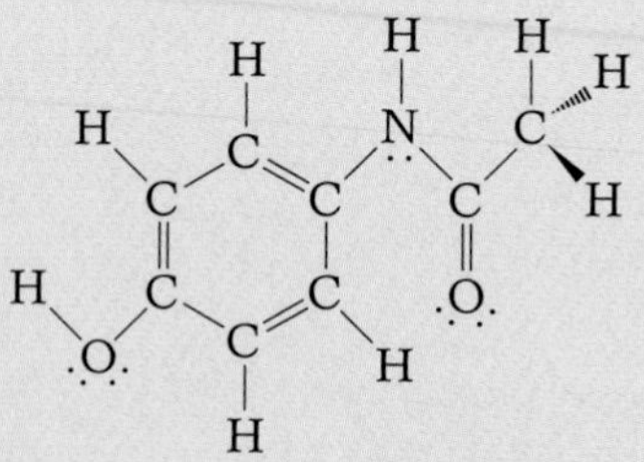

Acetaminophen

MALACHITE GREEN

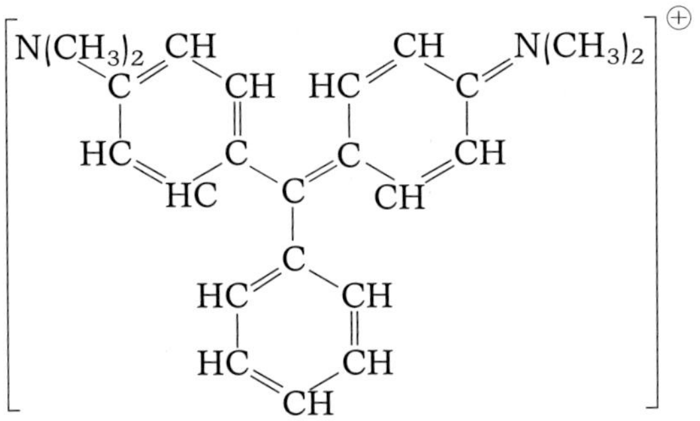

acidic form

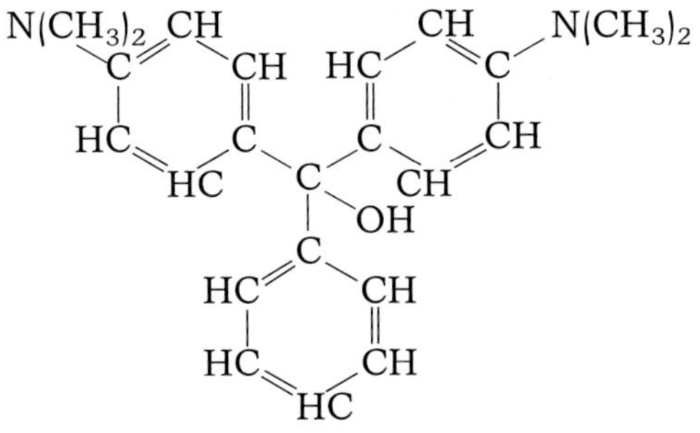

basic form

CURCUMIN (An extract of the plant turmeric)

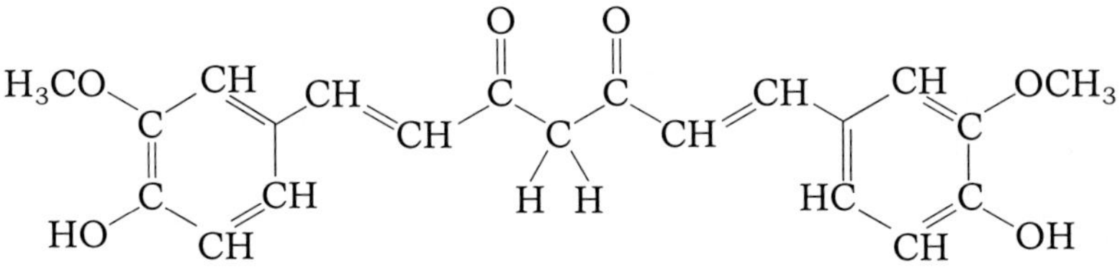

acidic form

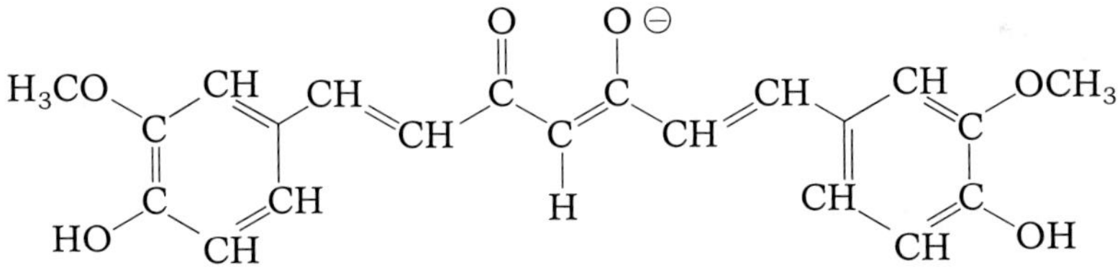

basic form

Thinking Through the Problem To convert these full structures to compressed structures we eliminate all C—H bonds and rewrite the C with the correct number of H's.

Working Through the Problem We get CH, CH_2, and CH_3 groups from carbons with bonds to one, two, and three hydrogens.

Answer

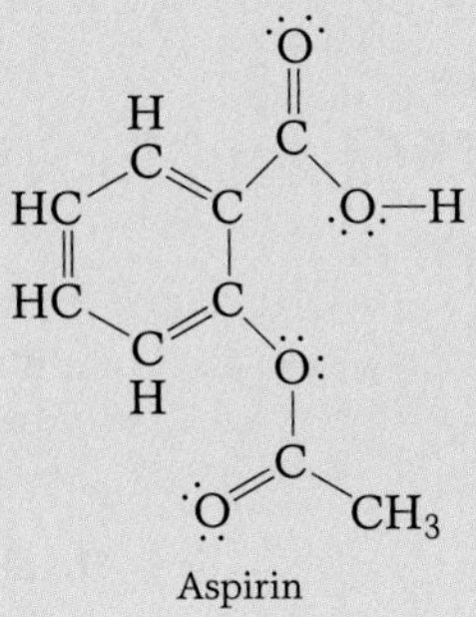

Aspirin

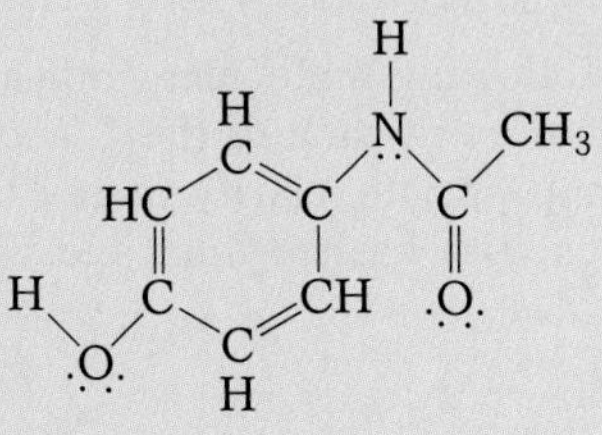

Acetaminophen

How are you doing? **13.4**

Draw the compressed structure for this drawing of the aspirin replacement ibuprofen.

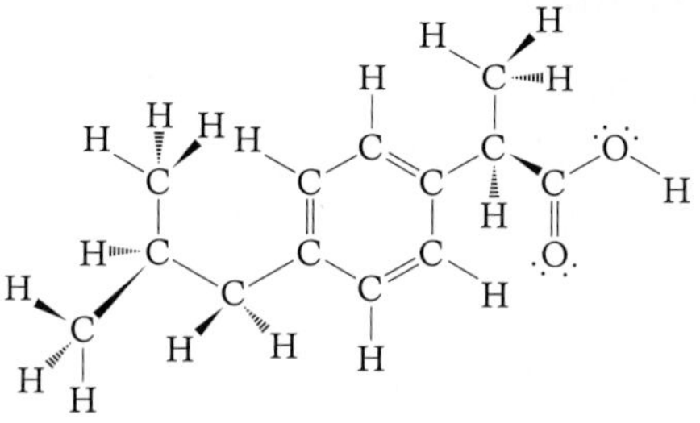

PROBLEMS

1. Draw the Lewis structure and predict the expected electron domain and molecular structure for each of the following molecules.

(a) Cl_2O (O in center)
(b) PF_3
(c) $NOCl$ (N in center)

2. Draw the Lewis structure and predict the expected electron domain and molecular structure for each of the following ions.

(a) ClO_3^- (b) PF_4^+ (c) NO_2^-

3. (a) Draw the Lewis structure and analyze the electron domain and molecular structure about the central atom in the molecules IF_3 and IF_5.
(b) The molecule IF_7 exists. What can you predict about its Lewis structure and the arrangement of electron domains about the central I?

4. Predict the expected geometries of the electron domains and bonded atoms about the central C in CH_3^-, $COCl_2$, and CBr_4.

5. Discussion question: Predict the expected geometries of the electron domains and the molecular arrangement in each of the following sulfur-containing substances and ions: SF^-, $SOCl_2$, SF_4, H_2S, SF_6, and SO_4^{2-}. Would it be possible to create a table like Table 13-1 for S? Why or why not?

6. For the following molecules, indicate the structure around each of the C and N atoms. Where appropriate, clarify three-dimensional structures with hashed lines and wedges.

(a)

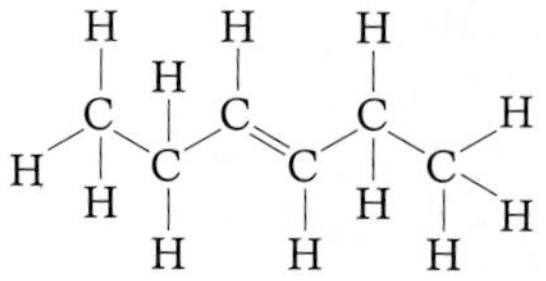

(b)

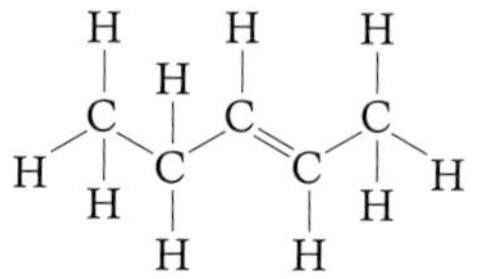

(c)

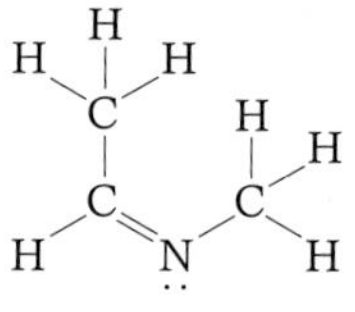

7. Draw the compressed structure of the molecules shown in the previous problem.

8. Analyze the electron domain and atom arrangements around each carbon and oxygen in the following molecules. Which atoms are trigonal and which are tetrahedral?

(a)

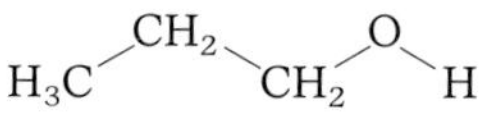

(b)

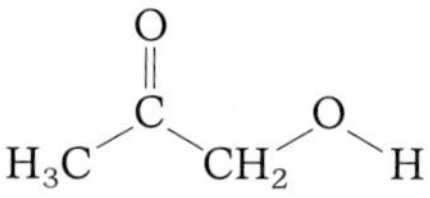

(c)

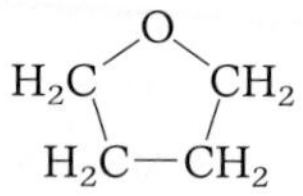

(d)

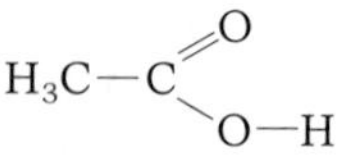

(e)

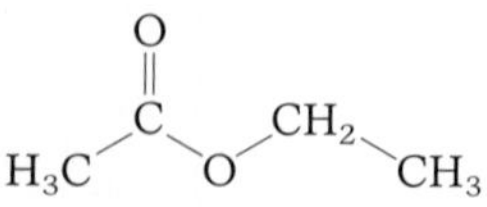

9. Draw the full structure of the molecules shown in the previous problem.

13.2 The Physical Origin of Valence Electron Behavior

SECTION GOALS

✔ Describe how light and light energy are characterized by chemists.

✔ Understand the evidence for quantized energy in the electrons of the atom.

✔ Use descriptions of atomic orbitals for the positions of the electron in one-electron atoms.

The Measurement of Color

The human eye and brain are extremely sensitive to color. When the eye and brain sense the entire spectrum of visible light that is the same as that of the sun, they perceive the color white or a shade of gray. When the eye and brain detect only a section of the spectrum, then a different color is perceived.

There are two ways an object can have color. First, the object can be a source of light from some portion of the visible spectrum. Under these circumstances, the observer experiences the effect of light coming from the object. Think of the bright yellow lights that are often used for exterior lighting. These use the emission of light by sodium vapor to give a bright light that is tightly restricted to the region at about 590 nm (one nanometer is 1×10^{-9} m), at the center of the yellow region of the spectrum. (You can tell that this light is of only a narrow wavelength range because if you look at a color picture illuminated by such a sodium vapor light the picture appears tan!)

Color can also arise when an object absorbs some wavelengths but reflects or transmits other wavelengths. An observer perceives the *complement* of the light that is absorbed. A color wheel, shown in Figure 13-10, helps us follow and predict the appearance of a simple

The color of a substance results from the interactions of light with its electrons. (Mary Kate Denny/PhotoEdit.)

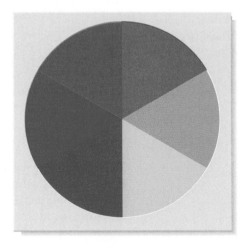

Figure 13-10 ▶ A color wheel is used to determine complementary colors.

Figure 13-11 ▶ The electromagnetic spectrum.

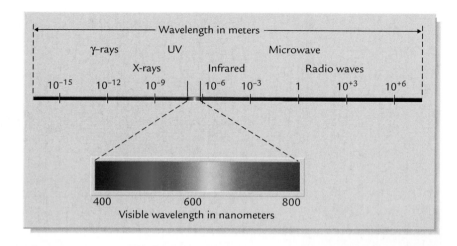

absorption. For example, when light of blue wavelengths is absorbed, then light of violet, green, yellow, orange, and red wavelengths is transmitted or reflected. This combination appears to our eyes as orange (the reason for this has to do with the physiology of vision, which is outside the scope of this course). Blue and orange are complementary colors (opposite each other on the color wheel).

Light energy is a form of electromagnetic energy. A diagram of the light spectrum and of other parts of the electromagnetic spectrum is given in Figure 13-11.

Visible light is primarily due to changes in the energy of electrons in atoms. When we discuss electrons and their energies in atoms and molecules we are mostly interested in the area of the spectrum in the range of 150 to 800 nm. This is the region of the spectrum where we find ultraviolet and visible light.

An important finding in the development of modern physics and chemistry was that light carries energy in the form of photons, or particles of light energy. The energy of a photon is inversely related to the wavelength of light. The longer the wavelength, the lower the energy. Figure 13-12 shows the electromagnetic spectrum in the range of 150 to 800 nm. In this figure, the light wavelengths are correlated with the energy that one mole of photons of light at that wavelength will carry.

✔ Meeting the Goals

Light is primarily characterized by its wavelength. Wavelengths in the range of 150–800 nm are associated with the electrons in atoms and molecules. The energy of the light is inversely related to its wavelength, with shorter wavelengths of light having more energy than longer wavelengths.

Figure 13-12 ▶ The energy of light in the ultraviolet and visible region.

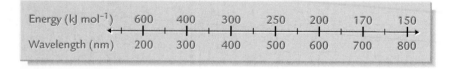

Energy (kJ mol^{-1})	600	400	300	250	200	170	150
Wavelength (nm)	200	300	400	500	600	700	800

Quantum Energy and the Properties of Atoms

The treatment of valence electrons that we have used so far is very helpful in describing bonds among atoms. But we need to look more closely at atoms if we are going to say anything else about how the electrons in atoms determine the bonding properties of atoms.

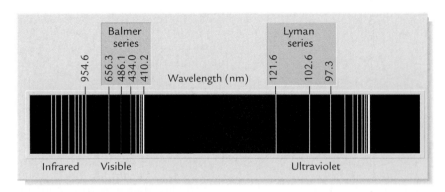

Figure 13-13 ▲ (a) Hydrogen discharge lamp and (b) spectrum produced by a diffraction grating. (a: Chip Clark.)

KEY IDEA 13.3

We begin by considering the experiment shown in Figure 13-13(a). Here, a tube is filled with hydrogen atoms and an electrical current flows through the tube, causing it to glow with a characteristic red light. Why does this happen? The answer is that the electrical current causes the electrons in the hydrogen atoms to acquire greater energy. An electron returns to a lower allowed level by giving off a photon in the red region of the visible light spectrum.

When we let light from a hydrogen tube fall on a device (called a spectroscope) that separates light into its wavelengths (as a prism does), we see that the light contains only particular wavelengths. These wavelengths, shown in Figure 13-13(b), are referred to as the **line spectrum** of hydrogen. The line spectrum of hydrogen is a physical property of the element. The fact that no energy is emitted in the regions between the lines points to a very important fact: the energy of the electron in the atom can only take on certain values, and transitions associated with light emission arise from transitions between these values.

This evidence for energy levels is explained in science by **quantum theory.** *"Quantum" means "amount" and the quantum theory states that energy levels of a system can only have certain values.* In atoms, we call these levels the energy levels of the atom. The most important energy levels for this course are those for the electrons in the atom.

When light is given off by an atom, the atom's electron structure changes. Because the atom can only have certain energies, it is reasonable that only certain energies of light will be observed. This is how the line spectrum of the atom arises: we only see light at wavelengths that correspond with "allowed" energy changes. This is a very unusual phenomenon for us, because we are accustomed to controlling energy changes to any chosen degree—we can warm ourselves a little from a large radiator or a lot from a very small, but very hot, piece of metal. But this experience is with huge assemblies of atoms, usually in condensed phases. We have to set aside our everyday experience and work with the observations of instruments that can detect individual atoms and their behavior.

The Concept of the Atomic Orbital

Throughout this book, we have approached the problem of covalent bonding from a valence electron standpoint. By now you are able to look at a group number on the periodic table and use it to determine the number of valence electrons in an atom. You can draw Lewis structures of simple molecules, and analyze those for much more complicated ones. And you can apply your knowledge of valence electrons to predict the formulas of many different ionic compounds. Finally, you know that the space occupied by valence electrons affects the three-dimensional structure of molecules.

The valence electron concept explains a lot. But why does it work? We can begin to explain this by considering a series of experimental observations for the simplest atom: hydrogen atoms (not molecules) alone in the gas phase. These atoms would often form molecules under normal conditions, but it is a relatively easy experiment to keep them in a monatomic, gaseous state.

According to quantum theory, the electrons in atoms can be present only in certain arrangements. This invites us to consider that the electrons may occupy only certain energy levels in the atom—a concept usually expressed by saying that *there are energy levels associated with individual electrons in the atom, with each electron assigned to a particular state.* This is exactly what is suggested by the line spectra of hydrogen and the other elements.

The states occupied by the electrons in atoms are called **atomic orbitals.** The noun *orbital* is one of the truly unfortunate terms in science, because it suggests incorrectly that the electron orbits the nucleus as a planet orbits the sun. This is simply not true, for the electron is best viewed as a cloud that cloaks the nucleus, much like a Nerf© ball might surround a marble inserted in its core.

What are the atomic orbitals, and how do we describe them? For this we will need to invoke just two more experiments from atomic physics related to waves and matter. We are familiar with the properties of waves in large bodies of water. For example, when water strikes a regularly spaced series of posts in a tank, the water rebounds in a characteristic wave pattern. When light shines on a microscopic series of slits, known as a diffraction grating, it gives a pattern of light on a screen.

But why are waves important for electrons? Electrons, after all, are particles. And they do not form a continuous thing like water in the ocean. This is true, but in the 1920s it was shown that electrons do travel in waves, allowing us to create an explanation of the electron levels in atoms. Because they are waves, electrons in atoms will follow patterns like those dictated by waves in a guitar string (Figure 13-14). If we pluck a string, we get a simple wave pattern that has a single hump. If we make it vibrate more quickly, we can set up the string to have a double-humped pattern. The fixed, or stationary, point in the string moves neither up nor down but rather becomes a *node* in space. If we put more energy into our string, we can induce other patterns, with three humps and two nodes.

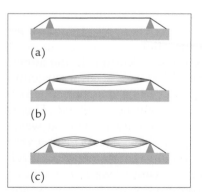

Figure 13-14 ▲ (a) A guitar string spans two contact points. When it vibrates, it can have (b) zero, (c) one, or more nodes.

We also realize that it is not possible to have a stable pattern with two and a half humps: it is either two or three, or some other integer value. A guitar string, like the electron in the atom, is an example of a standing wave, and it is restricted in the number of nodes it can have.

This wave analysis creates a simple picture of the energy levels of the electron in the atom and in all matter. The electron has to move in a confined region around the atom, creating boundaries that restrict the electron. Only certain levels, with 0, 1, 2, ... nodes, are allowed. We do not observe electrons in between these levels, just as we never see a stable guitar string vibrating with 2.5 nodes.

Each of these different levels represents a different energy level, designated by the variable n, where n is the **principal quantum number** of an energy level. The value of n can be any number in the series $n = 1, 2, 3,...$ (The values of n are chosen so that none of them is zero. This way we can say we have $n = 1$ for the "first quantum level.")

These observations allow us to summarize with the characteristics of atomic orbitals given in Table 13-2.

REMEMBER

Table 13-2 Characteristics of Atomic Orbitals

(a) Atomic orbitals represent energy levels for electrons in atoms.
(b) Atomic orbitals have an integer number of nodes associated with their descriptions; more nodes are associated with more energy.
(c) Atomic orbitals can accommodate $2e^-$ each.

Table 13-2 contains all the principles needed to set up the orbitals in an atom. They work with hydrogen and other atoms with a single electron and they are approximately correct for multi-electron atoms. The number of nodes of each orbital will define the shape of the orbital, which is important in determining how more complex atoms with two or more electrons behave.

We expect the lowest available energy level of an electron in an atom to have no nodes, just like the lowest energy state of any standing wave. Since an atom is a three-dimensional object, this orbital will have a shape that is spherical—just like the Nerf© ball. We call this orbital the 1s orbital. The 1s orbital is shown in Figure 13-15.

We expect only two electrons to occupy the 1s orbital. Beginning with the first period, hydrogen has only one electron and helium has two electrons. We state, then, that a hydrogen atom has a $1s^1$ arrangement and the He atom has a $1s^2$ arrangement, where the superscript number is the number of electrons in the s orbital. This completes the first row of the periodic table and the first energy level of the atom.

What about the second level? This will have *one* node, which was easy to picture for a wave moving in two dimensions, like a jump rope. But now we have three dimensions, and there are *two* ways to place a node in space. One is a sphere at a fixed radius from the nucleus. We call this a **radial node.** This gives rise to a 2s orbital, the

Figure 13-15 ▲ The shape of the 1s orbital.

PRACTICAL B — What does the length of a molecule tell us about the energy of the electrons in the molecule?

As we saw in Practical A, simply opening up the center of a molecule to multiple bonding can make a difference in the energy of light absorbed by that molecule. It is also possible to look at interesting trends in the light absorbed by different molecules as it compares to their length and composition. Consider the following trends:

Molecule and Energy of Absorbed Light	Molecule and Energy of Absorbed Light	Molecule and Energy of Absorbed Light
C_2H_4 655 kJ mol^{-1}		
C_4H_8 638 kJ mol^{-1}	C_4H_6 550 kJ mol^{-1}	
C_6H_{12} 667 kJ mol^{-1}	C_6H_{10} 526 kJ mol^{-1}	C_6H_8 445 kJ mol^{-1}

Figure 13-16 ▶ The shape of the 2s orbital.

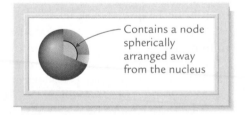

Contains a node spherically arranged away from the nucleus

shape of which is akin to a very primitive onion, with two layers. Figure 13-16 shows the 2s orbital.

But we can create a node in an angular direction, as a plane that slices the orbital in half. These are known as **angular nodes**. Of course, there are three unique planes in three-dimensional space—the xy plane, the xz plane, and the yz plane. As shown in Figure 13-17, this creates three other orbitals, which are pointed along the z, y, and x axes. We call these orbitals the $2p_x$, $2p_y$, and $2p_z$ orbitals.

Thus, we have four orbitals in the second level, distinguished by the kind of nodes they contain. In addition, there are two fundamen-

Describe the trends that you observe for these molecules. Compare the energies in the first column. These molecules differ from one another by their lengths. How much does the energy depend on the length alone? Now compare energies across the rows. All the molecules in one row contain a carbon chain of the same length. Compare the bonds present in these molecules. How much does the energy depend on the kinds of bonds in the molecule?

Next, consider what happens when we wrap a chain around itself. The molecule benzene, C_6H_6, has a ring of carbons, as shown below. It has a low energy absorp-

tion, 468 kJ/mole. How does this compare to the other C_6 molecules we have examined?

Carbon forms long chains very easily, and it is easy to relate these trends with other molecules. For example, the molecule *trans-β-carotene* is related to the substances used to absorb visible light in the eye, and, as the name implies, is present in a related form in carrots. The lowest energy transition for this molecule occurs at 247 kJ/mol, equivalent to blue-green light. Because we see the complementary color, we perceive the color of *trans-β-carotene* as orange.

benzene

β-carotene

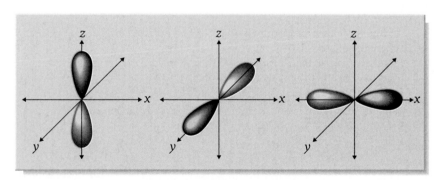

Figure 13-17 ▲ The shape of the three 2p orbitals.

✔ **Meeting the Goals**

Each of the atomic orbitals has a different shape, which determines the position of the electron that occupies the orbital. The shapes are distinguished by the different nodes—radial and angular.

tally different kinds of orbitals—s and p. We designate these as the *s* and *p* **subshells** of the $n = 2$ quantum level.

We are now ready for the third energy level. This will have two nodes. We can mix these among the radial and angular cases. When we have two radial nodes there can be no angular nodes, and we have a 3s orbital. An orbital in the third energy level with one radial node will have one angular node; there are three of these, the 3p orbitals. Finally, with the n = 3 level we also have orbitals with no radial nodes and two angular nodes. There are five ways to place two nodes in three-dimensional space, but we will not concern ourselves with their shape. These are designated as *d*-orbitals.

Table 13-3 summarizes the discussion of the first, second, and third energy levels, extending it to the fourth level. In all cases, the number of nodes is equal to one less than n, the principal quantum number.

Table 13-3 Quantum Levels, Nodes, and Orbitals for the First Four Quantum Levels

Principal Quantum Number	Nodes	Radial Nodes	Angular Nodes	Number of Orbitals	Subshell Designation
1	0	0	0	1	1s
2	1	1	0	1	2s
		0	1	3	2p
3	2	2	0	1	3s
		1	1	3	3p
		0	2	5	3d
4	3	3	0	1	4s
		2	1	3	4p
		1	2	5	4d
		0	3	7	4f

PROBLEMS

10. Quantum theory also explains why atoms and molecules absorb light only at certain energy values. Consider the discussion of the hydrogen emission spectrum and use it to explain the absorption spectrum.

11. Indicate the number and the distribution (radial and angular) of the nodes for the following orbitals:

(a) 4s   (b) 3p   (c) 4d

12. Look at Table 13-3. Count the number of orbitals in each energy level. Do you see a trend? How many orbitals do you expect in the fifth energy level?

13. In Figure 13-12, the energy of light is correlated with its wavelength. The energy amounts are given in terms of the amount of energy in one mole of light particles. Determine the

approximate light energy associated with one mole of light particles at 400, 500, and 600 nm. Also determine the light energy associated with one single light particle at each wavelength.

14. A firefly emits light energy obtained from a chemical reaction. If we assume that all of the energy in the reaction is released as light, how much energy is released per mole if this light is green?

15. Discussion question: An orbital has the following node structure. Identify the proper label for these orbitals.

(a) 2 radial nodes, 1 angular node
(b) 3 radial nodes, 0 angular nodes
(c) 0 radial nodes, 2 angular nodes

Can you ever tell what orbital you have from just the number of radial or angular nodes?

13.3 Electron Configuration and Atomic Properties

SECTION GOALS

✔ Understand the experimental meaning of ionization energy.

✔ Relate ionization energies to the shell structure of atoms.

✔ Write the electron configuration of an element.

Ionization Energy of Atoms and the Periodic Table

We will assume that electrons in atoms with two or more electrons (polyelectronic atoms) have the same orbitals available to them as do one-electron atoms. How these electrons fill the atomic orbitals will be discussed later in this section. But first, let's look at the energy levels and their different shapes.

Our next experiment involving electrons in atoms has to do with the origin of the familiar phenomenon of magnetism. When a material is magnetic, the force originates in the material's electrons. We know that the electron is charged, but how can a single particle give rise to magnetism? Experiments with larger objects say that a charged object that is spinning is magnetic. That is why we use the analogy of electron spin to explain electronic magnetism. The electron can spin in one of two directions, clockwise or counterclockwise, giving rise to a magnetic field that is "up" or "down."

If we shine light on an atom and the light has enough energy, then we find that we can get electrons to "pop off" the atoms, forming a positively charged ion. The **first ionization energy** of an atom is the smallest amount of energy needed to remove a single electron from the neutral atom in the gas phase. It represents the energy required for the process $A \rightarrow A^+ + e^-$. Table 13-4 lists the first ionization energies of the first eighteen elements.

The data in Table 13-4 are derived from experiments on neutral atoms in the gas phase. It is interesting to compare these data with the chemical properties of the elements. For example, if we graph the ionization energies of the first forty elements, we see that there is a

✔**Meeting the Goals**

Ionization energy is the energy, usually provided by light, that causes an electron to leave an atom or molecule.

Table 13-4 First Ionization Energies of the First Eighteen Elements, in kJ mol^{-1}

H 1310							He 2370
Li 520	Be 900	B 800	C 1090	N 1400	O 1310	F 1680	Ne 2080
Na 500	Mg 740	Al 580	Si 790	P 1010	S 1000	Cl 1250	Ar 1520

dramatic decrease in the ionization energy whenever we move from a group VIII element to a group I element.

Because ionization energy is an indication of how strongly an atom "holds onto" its electrons, we can use the ionization energy to suggest that an alkali metal atom holds onto its electrons less strongly than does a rare gas atom. Of course, we also know that the alkali metals are atoms that readily form ions with a +1 charge, since they lose one electron apiece.

A second trend is that the ionization energies of atoms generally increase as we go from left to right in a row of the periodic table. The largest ionization energies are those of the noble gases. As we saw in Chapter 2, noble gases are elements whose atoms have a full complement of valence electrons (two for helium, eight for the others). This, too, is further evidence that eight valence electrons (two for He) have particular stability. A set of eight electrons in the valence shell is called a **closed shell** of electrons.

CONNECT TO
SECTION 2.2

Of course, the first ionization energy might tell us that the electrons in certain atoms are *all* easy to remove. But this is not true. To get a closer look at the electron structure we can also examine data for the removal of a second and third electron. Consider the data for the elements Ne, Na, and Mg in Table 13-5. As we saw, it is hardest to pull the first electron out of Ne as compared with Na and Mg. But the element with the highest second ionization energy is Na, not Ne, and Mg has the highest third ionization energy.

Table 13-5 First, Second, and Third Ionization Energies, in kJ mol^{-1}

Element	First Ionization Energy $A \rightarrow A^+ + e^-$	Second Ionization Energy $A^+ \rightarrow A^{2+} + e^-$	Third Ionization Energy $A^{2+} \rightarrow A^{3+} + e^-$
Ne	2080	3950	6120
Na	500	4560	6910
Mg	740	1450	7732

✔ **Meeting the Goals**

Ionization energy indicates how tightly electrons are held. We observe that certain electron counts (for example, in the noble gases) are associated with high ionization energies, and consider that these electron counts indicate filled electron shells. Atoms with low ionization energies have shells with few electrons.

This suggests that while the first electron is easy to remove from Na, the other electrons are held very tightly. Similarly, it is relatively easy to remove one or two electrons from Mg. But Mg^{2+} has unusual stability. We recognize this trend as a familiar one, related to the formation of ions. Examining individual atoms in the gas phase confirms what we knew from compound formation: it is relatively easy to remove electrons from atoms until the number of electrons equals the number in a group VIII atom. Then the atom or ion has a closed shell of particular stability.

Valence Shells and Electron Cores

The ionization energy of an atom or ion does not tell us anything about the placement of the electrons in an atom. When we ionize Ne, for example, how do we know whether the first electron to be

See this with your **Web Animator** ▶

removed is a 1*s*, 2*s*, or 2*p* electron? The fact is that we have no way of knowing for sure—electrons do not come with little tags saying where they were before ionization. But a further study of ionization data does support the view that some orbitals are more likely to give up their electrons than others.

Table 13-4 includes the lowest energy that will cause an electron to come out of an atom. Higher energies will also remove an electron. We find that C, for example, has three peaks in its photoelectron spectrum, at 1,090, 1,720, and 28,600 kJ mol^{-1}. This is interpreted as evidence for electrons in 2*p*, 2*s*, and 1*s* levels. We see that ionizing some of the electrons, which we understand as the 1*s* electrons, requires much more energy than ionizing the others. But why aren't the 2*s* and 2*p* levels equal in energy? This has to do with the presence of more than one electron in the atom. The electrons interact with each other, not just with the nucleus. This makes the 2*p* orbital higher in energy than the 2*s* orbital.

Example 13.6   **Identifying trends in the first ionization energies**

Identify and discuss the trends in the first ionization energies as we move from one row of the periodic table to another.

Thinking Through the Problem We must examine the ionization energies given in Table 13-4. We are looking for trends.

Working Through the Problem A trend is noted as a continued increase or decrease in the numbers. First, we note that whenever we finish a full shell, as when we go from He to Li and from Ne to Na, there is a big drop in the first ionization energy. We also note that the first ionization energy for H is greater than that for Li, which is greater than that for Na.

Answer There are two major trends. First, there is a drop in ionization energies between the end of one period and the beginning of the next. This is interpreted as showing that the He and Ne have very stable electron structures, consistent with the filling of the $n = 1$ level or the $n = 2$ level. Secondly, ionization energies decrease as we go down a group of elements. This is consistent with higher shells (1, 2, and 3, respectively) being occupied.

How are you doing?   **13.5**

Look at the difference between Be and B and between Mg and Al in Table 13-4, then comment on how these differences reflect the shell structure of the atom.

How are you doing?   **13.6**

Look at the data for the first ionization energy for the third period elements. Do we see the same trends as in the second period?

It is surprising that all the electrons do not just go to the lowest ($n = 1$) orbital. To explain why not, we invoke a principle that no two electrons can be identical. This means that only two electrons can occupy a single orbital: one electron is said to have "spin" that is "up," and one has a "spin" that is "down." After that, the next electron must go into a different orbital. Thus, there are only two electrons in the 1s orbital, two in the 2s orbital, and two in each of the three 2p orbitals, totaling six electrons in the 2p orbitals.

We are now ready to consider the third energy level. This will have two nodes. We can mix these among the radial and angular cases.

> 3s orbital: two radial nodes. This will accommodate two electrons.
>
> 3p orbitals: one radial node, one angular node. There are three of these, and they will accommodate six electrons.
>
> 3d orbitals: two angular nodes. There are five ways to place two nodes in three dimensional space. We can call these "double" orbitals because of the double nodes. The five 3d orbitals can accommodate ten electrons.

See this with your **Web Animator**

As with the second level, we expect that the 3s subshell will be lower in energy than the 3p, and that the 3d will be highest in energy. This is also borne out by experiment, with one last observation: in neutral atoms of an element, electrons in the 3d subshell are so high in energy that the 4s subshell is filled first. Thus, we fill the 3s subshell (Na, Mg), the 3p subshell (Al through Ar) and then the 4s subshell (K and Ca). Only after this do we fill the 3d shell, with the elements Sc through Zn. We are then ready for the 4p subshell.

Writing Electron Configurations

When we apply the *s, p, d,* and *f* labels to the location of electrons in an atom, we are using a model for the arrangement of the electrons in an atom. Such an arrangement is called the **electron configuration**. We can use what we have developed in this section to help us write the electron configuration for any element, given its position in the periodic table. We will label the periods as values of *n*, where $n = 1, 2, 3$. *A value of n denotes a principal energy level. Energy increases as the value of n increases.*

The alkali metals (group I) add the first *s* electron for any principal energy level; alkaline earth metals (group II) add the second *s* electron for any principal energy level. Thus, the first two elements in every period have the electron configuration, ns^1, ns^2, respectively. For example, in the third period, Na has the configuration $3s^1$. Mg has the configuration, $3s^2$. In the fourth period, K is $4s^1$ and Ca is $4s^2$.

The *p* orbitals can accommodate six electrons. The "*p*" section of the periodic table is on the far right side (groups III–VIII). The electron configuration for the elements in the third period are shown in Table 13-6.

Table 13-6 Electron Configurations of Third Period Elements

Group	I	II	III	IV
Element	Na	Mg	Al	Si
Configuration	$1s^2 2s^2 2p^6 3s^1$	$1s^2 2s^2 2p^6 3s^2$	$1s^2 2s^2 2p^6 3s^2 3p^1$	$1s^2 2s^2 2p^6 3s^2 3p^2$
Group	V	VI	VII	VIII
Element	P	S	Cl	Ar
Configuration	$1s^2 2s^2 2p^6 3s^2 3p^3$	$1s^2 2s^2 2p^6 3s^2 3p^4$	$1s^2 2s^2 2p^6 3s^2 3p^5$	$1s^2 2s^2 2p^6 3s^2 3p^6$

Figure 13-18 ▼ Periodic table of the elements showing trends in the order in which orbitals are filled.

We have seen a trend where we fill the orbitals in the order of the principal quantum number n and then the subshells s and p. The next thing we might expect, beyond the $1s^2 2s^2 2p^6 3s^2 3p^6$ configuration of Ar, is to begin filling the $3d$ orbitals. However, for reasons beyond the scope of this text, the $3d$ orbitals in a neutral atom are filled only *after* the $4s$ orbitals. Thus, the configuration of K is $1s^2 2s^2 2p^6 3s^2 3p^6 4s^1$, that of Ca is $1s^2 2s^2 2p^6 3s^2 3p^6 4s^2$, and that of Sc is $1s^2 2s^2 2p^6 3s^2 3p^6 4s^2 3d^1$. This is true whenever we move past a noble gas: the next two elements get s electrons before the d orbitals are filled. In general, transition elements have configurations $(n-1)d^x$, where x has values from 1 to 10 (some exceptions occur, but none are important to us in this text). A similar "delay" in the filling of the $4f$ orbitals occurs also. These are filled (to make the lanthanide series) only after we have filled the $5p$ and $6s$ orbitals. This is summarized in Figure 13-18.

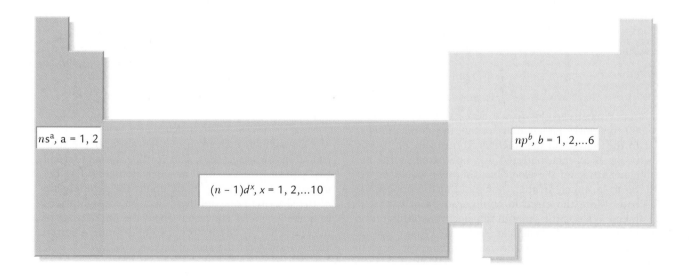

$ns^a, a = 1, 2$

$np^b, b = 1, 2, ...6$

$(n-1)d^x, x = 1, 2, ...10$

$(n-2)f^y, y = 1, 2, ...14$

PRACTICAL C   How do the "core" electrons affect the size of an atom?

In this chapter, we have learned a lot about how the valence electrons affect the behavior of atoms and the molecules they form. As we learned early in the book, valence electrons are involved in the interactions that form bonds and most ions. However, if valence electrons were all that mattered, we would expect all of the atoms in a column of the periodic table to be identical. But this is not the case. Be, Mg, Ca, Sr, and Ba are similar in many of their properties, but we find that while two (Mg and Ca) are essential to life, strontium is unimportant in biology, and both beryllium and barium are significant poisons.

Although there are many differences that account for the change in properties as we move down a column of the periodic table, perhaps none is as important as size. The size of an atom is often defined with respect to the distance between two of the atoms in the solid phase. The radius of the atom, which is half this distance, can be used as a measure of its size. (Note that there are some additional complications if the pure elements form molecules, like Cl_2, but we won't worry about those here.)

The table below shows radii for the nonradioactive alkaline earth elements.

Atom	Atomic Radius
Be	110 pm
Mg	160 pm
Ca	200 pm
Sr	215 pm
Ba	224 pm

What trend do we see here? What can account for this? First, let's remember that the atoms increase in the charge on the nucleus from +4 in Be to +56 in Ba. If all that mattered was how strongly the nucleus attracted the electrons, then we would expect Ba to be much smaller than Be. But the opposite is true. Write the electron configurations of each element. What is the same in the electron configuration of the outermost (valence) electrons? What changes? How can the similarities and differences explain these results?

The answers to these questions suggest that we can relate the configuration of valence electrons to the size of an atom. Let's next examine the size of the ions formed when these atoms are in compounds. As we know, only the +2 ions are important. Here are the sizes of these ions, in similar compounds.

Ion	Ionic Radius
Be^{+2}	27 pm
Mg^{+2}	72 pm
Ca^{+2}	100 pm
Sr^{+2}	116 pm
Ba^{+2}	136 pm

Is the same trend observed? If we consider that the formation of the ion is due to the removal of the valence electrons, what is the electron configuration of each of these ions? Comment again on the electron configuration of the ions and the influence of the principal quantum number on these atoms.

We have seen what happens as we move down a single column of the periodic table. Both the core and the valence electrons are important in the size of these atoms and ions. But the Group II elements are all metals. What happens when we look at a column of nonmetals? The following table has data for the halogens and for their −1 ions.

Atom/Ion	Atomic Radius	Ionic Radius
F / F⁻	64 pm	133 pm
Cl / Cl⁻	99 pm	181 pm
Br / Br⁻	114 pm	196 pm
I / I⁻	133 pm	220 pm

Consider the electron configuration of the neutral atoms and their ions. Can you see the same trend that you saw for the alkaline earth elements? Use the observations you have made about the relationship of the size of atoms and their electron configurations to make a general statement about the importance of nuclear charge, shell structure, and outer electron configuration on the size of an atom or ion.

It is a simple thing to write the electron configuration for any element, following this general scheme. (There are some exceptions to the exact ordering predicted by this scheme. However, you are not required to know these exceptions at this time.)

Example 13.8   **Writing electron configurations**

> Write the electron configurations of C, N, and O atoms.
>
> Thinking Through the Problem We must write the electron configuration for each of these elements.
>
> Working Through the Problem We locate the elements on the periodic table; these elements have 6, 7, and 8 electrons, respectively. We expect that they will each have 2 electrons in the $1s$ orbital and two electrons in the $2s$ orbital. The remaining electrons will be in the $2p$ orbital.
>
> Answer   C has a $1s^2 2s^2 2p^2$ configuration, N has a $1s^2 2s^2 2p^3$ configuration, and O has a $1s^2 2s^2 2p^4$ configuration.

Example 13.9   **Identifying atoms from electron configurations**

> Identify the atoms with these electron configurations:
> (a) $1s^2 2s^2 2p^6 3s^2 3p^5$ (b) $1s^2 2s^2 2p^6 3s^2 3p^6 3d^{10} 4s^2 4p^4$
>
> Thinking Through the Problem We must identify the elements with the above electron configurations.
>
> Working Through the Problem The first configuration belongs to an element in the third period, as the highest n value is 3. The second element belongs to the fourth period, as the highest n value is 4. We will count the s and p electrons in the highest energy orbitals to determine the group number for each element.
> (a) This configuration belongs to the element in the third period, group VII. (b) This configuration belongs to the element in the fourth period, group VI.
>
> Answer   (a) Cl (b) Se

How are you doing?   13.7

Determine the electron configuration for Sc and for In.

PROBLEMS

16. Identify the atoms with these electron configurations.

(a) $1s^2 2s^2 2p^5$ (c) $1s^2 2s^2 2p^6 3s^2$
(b) $1s^2 2s^2 2p^6 3s^2 3p^3$ (d) $1s^2 2s^2 2p^6 3s^2 3p^6 3d^{10} 4s^2 4p^6$

17. Indicate the number and the distribution (radial and angular) of the nodes for the following orbitals:

(a) $3s$ (b) $3d$ (c) $4f$

18. Identify the electron configurations you expect for the representative elements Ca, Rb, Sn, and Se.

19. Indicate the electron configurations you expect for the elements Cs, As, S, and I.

20. Consider the electron configurations for the elements Ne, Ar, Kr, and Xe. What is common? What is different, at least for the last electrons put into the atom? Use your answers to explain why the first ionization energies of these atoms vary as shown in the following table (all energies in kJ mol^{-1}).

Ne	Ar	Kr	Xe
2080	1520	1350	1170

21. Discussion question: Certain representative group metal atoms form more than one ion. For example, lead forms compounds with either the Pb^{4+} or the Pb^{2+} ion. Examine the electron configuration of Pb atoms and indicate why both of the ions may be formed.

22. Concept question: Consider the statement "The valence electrons in an atom are those in the subshell that is being filled for that atom." Is this true or false? Give an example as part of your answer.

23. The following atoms have a partially occupied subshell. Indicate how many electrons are in that subshell.

(a) Al (b) Br (c) N (d) Ti

24. We saw how the "fine structure" of the first ionization energy of an atom supports the presence of subshells in the atom. Subshells have some stability relative to partially filled shells. Examine the data in Table 13-4. What atoms will have an s^2 configuration? Which will have an s^2p^1 configuration? Compare the first ionization energies of these elements. Is there a pattern supporting the statement that filling just the s subshell gives particular stability?

25. When we add an electron to an atom, we get a monoanion, A^-, which has one electron more than the neutral atom. The energy required to remove an electron from a monoanion is the element's electron affinity.

(a) The electron affinities for O, F, Ne, and Na are shown in the following table. What is the expected electron configuration of each anion?

Element	Electron Affinity $A^- \rightarrow A + e^-$ (kJ mol^{-1})
O	141
F	328
Ne	0
Na	53

(b) For which anion is it easiest to remove an electron? For which is it hardest? How does this compare to the first ionization potential of these atoms? Include the expected electron configuration of each anion in your answer. (Note that the "0" value for Ne means that the A^- ion never forms; it loses an electron with no input of energy.)

Chapter 13 Summary and Problems

◀ Review with **Web Practice**

VOCABULARY

electron domain	The region of space around an atom occupied by a lone pair or a chemical bond to another atom (the bond may be single, double, or triple).
line spectrum	Only certain wavelengths of light are emitted from an excited atom; they appear as lines of light in a spectroscope.
quantum theory	A physical theory that explains the behavior of matter in terms of particular energy levels—quantum levels—within atoms and other species.
atomic orbital	A state occupied by an electron in an atom.
principal quantum number	The positive integer ($n = 1, 2, 3, \ldots$) that describes a quantum level for an electron in an atom.
radial node	A place where an atomic orbital has no amplitude; the word *radial* indicates that this node is a fixed distance from the nucleus, defining a sphere.
angular node	A kind of node defined by an angle relative to the position of the atom; this gives rise to the shapes of atomic orbitals.
subshell	The s, p, f, and d sub-divisions of the principal energy levels; in electrons with two or more electrons, these subshells are different for the same principal quantum number.

first ionization energy	The smallest energy required to completely remove an electron from a neutral atom in the gas phase.
closed shell	A situation in which all available atomic orbitals are filled with electrons; a set of eight electrons in the valence shell.
electron configuration	The assignment of electrons to particular principal energy numbers and type of atomic orbital.

SECTION GOALS

How does a Lewis structure relate to the three-dimensional structure of a molecule or polyatomic ion?	The Lewis structure of a molecule or polyatomic ion presents the arrangement of electrons in terms of chemical bonds and lone pairs. These occupy space, and the best spatial arrangement of these bonds and lone pairs determines the shape of the molecule or polyatomic ion.
How do electron domains predict the shape of a molecule or polyatomic ion?	We consider that each bond and each lone pair occupies an electron domain around an atom. These domains spread out in three dimensions so that they are as far apart as possible. Major cases are two domains (linear), three domains (triangular) and four domains (tetrahedral).
How do chemists characterize light and light energy?	Chemists understand that all light has characteristic wavelengths and frequencies associated with the wave motion of light. In addition, there is a direct relationship between the energy of the light and its frequency.
What observations give evidence for the quantized energy of electrons in the atom?	Several observations support the idea that electrons are in quantized levels in the atom. These include the presence of line spectra for atoms and the indication that certain numbers of electrons apparently fill subshells corresponding to quantum levels.
How do atomic orbitals describe the positions of electrons in atoms?	Atomic orbitals describe how three-dimensional space is divided up in atoms, so that different energy levels have different numbers of nodes in their orbitals and different orbitals can have different arrangements of the electrons in space.
What is involved in measuring an ionization energy?	An ionization energy is the energy needed to remove an electron from a species. The first ionization energy refers to a neutral atom, the second to an atom with a +1 charge, etc.
How do ionization energies tell us something about the shell structure of atoms?	There are dramatic changes in the ionization energy of atoms as we move through the periodic table. The atoms in Group VIII, for example, have the highest first ionization energy. This suggests that these atoms have a closed shell of electrons.
How do we write the electron configuration of an element?	We write the electron configuration following the energy of the atomic orbitals from lowest to highest. These configurations are written in the form "nx^y" where n is the orbital's principal quantum number, x indicates the subshell, and y indicates the number of electrons in that shell. s subshells can have two electrons, p subshells can have six, and d subshells ten.

PROBLEMS

26. Determine the number of valence electrons in the following atoms and ions.

(a) N^+
(b) C^{2-}
(c) N^-

27. In the first chapter, you were asked to remember which elements are found in normal conditions as diatomic elements. Draw the Lewis structure of each of these.

28. Determine the three dimensional structure of NF_4^+.

29. Draw the three dimensional structure of SO_4^{2-} ClO_3^-, and NO_2^-. Note any trends that you observe in the structures and in the number of electrons on each structure.

30. A few years ago, chemists determined that the anion AlF_4^- had a notable structure, something *you* can predict. What do you expect the structure of this anion is?

31. Electron domains can also be used to analyze larger structures. Describe the electron domains in the following organic molecules. You may have to fill in the number of lone pairs you expect on a certain atom. In the case of (a), what do we also know about the structure of the C=C bond?

(a)

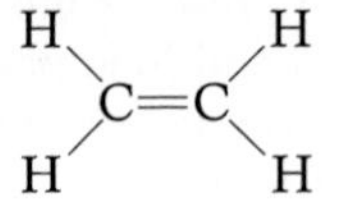

(b)

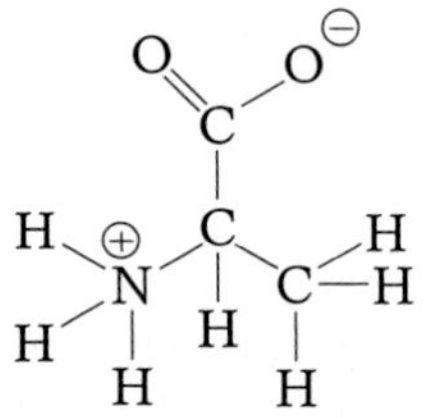

32. Examine the trends in the first ionization energies of the alkali metals and the halogens, shown below. What trend is present going down a column? How might this relate to the electron configuration?

Li	Na	K	Rb
520	500	410	400

F	Cl	Br	I
1680	1250	1140	1010

33. Determine the number of valence electrons in the following atoms and ions.

(a) Si^{4-}
(b) H^-
(c) Br^+
(d) B^{3-}

Solutions, Molarity, and Stoichiometry

Intravenous solutions used to stabilize patients are prepared using the concept of molarity. (Superstock.)

PRACTICAL CHEMISTRY Solution Concentrations in Medicine

Hospital patients often rely on sugar and salt solutions that are administered intravenously. The concentrations of these solutions are extremely important, and must be measured carefully beforehand. It is important in these stiuations to be able to relate volume amounts of a solution to moles. For example, how does a healthcare worker know what volume of solution contains a particular amount of glucose?

We have already seen how to relate mass amounts of a substance to moles. We can calculate the gram amount for one mole of any solid, liquid, or gas if we have the formula of that substance, because the chemical formula allows us to determine the molar mass. The molar mass ratio then allows us to convert from grams to moles or from moles to grams for any pure substance.

$$\text{mass} \xleftrightarrow{\text{molar mass}} \text{moles}$$

When we first discussed samples of pure substances in Chapter 6, we used the density ratio to relate mass amounts to volume amounts.

$$\text{mass} \xleftrightarrow{\text{density}} \text{volume}$$

Then, in a two-step calculation discussed in Chapter 8, we found mole amounts when given the density and the volume of a solution.

$$\text{volume} \xleftrightarrow{\text{density}} \text{mass} \xleftrightarrow{\text{molar mass}} \text{moles}$$

In this chapter, you will learn a shorter method to find mole amounts for solutions. This method uses a unit called *molarity* that directly relates solution volume to mole amounts of the pure substance dissolved.

This chapter will focus on the following questions:

PRACTICAL A How are intravenous solutions prescribed and administered?

PRACTICAL B How does intravenous "feeding" supply a patient's energy requirements?

PRACTICAL C How do we monitor the glucose in the blood with electron flow?

14.1 Solutions and Molarity

SECTION GOALS

✔ Define a solution in terms of its components.

✔ Understand the term *molarity*.

✔ Calculate the molarity of a solution from a mass or mole amount of solute and the volume of solution.

✔ Use molarity to convert between moles of solute and volume of solution.

Solutions Are Homogeneous Mixtures

A **solution** is a homogeneous mixture of two or more substances. Solutions exist in all physical phases of matter: solids, liquids, and gases. We make solutions by dissolving one substance (called a **solute**) in a second substance (called the **solvent**). A solvent is generally present in greater mole amounts than the solute. Table 14-1 lists some common examples of solutions in various phases.

We usually make solid solutions by melting the components and mixing them. Steel and other metal alloys are examples of solid solutions. The hot mixture becomes solid when it is cooled to room temperature.

The air that we breathe is a gas solution, primarily nitrogen, oxygen, water vapor, argon, and some trace gases. Liquid solutions are the most common type of solutions encountered in chemistry.

Lake and ocean water contain a number of dissolved substances that can comprise either a homogeneous or heterogeneous mixture. (Steve Kohls, Brainerd Daily Dispatch/AP.)

Table 14-1 Solutions in Different Phases

Physical Phase of Solution	Examples	Composition of the Solution
Solid	Steel	Fe and C, sometimes mixed with other metals such as Cr
	Dental amalgams	Hg and Ag
Liquid	Carbonated beverages	Flavored water solution and CO_2 (g)
	Vinegar	Acetic acid (l) and H_2O (l)
	Sea water	H_2O (l), dissolved air, and salts such as NaCl
Gas	Air	N_2, O_2, H_2O, CO_2, and traces of He and other noble gases
	Anesthetics	Gaseous drug, N_2, and O_2 (g)

Typical examples of liquid solutions are included in Table 14-1. These particular examples have water as the solvent in which the solute—either a solid, liquid, or gas—is dissolved. Solutions that use water as the solvent are called **aqueous solutions.**

Properties of Aqueous Solutions

Solutes such as table sugar ($C_{12}H_{22}O_{11}$) and table salt (NaCl) dissolve easily in water; we therefore say that they are **soluble** in water. Solutes that are very difficult to dissolve in water are called **insoluble.** Limestone ($CaCO_3$) and magnesium carbonate ($MgCO_3$) are examples of substances that are relatively insoluble in water. Carbonates such as $CaCO_3$ and $MgCO_3$ are responsible for the buildup of scale in water pipes.

Solubility is the maximum amount of a solute that dissolves in a given amount of solvent at a particular temperature. At room temperature and pressure, $MgCO_3$ has a solubility of 533.2 mg L^{-1}, while $CaCO_3$ has a solubility of 9.3 mg L^{-1}. This means that, under similar conditions of temperature and pressure, $MgCO_3$ is much more soluble in water than is $CaCO_3$.

Solubility is such an important property of solutions that we often refer to a solution in terms of its solubility properties. For example, solutions that have dissolved the maximum amount of solute possible at a specified temperature are called **saturated** solutions. Solutions that have less than this amount of dissolved solute and are able to dissolve additional solute are called **unsaturated** solutions. Figure 14-1 shows saturated and unsaturated solutions of NaCl.

Solubility generally has units of grams per liter (g L^{-1}) or moles per liter (mol L^{-1}). It is a property that often changes as the temperature of the solution changes. Some solutes, such as solids, usually dissolve more readily in heated solutions; other solutes, such as many gases, usually dissolve more readily in cooled solutions. Gases also dissolve more readily at higher pressure. This is why carbonated beverages are bottled under pressure.

Two factors that affect solubility, then, are temperature and, for gases, pressure. A third important factor is the choice of solvent.

See this with your Web Animator ▶

Figure 14-1 ▶ The solubility of NaCl in aqueous solution is 36 g per 100 mL of water. Solutions that contain less than this amount are called unsaturated solutions. Solutions that contain more than this amount are called saturated solutions.

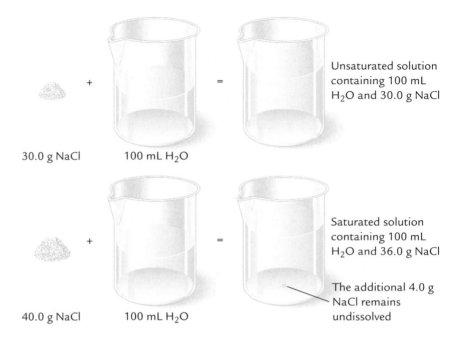

30.0 g NaCl   100 mL H₂O

Unsaturated solution containing 100 mL H_2O and 30.0 g NaCl

40.0 g NaCl   100 mL H₂O

Saturated solution containing 100 mL H_2O and 36.0 g NaCl

The additional 4.0 g NaCl remains undissolved

✔ **Meeting the Goals**

A solution is a homogeneous mixture of two or more components. One component, called the solute, is the substance that is dissolved in a second substance, which is called the solvent. When the solvent is water, the solution is called an aqueous solution.

Because solutions form when the solute and solvent particles mix freely with one another, the solubility of a particular solute depends strongly on the solvent used. For the remainder of this chapter, we will consider aqueous solutions only, where the solvent is water.

Much of what we know about solutions comes from experiments on electrical conductivity. Water itself does not conduct electricity. But some solutes form solutions that do conduct electricity.

Solutions are tested to determine whether or not they conduct electricity, using an apparatus similar to that in Figure 14-2. In a conductivity experiment, a light bulb and two wires are attached to a battery and the wires are inserted into a solution. Solutions that allow the bulb to light are called *electrolytes*, while those that do not allow the bulb to light are called *nonelectrolytes*. For example, the bulb remains dark when the wires are placed in a solution of sucrose ($C_{12}H_{22}O_{11}$); sucrose is a nonelectrolyte. However, the bulb glows brightly when the wires are inserted in a solution of NaCl and the

Figure 14-2 ▶
(a) Sucrose is a nonelectrolyte; the bulb is dark.
(b) Vinegar is a weak electrolyte; the bulb glows.
(c) NaCl is a strong electrolyte; the bulb glows brighter.
(Fundamental Photographs.)

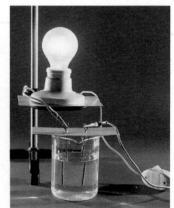

(a)   (b)   (c)

See this with your **Web Animator** ▶

Figure 14-3 ▶
Simple sugars such as glucose and sucrose exist in water solution as molecules surrounded by water molecules. (Water, as a solvent, surrounds solutes in a more complex way than shown here.)

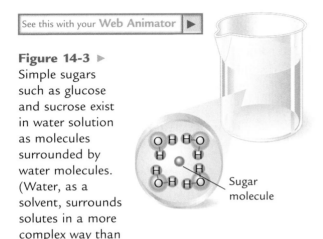

Sugar molecule

Figure 14-4 ▶
NaCl dissolves in water, releasing Na^+ and Cl^- ions into solution.

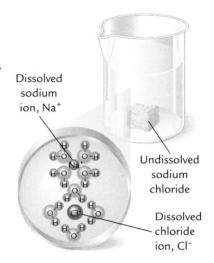

Dissolved sodium ion, Na^+

Undissolved sodium chloride

Dissolved chloride ion, Cl^-

bulb glows weakly in a solution of acetic acid $(C_2H_4O_2)$. Sodium chloride is a strong electrolyte and acetic acid is a weak electrolyte.

The difference in conductivity among the solutions shown in Figure 14-2 is due to the particles making up the solute. Sucrose is a molecular substance that dissolves in water as molecules. Figure 14-3 shows sucrose dissolving in water.

During dissolution (dissolving), sucrose molecules remain intact but become surrounded by water molecules in the aqueous phase. On the other hand, ionic substances **dissociate**, or separate, in water to form ions. NaCl is an ionic substance that forms Na^+ ion and Cl^- ion in aqueous solution, as shown in Figure 14-4. The acetic acid in vinegar is a molecular substance that dissociates very slightly in water, forming some H_3O^+ ions, some CH_3COO^- ions, and many CH_3COOH molecules. Weak electrolytes such as acetic acid will be discussed further in Chapter 16.

The Concentration of Solutes in Solution

Water is the most important solvent we use in general chemistry. Although we can dissolve different amounts of solute in water to make an aqueous solution, a solution is most useful to a chemist when he or she knows *how much* solute has been dissolved in a particular solution. This ratio, $\dfrac{\text{amount of solute}}{\text{amount of solution}}$, is one way of expressing the **concentration** of the solution, where "amount" refers to grams, moles, or liters. Solutions that contain a small amount of solute are called *dilute* solutions, whereas *concentrated* solutions contain a greater amount of solute.

In Chapter 6, we considered concentration units such as percentage by mass, parts per million, and parts per billion. In this chapter, we use the units *mole* for the solute and *liter* for the solution. This concentration unit is called the **molarity** of a solution.

CONNECT TO SECTION 6.3

$$\text{molarity} = \frac{\text{moles of solute}}{\text{volume (L) of solution}}$$

This ratio gives us the units of molarity, moles per liter, mol L^{-1}, or simply M. You will find it very helpful in problem solving if you "see" the M as $\dfrac{\text{moles of solute}}{\text{volume (L) of solution}}$. This gives us both units, moles and liters, which we need to consider in molarity calculations. Molarity is symbolized by M or by closed brackets, [ ]. A solution that contains $\dfrac{0.25 \text{ mole of } C_6H_{12}O_6}{\text{liter of } C_6H_{12}O_6 \text{ solution}}$ is labeled 0.25 M $C_6H_{12}O_6$, or $[C_6H_{12}O_6] = 0.25$ M.

To make a solution that is 0.25 M $C_6H_{12}O_6$, we first calculate grams of $C_6H_{12}O_6$ equivalent to 0.25 mole of $C_6H_{12}O_6$, using the molar mass ratio:

$$0.25 \text{ mol } C_6H_{12}O_6 \times \frac{180.1548 \text{ g}}{1 \text{ mol } C_6H_{12}O_6} \approx 45 \text{ g } C_6H_{12}O_6$$

Then we must dissolve this mass in enough water to end with one liter of solution. It is tempting to simply add one liter of water to 45 grams of $C_6H_{12}O_6$, but molarity refers to liters of solution, not solvent. Instead, we must use specially calibrated glassware, such as the volumetric flask shown in Figure 14-5. Volumetric flasks are designed to contain a particular volume of solution. To make a 0.25 M solution of $C_6H_{12}O_6$, then, we place 45 grams of $C_6H_{12}O_6$ in a 1-liter volumetric flask, dissolve it in a small amount of water, and then while stirring add more water to the mark on the volumetric flask. The same process is followed when the solute is a liquid, providing that the liquid solute dissolves in water.

Figure 14-5 ▶ Accurate solutions are made using volumetric glassware such as this 1-liter volumetric flask. The solute is weighed out, dissolved in a small amount of water, and then more water is added to the mark (line) on the flask.

(a) Solute

200 mL, 20 °C
(b) Solute and solvent

1000 mL, 20 °C
(c) Solution

Example 14.1 **Interpreting molarity labels**

A bottle is labeled 1.25 M $K_2Cr_2O_7$. What does this label mean?

Thinking Through the Problem The bottle certainly contains the chemical listed on the label, $K_2Cr_2O_7$, but how much? The "M" means $\dfrac{\text{moles of solute}}{\text{volume (L) of solution}}$. Here the solute is $K_2Cr_2O_7$, so the label, 1.25 M $K_2Cr_2O_7$, means that there are 1.25 moles of $K_2Cr_2O_7$ dissolved in enough water to make 1.00 liter of this solution.

Answer The label indicates that there are 1.25 moles of $K_2Cr_2O_7$ in enough water to make 1.00 liter of this solution.

Example 14.2 **Preparing solutions**

Describe how you would prepare 1.00 L of a 0.100 M solution of sucrose, $C_{12}H_{22}O_{11}$.

Thinking Through the Problem To make a 0.100 M solution of $C_{12}H_{22}O_{11}$, we must weigh out and dissolve 0.100 mole of $C_{12}H_{22}O_{11}$ in enough water to make 1.00 liter of solution. We first must calculate the grams of $C_{12}H_{22}O_{11}$ needed for 0.100 mole. The molar mass of $C_{12}H_{22}O_{11}$ is 342.295 g mol^{-1}.

Working Through the Problem First we calculate the grams required.

$$0.100 \ \text{mol } C_{12}H_{22}O_{11} \times \frac{342.295 \ \text{g } C_{12}H_{22}O_{11}}{1 \ \text{mol } C_{12}H_{22}O_{11}} \approx 34.2 \ \text{g}$$

Answer To make a 0.100 M solution of sucrose, $C_{12}H_{22}O_{11}$, weigh out 34.2 g $C_{12}H_{22}O_{11}$, dissolve in a small amount of water in a 1.00-liter volumetric flask. Add water until the volume reaches the 1.00-liter mark.

Example 14.3 **Determining molarity from mass and volume**

A solution is made by dissolving 3.25 grams of NaBr in enough water to make exactly 1250.0 mL solution. What is the molarity of the resulting solution?

Thinking Through the Problem To calculate molarity, we need moles of NaBr and the solution volume in liters. We convert grams of NaBr to moles of NaBr using molar mass (102.894 g mol^{-1}). We convert milliliters of solution to liters of solution (1 mL $\Leftrightarrow$ 1 $\times$ 10^{-3} L). We solve this problem in three steps:

1. Convert grams of NaBr to moles of NaBr.

2. Convert milliliters to liters of solution.

3. Calculate molarity.

The final answer will be rounded to three significant figures because of the quantity 3.25 grams given in the problem. But only the final answer is rounded to three significant figures.

Working Through the Problem

Step 1: Convert grams to moles using the molar mass ratio.

$$3.25 \text{ g NaBr} \times \frac{1 \text{ mol NaBr}}{102.894 \text{ g NaBr}} \approx 0.03158 \text{ mol NaBr}$$

Step 2: Convert the volume units from milliliters to liters.

$$1250.0 \text{ mL} \times \frac{1 \times 10^{-3} \text{ L}}{1 \text{ mL}} = 1.2500 \text{ L}$$

Step 3: Find molarity by dividing moles of NaBr by the liters of solution.

$$\frac{0.03158 \text{ mol NaBr}}{1.2500 \text{ L solution}} \approx 0.0253 \text{ M}$$

Answer The molarity of the solution is 0.0253 M or, because molar concentration can be indicated by brackets, we may write [NaBr] = 0.0253 M.

Molarity of Ions in Solution

Sometimes we need to find the molarity of an *ion* in solution. Ionic compounds dissociate into their ions in water solution, as shown in Figure 14-6. NaCl dissociates into its two ions, Na^+ and Cl^- in water solution.

Note that the formula of an ionic compound tells us the number of moles of each ion that will be produced by one mole of the compound. For example, one mole of $MgBr_2$ will dissociate into 1 mole of Mg^{2+} ion and 2 moles of Br^- ions. One mole of $Ca(C_2H_3O_2)_2$ will dissociate into 1 mole of Ca^{2+} ion and 2 moles of $C_2H_3O_2^-$ ions. You are familiar with this kind of "counting" from Chapter 1. How does this work for molarity problems?

To determine the molarity of an ion in solution, we multiply the molarity of the solution by the number of times that the ion appears in the formula of the dissolved compound.

Suppose that we have a 2.5 M NaCl solution. The molarity of Na^+ is also 2.5 M, as each mole of NaCl contains one mole of Na^+ ions.

$$\frac{2.5 \text{ mol NaCl}}{1 \text{ L solution}} \times \frac{1 \text{ mol } Na^+ \text{ ions}}{1 \text{ mol NaCl}} = \frac{2.5 \text{ mol } Na^+ \text{ ions}}{1 \text{ L solution}} = 2.5 \text{ M } Na^+$$

However, the molarity of Na^+ in a 2.5 M solution of Na_2SO_4 is twice 2.5 M because each mole of Na_2SO_4 contains two moles of Na^+ ions.

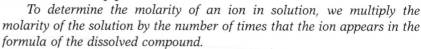

$$\frac{2.5 \text{ mol } Na_2SO_4}{1 \text{ L solution}} \times \frac{2 \text{ mol } Na^+ \text{ ions}}{1 \text{ mol } Na_2SO_4} = \frac{5.0 \text{ mol } Na^+ \text{ ions}}{1 \text{ L solution}} = 5.0 \text{ M } Na^+$$

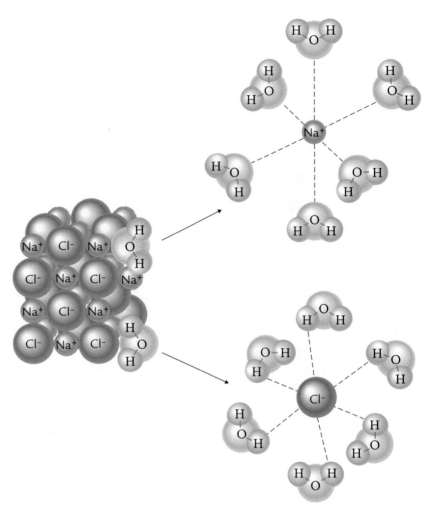

Figure 14-6 ▲ Ionic compounds dissociate in water solution. When sodium chloride dissociates, the Na^+ ion is attracted to the oxygen atoms in a water molecule, while the Cl^- is attracted to the hydrogen atoms.

Example 14.4 ⋯⋯ **Calculating molarity of ions**

Calculate $[NO_3^-]$ in a solution where $[Al(NO_3)_3] = 0.350$ M.

Thinking Through the Problem The square brackets indicate molarity and ⟶ (14.1) the key idea is to multiply the molarity of the solution by the number of times that the ion appears in the formula of the compound in solution.

Working Through the Problem

$$\frac{0.350 \text{ mol Al(NO}_3)_3}{1 \text{ L solution}} \times \frac{3 \text{ mol NO}_3^- \text{ ions}}{1 \text{ mol Al(NO}_3)_3} = \frac{1.05 \text{ mol NO}_3^- \text{ ions}}{1 \text{ L solution}}$$

Answer There are 1.05 moles of nitrate ions in one liter of this solution. Thus, $[NO_3^-] = 1.05$ M.

How are you doing? 14.1

Find $[K^+]$ for a solution made by dissolving 45.0 g of K_2SO_4 in enough water to make 500.0 mL of solution.

Volume-to-Mole Conversions Using Molarity

The most frequent use of the molarity ratio is to calculate either the moles of solute or the volume of solution. When the molarity ratio and either moles of solute or volume of solution are known, we can calculate the unknown variable. We use the known ratio and/or proportional reasoning or conversion factors, as we saw in Chapters 6–9.

For example, we calculate moles of KCl when given 25.0 mL of a 0.180 M KCl solution by solving the proportion:

$$\frac{0.180 \text{ mol KCl}}{1 \text{ L solution}} = \frac{x \text{ mol KCl}}{0.0250 \text{ L solution}}$$

$$x \text{ mol KCl} = 0.0250 \text{ L solution} \times \frac{0.180 \text{ mol KCl}}{1 \text{ L solution}}$$

In solving this problem, we see that the molarity ratio becomes a conversion factor:

$$\text{volume (L) of solution} \times \text{molarity ratio} = \text{moles of solute}$$

Figure 14-7 shows a general scheme for calculating moles from the volume and molarity of the solution.

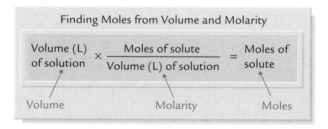

Figure 14-7 ▲ The molarity ratio is used to calculate moles of solute when the volume and molarity of a solution are given.

Example 14.5   **Finding moles when given volume and molarity**

How many moles of NaBr are contained in 125 mL of a 1.401 M solution of NaBr?

Thinking Through the Problem To calculate moles, we multiply volume (in liters) by the molarity ratio. We need to convert 125 mL to 0.125 L before doing the calculation.

Working Through the Problem

$$0.125 \text{ L solution} \times \frac{1.401 \text{ mol NaBr}}{1 \text{ L solution}} \approx 0.175 \text{ mol NaBr}$$

Answer There is 0.175 mole of NaBr in 125 mL of a 1.401 M solution of NaBr.

Example 14.6   **Finding moles of ions when given volume and molarity**

How many moles of Cl^- ions are there in 250.0 mL of a 0.500 M $MgCl_2$ solution? Assume that $MgCl_2$ completely ionizes in solution.

Thinking Through the Problem We must first find moles of $MgCl_2$ using the molarity ratio. We use the molarity of the $MgCl_2$ solution to find moles of $MgCl_2$ using the scheme shown in Figure 14-7. The number of moles of chloride ions will be twice the number of moles of $MgCl_2$, because there are 2 moles of Cl^- ions per mole of $MgCl_2$.

$$0.250 \; \text{L solution} \times \frac{0.500 \; \text{mol MgCl}_2}{1 \; \text{L solution}} = 0.125 \; \text{mol MgCl}_2$$

Now that we know the number of moles of $MgCl_2$, we can calculate the number of moles of Cl^- ion.

$$0.125 \; \text{mol MgCl}_2 \times \frac{2 \; \text{mol Cl}^-}{1 \; \text{mol MgCl}_2} = 0.250 \; \text{mol Cl}^-$$

Answer There is 0.250 mole of Cl^- ions in 250.0 mL of 0.500 M $MgCl_2$ solution.

✔ Meeting the Goals

The molarity ratio is used to calculate moles of solute from the volume and molarity of solution:

$$\begin{array}{l} \text{liters} \\ \text{of} \times \dfrac{\text{moles of solute}}{\text{volume (L) of solution}} \\ \text{solution} \end{array}$$

$$= \text{moles of solute}$$

How are you doing? **14.2**

Determine the number of moles of $Mg(C_2H_3O_2)_2$ in 0.250 L of a solution that contains 0.234 M $Mg(C_2H_3O_2)_2$. Also determine the moles of $C_2H_3O_2^-$ ions in this solution.

Mole-to-Volume Conversions Using Molarity

We also use the molarity ratio to calculate the volume of solution when given values of M and moles of solute. For example, suppose that we need to obtain 0.045 mole of HNO_3 by using a 0.015 M HNO_3 solution. We use molarity as the known ratio and solve for volume. The units work out if we remember to use mole liter^{-1} in place of the symbol M. We find the volume by solving the proportion:

$$\frac{0.105 \; \text{mol HNO}_3}{1 \; \text{L solution}} = \frac{0.045 \; \text{mol HNO}_3}{x \; \text{L solution}}$$

$$x \; \text{L solution} = 0.045 \; \text{mol HNO}_3 \times \frac{1 \; \text{L solution}}{0.105 \; \text{mol HNO}_3}$$

Here, the moles of HNO_3 divide out, leaving liters of solution. Notice that, to accomplish this, we multiplied by the reciprocal of the molarity ratio. This calculation is summarized in Figure 14-8.

Figure 14-8 ▶ The molarity ratio is used to find volume of solution from moles of solute.

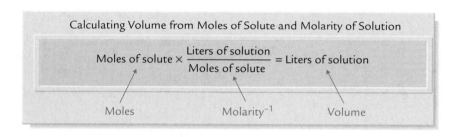

Example 14.7 **Finding volume from molarity and moles**

Determine the volume of 0.128 M ethanol solution required to provide 2.34×10^{-3} mole of ethanol.

Thinking Through the Problem We will use the molarity ratio to calculate the volume, in liters, because we are given the molarity of solution and the mole amount required.

Working Through the Problem Multiply moles of ethanol by the molarity ratio, as shown in Figure 14-8.

$$2.34 \times 10^{-3} \text{ mol ethanol} \times \frac{1 \text{ L solution}}{0.128 \text{ mol ethanol}} \approx 0.0182 \text{ L}$$

Answer The volume of ethanol solution required is 0.0182 L. For laboratory work, you may want to convert liters to milliliters. The volume, 0.0182 L, is equal to 18.2 mL.

Example 14.8 **Finding volume from molarity and moles**

A laboratory experiment requires 4.50×10^{-2} mole of HCl. What volume of HCl solution should you use if the solution is labeled 0.368 M?

Thinking Through the Problem We can use the molarity ratio as in example 14.7 to solve for the volume.

Working Through the Problem

$$x = \frac{4.50 \times 10^{-2} \text{ mol HCl} \times 1 \text{ L solution}}{0.368 \text{ mol HCl}} \approx 0.122 \text{ L solution}$$

Answer You will need to measure out 0.122 L (or 122 mL) of the 0.368 M HCl solution to get 4.50×10^{-2} mole of HCl.

✔**Meeting the Goals**

The reciprocal of molarity ratio can be used to calculate volume from the molarity of solution and moles of solute:

$$\begin{array}{c} \text{moles} \\ \text{of} \\ \text{solute} \end{array} \times \frac{\text{liters of solution}}{\text{moles of solute}}$$

$$= \text{liters of solution}$$

How are you doing? **14.3**

Calculate the volume of 0.962 M NaOH solution required in order to have 3.90×10^{-3} mole of OH^- ions.

PROBLEMS

1. **Concept question:** In Chapter 6, we used mass percentage to calculate the concentration of one substance in a mixture. Which do you think is more convenient—the mass percentage or the molarity? Might your answer depend on the situation?

2. What is a solution?

3. What are the names for the two components of a solution? Define each.

4. A solution is made by dissolving 13.9 grams of NaCl in 200.0 grams of water. What is the mass of the solution formed?

5. A solution is made by dissolving 25.0 grams of HCl in 500.0 grams of water. What is the mass of the solution?

6. A solution is made by dissolving 15.0 grams of $C_6H_{12}O_6$ in 100.0 grams of water. The density of the solution is 1.078 g mL^{-1}. What is the volume of the solution?

7. A solution is made by dissolving 22.5 grams of KCl in 250.0 grams of water. The density of the resulting solution is 1.023 g mL^{-1}. What is the volume of the solution?

8. A solution is labeled "1.92 M NaOH." What does this label tell us about the mole amount of NaOH in 500.0 mL of the solution?

9. A solution is labeled "0.760 M HCl." What does this label tell us about the mole amount of HCl in 250.0 mL of the solution?

10. Butane, C_4H_{10}, is used in cigarette lighters. As a liquid, it has a density of 0.74 g mL^{-1}. Determine the mass and the number of moles of butane in a lighter that contains 4.8 mL of butane. Determine the molarity of butane in this lighter.

11. Determine how many grams of each of the following solutes must be dissolved to make the indicated solution.
    (a) Dissolve C_2H_6O in water to make 500.0 mL of a solution that is 1.23 M C_2H_6O.
    (b) Dissolve H_3PO_4 in water to make 250.0 mL of a solution that is 8.76×10^{-3} M H_3PO_4.
    (c) Dissolve $Ca(OCl)_2$ in water to make 3.0 L of a solution that is 2.5 M $Ca(OCl)_2$.

12. **Discussion question:** Standardized solutions of the elements are important when chemists calibrate instruments. Determine what mass of each of the following compounds will dissolve in 1.00 liter of solution to give exactly a 0.0100 M solution.

    (a) $Ca(NO_3)_2$
    (b) K_3PO_4
    (c) $U_3(PO_4)_4$

    How will your calculations be different if we need 0.0100 M of the metal ion?

13. Determine the molarity of the following solutions.
    (a) Dissolve 25.0 g of NaCl in water to make 2.00 L of solution.
    (b) Dissolve 0.1052 g of oxalic acid, $H_2C_2O_4$ in water to make 1.00 L of solution.
    (c) Dissolve 0.025 g of PbI_2 to make 100. mL of solution.

14. Lead deposits in the home can be controlled in part through the use of a solution of "TSP," which stands for the chemical Na_3PO_4, often called "trisodium phosphate." A typical recipe is to dissolve 1 tablespoon—about 25 g—of TSP in 1 gallon of wash solution—about 3.6 L. Determine the molarity of TSP in such a solution.

15. Determine the volume of solution required to give the number of moles of solute indicated.
    (a) Use 3.25×10^{-3} M NaOH to give 0.024 mole of NaOH.
    (b) Use 2.50 M CH_3COOH to give 0.198 mole of CH_3COOH.
    (c) Use 0.094 M H_2SO_4 to give 1.00×10^{-3} mole of H_2SO_4.

16. Determine the number of moles of solute contained in each of the following solutions.
    (a) 250 mL of a solution of 0.0745 M KBr
    (b) 32.90 mL of a solution of 0.0953 M KOH
    (c) 50.00 mL of a solution of 0.108 M HNO_3

17. A solution is made by dissolving 3.088 grams of KCl in 25.0 grams of water. The density of the solution is 1.0092 g mL^{-1}. What is the volume of the solution? What is the molarity of the solution?

18. What volume of 1.250 M $Mg(OH)_2$ is required in order to provide 0.250 mole of OH^- ions? Assume complete ionization of $Mg(OH)_2$ in solution.

19. How many moles of Cl^- ion are there in 500.0 mL of a 0.750 M $BeCl_2$ solution? Assume complete ionization of $BeCl_2$ in solution.

20. How many moles of Cl^- ions are there in 200.0 mL of a 0.148 M $AlCl_3$ solution? Assume complete ionization of $AlCl_3$ in solution.

PRACTICAL A   How are intravenous solutions prescribed and administered?

Hospital patients often receive medications and nutrients intravenously (in solution, injected into the veins) in order to decrease physical stress and to stabilize the patient. So-called IV solutions are stored in sterile plastic bags of varying volumes, ready to be transferred to a patient in need. The most common solutions contain glucose, sodium chloride, or a combination of the two. Glucose, the most common simple sugar, is easily metabolized and provides some energy to the patient. It is easily transported by the blood and is often referred to as "blood sugar."

Suppose that we have 100.0 g of a 5.00% by mass glucose ($C_6H_{12}O_6$) solution. We know that 5.00% by mass means we have 5.00 grams of glucose in 100 grams of solution. So we can calculate moles of glucose. We already know how to determine the number of moles of glucose in 5.00 g of glucose:

$$5.00 \text{ g } C_6H_{12}O_6 \times \frac{1 \text{ mole } C_6H_{12}O_6}{180.1548 \text{ g } C_6H_{12}O_6} =$$

$$0.0278 \text{ mol } C_6H_{12}O_6$$

Now, to find the molarity of this solution, we need both moles of solute (which we have) and the volume of the solution in liters (we have grams instead). We need volume, but we have grams—this is a job for density! Suppose that the density of this solution is 1.03 g/mL. Using the density ratio, we can solve for the volume occupied by 100.00 grams of the solution:

$$\frac{1.03 \text{ g solution}}{1 \text{ mL solution}} = \frac{100.00 \text{ g solution}}{x \text{ mL}}$$

Solving for the unknown volume, we get:

$$x \text{ mL} = \frac{100.0 \text{ g solution} \times 1 \text{ mL solution}}{1.02 \text{ g solution}}$$

$$= 98.0 \text{ mL}$$

Finally, we calculate molarity:

$$\frac{0.0278 \text{ mol } C_6H_{12}O_6}{98.0 \text{ mL}} = \frac{0.0278 \text{ mol } C_6H_{12}O_6}{0.0980 \text{ L}}$$

$$\approx 0.284 \text{ M}$$

Saline solutions are salt solutions. Those used in medicine usually are very dilute solutions of NaCl that help the body maintain a healthy electrolyte balance. Calculate the molarity of a 0.9% by mass solution of NaCl. (NaCl has a molar mass of 58.443 g mol^{-1} and the density of 0.9% NaCl solution is 1.00 g mL^{-1}.)

How many 200. mL bags of saline are needed if the patient is to receive 0.05 mole of NaCl over a 24-hour period?

Suppose that a doctor prescribes 500.0 mL of a 10.% by mass $C_6H_{12}O_6$ IV solution. This has a density of 1.04 g mL^{-1}. How many moles of glucose are in this volume of solution?

How long will it take to deliver this amount if the drip rate is set at 8 drops per minute? Assume that there are 20 drops in 1 mL.

14.2 Molarity and Stoichiometry

SECTION GOALS

✓ Calculate the moles of either a reactant or a product in a chemical reaction by using molarity and volume information and the stoichiometry of the reaction.

✓ Calculate the volume of either a reactant or a product using molarity information and the stoichiometry of the reaction.

✓ Calculate the molarity of a product in a chemical reaction by using stoichiometry.

Solution Stoichiometry Calculations Using Molarity

The most important relationship in stoichiometry problems is the mole ratio that relates one substance to a second substance in a chemical reaction. The mole ratio is obtained using the coefficients of the balanced equation for the reaction. To do stoichiometry problems using data from laboratory measurements we often must convert measured values of mass and volume into moles. In Chapter 8, you learned how to use *mass* information to determine the number of moles of a substance in a chemical reaction. Mass and moles are related by the molar mass ratio:

$$\text{mass} \xleftrightarrow{\text{molar mass}} \text{moles}$$

Mass/mole relationships such as the one given by the molar mass ratio allow us to calculate the corresponding number of moles when we have measured the mass of various substances in a chemical reaction.

The second most common laboratory measurement is the solution volume. We use the mole/volume relationships defined by molarity to help solve stoichiometry problems involving solutions.

$$\text{volume of solution} \xleftrightarrow{\text{molarity}} \text{moles of solute}$$

In summary, to study the stoichiometry of the reaction we calculate moles from either mass or volume information.

- In a mass/mole calculation, we measure mass and convert to moles using the molar mass.
- In a volume/mole calculation, we measure volume and convert to moles using molarity.

Figure 14-9 summarizes how we solve solution stoichiometric problems involving mass amounts and volume amounts. This draws from strategies we discussed involving mass stoichiometry (Chapter 9) and gas stoichiometry (Chapter 10).

For solution stoichiometry, we must keep in mind the following steps.

Step 1. Molarity relates the volume of the reactant solution to the number of moles of that reactant.

Step 2. Using the stoichiometry of the equation, we can then relate number of moles of reactant to the number of moles of another reactant or a product.

Step 3. Finally, we can relate the number of moles of the second substance to *its* volume or mass, using the appropriate conversion factor as discussed above.

✔ **Meeting the Goals**

Stoichiometry problems can be solved for chemical reactions that include solutions, as the molarity ratio allows us to calculate moles.

The important thing to remember in stoichiometry problems is that the coefficients of the balanced equation give us the mole amounts of reactants and products. Therefore, no matter what infor-

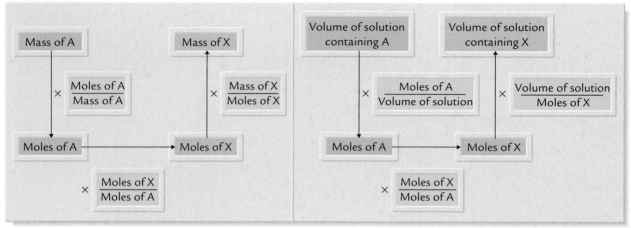

Mass/Mole Calculations   Volume/Mole Calculations

Figure 14-9 ▲ Stoichiometry problems rely on the mole ratio that relates two substances A and X in a chemical equation. To obtain moles, grams are converted using the molar mass ratio and volumes are converted using the molarity ratio.

mation we are given in the problem, *we must be able to convert it into moles.* We use the coefficients to form the mole/mole ratio needed to relate the amount of substance given to the amount of substance sought in the problem.

Calculating Product Information from Molarity

We use the molarity ratio, as indicated in Figure 14-9, to calculate moles of a product in solution stoichiometry problems. For example, suppose that we want to react solutions of $MgBr_2$ and $AgC_2H_3O_2$ and measure the mass of solid AgBr formed according to the equation:

$$MgBr_2\ (aq) + 2\ AgC_2H_3O_2\ (aq) \longrightarrow 2\ AgBr\ (s) + Mg(C_2H_3O_2)_2\ (aq)$$

We can calculate the mass of AgBr formed in this reaction if we know the volume amounts and molarities of the solutions reacting. The molarity ratio allows us to calculate mole amounts of solutes when we have some specified volume of solution, as we see in Example 14.9.

Example 14.9 | Calculating moles of product from volume and molarity of reactant

How many moles of AgBr will be formed by the complete reaction of 34.0 mL of a 0.692 M $MgBr_2$ solution? The equation for this reaction is:

$$MgBr_2\ (aq) + 2\ AgC_2H_3O_2\ (aq) \longrightarrow$$
$$2\ AgBr\ (s) + Mg(C_2H_3O_2)_2\ (aq)$$

Thinking Through the Problem ➤ (14.2) The key idea is to use information about $MgBr_2$ to calculate moles of AgBr. These two substances are related by the equation above and a mole ratio is written using the coefficients from the equation. To find moles of $MgBr_2$, we use the molarity and volume information given in the problem. The calculations we must do are:

$$\text{volume } MgBr_2 \xrightarrow{\text{molarity}} \text{mol } MgBr_2 \xrightarrow{\frac{2 \text{ mol AgBr}}{1 \text{ mol } MgBr_2}} \text{mol AgBr}$$

Working Through the Problem

Step 1: $0.0340 \text{ L solution} \times \dfrac{0.692 \text{ mol } MgBr_2}{1 \text{ L solution}} \approx$

$$0.02353 \text{ mole of } MgBr_2$$

Step 2: $0.02353 \text{ mol } MgBr_2 \times \dfrac{2 \text{ mol AgBr}}{1 \text{ mol } MgBr_2} \approx$

$$4.71 \times 10^{-2} \text{ mol AgBr}$$

Answer 4.71×10^{-2} mole of AgBr is formed.

Example 14.10 — Calculating grams of product from molarity and volume of a reactant

How many grams of $BaSO_4$ can be formed when 15.0 mL of 0.998 M Na_2SO_4 reacts with an excess of $Ba(NO_3)_2$ solution? The equation for this reaction is:

$$Na_2SO_4 \ (aq) + Ba(NO_3)_2 \ (aq) \longrightarrow BaSO_4 \ (s) + 2 \ NaNO_3 \ (aq)$$

Thinking Through the Problem This is a stoichiometry problem. Information is given for Na_2SO_4 but the question is about $BaSO_4$. These two substances are related by the equation. ◀━ (14.2) The key idea is to follow these steps:

Step 1: Use the molarity ratio to calculate moles of Na_2SO_4.

Step 2: Use the mole ratio to find moles of $BaSO_4$.

Step 3: Convert moles of $BaSO_4$ to grams of $BaSO_4$.

Working Through the Problem

Step 1: $0.0150 \text{ L solution} \times \dfrac{0.998 \text{ mol } Na_2SO_4}{1 \text{ L solution}} \approx$

$$1.497 \times 10^{-2} \text{ mol } Na_2SO_4$$

Step 2–3: $1.497 \times 10^{-2} \text{ mol } Na_2SO_4 \times \dfrac{1 \text{ mol } BaSO_4}{1 \text{ mol } Na_2SO_4} \times$

$$\dfrac{233.392 \text{ g}}{1 \text{ mol } BaSO_4} \approx 3.50 \text{ g } BaSO_4$$

Answer This process will result in the formation of 3.50 grams of $BaSO_4$. (The final answer is rounded to the 3 significant figures required by 0.0150 L.)

Examine the above calculation to make sure that you understand the steps. "Liters of solution" divides out to give "mol Na_2SO_4," which is converted to "mol $BaSO_4$" by using the coefficients of the equation. "Mol $BaSO_4$" is converted to "g $BaSO_4$" by using its molar mass.

How are you doing? 14.4

How many grams of $Mg(OH)_2$ can be formed by the complete reaction of 42.0 mL of 0.486 M NaOH? The equation is:
$$2\ NaOH\ (aq) + Mg(C_2H_3O_2)_2\ (aq) \rightarrow 2\ NaC_2H_3O_2\ (aq) + Mg(OH)_2\ (s).$$

Calculating the Volume of Reactants from Molarity

Sometimes we need to calculate the volume of solution to use in a chemical reaction. Suppose that we want to react a solution of NaCl with 0.500 L of a stock solution (0.123 M) of silver nitrate to form solid silver chloride, as shown in Figure 14-10. The equation is:

$$AgNO_3\ (aq) + NaCl\ (aq) \longrightarrow NaNO_3\ (aq) + AgCl\ (s)$$

What volume of 0.285 M NaCl solution is needed for this reaction?

We see from the balanced equation that the reactants are in a 1:1 ratio, 1 mol $AgNO_3$ to 1 mol NaCl. If we calculate the number of moles of $AgNO_3$ present in our solution, then we will also know the number of moles of NaCl needed for the reaction. Finally, we can use the molarity ratio to calculate the necessary volume of NaCl solution. We follow these steps:

Step 1: Use the volume and molarity of $AgNO_3$ to calculate moles $AgNO_3$.

$$0.500\ \text{L solution} \times \frac{0.123\ \text{mol AgNO}_3}{1\ \text{L solution}} = 0.0615\ \text{mol AgNO}_3$$

Step 2: Use a mole ratio made from the coefficients in the balanced equation to calculate moles of NaCl.

$$0.0615\ \text{mol AgNO}_3 \times \frac{1\ \text{mol NaCl}}{1\ \text{mol AgNO}_3} = 0.0615\ \text{mol NaCl}$$

Step 3: Use the moles of NaCl and the inverse of the molarity of NaCl to calculate the volume of NaCl solution required.

Figure 14-10 ▲ Silver chloride (AgCl) forms a precipitate when solutions of $AgNO_3$ and NaCl are mixed. (Chip Clark.)

$$0.0615\ \text{mol NaCl} \times \frac{1\ \text{L solution}}{0.285\ \text{mol NaCl}} \approx 0.216\ \text{L solution}$$

The three steps of this calculation can be put into a single calculator setup:

$$0.500 \text{ L solution} \times \frac{0.123 \text{ mol AgNO}_3}{1 \text{ L solution}} \times \frac{1 \text{ mol NaCl}}{1 \text{ mol AgNO}_3} \times$$
$$\frac{1 \text{ L solution}}{0.285 \text{ mol NaCl}} = 0.216 \text{ L}$$

Thus, 0.216 L, or 216 mL, of sodium chloride solution is required in the reaction.

The second step—relating moles of $AgNO_3$ to moles of NaCl—is very important. Do not forget it! Not every chemical reaction shows a 1:1 mole ratio, as we see in Example 14.11.

Example 14.11 **Calculating the volume of reactant using molarity**

Sulfuric acid and sodium hydroxide react together to give sodium sulfate and water, according to the equation:

$$H_2SO_4 \ (aq) + 2 \ NaOH \ (aq) \longrightarrow Na_2SO_4 \ (aq) + 2 \ H_2O \ (l)$$

Calculate the volume of 0.0215 M NaOH required to completely react with 100.0 mL of 0.0322 M sulfuric acid (H_2SO_4).

Thinking Through the Problem This is a stoichiometry problem, so we need to relate the two chemicals by their mole amounts. Molarity provides the relationship between mole and volume information.

$$\text{volume } H_2SO_4 \xrightarrow{\text{molarity}} \text{mol } H_2SO_4 \xrightarrow[1 \text{ mol } H_2SO_4]{2 \text{ mol NaOH}}$$
$$\text{mol NaOH} \xrightarrow{\text{molarity}} \text{volume NaOH}$$

Working Through the Problem ➥ (14.3) The key idea is to follow these steps:

Step 1: Calculate moles of H_2SO_4 solution from the volume and molarity given for H_2SO_4.

Step 2: Convert moles of H_2SO_4 to moles of NaOH using the stoichiometry of the equation.

Step 3: Convert moles of NaOH to volume of NaOH using molarity given for NaOH.

$$0.1000 \text{ L solution} \times \frac{0.0322 \text{ mol } H_2SO_4}{1 \text{ L solution}} \times \frac{2 \text{ mol NaOH}}{1 \text{ mol } H_2SO_4} \times$$
$$\frac{1 \text{ L NaOH solution}}{0.0215 \text{ mol NaOH}} \approx 0.300 \text{ L solution}$$

Answer 0.300 liter of the NaOH solution is required to react with 100.0 mL of 0.0322 M H_2SO_4.

✔Meeting the Goals

Using the molarity relationship, the needed volume for a reactant can be calculated if we know both the molarity of the reactant solution and the number of moles of that reactant needed.

$$L \text{ of solution} = \frac{\text{mol of solute}}{\text{molarity}}$$

✔Meeting the Goals

To find the molarity of a product in a chemical reaction, use stoichiometry to calculate moles of that product. Then use the molarity relationship,

$$M = \frac{\text{moles of solute}}{\text{liters of solution}}.$$

KEY IDEA 14.4

How are you doing? 14.5

The reaction of permanganate and iron (II) ions proceeds in acidic solutions according to the equation:

$$16\,H_3O^+\,(aq) + 2\,MnO_4^-\,(aq) + 5\,Fe^{2+}\,(aq) \longrightarrow$$
$$2\,Mn^{2+}\,(aq) + 5\,Fe^{3+}\,(aq) + 24\,H_2O\,(l)$$

Assuming that there is excess H_3O^+ present, what volume of 0.0200 M permanganate is needed to react with 32.5 mL of 0.0965 M Fe^{2+}?

Calculating the Molarity of Products

When two solutions are mixed and a chemical reaction occurs, the soluble products are present in a larger volume of solution than was present initially. This adds an additional layer of complexity to solution reaction calculations. It is not enough to specify the number of moles of the products. We must also calculate their molar concentration after mixing.

A problem arises in these cases because solution volumes are not necessarily additive. For example, if we mix 0.500 L of HCl solution and 0.500 L of another solution, the final volume may be more or less than 1.000 L. In such cases we have three options. We must *precisely determine the volume of the final solution, precisely adjust it to an exact amount, or assume that the final volume is the sum of the volumes of the solutions we mixed.* In all problems where the molarity is less than 0.500 M, this is a good assumption. We will always use this assumption.

Example 14.12   Calculating the molarity of products

Suppose that 24.2 mL of 0.186 M HCl solution completely reacts according to the following equation:

$$16\,HCl\,(aq) + 2\,Cr_2O_7^{2-}\,(aq) + C_2H_5OH\,(l) \longrightarrow$$
$$4\,Cr^{3+}\,(aq) + 16\,Cl^-\,(aq) + 11\,H_2O\,(l) + 2\,CO_2\,(g)$$

Calculate the molarity of Cr^{3+} ion formed if the final solution volume is 124.0 mL. ◀ (14.4) Assume HCl is the limiting reactant.

Thinking Through the Problem We calculate moles HCl from the molarity and volume given and then use stoichiometry to calculate moles Cr^{3+} produced. Molarity is calculated as the ratio, moles liter^{-1}.

Working Through the Problem

$$(0.0242 \,\text{L solution}) \times \frac{0.186 \,\text{mol HCl}}{1 \,\text{L solution}} \times \frac{4 \,\text{mol Cr}^{3+}}{16 \,\text{mol HCl}} =$$
$$1.1253 \times 10^{-3} \,\text{mol Cr}^{3+}$$

$$\text{molarity} = \frac{0.0011253 \, \text{mol Cr}^{3+}}{0.124 \, \text{L solution}} \approx 0.00908 \, \text{M Cr}^{3+}$$

Answer The Cr^{3+} solution formed in this reaction has a concentration of 0.00908 M.

How are you doing? 14.6

Find the molarity of the OH^- ion when 52.6 mL of a 0.0792 M CN^- solution react completely according to the equation:

$$2 \, H_2O \, (l) + 4 \, Ag \, (s) + 8 \, CN^- \, (aq) + O_2 \, (g) \longrightarrow$$
$$4 \, Ag(CN)_2^- \, (aq) + 4 \, OH^- \, (aq)$$

Assume that the final volume of the solution produced by this reaction is 93.4 mL.

PROBLEMS

21. Determine the volume of 0.231 M NaOH required to completely react with the following solutions of acids. The reaction in all cases is 1 mol NaOH + 1 mol acid.

(a) 25.00 mL of 0.982 M HCl
(b) 50.00 L of 0.982 M HCl
(c) 10.0 L of 2.3×10^{-3} M HNO_3

22. Determine the volume of 0.231 M NaOH required to completely react with 25.00 mL of 0.982 M H_2SO_4. The reaction is
$2 \, NaOH + H_2SO_4 \rightarrow Na_2SO_4 + 2 \, H_2O$

23. Determine the volume of 0.0221 M $AgNO_3$ required to completely react with all of the chloride in 100.0 mL of 0.00913 M NaCl. The reaction involved is:

$$Ag^+ \, (aq) + Cl^- \, (aq) \longrightarrow AgCl \, (s)$$

Now determine the mass of AgCl (s) formed.

24. Determine the volume of 0.0221 M $AgNO_3$ required to completely react with all of the chromate CrO_4^{2-} in 100.0 mL of 0.00619 M Na_2CrO_4. The reaction involved is:

$$2 \, Ag^+ \, (aq) + CrO_4^{2-} \, (aq) \longrightarrow Ag_2CrO_4 \, (s)$$

Now determine the mass of Ag_2CrO_4 (s) formed.

25. Calculate the molarity of OH^- ion present when 0.21 gram of Ca is reacted in enough water so as to make a 500.0-mL solution.

$$Ca \, (s) + 2 \, H_2O \, (l) \longrightarrow Ca(OH)_2 \, (aq) + H_2 \, (g)$$

26. Suppose that 6.92 grams of $Mg(OH)_2$ is reacted completely with 125.0 mL $HClO_4$ solution according to the following equation:

$$2 \, HClO_4 \, (aq) + Mg(OH)_2 \, (s) \longrightarrow$$
$$2 \, H_2O \, (l) + Mg(ClO_4)_2 \, (aq)$$

What is the molarity of the $HClO_4$ used in this reaction?

27. A solution is prepared by dissolving 0.812 gram of NaCl in water. How much 0.0221 M $AgNO_3$ is needed to completely react all of the chloride according to the reaction shown in Problem 23?

28. Calculate the molarity of an HCl solution if the following reaction requires 18.4 mL of the solution to react completely with 10.462 grams of $Ca(OH)_2$. The equation for this reaction is:

$$Ca(OH)_2 \, (s) + 2 \, HCl \, (aq) \longrightarrow$$
$$CaCl_2 \, (aq) + 2 \, H_2O \, (l)$$

29. The concentration of magnesium ion can be determined by reaction with $EDTA^{4-}$ (ethylene-diaminetetratacetate), according to the equation:

$$Mg^{2+} + EDTA^{4-} \longrightarrow Mg(EDTA)^{2-}$$

In one experiment, complete reaction of 50.00 mL of a Mg^{2+} solution required 23.95 mL of 0.1002 M EDTA.

(a) How many moles of EDTA were used?
(b) How many moles of Mg^{2+} were present?
(c) What was the molarity of Mg^{2+} in the original solution?

PRACTICAL B — How does intravenous "feeding" supply a patient's energy requirements?

Glucose metabolizes into carbon dioxide and water, releasing energy. This is a common way of providing energy through intravenous feeding. This essential reaction is:

$$C_6H_{12}O_6 \ (aq) + 6\,O_2 \ (g) \longrightarrow$$
$$6\,CO_2 \ (g) + 6\,H_2O \ (l) + energy$$

The energy yield on this reaction is approximately 2800 joules per gram of glucose. Let's assume that we need to provide a patient with the normal amount of energy required by an adult. This is approximately 7.5 megajoules per day. How many grams of glucose would the patient need to provide this amount of energy?

If we provided the glucose in the form of a 5.0% solution, what mass of solution is needed? If the density of such a solution is 1.02 g mL^{-1}, what volume of solution is this? What is the molarity of this solution?

Now, let's see how much oxygen this patient might require. Assuming that the oxygen is mostly used for the combustion of glucose in the patient's tissues, how many grams of oxygen is this? If the oxygen is provided as air (which is about 21.0% oxygen), how many liters of air will be needed to react with the amount of glucose just calculated, assuming a temperature of 298 K and a total pressure of 1.00 atm?

This kind of "solution energy" is not just provided in the hospital. Many drinks, especially sug-ared drinks, are used as a source of simple carbohydrates for energy. Suppose that you want to evaluate the energy content of a solution of Megasweet soda (not an actual brand). You know that it is sweetened with sucrose, $C_{12}H_{22}O_{11}$. When this is combusted by the body, 1350 J of energy are available per mole of sucrose. Determine the energy content of a 20.0% solution of sucrose in water, given a density of 1.08 g mL^{-1}. How many liters of this solution are needed to provide 7.5 megajoules? What is the energy content of a 20-ounce (600-mL) bottle of this stuff?

(Image Source/electraVision/PictureQuest.)

---

30. In a reaction similar to the one discussed in the previous question, the complete reaction of 25.00 mL of 0.0987 M EDTA required 15.90 mL of a solution of $Mg(NO_3)_2$.

  (a) How many moles of EDTA were used?
  (b) How many moles of Mg^{2+} were present?
  (c) What was the molarity of Mg^{2+} in the original solution?

31. Discussion question: The reaction of ammonia and hydrochloric acid proceeds according to the stoichiometry:

$$NH_3 \ (aq) + HCl \ (aq) \longrightarrow NH_4^+ \ (aq) + Cl^- \ (aq)$$

In one reaction of this type, 100.0 mL of 0.193 M HCl and 100.0 mL of 0.0865 M NH_3 were reacted.

  (a) How many moles of HCl were added to the

solution? How many moles of NH_3 were added?

  (b) Which reactant is the limiting reactant? Which is in excess?
  (c) How many moles of the excess reactant remained at the end? How many moles of NH_4^+ were formed?
  (d) Assuming a final volume of 200.0 mL, determine $[NH_4^+]$ and the concentration of the excess reactant in the final solution.

How much more of the limiting reactant is needed for complete reaction?

32. Determine the volume, in mL, of 0.443 M HCl required to react completely with 0.200 L of 0.0100 M $Na_2C_2O_4$, according to the equation:

$$Na_2C_2O_4 \ (aq) + 2 \ HCl \ (aq) \longrightarrow$$
$$H_2C_2O_4 \ (aq) + 2 \ NaCl$$

14.3 Aqueous Oxidation-Reduction Reactions

SECTION GOALS

- ✔ Be able to write the oxidation and reduction half reactions from a redox equation.
- ✔ Be able to write half reactions for reactions that occur at the anode and at the cathode in a galvanic cell.
- ✔ Identify from a redox equation the anode and the cathode in a galvanic cell.
- ✔ Balance redox equations using the half reaction method.
- ✔ Balance redox equations using the oxidation number method.

Electrochemistry

In Chapter 5, we saw several examples of chemical reactions between solid strips of metal and metal ions in aqueous solution. Depending on the metal and metal ions involved, either the chemical reaction occurred or it didn't. The reactions were single displacement reactions in which more active metals displaced less active metals in compounds. For example, using the activity series of metals, we found that zinc was less active than aluminum but was more active than copper. This led us to predict that zinc metal *would not* displace aluminum but *would* displace copper in these single displacement reactions. Figure 14-11 shows how zinc reacts with aluminum ions and with copper ions in solution.

$$Zn\ (s)\ +\ Al(NO_3)_3\ (aq)\ \longrightarrow\ \text{No reaction}$$
$$Zn\ (s)\ +\ Cu(NO_3)_2\ (aq)\ \longrightarrow\ Cu\ +\ Zn(NO_3)_2$$

(a)   (b)

Figure 14-11 ▲ (a) Zinc is less active than aluminum, so it will not react when placed in a solution containing Al^{3+}. (b) Zinc is more active than copper, so we do see a reaction when zinc is put into a solution containing Cu^{2+}. (a, b: Chip Clark.)

We classify the $Zn/Cu(NO_3)_2$ reaction as an oxidation-reduction reaction because of the change in oxidation numbers in both Zn and Cu^{2+}.

$$Zn^0 \, (s) + Cu^{2+} \, (aq) \longrightarrow Cu^0 + Zn^{2+}$$

When a redox reaction "works," we say that it occurs spontaneously. The spontaneous electron transfer produces electrical current that can be measured by an experiment using a galvanic cell. The type of galvanic cell with which you are most familiar is a battery. The energy produced by galvanic cells can be used to light lamps or do other kinds of work. What about redox reactions that do not occur spontaneously? Such reactions require an input of electricity in order for the chemical reaction to proceed. These are called *electrolytic* cells. Figure 14-12 shows an example of an electrolytic process in which a metal is plated onto a surface.

See this with your **Web Animator** ▼

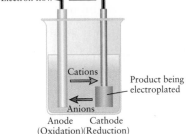

Figure 14-12 ▲ The picture on the left shows the process of electroplating. The picture on the right shows a final product that has been electroplated. (b: Comstock.)

In this section, we consider the relationship between chemical reactions and electrical energy. This area of study, called electrochemistry, involves the interconversion of chemical energy from chemical reactions and electrical energy.

Writing Half Reactions

We recall from Chapter 5 that an oxidation-reduction (redox) reaction is one in which electrons are transferred or moved from one element to another. Oxidation and reduction processes always occur together. Oxidation, the loss of electrons, can only occur when there is a corresponding reduction, or gain, of electrons. In a redox reaction, the element that is oxidized gives up electrons to a second element that is reduced. Remember that the oxidation number of an element standing alone is zero and that an increase in oxidation number indicates oxidation, whereas a decrease in oxidation number indicates reduction. In the Zn/Cu example, Zn^0 is oxidized to Zn^{2+} and Cu^{2+} is reduced to Cu^0.

We can follow the change in zinc's oxidation number by writing just the part of the reaction that involves zinc. This partial equation is called a half reaction. A **half reaction** shows the change in oxidation number for an element, along with the number of electrons involved in this change. Reactants and products in a half reaction are written in the same manner as the reactants and products in the full redox equation. Thus, Zn is a reactant and Zn^{2+} is a product in the half reaction equation. In the half reaction $Zn^0 \rightarrow Zn^{2+}$, we see that the oxidation number of Zn increases from $0 \rightarrow +2$. This tells us that Zn^0 loses 2 electrons when it forms Zn^{2+}. Because the loss of electrons signals an oxidation process, we know that Zn^0 is oxidized to Zn^{2+}. We show this change by writing in "$-2\ e^-$" for the zinc half reaction.

$$\overset{-2\ e^-}{Zn^0 \longrightarrow Zn^{2+}}$$

Finally, the half reaction equation for zinc is written

$$Zn^0 \longrightarrow Zn^{2+} + 2\ e^-$$

where we see the two electrons lost by Zn^0 on the product side of the equation. This allows us to check that the charges in the equation are balanced. To do this, look only at the charges. We see a total charge of zero on the reactant side of the equation. The charges appearing on the product side also add to zero; $0 = (2+) + 2\ (-1) = 0$. We conclude that the equation is charge-balanced.

For copper, we see that the oxidation numbers decrease from $+2 \rightarrow 0$. A decrease in oxidation number tells us that this is the reduction process; copper has gained 2 electrons.

$$\overset{+2\ e^-}{Cu^{2+} \longrightarrow Cu^0}$$

The copper half reaction is written

$$Cu^{2+} + 2\ e^- \longrightarrow Cu^0$$

where the two electrons gained by Cu^{2+} appear on the reactant side of the equation. Again, we verify that the charges are balanced: $(2+) + 2\ (-1) = 0$.

In this reaction, zinc has lost two electrons, which copper ion has gained. *The number of electrons lost in a chemical reaction must always equal the number of electrons gained in the reaction.* In this way, the chemical system preserves its electroneutrality.

✔ Meeting the Goals

Half reactions are obtained from an oxidation-reduction equation by using the change in oxidation number of the oxidized and the reduced species to determine the number of electrons lost or gained, respectively.

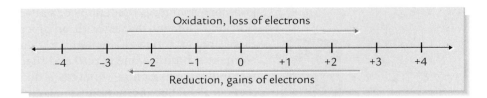

Figure 14-13 ▲ In a redox reaction, the oxidation number of the substance oxidized (or reduced) changes as the result of the number of electrons that have been lost (or gained).

Figure 14-13 shows how the change in oxidation number is related to the number of electrons gained or lost in a redox reaction. An increasing value in the oxidation number means that electrons have been lost; this is oxidation. A decreasing change in the oxidation number means that electrons have been gained; this is reduction.

Current in Simple Galvanic Cells

We can study the flow of electrons from zinc to copper by physically assembling the two half reactions in our Zn/Cu example and then making the proper electrical connections. This is easy to do.

The beaker shown on the left of Figure 14-14 contains the oxidation half reaction—a strip of Zn metal immersed in a solution of $Zn(NO_3)_2$. The beaker on the right side of the figure contains the reduction half reaction—a strip of Cu metal immersed in a solution of $Cu(NO_3)_2$. The two strips of metal act as electrodes and are connected by a wire that allows the electrons to flow from the oxidized substance to the reduced substance. The electrode at which oxidation occurs is called the **anode;** the electrode at which reduction occurs is called the **cathode.** In this reaction, Zn is the anode and Cu is the cathode. *Oxidation always occurs at the anode and reduction always occurs at the cathode in any redox reaction.* The oxidation and reduction half reactions are also shown in Figure 14-14. The two solutions

Figure 14-14 ▶ The oxidation of zinc produces electrons at the anode. The electrons travel through the external circuit to the cathode, where the reduction of Cu^{2+} to copper metal occurs.

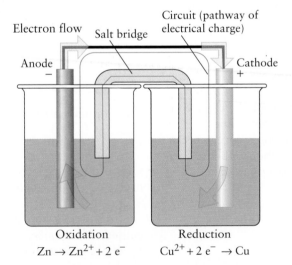

are connected by means of a salt bridge that prevents the buildup of positive or negative charge on the electrodes. Thus, the salt bridge is said to maintain the electroneutrality in the cell. Figure 14-14 shows a galvanic cell based on the Zn/Cu reaction discussed above.

How do we write equations for the half reactions that occur at both electrodes when we know the identities of the anode and the cathode? We use the following thought processes.

✔ **Meeting the Goals**

The oxidation half reaction occurs at the anode and the reduction half reaction occurs at the cathode.

- Anode $\xrightarrow{\text{means}}$ oxidation $\xrightarrow{\text{means}}$ the oxidation numbers must increase (electrons are lost).
- Cathode $\xrightarrow{\text{means}}$ reduction $\xrightarrow{\text{means}}$ the oxidation numbers must decrease (electrons are gained).

Example 14.13

Writing anode and cathode half reactions for a galvanic cell

A galvanic cell is made using a strip of Pb as the anode and a strip of Ag as the cathode. The Pb electrode is in a solution of $Pb(NO_3)_2$, and the Ag electrode is in a solution of $AgNO_3$. Write balanced half reactions for the processes that occur at the anode and at the cathode.

Thinking Through the Problem ◀📢 (14.6) The key idea is that oxidation always occurs at the anode, so Pb must be oxidized; that is, Pb must lose electrons in this reaction. How many electrons does it lose? We see that Pb is in a solution of $Pb(NO_3)_2$. Pb has an oxidation number of +2 in this compound, so we know that the change in oxidation number for Pb is from 0 to +2. Pb^0 loses 2 e^- to form Pb^{2+}. Reduction always occurs at the cathode, so copper ions must gain electrons in this process. We see that copper's oxidation number changes from +2 to 0, which tells us that Cu^{2+} gains two electrons.

Working Through the Problem Pb will lose electrons in the oxidation half reaction, so the electrons appear on the product side of the equation: $Pb\ (s) \rightarrow Pb^{2+}\ (aq) + 2\ e^-$. The electrons will be transferred from Pb to Cu^{2+} in the reduction reaction, so the electrons will appear on the reactant side of the equation: $Cu^{2+}\ (aq) + 2\ e^- \rightarrow Cu\ (s)$.

Answer Anode reaction: $Pb\ (s) \rightarrow Pb^{2+}\ (aq) + 2\ e^-$.
Cathode reaction: $2\ Ag^+\ (aq) + 2\ e^- \rightarrow 2\ Ag\ (s)$.

We can also do the reverse of this procedure and determine which metal is the anode and which is the cathode in a galvanic cell when we are given the equation for the spontaneous redox reaction. To do this, we assign oxidation numbers and find which elements undergo an increase or decrease in oxidation number during the reaction. *An element that shows a decrease in the value of its oxidation number during the course of the reaction is the element that is reduced. An element that shows an increase in the value of its oxidation number during the course of the reaction is the element that is oxidized.* Thus, we can

KEY IDEA 14.7

determine which element is oxidized and which is reduced in the redox reaction by looking to see where the oxidation numbers increase (oxidation) and where they decrease (reduction). Because oxidation occurs at the anode and reduction occurs at the cathode, we can identify the anode and the cathode in the reaction.

In many cases, we can see the electrodes changing as the redox reaction progresses—by focusing on the physical state of each substance in the equation. In our initial reaction, Zn^0 (s) + Cu^{2+} (aq) → Cu^0 (s) + Zn^{2+} (aq), we see that Zn^0 (s) is converted to Zn^{2+} (aq). This means that Zn is going into solution, so we expect the zinc electrode to become smaller as the reaction progresses. We also see that Cu^{2+} (aq) is converted to Cu^0 (s). This means that copper ions become copper atoms as they plate out on the electrode. We expect the copper electrode to become larger in this process.

How do we use a redox equation to find the anode and cathode of a galvanic cell? The thought process is as follows.

- Assign oxidation numbers to all elements in the equation.

- An increase in oxidation number $\xrightarrow{\text{means}}$ oxidation $\xrightarrow{\text{means}}$ anode.

- A decrease in oxidation number $\xrightarrow{\text{means}}$ reduction $\xrightarrow{\text{means}}$ cathode.

✔ **Meeting the Goals**

A redox equation identifies the oxidized and reduced species through the change in oxidation numbers. Hence the anode is identified as the location of the oxidation process and the cathode is identified as the location of the reduction process.

Example 14.14

Identifying from a redox equation the metal that is the anode and cathode

The reaction Mg (s) + Ni^{2+} (aq) → Mg^{2+} (aq) + Ni (s) is a spontaneous redox reaction. Determine which metal is the anode and which is the cathode. Then write the half reactions occurring at the anode and at the cathode, and predict which of the electrodes will become larger and which will become smaller during the redox reaction.

Thinking Through the Problem First we remember that oxidation always occurs at the anode and reduction at the cathode. Then we assign oxidation numbers to find which element is being oxidized and which is being reduced.

Working Through the Problem We see that magnesium has oxidation numbers that increase from 0 to +2, so this must be the oxidation half reaction. Because oxidation always occurs at the anode, Mg must be the anode. The other metal must be the cathode, but we should check to make sure we haven't made a mistake. The oxidation number of nickel decreases from +2 to 0, so this is the reduction half reaction; Ni must be the cathode.

Answer The oxidation half reaction is Mg^0 (s) → Mg^{2+} (aq) + 2 e^-; Mg is the anode. This electrode is slowly getting smaller as Mg atoms in the metal go into solution as Mg^{2+} ions.

The reduction half reaction is Ni^{2+} (aq) + 2 e^- → Ni^0 (s); Ni is the cathode. This electrode is getting larger as Ni^{2+} ions accept electrons and plate out on the nickel metal.

Example 14.15 **Labeling the anode and cathode in a galvanic cell**

Salt bridge

Anode   Cathode

See this with your **Web Animator** ▲

The oxidation reaction at the anode of a galvanic cell releases electrons that travel through the external wire to the cathode where reduction occurs.

According to the activity series for metals, tin is predicted to re-place copper in reactions such as: $Sn\ (s)\ +\ Cu(NO_3)_2\ (aq) \rightarrow$ $Sn(NO_3)_2\ (aq)\ +\ Cu\ (s)$. Determine the anode and cathode from this reaction and label the electrodes and the solutions in the gal-vanic cell drawing at left.

Thinking Through the Problem ◄═► (14.7) The key idea is that we can determine the anode and the cathode from the equation if we find which metal is oxidized and which is reduced. To do this, we need the oxidation numbers of Cu and Sn both when they stand alone and when they are in compounds. The metals standing alone have oxidation numbers of zero. Both copper and tin are written as nitrate compounds and we know that NO_3^- ion has an oxida-tion number of -1. This means that Cu in $Cu(NO_3)_2$ has an oxi-dation number of $+2$. This same reasoning shows that Sn in $Sn(NO_3)_2$ has an oxidation number of $+2$. When we write the two half reactions, we see that copper must gain electrons as its oxi-dation numbers decrease from $+2$ to 0; tin must lose electrons as its oxidation numbers increase from 0 to $+2$.

$$\text{Reduction half reaction: } Cu^{2+} + 2\ e^- \rightarrow Cu^0$$
$$\text{Oxidation half reaction: } Sn^0 \rightarrow Sn^{2+} + 2\ e^-$$

Working Through the Problem After analyzing the two half reactions, we see that copper is reduced and tin is oxidized. We know that reduction occurs at the cathode; this makes Cu the cathode in con-tact with Cu^{2+} ions in the solution. We know that oxidation oc-curs at the anode; this makes Sn the anode in contact with Sn^{2+} ions in the solution.

Answer The anode is labeled Sn; its solution is labeled Sn^{2+}. The cathode is labeled Cu; its solution is labeled Cu^{2+}. The NO_3^- ions are also present in both solutions but do not appear in the draw-ing, as they are spectator ions and are not part of the oxidation-reduction process. In this reaction, Sn is oxidized and Cu^{2+} is reduced.

How are you doing? **14.7**

For the reaction $Sb^{3+}\ (aq)\ +\ Fe\ (s) \rightarrow Fe^{2+}\ (aq)\ +\ Sb\ (s)$, determine which metal is the anode and which is the cathode. Then write the half reactions occurring at the anode and at the cathode, and predict which of the electrodes will become larger and which will become smaller during the redox reaction.

Balancing Simple Redox Equations by Half Reactions

To balance a redox equation, we must balance charges as well as atoms. As electrons are transferred from one reactant to another, we

✔Meeting the Goals

Half reactions are added to get a balanced redox equation after first making sure that the number of electrons lost equals the number of electrons gained.

must be careful to ensure that the number of electrons lost by one reactant is equal to the number of electrons gained by a second reactant. One method that is used to balance redox equations is called the **half reaction method.** In this method we use the oxidation and reduction half reactions to help us find out how many electrons are lost in the oxidation half reaction and how many are gained in the reduction half reaction. Although the oxidation reaction happens at the same time as the reduction reaction, it is often helpful to look at each half of the reaction alone.

Suppose we want to balance the equation $Mg\ (s) + Zn^{2+}\ (aq) \rightarrow Mg^{2+}\ (aq) + Zn\ (s)$. When we look closely at this equation, we can follow the conversion of Zn^{2+} ions to $Zn\ (s)$ and we can follow the conversion of $Mg\ (s)$ into $Mg^{2+}\ (aq)$. This gives us the beginning of two half reactions:

$$Zn^{2+}\ (aq) \longrightarrow Zn\ (s)$$
$$Mg\ (s) \longrightarrow Mg^{2+}\ (aq)$$

Both of these half reactions show a change of two in oxidation number. Mg^0 must lose two electrons to form Mg^{2+} (oxidation) and Zn^{2+} must gain two electrons to form Zn^0 (reduction).

$$\text{Oxidation half reaction: } Mg\ (s) \longrightarrow Mg^{2+}\ (aq) + 2\ e^-$$
$$\text{Reduction half reaction: } Zn^{2+}\ (aq) + 2\ e^- \longrightarrow Zn^0\ (s)$$

The overall redox reaction is the sum of these two half reactions. The balanced equation is $Mg\ (s) + Zn^{2+}\ (aq) \rightarrow Mg^{2+}\ (aq) + Zn\ (s)$. This equation was balanced as written because both the atoms (one Zn and one Mg on both sides of the equation) and the charges ($+2 = +2$) were balanced. Notice that the superscript zero on elements standing alone is omitted in the final equation. Also notice that the electron terms do not appear in the final equation, as they subtract out. *One requirement of a balanced redox equation is that the electrons in the oxidation half reaction must equal the electrons in the reduction half reaction.*

KEY IDEA 14.8

Table 14-2 Electrons Lost Must Equal Electrons Gained in a Balanced Redox Equation

Types of Half Reaction	Electrons in Half Reaction	Multiplication	Electrons Lost = Electrons Gained
Oxidation half reaction has	$2e^-$	Multiply by 1	$2e^-$ lost
Reduction half reaction has	$1e^-$	Multiply by 2	$2e^-$ gained
Oxidation half reaction has	$2e^-$	Multiply by 3	$6e^-$ lost
Reduction half reaction has	$3e^-$	Multiply by 2	$6e^-$ gained
Oxidation half reaction has	$2e^-$	Multiply by 2	$4e^-$ lost
Reduction half reaction has	$4e^-$	Multiply by 1	$4e^-$ gained

What do you do when the electron terms are not the same in the half reaction equations? You must find a multiplier for each half reaction that will make the electron terms the same. Let's look at a few examples. Notice that multiplication by 1 requires no action on your part. Sometimes you only need to find a multiplier for one of the half reactions to ensure that the final electron terms are the same. Note also that when the electron term is increased by multiplication, all elements in that half reaction are also increased by the same factor. Thus, if the half reaction $Fe^0 \rightarrow Fe^{3+} + 3\ e^-$ is multiplied by 2, we will get $2\ Fe^0 \rightarrow 2\ Fe^{3+} + 6\ e^-$. Table 14-2 shows examples of the process that makes the number of electrons lost equal to the number of electrons gained in redox reactions.

Example 14.16

Balancing a redox equation using the half reaction method

According to Table 5-3, nickel will displace silver ions from solution according to the (unbalanced) equation $Ni\ (s) + Ag^+\ (aq) \rightarrow Ag\ (s) + Ni^{2+}\ (aq)$. Balance this equation using the half reaction method.

Thinking Through the Problem We recognize this as an oxidation-reduction reaction because of the change in oxidation numbers of both nickel and silver.

$$\text{Oxidation half reaction: } Ni^0 \longrightarrow Ni^{2+} + 2\ e^-$$
$$\text{Reduction half reaction: } Ag^+ + 1\ e^- \longrightarrow Ag^0$$

In the oxidation half-reaction, the oxidation number of Ni increases from $0 \rightarrow +2$ and two electrons are added to the product side to make reactant charges equal to product charges. Ni^0 *loses* two electrons (oxidation) to form Ni^{2+}. In the reduction half reaction, the oxidation number of Ag decreases in value from $+1 \rightarrow 0$ and electrons are added to the reactant side to balance the charges. Ag^+ *gains* one electron (reduction) to form Ag^0.

Because each electron given up by Ni^0 must be accepted by the Ag^+ ion, we will need two Ag^+ ions in this reaction. Therefore, we will multiply the entire reduction half reaction by 2.

$$\text{Oxidation half reaction: } Ni^0 \longrightarrow Ni^{2+} + 2\ e^-$$
$$\text{Reduction half reaction: } 2\ Ag^+ + 2\ e^- \longrightarrow 2\ Ag^0$$

Now, when we add the two half reactions together we see that the number of electrons gained by the Ag^+ ions equals the number of electrons lost by Ni^0. Because coefficients are mole numbers, we say that two moles of electrons have been transferred in this process.

$$Ni\ (s) + 2\ Ag^+\ (aq) + 2\ e^- \longrightarrow 2\ Ag\ (s) + Ni^{2+}\ (aq) + 2\ e^-$$

The equation is now electrically balanced and, because the electrons appear in equal quantity on both sides of the equation, they are subtracted out from the final equation.

Answer The balanced equation is Ni (s) + 2 Ag$^+$ (aq) $\rightarrow$ 2 Ag (s) + Ni^{2+} (aq).

Example 14.17 **Balancing a redox equation using half reactions**

Balance the following reaction using the half reaction method.

$$\text{Al (s)} + \text{NiCl}_2 \text{ (aq)} \longrightarrow \text{Ni (s)} + \text{AlCl}_3 \text{ (aq)}$$

Thinking Through the Problem We recognize this as an oxidation-reduction reaction because of the change in oxidation numbers of both elements: Al^0 becomes Al^{3+} and Ni^{2+} becomes Ni^0. As the aluminum oxidation numbers increase from 0 to +3, aluminum must lose 3 e$^-$. This process is oxidation. As nickel oxidation numbers decrease from +2 to 0, nickel must gain electrons. This process is reduction. The two half reactions are:

$$\text{Oxidation half reaction: Al}^0 \longrightarrow \text{Al}^{3+} + 3 \text{ e}^-$$
$$\text{Reduction half reaction: Ni}^{2+} + 2 \text{ e}^- \longrightarrow \text{Ni}^0$$

From the two half reaction equations we see that we must multiply the oxidation half reaction by 2 and the reduction half reaction by 3 in order to get the same number of electrons (6) lost and gained.

Working Through the Problem After the appropriate multiplication to obtain 6 electrons in each half reaction, we will add the two half reactions to get the balanced equation. Notice that the e$^-$ terms subtract out in the final equation.

$$\text{Oxidation half reaction: 2 Al}^0 \longrightarrow \text{2 Al}^{3+} + 6 \text{ e}^-$$
$$\underline{\text{Reduction half reaction: 3 Ni}^{2+} + 6 \text{ e}^- \longrightarrow \text{3 Ni}^0}$$
$$\text{Balanced equation: 2 Al + 3 Ni}^{2+} \longrightarrow \text{2 Al}^{3+} + 3 \text{ Ni}$$

Answer The balanced equation is 2 Al + 3 Ni^{2+} $\longrightarrow$ 2 Al^{3+} + 3 Ni.

How are you doing? 14.8

Use the half reaction method to balance the redox equation Zn (s) + Au^{3+} (aq) $\rightarrow$ Au (s) + Zn^{2+} (aq).

Balancing Equations by the Oxidation Number Method

KEY IDEA 14.9

An alternate method of balancing redox equations is the **oxidation number method.** *In the oxidation number method, we find the number of electrons lost and gained by examination of the change in oxida-*

tion numbers of the oxidized and reduced species. The balanced equation results when we multiply each half reaction by some scalar that makes the number of electrons lost equal to the number of electrons gained. This is a shorter form of the half reaction method and works as follows. Suppose you want to balance the reaction:

$$Cu\ (s) + Ag^+\ (aq) \longrightarrow Ag\ (s) + Cu^{2+}\ (aq)$$

✔ Meeting the Goals

The oxidation number method builds on the original unbalanced equation. The oxidized and reduced species are identified from changes in oxidation number, and each half reaction is multiplied by some scalar that makes the number of electrons lost equal to the number of electrons gained.

The first thing to do is to assign oxidation numbers to each element: $Cu^0\ (s) + Ag^+\ (aq) \rightarrow Ag^0\ (s) + Cu^{2+}\ (aq)$. Then we look at the change in oxidation number for both Cu and Ag and construct a *mental* half reaction by drawing a bracket to connect Cu^0 with Cu^{2+}. Draw a second bracket to connect Ag^+ to Ag^0. Remember to include the electron terms for both half reactions.

Thinking this through, we see that the oxidation number for Cu increases from $0 \rightarrow +2$, so it must lose 2 e^-. We write in the electron term and label the half reaction as oxidation. The oxidation number of Ag^+ decreases from $+1 \rightarrow 0$, so Ag^+ must gain 1 e^-. We write in this electron term and label the half reaction as reduction.

$$\overset{\displaystyle -2\ e^-\ \text{(oxidation)}}{Cu^0\ (s) + Ag^+\ (aq) \longrightarrow \underset{\displaystyle +1\ e^-\ \text{(reduction)}}{Ag^0\ (s) + Cu^{2+}\ (aq)}}$$

Because the electrons lost must equal the electrons gained in the reaction, we multiply all terms in the silver half reaction by 2. This will give us 2 Ag^+ and 2 Ag in the equation.

$$\overset{\displaystyle -2\ e^-}{Cu^0\ (s) + 2\ Ag^+\ (aq) \longrightarrow \underset{\displaystyle +2\ (1\ e^-) = 2\ e^-}{2\ Ag^0\ (s) + Cu^{2+}\ (aq)}}$$

We have balanced the equation: Cu (s) + 2 Ag^+ (aq) → 2 Ag (s) + Cu^{2+} (aq).

Balancing redox equations using the oxidation number method

Example 14.18

Balance the following redox reaction using the oxidation number method. State which reactant is oxidized and which is reduced.

$$Al\ (s) + Ni^{2+}\ (aq) \longrightarrow Ni\ (s) + Al^{3+}\ (aq)$$

Thinking Through the Problem ◀━━▶ (14.9) The key idea is that, to use the oxidation number method, we must use the change in oxidation number for both Al and Ni^{2+} to find the number of electrons gained and the number of electrons lost in the unbalanced equation. Then we find the least common multiple of the electron coefficients and do the necessary multiplication to make the electron coefficients the same.

Working Through the Problem The aluminum half reaction shows the oxidation number increasing from 0 to +3. This indicates a loss of 3 electrons and tells us that aluminum is oxidized. The nickel half reaction shows the oxidation numbers decreasing from +2 to 0. This indicates a gain of 2 electrons and tells us that Ni^{2+} is reduced.

$$\overset{\overset{\displaystyle -3\ e^-}{\big\downarrow}}{Al^0\ (s)}\ +\ \underset{\underset{\displaystyle +2\ e^-}{\big\uparrow}}{Ni^{2+}\ (aq)}\ \longrightarrow\ Ni^0\ (s)\ +\ Al^{3+}\ (aq)$$

Now we see that if we multiply all of the aluminum terms by 2 and all of the nickel terms by 3, we will have 6 electrons lost (by 2 Al) equal to the 6 electrons gained (by 3 Ni^{2+}).

$$\overset{\overset{\displaystyle 2\ (-3\ e^-)}{\big\downarrow}}{2\ Al^0\ (s)}\ +\ \underset{\underset{\displaystyle 3\ (+2\ e^-)}{\big\uparrow}}{3\ Ni^{2+}\ (aq)}\ \longrightarrow\ 3\ Ni^0\ (s)\ +\ 2\ Al^{3+}\ (aq)$$

Answer   The balanced equation is 2 Al (s) + 3 Ni^{2+} (aq) → 3 Ni (s) + 2 Al^{3+} (aq). Al is oxidized and Ni^{2+} is reduced.

Example 14.19 **Balancing a redox equation using the oxidation number method**

Balance the following redox reaction using the oxidation number method. State which reactant is oxidized and which is reduced.

$$Ni\ (s)\ +\ Ag^+\ (aq)\ \longrightarrow\ Ag\ (s)\ +\ Ni^{2+}\ (aq)$$

Thinking Through the Problem To use the oxidation number method, we must use the change in oxidation number for both Ni and Ag^+ to find the number of electrons gained and the number of electrons lost in the unbalanced equation. Then we find the least common multiple of the electron coefficients and do the necessary multiplication to make the electron coefficients the same.

Working Through the Problem The nickel half reaction shows the oxidation numbers increasing from 0 to +2. This indicates a loss of 2 electrons and tells us that aluminum is oxidized. The silver half reaction shows the oxidation number decreasing from +1 to 0. This indicates a gain of 1 electron and tells us that Ag^+ is reduced.

$$\overset{\overset{\displaystyle -2\ e^-\ (oxidation)}{\big\downarrow}}{Ni^0\ (s)}\ +\ \underset{\underset{\displaystyle +1\ e^-\ (reduction)}{\big\uparrow}}{Ag^+\ (aq)}\ \longrightarrow\ Ag^0\ (s)\ +\ Ni^{2+}\ (aq)$$

Now we see that if we multiply all of the silver terms by 2, we will have 2 electrons lost (by Ni) equal to the 2 electrons gained (by 2 Ag^+ ions).

$$\overset{2\ e^-\ (oxidation)}{Ni^0\ (s)\ +\ 2\ Ag^+\ (aq)\ \longrightarrow\ 2\ Ag^0\ (s)\ +\ Ni^{2+}\ (aq)}$$

$$2\ (+1\ e^-)\ =\ 2\ e^-\ (reduction)$$

Answer The balanced equation is $Ni\ (s)\ +\ 2\ Ag^+\ (aq) \rightarrow 2\ Ag\ (s)\ +\ Ni^{2+}\ (aq)$. Ni is oxidized and Ag^+ is reduced.

How are you doing? 14.9

Use the oxidation number method to balance the redox equation $Mn\ (s)\ +\ Pb^{2+}\ (aq) \rightarrow Mn^{2+}\ (aq)\ +\ Pb\ (s)$.

PROBLEMS

33. In a galvanic cell, which reaction occurs at the anode? Which occurs at the cathode?

34. How do oxidation numbers help you decide whether a substance is oxidized or reduced?

35. Complete the following half reactions by writing in the electron term. Label the half reaction oxidation or reduction.

(a) $Cl_2 \rightarrow 2\ Cl^-$
(b) $Mn^{3+} \rightarrow Mn$
(c) $2\ H^+ \rightarrow H_2$

36. Complete the following half reactions by writing in the electron term. Label the half reaction oxidation or reduction.

(a) $NO_3^- \rightarrow NO_2^-$
(b) $SO_3^{2-} \rightarrow SO_4^{2-}$

37. Multiply each of the following half reactions by an appropriate factor to make the number of electrons gained equal to the number of electrons lost. Then add the two half reactions to get the balanced redox equation.

PRACTICAL C

How do we monitor the glucose in the blood with electron flow?

Glucose ($C_6H_{12}O_6$) is a small carbohydrate molecule that metabolizes easily to give quick energy to our cells. The average person has an amount of blood glucose (blood sugar) that falls within a particular range, called the "normal" range. Symptoms such as confusion, nausea, clammy skin, glazed eyes, and general lethargy may indicate that one's blood glucose is not within normal ranges. Such patients are given a course of treatment to bring their blood glucose levels to within the normal range.

As both higher-than-normal and lower-than-normal levels of blood glucose can be fatal, it is important for these patients to keep track of their glucose levels. Some over-the-counter glucose testing kits rely on electrochemical principles for their operation. A small current is produced when a droplet of blood is placed on the test strip. The blood dis-

solves some dry chemicals on the test strip and completes the circuit between two miniature electrodes. The detector is calibrated with standard glucose solutions so that the current is "read" as glucose concentrations.

One popular glucose testing kit measures the glucose concentration by monitoring the secondary half reaction involving iron ions.

What is the oxidation number of iron in $[Fe(CN)_6]^{3-}$? _____

What is the oxidation number of iron in $[Fe(CN)_6]^{4-}$? _____

Write the half reaction that converts $[Fe(CN)_6]^{3-}$ to $[Fe(CN)_6]^{4-}$ and determine if this is an oxidation or reduction half reaction. How many electrons are transferred?

(a) $Mg \rightarrow Mg^{2+} + 2\,e^-$ and $Cr^{3+} + 3\,e^- \rightarrow Cr$
(b) $Cu^+ + 1\,e^- \rightarrow Cu$ and $Al \rightarrow Al^{3+} + 3\,e^-$

38. How are the half reaction method and the oxidation number method of balancing redox equations similar? How are they different?

In Problems 39–42, balance the redox equations using either the half reaction method or the oxidation number method. Specify what is being oxidized and what is being reduced in each reaction.

39. $Ca\ (s) + Zn^{2+}\ (aq) \rightarrow Ca^{2+}\ (aq) + Zn\ (s)$

40. $Mg\ (s) + Fe^{2+}\ (aq) \rightarrow Mg^{2+}\ (aq) + Fe\ (s)$

41. $Al\ (s) + Co^{2+}\ (aq) \rightarrow Al^{3+}\ (aq) + Co\ (s)$

42. $Mn\ (s) + Cd^{2+}\ (aq) \rightarrow Mn^{2+}\ (aq) + Cd\ (s)$

43. The reaction $Zn\ (s) + Cu^{2+}\ (aq) \rightarrow Zn^{2+}\ (aq) + Cu\ (s)$ occurs in a galvanic cell. Write the equation for the reaction that occurs at the anode and at the cathode. Which electrode will gain in mass during the course of this reaction? Why?

44. The reaction $Mn\ (s) + Cd^{2+}\ (aq) \rightarrow Mn^{2+}\ (aq) + Cd\ (s)$ occurs in a galvanic cell. Write the equation for the reaction that occurs at the anode and at the cathode. Which electrode will gain in mass during the course of this reaction? Why?

45. Sketch a galvanic cell for the reaction $Co^{2+}\ (aq) + Zn\ (s) \rightarrow Co\ (s) + Zn^{2+}\ (aq)$ and label the electrodes and the solutions with which they are in contact.

46. Sketch a galvanic cell for the reaction $Ni^{2+}\ (aq) + Mn\ (s) \rightarrow Ni\ (s) + Mn^{2+}\ (aq)$ and label the electrodes and the solutions with which they are in contact.

Chapter 14 Summary and Problems

◀ Review with Web Practice

VOCABULARY

solution	A homogeneous mixture of two or more substances.
solute	The substance that is dissolved to make a solution.
solvent	The substance in which the solute is dissolved. Together, solute and solvent make up a solution.
aqueous solution	A solution in which the solvent is water.
soluble substance	A substance that dissolves in a solvent.
insoluble substance	A substance that does not dissolve in a solvent.
solubility	The maximum amount of solute that dissolves in a solvent at a specified temperature and pressure.
saturated solution	A solution that cannot dissolve any more solute to dissolve at a specified temperature and pressure.
unsaturated solution	A solution that is able to dissolve more solute at a specified temperature and pressure.
dissociation	The ionization that occurs when an ionic compound dissolves in water. Thus, NaCl dissociates into two ions, one Na^+ ion and one Cl^- ion, while $MgCl_2$ dissociates into three ions, one Mg^{2+} ion and two Cl^- ions.
concentration	The concentration of a solution tells the amount of solute that is dissolved in a specified amount of solution.
molarity	A concentration unit that tells us how many moles of solute are dissolved in 1 liter of solution: $\text{molarity} = \dfrac{\text{moles of solute}}{\text{volume (liters) of solution}}$.
half reaction	One of the two parts of an oxidation-reduction reaction. The oxidation half reaction shows the loss of electrons; the reduction half reaction shows the gain of electrons.

electrochemistry	The study of chemical systems and how they relate to electrical energy in oxidation-reduction reactions.
anode	The electrode at which oxidation takes place.
cathode	The electrode at which reduction occurs.
half reaction method	A method used to balance redox equations. Half reactions are written for both the oxidation and reduction processes, multiplied by an appropriate number to make the electron terms equal, and added to give the balanced equation.
oxidation number method	A method used to balance redox equations. This method uses the unbalanced equation to determine the number of electrons gained or lost. If the electron terms are not equal, the oxidized (or reduced) species are multiplied by some number that will make the electron terms equal. This gives the final balanced equation.

SECTION GOALS

What is the definition of a solution? What are the components of a solution?

A solution is a homogeneous mixture of two or more components. One component, called the solute, is the substance that is dissolved in a second substance, which is called the solvent. When the solvent is water, the solution is called an aqueous solution.

What is meant by molarity?

Molarity is a unit used to express the concentration of a solution. Molarity is the ratio of moles of solute to the volume of the solution in liters: $\text{molarity} = \dfrac{\text{moles of solute}}{\text{volume (liters) of solution}}$. The symbol for molarity is its first letter, M, where $\text{M} = \dfrac{\text{moles of solute}}{\text{liter of solution}}$.

How do we calculate the molarity of a solution?

The molarity ratio, $\text{M} = \dfrac{\text{moles of solute}}{\text{volume (liters) of solution}}$, contains three variables; M, moles of solute, and liters of solution. Given two of the three variables, we can solve for the third variable algebraically. An alternative method that uses a proportion can also be used.

How do we use molarity to convert between moles of solute and volume of solution?

There are two common ways to set up the calculation of moles of solute or volume of solution from molarity values. (1) We can substitute into the formula for molarity and solve it algebraically for the unknown quantity. (2) We can use the molarity ratio as a known value and set it equal to a similar ratio in which one of the variables is given and the other is unknown. The unknown is then calculated algebraically. You should use the method that makes the most sense to you.

How do we calculate moles when given either gram amounts or molarity and volume information for a solution?

We use the molar mass ratio to convert grams of a substance into moles of that substance (mass of substance A/molar mass of A = moles of A). The molarity ratio is used to calculate moles of a substance in a given volume of solution (molarity of solution containing A $\times$ volume of solution in liters = moles of A).

How do we calculate the moles of either a reactant or a product in a chemical reaction from molarity information and the stoichiometry of the reaction?

Solution molarity gives the numbers of moles of a substance per liter of solution. Given a volume of the solution, we can use the molarity relationship to calculate moles of solute. Once we know the moles of one of the substances in the chemical reaction, we can use the stoichiometry of the equation to determine the moles of any other substance in the reaction.

How can we calculate the volume of a solution of either a reactant or a product using molarity information and the stoichiometry of the reaction?

Using the molarity relationship, liters of solution $= \dfrac{\text{moles of solute}}{\text{molarity}}$, the needed volume of a solution of reactant or product can be calculated if we know both the molarity of the solution and the number of moles of the reactant or product needed. Use the stoichiometry of the reaction, if necessary, to calculate the number of reactant or product moles needed.

How can we calculate the molarity of a product in a chemical reaction from the stoichiometry of the reaction?

To find the molarity of a product in a chemical reaction, use stoichiometry to calculate moles of that product. Then use the molarity relationship,

$$M = \frac{\text{moles of solute}}{\text{volume (liters) of solution}},$$ to calculate the molarity.

How do we write the oxidation and reduction half reactions from a redox equation?

To write half reactions from an oxidation-reduction equation, we use the change in oxidation number of the oxidized and the reduced species to determine the number of electrons lost or gained, respectively. The substance that loses electrons is the oxidized species; the substance that gains electrons is the reduced species.

How do we write half reactions for reactions that occur at the anode and at the cathode in a galvanic cell?

The oxidation half reaction occurs at the anode and the reduction half reaction occurs at the cathode.

How do we identify from a redox equation the anode and the cathode in a galvanic cell?

A redox equation identifies the oxidized and reduced species through the change in oxidation numbers. Hence the anode is identified as the location of the oxidation process and the cathode is identified as the location of the reduction process.

How do we balance redox equations using the half reaction method?

Half reactions are added to get a balanced redox equation after first making sure that the number of electrons lost equals the number of electrons gained.

How do we balance redox equations using the oxidation number method?

The oxidation number method builds on the original unbalanced equation. The oxidized and reduced species are identified from changes in oxidation number and each half reaction is multiplied by some scalar that makes the number of electrons lost equal to the number of electrons gained.

PROBLEMS

47. Assume that each of the following compounds completely dissociates in water solution. How many moles of each ion will be formed from the following amounts?

(a) 1 mole of $K_2Cr_2O_7$
(b) 2 moles of $SnBr_2$
(c) 1.2 moles of $Co(NO_3)_3$

48. How many moles of each ion will be formed when the following compounds are dissolved in water? Assume complete dissociation.

(a) barium hydrogen carbonate
(b) ammonium phosphate
(c) copper (II) nitrate

49. Assume that each of the following compounds completely dissociates in water solution. How many moles of each ion will be formed from the following amounts?

(a) 3.5 moles of $MnCrO_4$
(b) 0.8 mole of $Ba(OH)_2$

50. How many moles of each ion will be formed when the following compounds are dissolved in water? Assume complete dissociation.

(a) silver permanganate
(b) iron (III) chloride

51. How many grams of PbI_2 will form when 125.0 mL of 0.200 M KI solution is reacted with an excess of $Pb(NO_3)_2$ solution? Assume that the reaction proceeds according to the equation:

$$2\,KI\;(aq) + Pb(NO_3)_2\;(aq) \longrightarrow$$
$$PbI_2\;(s) + 2\,KNO_3\;(aq)$$

52. Determine the molarity of a solution made by dissolving 4.58 grams of NaOH in enough water to make 50.0 mL of solution.

53. Determine the volume of 0.426 M NaOH required to completely react with 25.00 mL of a

0.982 M H_2SO_4 solution. Assume the reaction proceeds according to the equation:

$$2\,NaOH\ (aq) + H_2SO_4\ (aq) \longrightarrow$$
$$2\,H_2O\ (l) + Na_2SO_4\ (aq)$$

54. Determine the moles of OH^- ions present in a 36.8-mL sample of 0.098 M NaOH.

55. Determine the moles of OH^- ion present in a 36.8-mL sample of 0.098 M $Mg(OH)_2$, assuming complete dissociation.

56. Determine the total number of moles of OH^- ions present in a mixture of 34.0 mL of 0.487 M NaOH and 65.4 mL of 0.852 M KOH.

57. Determine the total number of moles of Na^+ ions present in a mixture containing 24.0 mL of a 1.024 M NaCl solution and 42.8 mL of 0.084 M NaOH solution.

58. Calculate the mass of Na_2S needed to react with 100.0 mL of a 0.984 M solution of $Ca(C_2H_3O_2)_2$ according to the equation:

$$Na_2S\ (s) + Ca(C_2H_3O_2)_2\ (aq) \longrightarrow$$
$$CaS\ (s) + 2\,NaC_2H_3O_2\ (aq)$$

59. Determine the volume of solution needed in order to provide 2.46×10^{-2} mole of $Ag_2Cr_2O_7$ from a solution of 0.982 M $Ag_2Cr_2O_7$.

Use the following equation for Problems 60–63.

$$2\,PbO_2\ (s) + 4\,HNO_3\ (aq) \longrightarrow$$
$$2\,Pb(NO_3)_2\ (aq) + 2\,H_2O\ (l) + O_2\ (g)$$

60. Calculate the volume of 0.062 M HNO_3 required to react with 6.98 grams of PbO_2.

61. Calculate the moles of $Pb(NO_3)_2$ produced when 34.0 mL of 0.096 M HNO_3 completely reacts.

62. Assume that a 24.8-gram sample of PbO_2 completely reacts with 25.0 mL of HNO_3 solution. What is the molarity of the HNO_3 solution?

63. A 74.2-mL sample of 0.880 M HNO_3 reacts with PbO_2 according to the given equation. Assuming complete reaction, what mass of $Pb(NO_3)_2$ is formed?

Balance the following redox equations using either the half reaction method or the oxidation number method. Specify what is being oxidized and what is being reduced in each reaction in Problems 64–67.

64. $Cd\ (s) + Pb^{2+}\ (aq) \rightarrow Cd^{2+}\ (aq) + Pb\ (s)$

65. $Zn\ (s) + Cu^{2+}\ (aq) \rightarrow Zn^{2+}\ (aq) + Cu\ (s)$

66. $Fe\ (s) + Ag^+\ (aq) \rightarrow Fe^{2+}\ (aq) + Ag\ (s)$

67. $Co\ (s) + Pt^{2+}\ (aq) \rightarrow Co^{2+}\ (aq) + Pt\ (s)$

68. The reaction $Fe\ (s) + Ag^+\ (aq) \rightarrow Fe^{2+}\ (aq) + Ag\ (s)$ occurs in a galvanic cell. Write the equation for the reaction that occurs at the anode and at the cathode. Which electrode will gain in mass during the course of this reaction? Why?

69. The reaction $Al\ (s) + Co^{2+}\ (aq) \rightarrow Al^{3+}\ (aq) + Co\ (s)$ occurs in a galvanic cell. Write the equation for the reaction that occurs at the anode and at the cathode. Which electrode will gain in mass during the course of this reaction? Why?

70. Sketch a galvanic cell for the reaction $Ti^{4+}\ (aq) + Mg\ (s) \rightarrow Ti\ (s) + Mg^{2+}\ (aq)$ and label the electrodes and the solutions with which they are in contact.

71. A 74.2-mL sample of 0.880 M HNO_3 reacts with PbO_2 according to the equation in Problem 59. Assuming complete reaction, what mass of $Pb(NO_3)_2$ is formed?

72. Two solutions are made: the first contains 5.00 grams $NaNO_3$ dissolved in 500.0 mL of solution; the second contains 8.00 grams Na_2SO_4 in 350.0 mL of solution.

(a) Determine the moles of Na^+ ion in each solution.

(b) Calculate the molarity of Na^+ ion in each solution.

73. A 30.0-mL portion of a 0.705 M solution of NaCl is added to 25.0 mL of 0.890 M $AgNO_3$. Assume that the reaction is $NaCl\ (aq) + AgNO_3\ (aq) \rightarrow AgCl\ (s) + NaNO\ (aq)$.

(a) Calculate the moles of AgCl precipitate formed.

(b) What is the limiting reactant in this reaction?

74. Determine how many grams of NaCl are needed to make 30.0 mL of a 0.705 M NaCl solution.

75. Determine the number of moles of solute contained in the following solutions:

(a) 75.0 mL of 0.0842 M K_2CO_3

(b) 125.0 mL of 0.648 M LiCl

76. Calculate $[Mg^{2+}]$ in a solution in which $[Mg_3(PO_4)_2] = 0.840$ M.

77. Determine the volume of 0.0464 M $Ba(OH)_2$ solution required to give 2.09×10^{-2} mole OH^-.

78. A solution is made by dissolving 6.25 grams KBr in 50.0 grams H_2O. The density of the solution is 1.074 g/mL.

(a) What is the volume of the solution?

(b) What is the molarity of the solution?

79. How many moles of I^- are in 15.4 mL of 1.012 M MgI_2 solution? Assume complete ionization of MgI_2 in solution.

80. How many moles of NO_2 are needed when 24.82 mL of 0.291 M HNO_3 are produced according to the following equation?

$$3\ NO_2\ (g)\ +\ H_2O\ (l)\ \longrightarrow\ 2\ HNO_3\ (aq)\ +\ NO\ (g)$$

81. How many moles of $Al(NO_3)_3$ are formed by the complete reaction of 20.10 mL of 0.948 M $Fe(NO_3)_2$ in the following reaction?

$$3\ Fe(NO_3)_3\ (aq)\ +\ 2\ Al\ (s)\ \longrightarrow$$
$$3\ Fe\ (s)\ +\ 2\ Al(NO_3)_3\ (aq)$$

82. Calculate the volume of 0.748 M $HClO_4$ solution needed to completely react with 2.482 grams $Cu(OH)_2$ in the following reaction:

$$Cu(OH)_2\ (s)\ +\ 2\ HClO_4\ (aq)\ \longrightarrow$$
$$Cu(ClO_4)\ (aq)\ +\ 2\ H_2O\ (l)$$

Acids and Bases

Monitoring a change in the properties of a system during the slow addition of a chemical substance is called titration. When enough base is added to an acid solution containing phenolphthalein, a color change results. (Richard Megna/Fundamental Photographs.)

PRACTICAL CHEMISTRY Titration

Many calculations allow us to determine the precise amounts of reactants needed for a complete reaction. This requires that we know the balanced equation and the amount of the other reactant. Sometimes, however, we do not know the balanced equation for the reaction or we do not know the amount of the other reactant. In these cases, we cannot calculate the amount of reactants needed, but we can use a technique called *titration* to add a reactant in small increments until we observe that the reaction is complete.

During a titration, we monitor a change in the properties of a system during the gradual addition of a chemical substance. There are many practical applications of titration, both inside and outside of the chemical laboratory. Consider three of them:

1. A patient's nerve response is monitored while an anesthetic is slowly administered. When the patient's nerves become deadened, the anesthesia is reduced to a much lower level, or is stopped altogether.

A standard amount of "chlorine" is added to pool water to keep it sanitary. (Shizuo Kambayashi/AP.)

2. Swimming pool water is kept sanitary by adding a small amount of "chlorine" in the form of hypochlorous acid, HOCl. A pool attendant determines whether or not there is enough hypochlorous acid in the pool by counting drops of another reactant.

3. A solution of NaOH is prepared. It is impossible that simple weighing determines the mass of NaOH used, because NaOH pellets absorb water from the atmosphere and we do not know how much of the mass of NaOH is due to water. Instead, we can determine the mass of NaOH in the solution by adding the solution to a precise mass of another reactant until the reaction is complete.

A successful titration has several requirements. First, the person carrying out the titration must have the manual skill to add the reactant in a carefully controlled way. Second, the change that occurs during the titration must be obvious to that person. Finally, the change must occur at a point that signifies something for the system under study. Titrations can be done for many different kinds of chemical reactions. In this chapter we will consider titrations of acids and bases.

The Practicals will link what you learn in this chapter about acids and bases to the skills and understanding needed to perform a good acid-base titration. The questions asked are:

PRACTICAL A   How do indicators signal a change in an acid-base titration?

PRACTICAL B   What chemical reactions do we encounter in an acid-base titration?

PRACTICAL C   How are chemical amounts determined by acid-base titration?

PRACTICAL D   How does pH change during acid-base titrations?

15.1 Acids and Bases in Water

SECTION GOALS

✔ Know what it means to say that a solution is acidic, basic, or neutral.

✔ Determine whether a particular solution is acidic, basic, or neutral.

Some Properties of Acidic and Basic Solutions

Acids, bases, and the solutions they form are a very important part of chemistry and of everyday life. Not only do they play a vital role in the chemical industry, but they are indispensable to life processes such as digestion.

Various substances can act as acids, but their solutions all share a set of properties we call "acidic." **Acidic** solutions are generally

(a)

(b)

(c)

Figure 15-1 ▲ Acids form solutions that have common properties. (a) Acids corrode metals with the formation of H_2; (b) acids dissolve carbonates with the formation of CO_2; (c) acids have a sour taste. (a: Ken Karp; b: Chip Clark; c: Corbis Images/PictureQuest.)

CONNECT TO
SECTION 3.2

corrosive—slowly dissolving many metals. Acids also dissolve compounds containing carbonate and some other oxyanions. Some common oxyanions are listed in Table 3-5. Acidic solutions can cause certain substances to change color and are responsible for the characteristic sour taste of some foods. Some of the properties of acids are shown in Figure 15-1.

Compounds known as bases also produce solutions with common properties. Many **basic** solutions feel slippery to the touch and can damage living tissue. They are also "caustic," which means that they often cause the decomposition of substances, especially biological ones. Basic solutions cause calcium solutions to become cloudy, cause a soap scum to form, and cause small amounts of grease to disperse. These are all properties that characterize and distinguish basic solutions from other substances.

Even before much of acid-base chemistry was understood, acids and bases fascinated chemists. When solutions of acids and bases were mixed, the characteristic properties of both solutions were diminished until these properties "disappeared" altogether! We call this property of acids and bases **neutralization**.

A second property that interested chemists was the ability of acidic and basic solutions to cause some organic substances to change color. Careful lists were compiled of the substances and the colors that "indicated" whether the solution was acidic or basic. These substances are known as **indicators**. Table 15-1 summarizes these properties of acids and bases.

Table 15-1 Properties of Common Acids and Bases

Properties of Acids	Properties of Bases
Corrosive	Caustic
Dissolves metals and carbonates	Causes calcium solutions to become cloudy
Sour taste	Slippery to touch
Causes indicators to change color	Causes indicators to change color
Neutralizes bases	Neutralizes acids

Acid-base properties are at the heart of some of today's most important environmental problems. "Acid rain" has been identified as a primary factor in the deterioration of our environment. Acid rain is caused by the burning of coal, fuel oil, and gasoline, which releases gaseous oxides of sulfur and nitrogen into the atmosphere. These nonmetallic oxides then react to form acidic solutions of sulfuric and nitric acids. Acidified rain or snow corrodes the materials that make up our buildings and monuments, and disrupts ecosystems when streams, rivers, and lakes become too acidified to support living things. Figure 15-2 shows the consequence of acid rain on a forested site in Germany.

Figure 15-2 ▶ The effect of acid rain on forestation. The photograph on the left was taken in 1970; the one on the right, in 1983. (Regis Bossu/ Sygma.)

Acids and bases have practical uses as well. Many of our household and industrial cleansers depend on the acidity or basicity of a solution. Basic solutions are the basis of many laundry detergents and soaps. Ammonia, an inexpensive base, is frequently used in glass cleaners. Acids are used in most toilet bowl cleaners, because acids dissolve mineral buildup on surfaces. In each of these cases, the cleaning agent acts in part through an acid-base reaction to make dirt dissolve in the cleaning solution. The acidic or basic properties of certain household products are shown in Figure 15-3.

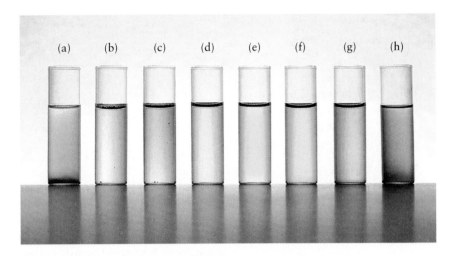

Figure 15-3 ▲ The acidities of various household products can be demonstrated by adding an indicator and noting the resulting color. (Here, pink is acidic and yellow is basic.) The common household products shown here are (a) lemon juice, (b) soda water, (c) 7-Up®, (d) vinegar, (e) ammonia, (f) lye, (g) milk of magnesia, and (h) detergent in water. (Ken Karp.)

✔ **Meeting the Goals**

This section identifies several of the tests and applications that were first used to identify acidic and basic properties in a practical manner. These properties define a solution as acidic or basic.

Two important questions we can ask about a solution at this point are: "Is this solution acidic or basic?" and "How acidic or basic is the solution?" The simplest answer to the first question comes from testing the solution with an acid-base indicator. This is a substance that shows different colors, depending on whether it is in an acidic or a basic solution. The most reliable and traditional acid-base indicator is litmus, a mixture of compounds obtained from lichens (mossy plants that often grow on rocks). The litmus mixture can be applied to paper to make "litmus paper." As shown in Figure 15-4, litmus paper is red when it is in contact with an acidic solution, and blue when it is in contact with a basic solution; this is therefore a simple and reliable test of whether a solution is acidic or basic. Other natural substances can be used as acid-base indicators, as shown in Figure 15-5. Note, however, that testing a solution with most acid-base

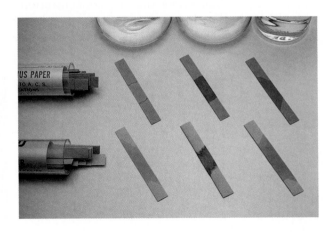

Figure 15-4 ▶ Acids such as the citric acid in lemons turn litmus paper red. Bases, such as that found in ammonia, turn litmus paper blue. (Chip Clark.)

Figure 15-5 ▲ Certain natural substances are acid-base indicators. The south Asian root spice turmeric, used as a component in curries, can be powdered. It has characteristic colors in acid, neutral, and basic solutions. (left: Tom McHugh/Photo Researchers; others: Tom Pantages.)

indicators does not answer the second question, "How acidic or basic is the solution?"

Strong Acids in Water

✔ **Meeting the Goals**

A fundamental part of tracking reaction products is knowing how solutions become acidic or basic.

Pure water, shown in Figure 15-6, contains a small amount of both hydronium ions and hydroxide ions. The reason for this is that water ionizes as shown by the following equation. The double arrow indicates that not all water molecules dissociate, just some of them.

$$2\ H_2O\ (l)\ \rightleftharpoons\ H_3O^+\ (aq)\ +\ OH^-\ (aq)$$

There are three common ways to describe how acids and bases work on a molecular level, and we will consider each in this chapter. The simplest is the Arrhenius definition of an acid or base, named for the insight of the Swedish chemist Svante Arrhenius. He suggested that a substance we now call an **Arrhenius acid** increases the concentration of hydronium ion, H_3O^+, in water. Conversely, **Arrhenius bases** increase the concentration of hydroxide ion, OH^-, in water. The link between the molecular (Arrhenius) definition and the practical definition of an acid or a base occurs because, regardless of the

▶ See this with your **Web Animator**

Figure 15-6 ▶ Water contains small amounts of hydronium ions and hydroxide ions.

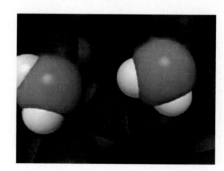

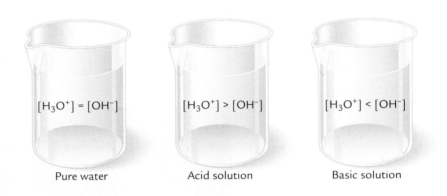

Figure 15-7 ▶ The first beaker contains pure water; because of the autoionization of water, pure water contains a small number of H_3O^+ and OH^- ions. The second beaker contains an acid solution; acid solutions contain more H_3O^+ than pure water. The third beaker contains a basic solution; basic solutions contain more OH^- than pure water.

$[H_3O^+] = [OH^-]$ $[H_3O^+] > [OH^-]$ $[H_3O^+] < [OH^-]$

Pure water   Acid solution   Basic solution

✔ Meeting the Goals

When a solution is acidic, it has more hydronium ion than pure water.

KEY IDEA 15.2

source, all acidic solutions show the effect of the hydronium ion, and all basic solutions show the effect of the hydroxide ion.

According to the Arrhenius definition, an acid *increases* the number of hydronium ions that are present in a sample of pure water. In aqueous solution, Arrhenius acids release H^+ ions that combine with H_2O to form H_3O^+ ions. Thus, the formula of an Arrhenius acid must contain hydrogen; to behave as an acid, the hydrogen must ionize in water solution. Figure 15-7 shows three beakers; one contains water, one contains an acid, and one contains a base.

There are many substances that undergo reaction in water to form hydronium ions. Some acids, called **strong acids**, undergo *complete* reaction when they dissolve. *This means that, for every molecule of the acid that enters the solution, one hydronium ion is formed.* Table 15-2 shows the correspondence between the initial number of acid molecules added to water and the resulting number of ions formed in solution. Given the general formula HA of an acid, its ionization in water is shown by the equation:

$$HA + H_2O \longrightarrow H_3O^+ + A^-$$

Of course, if we can say something about the behavior of the molecules of a substance, we can extend it to mole amounts, as shown by the final row in Table 15-2. So we can also say that when one mole of a strong acid is added to water, one mole of hydronium ions forms. As a specific example, consider the addition of the strong acid HCl to water:

$$HCl\ (g) + H_2O\ (l) \longrightarrow H_3O^+\ (aq) + Cl^-\ (aq)$$

Table 15-2 Ionization of a Strong Acid in Water

Initial Number of HA Molecules	Number of H_3O^+ Ions Formed	Number of A^- Ions Formed
1 molecule	1 ion	1 ion
5 molecules	5 ions	5 ions
1000 molecules	1000 ions	1000 ions
6.022×10^{23} molecules	6.022×10^{23} ions	6.022×10^{23} ions

Figure 15-8 ▶ Action of HCl on water to give hydronium ion.

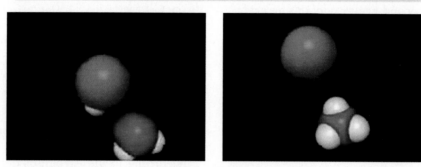

▶ | See this with your **Web Animator**

The reactions of all strong acids with water are very similar to the reaction of HCl. One hydronium ion and one anion are formed. The anion is formed from the original acid by the removal of one hydrogen ion. For the reaction of HCl, the anion is chloride, Cl^-. Examine the Lewis structures for this reaction, as shown in Figure 15-8. In particular, notice the transfer of hydrogen from HCl to H_2O to form H_3O^+.

✔ Meeting the Goals

When we see one of the seven compounds listed in Table 15-3 alone in a water solution, we know that the solution is an acidic solution formed by a strong acid.

The strong acids are very important substances and you should know the major ones by name and formula. The most common strong acids are shown in Table 15-3. Memorizing the acids in this table will help you recognize strong acids. Strong acids are expected to ionize completely in water solution, forming one hydronium ion for each molecule of acid initially present.

Table 15-3 Common Strong Acids

Hydrochloric acid	HCl
Hydrobromic acid	HBr
Hydroiodic acid	HI
Nitric acid	HNO_3
Sulfuric acid	H_2SO_4
Chloric acid	$HClO_3$
Perchloric acid	$HClO_4$

REMEMBER

Example 15.1

Writing equations for reactions of strong acids with water

Write an equation for the reaction of nitric acid, HNO_3, and water.

Thinking Through the Problem We know from Table 15-3 that HNO_3 is a strong acid, and that one mole of nitric acid will form one mole of hydronium ions in water solution. We begin with HNO_3 and H_2O as our reactants. The products are H_3O^+ and NO_3^-, the ion formed by the removal of H^+ from HNO_3.

Working Through the Problem We remove H^+ from HNO_3, leaving the ion NO_3^-. We add H^+ to H_2O, producing the hydronium ion, H_3O^+.

Answer $HNO_3\ (aq) + H_2O\ (l) \longrightarrow NO_3^-\ (aq) + H_3O^+\ (aq)$

Strong Bases in Water

KEY IDEA 15.3

Strong bases ionize completely in water to give hydroxide ions. Many compounds, especially the alkali metal (group I) hydroxides and barium hydroxide, are strong bases. *Strong bases give a stoichiometric amount of hydroxide ions when they dissolve in pure water.* This means that the salt dissolves completely and that every hydroxide ion in the salt becomes a hydroxide ion in the solution, no matter whether the salt contains one or two OH^- ions. (Water does not appear as a reactant in the following equations showing strong base ionization, because the hydroxide ion is already present in the salt and does not come from water.)

$$NaOH\ (s) \longrightarrow Na^+\ (aq) + OH^-\ (aq)$$
$$Ba(OH)_2\ (aq) \longrightarrow Ba^{2+}\ (aq) + 2\ OH^-\ (aq)$$

✔ **Meeting the Goals**

Not all metals form hydroxides that are strong bases that release all of their OH^- ions in water solution. Group I metals and barium form hydroxides that are strong bases.

In the first equation above, the formula $NaOH$ contains just one Na^+ and one OH^-. Ionization of one formula unit of $NaOH$ in water, then, results in the formation of one Na^+ and one OH^- ion. In the second equation, however, there are two OH^- groups in the formula $Ba(OH)_2$, as shown by the subscript number. Ionization of $Ba(OH)_2$ in water results in the formation of two OH^- ions, as shown by the coefficient 2 in front of the hydroxide ion in that equation.

Example 15.2

Writing equations for the dissolution of a strong base in water

Write the equation for the dissolution of $LiOH$ in water.

Thinking Through the Problem What ions enter into solution when lithium hydroxide dissolves in water? Lithium is a Group I metal, so lithium hydroxide is a strong base.

Working Through the Problem ➤ (15.3) The ions formed will be Li^+ and OH^-. We show they are in water solution by writing "(aq)" after their formulas.

Answer $LiOH\ (s) \longrightarrow Li^+\ (aq) + OH^-\ (aq)$

How are you doing? 15.1

Determine the chemical reaction that occurs when the following substances are dissolved in water. Write an equation for each reaction and identify the reactant as an acid or a base.

(a) $HClO_4$ (b) HI (c) $RbOH$

Lewis Picture of Acids and Bases

Before we consider the quantitative aspects of acid-base chemistry, it is important to have a molecular picture in mind. There are three common ways to describe the molecular behavior of acids and bases. We have considered the Arrhenius view of acids and bases, which is that acids produce hydronium ions and bases produce hydroxide ions in water. Now let's consider the second picture: the Lewis picture of acids and bases, which emphasizes that the properties of a species depend on its lone pair or lone pairs of electrons. *A **Lewis acid** accepts a lone pair of electrons from a Lewis base; a **Lewis base** donates a lone pair of electrons to a Lewis acid.*

Let's look at the specific example of HCl in water discussed earlier. In this reaction, shown in Figure 15-8, a hydrogen ion is transferred from HCl to H_2O, forming H_3O^+ and Cl^- ions. Let's analyze the same reaction from the Lewis viewpoint, in terms of electron lone pairs. We see that water has a lone pair of electrons available to donate to the hydrogen ion from HCl. Water is a Lewis base; it donates an electron pair to hydrogen ion. H^+ is a Lewis acid; it accepts an electron pair from oxygen in water.

Water is not the only species that can accept a hydrogen ion. In fact, according to the Lewis definition, any base can do so. Consider the reaction of hydrogen chloride gas with ammonia gas. When HCl and NH_3 vapors come into contact, they react to form solid NH_4Cl, as shown in Figure 15-9(a).

In this reaction, hydrogen ion from HCl is transferred to NH_3, leaving Cl^- ion. H^+ (a Lewis acid) is attracted to the lone pair of electrons on N (the Lewis base), forming NH_4^+ ion. NH_4Cl is formed when the Cl^- and NH_4^+ ions combine. The Lewis picture of this reaction between HCl and NH_3 is shown in Figure 15-9(b). A similar reaction is expected to occur between aqueous solutions of HCl and NH_3.

Examination of the Lewis dot picture shown in Figure 15-9(b), gives a clearer picture of an important concept in acid-base chemistry. Notice that there is a lone pair on N in the ammonia molecule, NH_3. *Compounds that contain anions from groups V through VII are*

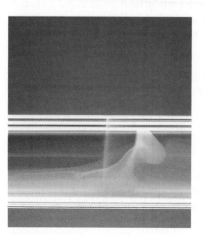

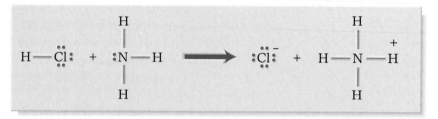

Figure 15-9 (a) ◄ Vapors of HCl and NH_3 react to form tiny particles of NH_4Cl powder that appear in this picture as a stream of dust. (Richard Megna/Fundamental Photographs.) (b) ▲ When NH_3 reacts with HCl, H^+ transfers from HCl to NH_3. In this reaction, ammonia acts as a Lewis base by sharing a pair of electrons with H^+, a Lewis acid. The bond formed is called a coordinate covalent bond.

We depend on the purity of our water for many uses. (Elyse Lewin/Brand X Pictures/PictureQuest.)

often good Lewis bases, because the negative charge associated with their lone pairs easily attracts H^+ ions. Neutral nitrogen atoms are also Lewis bases.

There may be several lone pairs on a species that acts as a Lewis base. Typically, these lone pairs are not equal to one another, and a particular site acts as the Lewis base. For example, hypochlorite, ClO^-, has the Lewis structure shown on the left in the following equation:

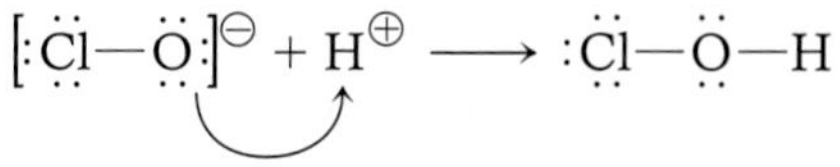

$$\left[:\ddot{\underset{..}{Cl}}-\ddot{\underset{..}{O}}:\right]^{\ominus} + H^{\oplus} \longrightarrow :\ddot{\underset{..}{Cl}}-\ddot{\underset{..}{O}}-H$$

The electrons in the Cl—O bond are not equally shared because oxygen has a different attraction for shared electrons than does chlorine. One of the oxygen lone pairs is a more attractive bonding site for H^+. When a hydrogen ion is added to hypochlorite, it bonds with one of the lone pairs on oxygen, forming hypochlorous acid, HOCl.

Example 15.3

Writing acid-base equations using Lewis structures

Use Lewis structures of acids and bases to indicate how a fluoride ion can react with a hydrogen ion to make HF.

Thinking Through the Problem ➧ (15.4) We need to draw the reaction of F^- with H^+. We start with the Lewis structure of F^-.

Working Through the Problem The Lewis structure of a base should contain a lone pair that can react with a Lewis acid.

Answer

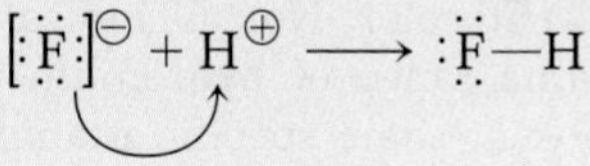

$$\left[:\ddot{\underset{..}{F}}:\right]^{\ominus} + H^{\oplus} \longrightarrow :\ddot{\underset{..}{F}}-H$$

Example 15.4

Using Lewis structures to identify Lewis acids and bases

Use a Lewis structure to explain why methylamine, CH_3NH_2, is a base. This molecule has a C—N bond, three H's on the C, and two H's on the N.

Thinking Through the Problem We need to draw a valid Lewis structure that explains the basicity of methylamine.

We draw the Lewis structure of methylamine, with three H's on C, a C–N bond, and two H's on N.

Working Through the Problem The Lewis structure contains a lone pair on N.

Answer

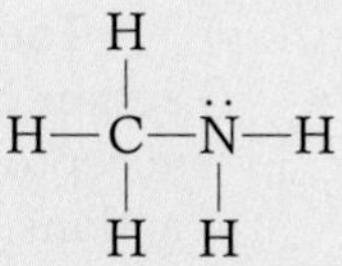

$$\begin{array}{c} H \\ | \\ H-C-\overset{..}{N}-H \\ | \quad | \\ H \quad H \end{array}$$

The N lone pair can react with a Lewis acid, accepting a hydrogen ion. Thus, methylamine is a base.

PRACTICAL A   How do indicators signal a change in an acid-base titration?

Acid-base indicators are substances that change color depending on the relative amounts of hydronium and hydroxide ion present in the solution. Generally, an indicator exhibits one color in acid solution and a different color in basic solution. Because of this difference, we can say that an indicator has an "acid form" and a "base form."

In acid-base titrations we add a solution of an acid to a solution of a base, or vice versa. There are many ways to follow these reactions, but the easiest way is to use an acid-base indicator. A color change is related to a particular concentration of hydronium ion. Thus, careful monitoring of the titration for the first sustained hint of a color change gives us a way to determine the hydronium ion concentration present.

The majority of indicators change color because of a reaction that changes the structure of the indicator by, for example, removing a hydrogen ion or by adding a hydroxide ion to the indicator. The change in structure corresponds to a change in color that helps us determine whether a solution is acidic or basic. If a solution becomes acidic, what very often happens is that either a hydrogen ion is put back onto the indicator molecule or the hydrogen ion removes a hydroxide ion from the indicator molecule. In either case, the indicator is converted into a different substance with a different color. Not all indicators change color in the same range of hydronium ion concentration. Let's look at some specific examples of three different acid-base indicators. Each has a different structure and exhibits a different color change when it converts from its acid form to its base form.

The acid form of the indicator *para*-nitrophenol $(C_6H_5NO_3)$ is shown at top right. This indicator goes from colorless to yellow when a hydrogen ion is removed. Redraw the Lewis dot picture of this compound with the lone pairs on the oxygen atoms. (Consult Section 2.3 if necessary.) In basic solution this molecule loses the hydrogen ion bonded to oxygen to form the yellow anion $C_6H_4NO_3^-$. Circle this hydrogen atom in the drawing below.

The acid form of the indicator malachite green does *not* have a hydrogen ion to release in basic solution. Compare the structure of malachite green in the acid and the base form (opposite page), and circle the part of the compound where hydroxide adds to convert the acidic form to the basic form.

In Figure 15-10 we see the effect of adding a solution of a strong base to a solution of a strong acid. The indicator phenophthalein is present. Describe what happens to the acid-base properties of the solution when we go from (a) to (b) and from (b) to (c). What can we say about the acidity or the basicity of these three solutions? Where is acid in excess? Where is base in excess? What is responsible for the color, and why does the color change? When would you say that the reaction of the strong acid and base is "complete" stoichiometrically?

How are you doing?   15.2

Draw a Lewis structure of one of the resonance structures of hydrogen carbonate, HCO_3^-. Which lone pair is expected to react first as a Lewis base?

The Lewis picture of acids and bases, which describes the interaction between a lone pair on an atom or molecule and a lone pair acceptor, is the general way that chemists analyze all acid-base reactions. If you become comfortable with analyzing acid-base reactions in this way, then you will have mastered a concept that will be useful far beyond this course!

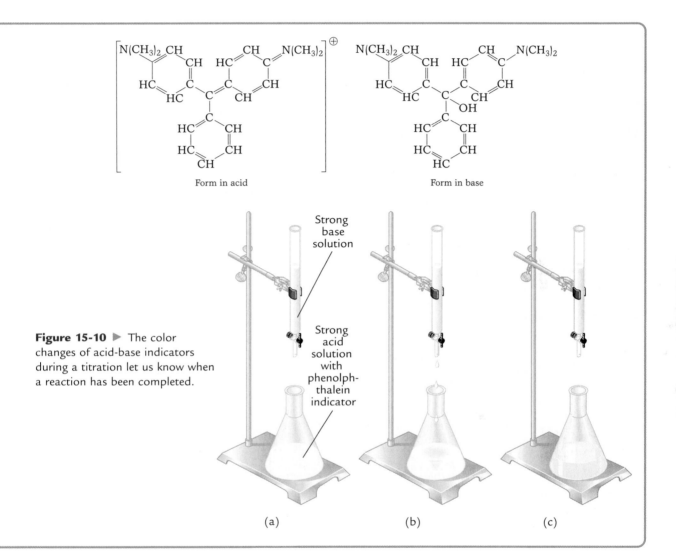

Form in acid   Form in base

Figure 15-10 ▶ The color changes of acid-base indicators during a titration let us know when a reaction has been completed.

Strong base solution

Strong acid solution with phenolph-thalein indicator

(a)   (b)   (c)

PROBLEMS

1. Concept question: A chemist observes bubbles when a piece of limestone is added to a solution. Is this solution an acid or a base?

2. Concept question: We have an acidic solution that turns litmus paper red. A second solution turns litmus paper blue. When equal volumes of the solutions are mixed, the mixture turns litmus paper blue. Indicate which of the following statements is true. If the statement is not true, then rewrite it to be true.

(a) The second solution contains a base.

(b) The mixture contains some acid from the first solution, but it can no longer affect the litmus paper.

(c) The mixture will cause grease to dissolve.

(d) The mixture will do a good job at removing soap scum from surfaces.

3. What is an acid? List four properties of acids.

4. What is a base? List four properties of bases.

5. Complete the equations for the complete ionization of the following strong acids and bases in water:

(a) $HClO_4 + H_2O \rightarrow$

(b) $HBr + H_2O \rightarrow$

(c) $KOH \rightarrow$
(d) $H_2SO_4 + H_2O \rightarrow$

6. Indicate which of the following substances are expected to be strong acids.

$$H_3PO_4 \qquad HClO_4 \qquad H_2CO_3$$

7. Explain what is meant by an indicator.

8. Describe how litmus acts as an indicator for acid and basic solutions.

9. How is acid rain formed?

10. What is a Lewis acid? A Lewis base?

11. Draw a Lewis structure for chloride ion. Discuss how Cl^- can act as a Lewis base.

12. Draw a Lewis structure for water. Discuss how H_2O can act as a Lewis base.

13. Use Lewis structures to write an equation for the reaction of HBr with water.

14. **Discussion question:** Draw the Lewis structure for perchlorate and phosphate. Comment on how both can act as bases. What differences do you see between them?

15. If we have 25 HNO_3 molecules, how many H_3O^+ will form according to the following reaction?

$$HNO_3\ (aq) + H_2O\ (l) \longrightarrow H_3O^+\ (aq) + NO_3^-\ (aq)$$

16. If we have 2.5×10^4 moles of KOH, how many moles of OH^- will form according to the following reaction?

$$KOH\ (aq) \longrightarrow OH^-\ (aq) + K^+\ (aq)$$

17. If we have 5.5×10^2 moles of H_2SO_4, how many moles of H_3O^+ will form according to the following reaction?

$$H_2SO_4\ (aq) + H_2O\ (l) \longrightarrow$$
$$H_3O^+\ (aq) + HSO_4^-\ (aq)$$

18. If we have 1.3×10^3 moles of NaOH, how many moles of OH^- will form according to the following reaction?

$$NaOH\ (aq) \longrightarrow OH^-\ (aq) + Na^+\ (aq)$$

19. Although weak acids ionize only slightly in water solution, the equations for their ionization are written the same way as the equations for strong acid ionization are written. We use a double arrow in the equation for the ionization of weak acids, however, to indicate the coexistence of both the ions and the original acid. The following substances are all weak acids. Write an equation for the reaction that occurs when these weak acids dissolve in water.

(a) H_3PO_4 (phosphoric acid)
(b) C_6H_5COOH (benzoic acid)
(c) H_2CO_3 (carbonic acid)
(d) HF (hydrogen fluoride)

15.2 Reactions of Acids and Bases

SECTION GOALS

✔ Describe how acids and bases react with each other.
✔ Use the Brønsted-Lowry description of acid-base reactions.
✔ Describe what happens when acids are brought into contact with substances or solutions containing carbonate and hydrogen carbonate.

Conjugate Acids and Bases

In Section 15.1, we discussed the reaction of acids and bases with water. This usually involves the transfer of a hydrogen ion to or from water or the direct release of hydroxide ion into the solution. But these are not the only reactions of acids and bases. In this section, we will discuss how acids and bases react not only with water, but with each other.

Recall Figures 15-8 and 15-9, which use Lewis structures to depict the action of HCl on water and ammonia. We can see that the acid HCl releases the hydrogen ion, which is transferred to the other molecule. The transfer of hydrogen ions from one species to another is summarized by the Brønsted-Lowry definitions of an acid and a base, our third way of describing acids and bases. A **Brønsted-Lowry acid** releases a hydrogen ion. The hydrogen ion is accepted by a **Brønsted-Lowry base.** We summarize this transfer of the hydrogen ion by calling the reacting acid "HA" and the reacting base "B." The products are the base A$^-$ and the acid BH$^+$:

$$HA + B \longrightarrow A^- + BH^+$$

There is a special relationship between the acid HA and the base A$^-$. They are linked to each other through the transfer of a hydrogen ion. We can define a one-to-one chemical relationship that transforms a base to an acid and an acid to a base. Two substances that exist in this kind of relationship are known as a **conjugate acid-base pair.** The word conjugate means "connected." Thus, two substances that are connected as a conjugate acid-base pair are connected to one another through the transfer of a hydrogen ion, as shown in Figure 15-11.

Figure 15-11 ▶ Relationship of conjugate acid-base pair HA and A$^-$.

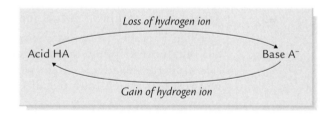

Loss of hydrogen ion

Acid HA   Base A$^-$

Gain of hydrogen ion

The difference between a conjugate acid and its conjugate base is one hydrogen ion. The acid has one more H$^+$ ion than does its conjugate base. The formulas for some conjugate acid-base pairs are given in Table 15-4. Compare the formulas of each conjugate acid with the formula of its conjugate base. Look specifically at the number of hydrogens in the formulas and the charge on each.

Table 15-4 Examples of Conjugate Acid-Base Pairs

Acid (contains 1 more H$^+$ than the base form)	Base (contains 1 fewer H$^+$ than the acid form)
$HClO_4$	ClO_4^-
NH_4^+	NH_3
HNO_2	NO_2^-
CH_3COOH	CH_3COO^-

Remember that the addition or removal of H^+ will affect the charge of the resulting conjugate acid or base. For example, we see in Table 15-4 that the removal of H^+ from a neutral molecule of $HClO_4$ results in a negatively charged conjugate base, ClO_4^-. Likewise, the removal of H^+ from NH_4^+ results in a neutral molecule of NH_3.

Example 15.5 Writing formulas for conjugate bases

What are the conjugate bases for the acids HI and HNO_3?

Thinking Through the Problem A conjugate base contains one fewer H^+ than its acid form. We start with the formulas of the acids, both of which are neutral and contain a single H atom. We expect the conjugate base to have a single negative charge. ➡️ (15.6) The key idea is that a conjugate acid-base pair is linked by the transfer of a hydrogen ion.

Working Through the Problem We remove H^+ from the acid, leaving a base with a negative charge and one fewer H atom.

Answer The conjugate base for the acid HI is I^-. The conjugate base for the acid HNO_3 is NO_3^-.

Example 15.6 Writing formulas for conjugate acids

What are the conjugate acids for the bases OH^- and F^-?

Thinking Through the Problem We remember that a conjugate acid has one more H^+ than its conjugate base. Each of these bases carries a single negative charge; we expect that the addition of H^+ will result in a neutral molecule with one more H than the initial formula for the base.

Working Through the Problem We add one H^+ to each base, to form the conjugate acid. Because both bases have a single negative charge, both conjugate acids will be neutral.

Answer The conjugate acid for OH^- has one oxygen and two hydrogens. It is "HOH," or H_2O. The conjugate acid for F^- is HF.

A special situation arises when a compound has more than one hydrogen atom or when it can accept more than one hydrogen ion from an acid. Such compounds are called **polyprotic acids** and **polyprotic bases.** The most important thing about them is that they always undergo acid-base reaction *one hydrogen ion at a time.* For example, sulfuric acid (H_2SO_4) can lose one hydrogen ion to form the conjugate base HSO_4^-.

$$H_2SO_4 \longleftrightarrow HSO_4^-$$
conjugate acid — conjugate base

The hydrogen ion in HSO_4^- is also available for acid-base reactions. Therefore, HSO_4^- can also be a Brønsted-Lowry acid. Its conjugate base is SO_4^{2-}.

$$HSO_4^- \longleftrightarrow SO_4^{2-}$$
$$\text{conjugate acid} \qquad \text{conjugate base}$$

Substances or ions such as HSO_4^- have a special classification because they can react both with acids and with bases. They are called **amphoteric compounds.**

Writing formulas for conjugate bases of polyprotic acids

Example 15.7

Carbonic acid, H_2CO_3, is a polyprotic acid with two hydrogen atoms. Indicate the conjugate base of carbonic acid.

Thinking Through the Problem The conjugate base is found by removing one H^+ from H_2CO_3. We start with the formula for carbonic acid, which is neutral and has two hydrogen atoms. We expect the conjugate base to have a charge of -1 after H^+ is removed.

Answer The conjugate base of carbonic acid is HCO_3^-. (This base is said to be amphoteric because HCO_3^- can also act as an acid by releasing a hydrogen ion.)

How are you doing? 15.3

(a) Write the formula of the conjugate acid for I^-, HCO_3^-, and NO_3^-.
(b) Write the formula of the conjugate base for HS^-, HF, and HCN.

Writing Acid-Base Reactions

According to the Brønsted-Lowry definitions of acids and bases, an acid gives up a hydrogen ion to a base that accepts it. What remains after the transfer of H^+ from acid to base are the conjugates of each. To write an equation for an acid-base reaction, we must write products that are the conjugate acid or conjugate base of the reactant acid and base. In general, these reactions look like this:

$$\overbrace{\text{acid}_1 + \text{base}_2}^{} \longrightarrow \text{base}_1 + \text{acid}_2$$

conjugate pair (top)
conjugate pair (bottom)

✔ **Meeting the Goals**

The first step in writing a Brønsted-Lowry reaction is the identification of the conjugate acid and conjugate base for each reactant. The conjugate pairs are written as products.

In this equation, the subscript numbers identify the acid-base conjugate pairs. We can predict the products of a simple acid-base reaction by writing, as products, the conjugate base of the reacting acid and the conjugate acid of the reacting base.

Example 15.8   **Analyzing equations for reaction of a strong acid with water**

Analyze from a Brønsted-Lowry point of view the reaction that occurs when hydrobromic acid is mixed with water.

Thinking Through the Problem We expect that an acid, in this case hydrobromic acid, will transfer a hydrogen ion to water. The formula for hydrobromic acid is HBr (see Table 15-3). Water acts as a base in this reaction. To complete the equation, we will need to determine the conjugate base of HBr and the conjugate acid of water.

Working Through the Problem The conjugate base for HBr is Br^-. Water has the conjugate acid H_3O^+. We write the reaction with HBr and H_2O as reactants and Br^- and H_3O^+ as products.

Answer $HBr\ (aq) + H_2O\ (l) \longrightarrow Br^-\ (aq) + H_3O^+(aq)$

Example 15.9   **Writing equations for acid-base reactions**

Write the formulas for the initial products of the reaction of sulfuric acid with ammonia.

Thinking Through the Problem Sulfuric acid is a strong acid (see Table 15-3). We expect a strong acid to transfer one H^+ ion to a base, ammonia in this case. The formula for sulfuric acid is H_2SO_4. It is a diprotic acid, containing two hydrogens. The equation will show the transfer of the first hydrogen ion to ammonia. The reactants are H_2SO_4 and NH_3. The products are the conjugate base of H_2SO_4 and the conjugate acid of NH_3.

Working Through the Problem The conjugate base of H_2SO_4 is HSO_4^-. Ammonia will accept H^+ to form its conjugate acid, NH_4^+.

Answer $H_2SO_4\ (aq) + NH_3\ (aq) \longrightarrow NH_4^+\ (aq) + HSO_4^-\ (aq)$

How are you doing? 15.4

Write the equation for the reaction that occurs when hydrogen carbonate ion acts as a base in a reaction with hydrochloric acid.

Naming Acids

CONNECT TO
SECTION 3.1

The conjugate acid-base concept, along with our understanding of how to name anions, discussed in Chapter 3, makes it relatively easy to name acids because the name of the acid is derived from the name of its conjugate base.

Acids contain hydrogen atoms that can be removed to give an anion. The most important acids are those that will act upon water to make hydronium ion. There are two kinds of substances that will do this: binary acids and oxyacids.

Binary acids are substances containing H and one other element. These have the general formula H_xA, where H is hydrogen and A is some other element. Binary acids are named by prefixing the anion name with *hydro-* and converting the last syllable of the anion name to *-ic.* Table 15-5 lists the formulas and names of three binary acids. Because these are aqueous acids, we name them in this way, not as simple molecules, which we did in Chapter 2.

Table 15-5 Formulas and Names of Some Binary Acids

Formula	Anion	Name of Anion	Name of Acid
HF	F^-	Fluoride	Hydrofluoric acid
HCl	Cl^-	Chloride	Hydrochloric acid
H_2S	HS^-	Hydrogen sulfide	Hydrosulfuric acid

Oxyacids are substances containing H, O, and one or (sometimes) more other atoms. They usually have the formula H_nXO_m and the acidic hydrogens are always attached to an O atom. The name used for an oxyacid is derived from the name of the oxyanion it contains, XO_m^{n-}. If the oxyanion ends in *-ate,* the suffix for the acid is *-ic.* If the oxyanion ends in *-ite,* the acid ends in *-ous.* As we saw in Chapter 3, the name of the oxyanion depends on how many oxyanions that element can form. For example, nitrogen forms two important oxyanions, nitrate, NO_3^-, and nitrite, NO_2^-. There are therefore two acids formed for nitrogen. Nitrate has the conjugate acid HNO_3, nitric acid, and nitrite has the conjugate acid HNO_2, nitrous acid. The rules for naming binary acids and oxyacids are summarized in Table 15-6.

Table 15-6 Rules for Naming Acids

Binary acids: Use prefix *hydro-* and suffix *-ic* with root of anion name.
  *hydro-*root of anion name-*ic*   Example: HBr is hydrobromic acid

Oxyacids: Use the name of the oxyanions (Table 3-4); convert the oxyanion endings as follows:

	Oxyanion Ending		Oxyacid Ending
Change	*-ate*	to	*-ic*
Change	*-ite*	to	*-ous*
Example: BrO^- is hypobromite ion			HBrO is hypobromous acid

The last consideration concerns the need for smooth-sounding names. This means that chemists choose to turn the *sulf-* root into *sulfur-* when naming the acids related to sulfur oxyanions. Thus, H_2SO_4 is sulfuric acid and H_2SO_3 is sulfurous acid. We also do the same thing with *phos-* by making *phosphor-* the root for the naming of acids.

Example 15.10 Naming oxyacids

Phosphorus forms two important oxyacids, H_3PO_4 and H_3PO_3. What are their names?

Thinking Through the Problem We will use the rules listed in Table 15-6 to determine the names of these acids. We start from the formula of the related oxyanions. For H_3PO_4 this is PO_4^{3-}, phosphate. For H_3PO_3 this is PO_3^{3-}, phosphite.

Working Through the Problem We will take the oxyanion names and change the ending to reflect that the substance is now an acid. The ending *-ate* is converted to *-ic*; the ending *-ite* is converted to *-ous*.

Answer H_3PO_4 is phosphoric acid; H_3PO_3 is phosphorous acid.

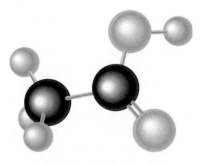

Acetic acid, CH_3COOH

Figure 15-12 ▲ Acetic acid is an organic acid found in vinegar. Organic molecules that contain the —COOH group are acidic.

How are you doing? 15.5

Name the acids $HClO_3$, $HBrO_4$, and HIO_2.

Most compounds containing carbon are known as organic compounds. Some organic compounds are acids, and these organic acids usually contain a —COOH group. The hydrogen in —COOH ionizes in water solution to form H_3O^+. An example of an organic acid is acetic acid, H_3C—COOH. The acetic acid molecule is shown in Figure 15-12. The C—H bonds in acetic acid and other organic compounds are not acidic in water.

Characteristic Reactions of Acids

When we take an egg and place it in a solution of vinegar, we see that small bubbles start to form on the surface of the shell as shown in Figure 15-13. If the vinegar is concentrated enough and we wait a day, then the eggshell will become very thin and will eventually dissolve, leaving behind the egg membrane surrounding the egg white and yolk.

This is an example of an acid-base reaction where one of the reactants, acetic acid, is a soluble acid and the other, calcium carbonate, is an ionic compound that is a base. The reaction that we see as the eggshell dissolves is the acid-base reaction giving calcium acetate and carbonic acid:

$$2\ CH_3COOH\ (aq) + CaCO_3\ (s) \longrightarrow$$
$$Ca(CH_3COO)_2\ (aq) + H_2CO_3\ (aq)$$

When the carbonic acid reaches a certain concentration in aqueous solution, it breaks down into water and carbon dioxide:

$$H_2CO_3\ (aq) \longrightarrow H_2O\ (l) + CO_2\ (g)$$

Figure 15-13 ▲ Vinegar contains acetic acid, which will react with calcium carbonate in an eggshell. Here we see bubbles forming on the surface of a broken shell. (Tom Pantages.)

This carbon dioxide forms the bubbles that we see on the surface of the eggshell.

The breakdown of calcium carbonate by acidic solution is responsible for the destruction of limestone and marble surfaces by acid rain. The most common components of acid rain are sulfuric acid and nitric acid.

Example 15.11

Figure 15-14 ▲ This limestone gargoyle is corroded by acid rain. (Richard Megna/Fundamental Photographs.)

Determining products of an acid-base reaction

Sulfuric acid in rainwater is responsible for damage to limestone and marble buildings, as shown in Figure 15-14. What are the products of the complete reaction of calcium carbonate (in limestone) with sulfuric acid?

Thinking Through the Problem We need to decide what the products of this reaction will be, and then write the relevant chemical equation. In a complete reaction, both hydrogens will be transferred. Sulfuric acid, H_2SO_4, can give up two hydrogen ions. The formula for calcium carbonate is $CaCO_3$. This is a base capable of accepting two hydrogen ions to form carbonic acid.

Working Through the Problem $CaCO_3$ will ionize to form Ca^{2+} and CO_3^{2-} ions. The transfer of both hydrogen ions from H_2SO_4 to CO_3^{2-} ion will leave SO_4^{2-} ion and will form H_2CO_3. The carbonic acid initially formed from carbonate will decompose to water and carbon dioxide. The overall reaction shows that solid $CaCO_3$ is destroyed.

Answer $CaCO_3\ (s) + H_2SO_4\ (aq) \longrightarrow$
$$Ca^{2+}\ (aq) + SO_4^{2-}\ (aq) + H_2O\ (l) + CO_2\ (g)$$

PRACTICAL B — What chemical reactions do we encounter in an acid-base titration?

Titration is a process in which one reactant is added gradually to a second reactant until the reaction between them is complete. Figure 15-16 shows the laboratory setup for a titration. The solution we wish to analyze is called the *analyte;* it contains an unknown amount of reactant. The solution added to the analyte is called the *titrant;* it contains a known amount of a reactant. Titration is used to determine the amount of analyte present in a sample.

A good titration reaction is quick and easily goes to completion. For titration to be useful, we must know the chemical reaction that occurs between the titrant and the analyte. When the titrant is added to the analyte solution, the two substances react. The titration continues until all analyte has reacted. In general, some device or chemical is used to indicate when the analyte is completely consumed. This is called the *end point* of the titration. Appropriate indicators should be used so that the end point of the titration is easy to detect.

The point at which actual stoichiometric amounts of the two reactants have been mixed and reacted is called the *equivalence point* of the titration. In a good titration, the indicator signals that the reaction is complete (end point) when the analyte is exactly used up (equivalence point).

Most acid-base titrations involve a solution of an acid and a solution of a base. In the simplest case, the acid and the base react in a 1:1 ratio to give the corresponding conjugate base and conjugate acid.

In all practical titrations in aqueous solution, either the starting acid or the starting base is strong. A strong acid or base will immediately give its full amount of either H_3O^+ or OH^-. For example, sup-

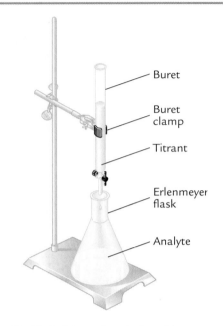

Figure 15-16 ▲ A buret is a device with a stopcock that can offer fine control of the addition of one solution to another.

pose that we add a strong acid titrant to a base. As soon as the strong acid enters the solution, it provides the full amount of available hydronium ion and the neutralization reaction with the base is rapid. Then, when the base is exhausted, even a small amount of added strong acid provides a big jump in hydronium ion, easily detected by many indicators.

Figure 15-15 ▲ Acids corrode some metals, forming hydrogen gas. (Ken Karp.)

How are you doing? 15.6

What reaction do you expect to occur between nitric acid and magnesium carbonate?

Acidic solutions don't destroy only limestone and marble. They also corrode certain metals, as shown in Figure 15-15.

$$2\ Al\ (s) + 3\ H_2SO_4\ (aq) \longrightarrow 3\ H_2\ (g) + 2\ Al_2(SO_4)_3\ (aq)$$

Corrosion is a major problem with aluminum cooking utensils, because acidic foods such as vinegar and tomato sauce can react with the aluminum.

This description allows us to look at combinations of reactants and decide how to carry out a useful titration. Imagine that you have available the solutions shown in Figure 15-17.

Now, decide how you will carry out the following tasks using these reactants. In each case, write the reaction that will occur in the titration, then describe which reactant will be in the buret and which will be in the beaker. The first one is done for you.

Problem 1: An unknown solution is tested with litmus paper; the paper turns red. Is this an acidic or basic solution? What titrant should we use to determine the solution's concentration? Answer: Because of the reaction with litmus paper, we know that this solution contains an unknown acid. We should titrate with a strong base, so we will choose NaOH for the titration. We write the unknown acid using the general formula HA. The reaction that will occur in the titration is:

$$HA + NaOH \longrightarrow NaA + H_2O$$

The NaOH is the titrant, and is in the buret; the unknown acid is the analyte, and is in the beaker.

Problem 2: You are given a mixture of solid sodium chloride and solid sodium benzoate, $C_7H_5O_2Na$. Sodium benzoate is a weak base that is completely soluble in water. The amount of sodium benzoate present in the mixture must be determined.

Problem 3: You know that a solution contains some ammonia, NH_3. How can you determine the concentration of the solution?

Figure 15-17 ▶ Some solutions you might use in an acid-base experiment.

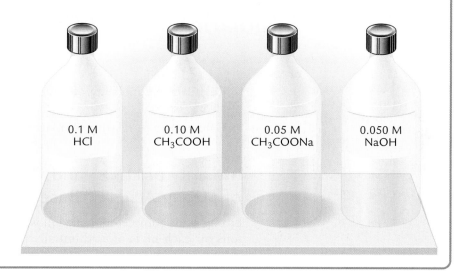

0.1 M HCl

0.10 M CH$_3$COOH

0.05 M CH$_3$COONa

0.050 M NaOH

PROBLEMS

20. Concept question: Look up the word "conjugate" in the dictionary and see how the definitions fit the Brønsted-Lowry view of acid-base behavior.

21. Give the formulas of the conjugate bases of the following acids:
(a) H_2SO_3
(b) H_2O
(c) H_3O^+
(d) $H_2PO_4^-$
(e) HS^-
(f) NH_4^+

22. Give the formulas of the conjugate acids of the following bases:
(a) H_2O
(b) ClO_3^-
(c) $HC_2O_4^-$
(d) HS^-
(e) SeO_4^{2-}
(f) $H_2PO_4^-$

23. Name each of the acids in Problem 22.

24. Write the formulas of the conjugate acids of the following bases:
(a) $CH_3CO_2^-$
(b) ClO^-

25. Write the formulas of the conjugate bases of the following acids:
(a) HSO_3^-
(b) H_2O
(c) HF

26. Name each conjugate base formed by the acids in Problem 25.

27. Sulfate has the Lewis structure shown below.

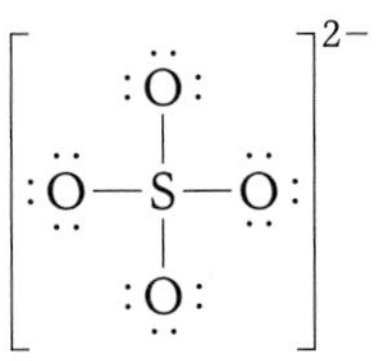

$$\left[\begin{array}{c} \ddot{\underset{\cdot\cdot}{O}}{:} \\ | \\ :\ddot{\underset{\cdot\cdot}{O}}-S-\ddot{\underset{\cdot\cdot}{O}}: \\ | \\ :\ddot{\underset{\cdot\cdot}{O}}: \end{array}\right]^{2-}$$

Draw the Lewis structure for its conjugate acid.

28. Discussion question: Indicate the acid-base conjugate pairs in the following acid-base reactions:

(a) $H_2O + HCl \longrightarrow H_3O^+ + Cl^-$
(b) $NH_3 + HI \longrightarrow NH_4^+ + I^-$

(c) $HClO + CO_3^{2-} \longrightarrow HCO_3^- + ClO^-$
(d) $H_2SO_4 + HCO_3^- \longrightarrow H_2CO_3 + HSO_4^-$

Write each reaction backwards. Does this change whether a given species is an acid or a base?

29. Write the likely products of the following reactions:

(a) $H_2SO_4 + NH_3 \rightarrow$
(b) $HSO_4^- + CO_3^{2-} \rightarrow$
(c) $H_3PO_4 + NaOH \rightarrow$
(d) $H_2SO_4 + Na_2CO_3 \rightarrow$

30. Lead (II) carbonate is a component of many old white paints. It is very resistant to weathering with normal rain, but it can dissolve in the presence of strong acid. Write a reaction for the dissolution of lead (II) carbonate by sulfuric acid.

15.3 Molarity of Acid-Base Mixtures

SECTION GOALS

✔ Calculate the molarity of aqueous solutions of strong acids or strong bases.

✔ Calculate the molarity of mixtures of strong acids and bases.

Molarity of H_3O^+ and OH^- in Mixtures of Strong Acids or Bases

Strong acids completely ionize in water, producing one mole of H_3O^+ ions for each mole of acid initially present. This makes it easy to determine the molarity of H_3O^+ present in an aqueous solution of a strong acid. For example, if we know that [HCl] = 0.500 M, then we also know that $[H_3O^+]$ = 0.500 M.

Likewise, strong bases ionize completely in water producing one mole of OH^- ions for each mole of OH^- present in the initial base. Thus, if [NaOH] = 0.400 M, then $[OH^-]$ = 0.400 M; and if $[Ba(OH)_2]$ = 0.400 M, then $[OH^-]$ = 0.800 M, as there are two hydroxides for every $Ba(OH)_2$.

For example, if we determine that a solution has 0.175 M HCl as the only acid present, then we expect that the concentration of hydronium ion will also be 0.175 M.

✔**Meeting the Goals**

For strong acids, $[H_3O^+]$ = [acid]. For strong bases, $[OH^-]$ = [base] $\times$ n, where n is the number of OH^- ions per unit of base.

Example 15.12   Calculating $[OH^-]$

Calculate $[OH^-]$ for a 0.219 M solution of CsOH.

Thinking Through the Problem Cs (cesium) is a group I metal, so CsOH is a strong base and will completely ionize in water.

Working Through the Problem We simply set the hydroxide concentration equal to the concentration of the CsOH.

Answer $[OH^-] = [CsOH] = 0.219$ M

Example 15.13 Calculating $[H_3O^+]$

Find $[H_3O^+]$ for a solution of 6.38 grams of HCl in exactly 1 liter of solution.

Thinking Through the Problem HCl is a strong acid, so $[HCl] = [H_3O^+]$.

Working Through the Problem We determine the molarity of HCl from the mass and volume:

$$6.38 \text{ g HCl} \times \frac{1 \text{ mol HCl}}{36.461 \text{ g HCl}}$$

$$\approx 0.175 \text{ mol HCl}$$

$$[HCl] = \frac{0.175 \text{ mol HCl}}{1 \text{ L solution}}$$

$$= 0.175 \text{ M HCl}$$

Answer $[HCl] = [H_3O^+] = 0.175$ M

How are you doing? 15.7

Determine the molarity of hydroxide in a solution containing 32.0 grams of KOH in 0.500 liters of water.

Many times we must consider what happens when we mix a strong acid with a strong base in a solution. Mixtures of H_3O^+ and OH^- in solution neutralize each other; they react in a 1:1 mole ratio to form water. Because of this, there are three possibilities to consider in an acid-base mixture.

KEY IDEA 15.7

1. If moles H_3O^+ = moles OH^-, the mixture will be neutral.
2. If moles H_3O^+ > moles OH^-, the mixture will be acidic; it will contain unneutralized H_3O^+.
3. If moles OH^- > moles H_3O^+, the mixture will be basic; it will contain unneutralized OH^-.

It is clear, then, that we must calculate both the moles of H_3O^+ and moles of OH^- to determine which reactant is used up and which reactant remains. To determine what remains unneutralized in solution, then, we must subtract the quantity of the ion present in the lesser amount from the quantity of the ion that is present in excess. Molarity is found by dividing moles by the *total* volume of the solution at the end.

PRACTICAL C How are chemical amounts determined by acid-base titration?

Titration is most often used to determine an unknown number of moles of a reactant by adding another reactant whose concentration *is* known. Determining the number of moles in a sample may provide the information we need to get the concentration of a solution, the molar mass of a solid acid or base, or the purity of a sample.

Determination of an Unknown Concentration

Suppose that the concentration of acetic acid in vinegar needs to be determined before the vinegar is sold. A titration is done with a strong base like NaOH, using the following reaction:

$$CH_3COOH + NaOH \longrightarrow NaCH_3COO + H_2O$$

We need to measure out a precise volume of vinegar. Then we measure the volume of NaOH used to complete the titration reaction. This volume of NaOH allows us to calculate (a) the number of moles of NaOH used (molarity × volume = moles), (b) the number of moles of CH_3COOH in the vinegar (using stoichiometry, this is a 1:1 mole ratio), and (c) the molarity of the CH_3COOH (moles CH_3COOH divided by total solution volume in liters = molarity). We follow these steps:

1. Measure a known volume of CH_3COOH into a beaker.
2. Add an indicator to the solution of CH_3COOH. (Here we can use the indicator phenolphthalein, which turns pink in its base form.)

Because the reaction ends when all of the CH_3COOH is consumed, the end point will be signalled when a small excess of NaOH is present. Use a buret to add a solution of NaOH with a known concentration to the solution of CH_3COOH.

Typical results might be the following: We measure 50.00 mL of the vinegar into a clean beaker. We add a drop of phenophthalein indicator; the solution remains colorless, as we expect for an acid solution. We titrate with 0.0654 M NaOH and observe that the solution turns pink (the endpoint) after the addition of 25.60 mL. Use this information to determine (a) the number of moles of NaOH added, (b) the number of moles of CH_3COOH present, and (c) the molarity of CH_3COOH.

Molar Mass Determination of a Crystalline Acid

In this experiment, we want to determine the molar mass of an unknown solid compound which we know is an acid. We know that this acid has one acidic hydrogen per molecule, which means that we can plan a titration using the reaction:

$$HA + NaOH \longrightarrow H_2O + NaA$$

Here HA is the unknown acid and A^- is its conjugate base. If we know the number of moles of NaOH added during the titration, then we will know the number of moles of HA present. Molar mass is

For example, let's say that we mix 50. mL of 0.025 M LiOH with 75. mL of 0.035 M HNO_3. Is the resulting solution acidic or basic? What is the molarity of the resulting mixture? We first calculate the number of moles of each reactant and the total solution volume (note that this assumes that volumes are additive, which is typical in these cases).

$$0.050 \text{ L LiOH solution} \times \frac{0.025 \text{ mol LiOH}}{1 \text{ L LiOH solution}} = 0.00125 \text{ mol LiOH}$$

$$0.075 \text{ L } HNO_3 \text{ solution} \times \frac{0.035 \text{ mol } HNO_3}{1 \text{ L } HNO_3 \text{ solution}} = 0.002625 \text{ mol } HNO_3$$

$$0.050 \text{ L} + 0.075 \text{ L} \approx 0.125 \text{ L total solution}$$

found by dividing the mass of HA by the moles of HA used in the experiment.

In a typical titration experiment of this sort, we follow these steps:

1. Dissolve a known mass of HA in water. Note that we do *not* need to know how much water we use, because this will not affect the reaction, assuming that we have enough water to dissolve HA completely.

2. Add an indicator to the solution of HA. Since the titration concludes when we consume all of HA, the end point will show the presence of a small excess of NaOH. We use phenolphthalein to signal the end point.

3. Use a buret to add a solution of NaOH with a known concentration to the solution of HA.

Typical results might be the following: We dissolve 0.059 g of the acid in about 50 mL of deionized or distilled water. We add a drop of phenophthalein indicator and the solution remains colorless. Titration with 0.054 M NaOH requires 8.96 mL to turn the solution pink (the end point). Use this information to determine (a) the number of moles of NaOH added, (b) the number of moles of HA present, and (c) the molar mass of HA.

Purity of a Solid Base

The base known as "tris" has the formula $C_6H_{15}NO_3$. It is very important in biochemistry laboratories, but it has the tendency to pick up moisture from the air. Therefore, samples of tris are titrated with known concentrations of acid to determine the number of moles of tris in a sample. Tris accepts one hydrogen ion per molecule, so the titration reaction will be:

$$tris + HCl \longrightarrow Cl^- + trisH^+$$

In a titration experiment of this sort, we follow these steps:

1. Dissolve a known mass of tris in water. Again, we do *not* need to know how much water we use.

2. Add an indicator to the solution. Since the titration concludes when we consume all of the tris, the end point will show the presence of a small excess of HCl. We use methyl red to indicate the end point.

3. Use a buret to add a solution of HCl with a known concentration to the solution of HCl.

Typical results might be the following: We dissolve 0.081 g of tris in about 25 mL of deionized or distilled water. We add a drop of methyl red indicator and the solution turns yellow. Titration with 0.0841 M HCl requires 6.35 mL to turn the solution red (the end point). Use this information to determine (a) the number of moles of HCl added, (b) the number of moles of tris present, (c) the mass of pure tris present, and (d) the percentage purity of the tris (mass by titration/mass dissolved $\times$ 100).

Because the reaction stoichiometry requires one mole of LiOH and one mole of HNO_3, we know that the limiting reactant is the one in the smaller mole amount: LiOH in this case. The HNO_3 is in excess. In other words, we know that 0.00125 mole of LiOH reacts with 0.00125 mole of HNO_3. The number of moles of HNO_3 that remain after the reaction is:

0.002625 mol HNO_3 (at start) − 0.00125 mol HNO_3 (reacted)
$$= 0.001375 \text{ mol } HNO_3 \text{ (left over)}$$

The molarity of remaining HNO_3 is equal to the number of moles of "left-over" HNO_3 divided by the final solution volume:

$$[HNO_3] = \frac{0.001375 \text{ mol } HNO_3}{0.125 \text{ L solution}} = 0.011 \text{ mol L}^{-1} = 0.011 \text{ M}$$

Example 15.14   **Calculating molarity of acid-base mixtures**

A solution was prepared by mixing 14.5 mL of 0.340 M HCl and 15.9 mL of 0.400 M NaOH. Is the final solution acidic or basic? What is the molarity of the final solution?

Thinking Through the Problem We need to find out which reactant is present in excess. This is the reactant that will remain in the solution at the end of the reaction. We are given the concentration and the volume of each reactant. This acid-base reaction will have a 1:1 stoichiometry.

Working Through the Problem We calculate and compare mole amounts of the HCl and the NaOH. We then determine how much remains of the reactant that is in excess. This amount, divided by the total volume, gives the molarity of the solution.

Working Through the Problem

$$(15.9 \times 10^{-3} \text{ L})\left(0.400 \ \frac{\text{mol}}{\text{L}}\right) = 6.36 \times 10^{-3} \text{ mol NaOH}$$

$$(14.5 \times 10^{-3} \text{ L})\left(0.340 \ \frac{\text{mol}}{\text{L}}\right) = 4.93 \times 10^{-3} \text{ mol HCl}$$

The NaOH is in excess and the HCl is the limiting reactant. The total volume is 30.4×10^{-3} L, assuming that the volumes are additive. We know, then, that 4.93×10^{-3} mole of HCl neutralizes 4.93×10^{-3} mole of NaOH. This leaves 1.43×10^{-3} mole of excess NaOH. The molarity of the resulting solution is:

$$\frac{1.43 \times 10^{-3} \text{ mol NaOH}}{30.4 \times 10^{-3} \text{ L NaOH solution}} \approx 4.7 \times 10^{-2} \text{ M} = [\text{NaOH}].$$

Answer The solution is $[\text{NaOH}] = [\text{OH}^-] = 4.70 \times 10^{-2}$ M.

✔ **Meeting the Goals**

A mixture of strong acid with strong base will have one of three outcomes. It will be neutral when moles H_3O^+ = moles OH^-, it will be acidic when moles H_3O^+ > moles OH^-, and it will be basic when moles H_3O^+ < moles OH^-. The first step is to calculate moles of strong acid and strong base present.

Example 15.15   **Calculating molarity of an acid-base mixture**

What is the molarity of the solution resulting when 28.7 mL of 0.251 M HCl reacts with 29.2 mL of 0.240 M NaOH?

Working Through the Problem

$$(28.7 \times 10^{-3} \text{ L})\left(0.251 \ \frac{\text{mol}}{\text{L}}\right) \approx 7.204 \times 10^{-3} \text{ mol HCl}$$

$$(29.2 \times 10^{-3} \text{ L})\left(0.240 \ \frac{\text{mol}}{\text{L}}\right) \approx 7.008 \times 10^{-3} \text{ mol NaOH}$$

Total volume is 57.9×10^{-3} L. There is 1.96×10^{-4} mole of acid in excess. The molarity of the resulting solution is:

$$\frac{1.96 \times 10^{-4} \text{ mole}}{57.9 \times 10^{-3} \text{ L solution}} \approx 3.38 \times 10^{-3} \text{ M} = [\text{H}_3\text{O}^+]$$

Answer The solution is $[\text{HCl}] = [\text{H}_3\text{O}^+] = 3.38 \times 10^{-3}$ M.

Find the molarity of the resulting solution when 1.00 L of 0.128 M NaOH and 0.500 L of 0.128 M HCl are mixed.

PROBLEMS

31. Find the number of moles of H_3O^+ contained in 850.0 mL of 0.0562 M $HClO_4$.

32. What volume of 1.002 M HCl will provide 2.66×10^{-2} mole H_3O^+?

33. Find the molarity of hydroxide ions present when 0.010 g $Ba(OH)_2$ is dissolved in enough water to make 400.0 mL solution. (Molar mass of $Ba(OH)_2 = 171.344$ g mol^{-1})

34. Find the moles in excess (not neutralized) when these two solutions are mixed together:

42.2 mL of 0.242 M NaOH and 25.0 mL of 1.044 M HCl.

35. Discussion question: A solution containing 36.4 mL of 0.491 M HCl is added to 38.6 mL of 0.442 M NaOH. Calculate the number of moles of each substance and determine whether the final solution is acidic or basic. How much more

of the limiting reactant is needed for complete reaction?

36. What is the molarity of a solution resulting when 6.98 grams of KOH is dissolved in enough water to make 200.0 mL of solution?

37. What is the molarity of the solution formed when 125.2 grams of HBr is dissolved in enough water to make 15.0 L of solution?

38. A solution was prepared by mixing 20.5 mL of 0.50 M HCl and 15.9 mL of 0.50 M NaOH. What is the molarity of the final solution? Is it acidic or basic?

39. A solution was prepared by mixing 51.3 mL of 0.380 M HCl and 35.8 mL of 0.410 M NaOH. What is the molarity of the final solution? Is it acidic or basic?

15.4 The pH Function

SECTION GOALS

✔ Understand and use pH and pOH.

✔ Understand the relationship between the acid-base properties of water and pH.

MAKING IT WORK WITH MATH

Exponential and Logarithmic Functions

Chemists routinely deal with acid and base solutions that have very small molarity values. The molarity values of these solutions can vary by several powers of ten and are more conveniently considered as a base-10 logarithmic function of the concentration called pH. Before we can discuss the pH function itself, let us briefly review what we mean by functions in general and exponential and logarithmic functions in particular.

Exponential functions are an important class of functions in which the independent variable y appears in the exponent. In general, an exponential function has the form $x = a^y$ where a is positive and $a \neq 1$. (Exponential functions are *not* the same as power functions where the variable appears in the base of an exponential term, e.g., $y = x^2$ or $y = x^3$.)

The function $x = a^y$ is a one-to-one function and therefore has an inverse. The inverse of this function is the function that "undoes"

$x = a^y$. For now, we will define the **logarithmic function** to be the inverse of the exponential function. This definition, $y = \log_a x$, and the fact that the function is the inverse of the exponential function permits us to write it as $x = a^y$. *These are two forms of the same relationship.* Because these equations represent the same function, they are equivalent statements about the relationship of x, y, and a.

Logarithmic form: $y = \log_a x$

Exponential form: $x = a^y$

The first form is the "logarithmic form." The second is called the "exponential form." It is important to be comfortable converting from one form to the other. You are already familiar with equations given in exponential form. With practice, you can become equally familiar with the logarithmic form.

In the case of base-10 logarithms, this relationship will be:

Logarithmic form: $y = \log x$

Exponential form: $x = 10^y$

REMEMBER

The number or variable following the log term is called the *argument* of the log. The argument of the log is the number for which we take the log.

Example 15.16   Writing exponential equations in log form

Rewrite the following equations in their logarithmic form:

(a) $0.001 = 10^{-3}$ (b) $100,000 = 10^5$

Thinking Through the Problem We want to rewrite equations that look like $x = 10^y$ in the form $\log x = y$. We examine each equation for x and y, then rewrite the equation in its log form. The power of 10 becomes y and the decimal is x.

Working Through the Problem Comparing the general equation above to the equation in part (a), we see that $0.001 = 10^{-3}$ gives $y = -3$ and $x = 0.001$. In log form, then, we write $-3 = \log 0.001$.

Answer (a) $\log 0.001 = -3$ (b) $\log 100,000 = 5$

Example 15.17   Writing log equations in exponential form

Rewrite the following equations in their exponential form:

(a) $\log 1,000,000 = 6$ (b) $\log 0.00001 = -5$

Thinking Through the Problem We want to rewrite equations, $\log x = y$ in the form $10^y = x$.

Working Through the Problem The argument of the log function is x. The exponent is y.

Answer (a) $1,000,000 = 10^6$ (b) $0.00001 = 10^{-5}$

Evaluate the following logarithms without using a calculator:

(a) log 1000   (b) −log 0.100   (c) log 0.001

For the base ten number system that we use, a logarithm (or log) is defined as the power to which ten must be raised in order to equal (or "get back") the original number. Table 15-7 shows the relationship between some numbers and their logarithms. The steps used to find the log of each number are given in successive columns across the table. As the numbers are written in scientific notation, we use the property of logs that states that the log of a product is the sum of logs (column two). Remember that the log 1 = 0 in any base and that the log 10^x is x (column 3). The final answer is given in the last column of the table.

REMEMBER

pH

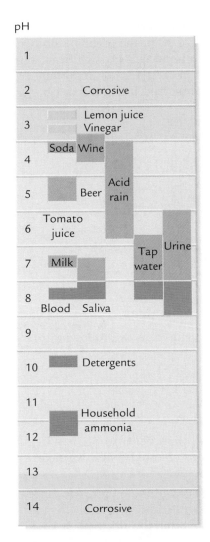

Figure 15-18 ▲ The pH of some common substances.

Table 15-7 Relating Numbers to Their Logs

$\log 1 \times 10^1 =$	$\log 1 + \log 10^1 =$	$0 + 1 =$	1
$\log 1 \times 10^2 =$	$\log 1 + \log 10^2 =$	$0 + 2 =$	2
$\log 1 \times 10^{-3} =$	$\log 1 + \log 10^{-3} =$	$0 + (-3) =$	-3
$\log 1 \times 10^{-6} =$	$\log 1 + \log 10^{-6} =$	$0 + (-6) =$	-6

The logarithm of each number is the exponent of 10 from the original number.

Calculating the pH of a Solution

Typical acid concentrations vary from very dilute solutions (1.0×10^{-14} M) to more concentrated solutions (10 M). In fact, the concentration of an acid can vary by factors as great as 1×10^{15}, or one quadrillion. Here we are faced with a choice: either become adept at manipulating and understanding very small numbers, or find a more convenient method to describe the acidity of a solution. Of course, the choice is simple. *Everyone* opts for convenience!

Chemists have found it much easier to express the acid content of a solution by using a logarithmic scale called the pH scale. Remember that a logarithm is an exponent.

Because most concentrations involve negative exponents (and most of us still find negative numbers difficult to think about), the pH function is defined as a *negative* logarithm. The **pH** of a solution is the negative base-10 log of the hydronium ion concentration. We often omit the "base-10" indication. Figure 15-18 shows the pH of some common substances.

$$\text{pH} = -\log_{10}[\text{H}_3\text{O}^+] \text{ or pH} = -\log[\text{H}_3\text{O}^+]$$

The pH function is more than a mathematical device. It is a very convenient way to keep track of the amount of hydronium ion in solution, as shown in Table 15-8.

Table 15-8 Relationship Between pH and $[H_3O^+]$

$[H_3O^+] = 3.21 \times 10^{-12}$ M	pH = 11.494
$[H_3O^+] = 3.21 \times 10^{-10}$ M	pH = 9.494
$[H_3O^+] = 3.21 \times 10^{-7}$ M	pH = 6.494
$[H_3O^+] = 3.21 \times 10^{-5}$ M	pH = 4.494
$[H_3O^+] = 3.21 \times 10^{-2}$ M	pH = 1.494

There are some notable similarities and differences among the log values contained in Tables 15-7 and 15-8. We notice that the part of the log value written *after* the decimal (called the mantissa of the log) is the same for all concentrations of H_3O^+ listed and must correspond to the "sameness" of the "3.21" in the original numbers. This leads us to the realization that the *number of significant figures* in 494 (three significant figures) equals the number of significant figures in 3.21 (three significant figures). We also notice that the number written *before* the decimal (called the characteristic of the log) corresponds to one less than the absolute value of the exponent of 10 in the original number. Just as the value of a power of 10 is not important for significant figures, so the characteristic of a logarithm like a pH is not important for significant figures. These points are summarized in Figure 15-19.

Figure 15-19 ▶ Only the numbers following the decimal point in a log number (the mantissa) are significant figures, as these come directly from the original number (before taking the log).

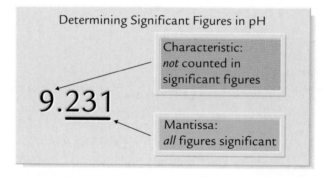

Determining Significant Figures in pH

Characteristic: *not* counted in significant figures

9.231

Mantissa: *all* figures significant

Example 15.18 Determining significant figures for pH values

Determine the number of significant figures in the following pH values.

(a) 7.231   (e) 12.32
(b) 8.432   (f) 0.22
(c) 1.2   (g) 7.0
(d) 9.000   (h) 0.323

Thinking Through the Problem We recall that, because pH is a log, only the numbers after the decimal are counted when we determine the significant figures in a pH value.

Working Through the Problem Counting is easy! In (a) there are three places after the decimal; there are 3 significant figures in this pH value. We count the number of places after the decimal in each pH value.

Answer (a, b, d, h) 3; (e, f) 2; (c, g) 1

Example 15.19   Converting $[H_3O^+]$ values to pH

Convert the following $[H_3O^+]$ values to pH; give the answer to the correct number of significant figures.

(a) 4×10^{-5} M
(b) 4.00×10^{-5} M
(c) 1.923×10^{-10} M
(d) 0.012 M
(e) 8.3×10^{-3} M

Thinking Through the Problem We need to carry out the pH calculation in each case. The answer should have the same number of significant figures as in the original concentration. We can apply the formula pH $= -\log [H_3O^+]$ to each. The number of significant figures in the original concentration is the number of places that we will include after the decimal in the value of the pH.

Working Through the Problem For the first, the pH calculation on the calculator gives:

$$-\log 4 \times 10^{-5} = 4.39794$$

But there is only one significant figure in 4×10^{-5}, so the answer must be rounded to one place after the decimal point: 4.4.

Answer (a) 4.4 (b) 4.398 (c) 9.7160 (d) 1.92 (e) 2.08

Calculating $[H_3O^+]$ from pH

What happens when we have the pH value but need $[H_3O^+]$? Suppose that we must find $[H_3O^+]$ for a solution that has pH = 3.4. We first notice that the pH has only one number after the decimal point, so we know that the $[H_3O^+]$ will also have only 1 significant figure. If we substitute into the expression for pH we have 3.4 $= -\log [H_3O^+]$. How do we solve this equation for $[H_3O^+]$? We can easily isolate log $[H_3O^+]$ by multiplying through by -1. To solve for $[H_3O^+]$, we must rewrite the log equation in its exponential form. The solution of this problem, then, is:

$$3.4 = -\log [H_3O^+]$$
$$-3.4 = \log [H_3O^+]$$
$$10^{-3.4} = [H_3O^+]$$
$$3 \times 10^{-5} \text{ M} = [H_3O^+] \text{ (1 s.f.)}$$

The inverse operation of log is 10^x. This is a second function key on many calculators and is accessed by pushing the keys labeled "2nd" and then "log." The 10^x operation allows us to find $[H_3O^+]$ when given the pH. Thus, for the function pH $= -\log_{10} [H_3O^+]$, the inverse function becomes:

$$[H_3O^+] = 10^{-pH}$$

We must also consider significant figures when converting from pH to $[H_3O^+]$. We look at the number of figures *after* the decimal place in the pH to determine the number of significant figures in $[H_3O^+]$. For example, a pH of 5.23 (2 s.f.) gives $[H_3O^+] = 5.9 \times 10^{-6}$ M and pH = 3.000 gives $[H_3O^+] = 1.00 \times 10^{-3}$ (3 s.f.).

Example 15.20   **Finding $[H_3O^+]$ from pH**

Find $[H_3O^+]$ for a solution with pH = 4.15.

Thinking Through the Problem We are given a pH (log) value and need to find the corresponding molarity, $[H_3O^+]$. We will solve this by using the exponential form for this situation, $[H_3O^+] = 10^{-pH}$. The pH, 4.15, has two significant figures; we will report $[H_3O^+]$ with two significant figures.

Working Through the Problem The direct answer is $10^{-4.15}$ but we need to convert this to standard scientific notation, which we do on the calculator. The answer will have two significant figures.

Answer $10^{-4.15} \approx 7.1 \times 10^{-5}$ (2 s.f.)

Example 15.21   **Finding $[H_3O^+]$ from pH**

Convert the following pH values to $[H_3O^+]$; give the answer to the correct number of significant figures.

(a) 7.902   (d) 1.00
(b) 3.22   (e) −0.13
(c) 12.9

Thinking Through the Problem This is the same as in the last example. Note that for the last case, we will wind up with $10^{-(-0.13)}$ as the calculator input.

Answer

(a) 1.25×10^{-8} M   (d) 1.0×10^{-1} M
(b) 6.0×10^{-4} M   (e) 1.3×10^{0} M = 1.3 M
(c) 1×10^{-13} M

How are you doing?   15.10

Supply the missing pH or $[H_3O^+]$ values for the following. Write each answer with the correct number of significant figures.

(a) $[H_3O^+] = 3.1 \times 10^{-7}$ M   pH = _____
(b) $[H_3O^+] = 3.05 \times 10^{-7}$ M   pH = _____
(c) $[H_3O^+] = 2.93 \times 10^{-7}$ M   pH = _____
(d) $[H_3O^+] =$ _____   pH = 6.25
(e) $[H_3O^+] =$ _____   pH = 9.593
(f) $[H_3O^+] =$ _____   pH = 11.079

Divers collect samples of lake water to test for acidity. (Robert Krueger/Photo Researchers.)

pH and the Acid-Base Characteristics of a Solution

As we saw in Section 15.1, all samples of liquid water contain some hydronium ion and some hydroxide ion because of a reaction called the *auto-ionization* of water:

$$2\ H_2O\ (l) \rightleftharpoons H_3O^+\ (aq) + OH^-\ (aq)$$

In pure H_2O, the amount of hydronium ion and the amount of hydroxide ion are equal to one another (Figure 15-7). At 25 °C, the concentration of hydronium ion and of hydroxide ion *in pure water* is 1.0×10^{-7} M. This is equal to a pH of 7.00. Water with a pH of 7.00 at 25 °C—even if it is not pure—is said to be *neutral*. Water with more hydronium ion than neutral water, with $[H_3O^+] > 1.0 \times 10^{-7}$ M, is said to be *acidic*. Water with less hydronium ion than neutral water, that is, with $[H_3O^+] < 1.0 \times 10^{-7}$ M, is said to be *alkaline* or *basic*. The correlation of pH with acidity and basicity is shown in Figure 15-20 for 25 °C.

KEY IDEA 15.8

Some values for $[H_3O^+]$ and pH are shown on the number line in Figure 15-20. Notice that, because of the nature of the pH function, the pH is *smallest* when $[H_3O^+]$ is greatest. Thus, *a solution with a smaller value for pH is more acidic than one with a larger value for pH.*

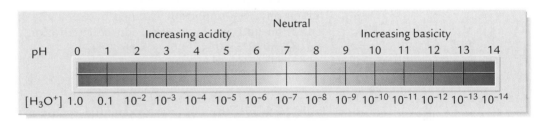

Figure 15-20 ▲ The pH of a solution at 25 °C indicates if the solution is neutral (pH = 7), acidic (pH < 7), or basic (pH > 7). Low values of pH indicate greater acidity than high values of pH.

Example 15.22   Using pH to assess acidity

Rank the following solutions in order of increasing acidity.
pH = 4.32; pH = 7.00; pH = 1.23; pH = 12.99; pH = −0.23

Thinking Through the Problem We need to write the values in order of lowest to highest acidity, (smallest $[H_3O^+]$ first). The pH scale is an inverse scale; larger numbers mean less hydronium.

Working Through the Problem We rewrite the values starting with the highest pH, because it has the smallest $[H_3O^+]$. The most acidic solution has the lowest pH: −0.23.

Answer   12.99, 7.00, 4.32, 1.23, −0.23

The pH provides a quick indication of whether a solution is acidic, basic, or neutral in relation to pure water. In pure water at 25 °C, pH = 7.00. A solution that is *acidic* has a pH *below* 7.00. A solution that is *basic* has a pH *above* 7.00. We can use this information to make rough predictions of the pH of acid and base solutions. An acid, when dissolved in pure water, yields more hydronium ion than is found in pure water alone. Therefore, an acid will give a pH *below* 7.00. Conversely, a base will give more hydroxide ion than is present in pure water. A solution of a base should have pH > 7.00.

Finding the pOH

The pH scale can be used to describe both acidic and basic solutions, as shown in Figure 15-20.

We can also measure basicity by calculating the **pOH**. The definition of pOH is similar to the definition of pH, except that pOH refers to the molarity of hydroxide ion, $[OH^-]$.

$$pOH = -\log [OH^-]$$

The relationship (at 25 °C) between pH and pOH is:

$$14 = pH + pOH$$

The precise value of the constant in this equation is 13.996, but for practical purposes the value 14 is quite adequate. All aqueous solutions have an important balance between the hydroxide ion concentration and the hydronium ion concentration. We can write the equation that relates $[H_3O^+]$ with $[OH^-]$:

$$1.0 \times 10^{-14} = [H_3O^+][OH^-]$$

Table 15-9 lists the log and exponential relationships you will need to solve pH and pOH problems.

Table 15-9 pH and pOH Relationships at 25 °C

$pH = -\log [H_3O^+]$	$pOH = -\log [OH^-]$	$pH + pOH = 14$
$[H_3O^+] = 10^{-pH}$	$[OH^-] = 10^{-pOH}$	$[H_3O^+][OH^-] = 1.0 \times 10^{-14}$

Example 15.23   **Finding pOH from pH**

Given a pH of −0.98, what is pOH?

Thinking Through the Problem We want the pOH. Because we don't need to say anything about concentrations, we start with the pH and the equation 14 = pH + pOH.

Working Through the Problem The equation can be solved to give pH = 14 − pOH

Answer   pOH = 14 − (−0.98) = 14.98

Example 15.24 **Finding pH and pOH from $[H_3O^+]$**

A solution has $[H_3O^+] = 3.2 \times 10^{-5}$ M. What are the pH and the pOH of the solution at 25 °C?

Thinking Through the Problem We need values for pH and pOH, beginning with $[H_3O^+] = 3.2 \times 10^{-5}$ M. It is easiest to use the initial hydronium ion concentration to get the pH, then solve for pOH.

Working Through the Problem We start by calculating pH; then we will find pOH by using the relationship $14 = pH + pOH$. The answer will have two significant figures because 3.2×10^{-5} has two significant figures. The pH and pOH will both have two numbers after the decimal.

$$pH = -\log(3.2 \times 10^{-5}) \approx 4.49 \ (2 \text{ s.f.})$$
$$pOH = 14 - pH = 9.51$$

Answer The pOH = 9.51 and the pH = 4.49 when $[H_3O^+] = 3.2 \times 10^{-5}$ M.

Example 15.25 **Finding $[H_3O^+]$ and $[OH^-]$ from pOH**

Given a pOH of 8.23, calculate both $[H_3O^+]$ and $[OH^-]$.

Thinking Through the Problem Given the pOH, we can easily find pH by subtracting from 14. Then we can use the exponential forms to find both $[H_3O^+]$ and $[OH^-]$.

Working Through the Problem To get $[OH^-]$ we use the relationship $[OH^-] = 10^{-pOH}$. We also use $14 = pH + pOH$ to get pH and finally we get $[H_3O^+] = 10^{-pH}$. Our answers all have two significant figures.

Answer

$$[OH^-] = 10^{-pOH} = 10^{-8.23} = 5.9 \times 10^{-9} \text{ M}$$
$$pH = 14 - pOH = 14 - 8.23 = 5.77$$
$$[H_3O^+] = 10^{-pH} = 10^{-5.77} = 1.7 \times 10^{-6} \text{ M}$$

How are you doing? **15.11**

Find the pH, pOH, $[H_3O^+]$, and $[OH^-]$ for a solution that is 0.0050 M $Ba(OH)_2$.

PROBLEMS

40. Logarithms and exponentials are different ways of writing the same relationship. This can be done with any function. Rewrite the following in exponential form.

(a) $3.00 = \log(1000)$
(b) $3.11 = \log(1290)$
(c) $-5.26 = \log(5.50 \times 10^{-6})$
(d) $8.26 = -\log(5.50 \times 10^{-9})$ (Hint: rearrange so that the logarithm is positive.)

PRACTICAL D   How does pH change during acid-base titrations?

The dramatic color changes that occur when we carry out a stoichiometry experiment using titration reflect some very large changes in pH. Looking at titrations with strong acids and strong bases only, we can identify three stages in a titration experiment, as shown in Figure 15-21. If we are adding a strong base to a strong acid:

STAGE I: As we begin the titration, we have primarily acid in solution. The pH is low.

STAGE II: At this point, we have added just enough base so that the acid and the base neutralize each other and there is no species dominating the pH. The pH is that of neutral water. This is a single point known as the equivalence point.

STAGE III: As we continue to add base, the base is in excess and, therefore, the solution is basic. The pH is high.

What is surprising in many cases is how difficult it is to get a titration to be exactly at the equivalence point. The points making up such a titration curve can be obtained by calculating the moles of excess reactant present in the acid-base mixture after various additions of base. The pH is calculated after finding the molarity of the mixture.

1. Complete the calculations in the table on the next page. We begin with 25.00 mL of 0.10 M HCl in the beaker, i.e., the initial amount of HCl is 2.5×10^{-3} mole. We add small volumes of 0.10 M NaOH and calculate the moles of acid or base in excess.

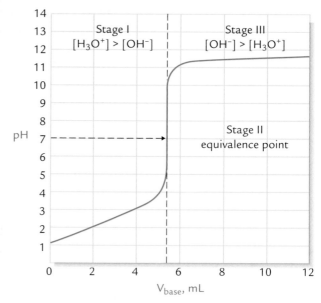

Figure 15-21 ▲ A titration curve showing the pH values as a strong base is added to a strong acid.

2. Use the completed table to answer the following questions:

(a) What happens when the moles of NaOH and HCl are equal?

(b) pH = 8.234 to $[H_3O^+]$
(c) $[H_3O^+] = 2.7 \times 10^{-9}$ M to pH
(d) $[H_3O^+] = 0.065$ M to pH
(e) $[OH^-] = 7.1 \times 10^{-10}$ to pOH

41. Concept question: In this section we said that logarithmic functions make it easier to write very large and very small numbers in a convenient form. Prepare a table with six numbers, three larger than 1×10^{10}, three smaller than 1×10^{-10}. Next to each, write their common logarithm (which you can determine using the "log" key on your calculator). Inspect your table, and comment on how the logarithmic version of the number can be considered "more convenient."

42. Carry out the following conversions:

(a) Convert pH = 1.23 to $[H_3O^+]$.
(b) Convert $[H_3O^+] = 1.4$ M to pH.

43. Carry out the following conversions:

(a) pH = 2.5 to $[H_3O^+]$

44. Concept question: A person discussing water pollution in a lake says, "The pH has only dropped from 6.3 to 5.6. That's only a ten percent change. So what's the worry?" Explain if the person is correct about the "ten percent" and if, in fact, this might be a worrisome change.

45. Discussion question: Indicate whether the following solutions can be expected to be acidic, basic, or neutral.

(a) A solution of laundry detergent having a pH of 10.2
(b) Vinegar with pH of 4.33

Initial Moles of HCl	Total Volume NaOH	Moles 0.10 M NaOH Added	Moles in Excess		Total Volume (L)	Molarity of Solution (M)	pH	pOH
			HCl	NaOH				
2.5×10^{-3}	0	0	2.5×10^{-3}		0.02500	0.10	1	13
2.5×10^{-3}	1.00 mL	1×10^{-4}	2.4×10^{-3}		0.02600	0.092		
2.5×10^{-3}	5.00 mL							
2.5×10^{-3}	10.00 mL							
2.5×10^{-3}	20.00 mL							
2.5×10^{-3}	22.50 mL							
2.5×10^{-3}	24.00 mL							
2.5×10^{-3}	24.50 mL							
2.5×10^{-3}	24.95 mL							
2.5×10^{-3}	25.00 mL							
2.5×10^{-3}	25.05 mL							
2.5×10^{-3}	25.50 mL	2.55×10^{-3}		5×10^{-5}	0.0505	9.9×10^{-4}	11	3
2.5×10^{-3}	26.00 mL							
2.5×10^{-3}	27.00 mL							

(b) How much of a pH change is there between 1.00 mL and 24.00 mL added NaOH?

(c) How much of a pH change is there between 24.00 mL and 24.50 mL added NaOH?

(d) How much of a pH change is there between 24.50 mL and 25.50 mL added NaOH?

(e) How much of a pH change is there between 25.50 mL and 26.00 mL added NaOH?

(f) What do you think will be the pH change in going from 27.00 mL to 50.00 mL added NaOH (3.0 units, 1.0 units, 0.25 units, or 0.05 units)?

(c) 0.19 M H_2SO_4
(d) 1.93×10^{-5} M NaOH

For (c) and (d), calculate the pH. Was your initial answer correct?

46. Determine the pH of the following solutions:
(a) A 2.1×10^{-2} M solution of HNO_3
(b) A solution of 0.25 mole of NaOH in 2.00 liters of solution
(c) A 2.1 M solution of $HClO_4$
(d) 15.0 g KOH dissolved in 300.0 mL solution

47. What is the pH of a solution of 0.25 mole of LiOH in 0.500 liter of solution?

48. The equation $1.01 \times 10^{-14} = [H_3O^+][OH^-]$ is correct only at 25 °C. At 40 °C, the correct equation is:

$$2.92 \times 10^{-14} = [H_3O^+][OH^-]$$

Use this expression to determine the value of the sum of pH and pOH at 40 °C.

49. Calculate $[H_3O^+]$, pH, pOH, and $[OH^-]$ for the following solutions:
(a) 0.035 M HCl
(b) 3.45×10^{-2} M RbOH
(c) 1.99×10^{-2} M H_2SO_4

50. What is the pH of the solution in which 20.0 mL of 0.10 M NaOH is added to 25.0 mL of 0.10 M HCl?

51. Calculate the pH of the solution made by mixing 35.0 mL of 0.15 M acid with 25.0 mL of 0.10 M base.

Chapter 15 Summary and Problems

VOCABULARY

acidic
An acidic solution has certain properties including the ability to cause characteristic color changes in indicators, corrosive character with metals and carbonates, and a pH less than 7. An acidic substance can cause a solution to become acidic.

basic
A basic solution has certain properties including the ability to cause characteristic color changes in indicators, causticity, and a pH greater than 7. A basic substance can cause a solution to become basic.

neutralization
The reaction that occurs when acidic and basic solutions are mixed, producing a solution that has neither basic nor acidic properties.

indicators
Substances or materials that exhibit characteristic color changes when the acid-base properties of a solution change; examples include litmus paper, phenolphthalein, and bromocresol green.

Arrhenius acid
A substance capable of donating hydrogen ion to water to form an acidic solution, or capable of neutralizing a basic solution.

Arrhenius base
A substance capable of forming hydroxide ion in water to form a basic solution, or capable of neutralizing an acidic solution.

strong acid
An Arrhenius acid that completely reacts in water to give hydronium ion and an anion; no un-ionized acid is left in the solution by a strong acid.

strong base
An Arrhenius base that completely reacts in water, usually involving the complete dissolution of a hydroxide salt.

Lewis acid
A species that can accept a lone pair of electrons.

Lewis base
A species that can donate a lone pair of electrons.

Brønsted-Lowry acid
A species capable of donating a hydrogen ion to another species in an acid-base reaction.

Brønsted-Lowry base
A species capable of accepting a hydrogen ion from another species in an acid-base reaction.

conjugate acid-base pair
Two species related to one another by the presence of a hydrogen ion (in the acid part of the pair) or the absence of a hydrogen ion (in the base part of the pair).

polyprotic acids
Species capable of losing more than one hydrogen ion in acid-base reactions; the hydrogen ions are lost one at a time.

polyprotic bases
Species capable of accepting more than one hydrogen ion in acid-base reactions; the hydrogen ions are accepted one at a time.

amphoteric
A species capable of acting as either a Brønsted-Lowry acid or a Brønsted-Lowry base, depending on the other reactants in the solution.

exponential functions
Mathematical functions of the form $y = a^x$, where $a > 0$.

logarithmic functions
Mathematical functions of the form $x = \log_a y$, where a is a positive number.

pH
A logarithmic function used to describe the concentration of hydronium ion in a solution. $pH = -\log_{10}[H_3O^+]$. The inverse function relates hydronium ion to pH, $[H_3O^+] = 10^{-pH}$.

pOH
A logarithmic function used to describe the concentration of hydroxide ion in a solution. $pOH = -\log_{10}[OH^-]$. The inverse function relates hydroxide ion to pOH, $[OH^-] = 10^{-pOH}$.

SECTION GOALS

What do we mean when we say that a solution is acidic, basic, or neutral?

Acidic properties include the ability to turn certain indicators to their acid form, and corrosive effects on substances such as carbonates and certain metals. It also means that the pH is less than 7 at 25 °C. Basic solutions are slippery and caustic, and cause solutions containing calcium ions to turn cloudy. They have a pH greater than 7. Neutral solutions possess neither property and have a pH of 7.

How do we determine whether a solution is acidic, basic, or neutral?

We can perform certain simple qualitative tests. The most simple of these involve acid-base indicators like litmus.

How do we describe how acids and bases react with each other and with water?

We describe acid-base behavior in any of three ways. According to the Brønsted-Lowry definition, acids donate hydrogen ion and bases accept hydrogen ion. Thus, acid-base reactions are hydrogen-ion transfer reactions. This allows us to write acids and bases as conjugate acid-base pairs. Another description, the Lewis picture, emphasizes that acids attach to lone pairs to form bonds while bases donate the lone pair. In the Arrhenius theory, acids increase the amount of hydronium ion in water and bases increase the amount of hydroxide ion in water.

What happens when acids are brought into contact with substances and solutions containing carbonate and hydrogen carbonate?

The acid will react with the hydrogen carbonate or the carbonate to give carbonic acid and then carbon dioxide.

How do we apply the unit of molarity to solutions of acids and bases?

The molarity of acid and base solutions refers to the molarity of a species that is capable of acting as an acid or a base.

How do we relate the molarity of a strong acid or base to the molarity of hydronium ion or hydroxide ion?

Because strong acids give one hydronium ion per molecule of strong acid, the molarity of hydronium ion in a strong acid solution is the same as the molarity of the strong acid. The molarity of a strong base is the molarity of the hydroxide ion it donates to solution.

What is the relationship between the acid-base properties of water and the pH function?

The conversion of hydronium ion and pH uses the function $pH = -\log [H_3O^+]$. This means that high concentrations of hydronium ion are found at low pH values and that low concentrations of hydronium ion are found at high pH values. The inverse function is $[H_3O^+] = 10^{-pH}$.

What is the relationship between pH and pOH?

At 25 °C the relationship between pH and pOH is $pH + pOH = 14$. This means that we can always calculate the pH from the pOH, and vice versa. It also means that we can convert $[OH^-]$ and $[H_3O^+]$ by pH and pOH conversions.

PROBLEMS

52. Find pH for the following values of $[H_3O^+]$:
  - (a) $1.0 = 1.0 \times 10^0$ M
  - (b) $0.0010 = 1.0 \times 10^{-3}$ M
  - (c) $0.0000010 = 1.0 \times 10^{-6}$ M
  - (d) $0.0000000010 = 1.0 \times 10^{-9}$ M
  - (e) $0.10 = 1.0 \times 10^{-1}$ M
  - (f) $0.00010 = 1.0 \times 10^{-4}$ M
  - (g) $0.00000010 = 1.0 \times 10^{-7}$ M
  - (h) $0.000000000010 = 1.0 \times 10^{-11}$ M

53. (a) Which of the pH values in Problem 52 has the largest concentration of hydronium ion?
  - (b) Which of the pH values in Problem 52 has the largest pH?
  - (c) According to your response to parts (a) and (b), as the pH becomes smaller, does $[H_3O^+]$ increase or decrease?

54. It is possible to carry out some acid-base reactions using a solid acid or base as one of the reactants. We have seen one example of this in the reaction of acids with carbonates. Many carbonates are not soluble in water, but they do dissolve when an acid is present. Explain why this may occur.

55. A solution must be neutralized before it can be disposed of. Which of the following questions must be answered before this can be done? If you think the question should be answered, suggest a way to answer it using the chemistry discussed in this chapter.

(a) Is the solution acidic or basic?
(b) If the solution contains an acid, is it a strong or a weak acid?
(c) If the solution contains a base, is it strong or weak?
(d) How many moles of the acid or base are present?
(e) What is the pH of the solution?

56. Fill in this table.

Conditions	pH	pOH	[OH$^-$]
2.3×10^{-2} M HI			
1.2 M LiOH			

57. What is the molarity of the solution formed when 3.05×10^{-2} mole of HI are dissolved in enough water to make 350.0 mL of solution? Determine the expected pH of this solution.

58. What is the molarity of the solution formed when 4.82 mg of NaOH is dissolved in enough water to make 25.0 mL of solution? Determine the expected pH of this solution.

59. What is the molarity of the solution resulting when 35.0 mL of 0.963 M HCl is added to enough water to make a final volume of 50.0 mL of solution? Determine the expected pH of this solution.

60. What is the molarity of the solution resulting when 27.5 mL of 1.025 M NaOH is added to enough water to make a final volume of 100.0 mL of solution? Determine the expected pH of this solution.

61. Determine the following values:

(a) A solution with [H$_3$O$^+$] = 9.54×10^{-4} M is diluted by a factor of 2. Determine [H$_3$O$^+$] in the resulting solution.
(b) A solution with [OH$^-$] = 4.2×10^{-5} M is diluted by a factor of 5. Determine [OH$^-$] and [H$_3$O$^+$] in the resulting solution.

(c) A solution with pH = 5.44 is diluted by a factor of 2. Determine pH in the final solution.
(d) A solution with pH = 12.42 is diluted by a factor of 5. Determine pOH and pH in the final solution.

62. Carry out the following conversions:

(a) Given pH = 7.42, find pOH.
(b) Convert pOH = 1.23 to [OH$^-$].

63. Carry out the following conversions:

(a) pOH = 5.33 to [OH$^-$]
(b) [OH$^-$] = 9.2×10^{-3} M to pH
(c) pH = 4.53 to [OH$^-$]
(d) pOH = 6.43 to pH
(e) pOH = 1.23 to [H$_3$O$^+$]

64. Calculate [H$_3$O$^+$], pH, pOH, and [OH$^-$] for the following solutions:

(a) 8.7×10^{-3} M HNO$_3$
(b) 2.1×10^{-1} M KOH
(c) 9.2×10^{-3} M Ba(OH)$_2$

65. (a) How many moles of HI are formed when 12.48 grams HI are completely dissolved in water?
(b) What volume of H$_2$O is needed for the solution above if we need [HI] = 0.560 M?

66. Consider the following substances in your answers to the questions that follow:

KOH, HClO$_4$, Al(OH)$_3$, HClO

(a) Classify these substances as acids or bases.
(b) Write the appropriate conjugate acid or base for these substances.
(c) Which of these substances is expected to completely dissolve in water?

67. Name the following acids:

(a) HI   (b) HIO$_4$ (c) HIO$_2$

68. An unknown solution is tested with litmus paper; the paper turns blue. Which of the solutions in Figure 15-17 should be used to determine the concentration of the unknown solution?

69. Alkaline solutions such as NaOH are stored in capped plastic bottles to limit exposure to air as CO$_2$ readily dissolves in bases, thus changing their concentrations.

$$OH^-\ (aq) + CO_2\ (g) \longrightarrow HCO_3^-\ (aq)$$

(a) Name the ion formed. What is its conjugate acid? What is its conjugate base?
(b) Draw the Lewis structures for this reaction.
(c) Which reactant is the Lewis acid in this reaction? Which is the Lewis base?

70. NaOH is a white solid that readily absorbs moisture from the air. Thus, solutions made from NaOH pellets have only an approximate molarity, as the absorbed water makes it impossible to measure the precise mass of NaOH.

(a) How many grams of NaOH are needed to make a 1-liter solution with a molarity about 0.100 M?

(b) A more precise molarity is often determined for a NaOH solution by titration with an acid solution of known molarity. Calculate [NaOH] if 25.00 mL of NaOH solution require 28.74 mL of 0.0998 M HCl to reach the endpoint.

71. Potassium acid phthalate ($C_8H_5O_4K$) is a very pure solid that is often used to find the molarity of sodium hydroxide solutions. The ionic equation for this reaction is:

$$K^+ (aq) + C_8H_5O_4^- (aq) + Na^+ (aq) + OH^- (aq) \longrightarrow$$
$$C_8H_4O_4^{2-} (aq) + H_2O + K^+ (aq) + Na^+ (aq)$$

(a) Write the net ionic equation for this reaction by eliminating any ion that appears in exactly the same way as a reactant and as a product.

(b) Which reactant is the Lewis acid and which is the Lewis base in this reaction?

(c) Which reactant is the Arrhenius acid and which is the Arrhenius base in this reaction?

(d) Calculate the molarity of NaOH if 20.00 mL of the NaOH solution is neutralized by 0.486 gram of $C_8H_5O_4K$.

72. Determine the moles of $C_8H_5O_4K$ in each of the following solutions:

(a) 0.800 g $C_8H_5O_4K$ in 500.0 mL solution

(b) 0.800 g $C_8H_5O_4K$ in 250.0 mL solution

(c) 0.800 g $C_8H_5O_4K$ in 100.0 mL solution

(d) Which measurement determines moles in these answers?

73. Find the moles of NaOH under the following conditions:

(a) 25.0 mL NaOH solution titrated with 23.82 mL of 0.205 M HCl solution

(b) 25.0 mL NaOH solution titrated with 23.82 mL of 0.103 M HCl solution

(c) 25.0 mL NaOH solution titrated with 23.82 mL of 0.410 M HCl

(d) Which measurement determines moles in these answers?

Equilibrium Systems

Chemical reactions such as acid-base, solubility, and equilibrium, can be used to test for, or remove, dissolved contaminants from water. Here, water is tested for the presence of pollutants.
(Ben Osborne/Tony Stone Images.)

PRACTICAL CHEMISTRY What's In Our Water Supply?

The purity and quality of our water supply have been the focus of national attention for many years for a very simple reason: Water is essential to life. Most of our water comes from lakes, rivers, and streams, which are fed by springs and by the runoff from rainfall and snow melt. While this may sound idyllic, waterways have been used since the beginning of history to carry off all types of waste products. Some of these waste products have profound effects on aquatic life and on our ability to use the water safely.

Fertilizers in agricultural runoff add unnatural concentrations of nitrogen and phosphorus to groundwater and streams. Pesticides are washed into the waterways by the rain. Industrial wastes such as oils, toxic chemicals, and acids pollute our water supply as well. Water pollution is not an easy problem to study or to solve.

Dissolved contaminants can be removed from water by using chemical reactions such as acid-base reactions, solubility reactions, complexation reactions, and equilibrium reactions. As we learn about equilibrium reactions in this chapter, we will see how they relate to the removal of dissolved water pollutants. (Numerical information in the Chapter Practicals is taken from *Environmental Chemistry,* 5th Ed.; Manahan, S.E.; Lewis, Chelsea MI, 1991.)

This chapter will focus on the following questions:

PRACTICAL A How are metal ions dissolved in our water?

PRACTICAL B How can dissolved lead be removed from drinking water?

PRACTICAL C Why does the pH of our water supply change?

PRACTICAL D Why does soap form a scum in hard water?

16.1 Equilibrium Reactions

SECTION GOALS

✔ Explain why an equilibrium system is called "dynamic" when, to the observer, the reaction appears to have stopped.

✔ Write an equilibrium expression.

✔ Explain how the equilibrium constant is made unitless.

Partial Reactions and Chemical Equilibrium

Throughout this book we have assumed that chemical reactions proceed to the extent allowed by the stoichiometry of the balanced equation. Such reactions are said to go *to completion.* There is a point when the reaction has "completely" converted the reactants into products. Even in the case of limiting reactants, one reactant is completely consumed in the reaction. The reaction stops only when it has completely used up the limiting reactant.

But there is an important kind of reaction that does *not* conform to this assumption. The reaction starts at some point, but when it has apparently stopped we find that a mixture of products and reactants is present. No matter how long we wait, the reaction is not completely converted to products.

Is there something wrong with such reactions? Not at all. It is just that they stop showing any change while there are still some reactants left. There is a special balance between reactants and products in this case, a balance we describe with the concept of **chemical equilibrium.**

As an example of equilibrium, consider the reaction of the ions Fe^{3+} and SCN^- (thiocyanate ion). The product is the dark red ion iron thiocyanate $Fe(SCN)^{2+}$. We call such an ion, where a metal ion is bound by one or more molecules or ions, a **metal complex** ion. Here, "complex" does not mean "complicated." Rather, it is a chemical term meaning "connected system," where the connection, or complexation, in this case is between the Fe^{3+} and SCN^- ions.

Figure 16-1 ▶ Iron (III) and thiocyanate ions are in equilibrium with a complex of the two ions.

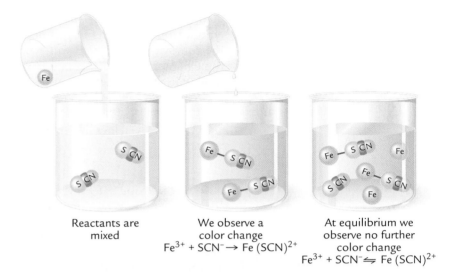

Reactants are mixed

We observe a color change
$$Fe^{3+} + SCN^- \rightarrow Fe(SCN)^{2+}$$

At equilibrium we observe no further color change
$$Fe^{3+} + SCN^- \rightleftharpoons Fe(SCN)^{2+}$$

See this with your **Web Animator** ▶

The equation for the reaction is:

$$Fe^{3+}\ (aq) + SCN^-\ (aq) \rightarrow Fe(SCN)^{2+}\ (aq)$$

The stoichiometry of the equation shows that if we mix 0.1 mole of Fe^{3+} ion with 0.1 mole of SCN^- ion, we will get 0.1 mole of $FeSCN^{2+}$ ion. However, when we do the experiment, we don't obtain this result. Upon analysis of the solution, we find that, although some of the $Fe(SCN)^{2+}$ complex is indeed formed, some Fe^{3+} and SCN^- ions still remain free in solution. Our initial reactant system has *not* been totally converted into product. Instead, we have what is called an **equilibrium system,** and we need to develop a more accurate description of this type of relationship.

Figure 16-1 illustrates the iron-thiocyanate equilibrium reaction. Notice that there is a double arrow in the figure, which tells us that the system is **dynamic.** This means that there is a tendency for the free Fe^{3+} and SCN^- ions to react, but there is also a tendency for the product, $FeSCN^{2+}$ ion, to react and form some Fe^{3+} and SCN^-. At equilibrium, the reaction appears to have stopped, as shown in Figure 16-1 (right beaker), but, on a molecular level, the reactants and products continue to interconvert. We indicate the dynamic nature of the equilibrium by using a double arrow in the equation:

$$Fe^{3+}\ (aq) + SCN^-\ (aq) \rightleftharpoons Fe(SCN)^{2+}\ (aq)$$

The description of this reaction includes both a macroscopic observation of a color change—which stabilizes after a period of time—and an atomic level observation of two ions and an ion complex that continually convert back and forth. Because the atomic level process occurs equally fast in the "forward" and "backward" directions, the net change is zero and we *observe* no change when the reaction is at equilibrium.

Accounting for Equilibrium Changes

Chemists use molarity and (for gases) pressure to describe the amounts of substances present in chemical equilibria. Thus, equilibrium does not require us to master new kinds of measurement. But the changes that occur from initial to ending conditions do invite us to use a method analogous to bookkeeping—careful measurement of initial amounts, the changes that occur, and the final amounts. We will use this bookkeeping throughout this chapter. We first introduce it with a completion reaction, where the limiting reactant is reduced to zero concentration at the conclusion of the reaction. We then show how it works in an equilibrium situation.

In the case of complete reactions, we know that one or more of the reactants will, at the completion of the reaction, be present in zero concentration. Consider the reaction of a strong acid HCl and a strong base NaOH. We know that the reaction involves the balanced chemical equation shown at the top of the table in Figure 16-2. Let's say that we have one liter of a reaction solution. If we mix 0.40 mole of HCl and 0.20 mole of NaOH, we "start" with 0.40 M of HCl and 0.20 M of NaOH. At the end, the NaOH is all gone and we have 0.20 mole of HCl remaining (with a concentration of 0.20 M) along with 0.20 mole NaCl and 0.20 mole of H_2O. The "bookkeeping" is summarized in Figure 16-2. Note that we don't need to "track" the concentration of water, because the concentration of water is essentially unaffected by the reaction.

✔ Meeting the Goals

A system at equilibrium is a dynamic system. At the microscopic level of ions and molecules, products continue to convert to reactants and reactants continue to convert to products even though we may observe no change in the system.

Bookkeeping for a Reaction that Goes to Completion			
Balanced equation:	$HCl\ (aq)$ + $NaOH\ (aq)$ $\rightarrow$	$NaCl\ (aq)$ +	$H_2O(l)$
Initial concentrations:	0.40 M 0.20 M	0.0 M	—
Change in concentrations:	– 0.20 M – 0.20 M	+ 0.20 M	—
Ending concentrations:	0.20 M 0.0 M	0.20 M	—

Figure 16-2 ▶ Bookkeeping of a reaction that goes to completion.

What happens when we keep track of an equilibrium reaction in this manner? We can start with the same approach—considering the initial concentrations. The final concentrations will not be the same as those we get for a completion reaction. Instead, the final concentrations are experimentally observed quantities. But the balanced equation is still useful to us as a guide to the relative concentration changes.

Let's compare the changes that result in a reaction between a metal ion and an anion that proceeds to an equilibrium of products and reactants. The reaction in this case is between zinc ion and an anion called *lactate*, $C_3H_5O_3^-$. A typical complexation reaction between lactate and zinc ions can be described in the following way:

$$Zn^{2+}\ (aq) + C_3H_5O_3^-\ (aq) \rightleftharpoons Zn(C_3H_5O_3)^+\ (aq)$$

An experiment in this system might start by preparing a solution that has $[Zn^{2+}] = 0.100$ M and $[C_3H_5O_3^-] = 0.100$ M. These concentrations are the initial concentrations. The system is allowed to react, forming some of the product, $Zn(C_3H_5O_3)^+$. When the system reaches an equilibrium state, the concentrations of all the species present are measured. The experiment would then show the equilibrium concentrations are $[Zn^{2+}]_{eq} = 0.031$ M, $[C_3H_5O_3^-]_{eq} = 0.031$ M, and $[Zn(C_3H_5O_3)^+]_{eq} = 0.069$ M. This is summarized in Figure 16-3. Note that we know the starting and final concentrations. In this case, the change in the concentrations is calculated from the difference in the initial and ending concentrations.

Figure 16-3 ▶ Bookkeeping of a reaction that forms an equilibrium mixture.

Bookkeeping for a Reaction that Forms an Equilibrium Mixture			
Balanced equation:	$Zn^{2+} (aq)$ +	$C_3H_5O_3^- (aq)$ ⇌	$Zn(C_2H_5O_3)^+ (aq)$
Initial concentrations:	0.100 M	0.100 M	0.0 M
Change in concentrations:	– 0.069 M	– 0.069 M	+ 0.060 M
Ending or Equilibrium concentrations:	0.031 M	0.031 M	0.069 M

The change in the concentrations of reactants and products in the table depends on the stoichiometry. In this chapter, we perform calculations only on reactions in which the stoichiometric coefficients for all species is 1. You may encounter more complicated equilibria in later courses.

Equilibrium Expressions

When we consider reactions that go to completion we know that stoichiometric calculations allow us to predict the amount of products and excess reactants present at the end of the reaction. Similarly, there is a relationship between the concentrations of reactants and products in an equilibrium system. This relationship is called an equilibrium expression and follows simple rules that were developed from an understanding of thousands of chemical reactions. The **equilibrium expression** is a ratio of product concentrations to reactant concentrations. It expresses the relationships among products and reactants at equilibrium as follows:

REMEMBER

- Product concentrations are multiplied together in the numerator of the ratio; reactant concentrations are multiplied together in the denominator of the ratio. Each concentration is raised to a power equal to the coefficient of that substance in the balanced equation.

- The concentration of aqueous solutions is given by their molarities, where square brackets, [ ], indicate molarity. Thus, CH_3COOH (aq) appears in the equilibrium expression as $[CH_3COOH]$.

- The concentration of gases is given by their pressure, P. Thus, $2 NH_3 (g)$ appears in the equilibrium expression as $(P_{NH_3})^2$.
- Pure solids, liquids, and water solvent species are given a value of 1 and therefore do not explicitly appear in an equilibrium expression.

Before writing an equilibrium expression, we should carefully *note the physical states of all reactants and products.* We now have a working definition of an equilibrium expression. *An equilibrium expression is the product of the products divided by the product of the reactants, where each pressure or molarity is raised to a power equal to its coefficient in the balanced equation.* Let's now write an equilibrium expression for the iron-thiocyanate reaction:

$$Fe^{3+} (aq) + SCN^- (aq) \rightleftharpoons Fe(SCN)^{2+} (aq)$$

All three species in this reaction are aqueous, so we will use molarity values (square brackets) in the equilibrium expression. There is only one product concentration, $[Fe(SCN)^{2+}]$; this goes in the numerator of the equilibrium expression. There are two reactant concentrations. $[Fe^{3+}]$ is multiplied by $[SCN^-]$ in the denominator of the equilibrium expression:

$$\frac{[Fe(SCN)^{2+}]}{[Fe^{3+}][SCN^-]}$$

Because only coefficients of 1 appear in the equation, each concentration is raised to the first power. *When molarity values are substituted into the equilibrium expression, the ratio equals a constant, called the* **equilibrium constant,** K. Thus, the complete equilibrium expression is:

$$K = \frac{[Fe(SCN)^{2+}]}{[Fe^{3+}][SCN^-]}$$

KEY IDEA 16.1

✔ **Meeting the Goals**

An equilibrium expression is a ratio in which the product of the concentrations of the products is divided by the product of the concentrations of the reactants. Each concentration is raised to a power equal to its coefficient in the balanced equation. Pure solids and liquids are given a value of 1 and do not appear in the equilibrium expression unless they are in the numerator.

KEY IDEA 16.2

Example 16.1   **Writing equilibrium expressions**

Write an equilibrium expression for each of the following reactions.

(a) $N_2 (g) + 3 H_2 (g) \rightleftharpoons 2 NH_3 (g)$
(b) $2 KNO_3 (s) \rightleftharpoons 2 KNO_2 (s) + O_2 (g)$
(c) $HClO (aq) + H_2O (l) \rightleftharpoons H_3O^+ (aq) + ClO^- (aq)$

Thinking Through the Problem ➡ (16.1) A key idea is to consider the physical state of matter for each substance. We use partial pressure for gases and molarity for aqueous solutions. Solids and liquids have a concentration value of 1 and therefore do not explicitly appear in the equilibrium expression. Each concentration is raised to a power equal to its coefficient; product concentrations go into the numerator and reactant concentrations go into the denominator.

Working Through the Problem

(a) All substances in this reaction are gases. The pressure of the product is squared and goes in the numerator: $(P_{NH_3})^2$. The pressures of the reactants are multiplied together in the denominator of the equilibrium expression: $(P_{N_2})(P_{H_2})^3$. Each pressure is raised to a power equal to its coefficient in the equation. This gives us a product/reactant ratio of $\dfrac{(P_{NH_3})^2}{(P_{N_2})(P_{H_2})^3}$.

(b) Only O_2 will appear in the equilibrium expression because the other two substances are solids. This gives us a product/reactant ratio $\dfrac{(1)^2(P_{O_2})}{(1)^2} = P_{O_2}$.

(c) The aqueous substances are written as molarity values using square brackets, and the concentration term of the water is given the value of 1. This gives us the product/reactant ratio, $\dfrac{[H_3O^+][ClO^-]}{[HClO]1}$.

Answer (a) $K = \dfrac{(P_{NH_3})^2}{(P_{N_2})(P_{H_2})^3}$ (b) $K = P_{O_2}$ (c) $K = \dfrac{[H_3O^+][ClO^-]}{[HClO]}$

How are you doing? 16.1

(a) Write the equilibrium expression for the zinc-lactate reaction described in Figure 16-3.

(b) Lactate also forms a complex with calcium ion. The stoichiometry is 1:1 for the metal:ligand mole ratio. Write the chemical equation and the equilibrium expression for this equilibrium reaction.

In both (b) and (c) of Example 16.1, the value of 1 for solids and liquids is usually omitted altogether. The "1" must be written, however, when it is the *only* substance in the numerator. For example, the system

$$Ag^+ (aq) + Cl^- (aq) \rightleftharpoons AgCl (s)$$

has the equilibrium expression, $K = \dfrac{1}{[Ag^+][Cl^-]}$. It is important to realize that the numerator is not zero, but 1.

Each of the equilibrium expressions in Example 16.1 could be solved for a numerical value of K if we knew the values for every substance in the ratio. This brings us to an additional requirement for an equilibrium expression. The equilibrium constant, K, cannot have units. We will divide out any molarity or pressure units in the following ways.

• To avoid molarity units, all molarity values in an equilibrium expression are divided by a standard value, 1 M.

• To avoid pressure units, all pressures in an equilibrium expression must be expressed in atmospheres.

Example 16.2

Calculating an equilibrium constant

Calculate the equilibrium constant for the reaction shown in Figure 16-1:

$$Fe^{3+} (aq) + SCN^- (aq) \rightleftharpoons Fe(SCN)^{2+} (aq)$$

Assume that the equilibrium concentration of both Fe^{3+} and SCN^- is 0.0003 M and the equilibrium concentration of $Fe(SCN)^{2+}$ is 0.010 M.

Thinking Through the Problem All substances are aqueous; all coefficients are 1. This means that $[Fe^{3+}]_{eq} = [SCN^-]_{eq} = 0.00030$ M, and $[Fe(SCN)^{2+}]_{eq} = 0.010$ M. To get K, we divide each molarity by 1 M and substitute the resulting values into the equilibrium expression.

Working Through the Problem

✔**Meeting the Goals**

The concentration unit for aqueous solutions is molarity; the concentration unit for gases is atmosphere. The numerical value of the equilibrium expression is constant at a given temperature. It is called the equilibrium constant for that system and is symbolized by K. Because the equilibrium constant must be unitless, each molarity is divided by a standard 1 M and each pressure is divided by a standard 1 atmosphere when numerical values are substituted into the equilibrium expression.

$$K = \frac{\dfrac{0.010 \text{ M}}{1 \text{ M}}}{\dfrac{0.00030 \text{ M}}{1 \text{ M}} \times \dfrac{0.00030 \text{ M}}{1 \text{ M}}}$$

Answer $K = \dfrac{(0.010)}{(0.00030)(0.00030)} = 1.1 \times 10^5$. The final answer is unitless.

How are you doing? **16.2**

(a) Calculate the equilibrium constant K for the zinc-lactate reaction described in Figure 16-3.
(b) For the calcium ion-lactate equilibrium, discussed in the previous *How are You Doing?* exercise, values of $[Ca^{2+}] = 0.059$ M, $[C_3H_5O_3^-] = 0.059$ M, and $[Ca(C_3H_5O_3)^+] = 0.041$ M are observed when the system is at 25 °C and at equilibrium. Calculate the value of K at this temperature.

The reaction between iron and thiocyanate ions represents only one type of equilibrium. However, chemical equilibrium can be encountered in any chemical transformation. The three types of equilibria that are most important to us are: **ionic solution equilibria, solubility equilibria,** and **gas phase equilibria.** They are described in Table 16-1, on page 560.

In this chapter, we will consider only ionic solution equilibria (both metal-ligand and acid-base examples) and solubility equilibria.

PRACTICAL A   How are heavy metal ions dissolved in water?

There are many metal ions present in our water because of natural processes. Even fresh water contains some salt, and other ions such as calcium and iron are present in our water supply because of the natural dissolving action of water on rocks and mineral deposits.

Because water is a polar molecule, it orients, or *coordinates,* itself around the positively charged metal ions. This forms a *coordination sphere* of water molecules around the metal ion in which the attractive force is between the lone pair electrons on oxygen and the positively charged metal ion. A metal ion surrounded by a number of water molecules is called a *hydrated metal ion.* $Fe(H_2O)_2^{3+}$, shown in Figure 16-4, is an example of a hydrated metal ion. The formula indicates that there are two H_2O molecules around one Fe^{3+} ion. The +3 charge is assigned to Fe because we know that water is a neutral substance.

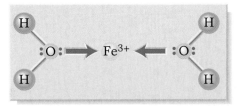

Figure 16-4 ▲ A complex between iron (III) ion and two water molecules.

Any material in water can be a problem, depending on how we want to use the water. Salty water is fine for the ocean, but not to quench our thirst. There are many metal ions, generally in the fifth row of the periodic table or below, that are almost always a problem. These are often called "heavy metal" ions because they are among the metals with the highest molar mass.

Heavy metal ions sometimes enter our water supply as the result of poor industrial, agricultural, or municipal waste management. Metal ions in soil or rock also can enter the water when molecules and ions form strong complexes with them. Metal-ligand

complexation is an equilibrium process. A simple example of a metal complex is $[Cd(CN)_4]^{2-}$. The brackets around the formula "group" the cadmium and cyanide ions and identify them as a metal complex. The Lewis structure of cyanide ion, CN^-, shows a triple bond between C and N and a lone electron pair on both C and N. Iron thiocyanate, which we saw in Figure 16-1, is another example of a metal-ligand complex.

Typical ligands contain lone pairs. By providing a lone pair of electrons to the metal ion, the ligand acts as a Lewis base; the metal ion, in accepting the lone pair from the ligand, acts as a Lewis acid. As shown in Figure 16-5, a metal complex forms between Cd^{2+} and CN^- because of the attraction between the lone pair on each C and the positively charged cadmium ion. In a metal complex, the metal ion is always the Lewis acid (electron pair acceptor) and the ligand is always the Lewis base (electron pair donor).

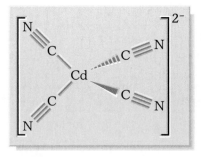

Figure 16-5 ▲ The complex ion formed from Cd^{2+} and four cyanide ions, $Cd(CN)_4^{2-}$.

Identify the Lewis acid and Lewis base in the reaction in which metal ligand complex $[Zn(CN)_4]^{2-}$ is formed from Zn^{2+} and CN^-. The equation for the equilibrium process is:

$$Zn^{2+} (aq) + 4\ CN^- (aq) \rightleftharpoons [Zn(CN)_4^{2-}] (aq)$$

What is the equilibrium expression for this system?

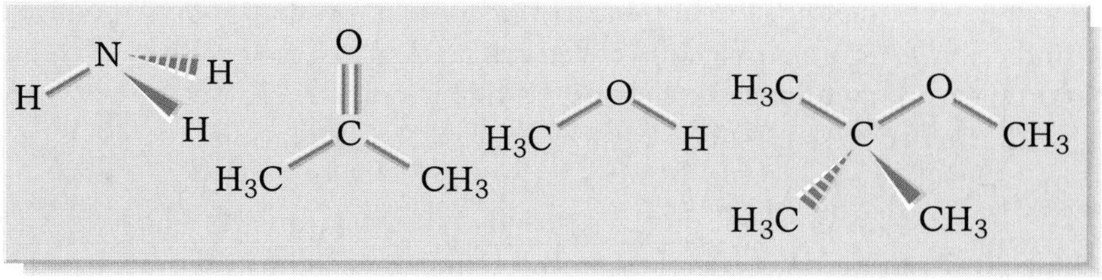

Figure 16-6 ▲ Some molecules and ions that form metal complexes.

Many other small molecules and ions can form strong complexes by acting as Lewis bases. A few are presented in Figure 16-6. Use your understanding of common bonding patterns (Chapter 2) to complete the atoms that have lone pairs. These are the probable sites of the complexation.

Other ligands contain two or more atoms with lone pairs (bonding sites) available to the metal ion. Ligands containing more than one bonding site form *chelates* because they act like big claws that surround and hold aqueous metal ions. (The word *chelate* comes from a Greek word meaning "claw." Metal complexes with which you may be familiar are the Fe-hemoglobin and the Mg-chlorophyll complexes. Both hemoglobin and chlorophyll are chelating agents in these complexes.) Strong chelating agents

play an important role in reducing the concentration of potentially hazardous hydrated metal ions in our water. Nitrilotriacete ion (NTA) is a strong chelating agent. The structure of NTA appears on the left in Figure 16-7. NTA has four binding sites: one on the N and one on one oxygen in each of the three COOH groups. Locate the binding sites and mark them with arrows.

Sodium **e**thylene**d**iamine**te**tra**a**cetate, or EDTA, is a chelating agent commonly added to water supplies during industrial water treatment, or to treat contaminants arising from detergent use and food preparation. EDTA has six bonding sites, located on each of the two N atoms and on one oxygen atom in each of the four COOH groups. Mark these positions on the drawing of EDTA shown in Figure 16-7.

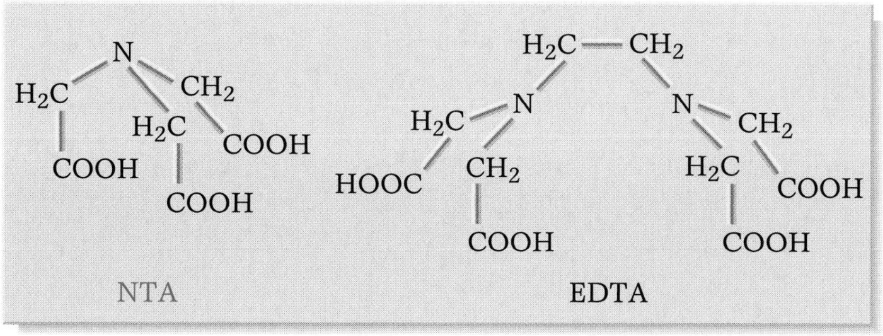

Figure 16-7 ▲ Chelating ligands for forming metal complexes.

Table 16-1 Types of Chemical Equilibria

1. Ionic solution equilibria comprise two types:
    (a) **Metal-ligand equilibria** These occur when a metal ion in solution and another ion or molecule, called a ligand, react to form a metal complex. Many of these solutions are highly colored. The easiest way to track the progress of such a system toward an equilibrium state is to monitor the color.
    (b) **Acid-base equilibria** These equilibria are among the most important in modern chemical research and especially in biomedical research. The easiest way to track the progress of such a system toward an equilibrium state is to monitor the concentration of H_3O^+.
2. Solubility equilibria involve the reactions of solids that dissolve in solution to a *very limited extent*. The easiest way to track the progress of such a system toward an equilibrium state is to monitor the ionic concentration or, if present, the color.
3. Gas phase equilibria are treated in the same manner as other types of equilibria. In the gas phase, the easiest way to track the progress of a system toward an equilibrium state is to monitor pressure changes.

PROBLEMS

1. How is an equilibrium system different from a chemical system not at equilibrium?

2. Why is an equilibrium system called a dynamic system?

3. Name some macroscopic properties one might observe when monitoring a chemical reaction.

4. An equilibrium expression is a ratio. Do product or reactant concentration terms go into the numerator of the ratio? What goes into the denominator of the ratio?

5. What unit is used to express solution concentration in an equilibrium expression?

6. What is the name of the constant that is obtained when the equilibrium expression is solved?

7. Write the equilibrium expression for each of the following equilibrium reactions.
    (a) $HF\ (aq) + H_2O\ (l) \rightleftharpoons H_3O^+\ (aq) + F^-\ (aq)$
    (b) $CaC_2O_4\ (s) \rightleftharpoons Ca^{2+}\ (aq) + C_2O_4{}^{2-}\ (aq)$
    (c) $Pb_2(CO_3)_3\ (s) \rightleftharpoons 2\ Pb^{3+}\ (aq) + 3\ CO_3^{2-}\ (aq)$

8. Calculate the equilibrium constant for the system $HF\ (aq) + H_2O\ (l) \rightleftharpoons H_3O^+\ (aq) + F^-\ (aq)$ when $[HF]_{eq} = 0.001$ M, and $[H_3O^+]_{eq} = [F^-]_{eq} = 2.0 \times 10^{-3}$ M.

9. Calculate the equilibrium constant for the system $CaC_2O_4\ (s) \rightleftharpoons Ca^{2+}\ (aq) + C_2O_4^{2-}\ (aq)$ when $[Ca^{2+}]_{eq} = [C_2O_4^{2-}]_{eq} = 3.0 \times 10^{-6}$ M.

10. Calculate the equilibrium constant for the system $Pb_2(CO_3)_3\ (s) \rightleftharpoons 2\ Pb^{3+}\ (aq) + 3\ CO_3^{2-}\ (aq)$ when $[Pb^{3+}]_{eq} = 4.6 \times 10^{-5}$ M and $[CO_3^{2-}]_{eq} = 6.9 \times 10^{-5}$ M.

16.2 Metal-Ligand Equilibria

SECTION GOALS

✔ Solve the quadratic formula for the roots of a quadratic equation.

✔ Solve an equilibrium expression for K, given equilibrium concentrations.

✔ Calculate the equilibrium concentrations of reactants and products when given only some initial concentrations and the value of K.

The Quadratic Formula

In Chapter 10, we applied the general linear equation $y = mx + b$ to chemical data. This is the general form of a linear equation in x. The value of x where the line crosses the x-axis is known as the x-intercept of the equation. The x-intercept of an equation is a very important point, known as the zero of the function.

We return to the idea of the x-intercept because it is important in calculating how a system will reach equilibrium in many cases. As we will see shortly, many equilibrium problems involve the solution of quadratic equations that have one or two x-intercepts, or roots. A quadratic equation is of the sort $y = ax^2 + bx + c$. The x-intercepts of this equation are when $0 = ax^2 + bx + c$. The values of x that satisfy this equation are called the *roots* of the quadratic equation. Unfortunately, compared with a linear expression, this equation is not as easy to solve for x, except in special cases. In most cases, we use a very helpful tool called the **quadratic formula.**

The quadratic formula is used to solve many equilibrium problems where a small change in some initial chemical concentration leads to a quadratic equation in x. The quadratic formula is the solution in terms of x of the general quadratic equation, $ax^2 + bx + c = 0$. The quadratic formula is:

REMEMBER

$$x = \frac{-b \pm \sqrt{b^2 - 4ac}}{2a}$$

KEY IDEA 16.3

where a, b, and c represent the coefficients in the quadratic equation. *The significance of the quadratic formula is that you can solve a quadratic equation using only the coefficients a, b, and c.* Remember that one side of the quadratic equation *must be set equal to zero* to ensure that the coefficients are correct for the quadratic formula.

The roots of a quadratic equation can be positive, negative, or zero. The numerical values obtained can be integers or fractions.

Example 16.3

Finding the roots of a quadratic equation

Find the roots of the equation $0 = x^2 + 3x + 2$.

Thinking Through the Problem To find the roots or values of x that make the equation equal to zero, we need the values of a, b, and c in the quadratic formula: $a = 1$, $b = 3$, and $c = 2$.

Working Through the Problem Substitute these values into the quadratic formula and solve.

$$x = \frac{-3 \pm \sqrt{3^2 - (4 \times 1 \times 2)}}{2(1)}$$

$$x = \frac{-3 \pm \sqrt{9 - 8}}{2} = \frac{-3 \pm \sqrt{1}}{2}$$

$$x = \frac{-3 \pm 1}{2}$$

Answer There are two roots to this equation: $x = -2$, $x = -1$.

The quadratic formula is

$$x = \frac{-b \pm \sqrt{b^2 - 4ac}}{2a}, \text{ where}$$

a, b, and c represent the coefficients in the quadratic equation. The significance of the quadratic formula is that we can solve a quadratic equation using only the coefficients. Remember that one side of the equation *must be set equal to zero* to ensure that the coefficients are correct.

The quadratic formula can be used to solve any quadratic equation. The results may be rational, irrational, or complex. For rational coefficients, the expression under the radical in the quadratic formula determines which type the solutions will be. For this reason, $b^2 - 4ac$ is called the *discriminant*. If $b^2 - 4ac = 0$, then $\sqrt{0} = 0$, so there will be *one root*. If $b^2 - 4ac$ is negative, the formula contains the square root of a negative number that yields two complex solutions. This is considered a *double root* (these will not be considered further in this book). If $b^2 - 4ac$ is positive, the square root will be a real number. This yields two real roots.

Example 16.4   **Solving a quadratic equation**

Solve the equation $0 = -9x^2 + 4$.

Thinking Through the Problem ➽ (16.3) To solve this quadratic equation for the values of x that will make the equation equal to zero, we must find its roots using the quadratic formula. We find the numerical values of a, b, and c to substitute into the quadratic formula. Here, $a = -9$, $b = 0$, and $c = 4$.

Working Through the Problem Substitute these values into the quadratic formula and solve for x.

$$x = \frac{-0 \pm \sqrt{0^2 - (-9 \times 4 \times 4)}}{2(-9)}$$

$$x = \frac{-0 \pm \sqrt{144}}{-18} = \frac{\pm 12}{-18}$$

Answer The two roots of this equation are: $x = \frac{2}{3}$, $x = -\frac{2}{3}$.

Example 16.5   **Solving a quadratic equation**

Solve $3x^2 - 5 = -2x$.

Thinking Through the Problem We must first rewrite this equation in proper form, setting it equal to zero. Then we must evaluate the coefficients a, b, and c and solve the quadratic formula for its roots.

Working Through the Problem $3x^2 + 2x - 5 = 0$; $a = 3$, $b = +2$, and $c = -5$.

$$x = \frac{-2 \pm \sqrt{4 - (4 \times 3 \times -5)}}{2(3)}$$

$$x = \frac{-2 \pm \sqrt{4 + 60}}{6} = \frac{-2 \pm 8}{6}$$

$$x = 1, -1\frac{2}{3}$$

Answer There are two solutions because the discriminant is positive.

How are you doing? 16.3

Determine the types of solutions for each equation.

(a) $3x^2 - 2x + 5 = 0$ (b) $2x^2 + 5 = 7x$

$a =$ _____ $a =$ _____

$b =$ _____ $b =$ _____

$c =$ _____ $c =$ _____

discriminant = _____   discriminant = _____

type of solution = _____   type of solution = _____

How do you make sense of your answers? In mathematics, the roots of a quadratic equation may be positive, negative, integral, fractional, or complex. In chemistry, however, we are only interested in the results that are chemically reasonable. Thus, values of x that result in a negative concentration, volume or pressure, are meaningless to a chemist. In equilibrium problems, we often must evaluate algebraic expressions that contain some number $\pm x$. If x is larger than the number we subtract it from, we end with a negative concentration or molarity, which is nonsense. Some roots may also result in a greater amount of starting material than we actually had. This situation also is nonsense. In cases such as these, we say that the solution is *chemically unreasonable* and we discard it.

Let's say that an equilibrium starts with a concentration M of a substance. At equilibrium, we find that the concentration of this substance will be $M - x$. In that case, if $x > M$, then the final concentration would be negative. This value of x is deemed chemically unreasonable. But a value of $x < M$ gives a final concentration that is positive. This value of x is acceptable.

Example 16.6   **Evaluating roots of a quadratic equation in chemistry**

Suppose that you have an initial concentration of 0.100 M for some substance. Over the course of an experiment, this initial value changes to some final equilibrium value of $0.100 - x$. Evaluate the $0.100 - x$ when $x = 0.102$ and when $x = +0.002$.

Thinking Through the Problem Substitute each value of x into the expression $0.100 - x$ and evaluate the result. Check to see whether or not the results are physically possible. If not, discard them.

Working Through the Problem For $x = 0.102$ we get $0.100 - 0.102 = -0.002$ M. A negative concentration is nonsense so we discard this value of x. For $x = +0.002$ we get $0.100 - (+0.002) = 0.098$ M. This is a positive concentration value and it is reasonable.

Answer Only the value of $x = 0.002$ results in an answer that is reasonable. There is one reasonable solution.

How are you doing? 16.4

Suppose that you have an initial concentration of 0.06 M for some substance. Over the course of an experiment, this initial value decreases to some final equilibrium value of $0.060 - x$. Evaluate the concentration $0.060 - x$ when $x = 0.0061$ and when $x = -0.004$. Determine which of the roots gives a physically reasonable result. Explain your decision.

STUDY TIPS   How To Use a Calculator

Many calculators are preprogrammed for quick solution of the quadratic formula. If your calculator does not have this option, you may program it yourself or you may simply learn to correctly enter the numbers as you do with any other type of mathematical problem. If you do not use a programmed form of the quadratic equation, here are two shortcuts that may help you save time.

1. A graphing calculator will allow you to type in the problem first using the "plus" sign before the square root, to obtain that solution. Then use the cursor to change the "plus" sign to a "minus" sign to obtain the second solution. A graphing calculator may also be used to graph the equation and the roots can be evaluated from the graph.

2. A non-graphing calculator can also be used in a way that will save you time. First, evaluate the expression under the square root symbol, take the square root, and save the result. Then the problem simplifies to:

$$x = \frac{-b \pm \text{saved result}}{2a}$$

3. We solve for both roots and evaluate them to see if physically reasonable concentrations result for all species.

Calculating Equilibrium Concentrations From Initial Conditions

If equilibrium concepts were used only to determine the K from experimental values of concentrations, then equilibrium would not be very helpful in chemistry. However, we can do the *opposite* calculation: given K and a set of initial concentrations, we *can predict the final equilibrium concentrations* of the substances.

While we can predict final concentrations of substances when a reaction goes to completion, as shown in Figure 16-2, we cannot assume that an equilibrium reaction goes to completion. Instead, we calculate the final amounts of reactants and products by using K.

Most equilibrium calculation problems begin with *initial* concentrations of reactants and then ask us to calculate all equilibrium concentrations. If we begin with initial concentrations of reactants only, with no product present, the reaction *must* begin to form some products, because all substances in the equilibrium will be present to some extent at any given time. We describe this by saying that the reaction must go from left to right. In the process, the reactant concentrations become smaller and the product concentration becomes larger as the reaction proceeds to an equilibrium state. Because equilibrium reactions have no "preferred" direction, it is also possible to start them with only the "products" present and no initial "reactant" concentration. In this case, the reaction must proceed in the opposite direction, from right to left, where the concentration of products must decrease and the concentration of reactants must increase. Finally, there are instances in which we start with a mixture of reactants and products. There may still be a reaction, but it takes a full equilibrium calculation to figure it out.

Following Changes In Equilibrium Concentrations: ICE Tables

The tables we used in Figures 16-2 and 16-3 prepare us to lay out equilibrium calculations in an orderly fashion. *These tables follow the changes in concentration from the start to the finish of a chemical reaction.* Notice that we list three concentrations: initial concentrations (I), change in concentration (C), and ending or equilibrium concentrations (E). We sometimes call this an ICE table as a reminder of the three types of concentration listed.

Using the information in Section 16.1, we can determine that $K = 72$ for the system

$$Zn^{2+}\ (aq) + C_3H_5O_3^-\ (aq) \rightleftharpoons Zn(C_3H_5O_3)^+\ (aq) \qquad K = 72$$

Now that we know the value of K for this system, we can use K to calculate equilibrium concentration values from any initial concentrations. This is summarized in Figure 16-8.

"ICE" Table for an Equilibrium Reaction–Starting with Products Only			
Balanced equation:	$Zn^{2+}\ (aq)$ +	$C_3H_5O_3^-\ (aq)$ $\rightleftharpoons$	$Zn(C_3H_5O_3)^+\ (aq)$
Initial concentrations:	0.020 M	0.020 M	0.0 M
Change in concentrations:	$-x$ M	$-x$ M	$+x$ M
Ending or Equilibrium concentrations:	$0.020 - x$ M	$0.020 - x$ M	x M

Figure 16-8 ▸ ICE table layout of equilibrium calculation—starting with reactants only.

Suppose that we start with initial concentrations of $Zn^{2+} = 0.020$ M and $C_3H_5O_3^- = 0.020$ M but with no $Zn(C_3H_5O_3)^+$ ions present. We write these values on the first line of an ICE table as shown in Figure 16-8. What will happen as the reaction proceeds toward equilibrium? The Zn^{2+} and $C_3H_5O_3^-$ concentrations will decrease by some amount (call the change $-x$) while the $Zn(C_3H_5O_3)^+$ concentration will increase by some amount (call the change $+x$). We write these values on the line labeled "change" in the ICE table. The equilibrium values are the algebraic sums of the first two lines in the table.

Now we see that the equilibrium concentrations are in terms of x. To find the numerical equilibrium concentrations, we must first calculate the value of x by substituting the equilibrium values (in terms of x) into the equilibrium expression, making the expression equal to 72, and solving the resulting mathematical expression for the variable, x. Thus,

$$K = \frac{[Zn(C_3H_5O_3)^+]}{[Zn^{2+}][C_3H_5O_3^-]} = \frac{x}{(0.020 - x)(0.020 - x)} = 72$$

$$72(0.020 - x)^2 = x$$
$$0 = 72x^2 - 3.88x + 0.0288$$

This is a quadratic equation, which we can solve by using the quadratic formula. In this case, $a = 72$, $b = -3.88$, and $c = 0.0288$. Using the quadratic formula, we get two real roots to this equation: $x = 0.0450$ and $x = +0.00888$. The first value of x is too large and gives a negative concentration for $[Zn^{2+}]$ and $[C_3H_5O_3^-]$ when we subtract it from 0.020. Negative concentrations do not make any sense chemically, so we discard them. The second value of x gives us reasonable values for all equilibrium concentrations. We place this value of x in our ICE table to find the equilibrium concentrations. $[Zn^{2+}] = [C_3H_5O_3^-] = 0.0011$ M and $[Zn(C_3H_5O_3)^+] = 0.0088$ M.

Now let's look at another alternative. Suppose that we instead start from an initial concentration of $Zn(C_3H_5O_3)^+ = 0.010$ M but with *no zinc or lactate* ions present. Now what will happen as the

"ICE" Table for an Equilibrium Reaction—Starting with Products Only			
Balanced equation:	Zn^{2+} (aq)	+ $C_3H_5O_3^-$ (aq) $\rightleftharpoons$	$Zn(C_3H_5O_3)^+$ (aq)
Initial concentrations:	0 M	0 M	0.010 M
Change in concentrations:	$+x$ M	$+x$ M	$-x$ M
Ending or Equilibrium concentrations:	$+x$ M	$+x$ M	$0.010 - x$ M

Figure 16-9 ▲ ICE table layout of equilibrium calculation—starting with products only.

✔ **Meeting the Goals**

Equilibrium concentrations can be calculated by using an ICE table. This way we can (1) substitute the resulting equilibrium concentrations in terms of x into the equilibrium expression, (2) simplify the equilibrium expression to obtain a quadratic equation equal to zero, (3) find the roots of the equation with the quadratic formula, and (4) determine which root yields a physically reasonable result.

reaction proceeds toward equilibrium? The $Zn(C_3H_5O_3)^+$ concentration will *decrease* by some amount (call the change $-x$) and the concentration of Zn^{2+} and $C_3H_5O_3^-$ will *increase* by the same amount because of the 1:1 stoichiometry (call this change $+x$). The equilibrium values listed in our ICE table are the algebraic sums of the first two lines in the table, as shown in Figure 16-9.

Again, the equilibrium concentrations are in terms of x and we must calculate the value of x. To find the numerical equilibrium concentrations, we must first calculate the value of x by substituting the equilibrium values (in terms of x) into the equilibrium expression, making the expression equal to 72, and solving the resulting mathematical expression for the variable, x. Thus,

$$K = \frac{[Zn(C_3H_5O_3)^+]}{[Zn^{2+}][C_3H_5O_3^-]} = \frac{0.010 - x}{(x)(x)} = 72$$
$$72x^2 = 0.010 - x$$
$$0 = -72x^2 - x + 0.010$$

Once again, we solve this quadratic equation using the quadratic formula. In this case, $a = -72$, $b = -1$, and $c = 0.010$. Solving, we get two values for x:

$$x = \frac{-b \pm \sqrt{b^2 - 4ac}}{2a} = \frac{1 \pm \sqrt{1 - 4(-72)(0.010)}}{2(-72)}$$
$$= \frac{1 \pm \sqrt{3.88}}{-144} = \frac{1 \pm 1.97}{-144} = -0.0206 \text{ and } +0.00674$$

Again we find two real roots to this equation: $x = -0.0206$ and $x = +0.00674$. But only the positive value makes sense chemically, because x is the concentration of $[Zn^{2+}]$ and of $[C_3H_5O_3^-]$. We cannot have a negative concentration. Thus, only $x = 0.00674$ is chemically reasonable. We substitute this value of x in the expressions in our ICE table and find the equilibrium concentrations $[Zn^{2+}] = [C_3H_5O_3^-] = 0.0067$ M and $[Zn(C_3H_5O_3)^+] = 0.0033$ M.

How are you doing? **16.5**

Calcium metal ions complex with lactate according to the following equation:

$$Ca^{2+} (aq) + C_3H_5O_3^- (aq) \rightleftharpoons Ca(C_3H_5O_3)^+ (aq) \qquad K = 12$$

Determine the equilibrium values of $[Ca^{2+}]$, $[C_3H_5O_3^-]$, and $[Ca(C_3H_5O_3)^+]$ if a solution is prepared starting with $[Ca(C_3H_5O_3)^+] = 0.010$ M. (The equilibrium constant is given following the equation.)

PRACTICAL B   How is a metal salt dissolved into drinking water?

Chelating agents are ligands with more than one binding site, as we saw in Practical A. But the extent to which dissolving of heavy metals takes place in water depends not only upon whether the ligand can form a strong complex with the metal, but on the stability of the complex under specific conditions and on the pH of the water.

Suppose that we want to determine whether or not the lead in $Pb(OH)_2$, found in old pipes, will dissolve into the water running through the pipes. Chelating agents can cause lead to dissolve under certain circumstances. Let's assume that the chelating agent NTA is present in the water and that the pH of the water is 8.00. In water solution, the predominant form of NTA contains one acidic hydrogen and carries a -2 charge. For simplicity we will use the abbreviated formula HT^{2-} for the predominant form of NTA in water. The equation for the complexation of Pb^{2+} by HT^{2-} is:

$$Pb(OH)_2 \ (s) + HT^{2-} \ (aq) \rightleftharpoons$$
$$PbT^- \ (aq) + OH^- \ (aq) + H_2O \ (l)$$

The equilibrium expression is:

$$K = \frac{[PbT^-][OH^-]}{[HT^{2-}]} = 2.07 \times 10^{-5}$$

Recall that solids such as $Pb(OH)_2$ and liquids such as H_2O do not appear in the equilibrium expression. We use the value of K as an indicator of the magnitude of the product-to-reactant ratio in the equilibrium expression. Decimal numbers such as K have an understood denominator of one. If we rewrite the equilibrium expression and include 1 as the denominator of K, we see a numerical value (ratio on the right) that corresponds with formulas for products and reactants (ratio on left):

$$K = \frac{[PbT^-][OH^-]}{[HT^{2-}]} = \frac{2.07 \times 10^{-5}}{1} = \frac{products}{reactants}$$

We can get a better understanding of how well HT^{2-} complexed the Pb if we directly compare the two concentrations, $\frac{[PbT^-]}{[HT^{2-}]}$.

The numerator gives the concentration of metal complex; the denominator gives the concentration of uncomplexed chelate. We obtain this ratio by algebraically rearranging the equilibrium expression to solve for this ratio:

$$\frac{[PbT^-]}{[HT^{2-}]} = \frac{K}{[OH^-]}$$

We can evaluate the ratio $\frac{[PbT^-]}{[HT^{2-}]}$ if we evaluate the ratio it equals, namely, $\frac{K}{[OH^-]}$. We have the value of K but not of the $[OH^-]$. Do you remember how $[OH^-]$ is related to the pH?

The pH is given as $pH = 8.00$. Then, $pOH = 6.00$, and $[OH^-] = 10^{-pOH} = 1.0 \times 10^{-6}$ M. Substitute into the problem this value and the value of K given:

$$\frac{[PbT^-]}{[HT^{2-}]} = \frac{K}{[OH^-]} = \frac{2.07 \times 10^{-5}}{1.0 \times 10^{-6}} = \frac{20.7}{1}$$

At a pH of 8.00, twenty times more of the NTA is complexed with Pb than is left uncomplexed. Ratios and proportions are very powerful mathematical tools. In this example, we *mentally* exclude everything but the relationship, $\frac{[PbT^-]}{[HT^{2-}]} = \frac{20.7}{1}$. This gives us a very clear idea of the relative concentrations of the quantity in the numerator compared to the quantity in the denominator.

How will the ratio of $\frac{[PbT^-]}{[HT^{2-}]}$ change if the pH is changed from 8.00 to 5.00? Will the numerical value of this ratio increase or decrease in value at $pH = 5.00$? (Hint: What is the pOH when $pH = 5.00$?)

Finally, because complexing the Pb^{2+} results in more of the lead being dissolved, then which pH value, 8.00 or 5.00, will help to keep more lead salt in the solid form? In other words, which pH value will result in less lead in the water?

PROBLEMS

11. What is hydrated metal ion?

12. Define a Lewis acid and a Lewis base.

13. What is the Lewis acid and what is the Lewis base in the metal complex $Cu(NH_3)_4^{2+}$?

14. Draw the Lewis structure of ammonia. How many lone pairs of electrons are present on N in the structure?

15. Draw the Lewis structure of water. How many lone pairs of electrons are present on O in this structure?

16. Use the quadratic formula to solve the following quadratic equations. If there are no real roots, write "no real roots" for your answer. Note that some answers may best be expressed as decimals

(a) $0 = -4x^2 + 3x + 4$
(b) $0 = x^2 - 10x$
(c) $0 = 20x^2 + 12x + 1$

17. Concept question: Make a table of quadratic functions and their possible roots. What have we learned about the possible solutions provided by the quadratic formula?

18. When the quadratic formula is used to solve the equation $3x^2 + 6x - 4 = 0$, what is the discriminant? What type of solution do you expect on the basis of this discriminant?

19. Assume the ratio has the value:

$$K = \frac{[HA]}{[A^-]} = 5.0$$

Do you expect to have more HA or A^- present in solution? Why?

20. Assume the ratio has the value:

$$K = \frac{[H_3O^+]}{[HA]} = 0.005$$

Do you expect to have more H_3O^+ or HA in solution? Why?

21. What are the units of K in an equilibrium calculation in which concentration is expressed in molarity units? Why?

22. Write the equilibrium expression for each of the following equilibrium systems, in which a metal-ligand is formed.

(a) $Ag^+ (aq) + 2\ NH_3\ (aq) \rightleftharpoons Ag(NH_3)_2^+ (aq)$
$K = 8.32 \times 10^3$
(b) $Tl^{3+} (aq) + 2\ CN^- (aq) \rightleftharpoons Tl(CN)_2^+ (aq)$
$K = 1.95 \times 10^{13}$

(c) $Ni^{2+} (aq) + 6\ NH_3\ (aq) \rightleftharpoons Ni(NH_3)_6^{2+} (aq)$
$K = 8.13 \times 10^{-1}$

23. Magnesium forms a complex with lactate. Data indicate that equilibrium can be achieved when $[Mg^{2+}] = 0.105$ M, $[C_3H_5O_3^-] = 0.105$ M, and $[Mg(C_3H_5O_3)^+] = 0.095$ M. Determine K for the reaction:

$$Mg^{2+} (aq) + C_3H_5O_3^- (aq) \rightleftharpoons Mg(C_3H_5O_3)^+ (aq)$$

24. Determine the equilibrium concentrations of Mg^{2+}, $C_3H_5O_3^-$, and $Mg(C_3H_5O_3)^+$ if a solution is prepared with initial concentrations of $[Mg^{2+}] = 0.300$ M, $[C_3H_5O_3^-] = 0.300$ M, and $[Mg(C_3H_5O_3)^+] = 0.000$ M. Use the K calculated in Problem 23 for magnesium lactate formation.

25. Discussion question: Determine the equilibrium concentrations of Mg^{2+}, $C_3H_5O_3^-$, and $Mg(C_3H_5O_3)^+$ if a solution is prepared with initial concentrations of $[Mg^{2+}] = 0.000$ M, $[C_3H_5O_3^-] = 0.000$ M, and $[Mg(C_3H_5O_3)^+] = 0.200$ M. Use the K calculated in Problem 23 for magnesium lactate formation. If we could add some Mg^{2+} to this solution, what would happen? Discuss qualitatively and quantitatively.

26. Tin (II) forms complex with lactate. Data indicate that equilibrium can be achieved when $[Sn^{2+}] = 0.165$ M, $[C_3H_5O_3^-] = 0.165$ M, and $[Sn(C_3H_5O_3)^+] = 0.134$ M. Determine K for the reaction:

$$Sn^{2+} (aq) + C_3H_5O_3^- (aq) \rightleftharpoons Sn(C_3H_5O_3)^+ (aq)$$

27. Determine the equilibrium concentrations of Sn^{2+}, $C_3H_5O_3^-$, and $Sn(C_3H_5O_3)^+$ if a solution is prepared with initial concentrations of $[Sn^{2+}] = 0.200$ M, $[C_3H_5O_3^-] = 0.200$ M, and $[Sn(C_3H_5O_3)^+] = 0.000$ M. Use the K calculated in Problem 26 for the formation of tin (II) lactate.

28. Determine the equilibrium concentrations of Sn^{2+}, $C_3H_5O_3^-$, and $Sn(C_3H_5O_3)^+$ if a solution is prepared with initial concentrations of $[Sn^{2+}] = 0.000$ M, $[C_3H_5O_3^-] = 0.000$ M, and $[Sn(C_3H_5O_3)^+] = 0.300$ M. Use the K calculated in Problem 26 for the equilibrium involving tin (II) lactate.

29. The chelating agent NTA is used to solubilize Pb from $PbCO_3$ at pH = 7.00. The equation for this reaction is:

$$PbCO_3\ (s) + HT^{2-} (aq) \rightleftharpoons$$
$$PbT^- (aq) + HCO_3^- (aq)$$

The equilibrium constant is $K = 4.06 \times 10^{-2}$.

(a) Write the equilibrium expression for this system.

(b) Calculate the ratio of $\dfrac{[\text{PbT}^-]}{[\text{HT}^{2-}]}$ when $[\text{HCO}_3^-] = 1.00 \times 10^{-3}$ M.

(c) Will the ratio in part (b) increase or decrease as $[\text{HCO}_3^-]$ decreases? Justify your answer by finding $\dfrac{[\text{PbT}^-]}{[\text{HT}^{2-}]}$ when $[\text{HCO}_3^-] = 1.00 \times 10^{-5}$ M.

30. NTA also forms complexes with dissolved calcium, Ca^{2+}. At pH = 7.00, the equation for this system is:

$$Ca^{2+} (aq) + HT^{2-} (aq) + H_2O (l) \rightleftharpoons CaT^- (aq) + H_3O^+ (aq)$$

The equilibrium constant for the reaction is $K = 7.75 \times 10^{-3}$.

(a) Write the equilibrium expression for this system.

(b) Calculate the ratio of $\dfrac{[\text{CaT}^-]}{[\text{HT}^{2-}]}$ when pH = 7.00 and $[Ca^{2+}] = 0.001$ M. (See Practical B for an example of this calculation.)

31. Hypochlorite ion is a weak base that forms its conjugate acid, HClO, when dissolved in water. The equation for this reaction is:

$$ClO^- (aq) + H_2O (l) \rightleftharpoons HClO (aq) + OH^- (aq) \quad K_b = 3.3 \times 10^{-7}$$

(a) Write the equilibrium expression for this system.

(b) Solve the equilibrium expression for the ratio $\dfrac{[\text{ClO}^-]}{[\text{HClO}]}$.

(c) Evaluate this ratio when the pH = 10.00. (Refer to Practical B.)

(d) Do you expect a greater or smaller concentration of ClO^- when the pH is changed from 10.00 to 3.00? Do a calculation to support your expectation.

(e) Explain why you expect hydroxide ion concentration to be greater at pH = 10 and less at pH = 3.

32. Formic acid is a weak acid that reacts with water to form its conjugate base:

$$HCOOH (aq) + H_2O (l) \rightleftharpoons H_3O^+ (aq) + HCOO^- (aq) \quad K_a = 1.8 \times 10^{-4}$$

(a) Write the equilibrium expression for this system.

(b) Solve the equilibrium expression for the ratio $\dfrac{[\text{HCOOH}]}{[\text{HCOO}^-]}$.

(c) Do you expect greater concentrations of $HCOO^-$ at pH = 11.0 or at pH = 2.0? Justify your expectations with a calculation.

(d) Explain why you expect hydronium ion concentration to be greater at pH = 2.0 than at pH = 11.0.

16.3 Weak Acid-Base Equilibria

SECTION GOALS

✔ Explain, in terms of ionization, the primary difference between an aqueous solution of a strong acid and an aqueous solution of a weak acid.

✔ Calculate the pH of an aqueous solution of a weak acid.

✔ Find the fractional ionization and the percent ionization of an aqueous solution of a weak acid.

✔ An approximation is sometimes used to avoid using the quadratic formula. Know when this approximation is valid.

The Equilibrium Expression for a Weak Acid in Water

The acid reactions that we discussed in Chapter 15 involved strong acids—those that ionize completely in aqueous solution. In the case of a strong acid, the amount of hydronium ion in the solution is equal to the amount of acid that is added. This makes it easy to determine the amount of hydronium ion. Then, given $[H_3O^+]$, we can calculate the pH:

Strong acid: $HA\ (aq) + H_2O\ (l) \longrightarrow A^-\ (aq) + H_3O^+\ (aq)$

$$[H_3O^+] = [HA]_{init}$$

$$-\log [H_3O^+] = pH$$

With weak acids, however, there is incomplete ionization. Some of the acid remains intact, and equilibrium is established among the acid, the water, the conjugate base, and the hydronium ion. As a result, the amount of hydronium ion is less than the amount of acid added to the solution.

Weak acid: $HA\ (aq) + H_2O\ (l) \rightleftharpoons A^-\ (aq) + H_3O^+\ (aq)$

$$[H_3O^+] < [HA]_{init}$$

✔ **Meeting the Goals**

In an aqueous solution of a strong acid, we assume that all the hydrogen present is ionized. For a strong acid, $[H_3O^+] = [HA]_{init}$. In an aqueous solution of a weak acid, only a fraction of the hydrogen available is ionized. For weak acids, $[H_3O^+] < [HA]_{init}$ and $[H_3O^+]$ must be calculated using an equilibrium expression.

Predicting the amount of hydronium ion in a solution of a weak acid thus requires an equilibrium expression. Following the rules for writing equilibrium expressions, we have:

$$HA\ (aq) + H_2O\ (l) \rightleftharpoons A^-\ (aq) + H_3O^+\ (aq)$$

$$K_a = \frac{[A^-][H_3O^+]}{[HA](1)} = \frac{[A^-][H_3O^+]}{[HA]}$$

As before, the concentration of water is dropped. Note the subscript "a" after the K. This is a reminder that this expression describes *acid* ionization equilibrium.

Using an ICE Table for Weak Acid Equilibria

As an example for our calculations, we will work with data for formic acid, HCOOH. This weak acid was first isolated from ants, and its name comes from the Latin name for ant, *formica*. It is the substance responsible for the burning sensation of many ant bites. Formic acid ionizes according to the reaction:

$$HCOOH\ (aq) + H_2O\ (l) \rightleftharpoons H_3O^+\ (aq) + HCOO^-(aq)$$

$$K_a = \frac{[HCOO^-][H_3O^+]}{[HCOOH]}$$

The measured K_a for formic acid is 1.77×10^{-4}. We use an ICE (initial-change-equilibrium) table to calculate the concentration of

"ICE" Table for a Weak Acid Equilibrium Reaction			
Balanced equation:	$HCOOH\ (aq) + H_2O\ (l) \rightleftharpoons H_3O^+\ (aq) + HCOO^-\ (aq)$		
Initial concentrations:	0.010 M —	0 M	0 M
Change in concentrations:	$-x$ M —	$+x$ M	$+x$ M
Ending or Equilibrium concentrations:	$0.010 - x$ M —	$+x$ M	$+x$ M

Figure 16-10 ▲ ICE table layout for a weak acid equilibrium calculation.

important species in a solution of 0.010 M formic acid. This is shown in Figure 16-10.

The initial solution (neglecting the very small amount of hydronium ion and hydroxide ion that exist in pure water) has no formate ($HCOO^-$) ion or hydronium ion. So initially, we have just 0.010 M formic acid. Some of this (x moles per liter) will ionize. The change in concentration for HCOOH, therefore, is $-x$. In the process, x moles per liter of formate and x moles per liter of hydronium ion will form. The change in concentration for both $HCOO^-$ and H_3O^+ is $+x$. This is summarized in Figure 16-10. Note that we do not concern ourselves with the concentration of water, because it is the solvent.

The last line of the table in Figure 16-10 is the sum of the first two lines. It lists the equilibrium values for the three substances in the K_a expression. We substitute the equilibrium concentrations into the K_a expression and solve for x using the equilibrium expression:

$$K_a = \frac{[HCOO^-][H_3O^+]}{[HCOOH]} = \frac{(x)\,(x)}{0.010 - x}$$

$$0.010\,K_a - K_a x = x^2$$

$$0 = x^2 + K_a x - 0.010\,K_a$$

✔ Meeting the Goals

For acid ionization equilibria, a quadratic equation is obtained from the equilibrium expression and solved using the quadratic formula. Then $x = [H_3O^+]$ and pH $= -\log\,[H_3O^+]$.

Substituting the given K_a value, this equation becomes:

$$0 = x^2 + 1.77 \times 10^{-4}\,x - 1.77 \times 10^{-6}$$

We can use the quadratic formula to solve for x, with $a = 1$, $b = 1.77 \times 10^{-4}$, and $c = -1.77 \times 10^{-6}$ (don't forget the negative sign with c in this case). The formula yields two roots: $x = +0.0012$ and $x = -0.0014$. The second root is rejected because it gives a negative value for the hydronium concentration: $[H_3O^+] = -0.0014$ M.

When we use the root $x = 0.0012$, we get:

$$[HCOOH] = 0.0088 \text{ M} \qquad [HCOO^-] = [H_3O^+] = 0.0012 \text{ M}$$

Because we now have $[H_3O^+]$, we can calculate pH:

$$pH = -\log\,[H_3O^+] = -\log(0.0012) = 2.92$$

✔ Meeting the Goals

The fraction of acid molecules (HA) ionized in water solution is $\dfrac{[A^-]}{[HA]}$. The percentage of ionization is obtained by multiplying the fractional amount by 100.

Another interesting—and important—value we can calculate from these results is the *fraction of formic acid that ionizes*. This is expressed as:

$$\text{fraction ionized} = \frac{[HCOO^-]}{[HCOOH]_{init}}$$

In this equation, $[HCOOH]_{init}$ is the initial concentration of formic acid. Then we have:

$$\text{fraction ionized} = \frac{0.0012 \text{ M}}{0.010 \text{ M}} = 0.12$$

$$\text{percentage ionization} = 12\%$$

Example 16.7   Calculating the pH of a weak acid solution

What is the pH of a solution that has $[HCOOH]_{init} = 0.10$ M? Calculate the fraction of molecules that will ionize.

Thinking Through the Problem There are three calculations required here. First, we must calculate $[H_3O^+]$ given $[HCOOH]_{init} = 0.10$ M. Then we must find the pH of HCOOH. Finally, we must calculate the fraction of HCOOH that ionizes. The measured K_a is 1.77×10^{-4}. ➤ (16.5) The key idea is to start with the acid ionization equation and the equilibrium expression for HCOOH, and create an ICE table.

Working Through the Problem

$$HCOOH \text{ } (aq) + H_2O \text{ } (l) \rightleftharpoons H_3O^+ \text{ } (aq) + HCOO- \text{ } (aq)$$

I	0.10 M	—	0 M	0 M
C	$- x$ M	—	$+ x$ M	$+ x$ M
E	$0.10 - x$ M	—	x M	x M

$$K_a = \frac{[HCOO^-][H_3O^+]}{[HCOOH]} = \frac{(x)(x)}{0.1 - x} = 1.77 \times 10^{-4}$$

The quadratic equation that results is $0 = x^2 + 1.77 \times 10^{-4} \, x - 1.77 \times 10^{-5}$, where $a = 1$, $b = 1.77 \times 10^{-4}$, and $c = -1.77 \times 10^{-5}$. Solving the quadratic formula gives two roots, $x = -0.00429$ and $x = 0.0041$. The only reasonable root in this case is $x = 0.0041$. Then $[HCOO^-] = [H_3O^+] = 0.0041$ M; pH $= -\log(0.0041) = 2.39$. Once we have calculated x, the fraction of ionized molecules is $\dfrac{[HCOO^-]}{[HCOOH]_{initial}} = \dfrac{0.0041}{0.10} = 0.041$, or 4.1%.

Answer $[H_3O^+] = 0.0041$ M and pH $= 2.39$. 4.1% of the HCOOH molecules have ionized.

How are you doing? **16.6**

Determine the pH of a solution that has [HCOOH] = 0.0010 M. Then calculate the fraction of ionized formic acid molecules. Use the equation and K_a value given in Example 16.7.

Trends and Approximations in Weak Acid Equilibria

Table 16-2 lists the K_a's for several acids. Note that the K_a values are for the ionization of a single hydrogen ion. Diprotic acids such as carbonic acid, H_2CO_3, contain two ionizable hydrogens and, for these acids, there is a distinct K_a value for each ionization. Ionization of the first hydrogen from H_2CO_3 is designated by (1) after the name carbonic. Ionization of the hydrogen from $H_2CO_3^-$ is designated by (2) after the name carbonic. Can you find the three K_a values in Table 16-2 for the ionization of the three hydrogens from phosphoric acid?

The strength of an acid depends on the amount of hydronium ion formed. For any general acid, HA:

$$HA\ (aq) + H_2O\ (l) \rightleftharpoons H_3O^+\ (aq) + A^-\ (aq) \quad \text{and} \quad K_a = \frac{[A^-]\,[H_3O^+]}{[HA]}$$

KEY IDEA 16.6

Acids that ionize to form larger amounts of A^- and H_3O^+ have larger values of K_a. K_a, then, is a good measure of acid strength.

Table 16-2 also lists the pH of a 0.10 M solution of each of the acids. Recall that a neutral solution at 25 °C has pH = 7, whereas the solution becomes increasingly acidic as pH approaches 0. *Note that a larger K_a indicates that a substance is a stronger acid.*

Table 16-2 Acid Ionization Constants

Name of acid	Acid	Conjugate base	K_a	pH of 0.10 M solution
Chlorous	$HClO_2$	ClO_2^-	1.1×10^{-2}	1.55
Phosphoric (1)	H_3PO_4	$H_2PO_4^-$	7.5×10^{-3}	1.62
Hydrofluoric	HF	F^-	6.6×10^{-4}	2.11
Formic	HCOOH	$HCOO^-$	1.8×10^{-4}	2.39
Acetic	CH_3COOH	CH_3COO^-	6.5×10^{-5}	2.59
Carbonic (1)	H_2CO_3	HCO_3^-	4.3×10^{-7}	3.68
Phosphoric (2)	$H_2PO_4^-$	HPO_4^{2-}	6.2×10^{-8}	4.10
Hypochlorous	HClO	ClO^-	3.0×10^{-8}	4.26
Hydrocyanic	HCN	CN^-	6.2×10^{-10}	5.10
Ammonium	NH_4^+	NH_3	5.6×10^{-10}	5.12
Carbonic (2)	HCO_3^-	CO_3^{2-}	4.8×10^{-11}	5.66
Phosphoric (3)	HPO_4^{2-}	PO_4^{3-}	2.2×10^{-13}	6.83

If we were to carry out a calculation for the pH of a solution of hypochlorous acid, as shown earlier, we would come to a very important conclusion about the extent to which this molecule ionizes. In Table 16-2, for example, we would determine that for $[HClO] = 0.100$ M, $[ClO^-] = [H_3O^+] = 0.000055$ M.

$$K_a = \frac{[ClO^-][H_3O^+]}{[HClO]} = \frac{(x)(x)}{0.100 - x}$$

$$x = 0.000055$$

Fewer than 1 in every 1,000 molecules are ionized at a given moment. This suggests that we will obtain a good approximation to the answer if we neglect the x compared to 0.100:

$$0.100 - x \approx 0.100$$

This approximation greatly simplifies our solution of the quadratic equation:

$$K_a \approx \frac{(x)(x)}{0.100} = \frac{x^2}{0.100}$$

$$x \approx \sqrt{K_a(0.100)}$$

Whenever we see that x is likely very much smaller than the initial concentration of the acid, we may invoke this approximation, neglect x in the denominator of the equilibrium expression, and avoid the quadratic formula altogether! The approximation is justified whenever the percentage of ionization is less than 5%.

✔ **Meeting the Goals**

It is reasonable to neglect the x in the denominator when the acid ionizes to a very small extent—about 5% or less of the total initial concentration of the acid before ionization. Under those circumstances we may use

$$[H_3O^+] = x \approx \sqrt{K_a[HA]_{init}}$$

Example 16.8   Calculating the approximate pH of a weak acid solution

Determine the pH of a 0.050 M solution of acetic acid, CH_3COOH. Verify any assumptions that are made.

Thinking Through the Problem Weak acids form equilibrium systems in solution. Find the K_a from Table 16-2, create an ICE table of values, write the equilibrium equation and equilibrium expression.

Working Through the Problem K_a for acetic acid is 6.5×10^{-5}. We will assume that there is very little ionization of acetic acid. The equation for the ionization of acetic acid in water is:

$$CH_3COOH\ (aq) + H_2O\ (l) \rightleftharpoons CH_3COO^-\ (aq) + H_3O^+\ (aq)$$

We will use the approximation that $x < 0.050$, and solve for $[H_3O^+]$:

$$[H_3O^+] \approx \sqrt{K_a[CH_3COOH]_{init}}$$

Substituting the appropriate values, we obtain:

$$[H_3O^+] \approx \sqrt{3.25 \times 10^{-6}} \approx 0.0018\ M$$

Vinegar contains acetic acid, a weak acid. (Stephen Frisch/Stock Boston.)

Note that 0.0018 is much smaller than $[CH_3COOH]_{init}$; only 3.6% of the molecules ionize. The assumption is justified in this case.

Answer The pH of a 0.050 M solution of acetic acid = 2.74.

REMEMBER

Note that it is always necessary to verify the assumption to neglect the subtracted x. This assumption is not valid when the value of x is close in magnitude to the initial concentration of the acid. Remember that the magnitude of the K_a gives an indication of the ratio of product to reactant. Hence the K_a value also indicates the extent of acid ionization expected. See what happens to this assumption in the following example.

Example 16.9

Calculating the pH of a solution

Determine the pH of a 0.050 M solution of phosphoric acid. Verify any assumptions that are made.

Thinking Through the Problem K_a for phosphoric acid is 7.5×10^{-3}. We will assume that there is very little ionization of phosphoric acid. We proceed as above to get a trial value for $[H_3O^+]$.

Working Through the Problem

$$[H_3O^+] \approx \sqrt{K_a[H_3PO_4]_{init}} = \sqrt{3.75 \times 10^{-4}} = 0.019 \text{ M}$$

In this case, the assumption is *not* verified; % ionization = $\dfrac{0.019}{0.05} \times 100 = 38\%$. We must find the exact solution provided by the quadratic formula.

Answer $[H_3O^+] = 0.016$ M, pH = 1.80.

How are you doing? 16.7

Determine the pH of a 0.02 M solution of CH_3COOH. Verify any assumptions made.

Equilibria Involving Weak Bases

Weak bases are capable of reacting in an equilibrium with water to give hydroxide ion. An example is the weak base hypochlorite:

Weak base: $ClO^- (aq) + H_2O (l) \rightleftharpoons HClO (aq) + OH^- (aq)$

$$[OH^-] < [ClO^-]$$

For this reaction, we designate an equilibrium expression that includes all of the reactants and products *except* water. The equilibrium

PRACTICAL C   Why does the pH of our water change?

Contaminated water seldom has a pH of 7.00. When acidic contaminants are present, the pH of the water is below 7; when basic contaminants are present, the pH of the water is greater than 7. For example, $H_2PO_4^{2-}$, dissolved CO_2, H_2S, proteins, and acidic metal ions such as Fe^{3+} or Al^{3+} are all substances that increase the acidity of water.

Some amount of dissolved CO_2 is present in all natural waters. In fact, even unpolluted air contains 350 ppm (by volume) of CO_2, and it is calculated that water in equilibrium with unpolluted air has a dissolved CO_2 concentration, $[CO_2 \ (aq)]$, of 1.146×10^{-5} M. (Recall that ppm is "parts per million"; in this context it means 350 liters CO_2 to 1,000,000 liters dry air.) Carbon dioxide is a weak Lewis acid. Dissolved CO_2 in water initially forms carbonic acid. A second equilibrium reaction occurs when the carbonic acid, a weak Brønsted-Lowry acid, forms hydrogen carbonate ions. The equation for the formation of carbonic acid is:

$$CO_2 \ (aq) + H_2O \ (l) \rightleftharpoons H_2CO_3 \ (aq)$$

with an equilibrium constant of $K = 2 \times 10^{-3}$ at 25 °C. What is the equilibrium expression for this system?

Carbonic acid partially dissociates to HCO_3^- so that the equilibrium equation is:

$$CO_2 \ (aq) + H_2O \ (l) \rightleftharpoons HCO_3^- \ (aq) + H^+ \ (aq)$$

with an equilibrium constant of $K = 4.45 \times 10^{-7}$.

Write the equilibrium expression for this second system; rearrange to solve for the ratio $\dfrac{[HCO_3^-]}{[CO_2]}$ and calculate this ratio when pH = 6.2.

As we saw in Practical B, the pH of the water supply has an important effect on the ability of chelating agents to solubilize metal ions. Chelating agents are generally conjugate bases of Brønsted-Lowry acids. Recall that a base accepts H^+ to form its conjugate acid. For example, ammonia, NH_3, is the conjugate base of NH_4^+. Ammonia is also a simple ligand that can form complexes with metal ions. Thus, ammonia in water can act in two different ways; it can form a metal complex such as $Cu(NH_3)_4^{2+}$ or it can accept H^+ to form its conjugate acid, NH_4^+. The pH of the water determines whether the ligand accepts H^+ or a metal ion. At pH ≤ 7, H^+ "wins" the competition for the ligand, and the acid form of ligand predominates. The ability of complexing agents to solubilize metal ions in this pH range is very weak. At pH > 7, ligands are mainly in their base form. Strong chelation therefore is seen at higher pH values.

constant in this case is K_b, the base hydrolysis (water splitting) constant.

$$K_b = \frac{[OH^-][HClO]}{[ClO^-]}$$

K_b problems are handled in a similar manner to K_a problems.

PROBLEMS

33. For each of the following acids, write the equation for its ionization in water. Then write the equilibrium expression. Use Table 16-2 for the K_a values.

(a) CH_3COOH (c) NH_4^+
(b) HF (d) $H_2PO_4^-$

34. Determine the pH of the following solutions. You may use the approximation method where appropriate.

(a) 0.100 M CH_3COOH
(b) 1.00×10^{-3} M CH_3COOH
(c) 0.100 M HF

(d) 1.00×10^{-3} M HF
(e) 0.100 M NH_4Cl
(f) 1.00×10^{-3} M NH_4Cl

35. A useful way to display the K_a of acids is through their pK_a. This is defined as:

$$pK_a = -\log_{10}K_a$$

Determine the pK_a for the acids in Table 16-2. Generally speaking, would the pH of a solution of a pure weak acid be less than, equal to, or greater than the pK_a?

36. Discussion question: The K_a expression can be used to express the *ratio* of the acid and the base form of a conjugate acid-base pair. The general formula for the conjugate acid is HA; HA stands in the place of each specific formula, for example, HCOOH. The general formula for the conjugate base is A^-, which stands in the place of each specific conjugate base, for example, $HCOO^-$.

(a) Write the K_a expression for the general equation:

$$HA\,(aq) + H_2O\,(l) \rightleftharpoons H_3O^+\,(aq) + A^-\,(aq)$$

Rearrange the K_a expression to solve for $\dfrac{[HA]}{[A^-]}$.

(b) Fill in the following table. Note that, for each acid, the ratio $\dfrac{[HA]}{[A^-]} = \dfrac{[H_3O^+]}{K_a}$.

Find the K_a values given for each acid in Table 16-2. Calculate the pK_a and the hydronium ion concentration from each pH given. Remember that $[H_3O^+] = 10^{-pH}$. Finally, calculate the ratio $\dfrac{[H_3O^+]}{K_a}$ because this is equal to the ratio $\dfrac{[HA]}{[A^-]}$.

Acid	HCOOH	HF	HClO	NH_4^+
K_a				
$pK_a = -\log K_a$				
$\dfrac{[HA]}{[A^-]}$ at pH = 2.0				
$\dfrac{[HA]}{[A^-]}$ at pH = 4.0				
$\dfrac{[HA]}{[A^-]}$ at pH = 6.0				
$\dfrac{[HA]}{[A^-]}$ at pH = 8.0				
$\dfrac{[HA]}{[A^-]}$ at pH = 10.0				

Circle all values for which the ratio is greater than 1. In such cases, there is more of the substance in the acid form HA than in the base form A^-. State the relationship between the pK_a and the amount of the substance in the acid form, HA.

16.4 Solubility Equilibria

SECTION GOALS

✔ Calculate the value of K_{sp} when given the solubility of a salt.
✔ Calculate the solubility of a salt when given its formula and K_{sp}.

Saturated Solutions of Salts and Solubility Product Expressions

Not all substances dissolve equally well in a given solvent at a specified temperature. Indeed, there is a whole spectrum of solubilities possible and substances can range from very soluble to very insoluble. Let us look again at the process of dissolution, restricting our discussion to that of salts in aqueous solution. Because solubilities change as temperature changes, we will consider each system we discuss to be at the same constant temperature.

As we saw in Chapter 14, when a small amount of table salt is added to water, the salt dissolves. Solutions that are able to dissolve additional salt are called unsaturated solutions. Sooner or later, depending on the salt, no more dissolution occurs, and we see the salt begin to accumulate on the bottom of the container. Solutions that are not able to dissolve additional solute are called saturated solutions. Here again, with saturated solutions, we have the conditions necessary for equilibrium: some salt continues to dissolve, but because the solution is saturated, some dissolved salt comes out of solution and redeposits as the solid salt. The two processes of "going into" and "coming out of" solution occur simultaneously so that, at a given temperature, the amount of salt in the saturated solution, expressed in grams or moles per liter, remains unchanged. A state of equilibrium is established between the dissolution and recrystallization of a salt that only dissolves to an extent by limited equilibrium. The equilibrium system is described by the general equation:

solid salt $\rightleftharpoons$ ions (forming the salt) in aqueous solution

Consider a saturated solution of silver bromide. The equation for this system is:

$$\text{AgBr } (s) \rightleftharpoons \text{Ag}^+ (aq) + \text{Br}^- (aq)$$

The amount of Ag^+ and Br^- ions in solution is determined by the *solubility* of AgBr. If we require the solubility to have units of mol L^{-1}, then the amount of Ag^+ and Br^- ions in solution is given by $[\text{Ag}^+]$ and $[\text{Br}^-]$, and the solubility of AgBr is given by either $[\text{Ag}^+]$ or $[\text{Br}^-]$. The equilibrium expression, then, is written as:

$$K_{sp} = [\text{Ag}^+][\text{Br}^-]$$

where the general equilibrium constant K has been replaced by K_{sp} for this specific case in which K_{sp} is *the product of the equilibrium concentration of the dissolved species (also known as the solubility product constant)*, each raised to the appropriate power.

In this section, we will consider only the simplest cases in which the soluble salt contains one positive and one negative ion.

Example 16.10 Writing K_{sp} equilibrium expressions

Write the K_{sp} expression for the reaction.

$$\text{PbCrO}_4 (s) \rightleftharpoons \text{Pb}^{2+} (aq) + \text{CrO}_4^{2-} (aq)$$

Thinking Through the Problem ◀◁ (16.7) The key idea is that a K_{sp} expression is an equilibrium expression in which K_{sp} is equal to the ratio of product to reactant concentrations, each concentration being raised to the appropriate power. For aqueous solutions, concentration is expressed in units of molarity and indicated by the square brackets, [ ]. Pure solids and liquids do not appear in an equilibrium expression.

Working Through the Problem The products in the equation are Pb^{2+} (aq) and CrO_4^{2-} (aq). These are written as $[Pb^{2+}]$ and $[CrO_4^{2-}]$. The reactant is $PbCrO_4$ (s); it does not appear in the K_{sp} expression.

Answer The equilibrium expression is $K_{sp} = [Pb^{2+}][CrO_4^{2-}]$.

How are you doing? 16.8

Write the equilibrium equation and the K_{sp} expression for each of the following salts:

(a) $CuCO_3$ (b) $CaSO_4$ (c) CdC_2O_4

Solubility Product Constant (K_{sp}) Values

REMEMBER

Values for K_{sp} can be calculated for any salt when its solubility is known. To fit the K_{sp} expression, we need solubility in moles per liter (mol L^{-1}). The **solubility** of a salt is the chemical amount in solution that is in equilibrium with undissolved solid at the specified temperature, where "amount" is the moles of substance per liter. Recall that one mole of NaCl gives one mole of Na^+ and one mole of Cl^- when it ionizes. This means that, knowing the solubility of a salt, we know the concentration of its ions in the K_{sp} expression.

Example 16.11 Calculating K_{sp} from solubility

Barium carbonate has a solubility of 1.1×10^{-5} mol L^{-1} at 25 °C. Calculate the value of its K_{sp} at 25 °C.

Thinking Through the Problem Write the equation for the dissolution reaction. Barium ion is Ba^{2+} and carbonate is CO_3^{2-}. The formula of the salt is $BaCO_3$. It is expected to dissociate in water solution.

$$BaCO_3\ (s) \rightleftharpoons Ba^{2+}\ (aq) + CO_3^{2-}\ (aq)$$

Working Through the Problem The K_{sp} expression is $K_{sp} = [Ba^{2+}][CO_3^{2-}]$. Substituting the solubility values for each ion, we have $K_{sp} = (1.1 \times 10^{-5})(1.1 \times 10^{-5}) = (1.1 \times 10^{-5})^2 \approx 1.2 \times 10^{-10}$.

Answer 1.2×10^{-10}

✔ Meeting the Goals

To find the value of K_{sp}, write an equilibrium expression for the dissolution of the salt, substitute equilibrium concentrations (solubility values), and solve for K_{sp}.

Now that we see how solubility values are related to the K_{sp}, we can use K_{sp} to calculate the solubility of the salt. It is the same problem but in reverse!

Because K_{sp} values are derived from solubility data, we have a "handle" on the solubility (or insolubility) of any salt for which we have a K_{sp} value. K_{sp} values like those in Table 16-3 can be used to determine the *relative* solubility of a salt.

Table 16-3 K_{sp} Values For Some Salts

Halides	K_{sp}	Carbonates	K_{sp}
AgBr	5.4×10^{-13}	$BaCO_3$	1.2×10^{-10}
AgCl	1.8×10^{-10}	$CaCO_3$	5.0×10^{-9}
AgI	8.5×10^{-17}	$CuCO_3$	1.4×10^{-10}
CuI	1.3×10^{-12}	$FeCO_3$	3.2×10^{-11}
CuCl	1.7×10^{-7}	$MgCO_3$	6.8×10^{-6}
CuBr	6.3×10^{-9}	$PbCO_3$	7.4×10^{-14}

Example 16.12 | **Calculating solubility from K_{sp}**

The K_{sp} for silver chloride is 1.8×10^{-8}. Calculate the solubility of AgCl.

Thinking Through the Problem   The equilibrium equation is $AgCl (s) \rightleftharpoons Ag^+ (aq) + Cl^- (aq)$; the equilibrium expression for this process is $K_{sp} = [Ag^+][Cl^-]$, where $[Ag^+] = [Cl^-] =$ unknown value. Call the unknown solubility s. The numerical value of K_{sp} is given.

Working Through the Problem $K_{sp} = [Ag^+][Cl^-]$; substituting the known and unknown values, we get $1.8 \times 10^{-8} = s^2$, and $s = \sqrt{1.8 \times 10^{-8}} = 1.3 \times 10^{-4}$. Remember that solubility has units of mol L^{-1}.

Answer   The solubility of AgCl is 1.3×10^{-4} mol L^{-1}.

How are you doing? 16.9

✔ **Meeting the Goals**

For a salt that dissociates to yield one cation and *one* anion, we calculate the solubility of a sparingly soluble salt by taking the square root of the K_{sp}.

(a) What is the solubility of cadmium sulfide, CdS, at 25 °C? (K_{sp} for CdS is 8.0×10^{-27}.)
(b) The solubility of magnesium carbonate is 2.6×10^{-3} mol L^{-1} at 25 °C. Calculate the K_{sp} for $MgCO_3$.
(c) Determine the solubility of copper (I) iodide, CuI, at 25 °C. (K_{sp} for CuI is 1.3×10^{-12}.)

Example 16.13 | **Determining relative solubility**

On the basis of the K_{sp} values contained in Table 16-3, rank the silver halides according to solubility from most soluble to least soluble.

Thinking Through the Problem K_{sp} is determined from solubility expressed as moles per liter. As solubility increases, the number of moles per liter increases; as solubility decreases, the number of moles per liter decreases.

> **Working Through the Problem** The most soluble silver halide will have the largest K_{sp} value, and the least soluble halide will have the smallest K_{sp} value.
>
> Answer
>
> | Most soluble | AgCl | $K_{sp} = 1.8 \times 10^{-10}$ | Largest magnitude |
> |---|---|---|---|
> | | AgBr | $K_{sp} = 5.4 \times 10^{-13}$ | |
> | Least soluble | AgI | $K_{sp} = 8.5 \times 10^{-17}$ | Smallest magnitude |

How are you doing? 16.10

Using the K_{sp} values in Table 16-3, rank $CaCO_3$, $BaCO_3$, and $MgCO_3$ according to solubility from the most to the least soluble.

PRACTICAL D   What causes soap to form scum in hard water?

Of the many dissolved metal ions in fresh water, calcium ions and magnesium ions are among the most common. The concentration of dissolved Ca in water is influenced by several factors, such as pH, but in fresh water the concentration is typically 1.0×10^{-3} M. Water containing Ca^{2+} and Mg^{2+} is called "hard water." In hard water, soaps form insoluble salts with Ca^{2+} and Mg^{2+} that appear as soap scum or curds. Most cleaning products now contain polyphosphates, which can complex Ca^{2+} and Mg^{2+} ions into soluble or suspended forms. The use of polyphosphates prevents precipitation of $CaCO_3$ in pipes and eliminates the effects of hard water.

Much of the dissolved calcium in fresh water comes from deposits of calcium carbonate, $CaCO_3$. Calcium carbonate is a sparingly soluble salt with a K_{sp} value of 4.47×10^{-9}. The K_{sp} expression, then, is:

$$K_{sp} = [Ca^{2+}][CO_3^{2-}] = 4.47 \times 10^{-9}$$

Calculate the solubility of $CaCO_3$.

However, the solubility equilibrium of $CaCO_3$ is not the only factor to consider as natural waters in contact with the atmosphere also contain dissolved CO_2. A more typical equilibrium reaction, then, is:

$$CaCO_3\,(s) + CO_2\,(aq) + H_2O\,(l) \rightleftharpoons$$
$$Ca^{2+}\,(aq) + 2\,HCO_3^-\,(aq) \qquad K = 4.24 \times 10^{-5}$$

Because HCO_3^- is the conjugate acid of CO_3^{2-}, the $\dfrac{[CO_3^{2-}]}{[HCO_3^-]}$ ratio is easily disturbed by the addition or removal of hydrogen ion. This makes the concentration of dissolved calcium strongly dependent on the pH of the water. The following equation relates HCO_3^- to CO_3^{2-}. Write the equilibrium expression for this reaction and solve it for the ratio $\dfrac{[CO_3^{2-}]}{[HCO_3^-]}$.

$$HCO_3^-\,(aq) + H_2O\,(l) \rightleftharpoons H_3O^+\,(aq) + CO_3^{2-}\,(aq)$$
$$K_a = 4.8 \times 10^{-11}$$

Do you expect $[HCO_3^-] > [CO_3^{2-}]$ at high or low values of pH?

(Richard Hutchings/PhotoEdit.)

PROBLEMS

37. Predict the likely products of the following double displacement reactions. Consult Table 5-3 and write the physical state, (aq) or (s) for each predicted product.

(a) $AgNO_3$ (aq) + Na_2CO_3 (aq) →
(b) CaF_2 (aq) + H_2SO_4 (aq) →
(c) KI (aq) + $Pb(NO_3)_2$ (aq) →
(d) $CuSO_4$ (aq) + Na_3PO_4 (aq) →
(e) H_2SO_4 (aq) + $PbCl_2$ (aq) →
(f) H_2SO_4 (aq) + $Ca_3(PO_4)_2$ (aq) →
(g) K_2SO_4 (aq) + $Ba(ClO_4)_2$ (aq) →

38. Predict the possible products if the following double displacement reactions occur.

(a) aluminum hydroxide (aq) + hydrogen chloride (aq) →
(b) sodium iodide (aq) + silver nitrate (aq) →
(c) water (l) + boron trifluoride (aq) →
(d) hydrogen fluoride (aq) + nickel (II) oxide (aq) →

39. Write net ionic equations for the formation of the following precipitates from their dissociated ions.

(a) $PbCl_2$ (c) $Mn(OH)_2$
(b) PbC_2O_4 (d) $Pb(IO_3)_2$

40. Write K_{sp} expressions for each of the following salts.

(a) CoS
(b) $SrCrO_4$
(c) MgC_2O_4
(d) $AgOH$

41. Calculate the solubility for each of the following salts. Values of K_{sp} (at 25 °C) are given in parentheses.

(a) HgS (2.0×10^{-53})
(b) $ZnCO_3$ (1.2×10^{-10})
(c) NiS (1.1×10^{-21})

42. Calculate the K_{sp} values for the following salts. The solubility of each salt is given in parentheses.

(a) FeS $(7.7 \times 10^{-10}$ mol $L^{-1})$
(b) SnS $(1.0 \times 10^{-13}$ mol $L^{-1})$
(c) PdS $(1.4 \times 10^{-29}$ mol $L^{-1})$

43. From the K_{sp} values you calculated in the last problem, rank these three sulfides from most soluble to least soluble.

44. What units does K_{sp} have?

Chapter 16 Summary and Problems

Review with **Web Practice**

VOCABULARY

chemical equilibrium	A dynamic system in which reactants continue to convert into products while the products also convert back into reactants. At any given time, all reactants and products are present in amounts determined by the rates of the forward and the reverse reactions.
metal complex	A chemical species formed when a metal ion is bound by one or more small molecules or ions. The species is usually bound by lone pairs on the molecules or ions interacting with the metal ion.
equilibrium system	Reactants and products in a chemical reaction that continue to interconvert.
dynamic relationship	Equilibrium systems have a dynamic relationship among the substances involved. This means that at the molecular/atomic level there is a constant conversion of products to reactants and reactants to products, even if there is no macroscopic change observable in the system.
equilibrium expression	An equation in which the ratio of product concentrations to reactant concentrations is equal to a constant value at a given temperature. The ratio gives the product of the concentrations of the products divided by the product of the concentrations of the reactants, each concentration raised to a power equal to its coefficient in the balanced chemical equation.

equilibrium constant	The equilibrium constant, K, is the numerical value of the equilibrium expression; K is unitless and valid at the particular temperature at which it is measured.
ionic solution equilibrium	An equilibrium system containing ions in solution; three important types of ionic solution equilibria are metal-ligand equilibria, acid-base equilibria, and solubility equilibria.
metal-ligand equilibrium	An equilibrium system composed of a metal ion, a ligand, and a complex formed of the two; a ligand is another ion such as Cl^- or CN^- or a neutral molecule such as H_2O or NH_3. The charge on the complex formed is determined by its component parts.
acid-base equilibrium	An equilibrium system containing an acid and its conjugate base.
solubility equilibrium	An equilibrium system containing a compound that is present as a solid and in solution.
gas-phase equilibrium	An equilibrium system that contains only gases.
quadratic formula	The quadratic formula is $x = \dfrac{-b \pm \sqrt{b^2 - 4ac}}{2a}$, where a, b, and c represent the coefficients in the quadratic equation; it is used to solve for the two roots (x) of a quadratic equation.
solubility	A measure of how well a salt dissolves in water. Salts that are soluble dissolve easily while those that are insoluble or sparingly soluble dissolve to a very limited extent.

SECTION GOALS

Why is an equilibrium system called "dynamic" when, to the observer, the reaction appears to have stopped?	A system at equilibrium is a dynamic system. At the microscopic level of ions and molecules, products continue to convert to reactants and reactants continue to convert to products even though we may observe no change in the color intensity, pressure, or other macroscopic properties.
How do we write an equilibrium expression?	An equilibrium expression is a ratio in which the product of the concentrations of the products is divided by the product of the concentrations of the reactants, and each concentration is raised to a power equal to its coefficient in the balanced equation. Pure solids and liquids are given a concentration value of 1 and usually do not appear in the equilibrium expression.
How does the equilibrium constant become unitless?	The concentration unit for aqueous solutions is molarity; if the concentration of a gas is expressed as partial pressure, it has units of atmospheres. The numerical value of the equilibrium expression is a constant at a given temperature. It is called the equilibrium constant for that system and is symbolized by K. Because the equilibrium constant must be unitless, each molarity is divided by a standard 1 M and each pressure is divided by a standard 1 atmosphere when numerical values are substituted into the equilibrium expression.
How is the quadratic formula solved for the roots of a quadratic equation?	The quadratic formula is $x = \dfrac{-b \pm \sqrt{b^2 - 4ac}}{2a}$, where a, b, and c represent the coefficients in the quadratic equation. The significance of the quadratic formula is that we can solve a quadratic equation using only the coefficients. Remember that one side of the equation *must be set equal to zero* to ensure that the coefficients are correct.
How do we solve an equilibrium expression for K when given equilibrium concentrations?	Equilibrium concentrations of all species must be used to solve for K. These are substituted into the equilibrium expression and K is evaluated. Values of K that are greater than 1 favor product formation; values of K less than 1 favor reactant formation.

How do we calculate the equilibrium concentrations of reactants and products when given only some initial concentrations and the value of K?	Equilibrium concentrations can be calculated by using an ICE table. With an ICE table, you can formulate an algebraic expression for the equilibrium concentrations in terms of initial concentrations as they are changed by slight increases $(+ x)$ or decreases $(- x)$ of reactant concentrations. Then, (1) substitute the resulting equilibrium concentrations in terms of x into the equilibrium expression, (2) simplify the equilibrium expression to obtain a quadratic equation equal to 0, (3) find the roots of the equation with the quadratic formula, and (4) determine which root yields a physically reasonable result.

How do we explain, in terms of ionization, the primary difference between an aqueous solution of a strong acid and that of a weak acid?

In an aqueous solution of a strong acid, we assume all the hydrogen present is ionized. Thus, for a strong acid, $[H_3O^+] = [HA]_{init}$. However, in an aqueous solution of a weak acid, only a fraction of the hydrogen available is ionized. For weak acids, $[H_3O^+] < [HA]_{init}$ and $[H_3O^+]$ must be calculated.

How do we calculate the pH of an aqueous solution of a weak acid?

For acid ionization equilibria, a quadratic equation is obtained from the equilibrium expression and solved using the quadratic formula. Then $x = [H_3O^+]$ and $pH = -\log [H_3O^+]$.

How do we find the fraction ionized and the percentage of ionization of an aqueous solution of weak acid?

The fraction of acid molecules (HA) ionized in water solution is $\dfrac{[A^-]}{[HA]}$. The percentage of ionization is obtained by multiplying the fractional amount by 100.

An approximation is sometimes used to avoid using the quadratic formula. When is this approximation valid?

It is reasonable to neglect the x in the denominator when the acid ionizes to a very small extent—about 5% or less of the total initial concentration of the acid before ionization. Under *those* circumstances we may use:

$$[H_3O^+] = x \approx \sqrt{K_a[HA]_{init}}$$

How do we calculate the value of K_{sp} when given the solubility of a salt?

To find the value of K_{sp}, write an equilibrium expression for the dissolution of the salt, substitute equilibrium concentrations (solubility values), and solve for K_{sp}.

How do we calculate the solubility of a salt when given its formula and K_{sp}?

For a salt that dissociates to yield one cation and *one* anion, we calculate the solubility of a sparingly soluble salt by taking the square root of the K_{sp}.

PROBLEMS

45. Write the K_{sp} expression for $CaCO_3$ (s). Begin with the ionization equation.

46. Calculate the solubility of $CaCO_3$ (s). The $K_{sp} = 4.47 \times 10^{-9}$.

47. At a pH < 4, aluminum ions form hydrated metal complexes with six water molecules. Write the equilibrium equation for this reaction.

48. At a pH > 10, aluminum ions form a metal-ligand complex with four hydroxide ions. Write the equilibrium equation for this reaction.

49. Write the equilibrium expression for

$$ZnS\ (s) + H_2O\ (l) \rightleftharpoons$$
$$HS^-\ (aq) + OH^-\ (aq) + Zn^{2+}\ (aq)$$

50. Calculate the solubility of ZnS in the previous problem when $K_{sp} = 2 \times 10^{-25}$.

51. When ZnS is dissolved in an acidic solution, the equilibrium constant for the reaction is $K = 2 \times 10^{-4}$. Is ZnS more or less soluble in acid solution than it is in plain water?

52. Fluoride ion forms insoluble salts with Ca^{2+}, Ba^{2+}, Sr^{2+}, and Pb^{2+}. Write the equilibrium expression for the formation of each of these insoluble salts.

53. Some aqueous metal nitrate solutions are acidic in nature. Calculate the hydronium ion concentration of $Ca(NO_3)_2$ when the pH $= 6.7$. Is this solution more or less acidic than a solution of $Pb(NO_3)_2$ that has pH $= 3.6$?

54. Water in equilibrium with $CaCO_3$ (s) and atmospheric CO_2 has a pH around 8.30. Calculate the $[H_3O^+]$ when the pH is 8.30.

55. Write the equilibrium expression for the reaction, HA (aq) + H_2O (l) $\rightleftharpoons$ A^- (aq) + H_3O^+ (aq). Then solve it for the $\dfrac{[H_3O^+]}{[HA]}$ ratio.

56. Sulfuric acid is a common strong acid, forming hydrogen sulfate ion when it ionizes in water. Hydrogen sulfate is an acid. Write the equilibrium equation for the ionization of hydrogen sulfate in water.

57. Concept question: Explain why it is wrong to think that, because sulfuric acid is a strong acid, hydrogen sulfate must be a strong acid also.

58. Phosphate ion is a weak base. Write an equilibrium equation for the reaction of phosphate and water.

59. Concept question: We have a solution containing 0.50 mole of a strong acid in one liter. Another solution contains 1.00 mole of a weak acid in one liter. Which solution contains more moles of acid? Which solution contains more hydronium ion?

Organic Chemistry and Biochemistry

Many medicines are based on organic chemistry and biochemistry. (Andrew Brookes/Corbis.)

PRACTICAL CHEMISTRY How Do Molecules Work?

The topic of organic chemistry and biochemistry invites us to look at the way arrangements *within* molecules are important for their properties and their identity. We will see that the way molecules behave depends on the order of the atoms in a backbone and the arrangement of atoms relative to one another in space. Ultimately, we will discuss how the presence or absence of different groups influences the ability of a bacteria to defend itself against antibiotics.

This chapter will focus on the following questions:

PRACTICAL A What are the bonding patterns that we associate with the odor of certain molecules?

PRACTICAL B How does the presence of a double bond matter in fats and oils?

PRACTICAL C What factors in molecular structure affect solubility?

PRACTICAL D How do amino acids affect antibiotic resistance?

17.1 Common Bonding Patterns of Second Period Elements

SECTION GOALS

✓ Be able to apply the predicted bonding patterns for elements in groups IV–VII.

✓ Know how to use bonding patterns as an aid to drawing Lewis structures.

Using Patterns in Bonding for First and Second Period Elements

Using chemistry in a practical way requires that you know what things are important in determining the properties of different chemical substances. In previous chapters, we have discussed how chemical substances can differ in whether they are elements or compounds, whether they have an extended or molecular structure, and also by the formula that we find for the substance. In this chapter, we will take a look at the significance of how the atoms in some important molecules, those containing carbon, are connected. **Organic chemistry** is the area of chemistry devoted to the study of molecules containing carbon. This recognizes the importance of carbon-containing molecules in nature, medicine, and industrial chemistry. **Biochemistry** is a more focused area, dealing with the way molecules interact in living systems.

For many of you, organic chemistry will constitute one, two, or more courses later on in your studies. But there are some simple principles and concepts to be presented here that will demonstrate that molecular structure is important, and that this can be the basis for classifying molecules.

The chemistry of organic compounds is dominated by hydrogen and other elements from the second period of the periodic table. In Chapter 2, we learned how to build Lewis structures from scratch, accounting for every valence electron and thinking carefully about each individual atom. In some ways this is a good thing, because it prompts us to consider each atom in turn. But it is also time consuming. We can save a lot of effort if we arrange a molecule according to *regular patterns* that arise time and time again. This, in turn, helps bring us back to the original goal of Lewis: to explain why carbon forms a particular number of bonds in many of its compounds.

We start with the hydrogen atom, which has one valence electron when neutral, and therefore needs to form one bond to make a two-electron valence configuration, equal to He. We expect hydrogen to form one bond. Next, let's examine the nonmetals in the second period, where an atom will need to acquire at least eight electrons around its core. Carbon, for example, is in group IV, so a neutral C atom has

Figure 17-1 ▶ Common
bonding schemes for C and O.

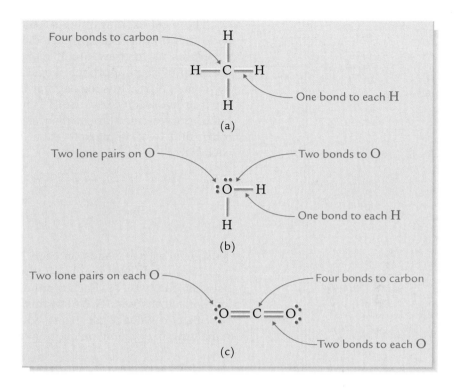

four electrons. It will need four more electrons to make eight, which can be done only by forming four bonds. We see this with methane, CH_4, in which carbon makes four single bonds, one to each H. This gives the hydrogen atoms their single bond (Figure 17-1a).

What about an element like oxygen? Oxygen is in group VI, so it has six valence electrons; it will tend to form two bonds, keeping four electrons in lone pairs. We see this bonding pattern for oxygen in water (Figure 17-1b), but oxygen can bond with elements other than hydrogen. Carbon dioxide, for example (Figure 17-1c), will have two double bonds, one between each of the oxygen atoms and the carbon atom. This gives C its preferred four bonds while each O atom has the preferred two bonds and two lone pairs.

How many bonds do you expect a neutral N atom to form? Table 17-1 gives expected bonding patterns for second period nonmetals.

Table 17-1 Bonding Patterns for Neutral Second Period Nonmetals

Element	Group Number/ Valence Electrons	Number of Bonds	Electrons in Lone Pairs
C	IV / 4	4	0
N	V / 5	3	2
O	VI / 6	2	4
F	VII / 7	1	6

The most exciting thing about this approach, and one that leads us to the kind of thinking that most chemists use in practice, is that it permits us to draw much more complicated molecules, whereas the step-by-step approach bogs us down. For example, consider hydrazine, N_2H_4. It is logical that the two N atoms are surrounded by the four hydrogens, because H is not expected to form more than one bond. So we can start with a skeleton with two N's next to each other, and two H's next to each N. We then form a single bond to each H:

$$\begin{array}{cc} H & H \\ \diagdown & \diagup \\ N \quad N \\ \diagup & \diagdown \\ H & H \end{array}$$

This gives us two bonds on each N. According to Table 17-1, we expect N to form three bonds, so it makes sense to put one single bond between the two N's. Each atom of N, then, shares one of its five valence electrons with two H atoms and one N atom. This leaves each N with two valence electrons not shared in bonds. Finally, put the remaining two valence electrons as a lone pair on each N.

$$\begin{array}{cc} H & H \\ \diagdown & \diagup \\ \ddot{N}\!-\!\ddot{N} \\ \diagup & \diagdown \\ H & H \end{array}$$

We now must carry out a very important check of the structure: have we got the right number of electrons on the molecule? We have four H's (one valence electron each) and two N's (five each), for a total of fourteen valence electrons. When we count the number of valence electrons in our final structure, we find five bonding pairs and two lone pairs, or fourteen electrons. The structure is correct.

Example 17.1   **Sketching structures using bonding patterns**

Sketch the Lewis structure of methanol, CH_3OH.

Thinking Through the Problem Methanol is not a simple diatomic molecule. ◀ (17.1) The key idea is that we use our knowledge of bonding patterns to help us set up the Lewis structure. We need a reasonable skeleton. The hydrogen atoms should be on the outside, because they can only form one bond. This leaves C and O next to each other in the "center" of the molecule. We expect a bond between C and O. The oxygen is expected to form two bonds, so it should also bond to one of the H atoms; the carbon is expected to form four bonds, so it should get three H's.

$$\begin{array}{ccc} & H & \\ H & C & O \quad H \\ & H & \end{array}$$

Working Through the Problem We add bonds to the structure next, four to the carbon and two to the oxygen. We expect the oxygen to have two lone pairs also. Oxygen has a total of six valence electrons, two of which are shared in the bonds with C and H. The remaining four valence electrons make up the two lone pairs on oxygen. Our final structure contains five bonding pairs and two lone pairs, for a total of fourteen valence electrons. This is the number we expect based on four H's (four valence electrons), one C (four valence electrons) and one O (six valence electrons).

Answer

$$
\begin{array}{c}
\text{H} \\
| \\
\text{H}-\text{C}-\overset{..}{\text{O}}-\text{H} \\
| \\
\text{H}
\end{array}
$$

Example 17.2   Sketching structures using bonding patterns

Sketch the Lewis structure of formaldehyde, CH_2O, using common bonding patterns.

Thinking Through the Problem We will use our knowledge of bonding patterns to help us set up the structure. As before, it is worth thinking for a minute before laying out the skeleton. With only two hydrogens, we do *not* expect the oxygen to be attached to a hydrogen. Why? Because to do so would mean we would have only one C—H bond, forcing the C to form three bonds to O. This would make the O have four bonds overall, which is not what we expect for O. So we start with the oxygen attached only to C, and with each H attached to C as well. We wind up with a C=O double bond and two lone pairs on O.

Answer

$$
\begin{array}{c}
\text{H} \\
\diagdown \\
\quad\text{C}=\overset{..}{\text{O}}\!: \\
\diagup \\
\text{H}
\end{array}
$$

How are you doing? 17.1

Sketch the Lewis structures of hydroxylamine (NH_2OH) and of hydrogen cyanide (HCN).

This approach has worked well for neutral molecules. How does it apply to polyatomic ions? This will require some additional thinking. If a polyatomic ion has a negative charge, we would like to put its charge on an atom that stabilizes negative charge well. The atoms that do this well are at the top right of the periodic table, especially oxygen and, to a lesser extent, sulfur and nitrogen.

Suppose we are working with the polyatomic ion hydroxide, OH^-. Here we expect the oxygen atom to have a negative charge and the hydrogen atom to be neutral. The H atom should still form one single bond, but the oxygen atom now has seven valence electrons—just like the neutral F atom. This means that the oxygen in the hydroxide ion is expected to form one bond and to have six electrons in three lone pairs. We check the final structure for hydroxide and find that it has eight valence electrons—exactly right for 1 (from H) + 7 (from O with a negative charge).

Example 17.3 | **Sketching polyatomic ion structures using bonding patterns**

> Sketch the Lewis structure of the amide ion, NH_2^-.
>
> Thinking Through the Problem We want to sketch the Lewis structure of a polyatomic ion. We will use bonding patterns to help us with this structure, beginning with the N in the middle of two H atoms. The "extra electron" should be assigned to the atom that has a position farthest to the right on the periodic table. Here the right-most element is N, so we expect it to have the negative charge. This would give the N atom six valence electrons, and it will need to form two single bonds, just as neutral O does.
>
> Answer
>
> $$H-\ddot{N}-H^-$$
>
> N in this case has **2** bonds and **2** lone pairs because it has a negative charge

Our check of the structure in the above example shows that we have eight valence electrons, which we expect based on one negatively charged N atom and two neutral H atoms.

We use similar strategies when a *positive* charge is present. A positive charge probably resides on the central atom in the structure, because the central atom is the one most likely to make more bonds, which requires fewer valence electrons on that atom.

Example 17.4 | **Sketching polyatomic ion patterns**

> Sketch the Lewis structure of the ammonium ion, NH_4^+, using preferred bonding patterns.
>
> Thinking Through the Problem We will lay out this molecule with N in the middle, as it can form more bonds than H can. We will assign the positive charge to N. This makes the N have four electrons—*so the N is like C, and we expect four bonds to N and no lone pairs.*
>
> Answer We expect a structure with four single bonds between N and the H's.

Figure 17-2 ▲ Molecules of oxygen and nitrogen illustrating double and triple bonds.

Figure 17-3 ▶ Molecules of ethane, ethylene, and acetylene illustrating single, double, and triple bonds.

How are you doing? 17.2

Sketch the Lewis structures for (a) methylamine, CH_3NH_2, and (b) methoxide ion, CH_3O^-.

The three-dimensional representations used in this book and in the WebLab Connections all include information on the number of bonds between two atoms. For example, in Figure 17-2 are shown the drawings for oxygen and nitrogen molecules. The molecules ethane (one C—C single bond), ethylene (one C=C double bond) and acetylene (one C≡C triple bond) are shown in Figure 17-3.

We can now combine bonding trends with resonance, a topic we also discussed in Chapter 2, to explore some other molecules. For example, what do we do when we construct the Lewis structure of formate ion, $HCOO^-$? We note that we have two oxygen atoms, so one of them will be negatively charged (with seven valence electrons and therefore one bond to make). The other will be neutral, so there will be a double bond. Then there is the carbon, which will make four bonds, one to H and three to the oxygen atoms. That matches very nicely with the requirements for the oxygen atoms, one bond to the negative oxygen (seven valence electrons) and two bonds to the neutral oxygen (six valence electrons). The Lewis structure for $HCOO^-$ is shown in Figure 17-4.

Figure 17-4 ▶ Analysis of bonding in formate ion.

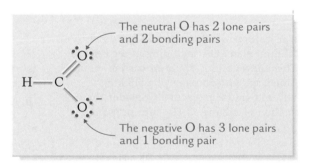

Of course, the construction of formate ion requires us to make a choice between the two oxygen atoms, which is a sign that there will be resonance between them. The other resonance structure will have the roles of the O atoms switched.

Is this method going to work in all cases? Alas, chemical substances can do some very unusual things, and we must understand why there are exceptions that fit in a wider picture. Consider the case of ozone that we discussed earlier. If we look again at this molecule, we see that one of the oxygen atoms has a neutral configuration, associated with two lone pairs and two bonds. The central oxygen, with one lone pair and three bonds, must have a positive charge, while the other outer oxygen will have a negative charge.

PROBLEMS

1. Draw the Lewis structure for the following molecules. The carbons are in the center.

(a) C_2H_6 (b) C_2H_4 (c) C_2H_2

2. In Chapter 2, you were asked to draw the Lewis dot structures for the following carbon-containing substances and ions: CH_3^-, HCN, $COCl_2$, CO, and CBr_4. Do these fit the expected pattern? Which are the exceptions?

3. Draw the structures you expect for the following molecules.

(a) $N(CH_3)_3$ (N in center, surrounded by three carbons)
(b) $CH_3CH_2NH_2$ (chain of C—C—N)

4. What is the structure you expect for the following compound?

$$H_3C—NH—CH_3$$

5. For the following structures, add the number of hydrogen atoms you expect to get in a neutral molecule.

$$C—C—O \qquad C—O—C$$
$$C—C{=}O \qquad C—C—N$$

6. An atom of carbon will form strong bonds with many different elements, including other carbon atoms. This allows a long chain of carbon atoms to be formed, at least on paper, by sticking additional units into a molecule. Consider the transformation of ethane, C_2H_6, into propane, C_3H_8:

Describe, in words, what has happened during this transformation.

Now predict what happens when a "CH_2" group is added to other molecules.

In some cases, the insertion can occur in more than one place. Predict two different products for this reaction:

7. The presence of preferred bonding patterns, especially for the second row elements, does not preclude having to make some choices in drawing a structure. Consider compounds with the formula C_3H_6. One structure has a simple chain of carbon atoms, whereas another has three carbon atoms in a ring.

$$C—C—C \qquad\qquad C{\overset{\displaystyle\diagdown\ \diagup}{}}C$$
$$C$$

Use your understanding of the preferred bonding pattern of C and H to complete these structures.

8. Web Question: For the illustrations in WebLab Connection Problem 17-1, determine the formula of the compounds shown. Redraw the molecules as Lewis structures. C, H, and O are present. Use the rules for common bonding patterns to determine the location of any lone pairs.

9. Web Question: For the illustrations in WebLab Connection Problem 17-2, determine the formula of the molecules shown. C, H, and N are present. Redraw the molecules as Lewis structures. Use the rules for common bonding patterns to determine the location of any lone pairs.

PRACTICAL A — What are the bonding patterns that we associate with the odor of certain molecules?

You have now reached a point where you can draw certain structures based on common bonding patterns. This means that you can also "fill in" the structures of certain molecules. In the process, you can discover what groups are associated with certain odors.

One group of consistent odors are associated with oxygen groups, as in the following examples. Each group has a special name, which you do not need to remember.

Name	Odor	Example in Common Experience	Molecular Example
Alcohols	Slightly sweet	Rubbing alcohol	$H_3C-\overset{\displaystyle OH}{\underset{\displaystyle H}{C}}-CH_3$
Ether	Pleasant and sweet	Methyl-*tert*-butyl ether, or MTBE (gasoline additive)	$H_3C-O-\overset{\displaystyle CH_3}{\underset{\displaystyle CH_3}{C}}-CH_3$
Carbonyl groups	Pungent; sometimes pleasant	Acetone (nail polish remover)	$H_3C-\overset{\displaystyle O}{\overset{\|}{C}}-CH_3$
Acid groups	Sour odor	Vinegar (solution of acetic acid)	$H_3C-\overset{\displaystyle O}{\overset{\|}{C}}-OH$
Esters	Fruit odor	Isoamyl acetate (found in banana)	$H_3C-\overset{\displaystyle O}{\overset{\|}{C}}-O$ … CH_3 $CH-CH_2-CH_2$ CH_3

For each of the structures, indicate the lone pairs needed to "fill" the normal bonding patterns of each atom. Characterize the environment around the oxygens in each case. Is there a difference in the number of bonds? Of lone pairs? What else may account for the differences?

17.2 An Introduction to Organic Chemical Structure

SECTION GOALS

✔ Determine what kinds of connections are important in an organic chemical structure.

✔ Determine some important structural differences between isomers of organic chemical structures.

Isomers of Hydrocarbons

One of the most interesting things that occurs in organic chemistry is that the consistent bonding patterns of C lead to a varied group of molecules. The reason for this is simple. We can "build" organic molecules in many different patterns based on different connections among the atoms. Let's start with a simple example: molecules containing just three carbons and hydrogen.

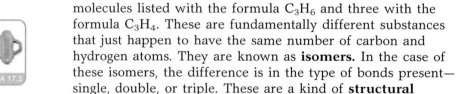

Figure 17-5 ▲ Hydrocarbon molecules with 3 carbons.

Examine each of the molecules in Figure 17-5. These diagrams do not simply list what is in the molecule; they also show how the atoms are *bonded* to one another. Each line represents a bond. By examining these molecules we can bring out some important questions in chemical structure.

1. *Formula and connections.* First, note that there are two molecules listed with the formula C_3H_6 and three with the formula C_3H_4. These are fundamentally different substances that just happen to have the same number of carbon and hydrogen atoms. They are known as **isomers.** In the case of these isomers, the difference is in the type of bonds present—single, double, or triple. These are a kind of **structural isomer.**

2. *Normal number of bonds.* There are single bonds (a line between two atoms), double bonds (a double line), and triple bonds (a triple line). Notice that there are four bonds to every carbon atom and one to each hydrogen. These are very important trends that are almost always observed for these atoms.

✔Meeting the Goals

Each of these points discusses a different issue in answering the question of what kinds of connections are important.

3. *Saturated molecules.* When a molecule contains only single bonds to carbon, then we say that the molecule is saturated. This means that the molecule contains as much hydrogen as it can. A molecule that contains one or more carbon atoms in a double or triple bond is known as an unsaturated molecule.

Example 17.5 **Finding isomers of molecules**

Examine each of these pairs of molecules. Indicate if they are isomers.

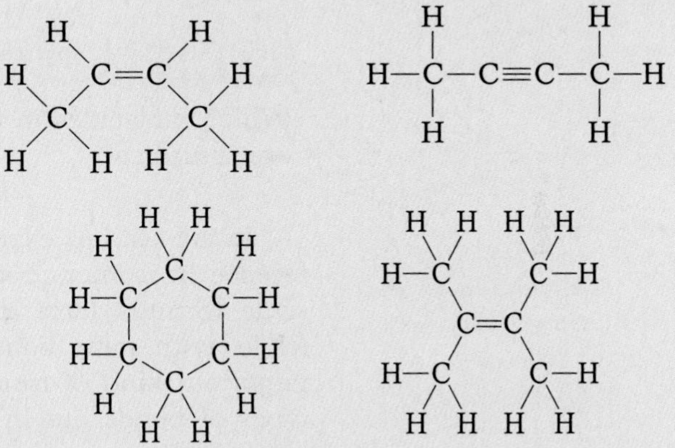

Thinking Through the Problem We need to understand if these molecules fit the definition of isomers. ➡ (17.2) The key idea is that isomers have the same formula but different connections of atoms. We can look at the formula and then the way the bonds are connected.

Answer The first molecule in the first pair has a formula of C_4H_8; the second, C_4H_6. They are not isomers. In the second pair, both molecules have the same formula, C_6H_{12}. One is a ring with only C—C single bonds; the other does not have a ring, and has a C=C double bond. They are isomers.

How are you doing? 17.3

Which of these molecules are isomers?

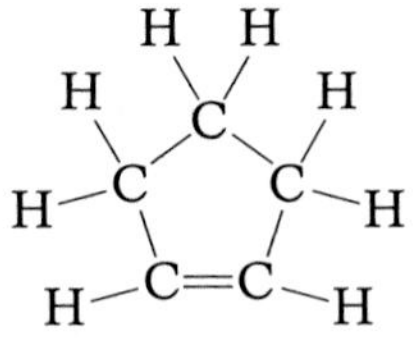

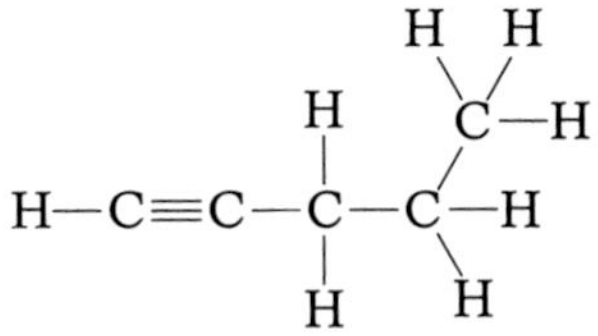

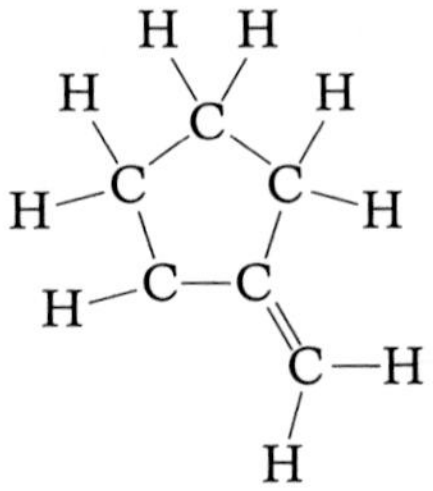

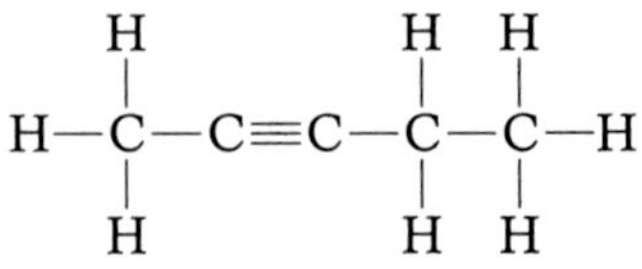

Example 17.6   **Finding saturation in a molecule**

> Which of the molecules in Example 17.5 are saturated?
>
> Thinking Through the Problem We need to understand if these molecules fit the definition of saturation. Saturated molecules have no double or triple bonds between carbon and another atom. We survey the molecules for those that contain double or triple bonds.
>
> Answer The only saturated molecule is the first molecule in the second pair.

How are you doing?   **17.4**

Which molecules in the previous "How are you doing?" exercise are saturated?

In the last few examples and exercises, we are able to detect a difference in isomers based on the fact that two substances with the same formula have different types of bonds—single, double, and triple, with some bonds arranged to form a ring. There is another important kind of isomer, created by having the same number and types of bonds but in different places in a molecule. Consider, for example, these two isomers of C_4H_8 in Figure 17-6.

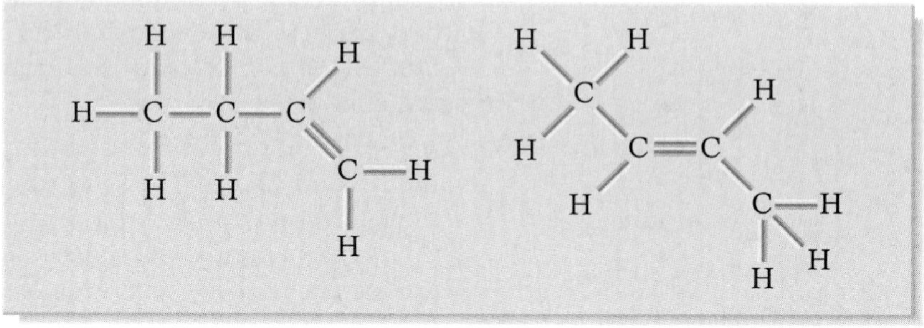

Figure 17-6 ▶ Two structural isomers of C_4H_8 based on a four-carbon chain.

Both have eight C—H bonds, two C—C bonds, and one C=C bond. In the molecule on the left, the C=C bond involves a C on the end of the molecule, whereas on the right we find the C=C bond in the middle. These molecules are another kind of structural isomer—a **geometric isomer,** based on the positions of the bonds, not on the number and types of the bonds.

Example 17.7   **Sketching isomers**

> The molecule C_5H_{10} has three structural isomers when the five carbons are in a chain with one C=C double bond. Draw them.
>
> Thinking Through the Problem We are looking for three structures, all with different positions of the bonds, and we know that all three structures will have five C's in a row. We can connect them with single bonds, then see if there are different positions for the

C=C double bond. After laying out the C skeleton, we can fill in each C with the required number of C—H bonds. We have ten H's to use.

Working Through the Problem

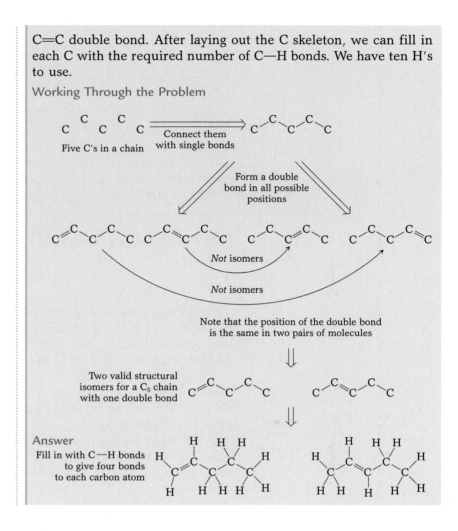

Five C's in a chain

Connect them with single bonds

Form a double bond in all possible positions

Not isomers

Not isomers

Note that the position of the double bond is the same in two pairs of molecules

Two valid structural isomers for a C_5 chain with one double bond

Answer
Fill in with C—H bonds to give four bonds to each carbon atom

How are you doing?   **17.5**

There is one structural isomer for the substance with the formula C_5H_8, in which all the C atoms are in a ring and there is one C=C bond. Determine the molecular structure of this substance. (Hint: You will have to show the different *possible* arrangements of a double bond.)

Heteroatoms in Organic Molecules

The definition of an organic molecule is any molecule that contains a carbon atom. Other atoms can be present, too. We have already discussed the presence of H in hydrocarbons. Any atom besides carbon and hydrogen in an organic molecule is called a "heteroatom," to show that it is not the same as C and H. In this section, we will discuss the two most important heteroatoms: oxygen and nitrogen. Others that are very important include halogens (F, Cl, Br, and I), sulfur, and phosphorus.

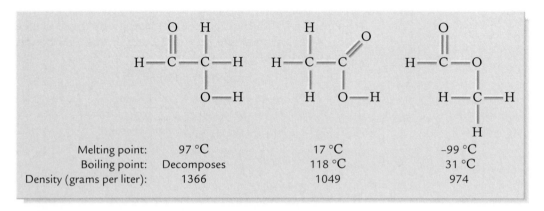

Melting point:	97 °C	17 °C	-99 °C
Boiling point:	Decomposes	118 °C	31 °C
Density (grams per liter):	1366	1049	974

Figure 17-7 ▲ Three isomers of $C_2H_4O_2$.

Figure 17-7 shows drawings of three different molecules that contain O, C, and H. They are all isomers of one another (they all have the formula $C_2H_4O_2$), but they have very different properties. The middle structure is that for acetic acid, the basis of common vinegar.

We can see that there are three ways that O can be present in these molecules. First, it can occur in a double bond to C. It can also occur with one bond to C and one bond to H. Finally, it can form two single bonds to C. This is true of *neutral* molecules, and it is the basis for how we expect O to be found in organic molecules, as shown in these neutral molecules. Later on we also consider what structures are available when charge is present in a molecule.

Example 17.8 Isomers and heteroatoms

Are these molecules isomers? If so, explain the differences.

Thinking Through the Problem We apply the same definition of isomers to these molecules that we used with hydrocarbons. The formula and the connections among the atoms are all we need. Once we determine that the molecules are isomers, we will document the difference in the connections.

Answer They both have the same formula, C_3H_6O. They are isomers, because the first molecule has a C=O double bond, and the second has two C—O single bonds.

Example 17.9   **Filling in heteroatom structures**

Which of the atoms in this structure is lacking one or more bonds, compared with the normal configuration for that atom? Draw a line to indicate where bonds can be formed to correct this.

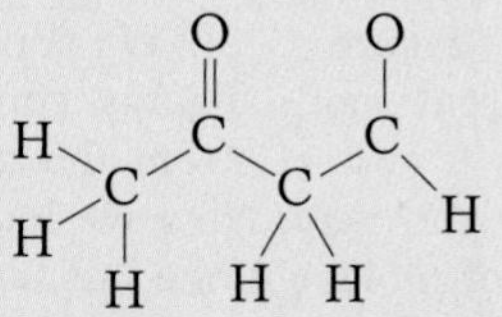

Thinking Through the Problem We will need to get a "correct" structure for this molecule. We start with what we expect for each atom: four bonds to C, one to H, and two to O. The atoms lacking enough bonds can be connected to give the correct structure.

Answer The rightmost oxygen and the carbon attached to it are both missing an essential bond. Forming a C=O double bond between them will fix the problem.

How are you doing? **17.6**

Determine which atoms in this molecule lack essential bonds. Draw in the bonds that will fix this situation.

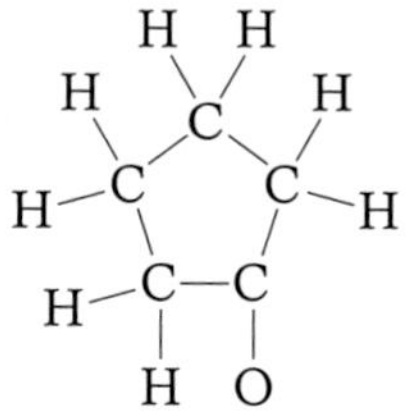

When we turn to nitrogen, we get a similar variety of bonding possibilities. In this case, we get three bonds formed to the N atoms when it has no charge, as suggested in Figure 17-8.

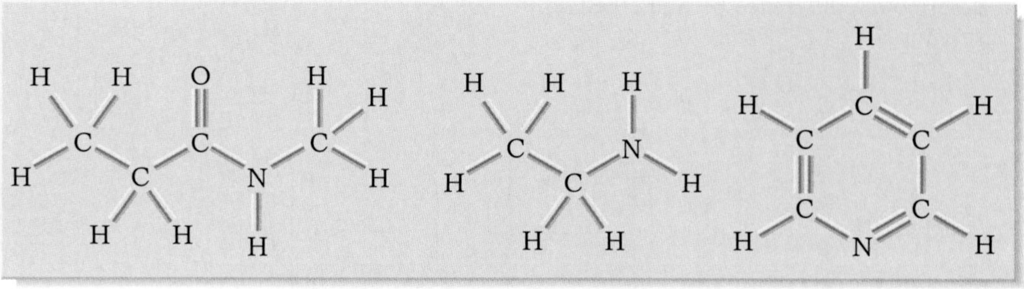

Figure 17-8 ▲ Some nitrogen-containing organic molecules.

How does the presence of a double bond matter in fats and oils?

We have discussed the organic chemistry of simple molecules. These are the building blocks of life, and some of them are literally that because they are the basis of our nutrition (Figure 17-9). We considered some aspects of different foods' energy content in Practicals in Chapters 9 and 11, from the point of view of stoichiometry and heat. Now we look a little closer at one type of molecule, the fats, and consider how their internal structure is important.

The molecules we call fats are rarely consumed as such, but a common pattern we do consume is the presence of three long chains of a small three-carbon backbone. This backbone is formed from glycerol, which has the formula $C_3H_8O_3$. Glycerol has three OH groups, one on each of the carbon atoms. Using your knowledge of common bonding patterns, predict the structure of glycerol.

Glycerol, as noted, forms larger molecules where the three chains are derived from one of the fatty acids. Fatty acids generally contain 14 to 22 carbons in a single chain that ends in an organic acid group, COOH. The rest of the fatty acid chain is a hydrocarbon, with only hydrogen and carbon present. When three fatty acids combine with glycerol, a molecule known as a triglyceride is formed. Figure 17-10 shows the triglyceride of the fatty acid palmitic acid. You should draw the $C_{15}H_{31}$ chains using your knowledge of common bonding patterns.

The different materials we call oils and fats are combinations of different triglycerides. The structure

Figure 17-9 ▲ Many foods are now labeled for the content of different types of fat and oils. The difference is often a simple matter of organic structure. (Bob Daemmrich/Stock Boston.)

(Arthur R. Hill/ Visuals Unlimited.)

PROBLEMS

10. Determine whether the following pairs of structures are isomers. If they are isomers, describe the difference between the two structures.

(a)

(b)

(c)

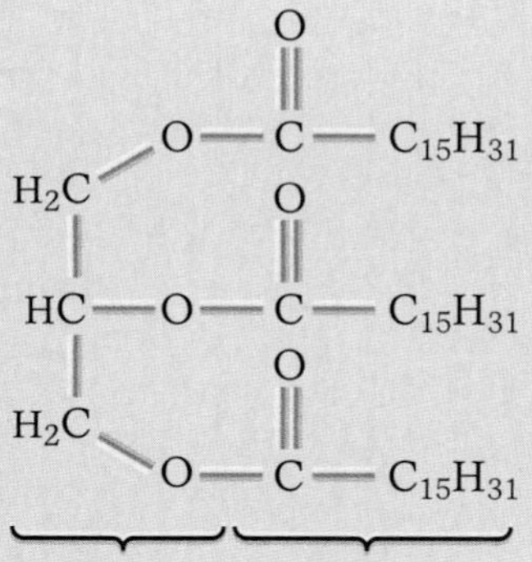

Figure 17-10 ▲ Drawing of the triglyceride of palmitic acid.

of the hydrocarbon chains have important effects on the properties of the triglycerides. For example, when the carbon-carbon bonds are all single bonds, then the triglyceride is called *saturated*, for there are no C=C double bonds that can be converted into C—C bonds by adding hydrogen. Saturated fats are usually solids at room temperature and are dominant in many animal-derived fats. The triglyceride of palmitic acid, for example, has a melting point of 66 °C. The solid nature of saturated fats is the reason why butter and cheese, which are concentrated fats derived from milk, are solids.

If we have one or more C=C bonds in the fatty acid side chains, then we have a site of unsaturation. If there is one C=C bond in each side chain, then the fat is monounsaturated. If there are two or more such bonds, it is polyunsaturated. Many plant-derived triglycerides are polyunsaturated. For example, oleic acid is a common monounsaturated fatty acid with 18 carbons in its chain. The C=C bond in oleic acid is located halfway down the chain, between the ninth and the tenth carbon from the COOH group. The triglyceride of oleic acid, which you should draw with the proper position of the C=C bond, has a melting point of only −4 °C.

There are relatively few sources of monounsaturated fats in nature, but there are many polyunsaturated fats. Corn oil, for example, is a mixture of triglycerides dominated by the fatty acid linoleic acid, which has eighteen carbons and C=C bonds in the position nine and thirteen carbons from the COOH group.

The differences in origin and physical properties of fats and oils have important implications in the food industry. Many humans are accustomed to the taste and consistency of saturated fats. These are more expensive than vegetable oils; therefore, the processing of vegetable oils to make solid foods that tasted more like saturated fats was done early in the history of industrial food processing. This process, where hydrogens are added to polyunsaturated fats, results in fewer C=C bonds and a greater tendency to form a solid. Such materials are well known to many of us as margarine.

11. Which of the structures in the preceding problem are saturated?

12. Examine the following three structures. They each lack one bond that will give all the atoms the proper number of connections. Indicate what that connection is.

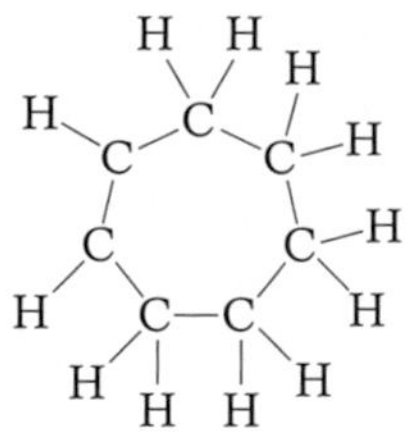

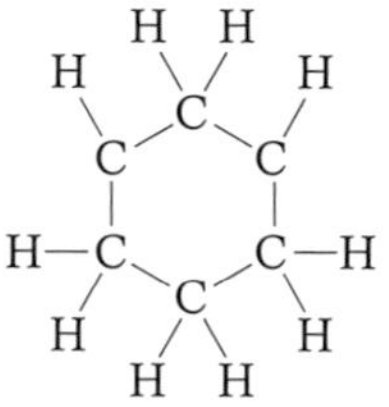

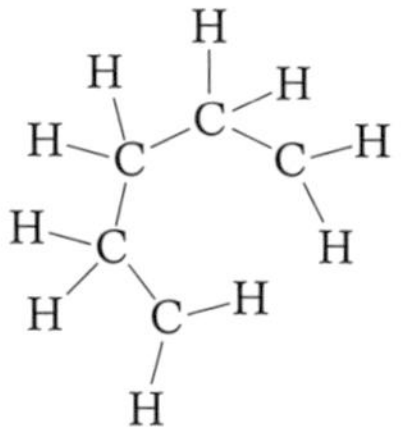

13. There are three pairs of isomers among the following six oxygen-containing substances. Indicate which are pairs with one another.

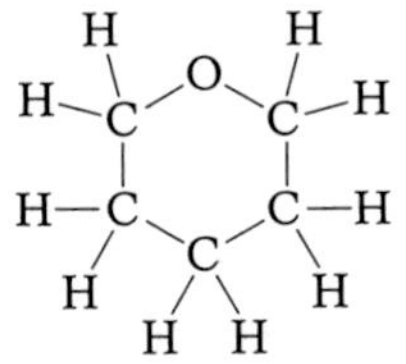

(a)

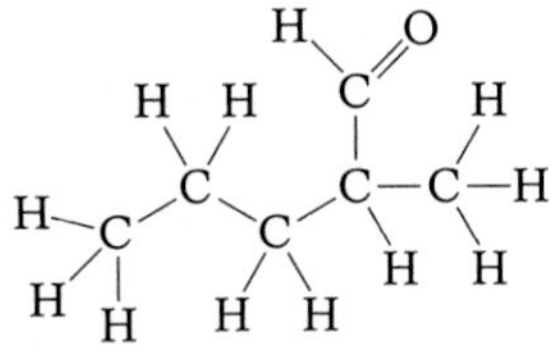

(b)

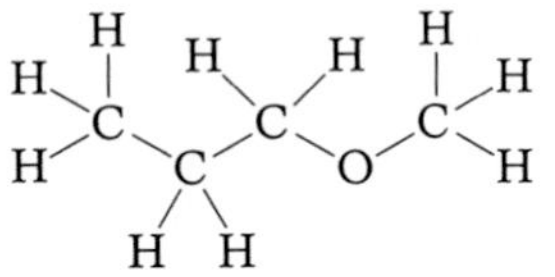

(c)

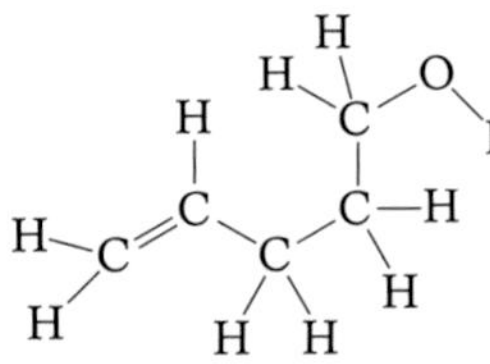

(d)

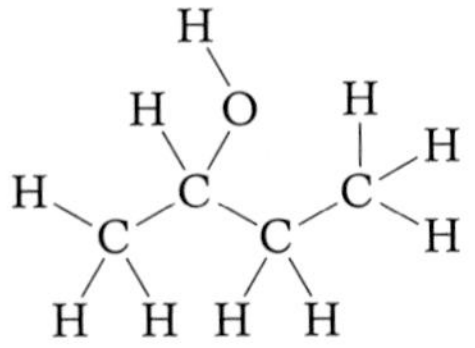

(e)

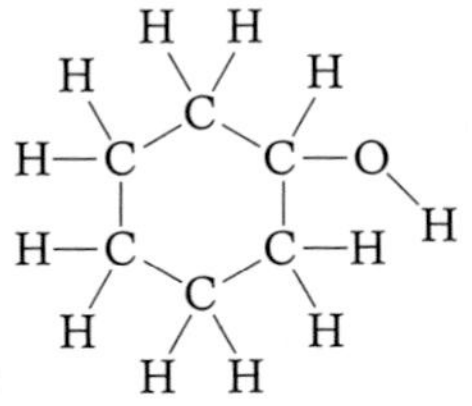

(f)

17.3 Similarity and Difference: Organic Functional Groups

SECTION GOALS

✔ Be able to see the key factors in making a molecule rigid.

✔ Examine a molecule and indicate important functional groups.

Rigid and Flexible Groups in Hydrocarbons

In the last section, we introduced several different molecules and emphasized that certain bonding patterns could be used to indicate whether two molecules, though they have the same formula, may actually be different substances. The arrangement of atoms implies a lot more than the connection. Unsaturated molecules have properties that imply that they are rigid, and this is very important in determining molecular properties.

Consider, first, the two molecules with the formula C_4H_8, shown in Figure 17-11. They all have the same formula, and both have a C—C bond flanking a central C=C bond in a chain of four carbon atoms. But they represent different substances with different properties. This suggests that the arrangement of atoms across a double bond is important in distinguishing isomers. When we compare the molecules, we also note that they differ by whether the two end C's are on the same or opposite sides of the central double bond. This suggests that the C=C bond is rigid, and that we cannot bring the end C's around to make the second molecule the same as the third.

You may note that the first molecule is also drawn in a way that suggests that the two end carbons are on the same side of the central C—C bond, just like in the second molecule. This is correct, but

✔ **Meeting the Goals**

Note that not all C=C bonds result in geometric isomers. Geometric isomers require that the groups on different sides of the C=C bond be different.

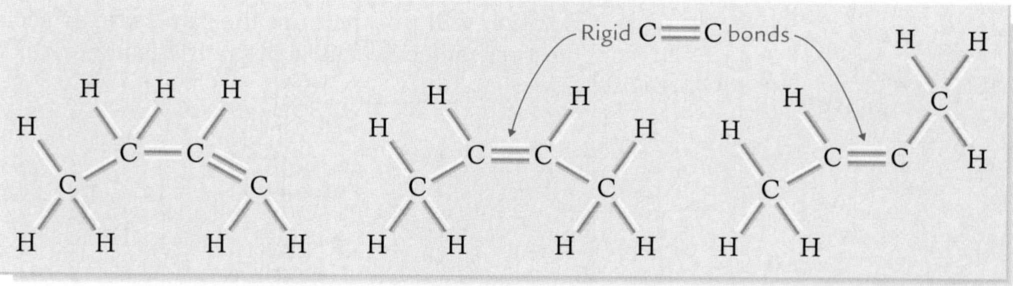

Figure 17-11 ▲ Some C_4H_8 isomers.

experimentally it is found that if the central C—C bond is rotated, as in Figure 17-12, then we get a molecule with the same properties as when we started. This is an important factor for the C—C single bond, or any single bond in a chain: it is relatively easy to rotate around that bond, so geometric isomers do not form with C—C bonds in a chain.

Figure 17-12 ▶ Rotation about a C—C single bond is easy.

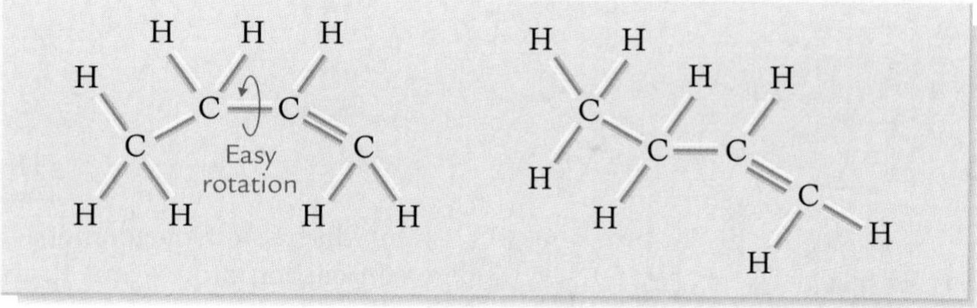

Example 17.10 **Geometric isomers**

Determine whether the following molecules have a geometric isomer. If they do, then draw it.

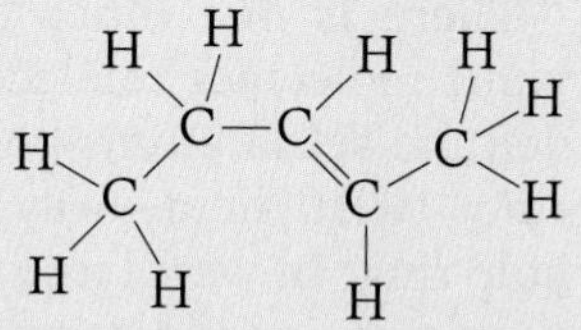

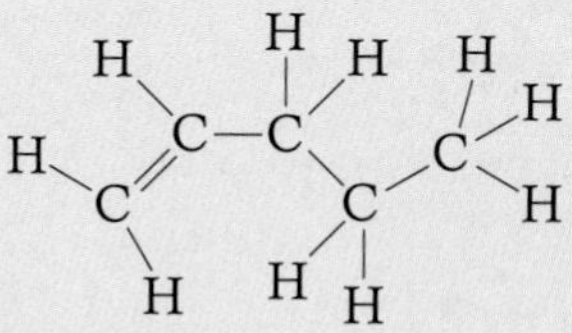

Thinking Through the Problem We need to determine whether either molecule has a geometric isomer. Molecules with a geometric isomer will have a C=C bond where there are different groups on different sides of the C=C bond.

Answer For the second molecule, the CH_2 group on the end, when rotated, gives the identical layout of atoms. For the first molecule, however, the two C's are on opposite sides of the double bond.

Rotation will put them on the same side as each other. Therefore, the first molecule has a geometric isomer. It is drawn below.

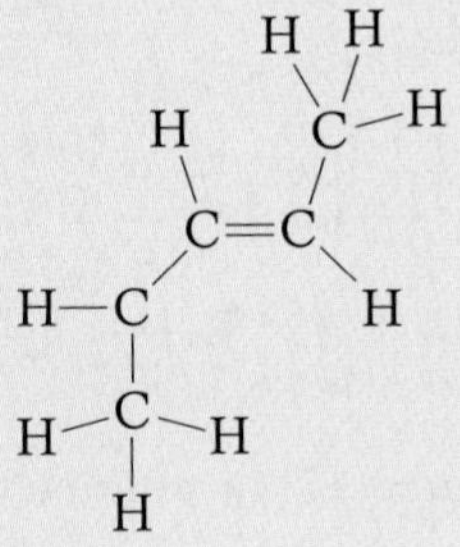

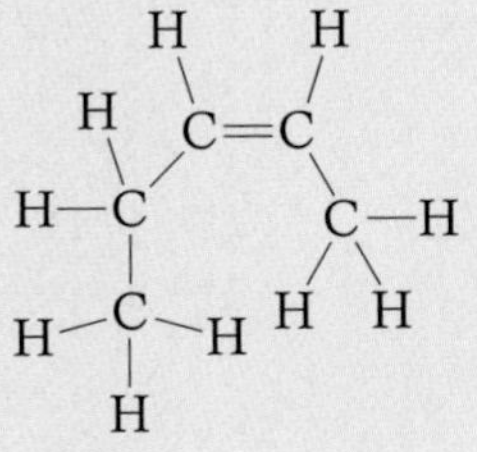

How are you doing? 17.7

If we have two carbons that are the same on the end of a double bond, then we do *not* get geometrical isomers. Show that this is the case for the molecule drawn here.

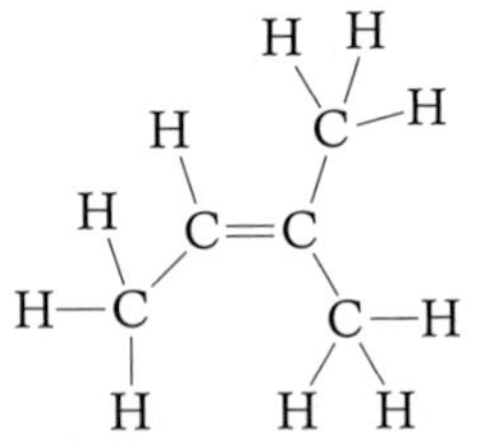

Our discussion of geometric isomers has focused on the bonds between carbon atoms, but it also allows us to conclude something that will make it *much* easier to draw organic molecules. We have discussed that the geometric isomers require the presence of a C=C bond. When there is a bond to H, on the other hand, it is always a single bond. That means, at least for this course, we never have to draw the bond to H! Instead, we can write a C—H bond in abbreviated form as a CH group. Figure 17-13 shows some examples of how we can simplify the writing of some of the structures we have seen.

There is one other kind of rigidity we need to watch for in organic molecules: molecules where there is a ring present. There is a definite set of atoms bonded to one another around a ring, and we cannot twist them away from one another without disrupting the whole ring. In very large rings, portions of the ring begin to look like a simple chain, and geometric isomers can form. But this is relatively uncommon.

Oxygen-Containing Functional Groups

As we saw with carbon, heteroatoms can have different bonding patterns associated with their position and the number of bonds they form. Consider, for example, three of the isomers with the formula C_4H_8O, shown in Figure 17-14. The bonding pattern for the oxygen changes from having a single bond to H and to C, to having two single bonds to carbons, and a double bond to a single carbon.

Figure 17-13 ▶ Writing C—H bonds as groups can simplify many drawings.

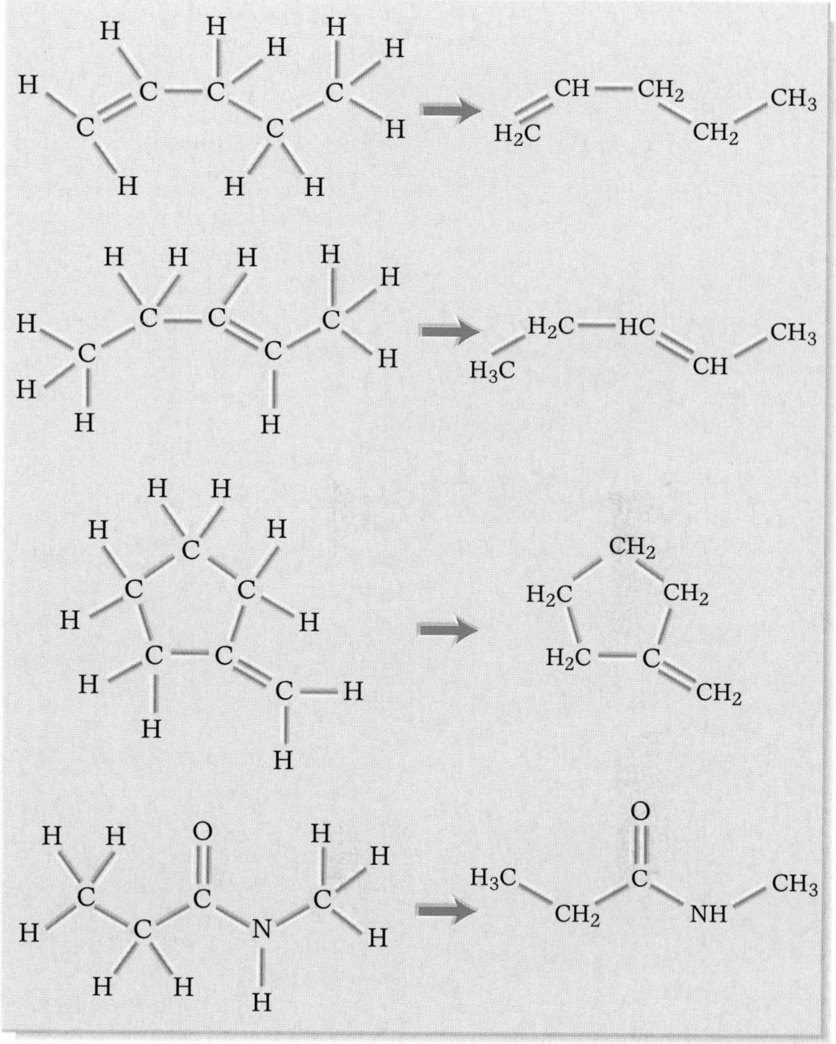

These three isomers all have very different properties, partly due to the difference in the C—C bonding between them. But the most important part is due to the presence of different bonding for the oxygen atoms. These different patterns are so important that they are recognized as having very different properties. The different patterns are named for their similarity to other important molecules and, because the bonding pattern controls the way the molecules function, the different groups are known as **functional groups.**

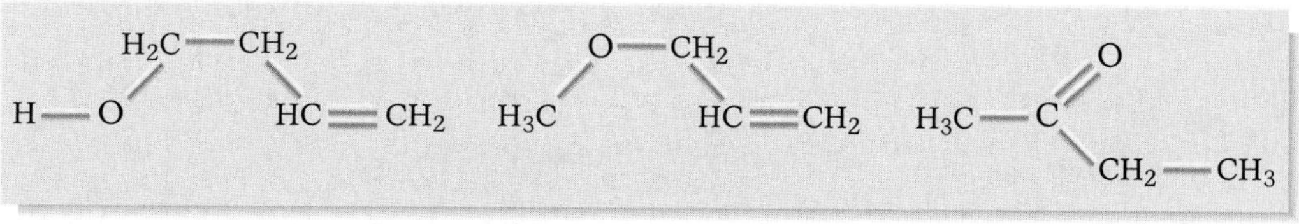

Figure 17-14 ▲ Three isomers of C_4H_8O.

There are five important functional groups involving C and O:

- Alcohol groups contain a C—O—H pattern.
- Ether groups contain a C—O—C pattern.
- Carbonyl groups contain a C=O bond with no other O's on the C.
- Carboxylic acid groups contain a COOH group, where one O is connected to the C by a double bond and the other is connected by a single bond. The oxygen in the C—O group is also bonded to an H.
- Ester groups are like acid groups, except that the oxygen in the C—O group is also bonded to another C.

Example `17.11`   **Organic functional groups**

Indicate the oxygen-containing functional groups in these molecules. Some molecules may contain more than one functional group.

Thinking Through the Problem We will list the oxygen-containing functional groups. The definitions of the different groups are used in categorizing the molecules. We start by looking at each O group and then determining into which group it fits.

Answer In molecule A, the only oxygen is in a C—O—H group; this is an alcohol. In molecule B, there are two oxygens, bonded to different C atoms; the C=O group is a carbonyl while the C—O—H is an alcohol. In molecule C, the only O is in a C—O—C group; this is an ether. Molecule D has the two O's bonded to the same C; because the oxygen in the C—O group also has a H, this is a carboxylic acid. Molecule E has the two O's bonded to the same C; because the oxygen in the C—O group also has an H, this is an ester.

`How are you doing?`   **17.8**

Indicate the oxygen functional groups in Figure 17-14 and Example 17.11.

(Stephen Frisch/Stock Boston.)

(PhotoDisc.)

(Ray Coleman/Visuals Unlimited.)

Nitrogen, Sulfur, and the Odor of Molecules

An old joke about science classes includes the line, "If it stinks, it's chemistry." Indeed, because the senses of smell and taste depend on the interaction of molecules with odor receptors, every unpleasant odor we encounter is usually due to a particular molecule or a group of molecules. Of course, every *pleasant* odor we encounter is just as dependent on the chemistry of the system. (In other words: "If it smells nice, it's chemistry, too!") The presence of certain functional groups are often readily detected by their odor.

For example, we have been discussing oxygen functional groups, as we introduced in Practical A. Here is what we can say about them, *in general*:

- Alcohol groups often have a slightly sweet odor. Example: rubbing alcohol, which contains the substance 2-propanol, $H_3C-CH(OH)-CH_3$.
- Ether groups have an odor that most people regard as pleasant and slightly sweet. Example: methyl-*tert*-butyl ether, or MTBE, is a gasoline additive.
- Carbonyl groups have an odor that depends on the other groups bonded to the $C=O$ group. Examples: the sweet odor of acetone in nail-polish remover (shown at top left), the almond odor in benzaldehyde.
- Carboxylic acid groups cause a sour and unpleasant odor. Example: acetic acid, $H_3C-COOH$, the acid in vinegar. This odor can change dramatically as the groups attached change. Butyric acid, $H_3C-CH_2-CH_2-COOH$, has a smell most people associate with vomit.
- Ester groups are generally pleasant, fruity smells. This is for good reason—lots of plants have esters in their odor. Example: isoamyl acetate, which is found in banana. This can make a big difference: ethyl butyrate, an ester formed from butyric acid, gives the pleasant odor to pineapple! (See photos at left.)

The odor of the oxygen-containing group, as suggested, is sometimes dependent on what is around it in a molecule. This is not as true with functional groups containing nitrogen and sulfur. Look at the group of substances in Figure 17-15.

Sulfur is an important example of how the bonding about a heteroatom can make a big difference in our perception of it. First, there is no preferred bonding pattern. Even in very common sulfur-containing molecules, there can be a different number of connections to the S. Figure 17-16 has some examples. Note, however, the presence of O in making those extra bonds. This is quite common for elements in the third row of the periodic table and below. The three compounds illustrated have very different odor properties. The first, as noted earlier, is the "smell of natural gas" and is related to the molecule H_2S, which also has a very unpleasant odor. The second molecule has an unpleasant odor, too; it is known as

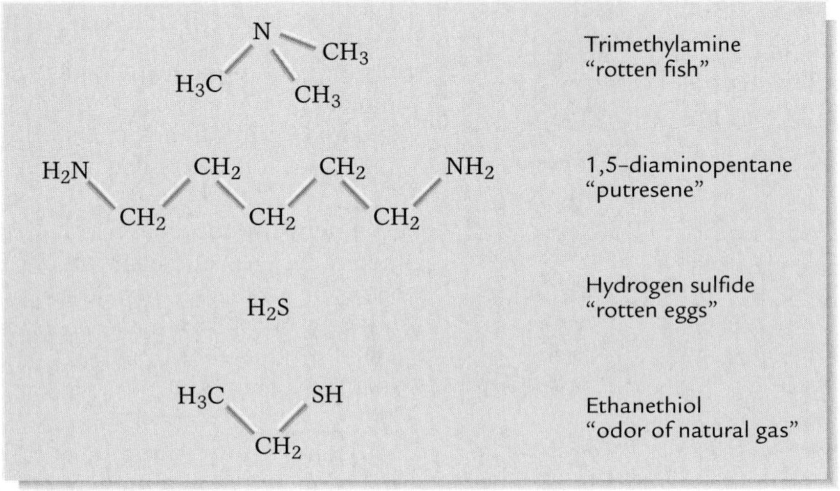
N
CH_3
H_3C
CH_3
Trimethylamine
"rotten fish"
H_2N
CH_2
CH_2
NH_2
CH_2
CH_2
CH_2
1,5-diaminopentane
"putresene"
H_2S
Hydrogen sulfide
"rotten eggs"
H_3C
SH
CH_2
Ethanethiol
"odor of natural gas"

Figure 17-15 ▶ Some nitrogen and sulfur-containing molecules.

"dimethylsulfoxide" and has powerful solvating abilities, partly because it is often contaminated with small amounts of compounds that have two bonds to an S atom. But it is still unpleasant. The third molecule, though it has a C—S bond, has no appreciable odor.

Why do we make such particular associations with certain kinds of functional groups? One speculation is that, when functional groups occur naturally, the environment where they are found is very hazardous. For example, partially decomposed foods are often sites of large quantities of dangerous bacteria. Do you think you should eat some meat before you find out if it is contaminated? Instead, the nose detects certain molecules that are emitted in the decomposition. *These* send a powerful message to the brain: "Do not eat this stuff!" Of course, certain pathogens are extremely aggressive and their presence in small amounts does not lead to a warning signal that we can perceive. The short and important lesson is: if you think it smells bad, it is bad.

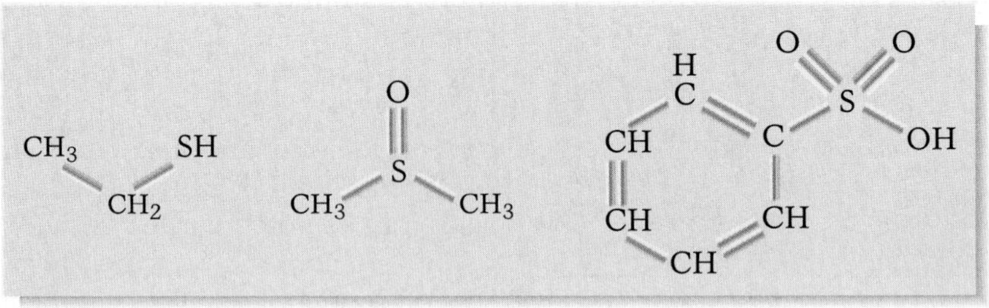
CH_3
SH
CH_2
O
S
CH_3
CH_3
H
C
O
O
S
CH
C
OH
CH
CH
CH

Figure 17-16 ▶ Some sulfur-containing organic molecules.

PRACTICAL C   What factors in molecular structure affect solubility?

In Chapter 5, we first encountered solubility as a property of certain ionic solids. We returned to this topic at the end of Chapter 16, when we saw that solubility is an equilibrium phenomenon. In both cases, we restricted our discussion to solubility in water, for that is the most common solvent in chemistry.

The relative solubility of substances in different solvents is important for many practical applications. A very simple one, shown in Figure 17-17, concerns oil and water.

With our knowledge of organic chemistry and organic functional groups, however, we can now expand the topic of solubility to the question of solubility involving molecules and to solvents other than water. Some data on the solubility of different substances in water and in the organic solvent benzene are given in Table 17-2 (for a picture of benzene, see Practical B in Chapter 13). Figure 17-18 shows this effect for NaCl.

Look through the structures and make a note of the functional groups that seem to be important for solubility in each case. Look carefully for exceptions to the trends you depict. Consider what happens when a change in one part of the molecule comes to "dominate" the solubility of a molecule in water or benzene. Also consider the effect of ion formation.

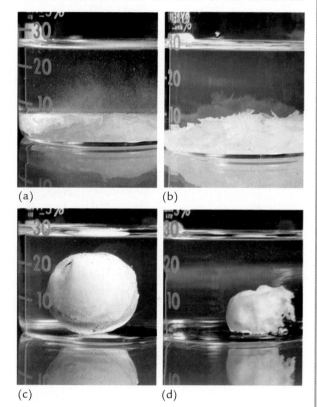

(a)   (b)

(c)   (d)

Figure 17-18 ▲ Solubility is dependent on the nature of the solvent.

(a)   (b)

Figure 17-17 ▲ Vegetable oil is a good example of an organic compound that is not soluble in water.

Table 17-2 Solubility of Molecules in Solvents

Compound	Solubility in Water	Solubility in Benzene
NaCl	Soluble	Insoluble
Ethanol, C_2H_5OH	Soluble	Soluble
Stearic acid, $C_{18}H_{37}COOH$	Insoluble	Soluble
Acetic acid	Soluble	Soluble

$$CH_3 - \overset{\overset{\displaystyle O}{\|}}{C} - O - H$$

Compound	Solubility in Water	Solubility in Benzene
Sodium acetate 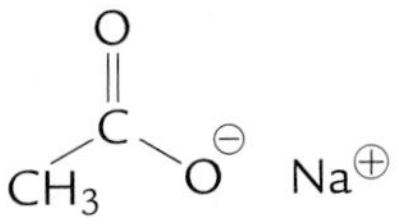	Soluble	Insoluble
Butyric Acid 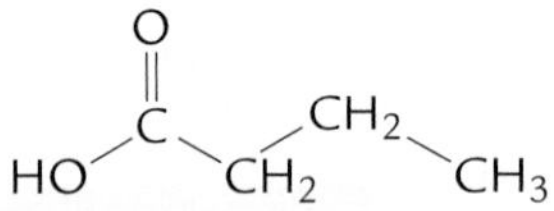	Soluble	Soluble
Ethyl acetate 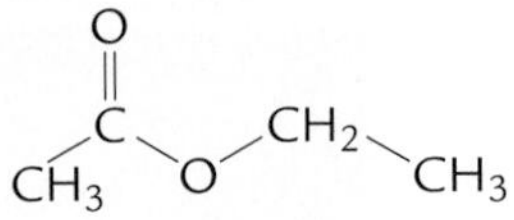	Insoluble	Soluble
Diethyl ether 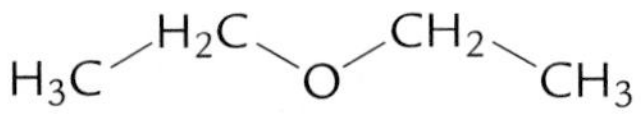	Insoluble	Soluble
Butanol 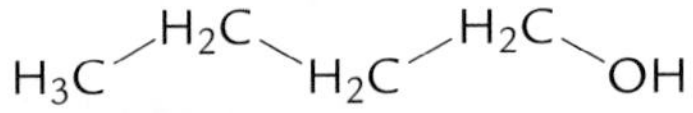	Soluble	Soluble

PROBLEMS

14. The following pairs of molecules are geometric isomers of one another. Indicate where the isomerism is present in each pair.

(a)

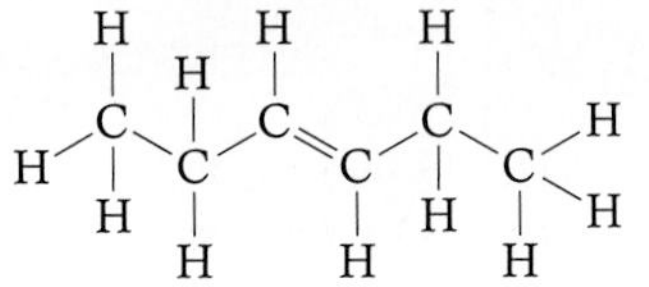

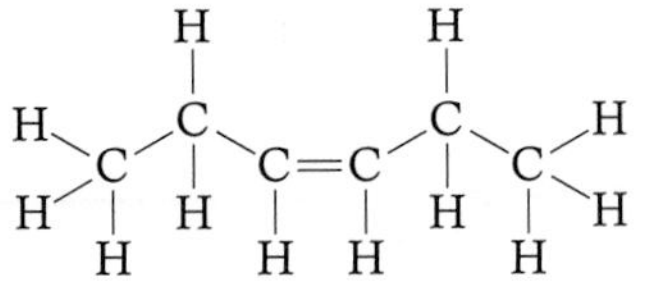

(b)

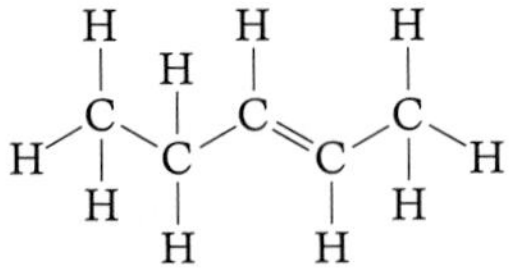

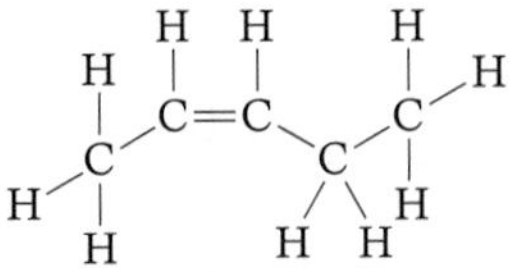

(c)

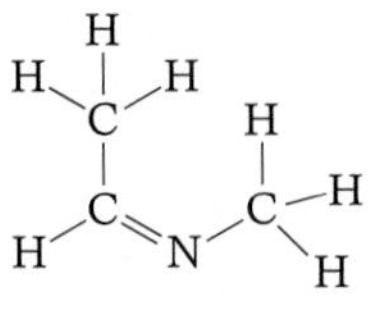

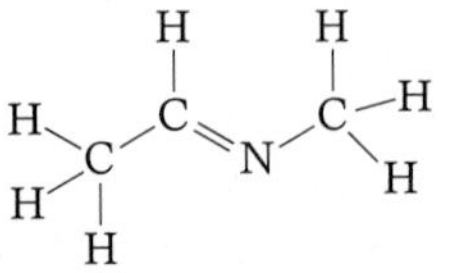

15. Redraw the molecules in the previous problem without any bonds to H. Instead, shorten the groups so that, for example, "C—H" becomes CH, etc.

16. The following drawing shows some molecules containing nitrogen functional groups. Circle the different bonding patterns you find for nitrogen. (Note: The meaning of the + and − charges on the molecule with two O atoms on the N will be discussed later in the chapter.)

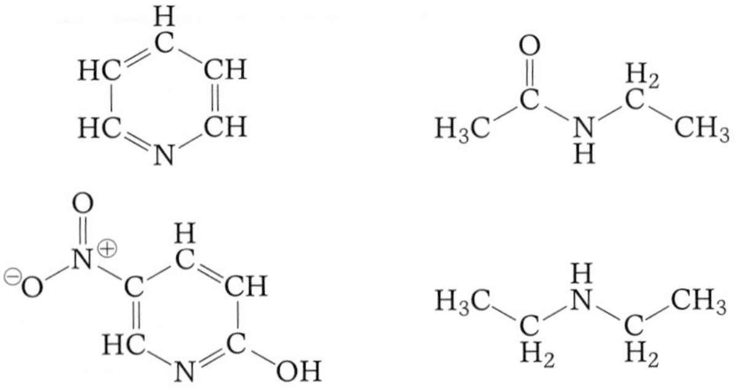

17. Phosphorus, which lies in the same row of the periodic table as sulfur, also has several different common bonding patterns. Look at the molecules below and indicate which (there may be more than one) is likely to have a strong and unpleasant odor. Explain your answer.

(a) 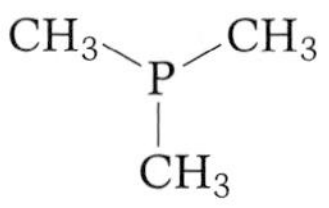 CH_3—P(—CH_3)—CH_3 with CH_3

(b) 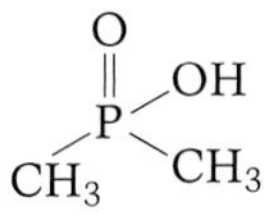 CH_3—P(=O)(—OH)—CH_3

(c) 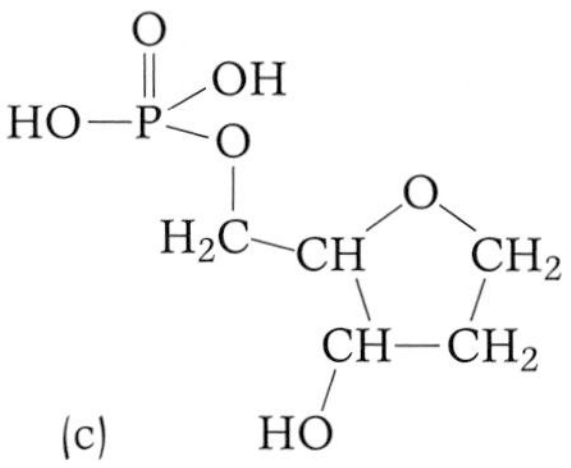 structure with HO—P(=O)(—OH)—O and ring H_2C—CH—O—CH_2, CH—CH_2, HO

(d) 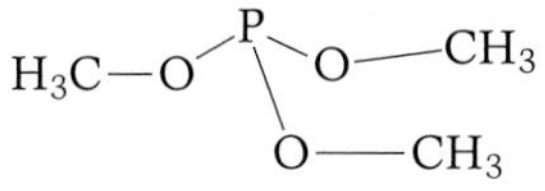 H_3C—O—P(—O—CH_3)(—O—CH_3) with O—CH_3

17.4 **Biochemical Structures**

SECTION GOALS

✓ Analyze the patterns of structure in carbohydrates, amino acids, and the building blocks of nucleic acids.

✓ Determine the sequence of amino acids in a peptide from its formula.

Organic Molecules in Biochemistry: Carbohydrates, Amino Acids, and Nucleic Acids

There are many organic molecules that are the major components of biological systems. Studying these and other chemical processes is the field of biochemistry. The rich reaction chemistry of these systems is beyond the scope of this text, but we do have enough understanding to proceed with a discussion of three of the most important structural systems in biochemistry. We begin with the structures of small molecules; in the next section, we take one of these structure classes and examine in more detail how it is used as the basis for much larger molecules.

In biochemistry, the three most important types of small molecules are carbohydrates, amino acids, and nucleic acids. **Carbohydrates** are molecules whose formula is, or is close to, that of $C_x(H_2O)_x$, where x is a small number. The most important carbohydrates in biochemistry have $x = 5$ or 6. They are most commonly in the form of a ring of five C atoms and one O atom (Figure 17-19).

The ring structure shown here fails to illustrate one very important thing: the arrangements of the OH and the H groups above and below the ring. Figure 17-20 shows both of these molecules from the side, where we can see the way the groups are arranged.

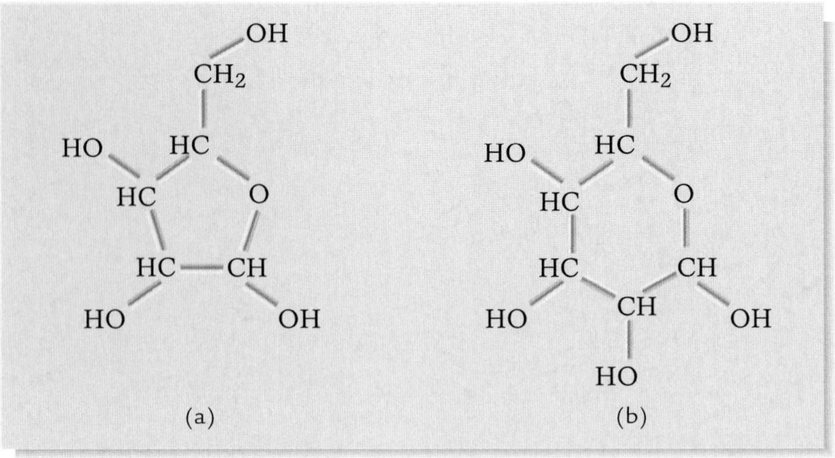

Figure 17-19 ▶ Basic ring structure of carbohydrates (a) ribose and (b) glucose.

This ring structure lets us see that the OH and other groups are arranged either on the same or on different sides of the five- or six-membered ring that is the basis of the structure. This, you may recognize, is like the *cis* and *trans* arrangements we discussed for molecules containing carbon-carbon double bonds. And, as with a double bond, the arrangement of the OH and other groups *cis* and *trans* about a ring cannot change without a chemical reaction. Therefore, the arrangement of groups around a carbohydrate ring (and other rings in organic chemistry) gives rise to isomers. In this case, the isomers are related to the three-dimensional arrangement of atoms, and this is referred to as **stereoisomerism.**

An example of a stereoisomer of a carbohydrate is galactose, which has one of the oxygen atoms on a different side of the ring (see Figure 17-21). The number of carbohydrate stereoisomers is finite, but it is very large. Looking at the glucose ring in Figure 17-20a, we see that there are five different ring carbons. If we consider the possible arrangements of groups, we get $2^5 = 32$ different molecules based on the glucose skeleton.

Not all carbohydrates have the formula $C_x(H_2O)_x$. An important example is the molecule deoxyribose. The prefix *deoxy-* indicates that there is one fewer oxygen in this molecule. While ribose has the

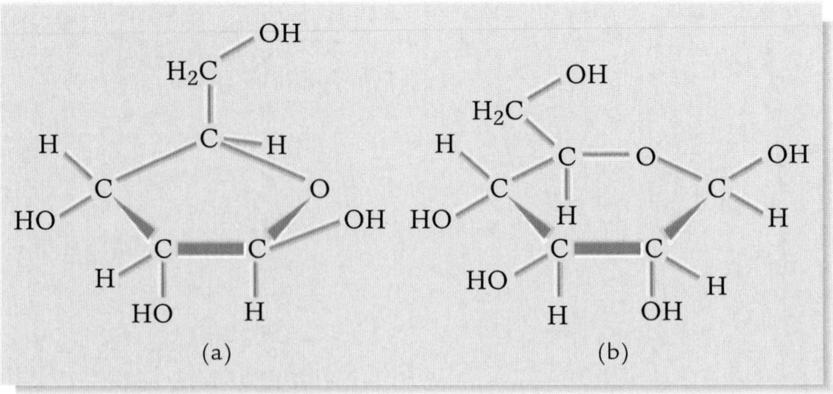

Figure 17-20 ▶ Side view of the structures of carbohydrates (a) ribose and (b) glucose.

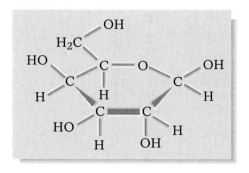

Figure 17-21 ▶ Side view of the structures of the carbohydrate galactose.

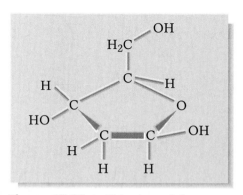

Figure 17-22 ▶ Side view of the structures of the carbohydrate deoxyribose.

molecular formula $C_5H_{10}O_5$, deoxyribose has the formula $C_5H_{10}O_4$. Different oxygen atoms can be removed from the structure shown in Figure 17-20a, but the most common molecule, called deoxyribose, has the structure shown in Figure 17-22.

Example 17.12   Carbohydrate stereoisomers

Draw at least two other stereoisomers of the glucose molecule.

Thinking Through the Problem The stereoisomers of glucose differ from glucose in the positions of the groups around the ring. If we take Figure 17-20a as a starting point, we can change the positions of any two oxygens.

Answer Counting clockwise from the O in the figure, we can, for example, do this for the second and third carbon. The names of these two other molecules are mannose (change in the 2-position) and allose (change in the 3-position).

How are you doing? 17.9

Present two other examples of molecules that can be described by the term *deoxyribose*.

Figure 17-23 ▶ Structure of
the amino acid glycine.

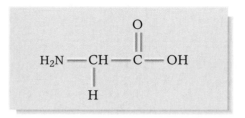

A second group of organic compounds important in biochemistry
are the **amino acids.** "Amino" refers to a functional group with an N
with three different single bonds and (in its neutral form) a lone pair.
Most commonly, the amino group is an NH_2 connected by a single
bond to a carbon. Some examples of amine groups were given earlier,
in Figure 17-15. The "acid" group in an amino acid refers to a carbon-
based acid, containing the COOH group.

Amino acids, in their name, can refer to any molecule containing
both an NH_2 group and a COOH group. But in biochemistry the most
important amino acids have both of these groups attached to the same
carbon. The simplest amino acid is glycine, shown in Figure 17-23.

The amino acids all have the same basic structure, but with one
variation: one of the hydrogens on the central carbon is substituted
by another group. There are twenty amino acids that occur with sufficient frequency that they are known as "natural amino acids" (other
amino acids are found in nature, but not in any general way). The different groups are referred to as the "side chains" of the amino acids.
Alanine, for example, has a CH_3 group in its side chain, while lysine
has a $CH_2CH_2CH_2CH_2NH_2$ group.

You may notice that some of the names of the amino acids are
familiar from nutritional guidelines. Nine of the amino acids cannot
be made by the human body and must therefore be obtained from
foods or nutritional supplements. A good diet provides these and
other amino acids, but a diet deficient in certain essential amino
acids can actually cause disease. For some, it is also possible that certain amino acids, if present in the diet, can cause disease as well.
Although such diseases are very rare, they have severe consequences
because of a problem in processing these amino acids. The most common disease of this kind is phenylketonuria, where the amino acid
phenylalanine is toxic to certain individuals.

Example 17.13   **Functional groups in amino acids**

List the amino acids (Table 17-3) that have an alcohol group.

Thinking Through the Problem We recall that the alcohol group occurs when an OH group is attached to a carbon that does *not* have
any other heteroatoms.

Answer Table 17-3 has three amino acids with an alcohol group—
serine, tyrosine, and threonine.

Table 17-3 The Naturally Occurring Amino Acids X—CH(NH$_2$)COOH

—X	Name	Abbreviation	—X	Name	Abbreviation
—H	Glycine	Gly	—CH$_2$(CH$_2$)$_2$NH—C—NH$_2$ (∥ NH)	Arginine	Arg
—CH$_3$	Alanine	Ala			
—CH$_2$⌬	Phenylalanine*	Phe	—CH$_2$ (imidazole ring N⟋⟍NH)	Histidine*	His
—CH(CH$_3$)$_2$	Valine*	Val	—CH$_2$ (indole ring NH)	Tryptophan*	Trp
—CH$_2$CH(CH$_3$)$_2$	Leucine*	Leu			
—CH(CH$_3$)CH$_2$CH$_3$	Isoleucine*	Ile			
—CH$_2$OH	Serine	Ser			
—CH(OH)CH$_3$	Threonine*	Thr	—CH$_2$CONH$_2$	Asparagine	Asn
—CH$_2$—⌬—OH	Tyrosine	Tyr	—CH$_2$CH$_2$CONH$_2$	Glutamine	Gln
—CH$_2$COOH	Aspartic acid	Asp	(NH⟋ ring ⟍COOH)	Proline†	Pro
—CH$_2$CH$_2$COOH	Glutamic acid	Glu			
—CH$_2$SH	Cysteine	Cys			
—CH$_2$CH$_2$SCH$_3$	Methionine*	Met			
—CH$_2$(CH$_2$)$_3$NH$_2$	Lysine*	Lys			

*Essential amino acids for humans.
†The entire amino acid is shown.

How are you doing?   **17.10**

Draw the structure you expect for the amino acid threonine. What functional groups are present? If the OH group in the side chain of threonine is converted to a CH$_3$ group, what amino acid results?

The last group of biochemically important molecules are the nucleic acids that are made from a set of small molecules called **nucleosides.** Nucleosides are the basis of the molecules that carry and communicate the code of life, because they are the basis of DNA and RNA. The nucleoside molecules in this case are derived from the substitution of an OH group in ribose or deoxyribose for another group. The four substituting groups are shown in Figure 17-24.

Figure 17-24 ► The four groups that are substituted for ribose or deoxyribose to make a nucleoside.

Adenine (A)

Guanine (G)

Thymine (T)

Cytosine (C)

Nucleosides are formed by adding one of the NH groups (shown at the bottom of the molecules in Figure 17-24) to the O on the carbon adjacent to the ring oxygen in ribose or deoxyribose. The nucleoside names are based on the side group. Figure 17-25 shows two nucleosides derived from adding thymine to deoxyribose and from adding adenine to ribose. The other nucleosides are named guanosine and cytidine. If they are built from deoxyribose, then the prefix *deoxy-* is added to the nucleoside.

A further modification of the ring is the addition of a phosphate group to the CH_2OH group of a nucleoside. This gives rise to a nucleotide. The nucleotide deoxyadenosine phosphate is shown in Figure 17-26.

Example 17.14 Nucleic acid names

Name the two molecules below.

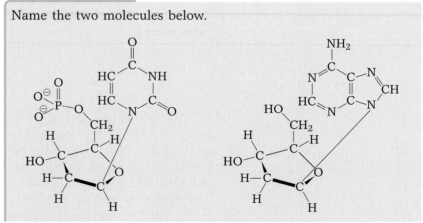

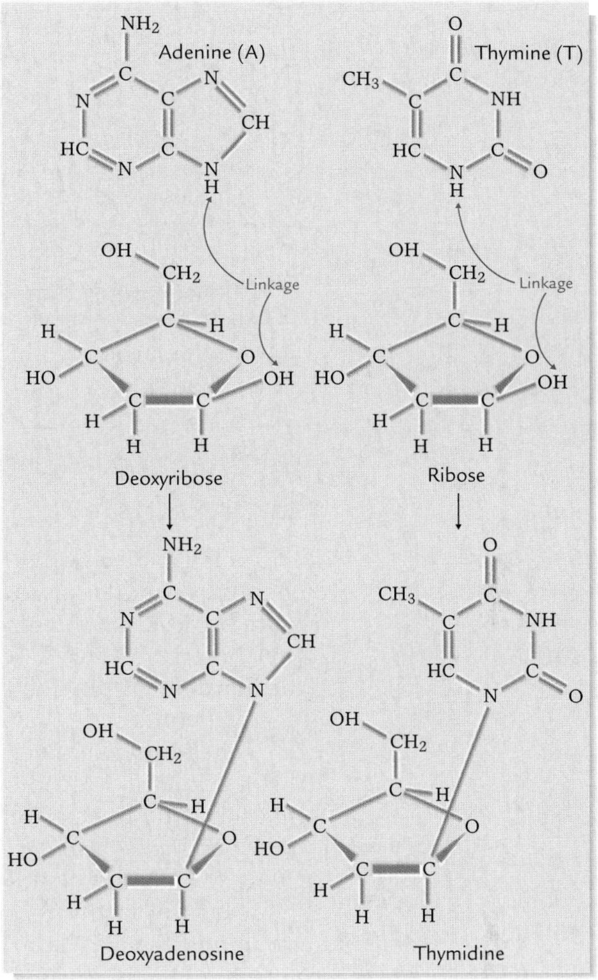

Figure 17-25 ▶ Formation and naming of two nucleosides.

Figure 17-26 ▶ A nucleotide with a single phosphate, a deoxyribose ring, and an adenine group.

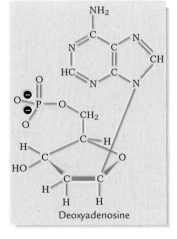
Deoxyadenosine

Thinking Through the Problem We can see that the first molecule has a phosphate group; it is a nucleotide. Because it is derived from thymine, the root for the name is thymidine. The second molecule has an adenine group and it is a nucleoside.

Answer The first molecule is deoxythymidine phosphate. The second molecule is adenosine.

How are you doing?   **17.11**

Draw the structures of deoxyguanosine and deoxyguanosine phosphate.

Biochemical Polymers: From Amino Acids to Proteins

The small molecules we have been discussing are important in biochemistry on their own, but they are responsible for something much more significant: the formation of large molecules that comprise the "machinery" of life. Carbohydrates are linked in long chains to form the starches of our food and the cellulose that is vital to plant life. The nucleotides are the basis of the long molecules of DNA that carry our genetic information. The amino acids, which we will focus on in this section, link together to create the proteins and other molecules that form the basis of our muscles, our cell's operating systems, and even our hair.

If we take two amino acids, say alanine and serine, we can link them by displacing the equivalent of water—an H from an amino group and the OH from the acid group. This is known as a conden-

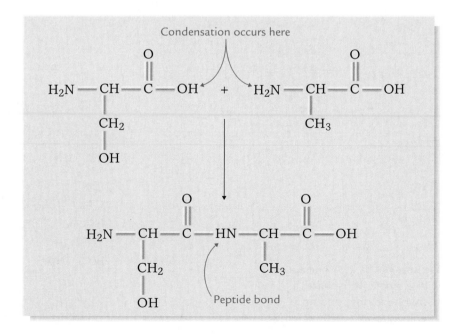

Figure 17-27 ▶ Formation of a dipeptide by condensation of two amino acids.

sation reaction. The bond that forms between the carbon and the nitrogen has a special name: a peptide bond. A molecule held together by one or more peptide bonds is known as a peptide, and we can name it based on the number of amino acids that are present—two amino acids make a dipeptide (shown in Figure 17-27), four make a tetrapeptide, and so on.

As we link several amino acids together, we can see the origin of the importance of these building blocks. A chain develops with a series of peptide bonds. If we have different amino acids, we begin to develop a structure in which there is a series of different side chains, some with different functional groups, all lined up. The name of a peptide is based on the order of the amino acids in its backbone, starting from the end with the NH_2 group. The dipeptide in Figure 17-27 is known as serylalanine, but we can use the abbreviations in Table 17-3 and instead write Ser-Ala.

Example 17.15   **Sketching structures of peptides**

Draw the structure of the tetrapeptide Leu-Glu-Met-Cys.

Thinking Through the Problem We know from the names that we have a leucine, a glutamic acid, a methionine, and a cysteine. The leucine is the NH_2 end of the chain, the cysteine is the COOH end.

Answer

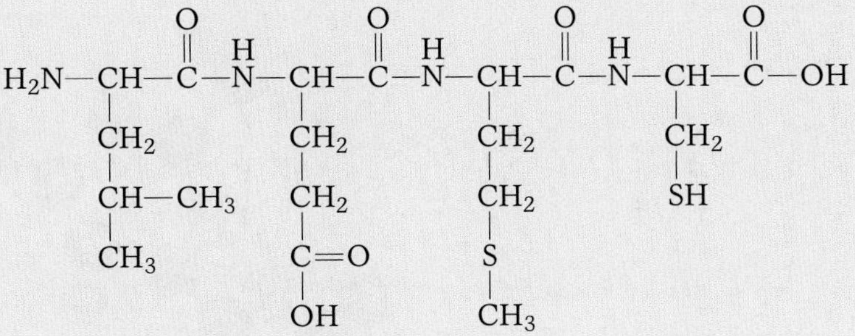

How are you doing?   **17.12**

Determine the shorthand notation for the pentapeptide shown below.

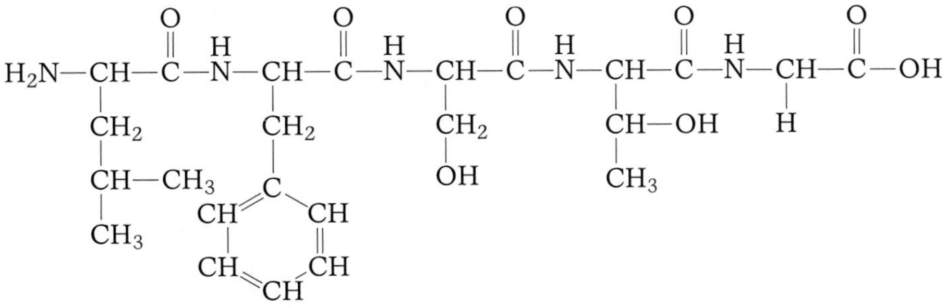

It is possible to build up peptide chains containing hundreds of amino acids. Molecules like this are known as **polypeptides,** based

on the prefix *poly-*, which means "many." Polypeptides are examples of molecules built from some small starting unit and assembled into very long chains. These long chain molecules are known as polymers. DNA and RNA are polymers of nucleotides. Cellulose is a polymer of glucose.

Polypeptides are the group in which we find the biochemically important molecules called *proteins*. For all proteins there is a primary structure that is described by the sequence of the amino acids. But very interesting effects occur, because the polypeptide rarely has the simple rod-like structure suggested in the previous figures. Rather, the molecule usually folds up in a very specific manner to produce the molecule's secondary structure. Finally, the secondary structure can fold up on itself, to give the final, tertiary structure. The marvelous thing about biochemistry is that this tertiary structure, still built on a simple peptide bond repeat unit, creates an intricate architecture upon which the side chains of the amino acids lie. These amino acids, especially where they have heteroatoms and functional groups, determine much of the activity of the protein.

Because of their large size, proteins are usually pictured with some abbreviations to the structure. Figure 17-28a shows such a structure, a snapshot of a winding structure known as a helix. You can see how the molecule wraps around on itself repeatedly. The side chains (the

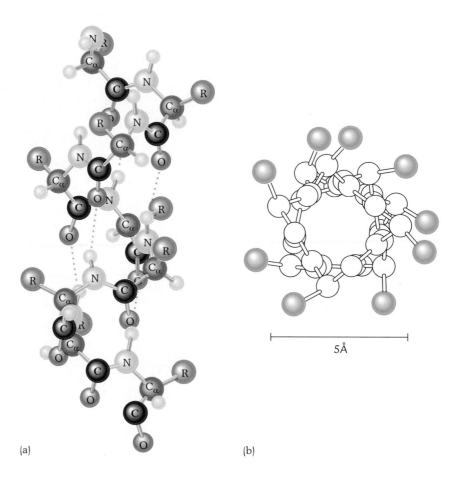

5Å

Figure 17-28 ▶ Two views of a polypeptide chain, emphasizing the secondary structure.

(a)   (b)

PRACTICAL **D** How do amino acids affect antibiotic resistance?

One characteristic of living systems is the presence of incredible diversity in structure. To be alive is to have to carry out a large number of molecular operations in an efficient manner. Similarly, the disruption of molecular processing systems can have a dramatic effect on an organism's ability to thrive and survive.

The control of bacterial pathogens is one of the great challenges of medicine and the drug industry. Almost all of us have taken an antibiotic at one time or another—perhaps the most famous of the antibiotics are the penicillins. Common penicillin, as with many related molecules, can prevent the spread of bacteria in a cell culture (Figure 17-29) and in the human body. The basic structure of a common penicillin is shown below. You should complete it with the proper positions of hydrogens and lone pairs on each of the atoms. The arrow indicates the critical reactive position.

The action of penicillin is rather simple. Penicillin reacts with and deactivates the machinery that the bacterium uses to form its cell wall. The cell wall consists of, among other components, a peptide chain, the bonds of which are formed by a series of peptide bond-forming reactions.

When we examine penicillin, we note that one of the bonds (highlighted by an arrow) is a peptide bond, with an amino group adjacent to a carbonyl group. This bond, known by the special name beta-lactam, is very reactive. It enters the enzyme that forms the cell wall of the bacterium and reacts with a serine residue that breaks open penicillin's reactive peptide bond to form an ester and an amine.

Not surprisingly, bacteria in an environment with many different antibiotics have been found to have a counter-strategy. They can protect themselves by reacting with the penicillin before it can disrupt the process of cell wall construction—an example of antibiotic resistance and a major concern in medicine today. Bacteria do this by destroying penicillins at the site of their greatest reactivity, the beta-lactam. The enzymes that destroy beta-lactams are known as beta-lactamases.

Scientists learn about the way beta-lactamases work through a combination of different techniques. In one case, powerful x-ray equipment was used to study the structure of a penicillin in the grip of a beta-latamase. Such a study resulted in the picture presented in Figure 17-30, where the portion in red shows the atoms of the penicillin attached through a covalent bond to a serine in the enzyme backbone.

The structure shown in 17-30 suggests that, in the absence of the serine that attacks the penicillin, the lactamase cannot work. This is confirmed by mutation studies, where different amino acids are substituted for that serine in the polypeptide chain of the lactamase. Only one other amino acid, tyrosine, will work in that role. Suggest why it makes sense that tyrosine can also act in a way similar to serine.

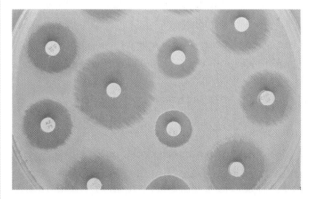

Figure 17-29 ▲ Antibiotics have dramatic effects on the growth of bacteria. (BBL/Visuals Unlimited.)

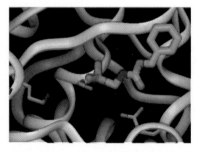

Figure 17-30 ▲ This illustration shows the detailed structure of a penicillin attached to a beta-lactamase. The atoms in red are the main atoms (not the hydrogens) of the opened-up penicillin along with the serine of the lactamase.

spheres marked with "R") stick out. It seems in this case that the side chains are arrayed randomly. But in Figure 17-28b, looking down the chain reveals that the peptide bonds form a core and the R groups (here shown as green spheres) coat the surface in specific ways.

PROBLEMS

18. The drawing below shows the ring structure of the common sugar fructose. Is fructose an isomer of glucose? Draw three other examples of molecules with the same ring structure that are stereoisomers of fructose.

$$HOCH_2 \quad O \quad CH_2OH$$

$$C\,H \quad HO\,C$$

$$H\,C = C\,OH$$

$$OH \quad H$$

19. Draw the structure of adenosine triphosophate, ATP.

20. Draw the structures of the amino acids glycine and valine. Draw the structure of the two different molecules that can be obtained when glycine and alanine combine.

21. Draw the structures of the amino acids alanine, phenylalanine, and tyrosine. Draw at least three different peptides that can be made from these three amino acids.

22. Draw the tetrapeptide Pro-Ile-Gly-Asp.

23. Determine the sequence of the peptide drawn below.

$$H_2N-CH-\overset{O}{\overset{\|}{C}}-\overset{H}{\overset{|}{N}}-CH-\overset{O}{\overset{\|}{C}}-\overset{H}{\overset{|}{N}}-CH_2-\overset{O}{\overset{\|}{C}}-\overset{H}{\overset{|}{N}}-CH-\overset{O}{\overset{\|}{C}}-\overset{H}{\overset{|}{N}}-CH-\overset{O}{\overset{\|}{C}}-OH$$

$$CH_2 \qquad CH_2 \qquad\qquad CH_2 \qquad CH_2$$

$$C=O \qquad \qquad\qquad\qquad OH \qquad SH$$

$$OH$$

Chapter 17 Summary and Problems

◄ | Review with **Web Practice**

VOCABULARY

organic chemistry	The branch of chemistry devoted to the study of carbon-containing molecules, including their structure and reactions.
biochemistry	The branch of chemistry that deals with molecules and reactions important in living things.
isomer	Substances that have the same chemical formula are isomers of each other.
structural isomers	Isomers that differ in the order of atoms.
geometric isomers	Isomers that differ in the arrangement of atoms around a point.
functional group	A special arrangement of atoms in an organic molecule, usually associated with multiple bonding or a heteroatom.
carbohydrates	Biologically important molecules with the empirical formula CH_2O.
stereoisomers	Isomers that differ based on the arrangements of groups around a tetrahedral carbon (one with four single bonds).
amino acid	Biologically important molecules that form the backbone of proteins.
nucleoside	One of a small number of molecules made of a nitrogen-containing base attached to a carbohydrate ring. These assemble into the DNA and RNA chains of the genetic code.
polypeptides	Molecules built from some small starting unit and assembled into very long chains, known as polymers.

SECTION GOALS

Is it possible to predict bonding patterns for elements in groups IV–VII?

For the second period elements it is possible. This depends on the number of valence electrons the atom "needs" to reach eight, and also on whether the atom has a charge or a lone pair.

Can bonding patterns be an aid to drawing Lewis structures?

We can be certain of some bonding patterns—for example, four bonds to C—and this makes it easy to inspect Lewis structures for validity.

What kinds of connections are important in an organic chemical structure?

When we look at bonds in organic chemical structure, we see single, double, and triple bonds. There are characteristic numbers of bonds for each element—one for H, four for C, two for O.

What are some important structural differences between isomers of organic chemical structures?

There are (for us) two important kinds of isomers. One has a different arrangement of bonds among the atoms. The other has a different geometrical layout of the bonds, something encountered only with double bonds to C.

Can you see the key factors in making a molecule rigid?

The key factors will be the presence of C=C double bonds, which cannot freely rotate around the bond.

Can you examine a molecule and indicate important functional groups?

To do this requires examining the type of bonds around an atom—for example, a C=O bond can be a carbonyl group, but when the C is also bonded to an —OH group, we get a carboxylic acid.

What are the patterns of structures in carbohydrates, amino acids, and the building blocks of nucleic acids?

The carbohydrates are chains of four or more carbons with an empirical formula CH_2O. The amino acids have an H_2N—C—COOH chain, where the central carbon has a variable side chain group that determines the identity of the amino acid. The four building blocks of amino acids are four specific nitrogen-containing ring structures.

How do we translate a formula of a sequence of amino acids into the structure?

The amino acid formula is a set of three-letter codes for the amino acids in a chain. We read along them to find the structure, with the first amino acid representing the H_2N— end of the chain.

PROBLEMS

24. The drawing below shows the common vitamin niacin. Analyze it for the common bonding patterns of each atom.

25. A molecule has a chain of four carbons attached to a ring of six carbons. All carbons have four single bonds. Draw the structure using common bonding patterns.

26. Analyze the following molecules for the presence of different organic chemical functional groups.

(a)

(b)

(c)

(d)

(e)

27. For the following molecules, one other isomer can be drawn that differs in the arrangement of the groups around the double bond. Draw the other isomer in each case.

(a)

(b)

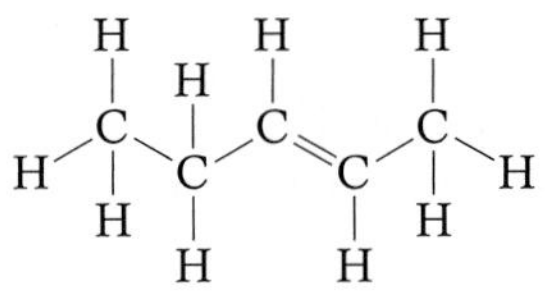

(c)

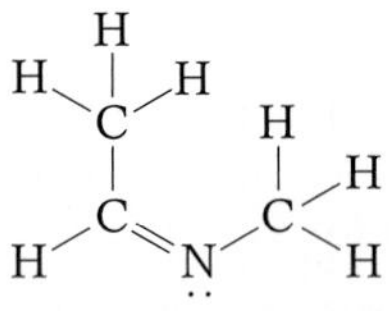

28. Draw the structure of guanosine triphosophate, ATP.

29. Draw the structures of the amino acids proline and glutamic acid. Draw the structure of the two different molecules that can be obtained when they combine.

30. Determine the sequence of the peptide drawn below.

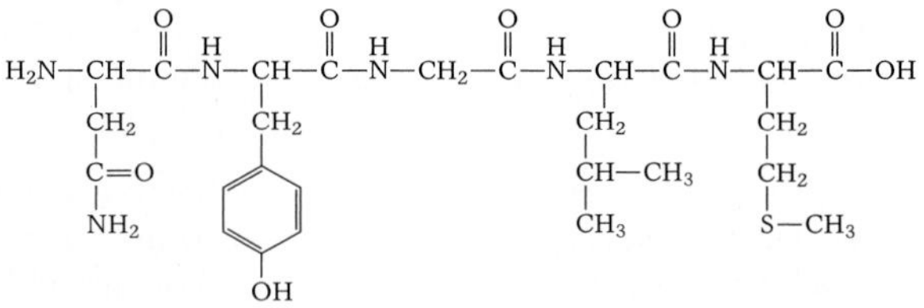

Answers to Practicals

Chapter 1

Practical B Both $(CaO)_3 \bullet Al_2O_3$ and $Ca_3Al_2O_6$ contain 3 Ca, 2 Al, and 6 O atoms. The dot indicates a weak connection between the $(CaO)_3$ and the Al_2O_3. $(CaO)_3 \bullet SiO_2$ contains 3 Ca, 5 O, and 1 S atoms. $CaSiO_3 \bullet H_2O$ contains 1 Ca, 1 Si, 4 O, and 2 H atoms.

Practical C The atomic number of uranium is 92, and a neutral atom of uranium contains 92 electrons. The atomic numbers of polonium and of radium are 84 and 88, respectively. $^{10}_5B$ is transmuted into $^{14}_7N$ by the addition of a particle with an atomic number of 2. This is ^4_2He. When ^4_2He adds to $^{27}_{13}Al$, in a bombardment process, the result is $^{31}_{15}P$.

Practical D The statement about H_2O describes a property of a compound. The statements about aluminum, iron in a ship's hull, and gold describe properties of an element. The statements about chlorine dioxide and iron in blood describe a property of an element in a compound.

Practical E On the basis of the listed properties, the compound that is most similar to NaBr is NaI. Most similar to Na_2S are Na_2Se, Na_2O, and NaF.

Chapter 2

Practical A Substances not likely to exist as a gas: CaS, KF, LiCl, and Br_2 (which we know from this chapter is a liquid). Substances that are likely to be gases are: H_2S (hydrogen sulfide), CF_4 (carbon tetrafluoride), CO (carbon monoxide), ICl (iodine monochloride), NCl_3 (nitrogen trichloride), and SO_3 (sulfur trioxide). C_2H_6 (dicarbon hexahydride, more commonly, ethane), HCl (hydrogen chloride), P_2H_4 (diphosphorus tetrahydride), NO_2 (nitrogen dioxide), and CO_2 (carbon dioxide) are all likely to be gases.

Practical B The presence of a lone pair on oxygen may indicate the gas is dangerous to humans as it "imitates" oxygen. NO^+ has a lone pair on oxygen; it may be dangerous. NO has 11 valence electrons and NO_2 has 17 in their respective Lewis structures. O_2^- has 13 valence electrons.

Practical C All structures contain $—N{=}O$, with two lone pairs on the oxygen.

Chapter 3

Practical A NaCl, KCl, and KI are the formulas for sodium chloride, potassium chloride, and potassium iodide.

Practical B $Ca_5(PO_4)_3OH$ is neutral, as adding up the changes shows: $5(+2) + 3(+5) + 12(-2) + (-2) + (+1) = 0$.

$CaCO_3$ is calcium carbonate, $NaC_2H_3O_2$ is sodium acetate, and $KHCO_3$ is potassium hydrogen carbonate.

The silicate ion contains Si and O; the selenate ion contains Se and O. The oxidation number of the silicate ion is -4 and that of the selenate ion is -2.

Practical C Mn in $MnSO_4$, manganese (II) sulfate, has an oxidation number of $+2$. Cu in CuO, copper (II) oxide, has an oxidation number of $+2$. Sn in SnF_2, tin (II) fluoride, has an oxidation number of $+2$. The balanced formulas are: $Fe_2(SO_4)_3$, $Mn_2(SO_3)_3$, Cu_2O, and SnF_4.

Chapter 4

Practical A Important substances in the striker are red phosphorus and powdered glass. In the match head, important substances for the reaction are sulfur, potassium chlorate, and other oxides.

Practical B An unbalanced equation for the reaction is:

$$P + KClO_3 + O_2 \rightarrow KCl + P_2O_5$$

The reactants are P, $KClO_3$, and O_2; the products are KCl and P_2O_5.

Practical C The balanced equations for the reaction of sulfur and potassium chlorate are:

$$3\ S_8 + 16\ KClO_3 \rightarrow 24\ SO_2 + 16\ KCl$$

and

$$3\ S + 2\ KClO_3 \rightarrow 3\ SO_2 + 2\ KCl$$

The balanced equation for the reaction of manganese (IV) oxide and sulfur is:

$$4\ MnO_2 + S \rightarrow 2\ Mn_2O_3 + SO_2$$

The balanced equation for the reaction of sulfur dioxide and calcium carbonate is:

$$SO_2 + CaCO_3 \rightarrow CaSO_3 + CO_2$$

The balanced equation for the reaction of tetraphosphorus trisulfide and potassium chlorate is:

$$3 P_4S_3 + 16 KClO_3 \rightarrow 6 P_2O_5 + 9 SO_2 + 16 KCl$$

Chapter 5

Practical A

(a) If the solution remains lightly colored, then either the original solution is very dilute or does not contain iron or cobalt ions.
(b) If cobalt (II) is present, then the solution should turn blue.

Sulfides present in the solutions:

(a) Pb^{2+}
(b) Hg^{2+}
(c) Sn^{4+} and/or Cd^{2+}
(d) Sb^{3+} and Pb^{2+}

The solution in part (c) has the most ambiguity.

Practical B The reaction of tin and oxygen is:

$$Sn\ (s) + O_2\ (g) \rightarrow SnO_2\ (s)$$

ZnO is the formula for zinc oxide.
The reactions that provide protective coating for aluminum and magnesium are:

$$2 Mg\ (s) + O_2\ (g) \rightarrow 2 MgO\ \text{(magnesium oxide)}$$
$$Al\ (s) + O_2\ (g) \rightarrow Al_2O_3\ (s)\ \text{(aluminum oxide)}$$

For the protection of zinc:

$$2 Zn\ (s) + 2 CO_2\ (g) + O_2\ (g) \rightarrow 2 ZnCO_3\ (s)$$

Practical C The unknown element X is less reactive than Fe but more reactive than Ni. Looking at Table 5-3, we conclude that X is either Cd or Co. We can narrow down the identity of X by reacting X with a solution of $Co(NO_3)_2$. If a displacement occurs, X is Cd. If not, X is Co.

In the second case, there is a wider range of possibilities for the identity of X (any element listed in Table 5-3 that is between Zn and H_2). We can narrow the range of possibilities for X by testing it against aqueous solutions of any of the elements with activities between Zn and H_2.

Practical D Concerning the first reaction:

(a) Pink to purple indicates a chemical reaction has occurred.
(b) Mn^{2+} is being oxidized to Mn^{7+}.
(c) Bi^{5+} is being reduced to Bi^{3+}.

Concerning the second reaction:

(a) Orange to green indicates a chemical reaction has occurred.
(b) Cr^{8+} is being reduced to Cr^{3+}.
(c) Carbon dioxide is evolved in this reaction.

1. (a) PbS
   (b) Lead (II) sulfide
   (c) Black

2. (a) CdS
   (b) Cadmium (II) sulfide
   (c) Yellow

3. (a) Sb_2S_3
   (b) Antimony (III) sulfide
   (c) Orange

4. Li, Na, K, Ru, Cs, and (NH_4) sulfides are soluble.

Chapter 6

Practical A Authors identify chemical substances wherever they are used. The authors may use a trivial name when the name is easy to recognize; for example, water.

When the authors measure a chemical substance they use the units mass (g), moles (mol), volume (mL or cm^3), and temperature (Celsius). More than one unit may be used for concentration and pressure, or when an extensive measurement is not important.

An example of an intensive quantity is 0.3 moles per liter and other concentrations.

The reason that these descriptions are provided is so that other scientists can repeat the experiments.

Practical B

Sample	Density, g/cm^3
1	1.341
2	1.285
3	1.256
4	1.302

Practical C

Writing the Ratio	Using the Ratio
2.5 g matter/25 g soil; 0.1 g matter/1 g soil	10%
0.250 mol/15 min; 0.0167 mol/minute	60 minutes
70% H_2O_2/100% solution; 0.7 g H_2O_2/1 g solution	8.6 g solution

Practical D

(a) Ratio of moles of oxygen to moles of barium is 2:1.
(b) NH_2
(c) $NaHSO_4$
(d) $C_{18}H_{38}O_2$
(e) Ratio of moles of CO to moles of palladium compound is 1:2.

Chapter 7

Practical A The systematic error of having wet sand can be corrected by drying the sand first. Spillage is a random error. A dirty container is a random error. These errors cannot be prevented but they can be minimized.

If the sand is not pure SiO_2 but a mixture of substances, then there will be a systematic error in all measurements of this sample.

The mass of the sand is 65.9 g and the volume of the sand is 25.6 mL, so the density of the sand is 2.57 g/mL. The calculated density of wet sand would be different from that for dry sand because we would be including the mass and volume of the water.

Practical B Plan A: The concrete patio has a volume of 0.24 m^3 with 2 significant figures. The volume of sand needed is 4.8 m^3 with 2 significant figures because 90 townhouses is an absolute count (no estimating).

Plan B: The paving brick patio needs 0.79 m^3 of sand. The number of significant figures is 2 because the trailing zero in 5.0 cm is significant. The total volume of sand needed is 71 m^3, which has a mass of 1.6×10^5 kg or 1.6×10^8 g.

Practical C The volume of a grain of sand is 5.24×10^{-4} cm^3 and the mass of a grain of sand is 1.38×10^{-3} g. The volume of sand on the beach is 2×10^5 m^3, the number of grains of sand on the beach is 3×10^{14}, and the number of moles of sand on the beach is 8×10^{-10}. In the Sahara, the number of grains of sand is 3×10^{23}.

Chapter 8

Practical A

Name	Molar Mass (g/mol)	Moles/pill
Thiamin	337.27	7.4×10^{-6}
Riboflavin	376.369	6.6×10^{-7}
Niacin	169.180	1.2×10^{-5}
Folic acid	441.403	4.5×10^{-7}
Biotin	244.31	8.2×10^{-8}
Pantothenic acid	424.472	1.2×10^{-6}

Practical B The molar mass of $C_6O_6H_8$ is 176.123 g/mol. There are 2.46×10^{20} molecules of vitamin C per glass of orange juice. The volume of the different brand that supplies the DV is 113 mL.

The volume of the solution of vitamin K that supplies the DV is 0.0832 mL.

Name	Molar Mass (g/mol)	Moles in 1.00 g	Molecules in 1.00 g
Vitamin E	416.685	2.40×10^{-3}	1.45×10^{21}
β-carotin	536.882	1.86×10^{-3}	1.12×10^{21}
Vitamin K	450.702	2.22×10^{-3}	1.34×10^{21}
Vitamin D_2	396.655	2.52×10^{-3}	1.52×10^{21}

Practical C The composition of $C_3H_4O_3$, $C_6H_8O_6$, and $C_{12}H_{16}O_{12}$ is 40.917% C, 4.578% H, and 54.504% O. The same ratio of C:H:O in $C_9H_{12}O_9$ gives the same percentage composition of these elements.

The composition of $C_{10}H_{12}NO$ is 74.046% C, 7.457% H, 8.635% N, and 9.63% O.

In 325 grams of quinine there are 240.6 g C (20.04 moles), 24.24 g H (24.04 moles), 28.064 g N (2.00 moles), and 32.05 g O (2.00 moles). The true formula is $C_{20}H_{24}N_2O_2$.

Practical D The subscript number of $C_9H_8O_4$ and of $C_7H_6O_3$ that causes the molecular formula to be the empirical formula is 9 and 7, respectively.

Chapter 9

Practical A 156.25 moles of O_2 are needed to produce 100 moles of CO_2 from octane, and 100 moles of O_2 are needed to produce the same amount of CO_2 from glucose. Octane needs more oxygen than glucose to burn completely. Octane has a higher energy content than glucose.

The equations for the combustion of oleic acid and alanine are:

$$2\ C_{18}H_{34}O_2 + 51\ O_2 \rightarrow 34\ H_2O + 36\ CO_2$$

and

$$C_3H_7NO_2 + 3\ O_2 \rightarrow 2\ H_2O + 3\ CO_2 + NH_3$$

Fats appear to have the highest energy content per molecule.

Practical B 651 liters of air are needed to make the orchid. The potato needs 35.8 g CO_2 and 12.2 g H_2O.

2160 grams of CO_2 are produced by the combustion of 700 grams of gasoline. 1330 grams of starch could be made from this amount of CO_2. This is equivalent to a small tree or about 30 pounds of potatoes.

Practical C A chemical equation for the formation of alanine is:

$$3 \ CO_2 + 2 \ H_2O + NH_3 \rightarrow C_3H_7NO_2 + 3 \ O_2$$

To obtain 1.23 grams of alanine, 1.82 grams of CO_2, 0.497 gram of H_2O, and 0.235 gram of NH_3 are needed.

$3 \ CO_2$	4.14×10^{-2} mol	1.82 g
$2 \ H_2O$	2.76×10^{-2} mol	4.97×10^{-1} g
NH_3	1.38×10^{-2} mol	2.35×10^{-1} g

If $(NH_4)_2SO_4 \rightarrow 2 \ NH_3 + H_2SO_4$, we need 0.912 g $(NH_4)_2SO_4$. The mass of N in the corn is 9.5 grams, and 0.339 mole of $(NH_4)_2SO_4$ is needed to produce this quantity of N, according to the following calculations:

Corn is 5% N
190 g corn $\rightarrow$ 9.50 g N

Assuming all N comes from $(NH_4)_2SO_4$:

$$9.50 \ \cancel{g \ N} \times \frac{1 \ \cancel{mol \ N}}{14.007 \ \cancel{g \ N}} \times \frac{1 \ mol \ (NH_4)_2SO_4}{2 \ \cancel{mol \ N}} = 0.339 \ mol \ (NH_4)_2 \ SO_4$$

Chapter 10

Practical A

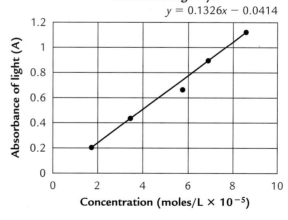

Absorbance versus Concentration of Carmine Indigo Dye

$y = 0.1326x - 0.0414$

We see in this graph that as concentration increases, absorption increases. This data fits the slope-intercept model; the slope of the line graphed is 0.1326 A/molarity $\times 10^{-5}$.

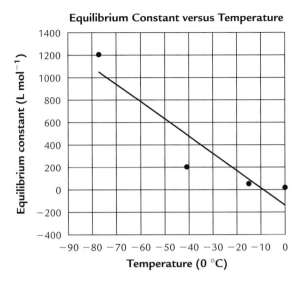

Equilibrium Constant versus Temperature

Data is not linear and will not fit the slope-intercept model. However, we can say that as the temperature decreases, the equilibrium constant increases.

Practical B In the graph in Figure 10-7, volume is the dependent variable and is graphed on the *y*-axis. Pressure is the independent variable and is graphed on the *x*-axis. A curve results when data points are smoothly connected. The graph shows that as pressure increases, volume decreases. The graph is consistent with Boyle's law.

The approximate values of volume at these pressures are 5.5 L, 2.2 L, and 0.8 L.

An equation for this curve is $V \propto 1/P$ or $V = k(1/P)$, where k is the slope of the line. This is a good example of a system that follows Boyle's Law.

Practical C The density of nitrogen gas at 0 °C, 25 °C, 50 °C, and 100 °C is 1.25 g/L, 1.15 g/L, 1.06 g/L, and 0.92 g/L, respectively. The density of hydrogen gas at 0 °C is 0.90 g/L.

Practical D The volume of the balloon shrinks, because the temperature is proportional to molecular motion.

The inflated balloon has a volume of 4.19×10^3 cm^3. This is equivalent to 4.19×10^3 mL or 4.19 L.

There is 0.182 mole of CO_2 in 8.00 grams of CO_2. This amount contains 1.09×10^{23} molecules of CO_2.

One molecule of CO_2 (gas phase) occupies 3.73×10^{-23} liter. One molecule of CO_2 (solid phase) occupies 5.41×10^{-23} liter, much less volume than in the gaseous phase. CO_2 molecules occupy more volume in the gas phase than in the solid phase as gas molecules move around more in the gas than they do in the solid.

Chapter 11

Practical A

	Natural gas	Gasoline	Coal	Wood	Ethanol
Portability	In gas cylinders (cumbersome)	Container (gas tank, safety gallon, etc.)	Easy; no special container required	Easy; no special container required	Container required
Safety	Gaseous (very easily ignited)	Volatile (easily ignited)	Safest	Safest	Pretty safe (easily ignited but not as volatile as gasoline)
Age and renewability	Old; nonrenewable	Old; nonrenewable	Old; nonrenewable	Old; not easily renewable	New; easily renewable
Relationship to the capture of the sun's energy at some point in the past	Ancient decayed plant material; photosynthesis	Ancient decayed plant material; photosynthesis	Ancient decayed plant material; photosynthesis	Old, undecayed plant material; photosynthesis	New material; photosynthesis later converted by yeast

Practical B

Substance	q (kJ/g)	q (L)	Equation
Toluene	-42.5	$-36,800$ kJ/L	$C_7H_8 + 9\,O_2 \rightarrow 7\,CO_2 + 4\,H_2O$
Heptane	-48.2	$-32,900$ kJ/L	$C_7H_{16} + 11\,O_2 \rightarrow 7\,CO_2 + 8\,H_2O$
Ethanol	-29.7	$-23,400$ kJ/L	$C_2H_6O + 3\,O_2 \rightarrow 2\,CO_2 + 3\,H_2O$
MTBE	-36.2	$-26,700$ kJ/L	$2\,C_5H_{12}O + 15\,O_2 \rightarrow 10\,CO_2 + 12\,H_2O$

Less O_2 is required for the combustion of oxygen containing organic compounds, but the energy per liter is less.

Chapter 12

Practical A A mass spectrum of iron would show that no iron-59 exists.

Percent Abundance	Isotopic Mass	Mass Number of Isotope
91.75	55.9349	56
2.20	56.9354	57
0.28	57.9333	58
5.84	53.9396	54

Thallium-201 is not a naturally-occurring isotope. The atomic mass for thallium is 204.3839.

Percent Abundance	Isotopic Mass	Mass Number of Isotope
29.5	202.9723	203
70.5	204.9745	205

Percent Abundance	Isotopic Mass	Mass Number of Isotope
93.26	38.9637	39
0.0118	39.974	40
6.73	40.9045	41

2.4×10^{-4} gram of ^{40}K. This is equivalent to 6.0×10^{-6} mole of ^{40}K or 3.6×10^{18} atoms ^{40}K.

Practical B The nuclear reactions are:

$$^{11}_{5}B + ^{4}_{2}He \rightarrow ^{15}_{7}N$$
$$^{196}_{78}Pt + ^{2}_{1}H \rightarrow ^{197}_{78}Pt + ^{1}_{1}P$$
$$^{197}_{78}Pt \rightarrow ^{0}_{-1}\beta + ^{197}_{79}Au$$

Practical C The number of neutrons given off in the first fission reaction is 3, and in the second fission reaction is 2.

Chapter 13

Practical A In the colorless form of phenolphthalein, all of the carbons have three electron domains except one carbon in the middle, which has four electron domains. In the colored form, all carbons have three electron domains.

In the colored form of malachite green (acidic form), all carbon atoms have three electron domains. In the basic form, there is a single carbon with four electron domains; all others have three electron domains.

The colorless form of curcumin has a four-electron domain on the carbon between the two oxygen atoms. This is the high-energy form. The colored form has all carbons with three electron domains. The colored form is the low-energy form.

Practical B No appreciable difference is seen in the energy values listed for the compounds in any column.

The energy decreases as the number of double bonds increases.

Comparing benzene with the other compounds containing six carbon atoms, we see that benzene (E = 468 kJ/mol) lies between C_6H_{10} (E = 526 kJ/mol) and C_6H_8 (E = 445 kJ/mol).

Practical C The atomic radius increases from Be to Ba. This is accountable to the fact that the d subshell is introduced for Sr (small increase from Ca), and there is increased shielding from the shells of electrons as the atoms get bigger (so the charge repulsion from the nucleus and electron cloud isn't as severe, and also because there are more electrons in the outer shell). All atoms have their highest s subshell filled; starting with Sr, the valence shell contains d electrons.

Atom	Atomic Radius	Electron Configuration	Ion	Ionic Radius	Electron Configuration
Be	110 pm	$1s^2 2s^2$	Be^{2+}	27 pm	$1s^2$
Mg	160 pm	$1s^2 2s^2 2p^6 3s^2$	Mg^{2+}	72 pm	$1s^2 2s^2 2p^6$
Ca	200 pm	$1s^2 2s^2 2p^6 3s^2 3p^6 4s^2$	Ca^{2+}	100 pm	$1s^2 2s^2 2p^6 3s^2 3p^6$
Sr	215 pm	$1s^2 2s^2 2p^6 3s^2 3p^6 4s^2 3d^{10} 4p^6 5s^2$	Sr^{2+}	116 pm	$1s^2 2s^2 2p^6 3s^2 3p^6 3d^{10} 4s^2 4p^6$
Ba	224 pm	$1s^2 2s^2 2p^6 3s^2 3p^6 4s^2 3d^{10} 4p^6 5s^2 4d^{10} 5p^6 6s^2$	Ba^{2+}	136 pm	$1s^2 2s^2 2p^6 3s^2 3p^6 3d^{10} 4s^2 4p^6 5s^2 5p^6$

The rate of increase for the corresponding ions roughly follows the same trend as the neutral atoms, except that the rate slows down from Mg to Ca, not Ca to Sr. This may be due to the $3d$ subshell being lower in energy than the $4s$ subshell in the ions.

Atom	Atomic Radius	Electron Configuration	Ion	Ionic Radius	Electron Configuration
F	64 pm	$1s^2 2s^2 2p^5$	F^-	133 pm	$1s^2 2s^2 2p^6$
Cl	99 pm	$1s^2 2s^2 2p^6 3s^2 3p^5$	Cl^-	181 pm	$1s^2 2s^2 2p^6 3s^2 3p^6$
Br	114 pm	$1s^2 2s^2 2p^6 3s^2 3p^6 4s^2 3d^{10} 4p^5$	Br^-	196 pm	$1s^2 2s^2 2p^6 3s^2 3p^6 3d^{10} 4s^2 4p^6$
I	133 pm	$1s^2 2s^2 2p^6 3s^2 3p^6 4s^2 3d^{10} 4p^6 5s^2 4d^{10} 5p^5$	I^-	220 pm	$1s^2 2s^2 2p^6 3s^2 3p^6 3d^{10} 4s^2 4p^6 4d^{10} 5s^2 5p^6$

For the halogens, there is a similar change in the size, but the anions are larger.

The importance of nuclear charge is that the smallest ions feel the largest effect. The importance of shell structure is that the size difference is much smaller when d electrons are introduced. The importance of outer electron configuration is that the anions are much larger while cations are much smaller than the isoelectronic neutral atoms.

Chapter 14

Practical A Because a 0.9% by mass solution of NaCl has a molarity of 0.154 M, 1.6 200-mL bags are needed.

A 10% by mass glucose solution has 0.289 mole of glucose in 500 mL of solution.

Twenty hours and 50 minutes are needed to deliver this amount of glucose at 8 drops per minute.

Practical B The patient needs 2,700 grams of glucose per day, or 54,000 grams of solution, which has a volume of 53 liters. The molarity of the solution is 0.28 M. This uses up 2,900 grams of oxygen, or 10,000 liters of air.

The energy content of Megasweet is 0.85 J/mL. The volume of Megasweet needed is 8,800 liters. A 20-ounce bottle contains 0.51 kilojoule.

Practical C In $[Fe(CN)_6]^{3-}$, Fe has an oxidation number of +3. In $[Fe(CN)_6]^{4-}$, Fe has an oxidation number of +2. In going from $[Fe(CN)_6]^{3-}$ to $[Fe(CN)_6]^{4-}$, Fe changes from Fe^{3+} to Fe^{2+}, a change of one electron. This may be written as:

$$[Fe(CN)_6]^{3-} + 1\ e^- \rightarrow [Fe(CN)_6]^{4-}$$

or, more simply, as:

$$Fe^{3+} + 1\ e^- \rightarrow Fe^{2+}$$

This is the reduction half-reaction.

Chapter 15

Practical A The indicator in the acid solution is colorless. It changes color to a dark pink when the acid is neutralized, leaving the base in excess. The color change occurs when the acid form of the indicator is converted to its basic form.

Practical B Sodium benzoate is a weak base so we should titrate with a strong acid. If we know both the molarity and the volume of acid used to neutralize the sodium benzoate, we can calculate the moles of acid. This is equal to the moles of sodium benzoate. We can convert moles to grams of sodium benzoate to determine the mass present.

The solution is basic as it contains hydroxide ions. Titrate the solution with a solution of strong acid. Determine the moles of acid needed to reach the end point of the titration (complete neutralization) by multiplying the acid molarity by its volume in liters. At the end point of this titration we know that moles of acid equals moles of base. Find molarity of the base by dividing moles of base by the volume of the base titrated.

Practical C In the determination of an unknown concentration, there were 1.67×10^{-3} mole of NaOH and CH_3COOH. The CH_3COOH had a molarity of 0.0221 M.

In the determination of the molar mass, there were 4.84×10^{-4} mole of NaOH and of HA. The molar mass of HA is 122 grams mol^{-1}.

In the determination of the purity of a solid base, there were 5.34×10^{-4} mole of HCl and tris. Tris has a purity of 98.4%.

Practical D

(a) pH = 7; neutral solution
(b) 1.65 pH units
(c) 0.30 pH unit
(d) 8.00 pH units
(e) 0.30 pH unit
(f) 1.0 unit

Chapter 16

Practical A Zn^{2+} is the Lewis acid and CN^- is the Lewis base in $Zn(CN)_4^{2-}$. The equilibrium expression for this reaction is:

$$K = \frac{[Zn(CN)_4^{2-}]}{[Zn^{2+}][CN^-]^4}$$

Nitrogen has one lone pair and the oxygen atoms in both structures have two lone pairs. The presence of lone pairs makes both N and O good Lewis bases.

Practical B At pH = 5, $\dfrac{[PbT^-]}{[HT^{2-}]} = \dfrac{2.07 \times 10^4}{1}$. There is less lead dissolved in water at pH = 8 than at pH = 5.

Practical C The equilibrium expressions are: $2 \times 10^{-3} = \dfrac{[H_2CO_3]}{[CO_2]}$ and $4.45 \times 10^{-7} = \dfrac{[HCO_3^-][H^+]}{[CO_2]}$.

When the pH = 6.2, $\dfrac{[HCO_3^-]}{[CO_2]} = \dfrac{0.705}{1}$.

Practical D The solubility of $CaCO_3$ is 6.69×10^{-5} mol L^{-1}.

The equilibrium expression is $K_a = \dfrac{[H_3O^+][CO_3^{2-}]}{[HCO_3^-]}$.

You would expect the acid form $[HCO_3^-]$ at low pH and the basic form $[CO_3^{2-}]$ at high pH.

Chapter 17

Practical A There are two lone pairs on every oxygen shown in the functional groups (in bold) in these compounds. In addition, these oxygens have two single bonds in alcohols and ethers, one double bond in ketones, and both single and double bonds in esters and acids.

Practical B

(a) Glycerol

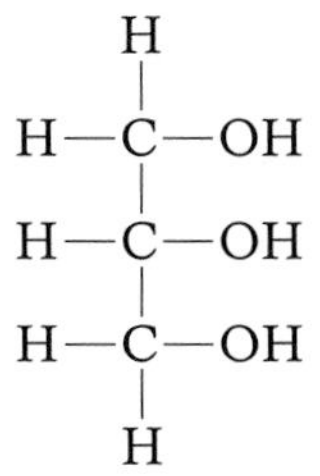

```
          H
          |
    H — C — OH
          |
    H — C — OH
          |
    H — C — OH
          |
          H
```

(b) A $C_{15}H_{31}$ chain

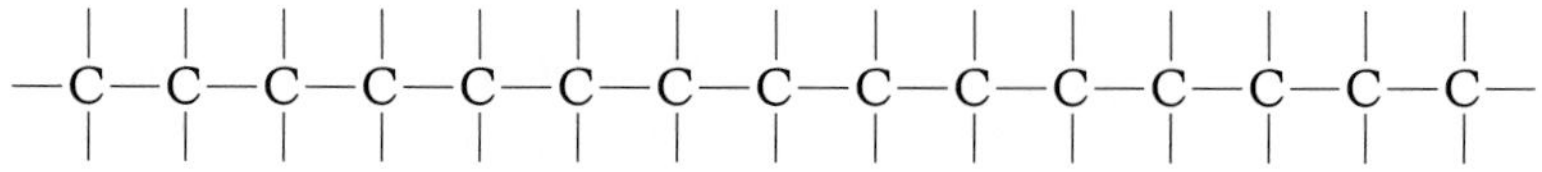

$$-\underset{|}{\overset{|}{C}}-\underset{|}{\overset{|}{C}}-\underset{|}{\overset{|}{C}}-\underset{|}{\overset{|}{C}}-\underset{|}{\overset{|}{C}}-\underset{|}{\overset{|}{C}}-\underset{|}{\overset{|}{C}}-\underset{|}{\overset{|}{C}}-\underset{|}{\overset{|}{C}}-\underset{|}{\overset{|}{C}}-\underset{|}{\overset{|}{C}}-\underset{|}{\overset{|}{C}}-\underset{|}{\overset{|}{C}}-\underset{|}{\overset{|}{C}}-\underset{|}{\overset{|}{C}}-$$

(c) Triglyceride of oleic acid, $C_{18}H_{34}O_2$

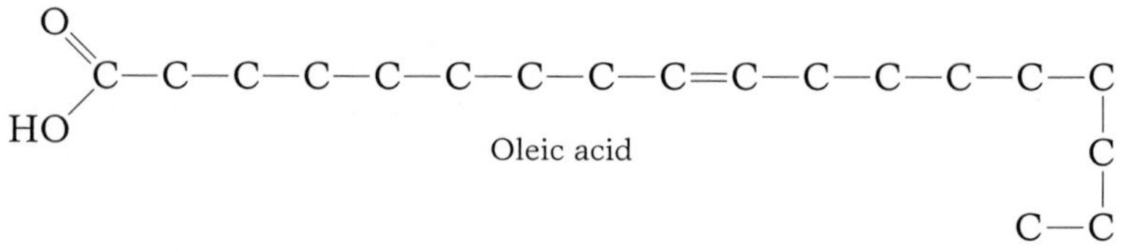

Oleic acid

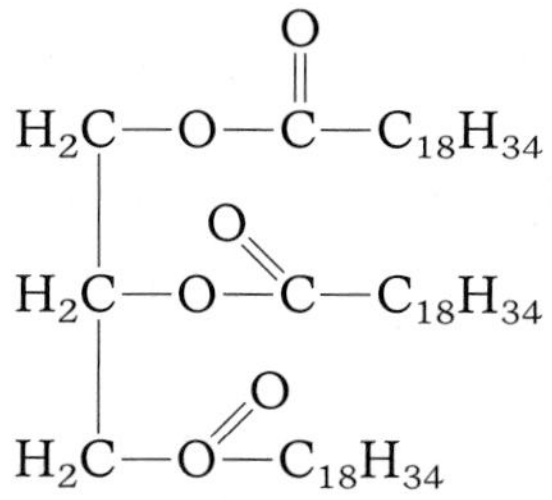

Triglyceride of oleic acid

Practical C The substances listed in Table 17-2 that are soluble in water either form ions in water solution (NaCl, sodium acetate) or contain —OH groups (ethanol, stearic acid, acetic acid). The substances that are soluble in benzene may contain —OH, —COOH, or —C—O—C— groups, but they also contain more —CH groups. These are larger molecules.

Practical D Both serine and tyrosine contain an —OH group at the end of their hydrocarbon chain.

Answers to How Are You Doing?

1.1 (a) Heterogeneous
(b) Homogeneous
(c) Heterogeneous

1.2 (a) 3   (b) 3   (c) 2   (d) 3

1.3

Symbol	At	As	Ar
Name	Astatine	Arsenic	Argon
Group Number	VII	V	VIII
Period	6	4	3

1.4 (1) CrAm
(2) OsMoNd
(3) Pd
(4) CaNe

1.5

Name	# of Protons	# of Neutrons	# of Electrons	Atomic #	Mass Number	Atomic Symbol
Oxygen	8	8	8	8	16	$^{16}_{8}O$
Copper	29	36	29	29	65	$^{65}_{29}Cu$
Sodium	11	12	11	11	23	$^{23}_{11}Na$
Aluminum	13	14	13	13	27	$^{27}_{13}Al$

1.6 Physical property; chemical property

1.7 Cu is solid.

1.8 Alkali and alkaline earth metals are both reactive with water, and both combine easily with the halogens, sulfur, and oxygen. These are chemical properties.

2.1 KBr is not molecular; PBr_3 is molecular.

2.2 Sulfur trioxide
Phosphorus tribromide
Nitrogen trichloride

2.3 5 and 8

2.4
$:\ddot{Cl}—\ddot{F}:$

2.5

$$[:N\equiv O:]^+$$

2.6

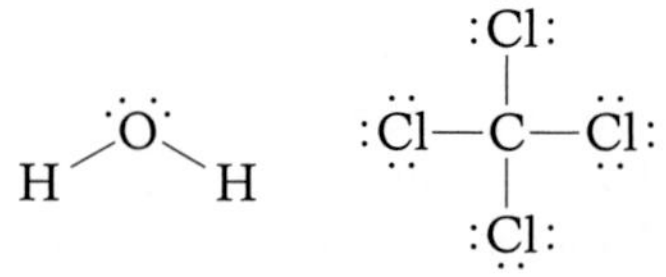

2.7

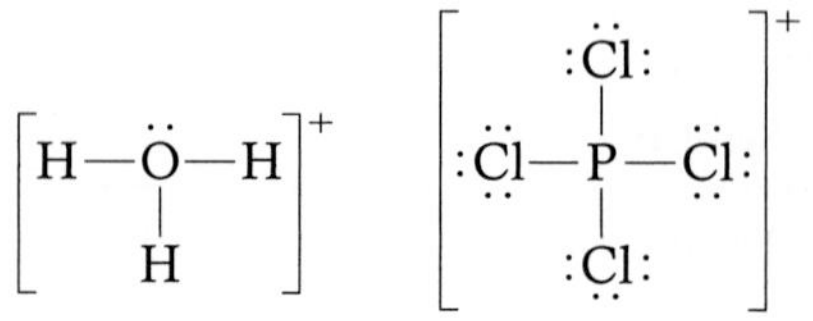

2.8

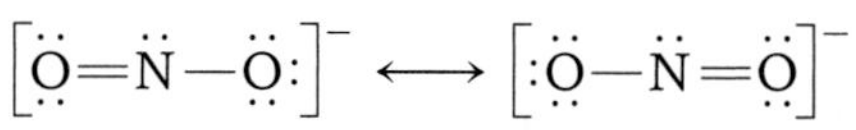

$$\left[\ddot{O}=\ddot{N}-\ddot{\underset{..}{O}}:\right]^- \longleftrightarrow \left[:\ddot{\underset{..}{O}}-\ddot{N}=\ddot{O}\right]^-$$

2.9

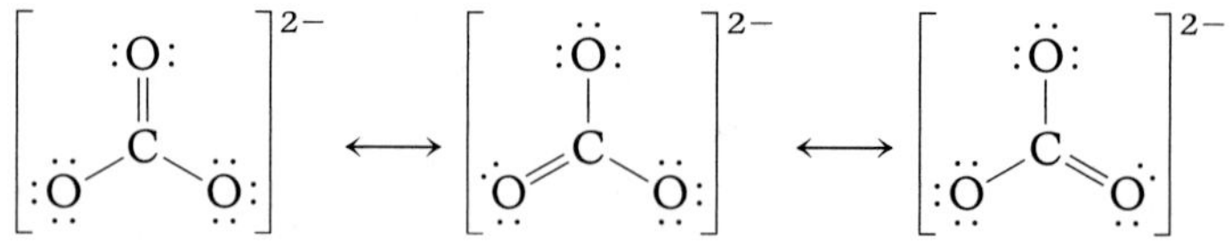

2.10

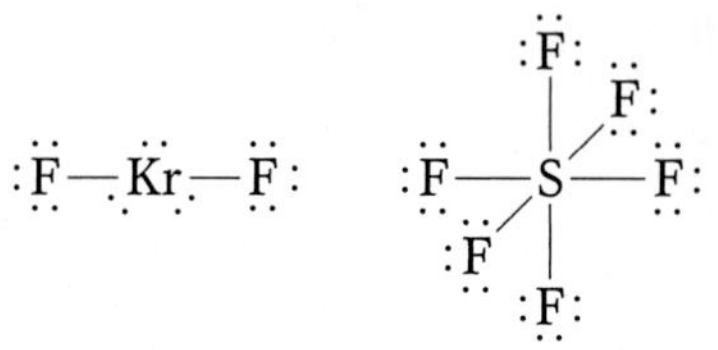

3.1

Compound	Al$_2$O$_3$	Li$_3$P	MgCl$_2$	NaF	PbO$_2$	K$_2$Se	BaO
Cation	Al^{3+}, 0	Li$^+$, 0	Mg^{2+}, 0	Na$^+$, 0	Pb^{4+}, 0	K$^+$, 0	Ba^{2+}, 0
Anion	O^{2-}, 8	P^{3-}, 8	Cl$^-$, 8	F$^-$, 8	O^{2-}, 8	Se^{2-}, 8	O^{2-}, 8

3.2 K$_3$P, BeF$_2$

3.3 (a) Calcium oxide
(b) Magnesium fluoride
(c) Barium nitride
(d) Aluminum oxide
(e) Lithium chloride
(f) Strontium iodide
(g) Cesium bromide
(h) Beryllium selenide
(i) Sodium fluoride
(j) Rubidium chloride

3.4 (a) Mg_3N_2
  (b) K_3P
  (c) Cs_2S

3.5

Name of Compound	Formula of Compound
Lithium acetate	$LiC_2H_3O_2$
Cadmium phosphate	$Cd_3(PO_4)_2$
Magnesium nitrate	$Mg(NO_3)_2$

3.6 (a) Rb^+, As^{3-}
  (b) Ca^{2+}, N^{3-}
  (c) Li^+, S^{2-}

3.7 (a) Ca^{2+}, C^-
  (b) Si^{4+}, O^{2-}

3.8 Mn^{7+}, O^{2-}

3.9 (a) Na^+, Mn^{7+}, O^{2-}
  (b) K^+, B^{3+}, O^{2-}
  (c) Li^+, H^+, C^{4+}, O^{2-}

3.10 (a) Cr^{6+}, O^{2-}
  (b) C^{3+}, O^{2-}
  (c) Mn^{7+}, O^{2-}

3.11 (a) V^{5+}, O^{2-}
  (b) Mn^{4+}, O^{2-}

3.12 Copper (II) sulfate

4.1 (a) One mole of ethane contains six moles of hydrogen atoms.
  (b) One mole of gold-containing arthritis drugs commonly contains one mole of gold and three moles of sulfur.
  (c) This sample of four molecules of calcium carbonate has four calcium ions.
  (d) Each molecule of the compound contains two atoms of nitrogen, three atoms of carbon, and ten atoms of hydrogen.

4.2 (a) 1 atom C per 3 atoms H and 1 mole of C per 3 moles of H
  (b) 5 atoms O per 1 molecule N_2O_5 and 5 moles of O per 1 mole of N_2O_5
  (c) 1 atom C per 3 atoms H and 1 mole of C per 3 moles of H

4.3 (a) CCl_4
  (b) C_2H_4O
  (c) C_4H_8

4.4 (a) Reactants: magnesium, HCl; products: hydrogen gas, $MgCl_2$
  (b) Reactant: hydrogen peroxide; products: water and oxygen gas

4.5 (a) $4\ Fe + 3\ O_2 \rightarrow 2\ Fe_2O_3$ (b) $2\ Al + 3\ Br_2 \rightarrow Al_2Br_6$

4.6 (a) $Pb + 2\ HCl \rightarrow PbCl_2 + H_2$
(b) $FeSO_4 + (NH_4)_2S \rightarrow FeS + (NH_4)_2SO_4$

5.1 (a) Physical   (b) Physical   (c) Chemical   (d) Chemical

5.2 (a) Change, chemical   (b) Property, physical

5.3 Decomposition, synthesis, synthesis

5.4 (1) $6\ Mg\ (s) + P_4\ (g) \rightarrow 2\ Mg_3P_2$ Ionic; salt formed
(2) $Si\ (s) + 2\ H_2\ (g) \rightarrow SiH_4$ Covalent; molecule

5.5 (a) $TiCl_4\ (s) + O_2\ (g) \rightarrow TiO_2 + 2\ Cl_2$
(b) $2\ NaI\ (aq) + Cl_2\ (g) \rightarrow 2\ NaCl\ (s) + I_2$
(c) $2\ Mg\ (s) + SiCl_4\ (l) \rightarrow MgCl_2 + Si$

5.6 $C\ (s) + 2\ PbO\ (s) \rightarrow 2\ Pb + CO_2$

5.7 (a) $2\ Al_2O_3\ (s) + 3\ C\ (s) \rightarrow 4\ Al\ (s) + 3\ CO_2\ (g)$
$Al^{3+} \rightarrow Al^0$ reduced
$C^0 \rightarrow C^{4+}$ oxidized
(b) $16\ Al\ (s) + 3\ S_8\ (s) \rightarrow 8\ Al_2S_3\ (s)$
$Al^0 \rightarrow Al^{3+}$ oxidized
$S_8{}^0 \rightarrow 8\ S^{2-}$ reduced

5.8 (1) N.R.
(2) $Zn\ (s) + 2\ AgNO_3\ (aq) \rightarrow 2\ Ag\ (s) + Zn(NO_3)_2$

5.9 (a) $3\ CuSO_4\ (aq) + 2\ Na_3PO_4\ (aq) \rightarrow$
$Cu_3(PO_4)_2\ (s) + 3\ Na_2SO_4\ (aq)$
(b) $Na_3SO_4\ (aq) + Ba(NO_3)_2\ (aq) \rightarrow 2\ NaNO_3\ (aq) + BaSO_4\ (s)$
(c) $MgBr_2\ (aq) + 2\ NaOH\ (aq) \rightarrow Mg(OH)_2\ (s) + 2NaBr\ (aq)$

6.1 (a) Extensive
(b) Intensive
(c) Extensive

6.2 CO; sublimation (solid to gas)

6.3 (a) Mercury is denser than manganese.
(b) Carbon tetrachloride and water do not mix and have different densities.

6.4 $V = nRT/P$ $T_2 = P_2\,V_2\,T_1/P_1V_1$
$h = 2A/b$ $V = m/d$

6.5 (a) 311 g   (b) 15.0 g

6.6 $2.16\ g/cm^3$

6.7 500 tons boron/5500 tons ore = 1/11 tons of boron per ton of ore

6.8 2,420,000 tons of ore

6.9 73.2 cm

6.10 (a) $\dfrac{1.9\ g\ nitrogen}{1\ g\ ammonia}$

(b) $\dfrac{3 \text{ g Se}}{1{,}000{,}000 \text{ g}}$

(c) $\dfrac{1 \text{ enrollment (online course)}}{50 \text{ enrollments (all courses)}}$

6.11 2.13 million tons of hydrogen

6.12 (1) 10 atoms C per 1 molecule indigo;
12 atoms H per 1 molecule indigo;
2 atoms N per 1 molecule indigo;
2 atoms O per 1 molecule indigo
(2) 5 atoms C per 6 atoms H;
5 atoms C per 1 atom N
6 atoms H per 1 atom O;
1 atom N per 1 atom O

6.13 (a) 90 atoms of O
(b) 152,500 molecules of BF_3
(c) 64 atoms of C

6.14 13 mol C per 1 mol DDT; 5 mol H per 1 mol DDT;
5 mol Cl per 1 mol DDT

6.15 0.0433 mole of glucose; 9.75 moles of H atoms

6.16 $\dfrac{3 \text{ mol Cl}}{1 \text{ mol AlCl}_3} = \dfrac{x \text{ mol Cl}}{5.21 \text{ mol AlCl}_3}$ $x = 15.6$ moles of Cl atoms

7.1 (a) 1   (b) 4   (c) 1

7.2 0.6

7.3 (a) $1/(9y^6)$
(b) $L^2 \text{ mol}^{-2} \text{ sec}^{-1}$

7.4 (a) 593,000
(b) 12.37
(c) 8.90×10^{-3}
(d) 1.256×10^2

7.5

Metric Unit	Length/Mass/Volume	Larger/Smaller than 1?	Exponential
Megaliter	Volume	Larger than 1	1×10^6 L
Nanogram	Mass	Smaller than 1	1×10^{-9} g
Millimeter	Length	Smaller than 1	1×10^{-4} m

7.6 (a) 4.1 nanometers   (b) 1.90×10^{-6} liter

7.7 (a) 7.98×10^{-9} g   (b) 1.8×10^{-3} km   (c) 3.45×10^3 μg

7.8 1.73×10^{26} molecules

7.9 1.63×10^{-4} mole

8.1 (a) 259.89329
(b) 76.0116
(c) 80.912

8.2 (a) 319.958 g/mol
(b) 373.640 g/mol
(c) 32.046 g/mol

8.3 (a) 0.08989 mol
(b) 1.09 mol
(c) 3.42 g
(d) 4,409 g

8.4 25.0 g $CaCl_2$ = 0.225 mol and 25.0 g $CaCO_3$ = 0.250 mol
Calcium carbonate has more calcium by mass.

8.5 0.368 g

8.6 1.810 L

8.7 N   21.21%
H   6.87%
P   23.46%
O   48.46%

8.8 (a) and (c) are correct empirical formulas.

8.9 When the sample mass is 100 g, % of element = grams of element. The only difference is the unit.

8.10 C_3H_6O

8.11 $C_6H_6Cl_6$ is the molecular formula.

9.1 (a) $\dfrac{3 \text{ mol } CO_2}{5 \text{ mol } O_2}$ (b) $\dfrac{1 \text{ mol } C_3H_8}{3 \text{ mol } CO_2}$ (c) $\dfrac{5 \text{ mol } O_2}{1 \text{ mol } C_3H_8}$

9.2 (a) 41.7 moles of O_2
(b) 6 moles of CO_2
(c) 4 moles of C_3H_8

9.3 (1) 300,000 moles of As
(2) 1.25 moles of NH_3
(3) 1.56 moles of O_2

9.4 0.270 mole of NO

9.5 (1) 19.97 grams of $MgCl_2$
(2) 3.05 grams of NaCl; 7.48 grams of AgCl

9.6 61.0 grams of $AlCl_3$, 7.67 grams of Al, 1.39 grams of H

9.7 35.0 grams of $WOCl_4$ is the actual yield.

10.1 The point (2,5) is located two spaces to the right and five spaces up from the origin.

10.2 The point (1,5) is on the line $y = 5x - 10$, and the point (6,10) is not.

10.3 $m = -2$

10.4 (a) "Different masses of Ca" is the independent variable; "measured volume of gas" is the dependent variable.

(b) "Limiting reactant $AgNO_3$" is the independent variable; "product AgCl" is the dependent variable.

10.5 (a) $PbNO_3$ is the independent variable; PbI_2 is the dependent variable.

(b) Graphed data:

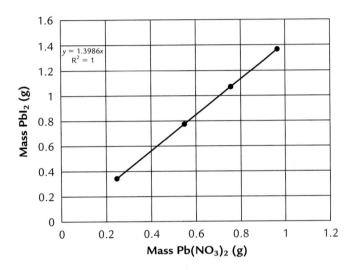

(c) $m = 10/7$ or 1.39 g PbI_2/g $Pb(NO_3)_2$

10.6 864.3 mm Hg = 1.14 atm

10.7 The initial volume is smaller than 2.90 L.

10.8 $P = 2.6$ atm; reasonable according to Boyle's law.

10.9 156 torr

10.10 373.15 K, 98.80 K, 273.15 K, -264.90 °C, -273.15 °C, -173.15 °C

10.11 $V_1 = 0.63$ L

10.12 $P = 3.73$ atm

10.13 $P = 671$ mm Hg

10.14 $P = 1.29$ atm

10.15 $V = 13.0$ L

10.16 $m = 5.99 \times 10^{-3}$ g

11.1 (a) Heat flows from the tray of water (system) into the freezer (surroundings).

(b) Heat flows from the wall (system) to you (surroundings).

11.2 (a) Iron, $q < 0$; water, $q > 0$

(b) Mold, $q > 0$; melted sugar, $q < 0$

11.3 Water must give off heat at zero degrees Celsius in order to go from liquid to solid. It undergoes a *phase change,* not a *temperature change*.

11.4 (a) 0.429 J
(b) 598 grams of water

11.5 To warm the acid requires 19.1 kJ. To vaporize the same sample requires 40.6 kJ.

11.6 H_2SO_4 + sugar → carbon + water
A chemical change evolves heat (exothermic). Heat is transferred from the reaction to the water, causing steam (heated water, phase change) to be emitted.

12.1 7

12.2 Isotope mass 133.0160: 65.11% abundance
Isotope mass 136.0998: 34.89%

12.3 $^{212}_{85}At \rightarrow\ ^{4}_{2}He +\ ^{208}_{83}Bi$

12.4 $^{190}_{75}Re \rightarrow\ ^{190}_{76}Os +\ ^{0}_{-1}\beta$

13.1 $:N\equiv N-\ddot{\underset{\cdot\cdot}{O}}:$

13.2

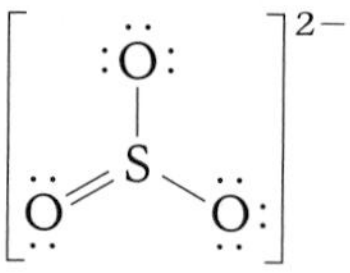

Trigonal, flat, 120 degrees

13.3 $H-C\equiv N$ Linear

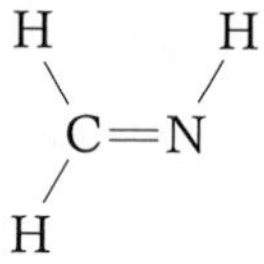

Trigonal planar

13.4

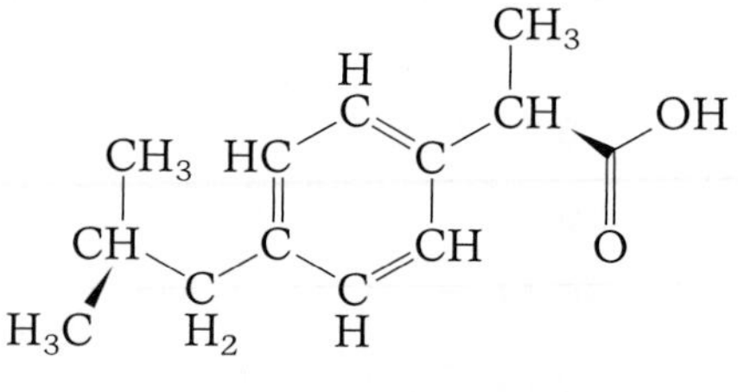

13.5 Be and Mg follow each other in group II; B and Al follow each other in group III. Mg and Al have lower ionization energies than Be and B, indicating a shell structure in the atom.

13.6 The pattern is the same, but with overall lower ionization energies.

13.7 Sc: $1s^2 2s^2 2p^6 3s^2 3p^6 4s^2 3d^1$

In: $1s^2 2s^2 2p^6 3s^2 3p^6 4s^2 3d^{10} 5s^2 4d^{10} 5p^1$

14.1 1.03 M

14.2 0.0585 mole of $Mg(C_2H_4O_2)_2$ and 0.117 mole of $(C_2H_4O_2)^-$

14.3 4.05 mL

14.4 0.595 g

14.5 62.7 mL

14.6 0.0223 M

14.7 $Sb^{3+} + 3\ e^- \rightarrow Sb^0$

This is the cathode reaction. Sb is expected to gain mass.

$Fe^0 \rightarrow Fe^{2+} + 2\ e^-$

This is the anode reaction. Fe is expected to lose mass.

14.8 $3\ Zn + 2\ Au^{3+} \rightarrow 3\ Zn^{2+} + 2\ Au$ (balanced)

14.9 $Mn + Pb^{2+} \rightarrow Mn^{2+} + Pb$ (balanced)

15.1 (a) $HClO_4 + H_2O \longrightarrow H_3O^+ + ClO_4^-$, acid
(b) $HI + H_2O \longrightarrow H_3O^+ + I^-$, acid
(c) $RbOH \longrightarrow Rb^+ + OH^-$, base

15.2

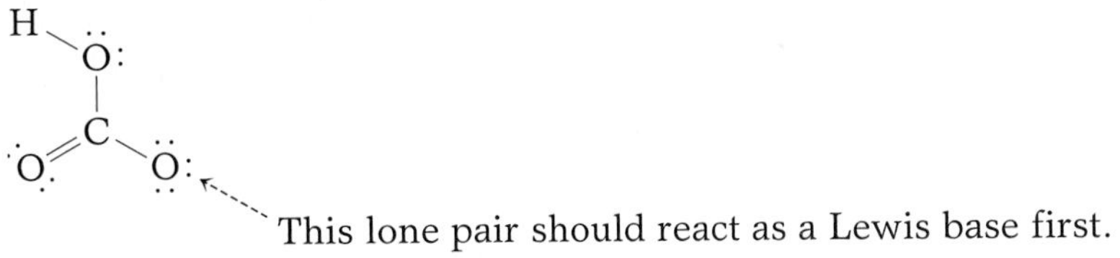

This lone pair should react as a Lewis base first.

15.3 (a) HI, H_2CO_3, HNO_3
(b) S^{2-}, F^-, CN^-

15.4 $HCO_3^- + HCl \longrightarrow Cl^- + H_2CO_3$

15.5 $HClO_3$: chloric acid
$HBrO_4$: perbromic acid
HIO_2: iodous acid

15.6 $2\ HNO_3\ (aq) + MgCO_3\ (s) \longrightarrow$
$CO_2\ (g) + Mg^{2+}\ (aq) + 2\ NO_3^-\ (aq) + H_2O$

15.7 $[OH^-] = 1.14$ M

15.8 $[OH^-] = 0.0427$ M

15.9 (a) 3
(b) 1
(c) -3

15.10

Concentration [H$_3$O$^+$]	pH
(a) 3.1×10^{-7} M	6.51
(b) 3.05×10^{-7} M	6.516
(c) 2.93×10^{-7} M	6.533
(d) 5.6×10^{-7} M	6.25
(e) 2.55×10^{-10} M	9.593
(f) 8.34×10^{-12} M	11.079

15.11 pH = 12, pOH = 2, [H$^+$] = 1×10^{-12} M, [OH$^-$] = 0.01 M

16.1 (a) $K = \dfrac{[Zn(C_3H_5O_3)^+]}{[Zn^{2+}][C_3H_5O_3^-]}$ (b) $K = \dfrac{[Ca(C_3H_5O_3)^+]}{[Ca^{2+}][C_3H_5O_3^-]}$

16.2 (a) 72 (b) 12

16.3 (a) The discriminant is negative and the solution contains complex numbers.
(b) The discriminant is positive and the solution contains real numbers.

16.4 The concentration is 0.064 M when $x = -0.004$; this is an unreasonable result as it gives a greater concentration than the initial concentration. When $x = 0.0061$, the concentration is 0.0539 M; this is a reasonable result.

16.5 9×10^{-3} M

16.6 pH = 3.47, 34% ionization

16.7 There is 5.7% ionization so the assumption of very little ionization is not verified. Using the quadratic formula, we get pH = 2.96.

16.8 $CuCO_3$ (s) $\leftrightarrow$ Cu^{2+} (aq) + CO_3^{2-} (aq); $K_{sp} = [Cu^{2+}][CO_3^{2-}]$
$CaSO_4$ (s) $\leftrightarrow$ Ca^{2+} (aq) + SO_4^{2-} (aq); $K_{sp} = [Ca^{2+}][SO_4^{2-}]$
CdC_2O_4 (s) $\leftrightarrow$ Ca^{2+} (aq) + $C_2O_4^{2-}$ (aq); $K_{sp} = [Cd^{2+}][C_2O_4^{2-}]$

16.9 (a) 8.9×10^{-14} mol L^{-1}
(b) $K_{sp} = 6.8 \times 10^{-6}$
(c) 1.1×10^{-6} mol L^{-1}

16.10 $MgCO_3 > CaCO_3 > BaCO_3$

17.1

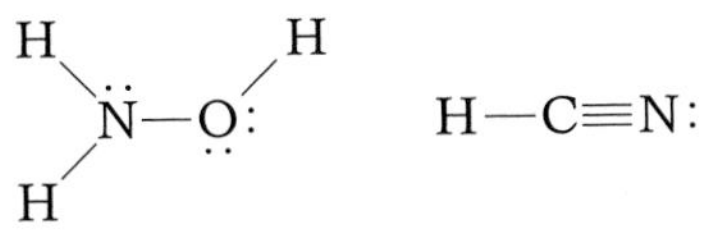

$$H-\overset{..}{\underset{..}{N}}-\overset{..}{\underset{..}{O}}: \qquad H-C\equiv N:$$

with H atoms on N.

17.2

$$H-\overset{\underset{|}{H}}{\underset{|}{C}}-\overset{..}{N}: \qquad \left[H-\overset{\underset{|}{H}}{\underset{|}{C}}-\overset{..}{\underset{..}{O}}: \right]^{-}$$

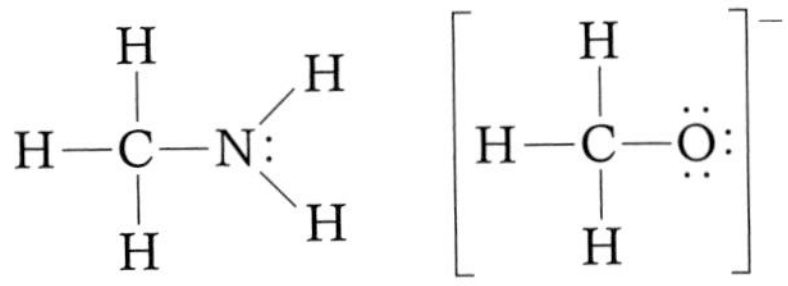

17.3 Molecules (a), (b), and (c) are isomers.

17.4 Only molecule (a) is saturated.

17.5 The carbons in the ring are each at the vertex point of a regular pentagon. As all of the carbons are equivalent, there is only one isomer when a single C=C is drawn.

17.6 All of the atoms in the structure will have the correct number of bonds if we draw a double bond between the C and O.

17.7 If we redraw the molecule and switch the positions of the final C atoms, we will end with the same drawing as when we began. Thus, redrawing the molecule in this way does not give us an isomer of the first molecule.

17.8 In Figure 17-14, the first molecule has an —OH (alcohol) group, the second molecule has a C—O—C (ether) group, and the third molecule has a C=O (carbonyl) group.

In Example 17.11: (a) alcohol group, (b) both a carbonyl and an alcohol group, (c) ether group, (d) acid group, (e) ester group.

17.9 We can arrive at two different drawings of deoxyribose by drawing the OH on different C atoms.

17.10

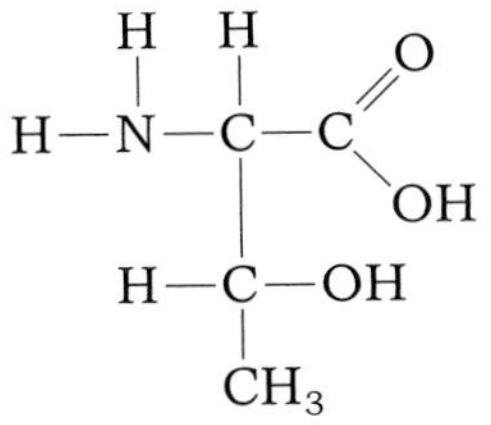

$$H-N-\overset{\underset{|}{H}}{\underset{|}{C}}-C\overset{\displaystyle O}{\underset{\displaystyle OH}{\diagup}}$$

The functional groups are —OH (alcohol) and —COOH (acid).

The amino acid valine results.

17.11 Deoxyguanonsine

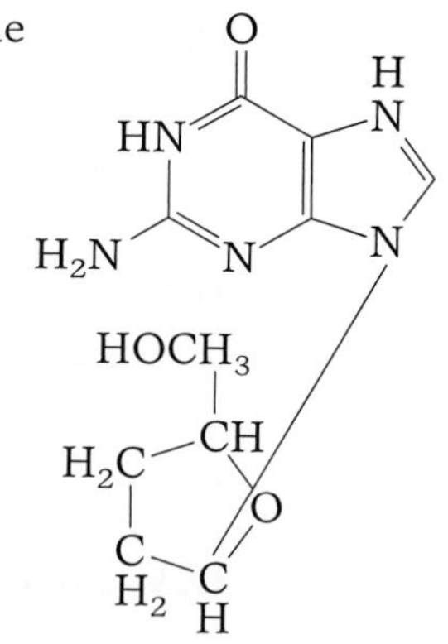

Deoxyguanosine phosphate

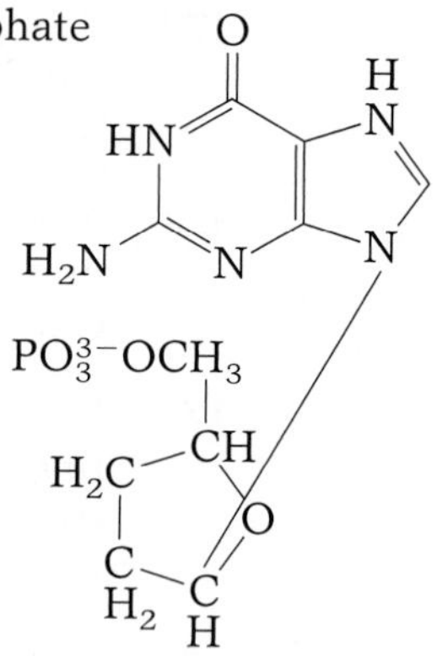

17.12 Leu-Phe-Ser-Thr-Gly

Answers to Odd-Numbered Problems

Chapter 1

1. Homogeneous substances have the same properties throughout the sample, while heterogeneous substances have different properties in various parts of the sample. For example, every sample of pure water will boil at 100 °C and freeze at 0 °C. The contaminants in a sluggish stream or lake may change the freezing and boiling points. The stream or lake represents a heterogeneous system.

3. A homogeneous sample of air is one in which a contaminant (a substance not usually found in air) is added to the air in such a manner that the air appears unchanged. Some possibilities are colorless gases, such as carbon monoxide and sulfur trioxide. A heterogeneous sample of air contains substances we can see, such as smoke or dust particles.

5. A mixture is usually heterogeneous; its properties will vary from place to place in the sample. The only mixture that is homogeneous is a solution. Mixtures, including solutions, can be separated by physical means, such as evaporation or boiling (distillation). A pure substance will have constant properties and cannot be broken down into smaller components by physical means.

7. (a) A pure substance that contains only one kind of atom
(b) A pure substance that contains more than one kind of atom in a fixed ratio and arrangement to one another

9. (a) Element (b) Compound (c) Element (d) Compound

11. (a) Be (b) O (c) He

13. (a) Nitrogen, hydrogen, phosphorus, oxygen
(b) Calcium, carbon, hydrogen, oxygen
(c) Aluminum, chlorine, oxygen

15. 14, Si, silicon   94, Pu, plutonium

17.

Element	Symbol	# of Protons	# of Neutrons	# of Electrons	Atomic Number	Mass Number
Lithium	Li	3	4	3	3	7
Cadmium	Cd	48	65	48	48	113
Iodine	I	53	74	53	53	127
Tungsten	W	74	111	74	74	185

19.

Symbol	Name	# of Protons	# of Neutrons	# of Electrons
$^{11}_{5}B$	Boron	5	6	5
$^{16}_{8}O$	Oxygen	8	8	8
$^{19}_{9}F$	Fluorine	9	10	9
$^{24}_{12}Mg$	Magnesium	12	12	12
$^{121}_{51}Sb$	Antimony	51	70	51
$^{202}_{80}Hg$	Mercury	80	122	80

21. The chemical formula of a compound gives information about the identity of the compound and its elemental makeup; sometimes we can infer whether the compound is a solid, a liquid, or a gas.

23. (a) No   (b) Yes; compound   (c) Yes; element

25. (a) Gas   (d) Gas
(b) Gas   (e) Liquid
(c) Gas

27. Nonmetals tend to be soft, dull, and crumble or fracture when pounded.

29. (a) $CaCl_2$ (b) BaI_2 (c) MgF_2

31. (a) True
(b) False: Some group VI elements are semimetallic.
(c) False: Almost all group VII elements are nonmetals, one is a semimetal.

33. Concept question: (a) When sugar is treated with H_2SO_4 to release steam and elemental carbon, carbon is not sweet because sweetness is a property of the compound sugar, not the elements of which it is made.   (b) Water is made up of hydrogen and oxygen but it cannot be used as a source of fuel because water is the *product* of combustion. We would have to reverse the process of combining H_2 and O_2, which would cost at least as much energy as the resulting fuel would provide. Again, the elemental properties of oxygen and hydrogen are not the same as the properties of the compound water.
(c) Same argument as above.

35. (a) Nonmetal   (b) Nonmetal   (c) Nonmetal   (d) Nonmetal

37. The only observation to make is that the product gas is less flammable than the reactant gas.

39. (a) Element   (f) Element
(b) Element   (g) Compound
(c) Compound   (h) Element
(d) Compound   (i) Element
(e) Element   (j) Compound

41. The moth balls are sublimating and the wax is melting. Only the water is undergoing a chemical change.

43. (a) A pure substance, either an element or a compound
(b) Compound
(c) A pure substance, either an element or a compound

45. 33

47. $^{115}_{48}$Cd

49. Chemical; chemical; physical; chemical

51. (a) Element   (b) Element   (c) Compound   (d) Compound

53. (a) Na   (b) P   (c) F

55.

Name	# of Protons	# of Neutrons	# of Electrons	Atomic Number	Mass Number	Atomic Symbol
Chlorine	17	18	17	17	35	$^{35}_{17}$Cl
Barium	56	81	56	56	137	$^{137}_{56}$Ba
Radium	88	138	88	88	226	$^{226}_{88}$Ra

57.

Symbol	Name	Period	Group
K	Potassium	4	1
N	Nitrogen	2	5
O	Oxygen	2	6
Cl	Chlorine	3	7
Ne	Neon	2	8
S	Sulfur	3	6

59. Physical; physical; chemical; chemical; chemical; chemical

61. (a) A molecule consists of two or more atoms covalently bonded together through sharing electrons. H_2O and NH_3 are examples of molecular substances, as they contain only nonmetals. A formula unit is the smallest whole number ratio in an extended (ionic) structure. Extended structures are typical of compounds that contain both a metal and a nonmetal. NaCl is an example of a formula unit.

(b) A compound is a pure substance if it contains all the same kind of molecules (if it is molecular) or formula units (if it has an extended structure).

63. (a) 1 magnesium, 1 nitrogen, 4 hydrogen, 1 arsenic, 4 oxygen
(b) 2 sodium, 4 boron, 17 oxygen, 20 hydrogen

65. (a) 15 (b) $^{41}_{20}$C (c) $^{80}_{35}$Br

67. Gold is easy to work with because it is easily shaped (malleable) and has a low melting point.

69. b

71. (a) Malleability (b) Ductility

73. (a) Gas (b) Gas (c) Solid (d) Liquid

75. c

77. b

79. (a) Cs (b) Se (c) I

Chapter 2

1. Three "official" names from outside of chemistry that are systematic, but that also might seem "strange," are President Pro Tempore, Your Highness, *T. rex*, for example.

3. SiH_4 PCl_3 HI SF_6

5. (a) NI_3 (b) P_2O_5 (c) PBr_3 (d) Cl_2O (e) SF_4

7. (a) Dinitrogen monoxide   (d) Dinitrogen tetraoxide
(b) Nitrogen monoxide   (e) Dinitrogen pentoxide
(c) Dinitrogen trioxide
These five compounds have different N : O ratios and different properties.

9. (a) Nitrogen trihydride (ammonia)
(b) Nitrogen trifluoride

11. See the *Practice of Chemistry* Web site.

13. $:\!\ddot{Br}\!-\!\ddot{Br}\!:$ $H\!-\!\ddot{Br}\!:$ $:\!\ddot{Br}\!-\!\ddot{F}\!:$

15.

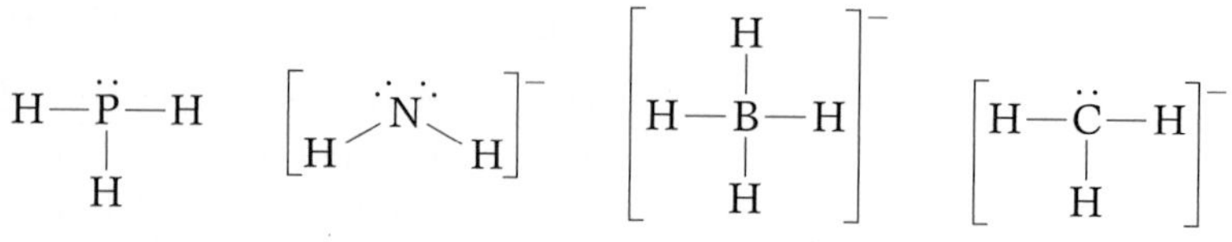

17. (a) $:C\equiv O:$ (b) $H-\overset{\cdot\cdot}{S}-H$ (c) $H-C\equiv N:$

19.

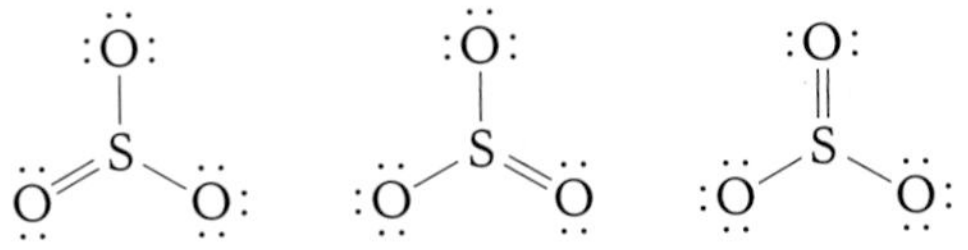

$$\overset{\cdot\cdot}{O}=\underset{O:}{\overset{:O:}{S}}-\overset{\cdot\cdot}{O}: \qquad \underset{:O:}{:O}=\underset{}{\overset{:O:}{S}}=\underset{}{\overset{}{O}}: \qquad \underset{:O:}{:O}-\underset{}{\overset{:O:}{S}}=\overset{\cdot\cdot}{O}:$$

21.

$$\left[\underset{:O:}{:O-\underset{:O:}{\overset{:O:}{Si}}-O:}\right]^{4-} \left[\underset{:O:}{:O-\underset{:O:}{\overset{:O:}{P}}-O:}\right]^{3-} \left[\underset{:O:}{:O-\underset{:O:}{\overset{:O:}{S}}-O:}\right]^{2-} \left[\underset{:O:}{:O-\underset{:O:}{\overset{:O:}{Cl}}-O:}\right]^{-}$$

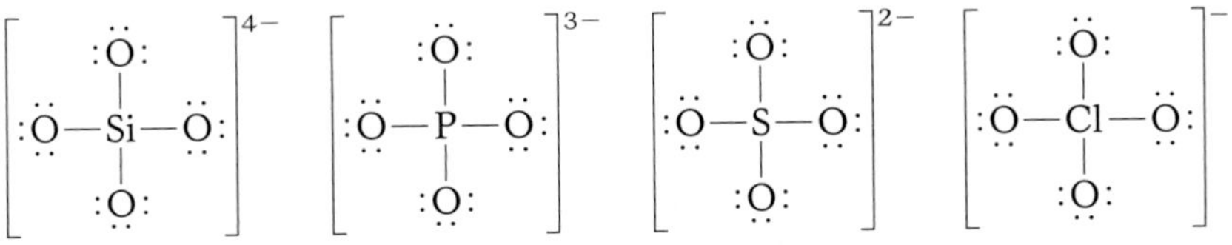

All structures above have 32 valence electrons (isoelectronic).

23. (a) $H-\overset{\cdot\cdot}{Se}-H$ (b) $\overset{\cdot\cdot}{O}=\overset{\cdot\cdot}{S}-\overset{\cdot\cdot}{O}:$

25. (a) $\left[:\overset{\cdot\cdot}{O}-\underset{:O:}{\overset{\cdot\cdot}{Cl}}-\overset{\cdot\cdot}{O}:\right]^{-}$ (b) $\left[:\overset{:F:}{F}-\underset{:F:}{\overset{}{P}}-\overset{\cdot\cdot}{F}:\right]^{+}$ (c) $\left[:\overset{\cdot\cdot}{O}-\overset{\cdot\cdot}{N}=\overset{\cdot\cdot}{O}\right]^{-}$

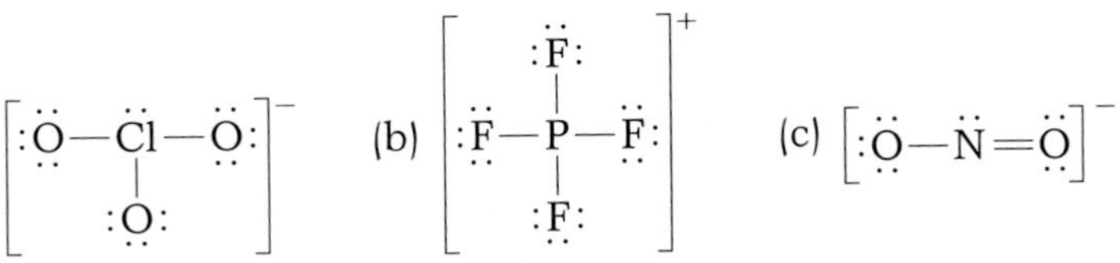

27. $\left[:\overset{\cdot\cdot}{Br}-\overset{\cdot\cdot}{O}:\right]^{-}$ $\left[:\overset{\cdot\cdot}{O}-\overset{\cdot\cdot}{Br}-\overset{\cdot\cdot}{O}:\right]^{-}$ $\left[:\overset{\cdot\cdot}{O}-\underset{:O:}{\overset{\cdot\cdot}{Br}}-\overset{\cdot\cdot}{O}:\right]^{-}$ $\left[:\overset{:O:}{O}-\underset{:O:}{\overset{}{Br}}-\overset{\cdot\cdot}{O}:\right]^{-}$

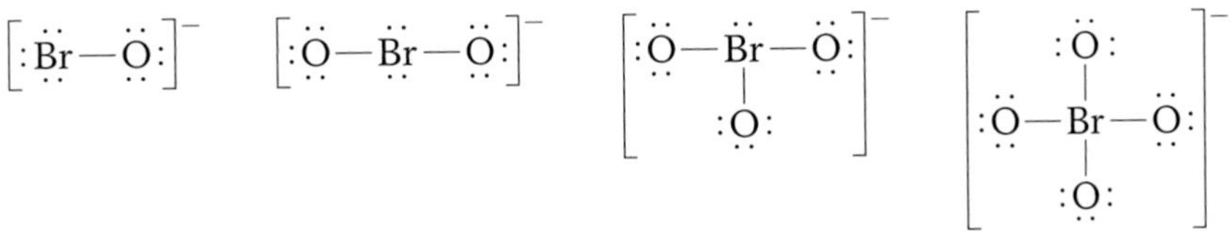

29. See the *Practice of Chemistry* Web site.

31. (a) 4   (b) 6   (c) 6   (d) 8   (e) 2   (f) 6   (g) 6

33. AlF_4^- $\left[:\overset{:F:}{F}-\underset{:F:}{\overset{}{Al}}-\overset{\cdot\cdot}{F}:\right]^{-}$

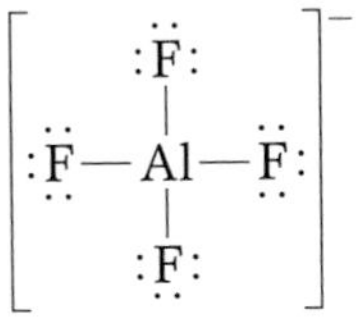

35.

Structure	Skeleton OK?	Electrons in Structure	Electrons from Atoms	Electrons on Outer Atoms	Electrons on Central Atom
XeO_4	Yes	32	32; 8 + 4(6)	8 on each	16
NF_3	No	24 (should have 26)	26; 5 + 3(7)	8 on each	8
CO_2	No	18 (should have 16)	16; 4 + 2(6)	8 on each	10
SO_2	Yes	18	18; 6 + 2(6)	S has 8 O has 8	8

37. 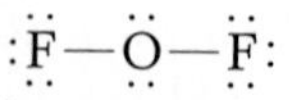 $:\!\ddot{F}\!-\!\ddot{O}\!-\!\ddot{F}\!:$ This structure should have 20 electrons; the incorrect one has only 18.

39. (a) Oxygen difluoride
(b) Dinitrogen tetraoxide
(c) Carbon disulfide

41. (a)

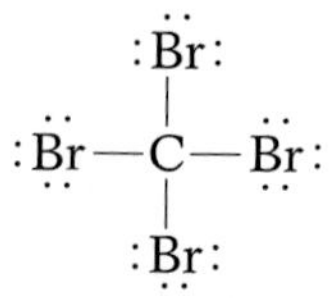

$$:\!\ddot{Br}\!-\!\underset{\underset{:\ddot{Br}:}{|}}{\overset{\overset{:\ddot{Br}:}{|}}{C}}\!-\!\ddot{Br}\!:$$

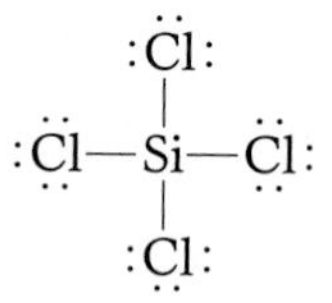

(b) $:\!\ddot{Cl}\!-\!\underset{\underset{:\ddot{Cl}:}{|}}{\overset{\overset{:\ddot{Cl}:}{|}}{Si}}\!-\!\ddot{Cl}\!:$ (c) $\ddot{O}\!=\!\ddot{S}\!-\!\ddot{O}\!:$

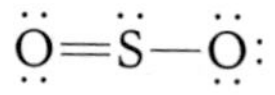

43. (a) SbF_5 (b) Sb_2O_3

45. Dichlorine heptaoxide; chlorine dioxide; dichlorine monoxide

47. Se: 6 valence electrons, 28 core electrons
I: 7 valence electrons, 46 core electrons

49. b

51. The components of a Lewis structure are:
(1) The skeleton showing connections between atoms
(2) Total number of valence electrons distributed among the atoms in the skeleton so that each atom (except H) has 8 electrons around it.

53. 10 electrons

55. ClO^- has 14 valence electrons; ClO_4^- has 32 valence electrons

57. 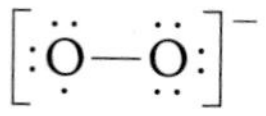 $\left[:\!\ddot{O}\!-\!\ddot{O}\!:\right]^-$

This structure has an odd number of valence electrons, resulting in one unpaired electron.

59. 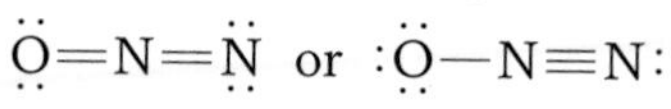 $\ddot{O}\!=\!N\!=\!\ddot{N}$ or $:\!\ddot{O}\!-\!N\!\equiv\!N\!:$

Chapter 3

1. Si is in group IV and has 4 valence electrons. Ga is in group III and has 3 valence electrons. As is in group V and has 5 valence electrons. In is in group III and has 3 valence electrons. P is in group V and has 5 valence electrons. Cd is a transition metal with 2 valence electrons. S is in group VI with 6 valence electrons. One pattern to see is that the atoms in each pair are "equidistant" from Si in group IV.

3. The ions of nonmetals tend to have *eight* valence electrons.

5.

Ion	Valence Electrons in the Ion
Sn^{2+}	$4 - 2 = 2$
Mg^{2+}	$2 - 2 = 0$
Ca^{2+}	$2 - 2 = 0$
C^{3-}	$4 + 3 = 7$
S^{2-}	$6 + 2 = 8$
P^{3+}	$5 - 3 = 2$

7. (a) OK   (b) $SrCl_2$ (c) KBr   (d) Al_2O_3 (e) AgI

9. (a) Potassium nitride   (b) Barium sulfide   (c) Cadmium oxide

11. (a) Na_2CrO_4 (b) $Ba(C_2H_3O_2)_2$ (c) $Al(NO_2)_3$

13. (a) $KMnO_4$ (b) $Cd(CN)_2$ (c) $AgNO_3$

15. (a) SO_4^{2-} (b) SO_3^{2-} (c) NO_3^- (d) NO_2^- (e) PO_4^{3-} (f) PO_3^{3-}

When two or more polyatomic ions contain oxygen and the same nonmetal, the *-ate* ending is used for the ion that contains more oxygen atoms; the *-ite* ending is used for the ion that contains fewer oxygen atoms. Hence, NO_3^- is named *nitrate,* and NO_2^- is named *nitrite.*

17. (a) Potassium oxalate
(b) Silver chromate
(c) Aluminum carbonate
(d) Magnesium perbromate

19. The convention that electrons sit in pairs around atoms to form bonds is not correct, but is the basis of the very useful Lewis dot theory. Also, the idea that ionic substances contain real, full negative and positive charges is a convention.

21. (a) Cs^+, $O^{1/2-}$ (b) Na^+, $S^{-1/5}$ (c) $Pb^{+8/3}$, O^{2-}

23. (a) Pb^{2+}
(b) Sn^{4+}
(c) Cu^+
(d) Fe^{3+}

25.

Compound	Charge on Ions	Number of Valence Electrons in Ions
PbO_2	$Pb^{4+}O^{2-}$	Pb^{4+}: 0; O^{2-}: 8
PbO	$Pb^{2+}O^{2-}$	Pb^{2+}: 2; O^{2-}: 8
In_2Se_3	$In^{3+}Se^{2-}$	In^{3+}: 0; Se^{2-}: 8
InI_3	$In^{3+}I^-$	In^{3+}: 0; I^-: 8
$GaCl_2$	$Ga^{2+}Cl^-$	Ga^{3+}: 0; Cl^-: 8

27. (a) Iron (II) nitrate
(b) Aluminum perchlorate
(c) Calcium hypochlorite
(d) Uranium (IV) sulfate

29. (a) $AgNO_3$
(b) $Pu(SO_4)_2$
(c) $PbSO_4$

31. (a) K^+, Cr^{6+}, O^{2-} (c) Co^{3+}, N^{5+}, O^{2-}
(b) Sn^{2+}, Br^- (d) Mn^{2+}, Cr^{6+}, O^{2-}

33.

Compound	Cation	Anion	Name
$Fe(NO_3)_3$	Fe^{3+} iron (III)	NO_3^- nitrate	Iron (III) nitrate
$CuSO_4$	Cu^{2+} copper (II)	SO_4^{2-} sulfate	Copper (II) sulfate
Li_2CO_3	Li^+ lithium	CO_3^{2-} carbonate	Lithium carbonate
K_3N	K^+ potassium	N^{3-} nitride	Potassium nitride
$Sn(OH)_2$	Sn^{2+} tin (II)	OH^- hydroxide	Tin (II) hydroxide

35.

	Cl^-	S^{2-}	N^{3-}	F^-	NO_3^-	SO_4^{2-}	PO_4^{3-}
Na^+	$NaCl$	Na_2S	Na_3N	NaF	$NaNO_3$	Na_2SO_4	Na_3PO_4
Mg^{2+}	$MgCl_2$	MgS	Mg_3N_2	MgF_2	$Mg(NO_3)_2$	$MgSO_4$	$Mg_3(PO_4)_2$
Al^{3+}	$AlCl_3$	Al_2S_3	AlN	AlF_3	$Al(NO_3)_3$	$Al_2(SO_4)_3$	$AlPO_4$
NH_4^+	NH_4Cl	$(NH_4)_2S$	$(NH_4)_3N$	NH_4F	NH_4NO_3	$(NH_4)_2SO_4$	$(NH_4)_3PO_4$
Cr^{3+}	$CrCl_3$	Cr_2S_3	CrN	CrF_3	$Cr(NO_3)_3$	$Cr_2(SO_4)_3$	$CrPO_4$
Cu^{2+}	$CuCl_2$	CuS	Cu_3N_2	CuF_2	$Cu(NO_3)_2$	$CuSO_4$	$Cu_3(PO_4)_2$

37. (a) Barium sulfite
(b) Calcium chlorite
(c) Cadmium nitrate
(d) Aluminum acetate

39. (a) $\overset{3+ \ -1}{CrCl_3}$ (b) $\overset{4+ \ 2-}{MnO_2}$ (c) $\overset{+ \ 2-}{Cu_2O}$ (d) $\overset{2+ \ 2-}{CuO}$ (e) Zn^{2+}; F_2^-

41. (a) Mn^{4+} (b) Co^{3+} (c) Ni^+ (d) Cr^{2+}

43. (a) Na^+ (b) Ti^{3+} (c) Mn^{2+} (d) K^+

45. (a) $CaSO_3$ (b) $FeCO_3$ (c) $Al_2(HPO_4)_3$

47. Both S and O are in group VI, so compounds of Al with group VI elements should be similar in formula: Ga_2O_3, Ga_2S_3.

49. AsO_4^{3-}, AsO_3^{3-}

51. (a) $Mg(IO)_2$ (b) $Mn_2(Cr_2O_7)_3$ (c) $Mg(MnO_4)_2$

53. (a) $\overset{2+\;5+\;2-}{Co(NO_3)_2}$ (b) $\overset{4+\;6+\;2-}{Pt(SO_4)_2}$

55. (a) $\overset{+\;6+\;2-}{Ag_2Cr_2O_7}$ (b) $\overset{2+\;6+\;2-}{BaCr_2O_7}$ (c) $\overset{3+\;6+\;2-}{Al_2(Cr_2O_7)_3}$

57. (a) $\overset{+\;-1/2}{CsO_2}$ (b) $\overset{+\;-1/2}{KO_2}$

Chapter 4

1. "One half of a mole" is acceptable because half a mole is half of a number; no division of an ethane molecule takes place.

3. (a) For a dozen cakes, you will need three dozen eggs.
(b) When you make sandwiches for six dozen students, you will need six loaves of bread.
(c) Building a dozen houses requires about twelve thousand pieces of wood.

5. (a) We find that there are three atoms of iron in one formula unit of iron (III) oxide.
(b) For one molecule of sulfur hexafluoride there are six atoms of F for every atom of sulfur.
(c) "White" phosphorus is a molecular substance with four P atoms per molecule of white phosphorus.

7. (a) 3 atoms Br per molecule of compound
(b) 1 atom P per molecule of compound
(c) 13 atoms H per molecule of compound
(d) 4 atoms Si per molecule of compound

9. (a) 1 mole Pt per molecule, 6 moles F per molecule: PtF_6
(b) 3 moles Ti per molecule, 4 moles PO_3 per molecule: $Ti_3(PO_4)_4$
(c) 1 mole NH_4 per molecule, 1 mole Fe per molecule, 2 moles SO_4 per molecule: $FeNH_4(SO_4)_2$

11. 1,200 moles

13. Reactants: Mg and O_2
Product: MgO

15. Reactants: Al and F_2
Product: AlF_3

17. Each side has 1 C, 4 H, and 4 O atoms.

$$H-\overset{\overset{\displaystyle H}{|}}{\underset{\underset{\displaystyle H}{|}}{C}}-H + \ddot{O}{=}\ddot{O} + \ddot{O}{=}\ddot{O} \longrightarrow \ddot{O}{=}C{=}\ddot{O} + \underset{H}{\overset{\ddot{O}\cdot}{\diagup}}\diagdown_H + \underset{H}{\overset{\cdot\ddot{O}}{\diagup}}\diagdown_H$$

19. $H-H + \ddot{\underset{..}{C}l}-\ddot{\underset{..}{C}l}{:} \longrightarrow H-\ddot{\underset{..}{C}l}{:} + H-\ddot{\underset{..}{C}l}{:}$

21. (a) To make a formula unit of copper (II) nitrate from copper requires two molecules of nitric acid, HNO_3.
(b) When we get two atoms of oxygen by decomposition of hydrogen peroxide we also form one molecule of water.
(c) Conversion of two molecules of methane to one molecule of ethanol also requires one molecule of oxygen.

23. (a) $2\ HgO \rightarrow 2\ Hg + O_2$
(b) $2\ H_2O_2 \rightarrow 2\ H_2O + O_2$

25. (a) $Cu + 2\ AgNO_3 \rightarrow 2\ Ag + Cu(NO_3)_2$
(b) $2\ Al + 3\ Sn(NO_3)_2 \rightarrow 2\ Al(NO_3)_3 + 3\ Sn$
(c) $3\ Ba(C_2H_3O_2)_2 + 2\ (NH_4)_3PO_4 \rightarrow Ba_3(PO_4)_2 + 6\ NH_4C_2H_3O_2$

27. (a) There are two atoms of carbon in one molecule of ethane.
(b) Decomposition of germane, GeH_4, gives two molecules of hydrogen gas for every molecule of germane.
(c) The destruction of a molecule of nitrogen dioxide in a car's catalytic converter gives one molecule of nitrogen gas for two molecules of nitrogen dioxide.

29. Four moles

31. (a) $2\ Fe(OH)_3 + 3\ H_2SO_4 \rightarrow Fe_2(SO_4)_3 + 6\ H_2O$
(b) $2\ NaCl + 2\ H_2O \rightarrow Cl_2 + H_2 + 2\ NaOH$
(c) $Na_2NH + 2\ H_2O \rightarrow NH_3 + 2\ NaOH$
(d) $3\ Al + 3\ NH_4ClO_4 \rightarrow Al_2O_3 + AlCl_3 + 3\ NO + 6\ H_2O$
(e) $3\ SCl_2 + 4\ NaF \rightarrow SF_4 + S_2Cl_2 + 4\ NaCl$
(f) $UO_2 + 4\ HF \rightarrow UF_4 + 2\ H_2O$
(g) $N_2O_5 + H_2O \rightarrow 2\ HNO_3$

33. (a) 4 N, 10 O, 12 H
(b) 10 C, 20 H, 30 O
(c) 2 S, 6 O

Drawn as follows:

$$\text{(a) }4\ \ \underset{\underset{\displaystyle H}{|}}{\overset{\overset{\displaystyle H\diagdown\ \diagup H}{\cdot\cdot}}{N}} + 5\ \ddot{O}{=}\ddot{O} \longrightarrow 4\ \ddot{N}{=}\ddot{O} + 6\ \underset{H\ \ \ H}{\overset{\cdot\ddot{O}\cdot}{\diagup\diagdown}}$$

(b) 2

$$\text{H-}\underset{\underset{\text{H}}{|}}{\overset{\overset{\text{H}}{|}}{\text{C}}}\text{-}\underset{\underset{\text{H}}{|}}{\overset{\overset{\text{H}}{|}}{\text{C}}}\text{-}\underset{\underset{\text{H}}{|}}{\overset{\overset{\text{H}}{|}}{\text{C}}}\text{-}\underset{\underset{\text{H}}{|}}{\overset{\overset{\text{H}}{|}}{\text{C}}}\text{-}\underset{\underset{\text{H}}{|}}{\overset{\overset{\text{H}}{|}}{\text{C}}}\text{-H} + 15\; \ddot{\text{O}}=\ddot{\text{O}} \longrightarrow$$

$$10\; \ddot{\text{O}}=\text{C}=\ddot{\text{O}} + 10\; \underset{\text{H}\quad\text{H}}{\overset{\cdot\ddot{\text{O}}\cdot}{\diagdown}}$$

(c) $2\; \ddot{\text{O}}=\ddot{\text{S}}=\ddot{\text{O}} + \ddot{\text{O}}=\ddot{\text{O}} \longrightarrow 2\; \underset{\ddot{\text{O}}\qquad\ddot{\text{O}}}{\overset{\overset{\cdot\ddot{\text{O}}\cdot}{\|}}{\text{S}}}$$

35. (a) Potassium bromide plus chlorine gas gives potassium chloride and bromine.

(b) Hydrogen chloride and ammonia gas give ammonium chloride.

(c) Magnesium sulfide plus sulfuric acid gives magnesium sulfate and hydrogen sulfide.

(d) Potassium carbonate and hydrogen chloride react to give water, carbon dioxide, and potassium chloride.

37. (a) $2\text{ KBr }(s) + \text{Cl}_2\ (g) \rightarrow 2\text{ KCl }(s) + \text{Br}_2\ (l)$

(b) $\text{HCl }(g) + \text{NH}_3\ (g) \rightarrow \text{NH}_4\text{Cl }(s)$

(c) $\text{MgS }(s) + \text{H}_2\text{SO}_4\ (aq) \rightarrow \text{MgSO}_4\ (s) + \text{H}_2\text{S }(s)$

(d) $\text{K}_2\text{CO}_3\ (s) + 2\text{ HCl }(l) \rightarrow \text{H}_2\text{O }(l) + \text{CO}_2\ (g) + 2\text{ KCl }(s)$

39. (a) C_5H_{12} (b) AgNO_3 (c) N_2O_4

41. $\text{Mn}_3\text{O}_4,\ \overset{8/3+}{\text{Mn}}$, manganese (II, III) oxide

43. $\text{Mn}_2\text{O}_4 \rightarrow \text{MnO}_2$, manganese (IV) oxide

45. $2\text{ C} + \text{O}{=}\text{O} \rightarrow 2\text{ C}{\equiv}\text{O}$

47. $6\text{ P} + 5\text{ KClO}_3 \rightarrow 5\text{ KCl} + 3\text{ P}_2\text{O}_5$

49. (a) $4\text{ Al} + 3\text{ O}_2 \rightarrow 2\text{ Al}_2\text{O}_3$

(b) $\text{Mg(NO}_3)_2 + 2\text{ NaCl} \rightarrow 2\text{ NaNO}_3 + \text{MgCl}_2$

(c) $2\text{ Al} + 6\text{ HCl} \rightarrow 2\text{ AlCl}_3 + 3\text{ H}_2$

51. (a) 1 K, 1 Cl, 3 O

(b) 2 Al, 3 S, 12 O

(c) 3 U, 4 P, 12 O

(d) 2 N, 8 H, 1 C, 3 O

53. $16/3\text{ KClO}_3 + \text{S}_8 \rightarrow 8\text{ SO}_2 + 16/3\text{ KCl}$

$16\text{ KClO}_3 + 3\text{ S}_8 \rightarrow 24\text{ SO}_2 + 16\text{ KCl}$

The ratio is 2:3.

55. $1\ P_4S_3\ +\ 16/3\ KClO_3\ \rightarrow\ 2\ P_2O_5\ +\ 3\ SO_2\ +\ 16/3\ KCl$
$3\ P_4S_3\ +\ 16\ KClO_3\ \rightarrow\ 6\ P_2O_5\ +\ 9\ SO_2\ +\ 16\ KCl$
(a) Tetraphosphorus trisulfide, potassium chlorate, diphosphorus pentoxide, sulfur dioxide, potassium chloride
(b) $P_4\overset{2-}{S_3}$; $\overset{4+}{S}O_2$
(c) KCl^- ; $K\overset{5+}{Cl}O_3$

Chapter 5

1. A physical property is one that can be described by one of the five senses and does not involve a change in the composition of the compound in question; a chemical property is described by the likelihood of a change to the composition of the compound.

3. One can conclude that sodium and chlorine react readily with oxygen because neither one is found free in nature, because sodium must be kept away from water or air (source of oxygen), and because chlorine combines readily with all elements.

5. Melting point, boiling point, density, physical phase at 25 °C, and appearance are all physical properties.

7. (a) $2\ C_2H_4O_2\ (aq)\ +\ CaCO_3\ (s)$
$\rightarrow Ca(C_2H_3O_2)_2\ (aq)\ +\ H_2O\ (l)\ +\ CO_2\ (g)$
(b) We expect the solid to dissolve, the gas to evolve, and expect to obtain a one-phase solution as the reaction proceeds to completion.
(c) If calcium carbonate dissolves in a clear colorless solution, you can draw the conclusion that some kind of acid must be present.

9. $2\ Cd\ +\ O_2\ \rightarrow\ 2\ CdO$

11. (a) Zn_3P_2 (b) BaF_2 (c) Sr_3N_2 (d) AlF_3

13. (a) $4\ Bi\ +\ 3\ O_2\ \rightarrow\ 2\ Bi_2O_3$ (synthesis)
(b) $Hg(OH)_2\ \rightarrow\ HgO\ +\ H_2O$ (decomposition)
(c) $N_2O_3\ +\ H_2O\ \rightarrow\ 2\ HNO_2$ (synthesis)

15. (a) $CuCl_2\ +\ 2\ Ag\ \rightarrow\ 2\ AgCl\ +\ Cu$
(b) $2\ KBr\ +\ Cl_2\ \rightarrow\ 2\ KCl\ +\ Br_2$
(c) $NiO\ +\ H_2\ \rightarrow\ Ni\ +\ H_2O$
(d) $2\ Cr_3O_3\ +\ 3\ C\ \rightarrow\ 3\ CO_2\ +\ 4\ Cr$

17. (a) $PbI_2\ +\ F_2\ \rightarrow\ PbF_2\ +\ I_2$ (single displacement)
(b) $ZnO\ +\ H_2\ \rightarrow\ H_2O\ +\ Zn$ (single displacement)
(c) $16\ K\ +\ S_8\ \rightarrow\ 8\ K_2S$ (synthesis)

19. (a) Fe in Fe_2S_3 has an oxidation number of +3.
(b) Iron (III) sulfide
(c) $16\ Fe\ +\ 3\ S_8\ \rightarrow\ 8\ Fe_2S_3$

(d) Fe is oxidized ($Fe^0 \rightarrow Fe^{3+}$); S is reduced ($S^0 \rightarrow S^{2-}$)

(e) $Fe_2S_3 + 2\,Al \rightarrow Al_2S_3 + 2\,Fe$

21. (a) $8\,Fe^0 + S_8^0 \rightarrow Fe^{2+}S^{2-}$ (Fe is oxidized, S_8 is reduced)

(b) $2\,Al^0 + 3\,F_2^0 \rightarrow 2\,Al^{3+}F_3^-$ (Al is oxidized, F_2 is reduced)

23. (a) Yes, a reaction takes place because a solid precipitates.

(b) There are solids on both sides of the equation; reaction might occur partially.

(c) There are solids on both sides of the equation; reaction might occur partially.

(d) Yes, a reaction takes place because a solid precipitates.

25. (a) $Mg(NO_3)_2\ (aq) + 2\,NaOH\ (aq)$
$$\rightarrow Mg(OH)_2\ (s) + 2\,NaNO_3\ (aq)$$

(b) $2\,Fe(C_2H_3O_2)_3\ (aq) + 3\,K_2SO_4\ (aq)$
$$\rightarrow Fe(SO_4)_3\ (s) + 6\,KC_2H_3O_2\ (aq)$$

(c) $2\,Na_3PO_4\ (aq) + 3\,Cd(NO_3)_2\ (aq)$
$$\rightarrow Cd_3(PO_4)_2\ (s) + 6\,NaNO_3\ (aq)$$

27. $Zn_3(PO_4)_2$, zinc phosphate

29. (a) $CH_4 + 2\,Cl_2 \rightarrow 4\,HCl + CCl_4$
$C^{4-} \rightarrow C^{4+}$ (oxidized); $Cl_2^0 \rightarrow 2\,Cl^-$ (reduced)

(b) $OF_2 + H_2O \rightarrow O_2 + 2\,HF$
$O^{2-} \rightarrow 1/2\,O_2^0$ (oxidized); $O^{2+} \rightarrow 1/2\,O_2^0$ (reduced)

(c) $N_2H_4 + 2\,H_2O_2 \rightarrow N_2 + 4\,H_2O$
$O^- \rightarrow O^{2-}$ (reduced); $N^{2-} \rightarrow N^0$ (oxidized)

31. (a) $4\,C_3H_5O_9N_3 \rightarrow 12\,CO_2 + 6\,N_2 + 13\,O_2 + 10\,H_2O$

(b) $(NH_4)_2Cr_2O_7 \rightarrow Cr_2O_3 + N_2 + 4\,H_2O$

33. (a) $2\,AgNO_3 + Na_2CO_3 \rightarrow Ag_2CO_3 + 2\,NaNO_3$

(b) $6\,HCl + Al_2O_3 \rightarrow 2\,AlCl_3 + 3\,H_2O$

(c) $CaF_2 + H_2SO_4 \rightarrow CaSO_4 + 2\,HF$

(d) $2\,KI + Pb(NO_3)_2 \rightarrow 2\,KNO_3 + PbI_2$

35. Water is a product of combustion. Its ability to extinguish a fire is one of its physical properties—a property not shared with oxygen, one of its components.

37. The product gas appears to be less flammable than the reactant gas.

39. Sr_3N_2

41. $V + H_2O$

43. K_3P

45. Three signs that a chemical reaction has taken place are 1) the formation of a precipitate, 2) the formation of a gas, and 3) a color change.

47. Combustion is the rapid reaction of a substance with oxygen. We recognize combustion when one of the two reactants is O_2. When a hydrocarbon (compound containing C, H, and possibly O) combines with O_2, the products are always CO_2 and H_2O.

49. (a) Solid carbon dioxide sublimes to form carbon dioxide gas.
(b) Solid sodium carbonate reacts with an aqueous solution of hydrogen chloride to form water, carbon dioxide gas, and an aqueous solution of sodium chloride.

51. (a) $2 \, AgCl + F_2 \rightarrow 2 \, AgF + Cl_2$
(b) $2 \, GaBr_3 + 3 \, Cl_2 \rightarrow 2 \, GaCl_3 + 3 \, Br_2$
(c) $2 \, Al_2O_3 + 3 \, C \rightarrow 3 \, CO_2 + 4 \, Al$

53. (a) The oxidation number of Pt in $PtCl_4$ is $+4$.
(b) Platinum (IV) chloride
(c) $Pt + 2 \, Cl_2 \rightarrow PtCl_4$
(d) Pt is oxidized ($Pt \rightarrow Pt^{4+}$); Cl_2 is reduced ($Cl^0 \rightarrow Cl^-$)
(e) $PtCl_4 + 2 \, F_2 \rightarrow PtF_4 + 2 \, Cl_2$

Chapter 6

1. Answers depend on the lengths of the broomstick, ballpoint pen, and paperclip in question as well as on the shoe box and kitchen dimensions. The ballpoint pen should be the best size for measuring the shoebox and the broomstick should be the best size for measuring the kitchen.

3. (a) Extensive   (b) Intensive   (c) Extensive   (d) Intensive

5. The remaining mass may have evaporated or burned; if we can collect the mass that evaporates and then weigh it, we may be able to determine whether the missing mass was evaporated or combusted.

7. Molten sulfur will crystallize or solidify upon contact with cold water; the impurities will not necessarily solidify.

9. (a) In the Si and P series, both boiling points and freezing points increase down each column. The Al series does not show a smooth trend.
(b) The freezing point decreases across the first row, but there are no particular freezing point trends across the second and third rows. If the first column is ignored, boiling points tend to increase in all three rows.
(c) Solids: $AlCl_3$, $GaCl_3$, $InCl_3$
Liquids: $SiCl_4$, $GeCl_4$, $SnCl_4$, PCl_3, $AsCl_3$, $SbCl_3$
No gases.

11. 15.7 cm^3

13. (a) Sinks   (b) Rises   (c) Rises   (d) Rises

15. Volume contracts at T_2 compared to T_1.

17. The new line should fall below the old line, as volume increases with heat.

19. 56 cm^3

21. 2.16 g/cm^3

23. 1.82 g/cm^3

25. (a) Miles and gallon
(b) Tons of Cl and tons of NaOH
(c) Atoms of C and molecules of methane
(d) Molecules of ammonia and molecules of hydrogen

27. 400 lb

29. 533 employees

31. 0.08 g Cu; 0.071 g Hg; 0.12 g Co; 0.58 g Ag

33. 225.0 g Fe

35. 30.1 g Ca

37. 28.4 g O$_2$

39. 250 liters of solution

41. (a) 52 cm
(b) The second photograph will have an area that is 4 times greater than the original.

43. 3 tagged alligators

45. 3,300 atoms of O

47. 0.40 mole of Cl atoms

49. 2,864 moles of S atoms

51. 88,000 atoms of Cl

53. 36,900 moles of S atoms

55. 0.02 g Cu; 0.10 g Ag; 0.098 g Hg; 0.15 g Co

57. 178.4 g

59. 0.264 g

61. 0.815 g/cm^3

63. (a) 22.2 cm^3
(b) d = 0.775 g/cm^3; this could be substance B
(c)

Substance	Density, g/cm^3	Float?	Sink?
Water	1.001	X	
Ethyl alcohol	0.791		X
Benzene	0.877	X	

65. 2,110 g

67. Sodium: 10.75 g
Glucose: 13.12 g
Potassium: 6.000 g

69. 1.36 ppm

71. 0.024 g

73. 1.50 g/mL

Chapter 7

1. Sometimes neither the guessed number nor the imprecise number seems satisfactorily close.

3. Significant figures are counted in order to determine the certainty of the measurement.

5. The cause of uncertainty in a measured number is due to estimating between demarcations of a measuring tool.

7. To determine the correct number of significant figures in a product or a quotient, we must use the lowest number of significant figures present in either divisor, dividend, or multipliers.

9. (a) 3   (b) 4   (c) 1   (d) 2   (e) 2

11.
$$a^2 + a^2 \neq (a^2)^2 = a^4 \qquad\qquad a^2 + a^2 = 2a^2$$
$$2^2 + 2^2 \neq (2^2)^2 = 2^4 \qquad\qquad 2^2 + 2^2 = 2 \cdot 2^2$$
$$4 + 4 \neq (4)^2 = 16 \qquad\qquad 4 + 4 = 2 \cdot 4$$
$$8 \neq 16 = 16 \qquad\qquad 8 = 8$$

13. A number in scientific notation is written as a number between 1 and 10 multiplied by 10 raised to some power.

15. (a) Gigagram   (c) Kilogram   (e) Microliters
(b) Decigram   (d) Milligram

17. (a) 1   (b) 2   (c) 5

19. (a) -1 (b) -1

21. $0.75x^{-5}$

23. $5z/3x^2y^3 = 5x^{-2}y^{-3}/3z^{-1}$

25. (a) 4.65×10^4
(b) 3×10^{-4}
(c) 3.61×10^6
(d) 4.17089×10^1
(e) 2.198×10^3

27. (a) 0.000163
(b) 231,700
(c) 7,300,000

29.

Measurement	Length, Volume, or Mass?	Number of s.f.
17.25 cm	Length	4
2.83 μg	Mass	3
1.0004×10^2 mL	Volume	5
32.400 kg	Mass	5
0.00068 pm	Length	2
1.48×10^{-2} dl	Volume	3
5.230×10^2 cm^3	Cubic length or volume	4

31. 38 g

33. 1.50 L/min

35. 2.65×10^{11}

37. 1070 km

39. 7.53×10^{15} μL

41. 2.03×10^5 pm

43. 18,900 dL

45. (a) 1.052×10^{-2} L   (e) 109.5 mg
(b) 1.50×10^3 mL   (f) 6.52×10^4 g
(c) 18.3 mL   (g) 5.837 kg
(d) 0.1250 mL

47. 0.5000 mole of CO_2

49. 46.6 cm^3

51. (a) 0.01236 L   (b) 0.01250 L $\geq x \geq 0.01225$ L   (c) 1.54 g

53. $42x^{-7}$

55. 74.7 mL

57. 9.347×10^{18} atoms Pb

59. 0.1 mole

61. (a) 4   (b) 4   (c) 7   (d) 3   (e) 5   (f) 2

63. (a) 8.73×10^{-7}
(b) 6.00835×10^5
(c) 5.26×10^9
(d) 9.14×10^{-6}
(e) 8.904×10^0

65. (a) 0.2500 L   (e) 0.168 g
(b) 0.0150 mL   (f) 4.975×10^6 μg
(c) 25.95 mL   (g) 4.027×10^{13} kg
(d) 9.87×10^{-3} L

Chapter 8

1. 1 molecule H_3PO_4 per 3 atoms H
4 atoms O per 1 molecule H_3PO_4
1 mol $C_6H_{12}O_6$ per 12 atoms H
30.9738 g P per 1 mol P

3. 2.00 mol

5. 2.67×10^{-4} mol

7. 329 g

9. 535 g

11. (a) 0.11 g   (b) 2340 g

13. (a) 0.382 mol C
(b) 3170 mol H

15. (a) 39.098 g K
(b) 109.876 g Mn
(c) 167.541 g Fe

17. 0.1902 g Fe, 0.2416 g Cl

19. 1.4 mol H_2O

21. (a) 4.2×10^{-5} (h) 5.1×10^{-3}
(b) 4.2×10^{-5} (i) 8.7×10^{-3}
(c) 4.2×10^{-5} (j) 5.5×10^{-2}
(d) 4.3×10^{-5} (k) 1.4×10^{-1}
(e) 1.4×10^{-2} (l) 1.0×10^{-1}
(f) 5.5×10^{-2} (m) 8.1×10^{-2}
(g) 1.3×10^{-2}

Gases have the same molar density regardless of mass. Nonmetals have approximately the same molar density to the order of magnitude; hydrocarbons are less dense. Metals do not follow a particular trend with respect to molar density.

23. (a) 40.002%   (d) 85.628%
(b) 40.002%   (e) 85.628%
(c) 40.002%   (f) 85.628%

The mass percentage of carbon is the same when the formulas have the same ratios of C, H, and O.

25. (a) 69.687% K, 28.516% O, 1.796% H
(b) 51.682% Rb, 48.318% Br
(c) 56.635% La, 48.365% Cl
(d) 62.194% Lu, 37.806% Cl

27. (a) 79.85 g Cu   (b) 60.500 g Cu   (c) 34.63 g Cu

Chalcocite (Cu_2S) has the highest mass percentage of copper.

29. (a) 44.87% K, 18.40% S, 36.73% O
(b) 30.02% Fe, 64.56% C, 5.42% H
(c) 53.77% Ba, 18.81% C, 2.37% H, 25.06% O

31. (a) 39.337% Na   (b) 74.504% Pb   (c) 62.6654% U

33. The metal is lithium. In both cases, molar mass must be used.

35. The empirical formula is based on measurements.

37. C_2H_5

39. $NaC_2H_3O_2$

41. Al_2S_3

43. $C_{10}H_{14}O$

45. C_2H_4

47. The molecular formula is H_2O_2. (The choice of method depends on the student. Some will prefer to find the empirical formula first and then the molecular, as this is a method they have already learned. Other students will prefer to find the molecular formula directly, as it is shorter and easier.)

49. P_4O_{10}

51. $Li_2C_2O_4$; $LiCO_2$

53. SnF_4

55. (a) 270.686 g/mol
(b) 195.265 g/mol
(c) 255.42 g/mol
(d) 133.0869 g/mol
(e) 120.06 g/mol

57. 4.9 mol

59. (a) 0.748 g Os   (b) 0.168 g Ca

61. 152 g Zn

63. (a) CH_2Cl
(b) $Hg(C_2H_3O_2)$
(c) NH_4S

65. $CaSO_4 \cdot 2\ H_2O$

67. PCl_5

69. Na_2O_2

71. N_2O_4

73. (a) 828 g   (b) 2.4×10^5 g

75. (a) Na_2SO_4 (b) $(NH_4)SO_4$

Chapter 9

1. This statement is true and shows the conservation of atoms.

3. (a) 104 molecules of CO_2 (b) 16 molecules of CO_2

5. (a) 63   (b) 42   (c) 42   (d) 42

7. (a) 157 moles of H_2 (c) 300 moles of H_2O
   (b) 108 moles of H   (d) 164 moles of H

9. 4,520 moles of CaO

11. 1,225 moles of Na_2CO_3 and 306 moles of Fe_3Br_8

13. (a) 0.101 mole of Li_3N (b) 1.01×10^2 moles of N_2

15. (a) 12.5 moles of Al_2S_3 (b) 5.17 moles of Al_2S_3

17. 5.06×10^{27} molecules of HCl

19. 88.01 grams of H_2O

21. 1.55 moles of CO_2

23. 17.0 moles of NO

25. 9.18×10^{24} molecules of N_2

27. 0.0476 mole of CO_2

29. One everyday example of a limiting reactant is the amount of hot water available for a shower. The length of the shower is limited by the amount of hot water available.

31. (a) CaO is the limiting reactant (0.8916 mol equivalent).
   (b) 57.2 grams of CaC_2
   (c) 84.0 g
   (d) 0 grams of CaO, 1.8 grams of C, 57.2 grams of CaC_2, 25.0 grams of CO

33. 0.960 gram of NO

35. (a) 12.6 grams of CO_2, LR = O_2
   (b) 7.72 g H_2O; LR = O_2
   (c) The limiting reactant in both cases is O_2.

37. 89.5% yield

39. 0.37 mole of O_2

41. 6.2×10^{-3} mole of H_2

43. 12.6 grams of ClO_2

45. (a) 0.220 mole of CO_2 (b) 0.132 mole of excess pentane

47. 9.45 grams of NaOH

49. 1.54 grams of $FeCO_3$

51. 0.334 mole, or 15.3 grams, of NO_2

Chapter 10

1. Path or squiggle

3. $x = -2$; $y = 1/7$

5. $K = °C + 273$

7. atm · L/mol

9. (a) Independent variable: moles of benzene; dependent variable: pressure
   (b) Pressure as a function of moles of benzene

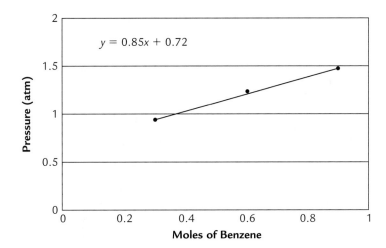

   (c) $m = 0.85$
   (d) atm/moles
   (e) $P = (0.85 \text{ atm/mol}) \, n_{\text{benzene}} + 0.72 \text{ atm}$
   (f) We can use the equation from part (e) to find P when $n_{\text{benzene}} = 0.40$ mole.

11. (a) 680 mm Hg   (b) 185 cm Hg   (c) 1.9 atm

13. An increase in pressure causes a proportional decrease in volume, or pressure is inversely proportional to volume when temperature and mole amount are held constant. $P_1V_1 = P_2V_2$

15. 1.6 L

17. Put it in a smaller cylinder.

19. 1.0×10^3 mm Hg

21. The pressure is 1.1 atm. To do this problem, first change 385 mL to L.

23. As the temperature approaches zero, the volume also approaches zero. We can't have a negative volume so we therefore can't have negative absolute temperature.

25. (a) 283.6 K   (c) −262.6 °C   (e) −272.90 °C
    (b) 262.6 K   (d) 90.6 K

27. Convert all Celsius temperatures into Kelvin.

V_1	T_1	V_2	T_2
220.0 L	-23.5 °C	271.6 L	35.0 °C
4.2 L	254.53 K	5.0 L	303.18 K
3.50 L	274.3 K	7.32 L	300.5 °C
25.0 L	285 K	0.500 L	5.7 K
25.30 L	50.7 °C	20.2 L	-15.0 °C
3.76×10^{-2} L	28.35 K	0.275 L	207.64 K

29. 51.2 K

31. 0.761 atm

33. The volume halves.

35.

P_1 (atm)	P_2 (atm)	V_1 (L)	V_2 (L)	T_1 (K)	T_2 (K)	n_1 (mol)	n_2 (mol)
1.00	1.00	0.250	5.00	273	5460	2.00	2.00
1.00	0.050	0.250	5.00	273	273	2.00	2.00
1.00	0.050	0.250	5.00	273	273	2.00	2.00
40.0	2.00	0.250	5.00	273	273	2.00	2.00
1.00	1.00	0.250	5.00	273	273	2.00	40.0
3.00	1.00	0.250	5.00	353	273	2.00	17.2
3.00	1.00	0.250	0.580	353	273	2.00	2.00

37. 16.0 L

39. (a) The pressure of H_2 is 15.8 atm.
(b) The pressure of CO is 7.90 atm.
(c) The total pressure is 23.7 atm. The pressure of H_2 is two times the pressure of CO, as the equation gives 2 mol H_2 and 1 mol CO.

41. (a) Pressure (P) is the independent variable, solubility (s) is the dependent variable.
(b) Moles dissolved vs. pressure:

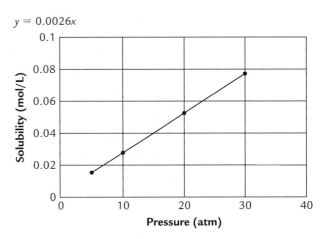

$y = 0.0026x$

(c) m = 2.6×10^{-3} mol/(L atm)
(d) mol/(L atm)
(e) s = 2.6×10^{-3} mol/(L atm)P(atm)

43. The height and arm span are equal. $H = ka$

45. x-intercept = 4; y-intercept = -3

47. (a) $V = (nR/P)T$
(b) nR/P equals the slope.
(c) The y-intercept is zero.
(d) Units of slope are L/°C.

49. $P = 1.13$ atm

51. $T = 317$ K

53. If only P changes, the value of the ratio will increase as P increases and decrease as P decreases.

55. $V = 27$ L

57. 103 g

59. 1680 g

61. 1.31 atm

63. (a) T and n vary directly with P.
(b) V varies inversely with P.
(c) R is a constant and does not change.

65. 1.21 atm

67. 4.71×10^{-4} mL

69. (a) 0.011 mol N_2
(b) 0.31 g N_2
(c) 1.2×10^{-3} g/mL

Chapter 11

1. (a) Campfire, $q < 0$; water, $q > 0$
(b) Cool teapot, $q > 0$; water, $q < 0$
(c) Steam, $q > 0$; coils, $q < 0$

3. (a) Sunlight
(b) Microwave
(c) Mechanical energy
(d) Nuclear energy

5. Stretching a rubber band, electrical current in wire, and addition of NH_4Cl to water

7. (a) The rock absorbs heat.
(b) q is positive for the rock.
(c) The heat comes from the sun.
(d) A reaction on the sun.

9. Endothermic

11. More

13. Endothermic

15. Decrease

17. Endothermic

19. 15,600 J

21. Assuming that no phase change occurs, ice undergoes the larger temperature change.

23.

Substance	Molar Mass (g/mol)	Specific Heat (J/gK)	Molar Heat Capacity (J/mol K)
C (diamond)	12.011	2.53	30.4
Si	28.0855	0.89	25
Ge	72.59	0.36	26
Sn	118.69	0.22	26
Pb	207.2	0.13	27

25.

Metal	q (kJ)	Moles of H_2O	H_2O (g)
Cr	404	9.94	179
Mo	292	7.18	129
Fe	247	6.07	109

27. You want a smaller heat capacity. Temperature change is inversely proportional to heat capacity.

29. (a) All of these are exothermic.

(b)

Oxide	q (kJ)
Cr_2O_3	−750
MoO_3	−524
Fe_2O_3	−515

31.

Food type	Q (kJ/g)	Calories
Carbohydrate	−15.28	−365
Protein	−15.46	−370
Fat/oil	−36.63	−875

33. (a) Heat is created by decomposition and flows from leaves to snow.

(b) Heat is created by crystallization and flows from sodium acetate to the heat pack.

35. Butane in the container is compressed; it must first expand (undergo a phase change) before it can ignite.

37. (a) 564 J   (b) −564 J   (c) −1130 J   (d) 5.4 K

39. −0.39 K

41. 127 J

43. The heat of vaporization is the heat required to cause one mole of a substance to change from the liquid to the gas phase. The heat of fusion is the amount of heat required to cause one mole of a substance to change from the solid to the liquid phase.

45. 2.01 kJ

47. (a) Ni: 4.75×10^5 J
    Ag: 1.32×10^5 J
    Au: 8.98×10^4 J

(b) Ni: 11.7 moles H_2O, 210. g H_2O
    Ag: 3.25 moles H_2O, 58.5 g H_2O
    Au: 2.21 moles H_2O, 39.8 g H_2O

49. (a) Au_2O_3 is formed in an endothermic reaction; the others are exothermic.

(b) Ag_2O: −13.2 kJ
    NiO: −321 kJ

(c) NiO provides more than 16 times the heat of either Ag_2O or Au_2O.

51. (a) 4.18×10^3 J
(b) 22.6 kJ
(c) 26.8 kJ

53. 1.74 J $(K \, g)^{-1}$

55. 72.8 kJ

57. 422 g H_2O

59. (a) Endothermic
(b) 11.8 kJ are needed

Chapter 12

1. Isotopes are atoms of the same element with the same number of protons but a different number of neutrons.

3. b and c

5. 64.923

7. (a) 24   (b) 41.998 g

9. 105.10 is too large and 103.82 is too small.

11. (a) $^{230}_{90}Th$ (b) $^{224}_{88}Ra$ (c) $^{131}_{53}I$ (d) $^{72}_{30}Zn$

13. Strontium

15. 52

17. (a) $^{4}_{2}He$ (b) $^{0}_{-1}\beta$ (c) γ

19. $^{39}_{18}Ar$

21. $^{4}_{2}He$

23. $^{0}_{-1}\beta$

25. $^{210}_{83}Bi \rightarrow \, ^{4}_{2}He + \, ^{206}_{81}Tl$

27. $^{88}_{35}Br \rightarrow \, ^{0}_{-1}\beta + \, ^{88}_{36}Kr$

29. $^{79}_{36}Kr$

31. $^{77}_{35}Br$

33. (a) $^{238}_{92}U \rightarrow \, ^{4}_{2}He + \, ^{234}_{90}Th$ (e) $^{230}_{90}Th \rightarrow \, ^{226}_{88}Ra + \, ^{4}_{2}He$
(b) $^{234}_{90}Th \rightarrow \, ^{234}_{91}Pa + \, ^{0}_{-1}\beta$ (f) $^{226}_{88}Ra \rightarrow \, ^{4}_{2}He + \, ^{222}_{86}Rn$
(c) $^{234}_{91}Pa \rightarrow \, ^{234}_{92}U + \, ^{0}_{-1}\beta$ (g) $^{222}_{86}Rn \rightarrow \, ^{4}_{2}He + \, ^{218}_{84}Po$
(d) $^{234}_{92}U \rightarrow \, ^{230}_{90}Th + \, ^{4}_{2}He$ (h) $^{218}_{84}Po \rightarrow \, ^{4}_{2}He + \, ^{214}_{82}Pb$

A one-step reaction is:
$^{238}_{92}U \rightarrow \, ^{214}_{82}Pb + \, ^{24}_{10}Ne$

35. "Half-life" means the length of time it takes for half of a sample of a radioactive element to decay.

37. 14.3 days

39. Half-life, type of particle emitted, ionization potential, amount of energy emitted

41. $^{38}_{17}Cl$

43. 69.723

45. 55.847

47. (a) Alpha   (b) Beta

49. (a) $^{214}_{82}Pb$ (b) $^{103}_{44}Ru$ (c) $^{230}_{90}Th$ (d) $^{234}_{91}Pa$

51. $^{230}_{90}Th \rightarrow {}^{226}_{88}Ra + {}^{4}_{2}He$

53. $^{214}_{82}Pb \rightarrow {}^{214}_{83}Bi + {}^{0}_{-1}\beta$

55. Alpha particle

Chapter 13

1. (a)

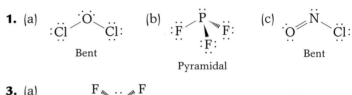

Bent   Pyramidal   Bent

3. (a)

F⟍ I ⟋F
F⟋ | ⟍F
    F

Square pyramidal (distorted)

For the sake of clarity, there are 3 pairs of electrons around each fluorine atom that are not drawn.

(b) A possible Lewis structure for IF_7 is a regular pentagon with two fluorine atoms on one side of the pentagon and one fluorine on the other side of the pentagon.

5.

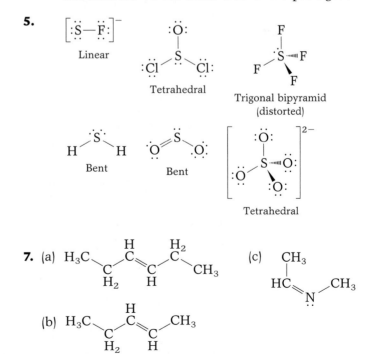

$[:S—F:]^-$

Linear

Tetrahedral

Trigonal bipyramid (distorted)

Bent   Bent   Tetrahedral

7. (a)

H₃C⟍C⟍C=C⟍CH₃
   H₂   H   H₂

(c)

CH₃
|
HC=N⟍CH₃

(b)

H₃C⟍C⟍C=C⟍CH₃
   H₂   H

All CH_2 and CH_3 groups shown are tetrahedral.

9. (a)

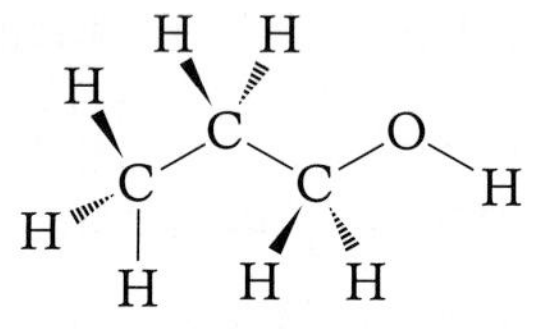

(d)

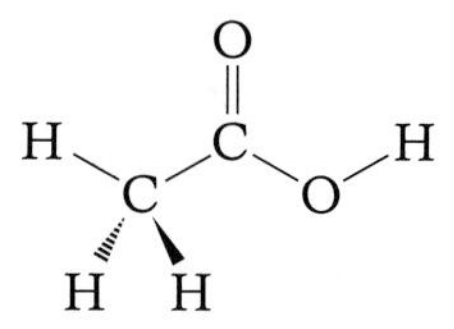

(b)

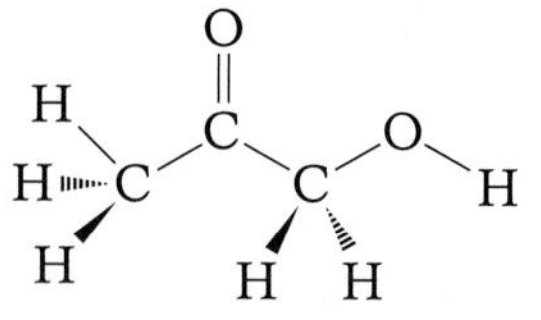

(e)

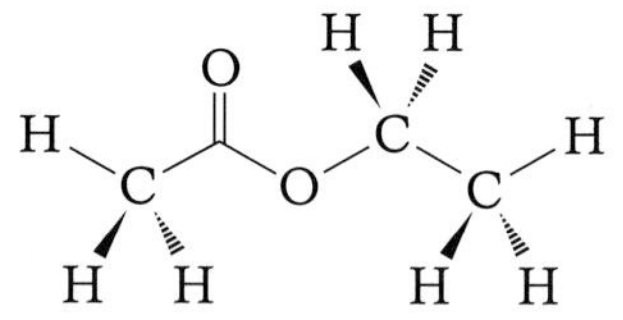

(c)

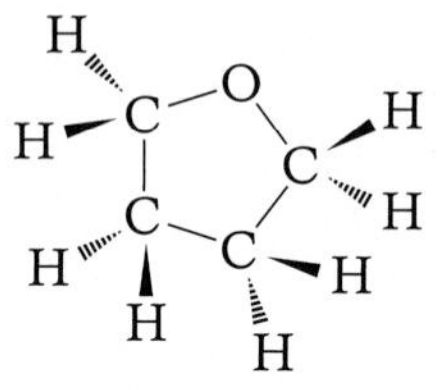

11. (a) 3 radial nodes
(b) 1 angular node, 1 radial
(c) 2 angular nodes, 1 radial

13. 400 nm: 300 kJ/mol e^- and 5×10^{-22} kJ/e^-
500 nm: 250 kJ/mol e^- and 4×10^{-22} kJ/e^-
600 nm: 200 kJ/mol e^- and 3×10^{-22} kJ/e^-

15. (a) $4p$ (b) $4s$ (c) $3d$

17. (a) 2 radial nodes
(b) 2 angular nodes
(c) 3 angular nodes

19. Cs: $1s^2 2s^2 2p^6 3s^2 3p^6 4s^2 3d^{10} 4p^6 5s^2 4d^{10} 5p^6 6s^1$
As: $1s^2 2s^2 2p^6 3s^2 3p^6 4s^2 3d^{10} 4p^3$
S: $1s^2 2s^2 2p^6 3s^2 3p^4$
I: $1s^2 2s^2 2p^6 3s^2 3p^6 4s^2 3d^{10} 4p^6 5s^2 4d^{10} 5p^5$

21. Pb has the electron configuration $6s^2 6p^2$, Pb^{2+} has $6s^2$, and Pb^{4+} has $6s^0$. Ions are formed by removing electrons to have either a filled outer shell or filled subshell (Pb^{2+}). This has more stability than having a partially filled subshell.

23. (a) 1   (b) 5   (c) 3   (d) 2

25. (a)

Monoanion	Electron Configuration of Anion
O^-	$1s^2 2s^2 2p^5$
F^-	$1s^2 2s^2 2p^6$
Ne^-	$1s^2 2s^2 2p^6 3s^1$
Na^-	$1s^2 2s^2 2p^6 3s^2$

(b) It is easiest to remove an e⁻ from Ne⁻; hardest to remove an e⁻ from F⁻. The electron affinity follows the same trend as the first ionization potential, but it is shifted to the left by one atom.

27. H—H   Ö=Ö   N≡N   :F̈—F̈:   :C̈l—C̈l: :B̈r—B̈r: :Ï—Ï:

29.

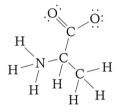

32 electrons
Tetrahedral

18 electrons
Bent

26 electrons
Trigonal pyramid

31. (a) Flat, trigonal
(b) CCO is trigonal; the other N and C domains are tetrahedral.

33. (a) 8   (b) 2   (c) 6   (d) 6

Chapter 14

1. The more convenient unit of measurement depends on what we are looking for. Mass percentage gives us grams of solute for any mass amount of solution; molarity gives us mole amounts of solute for any volume of solution.

3. A solute is a substance dissolved in another substance (the solvent), and a solvent is a substance which takes up another substance (the solute).

5. 525.0 g

7. 266 mL

9. 0.190 mole HCl

11. (a) 28.3 g   (b) 0.215 g   (c) 1.1×10^3 g

13. (a) 0.214 M   (b) 1.17×10^{-3} M   (c) 5.4×10^{-4} M

15. (a) 7.4 L   (b) 0.0792 L   (c) 0.011 L

17. 27.8 mL and 1.488 M

19. 0.750 mole

21. (a) 106 mL   (b) 0.213 mL   (c) 99.6 mL

23. 41.3 mL $AgNO_3$; 0.131 g AgCl

25. 0.02 M

27. 629 mL

29. (a) 2.40×10^{-3} mole   (b) 2.40×10^{-3} mole   (c) 0.0480 M

31. (a) 0.0193 mol of HCl; 8.65×10^{-3} mol of NH_3
(b) NH_3 is the limiting reactant; HCl is in excess.
(c) HCl is 0.0107 mole in excess; 8.65×10^{-3} mole of NH_4^+ is formed.
(d) 4.33×10^{-2} M NH_4^+; 5.35×10^{-2} M HCl

An additional 0.123 L, or 124 mL, are needed for complete reaction.

33. Oxidation always occurs at the anode; reduction always occurs at the cathode.

35. (a) $Cl_2 + 2 e^- \rightarrow 2 Cl^-$ (reduction)
(b) $Mn^{3+} + 3 e^- \rightarrow Mn^0$ (reduction)
(c) $2 H^+ + 2 e^- \rightarrow H_2$ (reduction)

37. (a) $3 Mg + 2 Cr^{3+} \rightarrow 3 Mg^{2+} + 2 Cr$
(b) $3 Cu^+ + Al \rightarrow Al^{3+} + 3 Cu$

39. $Ca + Zn^{2+} \rightarrow Ca^{2+} + Zn$

41. $2 Al + 3 Co^{3+} \rightarrow 2 Al^{3+} + 3 Co$

43. Anode: $Zn \rightarrow Zn^{2+} + 2 e^-$
Cathode: $Cu^{2+} + 2 e^- \rightarrow Cu$

We expect the Cu electrode to increase in mass, because Cu^{2+} is coming out of solution as solid Cu at that electrode.

45. Anode: $Zn \rightarrow Zn^{2+} + 2 e^-$

The anode is made of solid Zn and is in contact with a solution containing Zn^{2+} ions.

Cathode: $Co^{2+} + 2 e^- \rightarrow Co$

The cathode is made of solid Co and is in contact with a solution containing Co^{2+} ions.

47. (a) K^+: 2 mol; $Cr_2O_7^{2-}$: 1 mol
(b) Sn^{2+}: 2 mol; Br^-: 4 mol
(c) Co^{3+}: 1.2 mol; NO_3^-: 3.6 mol

49. (a) Mn^{2+}: 3.5 mol; CrO_4^{2-}: 3.5 mol
(b) Ba^{2+}: 0.8 mol; OH^-: 1.6 mol

51. 5.76 g

53. 115 mL

55. 0.00721 mole

57. 0.028 mole

59. 25.1 mL

61. 0.0016 mole

63. 10.8 g

65. $Zn \rightarrow Zn^{2+} + 2\ e^-$ (oxidation); $Cu^{2+} + 2\ e^- \rightarrow Cu$ (reduction)

67. $Co \rightarrow Co^{2+} + 2\ e^-$ (oxidation); $Pt^{2+} + 2\ e^- \rightarrow Pt$ (reduction)

69. Anode: $Al \rightarrow Al^{3+} + 3\ e^-$

Al decreases in mass as it goes into solution.

Cathode: $Co^{2+} + 2\ e^- \rightarrow Co^0$

Co increases in mass as it comes out of solution and plates onto the Co electrode as solid Co.

71. 10.8 g $Pb(NO_3)_2$

73. (a) 0.0212 mol AgCl
(b) NaCl is the limiting reactant.

75. (a) 6.32×10^{-3} mol K_2CO_3
(b) 8.10×10^{-2} mol LiCl

77. 0.225 L

79. 0.03116 mol I^-

81. 0.0127 mol $Al(NO_3)_3$

Chapter 15

1. The solution is an acid; the bubbles are caused by the hydrogen carbonate breaking down into water and carbon dioxide.

3. An acid releases hydronium ions (H^+). Properties include: sour taste, ability to dissolve metals and oxyanions, corrosive, turns litmus paper red.

5. (a) $HClO_4 + H_2O \longrightarrow H_3O^+ + ClO_4^-$
(b) $HBr + H_2O \longrightarrow H_3O^+ + Br^-$
(c) $KOH \longrightarrow K^+ + OH^-$
(d) $H_2SO_4 + H_2O \longrightarrow H_3O^+ + HSO_4^-$

7. An indicator is a chemical that will change color as the pH changes from acidic to basic solutions.

9. Acid rain is formed by molecules such as SO_3 or NO_2 that are emitted into the air and dissolved in rainwater to form an acid.

11. The chloride ion can act as a Lewis base by donating one of its four lone pairs.

13.

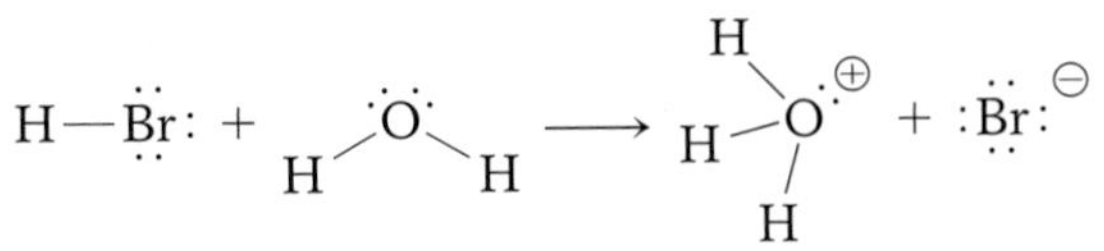

15. 25 molecules

17. 5.5×10^2 mol

19. $H_3PO_4 + H_2O \longleftrightarrow H_3O^+ + H_2PO_4^-$
$C_6H_5COOH + H_2O \longleftrightarrow H_3O^+ + C_6H_5COO^-$
$H_2CO_3 + H_2O \longleftrightarrow H_3O^+ + HCO_3^-$
$HF + H_2O \longleftrightarrow H_3O^+ + F^-$

21. (a) HSO_3^- (b) OH^- (c) H_2O (d) HPO_4^{2-} (e) S^{2-} (f) NH_3

23. (a) Hydronium ion
(b) Chloric acid
(c) Oxalic acid
(d) Hydrosulfuric acid
(e) Hydrogen perselenate ion
(f) Phosphoric acid

25. (a) SO_3^{2-} (b) OH^- (c) F^-

27.

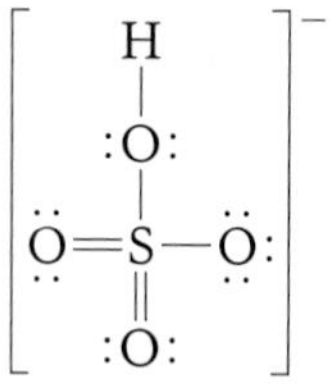

$$\left[\begin{array}{c} H \\ | \\ :O: \\ | \\ \ddot{O}=S-\ddot{O}: \\ \| \\ :O: \end{array} \right]^-$$

29. (a) $H_2SO_4 + NH_3 \longrightarrow HSO_4^- + NH_4^+$
(b) $HSO_4^- + CO_3^{2-} \longrightarrow HCO_3^- + SO_4^{2-}$
(c) $H_3PO_4 + NaOH \longrightarrow H_2O + NaH_2PO_4$
(d) $H_2SO_4 + Na_2CO_3 \longrightarrow NaHSO_4 + NaHCO_3$

31. 0.0478 mol

33. 2.92×10^{-4} M

35. There is 0.0179 mole of H^+ and 0.0171 mole of OH^-. The final solution is acidic, and an additional 1.8 mL of NaOH (limiting reactant) are needed for complete reaction.

37. 0.1032 M

39. 0.055 M in H^+, acidic

41.

Number	Log
4.2×10^{12}	12.62
2.2×10^{11}	11.34
3.0×10^{14}	14.48
4.2×10^{-12}	-11.38
2.2×10^{-11}	-10.66
3.0×10^{-14}	-13.52

43. (a) 3×10^{-3} M   (c) 8.57   (e) 9.15
(b) 5.83×10^{-9} M   (d) 1.19

45. (a) Basic   (c) Acidic, pH = 0.72
     (b) Acidic   (d) Basic, pH = 9.286

47. 13.70

49.

	$[H_3O]^+$	pH	pOH	$[OH^-]$
(a)	0.035	1.46	12.54	2.9×10^{-13}
(b)	2.88×10^{-13}	12.5384	1.462	3.45×10^{-2}
(c)	1.99×10^{-2}	1.701	12.299	5.03×10^{-13}

51. 1.34

53. (a) 0   (b) 11   (c) Increase

55. You should determine if it's an acid or a base by testing with litmus paper.

57. 0.0871 M, pH = 1.060

59. 0.674 M, pH = 0.171

61. (a) 4.77×10^{-4} M
     (b) 8.4×10^{-6} M OH^-, 1.2×10^{-9} M H_3O^+
     (c) pH = 5.74
     (d) pH = 13.12, pOH = 0.88

63. (a) 4.7×10^{-6} M
     (b) 11.96
     (c) 3.39×10^{-10} M
     (d) 7.57
     (e) 1.7×10^{-13} M

65. (a) 9.757×10^{-2} mol HI
     (b) 174 mL

67. (a) Hydroiodic acid   (b) Periodic acid   (c) Iodous acid

69. (a) HCO^-_3 is the hydrogen carbonate ion. Its conjugate acid is H_2CO_3, and its conjugate base is CO_3^{2-}.

(b)

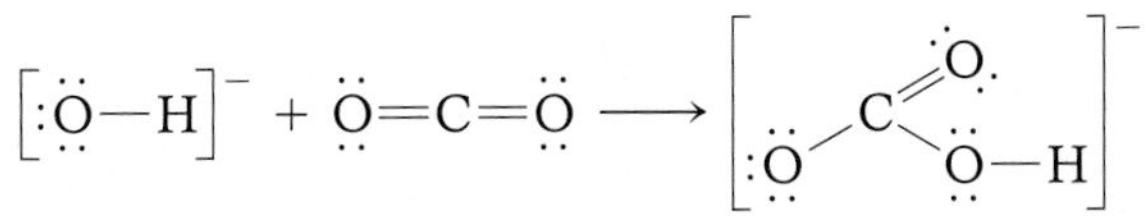

(c) OH^- is the Lewis base; CO_2 is the Lewis acid.

71. (a) $C_8H_5O_4^- + OH^- \longrightarrow C_8H_4O_4^{2-} + H_2O$
     (b) $C_8H_5O_4^-$ is the Lewis acid; OH^- is the Lewis base.
     (c) $C_8H_5O_4^-$ is the Arrhenius acid; OH^- is the Arrhenius base.
     (d) 0.119 M

73. (a) 4.88×10^{-3} mol NaOH
(b) 2.45×10^{-3} mol NaOH
(c) 9.77×10^{-3} mol NaOH
(d) Because the moles of NaOH = moles HCl, the moles of NaOH are determined by the volume and molarity of the HCl.

Chapter 16

1. An equilibrium system is different from a system not at equilibrium in that both forward and reverse reactions occur with equal rates.

3. Macroscopic properties include: color change, temperature change, or gas pressure.

5. mol L^{-1}

7. (a) $K = [H_3O^+][F^-]/[HF]$
(b) $K = [Ca^{2+}][C_2O_4^{2-}]$
(c) $K = [Pb^{2+}]^2[CO_3^{2-}]^3$

9. $K = 9 \times 10^{-12}$

11. A hydrated metal ion is a metal ion with water in its coordination sphere.

13. Cu^{2+} is the Lewis acid; NH_3 is the Lewis base.

15.

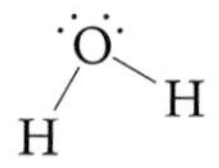

2 lone pairs

17. Using the quadratic formula, quadratic equations can be solved to give their two roots. However, not all solutions result in physically reasonable answers to a chemistry problem. For example, answers that result in negative volumes or pressures are unreasonable physical conditions and are discarded. Solutions that result in larger quantities than initially present are also discarded as unrealistic answers.

19. More HA, because K is greater than 1.

21. K has no units.

23. $K = 8.62$

25. $[Mg^{2+}] = [C_3H_5O_3^-] = 0.105$ M; $[MgC_3H_5O_3^+] = 0.095$ M

27. $[Sn^{2+}] = [C_3H_5O_3^-] = 0.124$ M; $[SnC_3H_5O_3^+] = 0.0759$ M

29. (a) $K = [PbT^-][HCO_3^-]/[HT_2^-]$
(b) $K/[HCO_3^-] = 40.6$
(c) Ratio should increase to 4,060.

31. (a) $\dfrac{[HClO][OH^-]}{[ClO^-]} = 3.3 \times 10^{-7}$

(b) $\dfrac{[OH^-]}{3.3 \times 10^{-7}} = \dfrac{[ClO^-]}{[HClO]}$

(c) At pH $= 10$, $\dfrac{[OH^-]}{K_b} = \dfrac{303}{1}$

(d) At pH $= 3$, $\dfrac{[OH^-]}{K_b} = \dfrac{0.0000303}{1}$

(e) At pH $= 10.0$, the solution is basic, so we expect the ClO^- ion concentration to be greater than at pH $= 3.00$ when the solution is acidic and the hydronium ion predominates.

33. (a) $CH_3COOH\ (aq) + H_2O\ (l) \longleftrightarrow H_3O^+\ (aq) + CH_3COO^-\ (aq)$;

$\dfrac{[H_3O^+][CH_3COO^-]}{[CH_3COOH]} = 6.5 \times 10^{-5}$

(b) $HF\ (aq) + H_2O\ (l) \longleftrightarrow H_3O^+\ (aq) + F^-\ (aq)$;

$\dfrac{[H_3O^+][F^-]}{[HF]} = 6.6 \times 10^{-4}$

(c) $NH_4^+\ (aq) + H_2O\ (l) \longleftrightarrow H_3O^+\ (aq) + NH_3\ (aq)$;

$\dfrac{[H_3O^+][NH_3]}{[NH_4^+]} = 5.6 \times 10^{-10}$

(d) $H_2PO_4^-\ (aq) + H_2O\ (l) \longleftrightarrow H_3O^+\ (aq) + HPO_4^{2-}\ (aq)$;

$\dfrac{[H_3O^+][HPO_4^{2-}]}{[H_2PO_4^-]} = 6.2 \times 10^{-8}$

35. In each case, the $pK_a = -\log K_a$. Each pK_a value has two places after the decimal because each K_a has two significant figures. For a 0.10 M solution of weak acid, the pH is generally less than the pK_a.

Name of acid	pK_a
Chlorous	1.96
Phosphoric (1)	2.12
Hydrofluoric	3.18
Formic	3.10
Acetic	4.19
Carbonic (1)	6.37
Phosphoric (2)	7.21
Hypochlorous	7.52
Hydrocyanic	9.21
Ammonium	9.25
Carbonic (2)	10.32
Phosphoric (3)	12.66

37. (a) $2 AgNO_3 (aq) + Na_2CO_3 (aq) \longrightarrow Ag_2CO_3 (s) + 2 NaNO_3 (aq)$
(b) $CaF_2 (aq) + H_2SO_4 (aq) \longrightarrow CaSO_4 (s) + 2 HF (aq)$
(c) $2 KI (aq) + Pb(NO_3)_2 (aq) \longrightarrow KNO_3 (aq) + PbI_2 (s)$
(d) $3 CuSO_4 (aq) + 2 Na_3PO_4 (aq) \longrightarrow$
$Cu_3(PO_4)_2 (s) + 3 Na_2SO_4 (aq)$
(e) $H_2SO_4 (aq) + PbCl_2 (aq) \longrightarrow 2 HCl (aq) + PbSO_4 (s)$
(f) $3 H_2SO_4 (aq) + Ca_3(PO_4)_2 (aq) \longrightarrow$
$3 CaSO_4 (s) + 2 H_3PO_4 (aq)$
(g) $K_2SO_4 (aq) + Ba(ClO_4)_2 (aq) \longrightarrow BaSO_4 (s) + 2 KClO_4 (aq)$

39. (a) $Pb^{2+} (aq) + 2 Cl^- (aq) \longrightarrow PbCl_2 (s)$
(b) $Pb^{2+} (aq) + C_2O_4^{2-} (aq) \longrightarrow PbC_2O_4 (s)$
(c) $Mn^{2+} (aq) + 2 OH^- (aq) \longrightarrow Mn(OH)_2 (s)$
(d) $Pb^{2+} (aq) + 2 IO_3^- (aq) \longrightarrow Pb(IO_3)_2 (s)$

41. (a) 4.5×10^{-27} mol/L
(b) 1.1×10^{-5} mol/L
(c) 3.3×10^{-11} mol/L

43. FeS > SnS > PdS

45. $CaCO_3 (s) \longrightarrow Ca^{2+} (aq) + CO_3^{2-} (aq)$ $K_{sp} = [Ca^{2+}][CO_3^{2-}]$

47. $Al^{3+} (aq) + 6 H_2O \longrightarrow Al(H_2O)_6^{3+} (aq)$

49. $K = [HS^-][OH^-][Zn^{2+}]$

51. More soluble

53. Less acidic; pH = 6.7, $[H_3O^+] = 2.0 \times 10^{-7}$ M; pH = 3.6,
$[H_3O^+] = 2.5 \times 10^{-4}$ M

55. $K = [A^-][H_3O^+]/[HA]$, $[H_3O^+]/[HA] = K/[A^-]$

57. HSO_4^- can act as an acid or a base, depending on conditions.

59. The weak acid contains more moles of acid but the strong acid will contain more H_3O^+ in a water solution.

Chapter 17

1. (a)

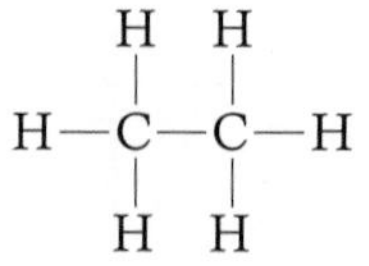

```
    H   H
    |   |
H — C — C — H
    |   |
    H   H
```

(b)

```
H           H
 \         /
  C = C
 /         \
H           H
```

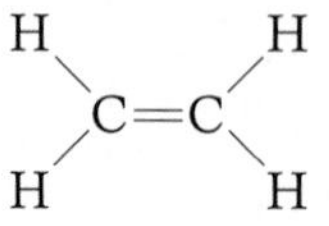

(c) $H-C{\equiv}C-H$

3. (a)

$$H-\underset{\underset{\displaystyle H-\underset{\displaystyle H}{\overset{\displaystyle |}{C}}-H}{\overset{\displaystyle |}{\underset{\displaystyle |}{C}}}}{\overset{\displaystyle H}{\overset{\displaystyle \diagdown}{C}}}-\ddot{N}-\underset{\displaystyle H}{\overset{\displaystyle H}{\overset{\displaystyle \diagup}{C}}}-H$$

(b)

$$H-\underset{\displaystyle H}{\overset{\displaystyle H}{C}}-\underset{\displaystyle H}{\overset{\displaystyle H}{C}}-\underset{\displaystyle H}{\ddot{N}}-H$$

5.

$$H-\underset{\displaystyle H}{\overset{\displaystyle H}{C}}-\underset{\displaystyle H}{\overset{\displaystyle H}{C}}-O-H \qquad H-\underset{\displaystyle H}{\overset{\displaystyle H}{C}}-\underset{\displaystyle H}{\overset{\displaystyle }{O}}-\underset{\displaystyle H}{\overset{\displaystyle }{O}}-H$$

$$H-\underset{\displaystyle H}{\overset{\displaystyle H}{C}}-\underset{\displaystyle H}{\overset{\displaystyle H}{C}}{=}O \qquad H-\underset{\displaystyle H}{\overset{\displaystyle H}{C}}-\underset{\displaystyle H}{\overset{\displaystyle H}{C}}-\underset{\displaystyle H}{\overset{\displaystyle }{N}}-H$$

7.

$$H-\underset{\displaystyle H}{\overset{\displaystyle H}{C}}-\underset{\displaystyle H}{\overset{\displaystyle H}{C}}-\underset{\displaystyle H}{\overset{\displaystyle H}{C}}-H \qquad \underset{\displaystyle H}{\overset{\displaystyle H}{\diagdown}}C\underset{\underset{\displaystyle H\diagup\overset{\displaystyle C}{\diagdown}H}{}}{\diagup}C\underset{\displaystyle H}{\overset{\displaystyle H}{\diagup}}$$

9. See the *Practice of Chemistry* Web site.

11. The only structure that is saturated is:

$$H-\underset{\displaystyle |}{\overset{\displaystyle H}{C}}-\underset{\displaystyle |}{\overset{\displaystyle H}{C}}-H$$
$$H-\underset{\displaystyle \underset{\displaystyle H}{|}}{\overset{\displaystyle |}{C}}-\underset{\displaystyle \underset{\displaystyle H}{|}}{\overset{\displaystyle |}{C}}-H$$

13. The pairs are (a) and (d), with formula $C_5H_{10}O$; (b) and (f), with formula $C_6H_{12}O$; and (c) and (e), with formula $C_4H_{10}O$.

15. (a)

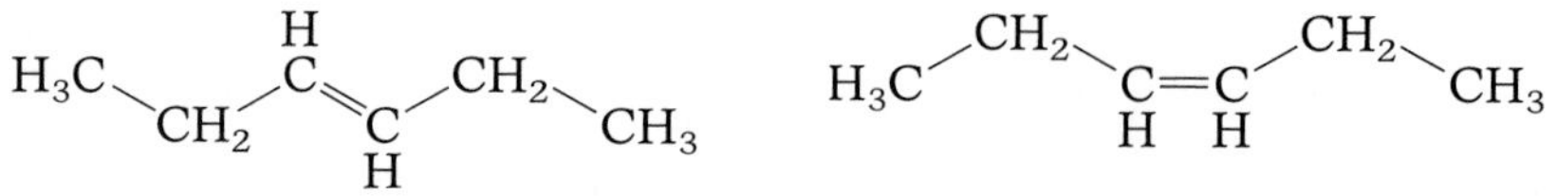

(b)

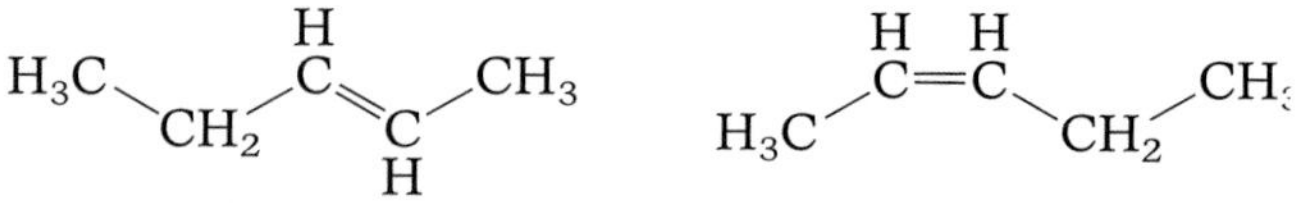

(c)

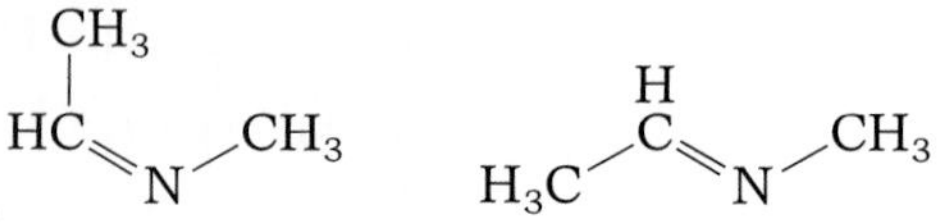

17. Molecule (a) might be similar to rotten fish, and molecule (b) might be of a sour or unpleasant odor.

19.

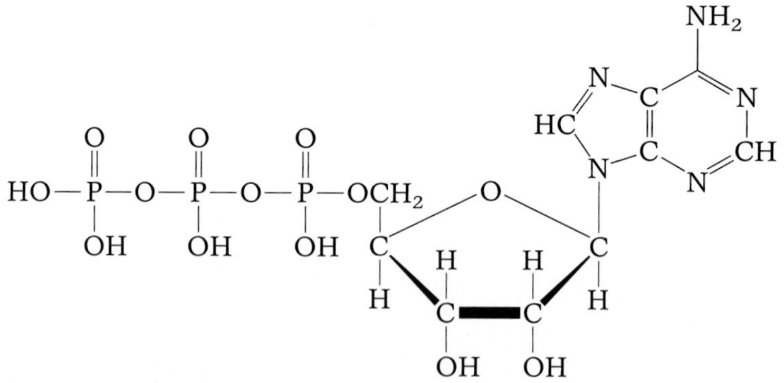

21.   Alanine

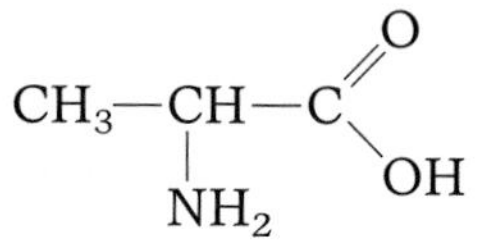

Phenylalanine

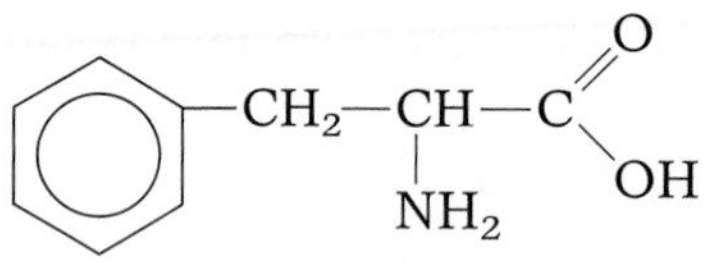

Tyrosine

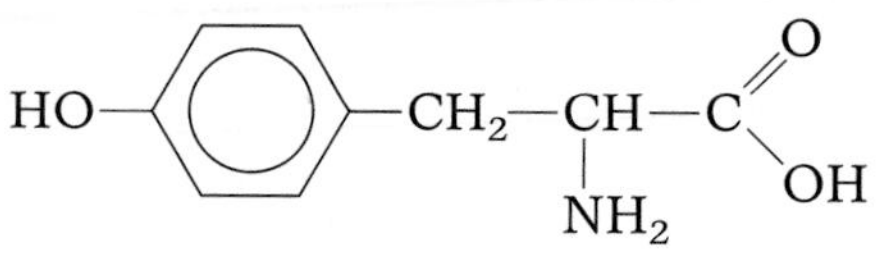

Peptide #1

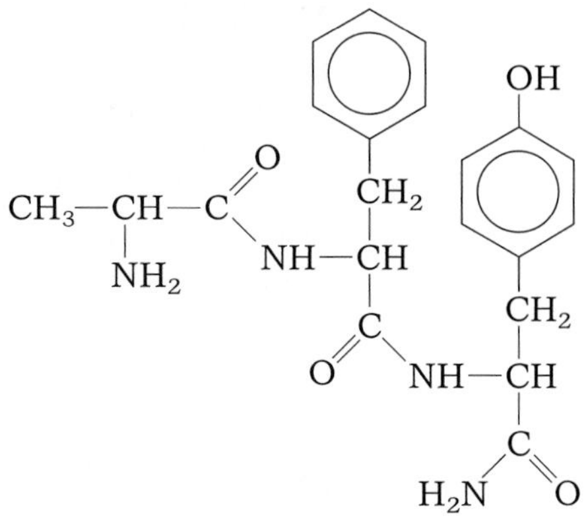

Ala–Phe–Tyr

Peptide #2

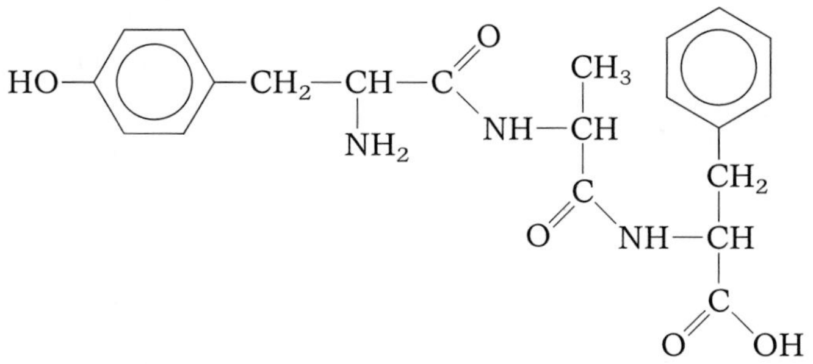

Tyr–Ala–Phe

Peptide #3

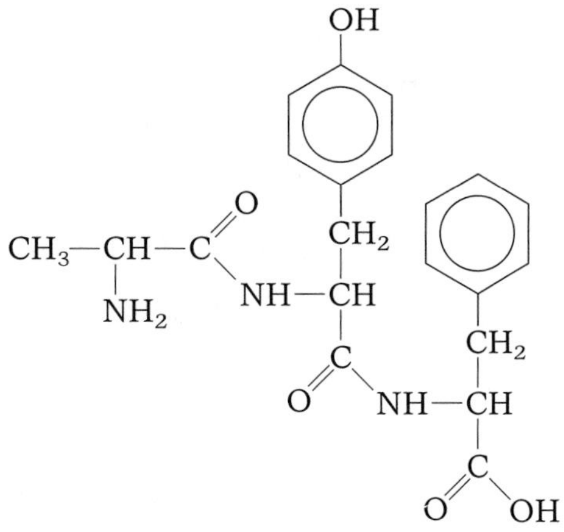

Ala–Tyr–Phe

23. This is a pentapeptide, with sequence Asp-Phe-Gly-Ser-Cys.

25.

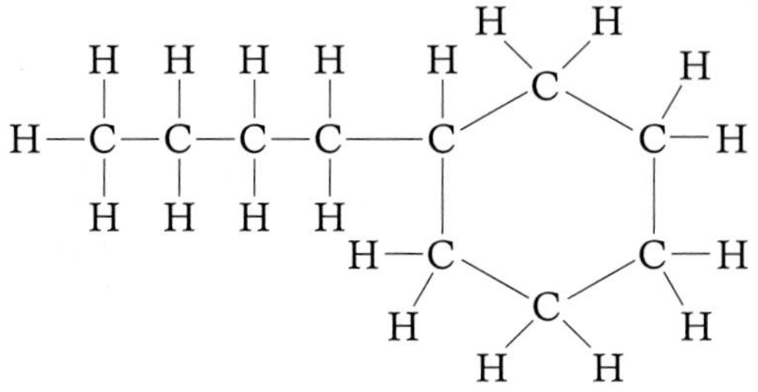

27. (a) 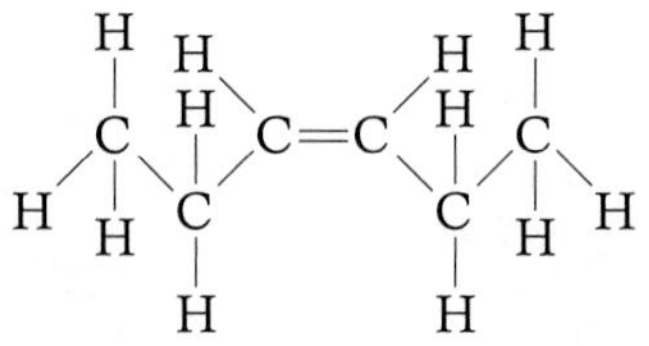   (b)

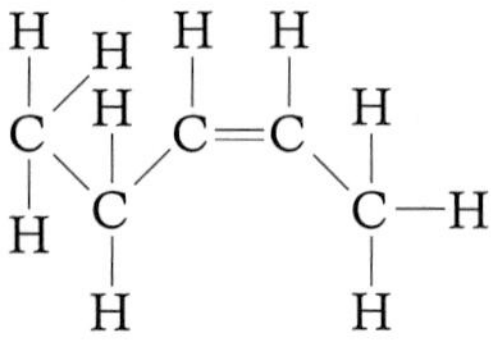

(c)

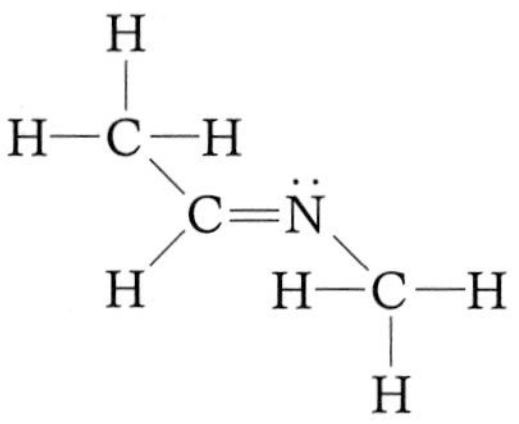

29.   Proline   Glutamic acid

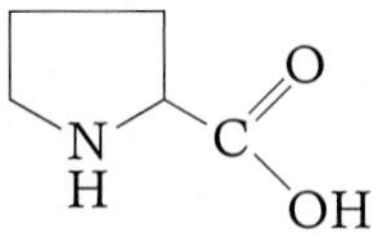

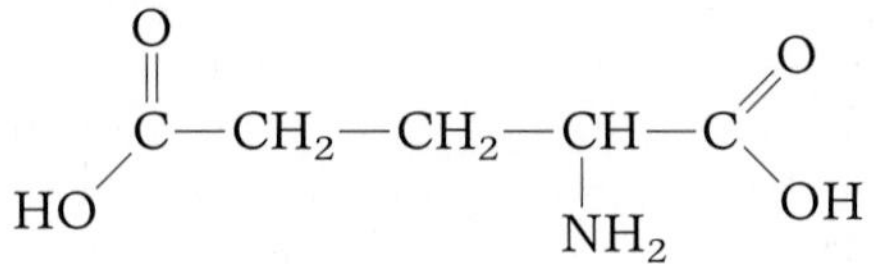

Pro–Glu 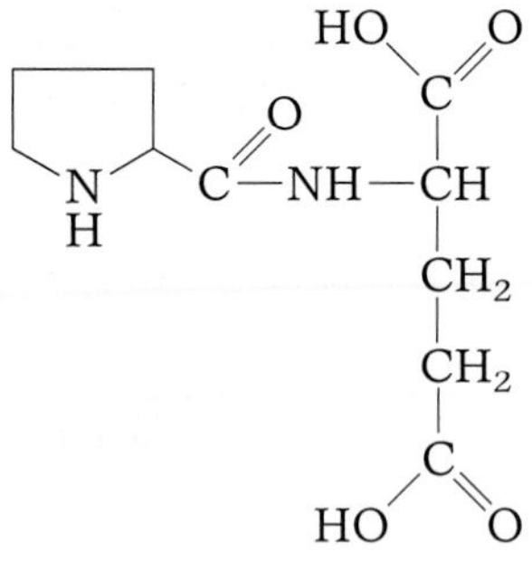   Glu–Pro

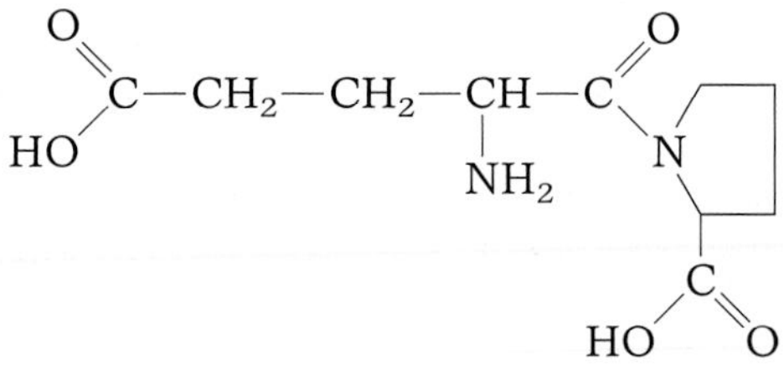

Index

ALPHABETICAL LISTING OF THE ELEMENTS

Element	Symbol	Atomic number	Atomic mass*	Element	Symbol	Atomic number	Atomic mass*
Actinium	Ac	89	(227)	Mendelevium	Md	101	(258)
Aluminum	Al	13	26.981538	Mercury	Hg	80	200.59
Americium	Am	95	(243)	Molybdenum	Mo	42	95.94
Antimony	Sb	51	121.760	Neodymium	Nd	60	144.24
Argon	Ar	18	39.948	Neon	Ne	10	20.1797
Arsenic	As	33	74.92160	Neptunium	Np	93	(237)
Astatine	At	85	(210)	Nickel	Ni	28	58.6934
Barium	Ba	56	137.327	Niobium	Nb	41	92.90638
Berkelium	Bk	97	(247)	Nitrogen	N	7	14.00674
Beryllium	Be	4	9.012182	Nobelium	No	102	(259)
Bismuth	Bi	83	208.98038	Osmium	Os	76	190.23
Bohrium	Bh	107	(262)	Oxygen	O	8	15.9994
Boron	B	5	10.811	Palladium	Pd	46	106.42
Bromine	Br	35	79.904	Phosphorus	P	15	30.973761
Cadmium	Cd	48	112.41	Platinum	Pt	78	195.078
Calcium	Ca	20	40.078	Plutonium	Pu	94	(244)
Californium	Cf	98	(251)	Polonium	Po	84	(209)
Carbon	C	6	12.0107	Potassium	K	19	39.0983
Cerium	Ce	58	140.116	Praseodymium	Pr	59	140.90765
Cesium	Cs	55	132.90545	Promethium	Pm	61	(145)
Chlorine	Cl	17	35.4527	Protactinium	Pa	91	231.03588
Chromium	Cr	24	51.9961	Radium	Ra	88	(226)
Cobalt	Co	27	58.933200	Radon	Rn	86	(222)
Copper	Cu	29	63.546	Rhenium	Re	75	186.207
Curium	Cm	96	(247)	Rhodium	Rh	45	102.90550
Dubnium	Db	105	(262)	Rubidium	Rb	37	85.4678
Dysprosium	Dy	66	162.50	Ruthenium	Ru	44	101.07
Einsteinium	Es	99	(252)	Rutherfordium	Rf	104	(261)
Erbium	Er	68	167.26	Samarium	Sm	62	150.36
Europium	Eu	63	151.964	Scandium	Sc	21	44.955910
Fermium	Fm	100	(257)	Seaborgium	Sg	106	(263)
Fluorine	F	9	18.9984032	Selenium	Se	34	78.96
Francium	Fr	87	(223)	Silicon	Si	14	28.0855
Gadolinium	Gd	64	157.25	Silver	Ag	47	107.8682
Gallium	Ga	31	69.723	Sodium	Na	11	22.989770
Germanium	Ge	32	72.61	Strontium	Sr	38	87.62
Gold	Au	79	196.96655	Sulfur	S	16	32.066
Hafnium	Hf	72	178.49	Tantalum	Ta	73	180.9479
Hassium	Hs	108	(265)	Technetium	Tc	43	(99)
Helium	He	2	4.002602	Tellurium	Te	52	127.60
Holmium	Ho	67	164.93032	Terbium	Tb	65	158.92534
Hydrogen	H	1	1.00794	Thallium	Tl	81	204.3833
Indium	In	49	114.818	Thorium	Th	90	232.0381
Iodine	I	53	126.90447	Thulium	Tm	69	168.93421
Iridium	Ir	77	192.217	Tin	Sn	50	118.710
Iron	Fe	26	55.845	Titanium	Ti	22	47.867
Krypton	Kr	36	83.80	Tungsten	W	74	183.84
Lanthanum	La	57	138.9055	Uranium	U	92	238.03
Lawrencium	Lr	103	(262)	Vanadium	V	23	50.9415
Lead	Pb	82	207.2	Xenon	Xe	54	131.29
Lithium	Li	3	6.941	Ytterbium	Yb	70	173.04
Lutetium	Lu	71	174.967	Yttrium	Y	39	88.90585
Magnesium	Mg	12	24.3050	Zinc	Zn	30	65.39
Manganese	Mn	25	54.938049	Zirconium	Zr	40	91.224
Meitnerium	Mt	109	(266)				

*Parentheses around an atomic mass indicate the most stable isotope of a radioactive element.
Source: Vocke, R. D. Jr., *Pure and Applied Chemistry*, 1999, Vol. 71, pp. 1593–1607.

Name and Formulas of Common Polyatomic Ions

NO_3^-	Nitrate ion	$C_2H_3O_2^-$	Acetate ion
NO_2^-	Nitrite ion	CN^-	Cyanide ion
NH_4^+	Ammonium ion	$C_2O_4^{2-}$	Oxalate ion
SO_4^{2-}	Sulfate ion	ClO_4^-	Perchlorate ion
SO_3^{2-}	Sulfite ion	ClO_3^-	Chlorate ion
HSO_4^-	Hydrogen sulfate ion	ClO_2^-	Chlorite ion
HSO_3^-	Hydrogen sulfite ion	ClO^-	Hypochlorite ion
PO_4^{3-}	Phosphate ion	BO_3^{3-}	Borate ion
HPO_4^{2-}	Hydrogen phosphate ion	H_3O^+	Hydronium ion
$H_2PO_4^-$	Dihydrogen phosphate ion	OH^-	Hydroxide ion
PO_3^{3-}	Phosphite ion	MnO_4^-	Permanganate ion
CO_3^{2-}	Carbonate ion	CrO_4^{2-}	Chromate ion
HCO_3^-	Hydrogen carbonate ion	$Cr_2O_7^{2-}$	Dichromate ion

SI Unit Prefixes

Tera (T)	10^{12}	pico (p)	10^{-12}
Giga (G)	10^9	nano (n)	10^{-9}
Mega (M)	10^6	micro (μ)	10^{-6}
kilo (k)	10^3	milli (m)	10^{-3}
hecto (h)	10^2	centi (c)	10^{-2}
deca (da)	10^1	deci (d)	10^{-1}